ATOM, MOLECULE AND SPECTRUM

ATOM, MOLECULE AND SPECTRUM

S.K. Dogra
Former Professor
Department of Chemistry
I.I.T. Kanpur, Uttar Pradesh

H.S. Randhawa
Former Assistant Professor
Department of Biochemistry and Chemistry
Punjab Agricultural University
Ludhiana, Punjab

PUBLISHING FOR ONE WORLD

NEW AGE INTERNATIONAL (P) LIMITED, PUBLISHERS
New Delhi • Bangalore • Chennai • Cochin • Guwahati • Hyderabad
Jalandhar • Kolkata • Lucknow • Mumbai • Ranchi
Visit us at **www.newagepublishers.com**

Published by New Age International (P) Ltd., Publishers
First Edition: 2011

Branches:

- 37/10, 8th Cross (Near Hanuman Temple), Azad Nagar, Chamrajpet, **Bangalore**-560 018. Tel.: (080) 26756823, Telefax: 26756820, E-mail: bangalore@newagepublishers.com
- 26, Damodaran Street, T. Nagar, **Chennai**-600 017. Tel.: (044) 24353401, Telefax: 24351463. E-mail: chennai@newagepublishers.com
- CC-39/1016, Carrier Station Road, Ernakulam South, **Cochin**-682 016. Tel.: (0484) 2377004, Telefax: 4051303. E-mail: cochin@newagepublishers.com
- Hemsen Complex, Mohd. Shah Road, Paltan Bazar, Near Starline Hotel, **Guwahati**-781 008. Tel.: (0361) 2513881. Telefax: 2543669, E-mail: guwahati@newagepublishers.com
- 105, 1st Floor, Madhiray Kaveri Tower, 3-2-19, Azam Jahi Road, Nimboliadda, **Hyderabad**-500 027. Tel.: (040) 24652456, Telefax: 24652457, E-mail:hyderabad@newagepublishers.com
- Adda Tanda Chowk, **Jalandhar City**. Telefax: (0181) 2221272. E-mail:jalandhar@laxmipubications.com
- RDB Chambers (Formerly Lotus Cinema)106A, 1st Floor, S.N. Banerjee Road, **Kolkata**-700 014. Tel.: (033) 22273773, Telefax: 22275247, E-mail:kolkata@newagepublishers.com
- 16-A, Jopling Road, **Lucknow**-226 001. Tel.: (0522) 2209578, 4045297, Telefax: 2204098. E-mail: lucknow@newagepublishers.com
- 142C, Victor House, Ground Floor, N.M. Joshi Marg, Lower Parel, **Mumbai**-400 013. Tel.: (022) 24927869. Telefax: 24915415, E-mail: mumbai@newagepublishers.com
- 22, Golden House, Daryaganj, **New Delhi**-110 002. Tel.: (011) 23262370, 23262368, Telefax: 43551305. E-mail: sales@newagepublishers.com

---

ISBN: 978-81-224-3073-8

₹ 795.00

C-10-11-5201

Printed in India at Glorious Printers, Delhi.
Typeset at Kalyani Computer Services, Delhi.

PUBLISHING FOR ONE WORLD
NEW AGE INTERNATIONAL (P) LIMITED, PUBLISHERS
4835/24, Ansari Road, Daryaganj, New Delhi-110002
Visit us at **www.newagepublishers.com**

Dedicated to

Dr. (Mrs.) Sulekha Dogra

and

Late Mrs. Mohinder Kaur Randhawa

Preface

From the cast subject of quantum theory we have chosen material which is of interest mainly to chemists. The concepts of quantum mechanics and Schrodinger wave theory are concerned in the body of this book.

The authors main purpose is to introduce to the students at early stage in their studies, a new set of ideas differing greatly from classical mechanics. It is necessary for the students not only to master the basic principles of quantum mechanics but also to develop an ability to apply these principles to their respective field of specialization they choose in their later career.

The first course in the first semester of all the M.Sc. programmes of all the Universities or the Institutes of the country involves elementary quantum mechanics. Thus, the contents of this book are mainly based on the notes we developed while teaching a course in quantum chemistry to the first semester students of M.Sc. programme. Although, we have regularly taught this material at the first year M.Sc. level but it can be easily taught to the final year students of B.Sc. (Hons) in all the universities. We have included more material that can be easily covered in one semester.

The most common apprehension among the students from Indian Universities or Institutes is that one requires sound background of mathematics before taking up Physical Chemistry in general and quantum mechanics in particular. For presenting this book we have taken special precaution not to include very sophisticated mathematics. The emphasis is laid more on the development of physical concepts rather than on mathematics. This book can be easily understood by the students if they have a mathematical background of one year of calculus, with no knowledge of differential equations. Most of the mathematical techniques are developed in the book.

Most of the text books on quantum mechanics meant for students of undergraduate and postgraduate levels generally have few solved examples which too are of an academic nature with more emphasis on theory than applications. In this book, an attempt has been made to present the basic concept in quantum mechanics with greater emphasis on applications in different areas of chemistry viz. physical, organic and inorganic chemistry.

The book comprises of five chapters. The chapter 1 entitled "Classical Mechanics and Evolution of Quantum Mechanics" deals with the concepts and limitations of classical mechanics. In this chapter, we have discussed topics, starting from the ideas developed by Planck, Bohr, Sommerfeld, de Broglie and Heisenberg, which led to the development of quantum mechanics and are required for any discussion on quantum mechanics. The chapter 2 entitled "Quantum Mechanics" deals with the description of matter waves and quantum mechanical laws. Concept of operators, eigenvalue, eigenfunction, expectation values, normalization, orthoganility relation etc. was also developed in this chapter. Having gained familiarity with these concepts, few problems, e.g. particle in one-dimensional and three-dimensional box, simple harmonic oscillator and rigid rotator, have been discussed so that the students acquire some concrete experience with quantum mechanical principles. Third chapter entitled "Atomic Structure", an extension of an exact solution the problem in three dimensional space, i.e. hydrogen atom. Here we have pointed out that it is not possible to have an exact solution beyond hydrogen atom and thus have discussed the approximate methods, like perturbation theory and variation method. Qualitative picture of multi-electron atoms as well as mat angular momentum is also given in chapter 3. Chapter 4 entitled "Molecular Structure" covers qualitatively the structure of molecules, based on valence bond and molecular orbital theories. Efforts have been made to cover the unsaturated systems also. The Chapter 5, entitled "Spectroscopy" presents the applications and results of previous chapters to molecular spectroscopy. Here the emphasis is made mostly to diatomic molecules. One of the good features of this book is that it contains lot of solved examples within each chapter as well as at the end of each chapter. This will provide the students an opportunity to solve some or all of these problems by themselves if they desire so. At the moment, we are not providing the answers to these unsolved problems, but if the demands come we will do in the next edition.

It has been always a dilemma for the chemists to use SI systems of units or cgs. We have tried to stress for the SI systems but the cgs system is also not neglected. For example in spectroscopy, especially in the vibrational frequencies etc. one cannot avoid the cm^{-1} unit.

This book is the outcome of the comments and criticism of the students whom we taught at our respective Institutes. We are really thankful to them for their sincere criticism which made us to revise our class notes regularly.

S. K. Dogra

H. S. Randhawa

Contents

1. Classical Mechanics and Evolution of Quantum Mechanics 1

1.1 Classical Mechanical Era 1

1.1.1 Generalized Coordinates 5

1.1.2 Example of a Motion Subject to Constraints 5

1.1.3 Generalized Momenta and Hamiltonian Equations 11

1.1.4 Field Concept 12

1.2 Post Classical Mechanics and Pre-quantum Mechanical Era 16

1.2.1 Black Body Radiations 16

1.2.2 Heat Capacity of Solids 22

1.2.3 Photoelectric Effect and Compton Effect 25

1.2.4 Compton Effect 27

1.2.5 Bohr's Theory 30

1.2.6 Bohr's Correspondence Principle 42

1.2.7 Bohr Model for Hydrogen Like Atoms 43

1.2.8 Sommerfeld's Model 44

1.2.9 Drawbacks of the Bohr Theory 50

1.3 Matter Wave or de Broglie Waves 52

1.3.1 Hydrogen Atom and de Broglie Hypothesis 56

1.4 Heisenberg's Uncertainty Principle 57

1.4.1 Diffraction Phenomenon and Heisenberg Principle 59

1.4.2 Bohr's Hydrogen Atom and Uncertainty Principle 61

Problems 71

2. Quantum Mechanics 73

2.1 Wave Mechanics 73
2.2 Schrödinger Wave Equation 74
2.3 Physical Significance of Wave Function 76
2.4 Algebraic and Quantum Mechanical Operators 83
2.4.1 Concept of Functions and Operators 83
2.4.2 Principle of Superposition 85
2.4.3 Algebra of Operators 87
2.4.4 Quantum Mechanical Operators 91
2.4.5 Commutation Rules for Angular Momentum Operators 105
2.4.6 Heisenberg Rule for the Product of Position and Momentum Operators 108
2.4.7 Hermiticity of Quantum Mechanical Operators 111
2.4.8 Properties of Angular Momentum from Commutation Rules 116
2.5 Postulates of Quantum Mechanics 119
2.6 Applications of Schrödinger Equation 122
2.6.1 Free Particle 122
2.6.2 Particle in a One Dimensional Box of Infinite Walls 123
2.6.3 Non-interacting Particles: (Theorem Regarding Separation of Variables) 132
2.6.4 Particle in a Three Dimensional Box 134
2.6.5 Particle on a Circular Ring 138
2.6.6 Power Series Method 140
2.6.7 Simple (or Linear) Harmonic Oscillator 143
2.6.8 Rigid Rotator 156
Problems 163

3. Atomic Structure 170

3.1 Hydrogen Atom (One Electron System) 170
3.1.1 Solution of Azimuthal Equation 176
3.1.2 Solution of Polar Equation (*i.e.,* Θ part) 180
3.1.3 Solution of Radial Equation 186
3.1.4 Quantum Numbers and Energy for Hydrogen Like Atoms 201

	3.1.5	Total Number of Energy States	204
	3.1.6	Wave Functions for Hydrogen Like Atoms	206
	3.1.7	Geometrical Description of Hydrogen Like Atomic Orbitals	212
	3.1.8	Quantum Mechanical Model of Hydrogen Atom	228
3.2.	Virial Theorem		229
3.3	Quantum Mechanical Tunneling		232
	3.3.1	Thin Potential Wall as Potential Barrier	233
3.4	Hamiltonian and its Symmetry		239
3.5	Approximate Methods		240
	3.5.1	The Perturbation Method	240
	3.5.2	Atom (ion) with Two Electrons	246
	3.5.3	Exchange Degeneracy	248
3.6	The Variation Method		250
	3.6.1	Ground State Energy of the Hydrogen Like Atoms by the Variation Method	255
	3.6.2	Helium Atom	258
3.7	Electron Spin or Spin Angular Momentum		265
3.8	Pauli Exclusion Principle		268
3.9	Multi-Electron Atoms		272
3.10	Screening Effect of the Inner Shell Electron Orbital		279
3.11	Auf-Bau Principle		282
3.12	Periodic Table and Periodicity in Properties		284
	3.12.1	Modern Periodic Table	284
	3.12.2	Ionization Energy	290
	3.12.3	Electron Affinity	291
	3.12.4	Atomic and Ionic Radii	292
3.13	Angular Momentum		293
	3.13.1	For One Electron	293
	3.13.2	Multi-Electron Atoms	300
3.14	State Symbol or Term Symbols		303
	Problems		315

4. MOLECULAR STRUCTURE **320**

4.1 Born Oppenheimer Approximation 322
4.2 Molecular Orbital Theory 324
4.3 Linear Combination of Atomic Orbitals (LCAO) 325
4.3.1 Conditions on a.o. to form m.o. 326
4.4 The H_2^+ Molecule Ion 328
4.5 Hydrogen Molecule (H_2) 337
4.6 Shapes and Energies of Various m.o.'s 341
4.6.1 Shapes of m.o.'s 341
4.6.2 Energies of m.o.'s 345
4.7 Electronic Configuration of A_2 Molecules 346
4.8 Other A_2 Types of Molecules 349
4.9 Molecular Orbitals Formed by the Overlap of Different Atomic Orbitals 350
4.10 Heteronuclear Diatomic Molecules 352
4.10.1 LiH 353
4.10.2 AB Molecules in General 355
4.10.3 Different Types of m.o.'s 357
4.11 Valence Bond Theory (v.b.t.) 358
4.11.1 Hydrogen Molecule (H_2) 358
4.11.2 Homonuclear Diatomic Molecules 367
4.11.3 Heteronuclear Diatomic Molecules 368
4.12 Comparison of V.B. and M.O. Theories 370
4.13 Molecular Term Symbols 375
4.14 Dipole Moment of AB Molecules 378
4.15 Electronegativity 380
4.15.1 Pauling's Method 380
4.15.2 Mulliken Method 383
4.16 Saturation of Chemical Forces 385
4.17 Triatomic Molecules 385
4.17.1 Triatomic Linear Molecules 386
4.17.2 Triatomic Bent Molecules 399
4.18 Carbon Compounds 404

	4.18.1	Saturated Hydrocarbons	404
	4.18.2	Unsaturated Hydrocarbons	407
	4.18.3	Carbonyl Group	411
4.19	Hybridization		413
	4.19.1	Wave Functions	416
	4.19.2	Angular Dependence of Hybrid Orbitals	423
	4.19.3	Strength of Hybrid Bonds	426
4.20	Hückel Molecular Orbital Theory (HMO)		427
4.21	Bonding in Transition Metals		454
	Problems		477
5. SPECTROSCOPY			**480**
5.1	Intensity		480
	5.1.1	Beer-Lambert's Law	480
5.2	Theories of Absorption or Emission Spectrum		485
5.2.1	Classical Theory		486
5.2.2	Quantum Theory		487
5.2.3	Quantum Mechanical Formulation		489
5.2.4	Selection Rules		492
5.3	Different Kinds of Spectroscopy		494
5.4	Rotational Spectrum		496
	5.4.1	Diatomic Molecules	496
	5.4.2	Quantum Restrictions	498
	5.4.3	Interaction of Radiation with Molecule	499
	5.4.4	Spectrum	501
	5.4.5	Thermal Distribution of Rotational Quantum States (Intensities of the Rotational Lines)	503
	5.4.6	Extension to Polyatomic Molecules	505
	5.4.7	Limitations of Simple Model	507
5.5	Vibrational Spectroscopy		513
	5.5.1	Vibrational Spectrum	513
	5.5.2	Vibrational Energy Levels of Diatomic Molecule	516

5.5.3	Mechanism of Absorption of Infrared Radiation by Molecules	517
5.5.4	Spectrum	518
5.5.5	Intensity or Thermal Population Distribution	519
5.5.6	Potential Energy and Asymmetry in Chemical Bond (An-harmonic Motion)	522
5.5.7	Energy levels	526
5.5.8	Spectrum	527
5.5.9	Rotational-Vibrational Spectrum	529
5.5.10	Vibration of Polyatomic Molecules	537
5.5.11	Applications of Infrared Spectroscopy	541
5.6	Raman Scattering	543
5.6.1	Classical Theory of Raman Scattering	544
5.6.2	Quantum Theory of Raman Scattering	547
5.6.3	Vibrational Raman Scattering	548
5.6.4	Rotational Raman Scattering	549
5.6.5	Rotational-Vibrational Raman Scattering	551
5.6.6	Vibrational Raman Spectrum of Polyatomic Molecules	552
5.7	Electronic Spectrum	553
5.7.1	Electronic Spectrum of Diatomic Molecules	553
5.7.2	Vibrational Coarse Spectrum	554
5.7.3	Intensity Distribution in Vibrational Structure	556
5.7.4	Wave Mechanical Description of Franck-Condon Principle	558
5.7.5	Selection Rules	562
5.7.6	Electronic Spectrum of Polyatomic Molecules	564
5.7.7	Effect of Substituents on Electronic Transitions	566
5.7.8	Transition Metal Complexes	569
5.8	Behaviour of Electronically Excited Molecules	572
5.8.1	Fluorescence	573
5.8.2	Phosphorescence	575
5.8.3	Delayed Fluorescence	576
5.8.4	Excimer Formation	576
5.8.5	Radiationless Transitions	577

5.9	Nuclear Magnetic Resonance	578
	5.9.1 Population of Different Spin States	582
	5.9.2 Nuclear Magnetic Resonance Spectrometer	583
	5.9.3 Chemical Shift	585
	5.9.4 Origin of Proton Chemical Shifts	589
	5.9.5 Spin-Spin Coupling and Fine Structure	596
	5.9.6 Coupling Between Equivalent Nuclei	599
	5.9.7 First Order Analysis	600
	Appendix 5.1: Evaluation of Transition Moment Integral for Rotational Spectrum	603
	Appendix 5.2 : Evaluation of Transition Moment Integral for the Vibrational Spectrum	604
	Appendix 5.3: Evaluation of Bohr nuclear magneton	605
	Problems	607
	Appendix A-1–A-6	611
	Index	**615**

CHAPTER

1

Classical Mechanics and Evolution of Quantum Mechanics

1.1 CLASSICAL MECHANICAL ERA

Classical mechanics is based on the concepts such as particle (an entity of matter), position, velocity, linear momentum, angular momentum, energy, time and even field. They are related to one another through certain laws of classical mechanics. The classical mechanics in its most elementary form is also called *Newtonian mechanics*, may be considered as based on Newton's three laws of motion.

(1) Law of inertia states that a free body is either at rest or moving with constant velocity, say v.

(2) Law of force states that the force acting on a particle of mass m equals the rate of change of momentum ($p = mv$) with time (t).

$$f = \frac{d}{dt}(mv) = m\frac{dv}{dt} = m\frac{d^2r}{dt^2} = ma \qquad ...(1.1)$$

where r is the distance traversed by the body in time t, *i.e.*, $r = v \times t$ and the quantity $a = \frac{d^2r}{dt^2} = \frac{dv}{dt}$, is known as *acceleration*. The motion of particle may be written as:

$$m\frac{d^2x}{dt^2} = f_x, m\frac{d^2y}{dt^2} = f_y, m\frac{d^2z}{dt^2} = f_z \qquad ...(1.2)$$

where x, y and z are the Cartesian coordinates of the particle and f_x, f_y and f_z are the three components of force f acting on the particle.

(3) Law of action and reaction states that when a body A exerts a force f_A on body B by any means, then body B in turn exerts an equal and opposite force f_B on body A, such that

$$f_A = -f_B \qquad ...(1.3)$$

The law of inertia defines the equilibrium condition for a body and implies the conservation of momentum. The law of conservation of momentum states that total momentum of a cluster is composed of the vector sum of all the momenta of the particles and that this sum remains constant even though individual particle momentum may change *i.e.*,

$$\sum_{i=1}^{n} m_i v_i = m_1 v_1 + m_2 v_2 + m_3 v_3 + \ldots + m_n v_n = \text{Constant} \quad \ldots(1.4)$$

where m_i's and v_i's are the mass and velocity of the respective particle.

A very important approach to classical mechanics which is most widely used in advanced classical dynamics is based on *Hamilton's principle*. The principle states that the real path followed by a particle between points A (t_1) and B (t_2) (such that $t = t_2 - t_1$) and slightly 'different path' connecting the same points have the same value of integral $\int_{t_1}^{t_2} (K - V)dt$ for both the paths, provided kinetic energy (K) and potential energy (V) are defined as function of time and position along the path as shown in Fig. 1.1. In other words we can say that the integral has stationary value. It can be a minimum or maximum. The quantity L = K – V is called *Langrangian function* or *kinetic potential*. Another quantity which is equal to the energy of system is called the *Hamiltonian* (H) *function*. Mathematically it is:

$$H = K + V \quad \ldots(1.5)$$

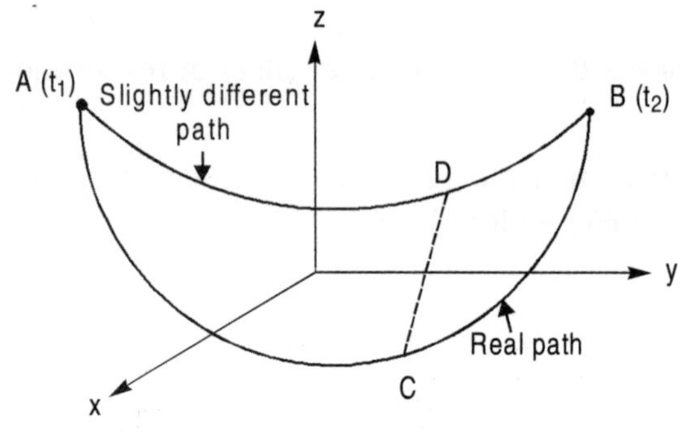

Figure 1.1: *A schematic diagram illustrating the Hamilton's principle ACB is the real path and ADB is any slightly different path*

All the laws of classical mechanics can be derived from Hamilton's principle, and it provides a system of mechanics based on energies rather than on vector quantities such as forces and momenta. Since energy seems to be the primary stuff, out of which all physical universe, is formed, perhaps a Hamiltonian approach to mechanics is the most fundamental one (and can be adapted to both relativistic and non-relativistic quantum mechanics).

Newton's equations of motion in the Langrangian form are deduced as follows: Let x_i, y_i and z_i are three Cartesian coordinates of the *i*th particle with mass m_i. Newton's equations for *n*-point particle are:

$$m_i \frac{d^2 x_i}{dt^2} = m_i \ddot{x}_i = X_i$$

$$\left.\begin{aligned} m_i \frac{d^2 y_i}{dt^2} &= m_i \ddot{y}_i = Y_i \\ m_i \frac{d^2 z_i}{dt^2} &= m_i \ddot{z}_i = Z_i \end{aligned}\right\} \quad ...(1.6)$$

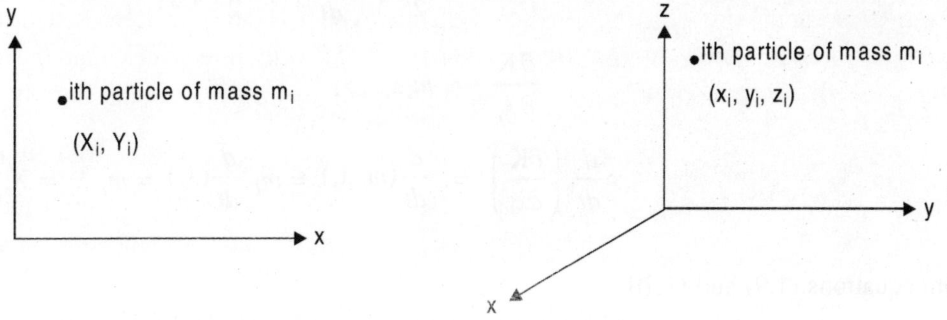

Figure 1.2: *Coordinates of a point in two and three dimensional space*

where $i = 1, 2, 3\ldots n$. X_i, Y_i and Z_i are the components of force along the three axes, acting on *i*th particle. There is a set of such equations for each particle. Dots (..) refer to differentiation with respect to time, *i.e.*,

$$\ddot{x}_i = \frac{d^2 x_i}{dt^2}, \quad \dot{x}_i = \frac{dx_i}{dt} \quad ...(1.7)$$

The equation (1.6) could be more useful if we introduce some familiar definitions. We define kinetic energy K (for Cartesian coordinates) as:

$$\begin{aligned} K &= \frac{1}{2} m_1 (\dot{x}_1^2 + \dot{y}_1^2 + \dot{z}_1^2) + \ldots + \frac{1}{2} m_n (\dot{x}_n^2 + \dot{y}_n^2 + \dot{z}_n^2) \\ &= \frac{1}{2} \Sigma m_i (\dot{x}_i^2 + \dot{y}_i^2 + \dot{z}_i^2) \end{aligned} \quad ...(1.8)$$

For conservative systems (a certain class of systems) it is possible to define another quantity, the potential energy V, which is a function of coordinates $x_1, y_1, z_1 \ldots x_n, y_n, z_n$, (but not of time) of all the particles such that the force acting on each particle is equal to the partial derivatives of the potential energy with respect to coordinates of the particles with negative sign, *i.e.*,

$$\left.\begin{aligned} X_i &= -\frac{\partial V}{\partial x_i} \\ Y_i &= -\frac{\partial V}{\partial y_i} \\ Z_i &= -\frac{\partial V}{\partial z_i} \quad i = 1, 2, 3\ldots n \end{aligned}\right\} \quad ...(1.9)$$

It is possible to find a function V which will, in this manner, express forces of types mechanical, electrostatic and gravitational. However the potential energy function cannot be set for electromagnetic type forces and this is also not important in chemical applications.

∴ Newton's equations of motion in terms of K and V are:

$$K = \frac{1}{2}m_i\left(\frac{dx_i}{dt}\right)^2 = \frac{1}{2}m_i\dot{x}_i^2$$

∴
$$\frac{\partial K}{\partial \dot{x}_i} = m_i\dot{x}_i$$

or
$$\frac{d}{dt}\left(\frac{\partial K}{\partial \dot{x}_i}\right) = \frac{d}{dt}(m_i\,\dot{x}_i) = m_i\frac{d}{dt}(\dot{x}_i) = m_i\,\ddot{x}_i = X_i \qquad ...(1.10)$$

From equations (1.9) and (1.10)

$$\left.\begin{aligned}\frac{d}{dt}\left(\frac{\partial K}{\partial \dot{x}_i}\right)+\frac{\partial V}{\partial x_i} &= 0\\ \frac{d}{dt}\left(\frac{\partial K}{\partial \dot{y}_i}\right)+\frac{\partial V}{\partial y_i} &= 0\\ \frac{d}{dt}\left(\frac{\partial K}{\partial \dot{z}_i}\right)+\frac{\partial V}{\partial z_i} &= 0\end{aligned}\right\} \qquad ...(1.11)$$

There are three such equations for each particle. These results are strictly restricted to Cartesian coordinates. The equations of motions in equation (1.11) can be written in terms of function L as:

$$\left.\begin{aligned}L &= K - V\\ \frac{\partial L}{\partial \dot{x}_i} &= \frac{\partial K}{\partial \dot{x}_i}\end{aligned}\right\}$$

(since V is a function of coordinates only and hence constant)

and
$$\left.\frac{\partial L}{\partial x_i} = -\frac{\partial V}{\partial x_i}\right\}$$

(since K is a function of velocity only and hence constant)

$$\left.\begin{aligned}\frac{d}{dt}\left(\frac{\partial L}{\partial \dot{x}_i}\right)-\frac{\partial L}{\partial x_i} &= 0\\ \frac{d}{dt}\left(\frac{\partial L}{\partial \dot{y}_i}\right)-\frac{\partial L}{\partial y_i} &= 0\\ \frac{d}{dt}\left(\frac{\partial L}{\partial \dot{z}_i}\right)-\frac{\partial L}{\partial z_i} &= 0\end{aligned}\right\} \qquad ...(1.12)$$

These are the Langrangian's equations of motion in Cartesian coordinates and there is always a set of three such equations for each particle. Similarly we can deduce the Hamilton's equations in Cartesian coordinates from the function H = K + V.

1.1.1 GENERALIZED COORDINATES

In many problems it is not convenient to use the Cartesian coordinates $x_1, y_1, z_1 \ldots x_n, y_n, z_n$, *e.g.*, the problem of planetary motion could not be treated in terms of rectangular coordinates since the equations of motion become quite complicated and hence difficult to solve. However, the expressions for forces are simple in polar coordinates. Similarly the expressions for forces in the problem of isotropic spatial harmonic oscillator are simple in terms of polar coordinates but the problem becomes *quite complicated in the treatment of system composed of two attracting particles in space, i.e., the motion is not free but subjected to constraint*. For example, when a particle is moving on a surface, then it is preferable to use a coordinate system such that the conditions of constraint take the forms of a requirement that *one or more of the coordinates remain constant. This type of set of coordinates for specifying the configuration of a system is called the set of generalized coordinates.*

1.1.2 EXAMPLE OF A MOTION SUBJECT TO CONSTRAINTS

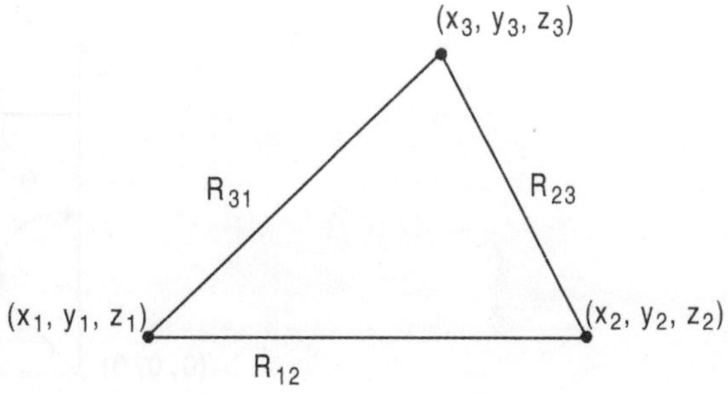

Let us consider the motion of a rigid body. Assume the rigid body as a system of particles moving in such a way that the distance between any two of them remains constant. If these are *n*-particles, their positions are specified by 3n rectangular coordinates, $x_1, y_1, z_1 \ldots x_n, y_n, z_n$. These variables are not independent since we allow them to vary in such a way that the distance between particles remains constant. If we fix the coordinates of three particles which do not lie in a straight line, the coordinates of all other particles can be determined. Among the nine coordinates (three for each particle) of the chosen three particles, there are three relations corresponding to the three distances between the particles so that the position of three particles and hence of the whole body are fixed by only six variables, (*i.e.*, $x_1, y_1, z_1; x_2, y_2, z_2; x_3, y_3, z_3; R_{12}, R_{23}, R_{31}$). The exception to this rule occurs when three particles lies in a straight line. In this case the position of the whole system is determined by five variables, ($x_1, y_1, z_1; x_2, y_2, z_2; x_3, y_3, z_3; R_{12}, R_{23}$). Thus *the number of variables necessary to specify the position of a system is called the number of degrees of freedom of the system*. A rigid body has six degrees of freedom.

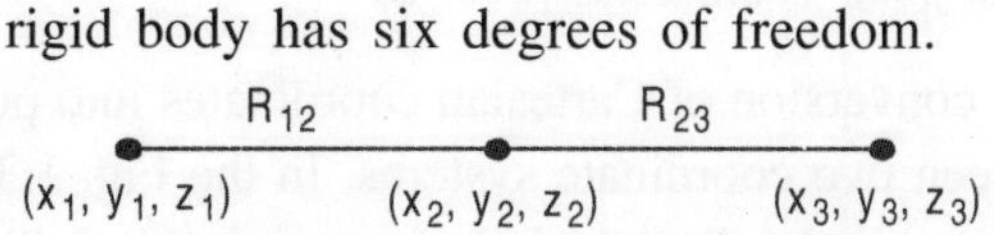

The position of a system of *n* particles with *f* degrees of freedom is determined by *f* coordinates denoted by $q_1, q_2, q_3 \ldots q_f$. The condition of constancy is expressed by $(3n - f)$ other variables, $q_{f+1}, q_{f+2}, q_{f+3} \ldots q_{3n}$. In terms of the *q's* the rectangular coordinates of the particles are given by the 3*n* functions, called the functions of transformations.

$$x_1 = \varphi_1 (q_1, q_2 \ldots q_{3n}),\ y_1 = \varphi_2 (q_1, q_2 \ldots q_{3n}),\ z_1 = \varphi_3 (q_1, q_2 \ldots q_{3n})$$
$$x_2 = \varphi_4 (q_1, q_2 \ldots q_{3n}),\ y_2 = \varphi_5 (q_1, q_2 \ldots q_{3n}),\ z_2 = \varphi_6 (q_1, q_2 \ldots q_{3n})$$
$$x_3 = \varphi_7 (q_1, q_2 \ldots q_{3n}),\ y_3 = \varphi_8 (q_1, q_2 \ldots q_{3n}),\ z_3 = \varphi_9 (q_1, q_2 \ldots q_{3n}) \quad \ldots(1.13)$$

....

....

$$x_n = \varphi_{3n-2} (q_1, q_2 \ldots q_{3n}),\ y_n = \varphi_{3n-1} (q_1, q_2 \ldots q_{3n}),\ z_n = \varphi_{3n} (q_1, q_2 \ldots q_{3n})$$

or we can write:

$$\left.\begin{aligned} x_i &= f_i (q_1, q_2 \ldots q_{3n}) \\ y_i &= g_i (q_1, q_2 \ldots q_{3n}) \\ z_i &= h_i (q_1, q_2 \ldots q_{3n}) \end{aligned}\right\} \quad \ldots(1.14)$$

Thus there is a set of three equations for each particle i. The functions f_i, g_i and h_i may be functions of any or all of the new coordinates q_n, such that these new variables do not necessarily split into sets which belong to the particular particle. For example, in the case of two particles, the six new coordinates may be the three Cartesian coordinates of the centre of mass together with the polar coordinates of the one particle in reference to the other particle as origin.

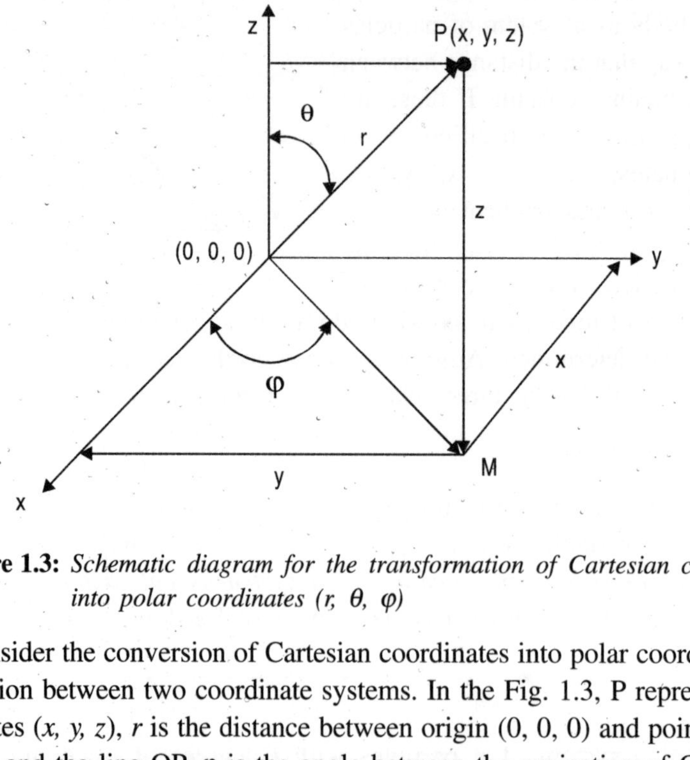

Figure 1.3: *Schematic diagram for the transformation of Cartesian coordinates (x, y, z) into polar coordinates (r, θ, φ)*

Let us first consider the conversion of Cartesian coordinates into polar coordinates. Figure 1.3 represents the relation between two coordinate systems. In the Fig. 1.3, P represents a point with Cartesian coordinates (x, y, z), r is the distance between origin (0, 0, 0) and point P, θ is the angle between the z-axis and the line OP, φ is the angle between the projection of OP in the xy-plane and x-axis. The rectangular coordinates are related to the spherical polar ones by:

$$\left.\begin{array}{ll} x = r \sin\theta \cos\varphi & r = (x^2 + y^2 + z^2)^{1/2} \\ y = r \sin\theta \sin\varphi & \theta = \cos^{-1}\left(\dfrac{z}{(x^2 + y^2 + z^2)^{1/2}}\right) \\ z = r \cos\theta & \varphi = \tan^{-1}(y/x) \end{array}\right\} \quad ...(1.15)$$

Another term which is very commonly used is the integration of some functions over a region of space. In mathematical sense, this is the summation of the function over infinitesimal small volume element, commonly denoted by dV or $d\tau$. In Cartesian coordinate system, it can be very easily expressed by $dx\ dy\ dz$, *i.e.*,

$$dV \text{ (or } d\tau) = dx\ dy\ dz \quad ...(1.16)$$

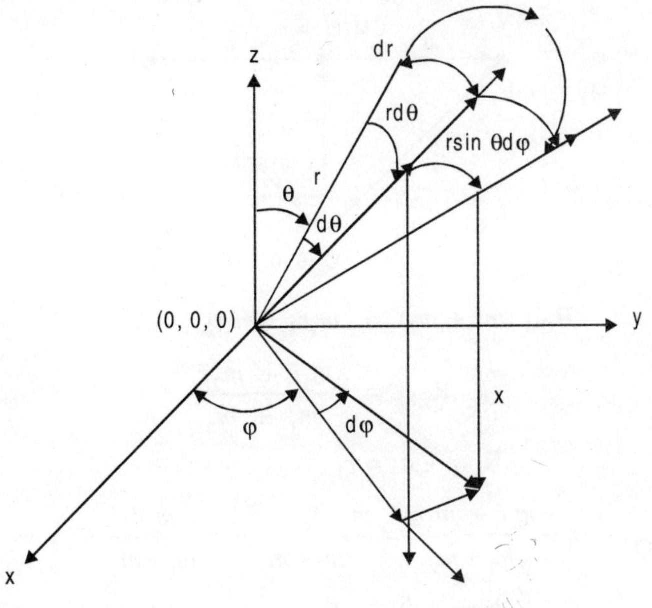

Figure 1.4: *Relation between the Cartesian and spherical polar coordinates for small volume element dτ*

and possesses the unit of volume. The corresponding expression for the volume in terms of spherical polar coordinates is not simply $dr\ d\theta\ d\varphi$, because this does not represents the unit of volume. The value of small volume element $d\tau$ or dV can be easily determined from Fig. 1.4. Let the small volume element can be defined by the space enclosed by r and $r + dr$; θ and $\theta + d\theta$; φ and $\varphi + d\varphi$. Thus the value of this length along the direction of radius is dr. The edge along which θ varies is defined by a circular arc $r\ d\theta$. Lastly the edge along which φ varies equals to its projection in the xy plane, which is circular arc of radius $r \sin\theta$ and angle $d\varphi$, resulting its value as $r \sin\theta\ d\varphi$. In the limits of small dimensions this irregular volume element will be the product of these three edges of length dr, $r\ d\theta$ and $r \sin\theta\ d\varphi$ respectively, *i.e.*

$$d\tau \text{ or } dV = dx\ dy\ dz = (r \sin\theta\ d\varphi)\ (r d\theta)\ (dr) = r^2 \sin\theta\ dr\ d\theta\ d\varphi \quad ...(1.17)$$

Let us consider a two particle system as shown below:

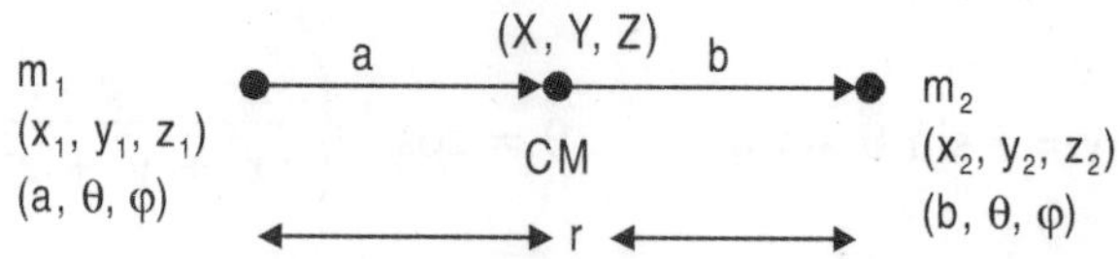

Here the coordinates of centre of mass (CM) of the system are (X, Y, Z). The distance between the two masses (r) is given by:

$$r = [(x_2 - x_1)^2 + (y_2 - y_1)^2 + (z_2 - z_1)^2]^{1/2} \qquad ...(1.18)$$

and the mass concentrated at CM = $m_1 + m_2$. Let CM be the origin and with respect to it a, θ, φ and b, θ, φ are the coordinates of m_1 and m_2.

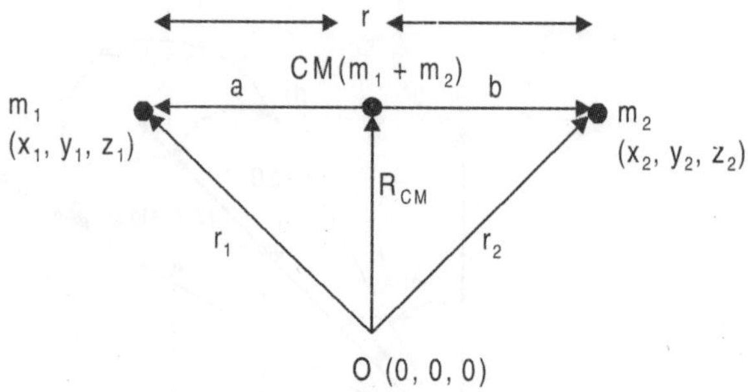

Now $\qquad \mathbf{R}_{CM}\,(m_1 + m_2) = m_1\mathbf{r_1} + m_2\mathbf{r_2}$

Or $\qquad \mathbf{R}_{CM} = \dfrac{m_1\vec{r}_1 + m_2\vec{r}_2}{m_1 + m_2}$

∴ $\qquad \mathbf{R}_{CM} + \boldsymbol{a} = \mathbf{r_1}$

$$\Rightarrow \quad \boldsymbol{a} = \mathbf{r_1} - \mathbf{R}_{CM} = r_1 - \frac{m_1\vec{r}_1 + m_2\vec{r}_2}{m_1 + m_2} = \frac{m_2\,(\vec{r}_1 - \vec{r}_2)}{m_1 + m_2} = \frac{m_2\vec{r}}{m_1 + m_2} \quad (\because r = \vec{r}_1 - \vec{r}_2)$$

Similarly $\qquad \mathrm{R}_{CM} + b = r_2$

$$\boldsymbol{b} = \mathbf{r_2} - \mathbf{R}_{CM} = r_2 - \frac{m_1\vec{r}_1 + m_2\vec{r}_2}{m_1 + m_2} = \frac{m_1\,(\vec{r}_2 - \vec{r}_1)}{m_1 + m_2} = -\frac{m_1\vec{r}}{m_1 + m_2} \quad (\because r = \vec{r}_1 - \vec{r}_2)$$

Thus $\qquad a = \dfrac{m_2 r}{m_1 + m_2}$ and $b = -\dfrac{m_1 r}{m_1 + m_2} \qquad ...(1.19)$

Now if coordinates of one of the particles, say m_1 are taken as origin and of the other particle are taken as polar coordinates r, θ, φ as shown below, then this amounts to the replacement of six Cartesian coordinates (x_1, y_1, z_1; x_2, y_2, z_2) by six new coordinates comprising three Cartesian and three polar coordinates (X, Y, Z, r, θ, φ). Taking moments about CM (in purely Cartesian coordinate system)

$$m_1\,x_1 + m_2\,x_2 = (m_1 + m_2)\,\mathrm{X}$$

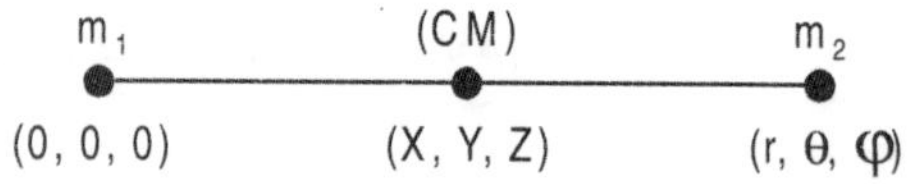

$$\therefore \quad \left.\begin{aligned} X &= \frac{m_1 x_1 + m_2 x_2}{m_1 + m_2} \\ Y &= \frac{m_1 y_1 + m_2 y_2}{m_1 + m_2} \\ Z &= \frac{m_1 z_1 + m_2 z_2}{m_1 + m_2} \end{aligned}\right\} \qquad ...(1.20)$$

In general the coordinate of CM of the system is given by:

$$q = \frac{\Sigma m_i q_i}{\Sigma m_i}$$

where m_i is the mass and q_i is the coordinate of ith particle respectively. The polar coordinates are:

$$\left.\begin{aligned} x_1 - x_2 &= r \sin\theta \cos\varphi \\ y_1 - y_2 &= r \sin\theta \sin\varphi \\ z_1 - z_2 &= r \cos\theta \end{aligned}\right\} \qquad ...(1.21)$$

Eliminating (x_2, y_2, z_2) between two equations (1.20) and (1.21), we obtain:

$$x_1 = x_2 + r \sin\theta \cos\varphi = X + \frac{m_2}{m_1 + m_2} r \sin\theta \cos\varphi$$

or

$$\left.\begin{aligned} x_1 &= X + \frac{\mu}{m_1} r \sin\theta \cos\varphi \\ y_1 &= Y + \frac{\mu}{m_1} r \sin\theta \sin\varphi \\ z_1 &= Z + \frac{\mu}{m_1} r \cos\theta \end{aligned}\right\} \qquad ...(1.22)$$

Elimination of (x_1, y_1, z_1) between the equations (1.20) and (1.21) yields:

$$\left.\begin{aligned} x_2 &= X - \frac{\mu}{m_2} r \sin\theta \cos\varphi \\ y_2 &= Y - \frac{\mu}{m_2} r \sin\theta \cos\varphi \\ z_2 &= Z - \frac{\mu}{m_2} r \cos\theta \end{aligned}\right\} \qquad ...(1.23)$$

Here $\mu = \dfrac{m_1 m_2}{m_1 + m_2}$ is called the *reduced mass*. Equations (1.22) and (1.23) express the rectangular coordinates of masses m_1 and m_2 in terms of new coordinate system. In this new coordinate system, the *Lagrangian* and *Hamiltonian* functions become:

$$H = \frac{m_1 + m_2}{2}\left[\left(\frac{dX}{dt}\right)^2 + \left(\frac{dY}{dt}\right)^2 + \left(\frac{dZ}{dt}\right)^2\right] + \frac{\mu}{2}\left[\left(\frac{dr}{dt}\right)^2 + r^2\left(\frac{d\theta}{dt}\right)^2 + r^2\sin^2\theta\left(\frac{d\varphi}{dt}\right)^2\right] \mp V \qquad ...(1.24)$$

Minus sign is for Lagrangian. These relations can be proved as follows. We know that:

$$x = r\sin\theta\cos\varphi; \qquad y = r\sin\theta\sin\varphi; \qquad z = r\cos\theta$$

Differentiate with respect to t and manipulate the resulting equations to obtain:

$$\left(\frac{dx}{dt}\right)^2 + \left(\frac{dy}{dt}\right)^2 + \left(\frac{dz}{dt}\right)^2 = \left(\frac{dr}{dt}\right)^2 + r^2\left(\frac{d\theta}{dt}\right)^2 + r^2\sin^2\theta\left(\frac{d\varphi}{dt}\right)^2 \qquad ...(1.25)$$

Differentiating equations (1.22) and (1.23), we get:

$$\frac{dx_1}{dt} = \frac{dX}{dt} + \frac{\mu}{m_1}\frac{dx}{dt}; \quad \frac{dx_2}{dt} = \frac{dX}{dt} - \frac{\mu}{m_2}\frac{dx}{dt}$$

$$\left.\begin{aligned}
\left(\frac{dx_1}{dt}\right)^2 &= \left(\frac{dX}{dt}\right)^2 + \frac{\mu^2}{m_1^2}\left(\frac{dx}{dt}\right)^2 + \frac{2\mu}{m_1}\frac{dX}{dt}\frac{dx}{dt}\\
\left(\frac{dy_1}{dt}\right)^2 &= \left(\frac{dY}{dt}\right)^2 + \frac{\mu^2}{m_1^2}\left(\frac{dy}{dt}\right)^2 + \frac{2\mu}{m_1}\frac{dY}{dt}\frac{dy}{dt}\\
\left(\frac{dz_1}{dt}\right)^2 &= \left(\frac{dZ}{dt}\right)^2 + \frac{\mu^2}{m_1^2}\left(\frac{dz}{dt}\right)^2 + \frac{2\mu}{m_1}\frac{dZ}{dt}\frac{dz}{dt}
\end{aligned}\right\} \qquad ...(1.26)$$

Similarly

$$\left.\begin{aligned}
\left(\frac{dx_2}{dt}\right)^2 &= \left(\frac{dX}{dt}\right)^2 + \frac{\mu^2}{m_2^2}\left(\frac{dx}{dt}\right)^2 - \frac{2\mu}{m_2}\frac{dX}{dt}\frac{dx}{dt}\\
\left(\frac{dy_2}{dt}\right)^2 &= \left(\frac{dY}{dt}\right)^2 + \frac{\mu^2}{m_2^2}\left(\frac{dy}{dt}\right)^2 - \frac{2\mu}{m_2}\frac{dY}{dt}\frac{dy}{dt}\\
\left(\frac{dz_2}{dt}\right)^2 &= \left(\frac{dZ}{dt}\right)^2 + \frac{\mu^2}{m_2^2}\left(\frac{dz}{dt}\right)^2 - \frac{2\mu}{m_2}\frac{dZ}{dt}\frac{dz}{dt}
\end{aligned}\right\} \qquad ...(1.27)$$

Multiply equation (1.26) by $m_1/2$ and equation (1.27) by $m_2/2$ and add:

$$H = \frac{m_1}{2}\left[\left(\frac{dx_1}{dt}\right)^2 + \left(\frac{dy_1}{dt}\right)^2 + \left(\frac{dz_1}{dt}\right)^2\right] + \frac{m_2}{2}\left[\left(\frac{dx_2}{dt}\right)^2 + \left(\frac{dy_2}{dt}\right)^2 + \left(\frac{dz_2}{dt}\right)^2\right] + V$$

$$= \frac{m_1 + m_2}{2}\left[\left(\frac{dX}{dt}\right)^2 + \left(\frac{dY}{dt}\right)^2 + \left(\frac{dZ}{dt}\right)^2\right] + \frac{\mu^2}{2}\left[\frac{1}{m_1} + \frac{1}{m_2}\right]\left[\left(\frac{dx}{dt}\right)^2 + \left(\frac{dy}{dt}\right)^2 + \left(\frac{dz}{dt}\right)^2\right] + V$$

or $$H = \frac{m_1 + m_2}{2}\left[\left(\frac{dX}{dt}\right)^2 + \left(\frac{dY}{dt}\right)^2 + \left(\frac{dZ}{dt}\right)^2\right] + \frac{\mu}{2}\left[\left(\frac{dx}{dt}\right)^2 + \left(\frac{dy}{dt}\right)^2 + \left(\frac{dz}{dt}\right)^2\right] + V \qquad ...(1.28)$$

Replacing the second term of equation (1.28) by equation (1.25) we get:

$$H = \frac{m_1 + m_2}{2}\left[\left(\frac{dX}{dt}\right)^2 + \left(\frac{dY}{dt}\right)^2 + \left(\frac{dZ}{dt}\right)^2\right] + \frac{\mu}{2}\left[\left(\frac{dr}{dt}\right)^2 + r^2\left(\frac{d\theta}{dt}\right)^2 + r^2 \sin^2\theta\left(\frac{d\varphi}{dt}\right)^2\right] + V$$

...(1.29)

This is the Hamiltonian function in the new coordinate system.

1.1.3 GENERALIZED MOMENTA AND HAMILTONIAN EQUATIONS

In many body problems it is convenient to express the kinetic energy in terms of momentum instead of velocities. For generalized coordinates, we define the momentum p_k associated with coordinate q_k as:

$$p_k = \left(\frac{\partial L}{\partial \dot{q}_k}\right), \qquad k = 1, 2, 3 ...3n \qquad ...(1.30)$$

Proof. In Cartesian coordinates the momentum related to direction x_k is $m_k\, x_k$. V is a function of coordinates only but not of time.

Now $$L = K - V = 1/2\ m_k\ \dot{x}_k^2 - V$$

$$\therefore \qquad \frac{\partial L}{\partial \dot{x}_k} = 1/2\ m_k\ 2\dot{x}_k = m_k \dot{x}_k = p_k$$

or $$p_k = \frac{\partial L}{\partial \dot{x}_k}, \qquad k = 1, 2, 3 ... 3n \qquad ...(1.30a)$$

Similarly in generalized coordinates we can write:

$$p_k = \frac{\partial L}{\partial \dot{q}_k}, \qquad k = 1, 2, 3 ... 3n \qquad ...(1.30b)$$

Thus the form taken by Lagranges equation (1.12) when the definition of p_k introduced is

$$p_k = \frac{\partial L}{\partial \dot{q}_k}, \qquad k = 1, 2, 3 ... 3n \qquad ...(1.31)$$

Equations (1.30) and (1.31) form a set of 6*n* first order differential equations, equivalent to 3*n* second order equations of equation (1.32)

$$\frac{d}{dt}\left(\frac{\partial L}{\partial \dot{q}_j}\right)-\left(\frac{\partial L}{\partial q_j}\right) = 0, \qquad j = 1, 2, 3 \ldots 3n \qquad \ldots(1.32)$$

$\frac{\partial L}{\partial \dot{q}_j}$, being the general function of both q's and $\dot{q}$'s. The definition of p_k given by equation (1.30) provides 3*n* relations between the variables q_k, $\dot{q}_k$ and p_k, permitting the elimination of 3*n* velocities ($\dot{q}_k$) such that the system can now be described in terms of 3*n* coordinates q_k and 3*n* conjugate momenta p_k. Hamilton in 1834 showed that the equations of motion can in this way be thrown into an especially simple form, involving a function of H of the p_k's and q_k's called Hamiltonian function.

1.1.4 FIELD CONCEPT

The field concept is less directly connected to our every day experience than the concept of a particle, but it can be easily realized by our senses. For example, if one feels the attraction of a piece of iron by a magnet, one obtains the immediate impression that there is something surrounding the magnet that acts on the iron. Infra-red rays emitted by a body heated to 60 to 70°C can be felt as heat if the hand is held at short distance. Further, fields are of two categories: scalar and vector. Examples of the former are temperature, pressure, density distribution etc. and that of latter are force of interactions. The relative strengths of vector fields are compared in Table 1.1.

Table 1.1: *Relative strengths of force of interactions*

Interaction	Relative Strength	Remarks
Gravitational	1	Macroscopic
Weak (nuclear)	10^{27}	Microscopic
Electromagnetic	10^{38}	Microscopic
Strong (nuclear)	10^{40}	Microscopic

The physical properties such as position, velocity, energy etc. are regarded as the measurable properties of any physical system and it is assumed that one can simultaneously determine as many properties of system as one desires. The macroscopic word of earth and heaven, *i.e.* planets, is governed by the laws of classical mechanics. When we describe various phenomena in terms of classical mechanics, we assume that physical processes are independent of the conditions of observation. Thus we can take it for granted that we always spy on the process and yet not interfere with it or influence it. If we spy on a physical process from different view points (and correspondingly use different frames of reference for its description) it will appear to us in different ways. For instance, the free fall of a body may proceed in a straight line in one frame of reference and in a parabola in another. But the dependence of the form of a phenomenon on the frame of reference has always been taken into consideration, namely transforming from the coordinates of one frame of reference to the coordinates of another frame of reference. Such a change in form introduces nothing new into the phenomenon. For this reason, in classical mechanics, we can speak of the independence of phenomenon from the manner of observation. On the other hand, quantum mechanics shows that in microprocesses this is not the case. Here the very possibility of observation presupposes definite physical conditions that may be related to the essence of the process. Specifying these conditions does not mean simply indicating a particular frame of reference but requires more detailed elaboration. Although classical mechanics is more than 425 years old and has been successful in solving variety of problems, but it is not universal in its applications, *e.g.*, the description of motion of a body given by classical mechanics has proved to be inadequate when the speed of the body approaches the speed of light. This limitation of classical mechanics leads to the evolution of relativistic mechanics. Some of the important deductions of relativistic mechanics are:

(1) Mass of a body depends on its relative velocity, *i.e.*,

$$m = m_0\left(1 - \frac{v^2}{c^2}\right)^{-\frac{1}{2}} = \gamma m_0 \qquad ...(1.33)$$

γ is called Lorentz constant, m_0 the rest mass is taken as the mass measured by an observer relatively at rest. It is the smallest possible value of m for any object; c is the velocity of light. When $v/c \to 0$, $m \to m_0$. The rest mass of some atomic particles is compared here in Table 1.2.

(2) Relativistic kinetic energy is given as:

$$\text{K} = m_0\left(1-\frac{v^2}{c^2}\right)^{-\frac{1}{2}} c^2 - m_0c^2 = (m-m_0)c^2 \qquad ...(1.34)$$

Table 1.2: *Comparison of rest mass of some atomic particles*

Particle	Rest mass (a.m.u.)	Rest mass (kg)	Rest energy (MeV)
Atomic mass unit	1	1.660×10^{-27}	931
Electron	0.00055	9.109×10^{-31}	0.511
Neutron	1.00867	1.675×10^{-27}	939.6
Proton	1.00729	1.673×10^{-27}	938.3

1 MeV = 10^6 eV, 1G eV = 10^9 eV

We can easily reduce this expression for the kinetic energy to classical form, K = $\frac{1}{2}m_0c^2$, when $v << c$. Applying the binomial expression to equation (1.34) we obtain:

$$\text{K} = m_0c^2\left[\left(1 - \frac{v^2}{c^2}\right)^{-\frac{1}{2}} - 1\right]$$

or
$$K = m_0c^2\left[1+\left(-\frac{1}{2}\right)\left(-\frac{v^2}{c^2}\right)+\frac{\left(-\frac{1}{2}\right)\left(-\frac{3}{2}\right)}{2!}\left(-\frac{v^2}{c^2}\right)^2+...-1\right]$$

$$= m_0c^2\left[1+\frac{v^2}{2c^2}+\frac{3v^4}{8c^4}+...-1\right]$$

$$= m_0c^2\left[\frac{v^2}{2c^2}+\frac{3v^4}{8c^4}+...\right]$$

As $\frac{v}{c} \to 0$, the higher power of $\frac{v}{c}$ can be neglected and then

$$K = m_0c^2\left(\frac{1}{2}\frac{v^2}{c^2}\right)=\frac{1}{2}m_0v^2 \quad ...(1.35)$$

This is also known as a statement of corresponding principle.

(3) Total energy is defined as:

$$E = E_0\ (= mc^2) + K \quad ...(1.36)$$

where K is the kinetic energy at a given point. The definition of total energy in relativistic mechanics does not include potential energy and is given by:

$$E = mc^2 \quad ...(1.36a)$$

i.e., this energy is made up of the rest energy m_0c^2 and the energy of motion. Without having any information about the structure of the body and the nature of interactions of its component particles, we can state that its internal energy is equal to m_0c^2.

The equivalence between mass and energy expressed by equation (1.36) is one of the most important consequences of the special theory of relativity. It gives a different and more insight into conservation of energy, also known as the principle of mass-energy conservation, which for an isolated system is stated as:

$$\Sigma\ (\text{rest energy + kinetic energy + potential energy}) = \text{constant} \quad ...(1.37)$$

Another useful relationship that includes the total energy E can be obtained directly from the relation

$$m_0\left(1-\frac{v^2}{c^2}\right)^{-\frac{1}{2}} = m \quad ...(1.33)$$

Multiplying both sides of equation (1.33) by c^2 and followed by squaring and simplifying we obtain:

$$m^2c^4 = m_0^2c^4 + m^2v^2c^2$$

Since $p = mv$, this expression can be written as:

$$E^2 = E_0^2 + p^2c^2 \qquad ...(1.38)$$

If the body is moving at very high speed, then E_0^2 can be neglected as compared to $p^2\ c^2$, consequently,

$$E^2 = p^2c^2 \text{ or } E = pc \qquad ...(1.39)$$

At high speeds E_0 is also small as compared to the kinetic energy K at a given point and equation (1.36) shows that:

$$E \approx K \text{ or from equation (1.39) } K \approx pc \qquad ...(1.40)$$

Particles moving at very high speeds for which equations (1.39) and (1.40) are useful, are said to be in the extreme relativistic region. Another interesting relationship involving the total energy is obtained by differentiating equation (1.39) with respect to p. This is:

$$2E\frac{dE}{dp} = 2c^2p \Rightarrow \frac{dE}{dp} = \frac{pc^2}{E}$$

or

$$\frac{dE}{dp} = \frac{pc^2}{mc^2} = v \qquad ...(1.41)$$

Now if the body is moving at the speed of light, *i.e.*, $v = c$, then

$$dE = c\ dp \qquad ...(1.42)$$

or

$$\int dE = c\int dp + \text{const.}$$

$$E = cp + \text{const.} \qquad ...(1.43)$$

For $p = 0$. $E = E_0$ and therefore const. $= E_0$

So equation (1.43) can be expressed as:

$$E - E_0 = pc \qquad ...(1.44)$$

But according to equation (1.38),

$$E^2 - E_0^2 = p^2c^2 \qquad ...(1.45)$$

Combining equations (1.44) and (1.45), we get:

$$E + E_0 = pc \qquad ...(1.46)$$

Comparison of equations (1.44) and (1.46) implies that $E_0 = 0$ or $m_0 = 0$, *i.e.*, if a body is moving at the speed of light, its rest mass and its rest energy must be zero. The reciprocal conclusion must also be true: *If an entity has neither rest mass nor rest energy, it must be travelling at the speed of light*. Although it does not make sense for a body to have a mass of zero from classical point of view, it is the correct description of a photon and of a neutrino. Experiments based on Mössbauer Effect showed that the mass of a photon moving at the speed of light (this is the only speed at which it moves) is given by $m = \frac{h\nu}{c^2}$, in agreement with the theoretical equation.

$$E = h\nu = mc^2 \qquad ...(1.46a)$$

Another limitation of the classical mechanics is that it failed to interpret the properties of matter of very small dimensions, *i.e.*, microscopic world. For example, classical mechanics fails to resolve phenomenon of black body radiation, the structure of atom, the properties of electrons and atomic nuclei, properties of other particles such as protons, quarks etc. The very stability of a system consisting of a positively charged nucleus and negatively charged electrons, radiation of light emitted by atoms and molecules, diffraction of electrons, stability of DNA helix etc.

1.2 POST CLASSICAL MECHANICS AND PRE-QUANTUM MECHANICAL ERA

The failure of classical mechanics to explain the properties and phenomenon associated with the microscopic physical systems led to the birth of *old quantum theory*. This theory has been able to resolve the discrepancies between the theoretical results based on classical mechanics and experimental observations to a great extent at least in simple physical systems. This theory stemmed from the laws of classical mechanics in conjunction with new concepts, based on the works of Max Planck, Albert Einstein and Neils Bohr.

According to wave theory, radiant energy can be compared to rippling stream of water, but according to quantum theory of Planck it is comparable to rain, the drops corresponding to quanta and just as the drops vary in size, so do the quanta vary in energy. Again according to quantum theory, radiation is emitted when the radiant system drops from one stable level to another. The theory was originally developed for black body radiation spectrum of an incandescent body. Einstein used it to explain the variations in specific heats of solids and gases with temperature and to explain the photoelectric effect. It has also been used to explain the photochemical reactions. Bohr used it to explain the atomic spectrum. The three new postulates of old quantum theory are:

1.2.1 BLACK BODY RADIATIONS

The first and fundamental postulate of the theory was proposed by Max Planck, in 1900. According to him radiation consists of particles and the energy of these particles is quantized. The classical oscillators (energy of a classical oscillator is k_b T, where k_b is Boltzmann's constant and T the temperature in absolute) in a black body do not emit radiation continuously as demanded by classical theory, but each black body harmonic oscillator emit light in discrete amounts called *photon* or *quantum* of radiation having the magnitude of $h\nu$ ($k_b T \ll h\nu$), where ν is the frequency of radiation and h is a constant (now called Planck's constant). This constant is also known as *action constant* since it has the dimensions of energy × time ($MLT^{-2}\ LT = ML^2T^{-1}$) *i.e.*, action.

This hypothesis of quantum oscillators led Planck to derive the energy distribution law, called Planck's energy distribution law of black body radiation.

$$E\lambda \cdot d\lambda = \frac{c_1}{\lambda^5} \frac{d\lambda}{e^{c_2/\lambda T} - 1} \qquad ...(1.47)$$

where $c_1 = 8\pi ch$, $c_2 = \dfrac{hc}{k_b}$ are constants, E_λ is the energy density per unit wavelength of radiation in a small wavelength range $d\lambda$, *i.e.*, lying between the limits of wavelength λ and $\lambda + d\lambda$. This relation agrees well admirably with experimental results for all wavelengths, *i.e.*, it explains the experimental and theoretical shape of curves of the power radiated from a black body as a function of wavelength to a great extent as it is evident from Fig. 1.5. The expression obtained by Planck to explain the energy distribution of black body is based on two points.

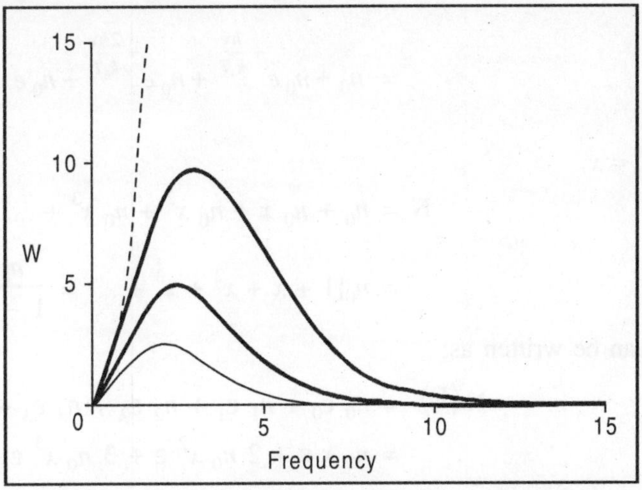

Figure 1.5: *The graph indicates how Planck's energy distribution law, which is based on the idea of quantized oscillators, agreed with experiment. Frequency is in the units of 10^{14} Hz and intensity in the units of 10^{-16} J per unit volume per unit frequency. Dotted line represents the Rayleigh-Jean law. Thick line represents at 5000 K, medium line at 4000 K and thin line at 3000 K. The maximum shifts towards higher frequency (energy) with increase of temperature*

(*i*) Calculation of average energy of the oscillator.

(*ii*) Number of vibrational modes of the oscillator over the wavelength range of λ and $\lambda + d\lambda$.

The first point is given by the equation (1.48).

$$\langle E \rangle = \frac{\Sigma n_i \varepsilon_i}{\Sigma n_i} \qquad ...(1.48)$$

where n_i is the number of molecules present in the energy level ε_i.

$$N = \Sigma n_i = n_0 + n_1 + n_2 + ... \qquad ...(1.49)$$

The number of molecules in an energy level can be calculated with the help of Maxwell-Boltzmann distribution law, *i.e.*,

$$n_i = n_0 \frac{e^{-\frac{\varepsilon_i}{k_b T}}}{e^{-\frac{\varepsilon_0}{k_b T}}} \qquad ...(1.50)$$

Assuming that energy is measured from zeroth level, *i.e.*, $\varepsilon_0 = 0$, and the energy difference in the successive levels is the same, *i.e.*, $hv = \varepsilon$, the equation (1.49) can be written as:

$$N = \Sigma n_i = n_0 e^{-\frac{\varepsilon_0}{k_b T}} + n_0 e^{-\frac{\varepsilon_1}{k_b T}} + n_0 e^{-\frac{\varepsilon_2}{k_b T}} + ...$$

$$= n_0 + n_0 e^{-\frac{hv}{k_b T}} + n_0 e^{-\frac{2hv}{k_b T}} + n_0 e^{-\frac{3hv}{k_b T}} + ...$$

Substituting $e^{-\frac{hv}{k_b T}} = x$

$$N = n_0 + n_0 x + n_0 x^2 + n_0 x^3 + ...$$

$$= n_0[1 + x + x^2 + x^3 + ...] = \frac{n_0}{1 - x} \qquad ...(1.51)$$

Equation (1.48) can be written as:

$$N\langle E\rangle = n_0 \varepsilon_0 + n_1 \varepsilon_1 + n_2 \varepsilon_2 + n_3 \varepsilon_3 + ...$$

$$= n_0 x \varepsilon + 2 n_0 x^2 \varepsilon + 3 n_0 x^3 \varepsilon$$

$$= n_0 x \varepsilon [1 + 2x + 3x^2 + ...] \qquad ...(1.52)$$

Now $1 + 2x + 3x^2 + ... = \frac{d}{dx}[1 + x + x^2 + x^3 + ...] = \frac{d}{dx}\left[\frac{1}{1 - x}\right] = \frac{1}{(1 - x)^2}$

Substituting this in equation (1.52):

$$N \langle E\rangle = \frac{n_0 x \varepsilon}{(1 - x)^2} \qquad ...(1.53)$$

Combining equations (1.51) and (1.53)

$$\langle E\rangle \frac{n_0}{1 - x} = \frac{n_0 x \varepsilon}{(1 - x)^2}$$

or

$$\langle E\rangle = \frac{x \varepsilon}{(1 - x)} = \frac{hv e^{-\frac{hv}{k_b T}}}{1 - e^{-\frac{hv}{k_b T}}} \qquad ...(1.54)$$

This hypothesis of Planck theory enables us to explain why a piece of iron becomes first red, then yellow, then blue and finally white when it is heated to higher temperature and provided it does not melt. At room temperature the thermal energy is less than the energy of the quantum of visible light. Iron at room temperature does emit infra-red radiations, whose quanta are of lower energy than that of visible radiation. We cannot see it but we can measure it and we can feel it as heat radiation. At higher temperature, thermal energy reaches the energy of a visible quantum. Red light has the lowest quantum in the visible range, and then comes yellow, followed by blue. Thus red is the first colour to be seen when the iron is heated. As the temperature is increased, iron becomes yellow-blue and then white, because the combination of red-yellow and blue is white. Due to this very reason carbon filament is not used (but tungsten is used) in incandescent lamps since carbon filament can be heated to 2110 K, while tungsten filament can be heated way up to 2500 K. Why are these 400 degrees so important? Because the purpose of an incandescent lamp is to provide visible light and not heat, and so the aim is to have the maximum of the curve located in the visible portion of the radiation (Fig. 1.6). It can be seen from the graph that the idea is to have a filament that can withstand temperature of sun's surface or 6000 K. But for all practical purposes even the filaments that can withstand temperatures of 2500 K degrees would suffice since the step from 2100 to 2500 degrees raises the portion of energy involving visible radiation from 0.5% to 1.6%. The various energy distribution laws of black body radiation based on classical theory of continuous oscillators can be deduced from Max Planck's law of energy distribution equation (1.47).

(*i*) When λ is small, $\frac{1}{\lambda}$ is large. Neglecting 1 in comparison to $e^{\frac{c_2}{\lambda T}}$, we obtain

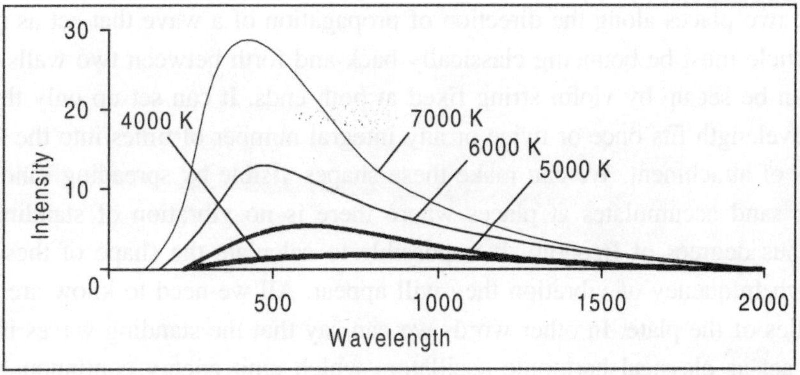

Figure 1.6: *The graph showing the Planck's energy distribution at different temperature*

From equation (1.47):

$$E_\lambda \, d\lambda = \frac{c_1 \, d\lambda}{\lambda^5 \, e^{\frac{c_2}{\lambda T}}} \qquad ...(1.55)$$

which is Wein's radiation law based on the principle that every atom/molecule in a black body behaves as an oscillator and hence light in the body is emitted by classical oscillators. The energy distribution according to this law has also been shown in Fig. 1.5. The formula agrees well with experimental results for waves of length shorter than that corresponding to the point of maximum emission, but it breaks down for longer wavelengths and for higher temperatures.

(*ii*) Again if λ is large, 1/λ is small. Expanding exp ($c_2/\lambda T$) and neglecting higher terms of exponential, we obtain

$$e^{\frac{c_2}{\lambda T}} = \left(1+\frac{c_2}{\lambda T}\right)-1=\frac{c_2}{\lambda T}$$

Accordingly equation (1.47) reduces to:

$$E_\lambda \, d\lambda = \frac{c_1 \, \lambda \, T \, d\lambda}{\lambda^5 \, c_2} = \frac{a \, T \, d\lambda}{\lambda^4} \qquad \text{...(1.56)}$$

which is Rayleigh-Jean's law and $a = c_1/c_2$.

$$E_\lambda \, d\lambda = k_b \, T \, dn = \frac{a \, T \, d\lambda}{\lambda^4}$$

where dn is the total number of oscillators at any particular range of wavelengths λ and λ + $d\lambda$. It is based on the principle of standing waves that are possible in the cavity of black body and energy is equally distributed among the frequencies.

Standing Waves: The wave is reflected back and forth, somehow, to set up standing waves. There must be two places along the direction of propagation of a wave that act as mirrors for the wave. The particle must be bouncing classically back and forth between two walls. For example, such waves can be set up by violin string fixed at both ends. It can set up only those vibrations whose half wavelength fits once or twice or any integral number of times into the space between the two points of attachment. We can make these shapes visible by spreading sand uniformly on the plate. The sand accumulates at places where there is no vibration of standing waves (Fig. 1.7), *i.e.*, various degrees of freedom. It is possible to calculate the shape of these patterns and predict at which frequency of vibration they will appear. All we need to know are the shape and elastic properties of the plate. In other words we can say that the standing waves in the cavity of a black body act as classical harmonic oscillators which emit energy continuously. The energy distribution according to this law is also shown in Fig. 1.5. The energy curve corresponding to this formula runs to infinity and shows no point of maximum emission. Again according to this formula, the energy radiated in any wavelength $d\lambda$ increases rapidly as λ decreases and approaches infinitely for very small wavelengths, which obviously is incorrect. This wrong prediction is called *ultraviolet catastrophe*. Further if the expression (1.56) be integrated over the whole range of wavelengths from 0 → ∞, the energy comes out to be infinity for all temperatures except at absolute zero and this is clearly absurd.

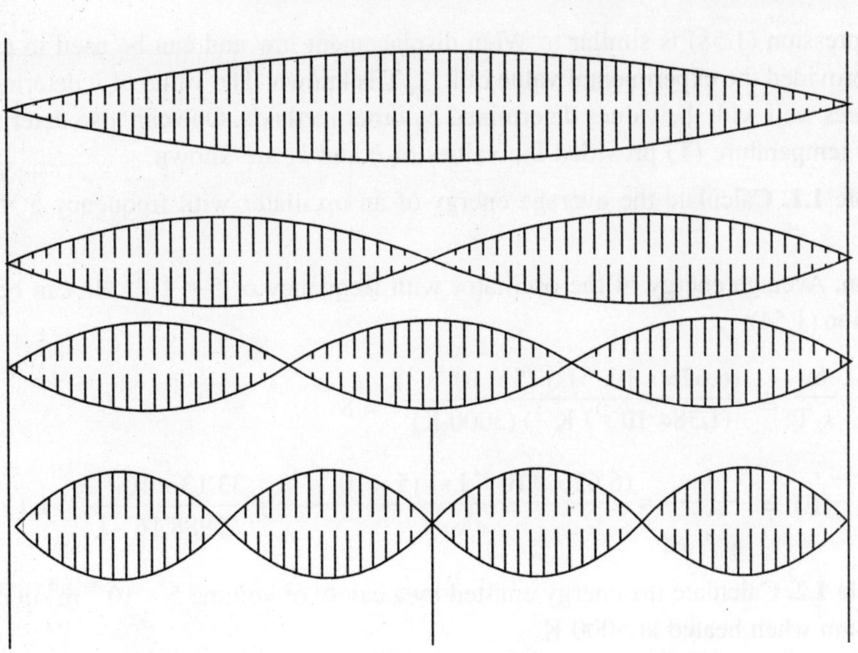

Figure 1.7: *Waveforms of a string fixed at both ends*

(*iii*) Further it is also possible to deduce Wien displacement law, according to which the wavelength at the point of maximum radiation, λ_{max} decreases as the temperature increases. (Fig. 1.6) or

$$\lambda_{max} \propto \frac{1}{T} \quad \text{or} \quad \lambda_{max}\, T = \text{constant} \qquad \text{...(1.57)}$$

Differentiating the expression (1.47) with respect λ and applying the condition of maximum, *i.e.*, equating to zero.

$$\frac{d[E_\lambda]}{d\lambda} = 8\pi hc\left[-\frac{5}{\lambda^6}e^{-\frac{hc}{\lambda k_b T}} + \frac{1}{\lambda^5}e^{-\frac{hc}{\lambda k_b T}}\left(-\frac{hc}{k_b T}\right)\left(-\frac{1}{\lambda^2}\right)\right]$$

$$0 = \frac{8\,\pi hc}{\lambda_{max}^6}e^{-\frac{hc}{\lambda_{max} k_b T}}\left[-5+\frac{hc}{k_b\, T\,\lambda_{max}}\right]$$

or
$$\frac{hc}{k_b\, T\,\lambda_{max}} = 5 \Rightarrow \lambda_{max}\, T = \frac{hc}{5\, k_b} = \text{constant} \qquad \text{...(1.58)}$$

or
$$\lambda_{max} = \frac{2886}{T} \qquad \text{...(1.59)}$$

provided λ is in micrometers and T is in Kelvins. At the lowest temperature the maximum lies in the infra-red region. That is precisely why infra-red radiation is also termed as thermal radiation. Some animals have a highly sensitive perception of the thermal rays because of which they are able to feel the presence of another body (who is also a source of radiation) in darkness.

The expression (1.58) is similar to Wien displacement law and can be used to calculate the value of h provided the experimental value of λ_{max} T is known. The value of h determined by this method agrees well with the values determined by other methods. One can also determine λ_{max} at a particular temperature (T) provided the values of h and k_b are known.

Example 1.1. Calculate the average energy of an oscillator with frequency 5×10^{15} Hz at 3000 K.

Solution. Average energy of the oscillator with frequency of 5×10^{15} Hz can be calculated using equation (1.54).

$$\frac{h\nu}{k_b T} = \frac{(6.626 \times 10^{-34}\,\text{Js})\,(5 \times 10^{14}\,\text{s}^{-1})}{(1.38 \times 10^{-23}\,\text{J K}^{-1})\,(3000\,\text{K})} = 8$$

Further $$\langle E \rangle = \frac{h\nu}{e^{\frac{h\nu}{k_b T}} - 1} = \frac{(6.626 \times 10^{-34}\,\text{J s})\,(5 \times 10^{14}\,\text{s}^{-1})}{e^{8} - 1} = \frac{33.13 \times 10^{-20}\,\text{J}}{2988.16 - 1} = 1.11 \times 10^{-22}\,\text{J}$$

Example 1.2. Calculate the energy emitted by a cavity of volume 5×10^{-6} m^3 in the range of 380 to 420 nm when heated at 5000 K.

Solution. The energy emitted in the range of 380 to 420 nm can be obtained by using equation (1.47). The mean value of wavelength can be obtained by taking the average of wavelength range, *i.e.*, $\frac{1}{2}$ (380 + 420) = 400 nm. The equation (1.47) can be solved by dividing it as:

$$\frac{hc}{k_b T \lambda} = \frac{(6.626 \times 10^{-34}\,\text{Js})\,(3 \times 10^{8}\,\text{ms}^{-1})}{(1.38 \times 10^{-23}\,\text{J K}^{-1})\,(5000\,\text{K})\,(400 \times 10^{-9}\,\text{m})} = 7.20$$

$$e^{\frac{hc}{k_b T \lambda}} - 1 = e^{7.2} - 1 = 1339.5 - 1 = 1338.5$$

$$\frac{8\,\pi hc}{\lambda^5} = \frac{(8)\,(22/7)\,(6.626 \times 10^{-34}\,\text{Js})\,(3 \times 10^{8}\,\text{ms}^{-1})}{(400 \times 10^{-9}\,\text{m})^5} = 0.488 \times 10^{9}\,\text{Jm}^{-4}.$$

$$d\varepsilon = \frac{8\,\pi hc}{\lambda^5} e^{-\frac{hc}{k_b T \lambda}} d\lambda = (0.488 \times 10^{9}\,\text{Jm}^{-4})\,(7.466 \times 10^{-4})\,(40 \times 10^{-9}\,\text{m})$$

$$= 1.45 \times 10^{-2}\,\text{Jm}^{-3}$$

Therefore the energy emitted by a cavity of volume 5×10^{-6} m^3 is

$(1.45 \times 10^{-2}\,\text{Jm}^{-3})\,(5 \times 10^{-6}\,\text{m}^3) = 7.25 \times 10^{-8}$ J.

1.2.2 HEAT CAPACITY OF SOLIDS

Einstein tried to explain the variation of heat capacity of solids with temperature on the concept of quantum oscillators put forward by Max Planck. For many solid elements and compounds heat capacity have been found to be dependent on temperature as shown in Fig. 1.8. At 0 K, heat capacity of all solids becomes zero and at higher temperatures it approaches a limiting value of 25 J K^{-1} mol^{-1}. But according to Dulong-Petit's empirical law (which is based on classical mechanics), the heat capacity of all solids should be 25 J K^{-1} mol^{-1}. Thus classical mechanics fails to explain the behaviour of heat capacity with temperature.

The theoretical justification for the independence of heat capacity of some solids from temperature is very simple and it can be explained as follows. A crystal is composed of orderly array (lattice) of atoms or ions. The equilibrium position of each atom is called lattice site. There is repulsion between each atom and thus crystal structure, where each atom occupies a lattice site, depicts one of the minimum potential energy. Now at any other temperature, except at absolute zero temperature where kinetic energy is zero, the atoms possess kinetic energy. Due to this, they move about their mean positions, collide with other atoms and get rebound. To a good approximation, this may be considered as atoms vibrating about their mean positions. Since each particle has three independent degrees of translational motion, one can assign three independent degrees of vibrational motion and let us assume that these are having simple harmonic motion. If there is Avogadro's number (N_{av}) of atoms in a crystal, there will be 3 N_{av} harmonic oscillators. According to equipartition principle each harmonic oscillator can be assigned energy of magnitude k_b T. The total energy for 3 N_{av} oscillators will be 3 N_{av} k_b T. Heat capacity, which is defined as the energy required to raise the temperature of a given amount of substance by one degree, will be 3 N_{av} k_b = 3 R = 24.943 J K^{-1} mol^{-1}. This agrees nicely with the value given by Dulong-Petit law.

Einstein was the first, who tried to explain these results on the basis of quantum theory. The assumptions involved in his derivation are:

(*a*) Each atom or ion at the lattice point can be considered as an oscillator and its behaviour is independent of the presence of other atoms or ions.

(*b*) Vibrational motion of each oscillator can be described by simple harmonic motion.

(*c*) Vibrational frequency of oscillation for each oscillator is ν.

Using these assumptions, the total energy of 3 N_{av} oscillators at any temperature will be:

$$E = 3\,N_{av}\,\langle E\rangle = 3\,N_{av}\,h\nu\,\frac{e^{-h\nu/k_b T}}{1-e^{-h\nu/k_b T}} = 3\,N_{av}\,h\nu\,\frac{1}{e^{h\nu/k_b T}-1} \qquad ...(1.60)$$

where $\langle E\rangle$ is the average energy of each oscillator. Heat capacity will be:

$$C_V = \left(\frac{\partial E}{\partial T}\right)_v = 3R\left(\frac{h\nu}{k_b T}\right)^2 \frac{e^{h\nu/k_b T}}{[e^{h\nu/k_b T}-1]^2} \qquad ...(1.61)$$

Later on Debye explained the variation of C_V with temperature exactly on the same lines as carried out by Einstein, except that these oscillators are not independent but their oscillations are affected by the presence of other oscillators. The vibrational frequencies of the oscillator vary from 0 to λ_{max}. That is some oscillators are not vibrating and the other do with different frequencies. The expression thus obtained is:

$$C_V = \frac{9\,RT^3}{\theta_D^3}\int_0^{\theta_D/T} x^4 e^x\,[e^x - 1]^{-2} dx \qquad \text{...(1.62)}$$

where $x = \dfrac{h\nu}{k_b T}$, θ_D is the Debye's characteristic temperature. The variation of C_V with temperature is thus shown in Fig. 1.8 and shows that the agreement with experimental curve is good. The values predicted by Debye's expression is little higher than that predicted by Einstein's equation (1.61). C_V has T^3 dependence at low temperatures. The physical meaning of this model or better agreement with experimental results can be explained on the similar lines as done for black body. At low temperature, the thermal energy present is small in comparison to the energy of each oscillator ($h\nu$). Thus, at low temperature, these oscillators are not populated and do not contribute to the total energy. Due to this the heat capacity will change with temperature, whereas on the basis of classical mechanics, the total energy will be a cumulative effect of addition of small amounts at low temperature. The difference between Einstein's and Debye's results is because of the presence of oscillators of different frequencies, *i.e.*, even at low temperature (low energy), there will be enough oscillators to absorb this energy and contribute to the total energy. Due to this very reason, the value of heat capacity predicted by Debye's model will be little more than that predicted by Einstein's model. But one thing is certain that both these expressions prove one point that the radiation or energy does not increase or decrease continuously but changes by absorbing or emitting quantum of value ($h\nu$).

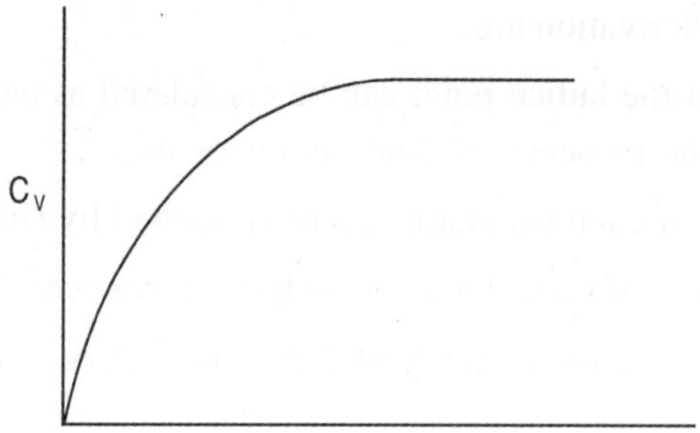

Figure 1.8: *Behaviour of heat capacity of solid with temperature*

Example 1.3. Using equation (1.61) show that the limiting value of heat capacity of metals agrees with that obtained by Dulong-Petit.

Solution. If the temperature is high, $h\nu/k_b T << 1$, the exponential term can be expanded.

$$e^{\frac{h\nu}{k_b T}} = 1 + \frac{h\nu}{k_b T} + \ldots \text{ Neglecting higher terms.}$$

$$C_V = 3R\left(\frac{h\nu}{k_b T}\right)^2 \frac{\left(1 + \dfrac{h\nu}{k_b T}\right)}{\left(1 + \dfrac{h\nu}{k_b T} - 1\right)^2} = 3R = 3 \times 8.314 \text{ J K}^{-1}\text{ mol}^{-1} = 24.942 \text{ J K}^{-1}\text{ mol}^{-1}$$

Dulong-Petit's value is at constant pressure, *i.e.*,

$$C_P = 6.4 \text{ cal K}^{-1} \text{ mol}^{-1} = 26.765 \text{ J K}^{-1} \text{ mol}^{-1}$$

$C_P - C_V$ for solids can be calculated using coefficient of expansion (α), coefficient of compression (β) and equation $C_P - C_V = \dfrac{\alpha^2 VT}{\beta}$ and is:

$$C_P - C_V \approx 0.4 \text{ cal K}^{-1} \text{ mol}^{-1} = 1.673 \text{ J K}^{-1} \text{ mol}^{-1}$$

$$C_V = 26.765 - 1.673 = 25.09 \text{ J K}^{-1} \text{ mol}^{-1}$$

Thus, the agreement is not bad.

Example 1.4. Show that at low temperature, heat capacity of solids at constant volume is proportional to cube of absolute temperature.

Solution. At low temperature $\dfrac{h\nu}{k_b T}$ is very large ($\rightarrow \infty$), the integral can be approximated to:

$$\int_0^\infty \frac{x^4 e^x}{[e^x - 1]^2} dx = 4/15\, \pi^4$$

So the equation (1.62) becomes:

$$C_V = 9R\left(\frac{k_b T}{h\nu}\right)^3 \left(\frac{4\pi^4}{15}\right) = \frac{36}{15}\pi^4 R \left(\frac{k_b}{h\nu}\right)^3 T^3$$

i.e., $C_V \alpha T^3$. The behaviour of C_V with temperature, as predicted by Einstein's expression, is also shown in Fig. 1.8. Though the agreement was much better than Dulong-Petit, but still differs at low temperatures.

Example 1.5. The Debye characteristic temperature for diamond is 1860 K. Calculate the heat capacity of diamond at 6 K.

Solution. Use the expression

$$C_V = \frac{36}{15}\pi^4 R \left(\frac{k_b}{h\nu}\right)^3 T^3, \quad \left[\theta_D = \frac{h\nu}{k_b}\right]$$

$$C_V = \left(\frac{36}{15}\right)\left(\frac{22}{7}\right)^4 (8.314 \text{JK}^{-1} \text{ mol}^{-1}) \left(\frac{1}{1860 \text{ K}}\right)^3 (6\text{K})^3$$

$$= 6.52 \times 10^{-5} \text{ JK}^{-1} \text{ mol}^{-1}.$$

1.2.3 PHOTOELECTRIC EFFECT AND COMPTON EFFECT

The second postulate of old quantum theory was proposed by Einstein in 1905 in order to explain the phenomenon of emission of electrons from a clean metal surface when light falls on it, *i.e., photoelectric effect* which was discovered by Hertz in 1887. The postulate states that radiation consists of bundles or quanta of localized energy ($E = h\nu$), called photon (the term photon has been introduced by G.N. Lewis in 1926) that travels with speed of light and that emission of a photoelectron occurs as a result of interaction of single quantum, *i.e.*, of the photon with a matter.

Classical theory, *i.e.*, electrodynamics demands that energy of light should be spread out through the wave, consequently: (*i*) the energy of emitted electrons should be directly proportional to the intensity of incident radiations and (*ii*) if the intensity of incident radiation is weak, time will be required to build up sufficient energy in metal to eject electrons. On the other hand, according to Einstein hypothesis, the observed maximum kinetic energy (K) of the ejected photoelectron varies linearly with frequency of the incident radiation and one may write

$$K = \frac{1}{2} m_e v^2 = h\nu - h\nu_0 = h\nu - \varphi \qquad ...(1.63)$$

where $h\nu_0 = \varphi$ is a constant, which is characteristic of the metal and is called work function of metal. Actually it is a measure of the work done to remove the electron. The frequency ν_0 is called the cut off or threshold frequency and for $\nu < \nu_0$ no photoelectron will be emitted. For example for sodium metal, $\nu_0 = 4.4 \times 10^{14}\ s^{-1}$. Thus equation (1.63) clearly explains the experimental facts in contradiction to classical mechanics, *i.e., the intensity of radiation determines the number of electrons and frequency of radiation determines the energy of the ejected photoelectron.*

Example 1.6. Calculate the velocity of the electron ejected from the copper surface when a photon of wavelength 250 nm strikes on it. The work function of copper is 4.653 eV.

Solution. Convert work function eV into energy units as: $4.653 \times 1.6 \times 10^{-19}\ J = h\nu_0 = 7.445 \times 10^{-19}\ J$

$$h\nu = \frac{(6.626 \times 10^{-34}\,J\ s)\ (3 \times 10^{8}\,m\ s^{-1})}{(250 \times 10^{-9} m)} = 7.95 \times 10^{-19} J$$

Using equation (1.63)

$$v = \left[\frac{2\ (h\nu_0 - h\nu)}{m_e}\right]^{\frac{1}{2}} = \left[\frac{2(7.95 \times 10^{-19} - 7.445 \times 10^{-19} J)}{9.31 \times 10^{-31} kg}\right]^{\frac{1}{2}}$$

$$= 2.32 \times 10^{5}\ ms^{-1}$$

Example 1.7. Voltage required to stop the ejection of an electron from platinum surface is 0.647 eV when the radiation of wavelength 220 nm strikes the surface. Calculate the work function of the platinum surface.

Solution.
$$\varepsilon = h\nu = \frac{hc}{\lambda} = \frac{(6.626 \times 10^{-34} J\ s)\ (3 \times 10^{8} m\ s^{-1})}{(220 \times 10^{-9} m)\ (1.6 \times 10^{-19} J\ eV^{-1})} = 5.647\ eV$$

$$K.E. = \frac{1}{2} m_e v^2 = 0.647\, eV$$

Using equation (1.63)

$$\varepsilon_0 = \varepsilon - \text{K.E.} = 5.647 - 0.647 = 5.0\ eV.$$

1.2.4 COMPTON EFFECT

In addition to photoelectric effect, the other phenomenon that throws light on the particle nature of radiation is the Compton's effect. When X-rays of wavelength λ_0 interacts with atom or molecule of low atomic mass, secondary X-rays and electrons are produced as shown in Fig. 1.9. The following experimental observations have been made on the properties of the secondary X-rays. (*i*) Scattered radiation consists of two wavelengths, the original λ_0 and additional wavelength λ_s. λ_s is always greater than λ_0. (*ii*) λ_s depends on the scattering angle θ, *i.e.*, the angle between the incident and scattered radiation, but not on the scattering medium. The wavelength of the scattered radiation is given by the expression (1.64).

$$\lambda_s = \lambda_0 + k \sin^2(\theta/2)$$

$$\Delta\lambda = \lambda_s - \lambda_0 = k \sin^2(\theta/2) \qquad ...(1.64)$$

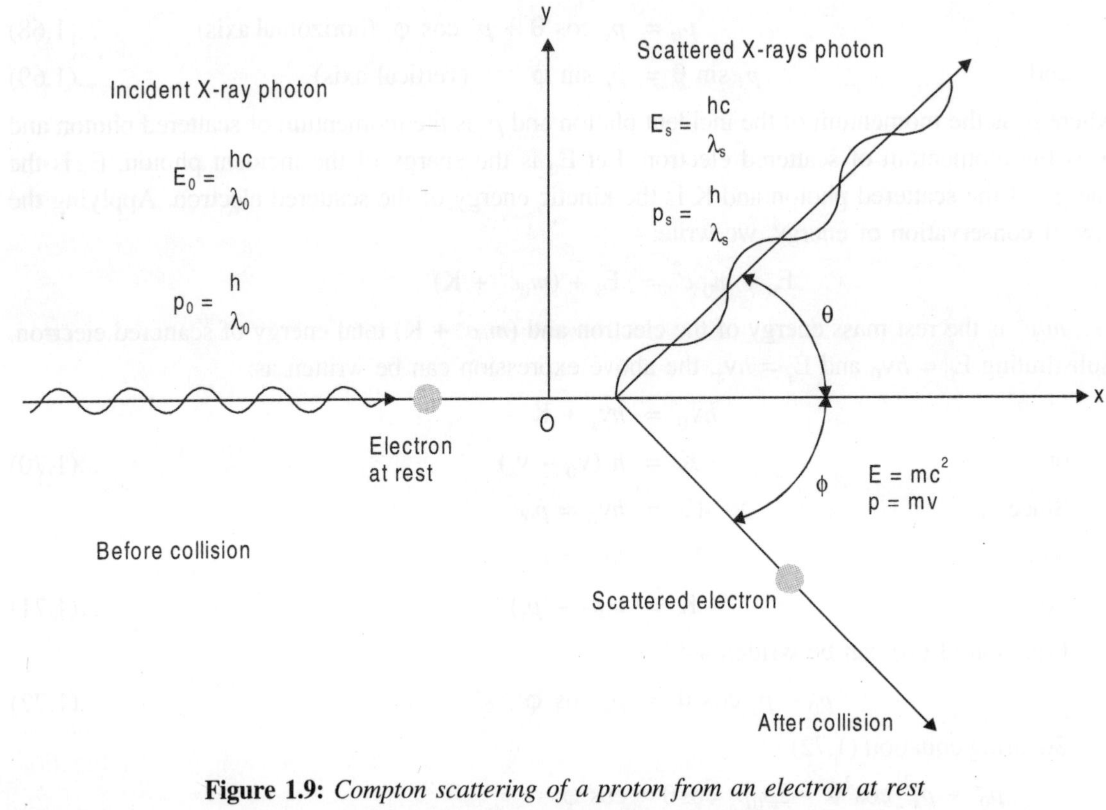

Figure 1.9: *Compton scattering of a proton from an electron at rest*

where k is a constant. The classical electrodynamics theory failed to explain these experimental observations. According to this theory when monochromatic radiation interacts with matter, the frequency of incident radiation should remain unaltered. However Compton in 1923 proposed that: (*i*) X-ray photons have a momentum and hence can be considered as particles, (*ii*) the scattering process is an elastic collision between a photon and electron. *The shift in wavelength of x-ray photons due to elastic scattering from electrons is called the Compton Effect.* This effect is also observed in the scattering of γ-rays as well.

Let us now try to explain the Compton Effect on the basis of particle nature of electromagnetic radiation. The energy of a photon, *i.e.*, a particle of rest mass zero is:

$$E = pc \qquad ...(1.65)$$

The energy of photon is also given as:

$$E = h\nu \qquad ...(1.66)$$

Thus, the momentum of photon is:

$$p = \frac{h\nu}{c} = \frac{h}{\lambda} \qquad ...(1.67)$$

The *x*-ray photon is scattered from stationary free electron (Fig. 1.9). According to law of conservation of momentum, we can write:

$$p_0 = p_s \cos\theta + p_e \cos\varphi \quad \text{(horizontal axis)} \qquad ...(1.68)$$

and

$$p_s \sin\theta = p_e \sin\varphi \quad \text{(vertical axis)} \qquad ...(1.69)$$

where p_0 is the momentum of the incident photon and p_s is the momentum of scattered photon and p_e is the momentum of scattered electron. Let E_0 is the energy of the incident photon, E_s is the energy of the scattered photon and K is the kinetic energy of the scattered electron. Applying the law of conservation of energy, we write:

$$E_0 + m_0 c^2 = E_s + (m_0c^2 + K)$$

i.e., m_0c^2 is the rest mass energy of the electron and $(m_0c^2 + K)$ total energy of scattered electron. Substituting $E_0 = h\nu_0$ and $E_s = h\nu_s$, the above expression can be written as:

$$h\nu_0 = h\nu_s + K$$

or

$$K = h(\nu_0 - \nu_s) \qquad ...(1.70)$$

Since

$$E_0 = h\nu_0 = p_0c$$

and

$$E_s = h\nu_s = p_sc$$

So

$$K = c(p_0 - p_s) \qquad ...(1.71)$$

Equation (1.68) can be written as:

$$p_0 - p_s \cos\theta = p_e \cos\varphi \qquad ...(1.72)$$

Squaring equation (1.72)

$$p_0^2 + p_s^2 \cos^2\theta - 2p_0p_s \cos\theta = \cos^2\varphi$$

or

$$p_0^2 + p_s^2 \cos^2\theta - 2p_0 p_s\cos\theta = (1 - \sin^2\varphi) \qquad ...(1.73)$$

Combining equations (1.69) and (1.73), equation (1.73) can be written as:

$$p_0^2 + p_s^2 \cos^2\theta - 2p_0 p_s \cos\theta = p_e^2 (1 - p_s^2 \sin^2\theta/p_e^2) = p_s^2 - p_e^2 \sin^2\theta$$

or $p_0^2 + p_s^2 (\sin^2\theta + \cos^2\theta) - 2p_0 p_s \cos\theta = p_e^2$

or finally $p_0^2 + p_s^2 - 2p_0 p_s \cos\theta = p_e^2$ ...(1.74)

According to special relativistic mechanics the energy of the electron is:

$$E^2 = (m_0c^2)^2 + p_e^2c^2 \quad ....(1.75)$$

Also the total energy of the scattered electron is:

$$E = m_0c^2 + K \quad ...(1.76)$$

Equating equations (1.75) and (1.76)

$$\frac{K^2}{c^2} + 2\,K\,m_0 = p_e^2 \quad ...(1.77)$$

Now combining equations (1.71), (1.74) and (1.77) to eliminate K:

$$\frac{c^2(p_0 - p_s)^2}{c^2} + 2\,m_0\,c(p_0 - p_s) = p_0^2 - 2p_0\,p_s\cos\theta + p_s^2$$

or $$p_0^2 + p_s^2 - 2p_0p_s + 2m_0c(p_0 - p_s) = p_0^2 - 2p_0p_s\cos\theta + p_s^2$$

$$2\,p_0\,p_s\,(1 - \cos\theta) = 2\,m_0c\,(p_0 - p_s)$$

or $$\frac{1 - \cos\theta}{m_0c} = \frac{1}{p_s} - \frac{1}{p_0} \quad ...(1.78)$$

Since $p_s = \frac{h\nu_s}{c}$ and $p_0 = \frac{h\nu_0}{c}$, equation (1.78) can be expressed as:

$$\frac{1 - \cos\theta}{m_0c} = \frac{c}{h\nu_s} - \frac{c}{h\nu_0}$$

or in terms of wave length

$$h\left(\frac{1 - \cos\theta}{m_0c}\right) = \lambda_s - \lambda_0$$

or $$\lambda_s - \lambda_0 = \frac{h}{m_0c}(1 - \cos\theta)$$

$$\Delta\lambda = (\lambda_s - \lambda_0) = \lambda_c\,(1 - \cos\theta) = 2\,\lambda_c\,\sin^2(\theta/2) \quad ...(1.79)$$

where $\lambda_c = \frac{h}{m_0c}$ = 0.0024 nm is defined as Compton wavelength. Equation (1.79) conforms to the equation (1.64) which is based on experimental results. The value of constant k in equation (1.64) is thus $2\frac{h}{m_0c}$.

In Compton's experiments on scattering of *x*-rays on graphite, use was made of rays with wavelength $\lambda = 0.0712605$ nm, to which there corresponds a quantum energy $h\nu = 17395$ eV. Since the ionization energy (IE) of a carbon atom is 11.217 eV, in this case we have $h\nu/\text{IE} = 1550$. Compton studied the scattering of γ rays on RaC on Al and paraffin. ($\lambda = 0.0029$ nm, $h\nu = 0.536$ MeV).

1.2.5 BOHR'S THEORY

The third postulate of the old quantum theory was proposed in 1913 by Neil's Bohr in order to explain the spectra of atoms. *This postulate states that in an atom only those orbits are allowed in which the angular momentum of the revolving electron about the positively charged nucleus is quantized, i.e.*, an integral multiple of $\frac{h}{2\pi}(\hbar)$;

$$mvr = n\frac{h}{2\pi} = n\hbar\,,\ n = 1, 2, 3,... \qquad ...(1.80)$$

and an electron in one of the allowed orbits does not radiate energy inspite of its centripetal acceleration. This is in contradiction with classical electrodynamic theory, which demands that any accelerating charged particle will radiate electromagnetic energy. Accordingly the allowed orbits are also known as *non-radiating orbits or stationary orbits.* The radiation will be emitted if the electron jumps from an initial orbit of energy E_n to a final orbit of energy E_m ($E_n > E_m$) and the frequency of the emitted photon will be:

$$h\nu = E_n - E_m$$

or

$$\nu = \frac{E_n - E_m}{h} \qquad ...(1.81)$$

Classical mechanics and this postulate of old quantum theory, if true, should enables one to calculate the quantities related to the structure of atom, such as spectra of atoms, magnetic moments of atoms etc. The following simple model will be considered to calculate these quantities.

Nucleus with infinite mass, being heavier than electron (mass of proton is 1836 times that of electron) can be assumed to be stationary, *i.e.*, at rest. Only electron is moving around the nucleus in circular orbits (later on for a nucleus of finite mass, both the motion of nucleus and the electron around the common centre will be taken into account in order to improve up on the magnitude of parameters governed by mass of the electron). Let us consider hydrogen like atoms in which the nuclear charge is Ze (for hydrogen the atomic number Z = 1), but where only one electron revolves around the nucleus with charge Ze (Fig. 1.10). These include such atoms as singly ionized helium He^+ (Z = 2), doubly ionized lithium Li^{2+} (Z = 3) and so on. Let the mass of the electron be m_e and it is moving in an orbit of radius r, with velocity v_e. Keeping these assumptions in view, we proceed as follows in order to determine the quantities related to the structure of atoms.

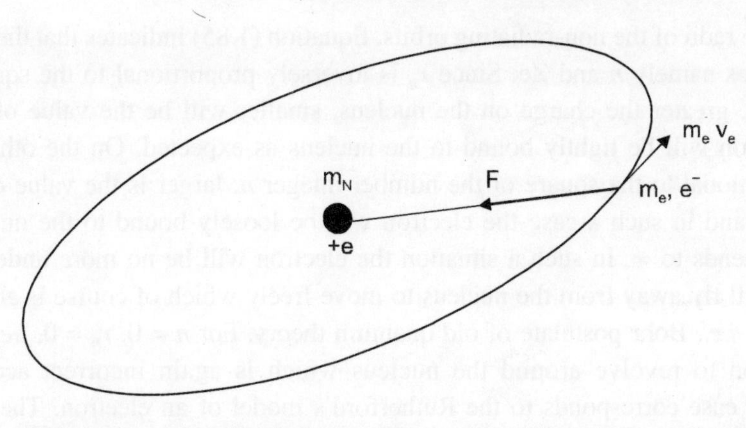

Figure 1.10: *Bohr model: The hydrogen atom consists of an electron of mass m_e and charge –e revolving about a very massive proton of mass $m_N >> m_e$ and charge +e*

The magnitude of electrostatic force (F) of attraction between the positively charged nucleus (+Ze) and negatively charged electron (–*e*) is:

$$F = \frac{Ze^2}{r^2} \quad \text{or} \quad F = \frac{Ze^2}{(4\pi\varepsilon_0)\, r^2} \qquad ...(1.82)$$

The term $4\pi\varepsilon_0$ is used to convert the cgs units to SI units and is known as permitivity constant. In order that the electron should remain moving in the orbit, this force should be balanced by an equal amount and opposite force acting on the electron. According to Newton's second law, the balancing force is the centripetal force acting on the electron,

$$\frac{Ze^2}{(4\pi\varepsilon_0)\, r^2} = \frac{m_e v_e^2}{r} \qquad ...(1.83)$$

where v_e^2/r is the centripetal acceleration. Solving equation (1.83) for *r*, we get:

$$r = \frac{Ze^2}{(4\pi\varepsilon_0)\, m_e v_e^2} \qquad ...(1.84)$$

According to third postulate of Bohr's quantum theory, angular momentum of electron in the stationary orbit is quantized, *i.e.*,

$$p = m_e\, v_e\, r = n\frac{h}{2\pi} \qquad ...(1.80)$$

Substituting the value of v_e from equation (1.80) into equation (1.84) and solving the resulting equation for *r*, we get

$$r = \frac{Ze^2}{(4\pi\varepsilon_0)\, m_e}\, \frac{1}{(nh/2\pi m_e r)^2}$$

or

$$r = r_n = \frac{(4\pi\varepsilon_0)\, n^2 h^2}{4\pi^2 m_e Ze^2},\ n = 1, 2, 3... \qquad ...(1.85)$$

which gives the radii of the non-radiating orbits. Equation (1.85) indicates that the radius r_n depends on two variables namely n and Ze. Since r_n is inversely proportional to the square of the charge on the nucleus, greater the charge on the nucleus, smaller will be the value of r_n and in such a situation electron will be tightly bound to the nucleus as expected. On the other hand since r is directly proportional to the square of the number integer n, larger is the value of n larger will be the value of r and in such a case the electron will be loosely bound to the nucleus. For infinite value of n, r_n tends to ∞. In such a situation the electron will be no more under the influence of nucleus and will fly away from the nucleus to move freely which of course is absurd according to third postulate, *i.e.*, Bohr postulate of old quantum theory. For $n = 0$, $r_n = 0$, *i.e.*, there is no orbit for the electron to revolve around the nucleus which is again incorrect according to Bohr's postulate. This case corresponds to the Rutherford's model of an electron. The model states that the atom consists of a very small nucleus (dimensions of the order of 10^{-14} m) in which all the positive charge and most of the mass is concentrated and the cloud of the negatively charged electron surrounds the nucleus. Since the dimension of the atom is of the order of 10^{-10} m, most of the space within an atom is empty. For neutral atoms, the charge on the electrons about the nucleus is equal to the positive charge of the nucleus. For the ground state, $n = 1$ and

$$r_1 = \frac{(4\pi\varepsilon_0)\, h^2}{4\pi^2 Ze^2 m_e} \qquad ...(1.86)$$

For hydrogen atom, Z = 1, equation (1.86) reduces to:

$$r_1 = \frac{(4\pi\varepsilon_0)\, h^2}{4\pi^2 e^2 m_e} = 0.053 \text{ nm} \qquad ...(1.87)$$

which is called the Bohr's radius and is denoted by a_0. This result agrees well with the radius of hydrogen atom obtained by other experimental techniques and also obtained from the planetary model of hydrogen atom. Based on the planetary model, the total energy of the system is:

$$E = K + V = -\frac{e^2}{2r(4\pi\varepsilon_0)} \qquad ...(1.88)$$

Substituting the experimental value of binding energy of the hydrogen atom (–13.6 eV) into equation (1.88), we get:

$$r = r_1 = \frac{e^2}{2\times 13.6\,(4\pi\varepsilon_0)} = 0.53 \times 10^{-10} \text{ m} = 0.053 \text{ nm}$$

From equation (1.85)

$$r_n = n^2\, r_1 \qquad ...(1.89)$$

which shows that the radii of the orbits for stationary states are also quantized and are given by r_1, $4r_1$, $9r_1$ and so on.

In order to obtain the velocity of the revolving electron, v_e, equations (1.84) and (1.85) are equated to eliminate r.

$$\frac{Ze^2}{(4\pi\varepsilon_0)\, m_e v_e^2} = \frac{(4\pi\varepsilon_0)\, n^2 h^2}{4\pi^2 Z\, e^2 m_e}$$

$$v_e^2 = \frac{4\,\pi^2 Z^2 e^4}{(4\pi\varepsilon_0)^2\, h^2 n^2}$$

or $$v_e = \frac{2\,\pi\, Ze^2}{(4\pi\varepsilon_0)\, hn} \qquad ...(1.90)$$

Equation (1.90) reveals that the velocity of the electron is directly proportional to the atomic number Z and inversely proportional to n. This also suggests that velocity is quantized. Greater the value of Z on the nucleus, greater will be the velocity, whereas greater is the value of n, smaller will be velocity. For $n = 0$, $v_e = \infty$, which is again impossible since electron cannot have velocity more than that of light particles.

Further since the radii in this model are also quantized, the linear velocity v_e is related to the frequency (f_r) of revolution of the electron in its orbit by:

$$v_e = wr_1 n^2 = 2\pi f_r\, r_1 n^2 \qquad ...(1.91)$$

Equating the equations (1.90) and (1.91) we have

$$2\pi f_r r_1 n^2 = \frac{2\,\pi\, Ze^2}{(4\pi\varepsilon_0)\, hn}$$

Thus $$f_r = \frac{Ze^2}{(4\pi\varepsilon_0)\, r_1\, hn^3} \qquad ...(1.92)$$

gives the number of revolutions per second made by the electron in n orbit. Since f_r is proportional to $1/n^3$ and r_n is quantized, frequency is also quantized.

For hydrogen atom, Z = 1 and for $n = 1$

$$f_r = \frac{e^2}{(4\pi\varepsilon_0)\, r_1\, h} \qquad ...(1.93)$$

Using the value of $r_1 = 0.053$ nm and the known value of e for the electron yields a value of $f_r = 6.44 \times 10^{15}\ \text{s}^{-1}$. This agrees well with the determination of the frequency of revolution obtained by other methods and planetary model of hydrogen atom. According to planetary motion the frequency of revolution is:

$$m_e\, (2\pi f_r r)^2 = \frac{Ze^2}{(4\pi\varepsilon_0)\, r}$$

or $$f_r = \frac{1}{2\pi}\left(\frac{Ze^2}{(4\pi\varepsilon_0) m_e r^3}\right)^{1/2} \qquad ...(1.94)$$

For hydrogen atom, equation (1.94) yields $f_r = 6.6 \times 10^{15}\ s^{-1}$. Note that frequency in planetary motion is not quantized.

Example 1.8. Calculate: (*i*) the radii, (*ii*) velocities and (*iii*) frequencies of revolution of electron in hydrogen atom in the first and second orbits.

Solution. (*i*) The radius of any orbit in hydrogen atom can be calculated using equation (1.85).

$$r_n = \frac{(4\pi\varepsilon_0)h^2n^2}{4\pi^2 m_e Ze^2} = \frac{(1.113\times10^{-10}\,\text{C}^2\text{N}^{-1}\text{m}^{-2})\,(6.626\times10^{-34}\,\text{Js})^2\,(n^2)}{(9.109\times10^{-31}\,\text{kg})\,(1.6\times10^{-19}\,\text{C})^2(1)\,(4\pi)^2} = 5.295 \times 10^{-11}\ n^2\text{m}.$$

So $r_1 = (52.95)\,(1^2) = 52.95$ pm; $r_2 = (52.95\text{ pm})\,(2^2) = 211.8$ pm.

(*ii*) The velocity of an electron in the hydrogen orbit can be calculated by using equation (1.90), *i.e.*,

$$v_e(n) = \frac{2\pi\, Ze^2}{4\pi\varepsilon_0\, hn} = \frac{(2)(22/7)\,(1)\,(1.6\times10^{-19}\,\text{C})^2}{(1.113\times10^{-10}\,\text{C}^2\ \text{N}^{-1}\ \text{m}^{-2})\,(6.626\times10^{-34}\,\text{Js})(n)} = \frac{2.182\times10^6\,\text{ms}^{-1}}{n}$$

$$v_e(1) = \frac{2.182\times10^6\,\text{ms}^{-1}}{1} = 2.182\times10^6\,\text{ms}^{-1},\ v_e(2) = \frac{2.182\times10^6\,\text{ms}^{-1}}{2} = 1.091\times10^6\,\text{ms}^{-1}$$

(*iii*) The frequency of revolution can be calculated by using either equation (1.92) or equation (1.94), *i.e.*,

$$f_r(n) = \frac{Ze^2}{(4\pi\varepsilon_0)\,(r_1\, hn^3)}$$

$$= \frac{(1)\,(1.6\times10^{-19}\,\text{C})^2}{(1.13\times10^{-10}\,\text{C}^2\ \text{N}^{-1}\text{m}^{-2})\,(6.626\times10^{-34}\,\text{Js})\,(n^3)\,(52.95\times10^{-12}\,\text{m})} = \frac{6.427\times10^{15}\,\text{Hz}}{n^3}$$

$$\text{So } f_r(1) = \frac{6.427\times10^{15}\,\text{Hz}}{n^3} = \frac{6.427\times10^{15}\,\text{Hz}}{1^3} = 6.427\times10^{15}\,\text{Hz}$$

$$f_r(2) = \frac{6.427\times10^{15}\,\text{Hz}}{2^3} = 0.8\times10^{15}\ \text{Hz}$$

$$f_r(1) = \frac{1}{2\pi}\left(\frac{Ze^2}{(4\pi\varepsilon_0)m_e r^3}\right)^{1/2}$$

$$= \frac{1}{2\pi}\left[\frac{(1)\,(1.6\times10^{-19}\,\text{C})}{1.113\times10^{-10}\,\text{C}^2\text{N}^{-1}\text{m}^{-2})\,(9.109\times10^{-31}\,\text{kg})\,(52.9\times10^{-12}\,\text{m})^3}\right]$$

$$= 6.6\times10^{15}\ \text{Hz}$$

The value of f_r for the first orbit obtained by both the equations are similar.

Total Energy: The next problem is the calculation of total energy which is sum of its kinetic and potential energies, *i.e.*, K + V. From equation (1.83), the kinetic energy comes out to be:

$$K = \frac{1}{2} m_e v_e^2 = \frac{Ze^2}{(4\pi\varepsilon_0)2r} \qquad ...(1.95)$$

or
$$K = \frac{Ze^2[4\pi^2 Ze^2 m_e]}{(4\pi\varepsilon_0)[(4\pi\varepsilon_0)2n^2h^2]} = \frac{2\pi^2 Z^2 e^4 m_e}{(4\pi\varepsilon_0)^2 h^2 n^2} \qquad ...(1.96)$$

According to classical mechanics, the potential energy of the electron will be the work done in bringing the electron from an infinite distance to the nucleus, where it is at rest and has zero potential energy, to a distance r from the nucleus. Accordingly:

$$V = \int_\infty^r F\,dr = \int_\infty^r \frac{Ze^2}{(4\pi\varepsilon_0)r^2}\,dr = -\frac{Ze^2}{(4\pi\varepsilon_0)r} \qquad ...(1.97)$$

The minus sign is an indication that the system is one of attraction type rather than of repulsion type, as the electron is attracted to the positive nucleus.

or
$$V = -\frac{Ze^2[4\pi^2 Ze^2 m_e]}{(4\pi\varepsilon_0)[(4\pi\varepsilon_0)n^2h^2]} = -\frac{4\pi^2 Z^2 e^4 m_e}{(4\pi\varepsilon_0)^2 h^2 n^2} \qquad ...(1.98)$$

Thus according to equation (1.88) the total energy of the system is:

$$E = K + V = \frac{Ze^2}{(4\pi\varepsilon_0)2r} - \frac{Ze^2}{(4\pi\varepsilon_0)r} = -\frac{Ze^2}{(4\pi\varepsilon_0)2r} \qquad ...(1.99)$$

or
$$E_n = \frac{2\pi^2 Z^2 e^4 m_e}{(4\pi\varepsilon_0)^2 h^2 n^2} - \frac{4\pi^2 Z^2 e^4 m_e}{(4\pi\varepsilon_0)^2 h^2 n^2} = -\frac{2\pi^2 Z^2 e^4 m_e}{(4\pi\varepsilon_0)^2 h^2 n^2} \qquad ...(1.100a)$$

By using the values of $m_e = 9.11 \times 10^{-31}$ kg and $e = 1.6 \times 10^{-19}$ C, we obtain:

$$E_n = -\frac{13.6\,Z^2}{n^2}\,\text{eV} \qquad ...(1.100b)$$

where negative charge again indicates a bound system. The binding energy of an electron is defined as the minimum energy required to remove the electron completely from the atom or in other words, to ionize the atom. Equations (1.100*a*) and (1.100*b*) also show that the energies of the orbits for stationary states are also quantized and are given by $-13.6Z^2, -\frac{13.6}{4}Z^2, -\frac{13.6}{9}Z^2$ and so on (in electron volts, eV). State of the lowest energy or ground state corresponds to $n = 1$ and its energy is $E_1 = -13.6$ eV. The ionization energy of hydrogen atom is 13.6 eV and is in excellent agreement with the values obtained by other experimental techniques. The ionization energies of few one electron systems are compiled in Table 1.3.

Table 1.3: *Comparison of ionization energies of some one electron systems*

System	H	He^+	Li^{2+}	Be^{3+}	B^{4+}	C^{5+}
Ionization energy (eV)	13.606	54.403	122.420	217.657	340.127	489.84

The other deductions that can be made from equation (1.100) are:

The absolute value of energy is inversely proportional to the square of the integral value of *n* and directly proportional to the square of the charge on the nucleus. For a particular nucleus, the energy of the electron is governed by *n*. For $n = 1$, it has the maximum stability. As *n* increases, the electron energy relative to the ground state increases and tends to zero as *n* tends to infinity. On the other hand, for fixed value of *n*, the stability of hydrogen like nuclei increases with increase of Z, *i.e.*, Be^{3+} (Z = 4) > Li^{2+} (Z = 3) > He^{2+} (Z = 2) > H (Z = 1).

The energy levels diagram as deduced from equation (1.100) for hydrogen atom has been shown in Fig. 1.11. It shows that all the states from $n = 1$ to $n = \infty$ are bound states, since they have negative energies. When *n* grows and approaches ∞, the energy states become closer and at a certain stage, the energy difference between two successive states becomes so small that a continuous spectrum results as expected on the basis of planetary model and correspondence principle. Above the line given by $n = \infty$, the energy states have positive energy, $E > 0$ and the spectrum of states is continuous. The system is unbound, meaning the electron is free.

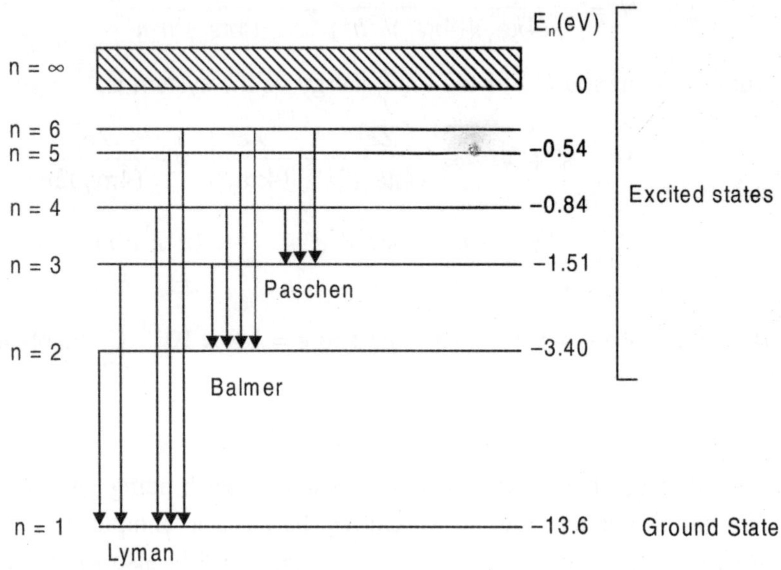

Figure 1.11: *Energy level diagram for hydrogen atom*

Spectral Series and Rydberg Constant

According to classical electrodynamics or planetary model such a system, *i.e.*, equation (1.100), should emit radiation with frequency equal to frequency of revolution of electron around the nucleus. But when an electron is in one of its allowed orbits it will not radiate energy inspite of centripetal acceleration. The radiation will be emitted only if the electron jumps from an orbit (*n*) of energy E_n to a final orbit (*m*) of energy E_m ($E_n > E_m$). The frequency of the emitted photon can be obtained using equation (1.81).

$$\nu = \frac{\Delta E}{h} = \frac{E_n - E_m}{h} \qquad \text{...(1.81)}$$

Substituting the energy expression from equation (1.100*a*), we obtain:

$$\nu = \frac{2\pi^2 Z^2 e^4 m_e}{(4\pi\varepsilon_0)^2 h^3}\left[\frac{1}{m^2} - \frac{1}{n^2}\right] \qquad \text{...(1.101)}$$

or $$\bar{\nu} = \frac{\nu}{c} = \frac{2\pi^2 Z^2 e^4 m_e}{(4\pi\varepsilon_0)^2 c\, h^3}\left[\frac{1}{m^2} - \frac{1}{n^2}\right] \qquad \text{...(1.102)}$$

This expression gives $$R = \frac{2\pi^2 e^4 m_e}{(4\pi\varepsilon_0)^2 ch^3} \qquad \text{...(1.103)}$$

the theoretical value of Rydberg constant. Equation (1.102) can now be written as:

or $$\bar{\nu} = RZ^2\left[\frac{1}{m^2} - \frac{1}{n^2}\right] \qquad \text{...(1.104)}$$

When parameters in equation (1.103) are replaced by their numerical values, the value of R comes out to be $1.097373 \times 10^7\ m^{-1} = 1.097373 \times 10^5\ cm^{-1}$, which is in agreement with the empirical value of $R = 1.0971136 \times 10^5\ cm^{-1}$ for $n = 3$, $\lambda = 656.3$ nm, which is identified as the red H_α line; for $n = 4$, $\lambda = 486.1$ nm, the blue H_β line, determined using the Rydberg empirical formula.

$$\frac{1}{\lambda} = R\left[\frac{1}{2^2} - \frac{1}{n^2}\right]$$

The uncertainty between the experimental and theoretical values of R is of the order of 0.02%. The atomic spectrum of hydrogen atom as deduced from relation (1.104) has been summarized in Table 1.4 and is also nicely in agreement with the results based on empirical formula in the various spectral regions.

To conclude we can say that Bohr model has been very fair to explain the discrete lines in the spectra of atoms and not continuous distributions of frequencies as demanded by planetary model. In planetary model, according to classical electrodynamics, (*a*) an accelerating charge must continuously emit electromagnetic radiation, and (*b*) the frequency of emitted radiation must be equal to the frequency of revolution. Accordingly the total energy of atom should decrease (become more negative), while the frequency of rotation (1.94) must continuously increase and in about 10^{-6} s complete collapse will take place. The small uncertainty, *i.e.*, 0.02% between the experimental and theoretical value of Rydberg constant may be to due the non consideration of finite mass of the nucleus. So in order to reduce the uncertainty, the correction for the same must be applied.

Table 1.4: *Atomic spectrum of hydrogen atom*

Series	Spectral region	Series formula	Series limit
Lyman (1906)	Ultraviolet	$\frac{1}{\lambda} = R\left[\frac{1}{1^2} - \frac{1}{n^2}\right]$, $n = 2, 3, 4...$	91.127 nm
Balmer (1885)	Visible	$\frac{1}{\lambda} = R\left[\frac{1}{2^2} - \frac{1}{n^2}\right]$, $n = 3, 4, 5...$	364.51 nm
Paschen (1908)	Infra-red	$\frac{1}{\lambda} = R\left[\frac{1}{3^2} - \frac{1}{n^2}\right]$, $n = 4, 5, 6...$	820.14 nm
Bracket (1922)	Infra-red	$\frac{1}{\lambda} = R\left[\frac{1}{4^2} - \frac{1}{n^2}\right]$, $n = 5, 6, 7....$	1458.0 nm
Pfund (1924)	Infra-red	$\frac{1}{\lambda} = R\left[\frac{1}{5^2} - \frac{1}{n^2}\right]$, $n = 6, 7, 8...$	2278.2 nm

Example 1.9. Calculate the energies of the second and third energy levels of the Bohr's orbits. Also calculate the transition energy in terms of eV, J and cm^{-1}.

Solution. The energies of the second and third orbits can be calculated using relation (1.100*b*), *i.e.*,

$$E_n = -\frac{13.6\,Z^2}{n^2}$$

$$E_2 = -\frac{13.6\,(1)^2}{2^2} = -3.4\text{ eV} ;\; E_3 = -\frac{13.6\,(1)^2}{3^2} = -1.511\text{ eV}$$

$$\Delta E = E_3 - E_2 = -1.511 - (-3.4)\text{ eV} = 1.889\text{ eV}$$

$$= (1.889\text{ eV})\,(1.6 \times 10^{-19}\,\text{JeV}^{-1}) = 3.022 \times 10^{-19}\text{J}$$

$$\nu = \frac{\Delta E}{h} = \frac{3.022\times10^{-19}\,\text{J}}{6.626\times10^{-34}\,\text{Js}} = 4.56\times10^{14}\,\text{s}^{-1}$$

$$\bar{\nu} = \frac{\nu}{c} = \frac{4.56\times10^{14}\,\text{s}^{-1}}{3\times10^{8}\,\text{ms}^{-1}} = 1.52\times10^{6}\,\text{m}^{-1} = (1.52\times10^{6}\,\text{m}^{-1})\,(10^{-2}\,\text{cm}^{-1}\text{m})$$

$$= 1.52 \times 10^{4}\,\text{cm}^{-1}$$

Correction for Finite Mass or Motion of Nucleus

Let us now assume that both nucleus of mass m_N and electron of mass m_e are moving in respective circles of radii r_N and r_e around an axis passing through the centre of gravity (C), as shown in Fig. 1.12. The above quantities are related to each other by the following equations. Taking moments around the centre of mass:

$$m_N\, r_N = m_e\, r_e \qquad ...(1.105)$$

and $$r = r_N + r_e \qquad ...(1.106)$$

values of r_e and r_N, in terms of r, can be obtained by solving equations (1.105) and (1.106).

$$r_N = \frac{m_e}{m_N + m_e} r, \; r_e = \frac{m_N}{m_N + m_e} r \qquad ...(1.107)$$

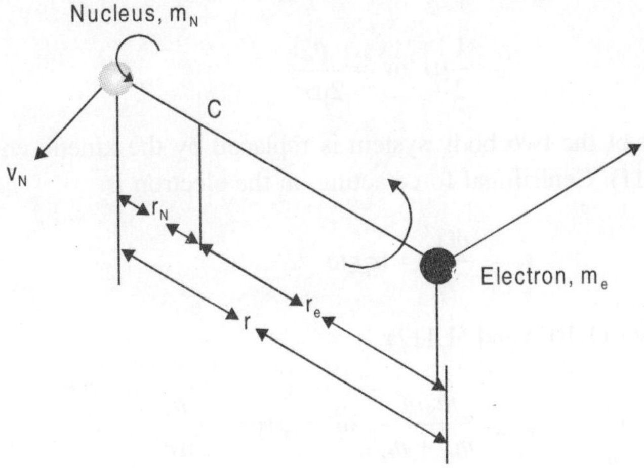

Figure 1.12: *The electron and the nucleus revolve around the common centre of mass*

The total angular momentum of the system will be the sum of the angular momentum of individual particle.

$$p_\varphi = m_N v_N r_N + m_e v_e r_e = m_N r_N^2 \omega + m_e r_e^2 \omega \qquad ...(1.108)$$

where ω is the angular velocity. Substituting the values of r_N and r_e in equation (1.108)

$$p_\varphi = m_N \left(\frac{m_e}{m_N + m_e} \right)^2 r^2 \omega + m_e \left(\frac{m_N}{m_N + m_e} \right)^2 r^2 \omega \qquad ...(1.109)$$

$$= \frac{m_N m_e}{(m_N + m_e)^2} (m_N + m_e) r^2 \omega = \frac{m_N m_e}{m_N + m_e} r^2 \omega = \mu r^2 \omega$$

Equation (1.109) shows that the angular momentum of two particle system can be reduced to one particle system, with the difference that the mass of two particles is equivalent to the mass $\frac{m_N m_e}{m_N + m_e}$. This is called *reduced mass*, defined as:

$$\mu = \frac{m_N m_e}{m_N + m_e} \quad \text{or} \quad \frac{1}{\mu} = \frac{1}{m_N} + \frac{1}{m_e} \qquad ...(1.110)$$

i.e., harmonic mass of the two masses. If the mass of one particle is very heavy (say nucleus, *i.e.*, m_N) as compared to that of electron (m_e), reduced mass will nearly be equal to that of the mass of the light particle. Thus the assumption made earlier for deriving Bohr's expression was totally justified. Better accurate picture can be obtained if we include the mass of the nucleus in deriving the expression for kinetic energy.

$$K = \frac{1}{2}m_e v_e^2 + \frac{1}{2}m_N v_N^2 = \frac{1}{2}m_e r_e^2\omega^2 + \frac{1}{2}m_N r_N^2\omega^2$$

$$= \frac{1}{2}m_e\left(\frac{m_N}{m_N + m_e}\right)^2 r^2\omega^2 + \frac{1}{2}m_N\left(\frac{m_e}{m_N + m_e}\right)^2 r^2\omega^2$$

$$= \frac{1}{2}\mu r^2\omega^2 = \frac{p_\varphi^2}{2\mu r^2} \qquad \text{...(1.111)}$$

The kinetic energy of the two body system is replaced by the kinetic energy of one particle given by equation (1.111). Centrifugal force acting on the electron is:

$$F = \frac{m_e v_e^2}{r_e} = m_e r_e \omega^2 \qquad \text{...(1.112)}$$

Combining equations (1.107) and (1.112)

$$F = \frac{m_N m_e}{m_N + m_e} r\omega^2 = \mu r\omega^2 = \frac{p_\varphi^2}{\mu r^3} \qquad \text{...(1.113)}$$

One can see that all the equations required in the derivations of Bohr's model, discussed earlier are similar, with the exception that m_e is replaced by the reduced mass μ. Proceeding on the same lines, the respective expressions for r_n, v_n and E_n can be derived and are:

$$r_n = \frac{(4\pi\varepsilon_0)n^2h^2}{4\pi^2\mu Ze^2} \qquad \text{...(1.114)}$$

$$v_n = \frac{2\pi Ze^2}{(4\pi\varepsilon_0)hn} \qquad \text{...(1.115)}$$

$$E_n\ (\text{cm}^{-1}) = -\frac{2\pi^2Z^2e^4\mu \times 10^{-2}}{(4\pi\varepsilon_0)^2h^3cn^2} \qquad \text{...(1.116)}$$

and the Rydberg constant

$$R\mu \text{ or } R_H = \frac{2\pi^2e^4\mu \times 10^{-2}}{(4\pi\varepsilon_0)^2ch^3} \qquad \text{...(1.117)}$$

The ratio of R_H and R is given as:

$$\frac{R_H}{R} = \frac{\mu}{m_e} \qquad \text{...(1.118)}$$

The difference between the two Rydberg constants, calculated by two methods differs through their use of mass of the electron or reduced mass of the electron and nucleus. Substituting the value of μ in equation (1.118), we get:

$$R_H = \frac{m_N}{m_N + m_e} R \qquad ...(1.119)$$

It is clear from equation (1.119) that R_H will be less than R because $m_N < (m_e + m_N)$. Substituting the values of m_e and m_N in equation (1.119)

$$R_H = \frac{1.67252\times10^{-27}}{9.1091\times10^{-31} + 1.67252\times10^{-27}}(109737.31 \text{ cm}^{-1}) = 109678 \text{ cm}^{-1}$$

The agreement between R_H and R, the experimentally determined value is very good. One thing is quite clear from equation (1.119) that as the mass of the nucleus increases, the value of R_H approaches the value of R and can be called R_∞. Moreover, for the same

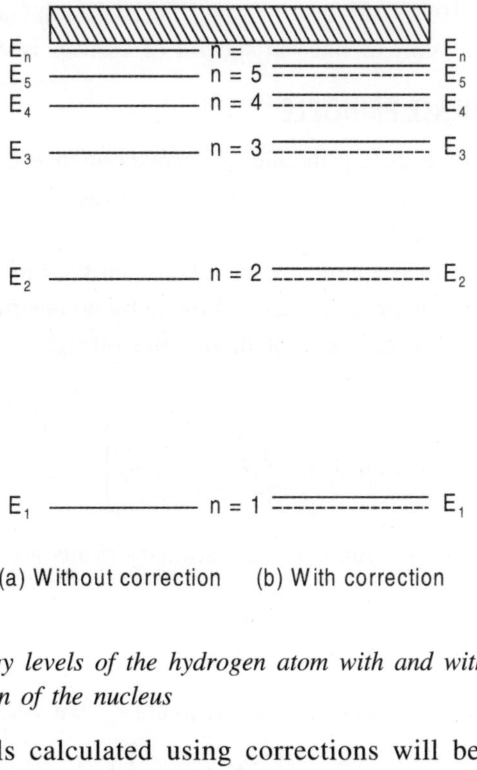

Figure 1.13: *Energy levels of the hydrogen atom with and without corrections due to the motion of the nucleus*

value of *n*, the energy levels calculated using corrections will be less negative than the corresponding levels without using corrections, *i.e.*, E_n (with corrections) > E_n (without corrections) Fig. 1.13. A comparison of equations for wavelengths without and with corrections also shows that:

$$1/\lambda \text{ (with correction)} < 1/\lambda \text{ (without correction)}$$

Accordingly the calculated wavelengths of the emitted photons (with corrections) will be slightly larger.

The deuterium atom ^{2}D, an isotope of hydrogen has a nucleus composed of proton and a neutron. The reduced mass of deuterium will be $\mu_D = \dfrac{m_e}{1+\dfrac{m_e}{2m_N}}$; which makes $\mu_D > \mu_H$. Since the Rydberg constant is directly proportional to the reduced mass, it is evident that $R(\mu_D) > R(\mu_H)$. This small discrepancy played an important role in the discovery of deuterium by Urey. For this discovery, Urey was awarded Nobel Prize in 1934 in chemistry. For tritium, $\mu_T = \dfrac{m_e}{1+\dfrac{m_e}{3m_N}}$. Thus the isotopes of the same elements will have different values of R. For example, for the isotopes of H, D and T, we have $R_H = 109677.58\ \text{cm}^{-1}$, $R_D = 109710.48\ \text{cm}^{-1}$ and $R_T = 109715.97\ \text{cm}^{-1}$. This difference leads to a certain shift of lines in spectra of various isotopes (isotopic shift).

1.2.6 BOHR'S CORRESPONDENCE PRINCIPLE

The basic philosophy of Bohr's principle of correspondence is that *when Bohr model, which is especially constructed for microscopic principle system, is applied to macroscopic system, the results obtained are identical to those obtained with classical mechanics.* Thus in an atom for large values of n and m, the frequency of the photon emitted in a transition from n to m should be equal to the number of revolutions per second made by an electron in an orbit, as demanded by classical mechanics. According to classical theory the orbital frequency given by the equation (1.94) by taking Z = 1, is:

$$f_r = \frac{1}{2\pi}\left(\frac{e^2}{4\pi\varepsilon_0 m_e r^3}\right)^{1/2} \qquad \text{...(1.94)}$$

According to Bohr model, the radii of the stationary orbits are given by:

$$r_n = \frac{(4\pi\varepsilon_0)n^2h^2}{4\pi^2 m_e e^2} \qquad \text{...(1.85)}$$

Combining equations (1.94) and (1.85) to eliminate r_n, we get:

$$f_r = \left[\frac{4\pi^2\ m_e\ e^4}{(4\pi\varepsilon_0)^2\ h^3}\right]\frac{1}{n^3} \qquad \text{...(1.120)}$$

Now the frequency of photon emitted in a transition from n to m is:

$$\nu = \frac{2\ \pi^2\ e^4 m_e}{(4\pi\varepsilon_0)^2\ h^3}\left[\frac{1}{m^2} - \frac{1}{n^2}\right] \qquad \text{...(1.102)}$$

Let us now manipulate the quantities with in bracket for very large values of m and n. It can be expressed as:

$$\frac{1}{m^2} - \frac{1}{n^2} = \frac{n^2 - m^2}{n^2\, m^2} = \frac{(n - m)\,(n + m)}{n^2\, m^2}$$

For very large values of n and m we can write $n - m = \Delta n$, $n + m \cong 2n$ and $m^2 n^2 \cong n^4$

Accordingly

$$\frac{1}{m^2} - \frac{1}{n^2} = \frac{(\Delta n)\,(2n)}{n^4} = \frac{2\,\Delta n}{n^3} \qquad \text{...(1.121)}$$

Thus we can express equation (1.102) as:

$$\nu = \frac{2\pi^2 e^4 m_e}{(4\pi\varepsilon_0)^2 h^3}\frac{2\Delta n}{n^3} = \frac{4\pi^2 e^4 m_e}{(4\pi\varepsilon_0)^2 h^3}\frac{\Delta n}{n^3} \qquad \text{...(1.122)}$$

For $\Delta n = 1$,

$$\nu = \frac{4\pi^2 e^4 m_e}{(4\pi\varepsilon_0)^2 h^3}\frac{1}{n^3} \qquad \text{...(1.123)}$$

which is identical to classical frequency f_r in equation (1.120). For $\Delta n = 2, 3, 4, \ldots$ we obtain the harmonics of frequency.

1.2.7 BOHR MODEL FOR HYDROGEN LIKE ATOMS

The expressions for hydrogen like atoms namely He^+, Li^{2+}, Be^{3+}, etc., are listed in Table 1.5. For the same value of n, the radius of electron orbit in hydrogen like atoms is smaller by a

Table 1.5: *Comparison of parameters of hydrogen and hydrogen like atoms according to Bohr model*

Parameter	Hydrogen	Hydrogen like
r_n	$\frac{(4\pi\varepsilon_0)\, n^2 h^2}{4\,\pi^2\, m_e\, e^2}$	$\frac{(4\pi\varepsilon_0)\, n^2 h^2}{4\,\pi^2\, Z m_e e^2}$
E_n	$-\frac{2\pi^2 e^4 m_e}{(4\pi\varepsilon_0)^2\, h^2 n^2}$	$-\frac{2\pi^2\, Z^2 e^4 m_e}{(4\pi\varepsilon_0)^2\, h^2 n^2}$
R	$\frac{2\pi^2\, e^4 m_e}{(4\pi\varepsilon_0)^2\, ch^3}$	$\frac{2\pi^2 Z^2 e^4\, m_e}{(4\pi\varepsilon_0)^2\, ch^3}$
$1/\lambda$	$R\left(\frac{1}{m^2} - \frac{1}{n^2}\right)$	$RZ^2\left(\frac{1}{m^2} - \frac{1}{n^2}\right)$

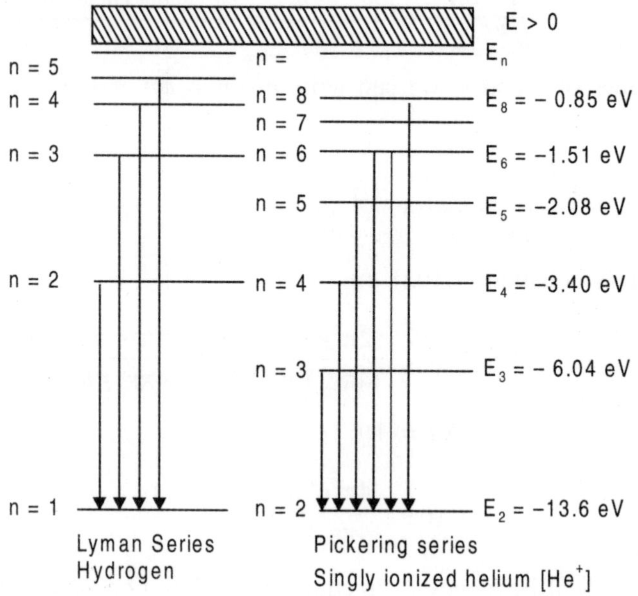

Figure 1.14: *A comparison of energy levels for H and He^+*

factor Z. The energy levels for the same *n* are more negative by a value of $1/Z^2$. The energy levels for H and He^+ are compared in Fig. 1.14. Rydberg constant R is Z^2 times greater in He^+ than that in H for any given transition (*n* to *m*).

1.2.8 SOMMERFELD'S MODEL

The Bohr's model of atom was modified in 1916 by Sommerfeld (the Ph.D. guide of Heisenberg) in order to match more precisely the theoretical spectral results with the experimentally observed data. According to him motion of the particle in a field of central forces in general leads to elliptical orbits and the circle is a special case of an ellipse whose major and minor axes are of equal length. He generalized the quantum condition of Bohr that $p = h/2\pi$ to the condition

$$\oint p_i\, dq_i = n_i h \qquad ...(1.124)$$

where n_i can take on all the integral values except zero and p_i is the momentum of the given generalized coordinate q_i. The symbol $\oint$ means that the integral is taken over complete cycle if the motion is periodic. Non-periodic motions are not quantized.

The particle moving in an elliptical orbit has two degrees of freedom, *i.e.*, both the angle φ, and the radius can vary against one another. For electrons moving in circular orbits, only the angle of revolution (φ) varies. (Fig. 1.15). Accordingly the kinetic energy of an electron moving in an elliptical orbit is a function of $\varphi = \frac{d\varphi}{dt}$ and $r = \frac{dr}{dt}$. Thus two quantum numbers corresponding to the two variables, *i.e.*, r and φ are required to describe completely the motions of the particle in elliptical orbits. According to Sommerfeld, quantum condition (1.124), two integrals are:

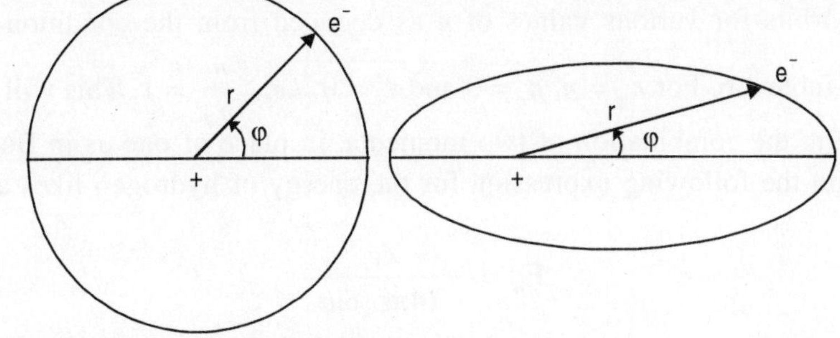

Figure 1.15: *Effect of an elliptical orbit on the variables r and φ*

$$\oint p_\varphi \, d\varphi = n_\varphi \, h \qquad ...(1.125)$$

$$\oint p_r \, dr = n_r \, h \qquad ...(1.126)$$

Since the angular momentum of an isolated system is constant, the integral equation (1.125) can be written as:

$$p_\varphi = n_\varphi \, h/2\pi \qquad ...(1.127)$$

Here p_φ is the angular momentum, p_r is the radial momentum, n_φ and n_r are the respective quantum numbers. According to law of conservation of momentum, the sum of these quantum numbers should be constant and equal to the number n of Bohr's theory and is called the principal quantum number.

$$n = n_r + n_\varphi \qquad ...(1.128)$$

Based on classical dynamics

$$\frac{n}{n_\varphi} = \frac{\text{length of major axis}}{\text{length of minor axis}} = \frac{a}{b}$$

The eccentricity of the electron orbit (ε_e) is determined by the relationship:

$$\varepsilon_e = \left[1 - \left(\frac{n_\varphi}{n}\right)^2\right]^{\frac{1}{2}} \qquad ...(1.129)$$

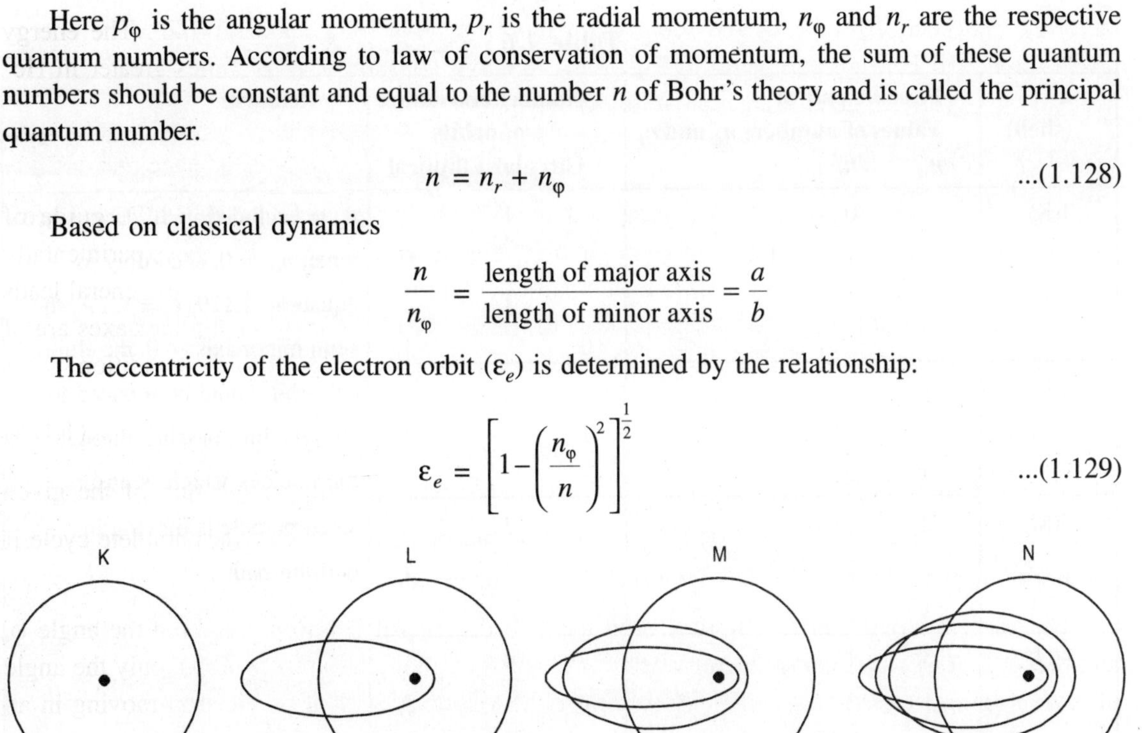

Figure 1.16: *Sommerfeld orbits of hydrogen-like atoms*

The Bohr Sommerfeld orbits of hydrogen like atoms for $n = 1$ (K), 2 (L), 3 (M) and 4 (N) are shown in Fig. 1.16. The value of '*a*' in Fig. 1.16 is the same for all orbits. Actually according to equation (1.128), '*a*' increases with increase of n as n^2. The number of circular and elliptical orbits for various values of n as deduced from the condition $n = n_r + n_\varphi$ has been listed in Table 1.6. For $n_\varphi = n$, $n_r = 0$ and $\varepsilon_e = 0$, *i.e.*, $\frac{n}{n_\varphi} = 1$. This will result in circular orbit. Now using the combination of two momenta, in place of one as in Bohr's treatment of atom, we obtain the following expression for the energy of hydrogen likes atoms.

$$E_n = -\frac{Ze^2}{(4\pi\varepsilon_0)2a} \qquad ...(1.130)$$

where '*a*' is the semi-major axis of ellipse.

$$a = \frac{(4\pi\varepsilon_0)h^2}{Zm_e e^2}(n_r + n_\varphi)^2 \qquad ...(1.131)$$

Substituting the value of '*a*' in equation (1.130) we get the same equation (1.116) as in the case of circular orbits.

$$E_n\ (\text{cm}^{-1}) = -\frac{2\pi^2 Z^2 e^4 \mu \times 10^{-2}}{(4\pi\varepsilon_0)^2 h^3 c n^2} \qquad ...(1.116)$$

Table 1.6

***n* (shell)**	**Possible pair of Values of numbers n_φ and n_r**			**Number and nature of orbits**			**Remarks**
	n_φ	n_r		**Circular**		**Elliptical**	
1(K)	1	0			1		n_φ is excluded in all cases since when $n_\varphi = 0$, according to equation (1.129) $\varepsilon_e = 1$, *i.e.*, the semi minor axis → 0, the elliptical orbit would be reduced to straight line passing through the nucleus which is absurd, since particle is moving in definite path.
		1	1	1		0	
	0	1			1		
				0		1	
2(L)	1	1			2		
			2	1		1	
	2	0					
3(M)	2	1			3		
	1	2	3	1		2	
	3	0					
4(N)	3	1			4		
	1	3		1		3	
			4				
	2	2					
	4	0					

Consequently for a given value of principal quantum number n, the energy of an atom (ion) is independent of the shape of the orbit. In general case the given value of n is obtained by means of n different combinations of numbers n_φ and n_r. $n_\varphi = 1, 2, 3, \ldots n$ and $n_r = 0, 1, 2, \ldots (n-1)$. Thus for each given value of n we have different states, *i.e.*, orbits of the atom with identical energy. Such states are called degenerate states.

Fine structure is observed in the high resolution spectra of hydrogen atom and this is attributed to splitting of degenerate levels. Hence the degeneracy of levels of a hydrogen atom and also the levels of ions, He^+, Li^{2+}, etc. is only apparent and these levels, due to slight degree of their splitting may be considered practically degenerate. The small degeneracy arises since the energy of an atom is not a function of n only. With different n_φ, levels will differ in energy. This is attributed to the continuous non-uniform motion of the electron in elliptical orbits. The velocity is maximum when the electron is nearest to the nucleus, which is situated at the focus of ellipse. It decreases continuously and becomes minimum when it is far away from the nucleus. This continuous change of velocity will cause difference in the effective mass of electron at different positions in the elliptical orbit. According to special relativistic mechanics, the mass of the moving particle is given by:

$$m = m_0\left(1-\frac{v^2}{c^2}\right)^{-\frac{1}{2}} \qquad \ldots(1.33)$$

Applying the mass correction, the equation (1.116) gets modified to:

$$E_{n,n_\varphi} = \frac{hcRZ^2}{n^2} + \frac{h\alpha^2RcZ^4}{n^4}\left(\frac{n}{n_\varphi}-\frac{3}{4}\right) \qquad \ldots(1.132a)$$

Here $\alpha = \dfrac{2\pi e^2}{hc} = 7.2977 \times 10^{-3}$ is called Sommerfeld structure constant. For the ground level of the hydrogen level ($n = n_\varphi = 1$), this correction amounts to 0.001% and decreases sharply as the principal quantum number increases. Inspite of this extremely small correction, the splitting of degenerate levels into components will take place. As already stated that the energy of an atom is now a function of not only of the number n, but also of the number n_φ The correction term in (1.132a) shows that the frequency separation varies with Z as Z^4 in hydrogen like atoms. For singly ionized helium, the frequency separation should be sixteen times that for hydrogen atom. This deduction conforms to the experimentally observed results.

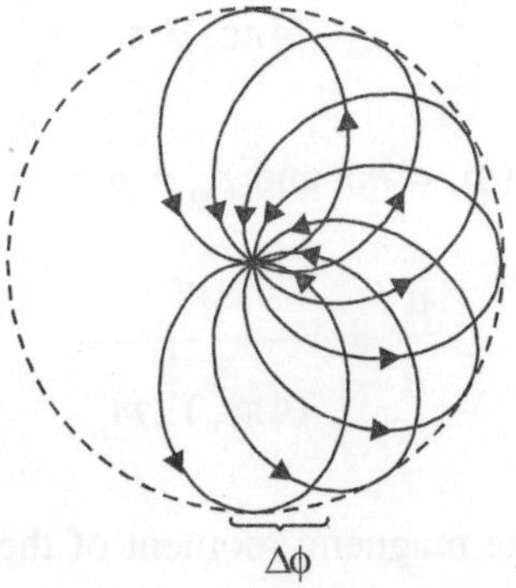

Figure 1.17: *Orbital precession (Sommerfeld rosette)*

Further it is to be noted that according to relativistic theory of the atom, the electronic orbits are no longer closed, but take the shape of a Sommerfeld rosette (Fig. 1.17). In this case, the motion of the electron may be pictured as motion in a closed flat orbit that precesses with certain angular velocity and during one circuit of the electron around the nucleus the orbit axis turns through an angle $\Delta\phi$, which does not exceed 0.01. The electron orbits of hydrogen atom may be considered to the extent that n-degeneracy of energy levels of H-atom is maintained. Accordingly equation (1.132a) modifies to:

$$E_{n,n_\varphi} = -\frac{hc\,RZ^2}{[n - f(n, n_\varphi)]^2} = -\frac{hcRZ^2}{n^{*2}} \qquad \text{...(1.132}b\text{)}$$

where $f(n, n_\varphi) = \dfrac{\alpha^2 h^2}{2}\left(\dfrac{1}{n_\varphi} - \dfrac{3}{4}\right)$ and $n^* = n - f(n, n_\varphi)$. In this form the expression for the energy of the atom, (or ion) with one electron is similar to Rydberg formula.

$E = -\dfrac{hcRZ^2}{n_{eff}^2}$ for the energy of atoms of the alkaline elements of corresponding ions with one extra electron, *e.g.*, Be^+, B^{2+}, C^{3+}, ... or Mg^+, Al^{2+}, Si^{3+} ... . With this picture, a reasonable agreement with the experimentally observed fine structure in the hydrogen spectrum is obtained.

In the presence of magnetic field, it is found that the spectral lines in the hydrogen atom are splitted further. To account for this splitting the magnetic properties of the atoms need to be deduced from the Bohr-Sommerfeld model of atoms. An electron revolving around a nucleus may be assumed as a certain closed electric current, $i = e/\tau$ (where τ is the orbital period of the electron). The idea that magnetism is due to certain hypothetical molecular currents was expressed by Ampere in 1822. The magnetic dipole moment associated with this current is given by:

$$\mu = iS \qquad \text{...(1.133)}$$

where $S = \pi\, a\, b$ is the area of the orbit (a and b are the sub-major and sub-minor axes of the ellipse). Thus it follows:

$$\mu = \frac{\pi abe}{(4\pi\varepsilon_0)^{\frac{1}{2}}\,\tau} \qquad \text{...(1.134)}$$

Since $\dfrac{\pi ab}{\tau} = \dfrac{1}{2} r^2 \dot{\varphi} = p/2m_e$, ($p_r = m\dot{r}$ and $p_\varphi = p = m_e r^2 \dot{\varphi}$), we have:

$$\frac{\mu}{p} = \frac{e^2}{(4\pi\varepsilon_0)^{\frac{1}{2}}\, m_e} \qquad \text{...(1.135)}$$

Substituting $p_\varphi = n_\varphi \dfrac{h}{2\pi}$, for the magnetic moment of the electron in the orbit we get:

$$\mu = \frac{ehn_\varphi}{4\pi(4\pi\varepsilon_0)^{1/2}\, m_2} = \mu_B n_\varphi \qquad ...(1.136)$$

It follows that the magnetic moment of the atom, like its angular momentum p_φ is a quantized quantity in multiples of a certain elementary magnetic moment.

$$\mu_B = \frac{eh}{4\pi\,(4\pi\varepsilon_0)^{\frac{1}{2}} m_e} \qquad ...(1.137)$$

This quantity is numerically equal to 9.27×10^{-21} erg orsted^{-1}. Since for ground state of hydrogen atom, $n_\varphi = n = 1$, the magnetic moment of the hydrogen atom should be equal to one Bohr magneton. Now considering the motion of the electron in three dimensional space, *i.e.*, in the coordinates r, θ, φ (φ is the angle between the normal to the plane of the electron orbit and the polar axis, θ is the angle reckoned on the equatorial plane from its line of interaction with the planes of the orbit), it is found that the projection of the orbit angular momentum p on the direction of the polar axis turns out to be a quantized quantity.

$$p = p\cos\varphi = m_\varphi \frac{h}{2\pi} \qquad ...(1.138)$$

where m_φ is an integer. We also know that

$$\cos\varphi = m_\varphi / n_\varphi \qquad ...(1.139)$$

Thus cos φ can take only definite discrete values, *i.e.*, the electron orbit is definitely and discretely oriented in space (space quantization). Equation (1.139) reveals that the maximum and minimum values of the quantum number m_φ must be equal to n_φ and $-n_\varphi$ respectively, *i.e.*,

$$m_\varphi = \pm n_\varphi,\ \pm(n_\varphi - 1),\ \pm(n_\varphi - 2),\ \ldots\ \pm 1,\ 0$$

i.e., m_φ can have $(2n_\varphi + 1)$ different values for each value of n_φ. For $n_\varphi = 2$, $m_\varphi = 2, 1, 0, -1, -2$.

Combining equations (1.136) and (1.139), it follows that:

$$\mu' = \mu\cos\varphi = \frac{ehm_\varphi}{4\pi(4\pi\varepsilon_0)^{\frac{1}{2}}\, m} = \mu_B m_\varphi \qquad ...(1.140)$$

Thus, the projection of the magnetic moment of the atom is also quantized. When an external magnetic (or electrical) field is applied, the direction of the polar axis becomes fixed in space, *i.e.*, the magnetic field is parallel to the polar axis (z-axis, Fig. 1.18).

Assuming $\varphi = \mu H_{ext} = \alpha$, the expression for the projection of the magnetic moment on the direction of magnetic field will be:

$$\mu' = \mu\cos\alpha = \mu_B\, m_\varphi \qquad ...(1.141)$$

Since the quantum number m_φ determines the magnitude of the projections of the magnetic moment on the direction of the magnetic field, it may be called the magnetic quantum (orbital) number. The normal Zeeman effect (to be studied later) has proved to be one of the most direct formulations for the concept of quantization of angular momentum and has also been able to introduce the third quantum number m_φ, the selection rule for the transition between magnetic states of the atom ($\Delta m_\varphi = 0, \pm 1$)

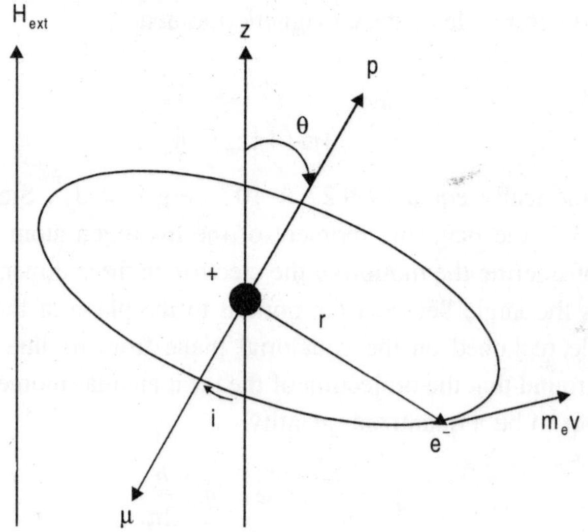

Figure 1.18: *A hydrogen atom placed in an external magnetic field that determines the z-direction. The orbital motion of the electron develops a magnetic dipole moment μ which interacts with H_{ext} to produce a potential energy $\Delta E_m = -\mu H_{ext}$. p is the orbital angular momentum of the electron*

1.2.9 DRAWBACKS OF THE BOHR THEORY

The postulates of Bohr model concerning the quantization of momentum (which is expressed by the quantum condition $\oint p_i dq_i = n_i h$) follows from the empirical expression for the energy of the atom. For this reason, coincidence with experimental values of the atom, calculated on the basis of this postulate, cannot be considered as experimental confirmation of Bohr model. The experimental verification of the Bohr model must be based only on those experimental properties of atom which are not used in the design of the model. Among others, these are magnetic properties of the atom. The results of the theory, both as regards the magnitude of the magnetic moment of atoms and the phenomenon of space quantization, are in close agreement with experiment. However, a more detailed examination of the magnetic properties of the atoms shows that this correspondence is only qualitative and, actually, apparent. Even the introduction of third quantum number in the Bohr-Sommerfeld model could not explain the anomalous Zeeman Effect. We know that orbit motion of the electron develops a magnetic dipole moment μ, which interacts with external magnetic field (H_{ext}) to produce a potential energy:

$$\Delta E_m = + \mu H_{ext} \cos \alpha = + \mu_B H_{ext} m_\varphi \qquad ...(1.142)$$

Thus the total energy of hydrogen atom in the presence of external magnetic field is now:

$$E_{n,n_\varphi,m_\varphi} = -\frac{me^2}{8\,\varepsilon_0^2\,h^2 n^2} + \mu_B H_{ext} m_\varphi$$

or
$$E = E_0 + \mu_B H_{ext} m_\varphi \qquad ...(1.143)$$

The first term is due to Coulombic interaction between the electron and proton, while the second term is due to interaction between external magnetic field and magnetic dipole moment of the circulating electron.

It follows from equation (1.143) that the separation between any two consecutive energy levels is constant and given by $\mu_B H_{ext}$. In the simplest case the Zeeman Effect causes a symmetrical splitting of the frequencies ν_+, ν_0, and ν_{-1} and the magnitude of splitting $\Delta\nu_0 = \nu_+ - \nu_0 = \nu_0 - \nu_{-1}$ is equal to $\mu_B H_{ext}/h$. From this value, μ_B can be obtained. *It is observed for instance, in the system of singlets of helium and the group of alkaline earth elements and also in the spectra of Zn, Cu and Hg. One of the single lines of Cd, 643.846 nm shows normal triplet.*

In order to deduce the Zeeman spectrum of atom, we write from equation (1.143)

$$E' = E_0' + m_\varphi' \mu_B H_{ext} \qquad \text{(for ground state)}$$
$$E'' = E_0'' + m_\varphi'' \mu_B H_{ext}$$

It follows that

$$h\nu = E'' - E' = E_0'' - E_0' + (m_\varphi'' - m_\varphi')\mu_B H_{ext}$$

or
$$\nu = \frac{E''-E'}{h} = \frac{E_0''-E_0'}{h} + \frac{m_\varphi''-m_\varphi'}{h}\mu_B H_{ext}$$

or
$$\nu = \nu_0 + (m_\varphi'' - m_\varphi')\frac{\mu_B H_{ext}}{h} \qquad ...(1.144)$$

or
$$\nu = \nu_0 + \Delta m_\varphi \frac{\mu_B H_{ext}}{h}$$

For $\Delta m_\varphi = 0$, *i.e.*, no external field is applied, $\nu = \nu_0$ and corresponding wavelength $\lambda = c/\nu_0$. The original spectral line at ν_0 or λ_0 will split into three components of frequencies/wavelengths for $\Delta m_\varphi = 0, \pm 1$. Hence for $n_\varphi = 1$, $n = 1$, the frequencies are

$$\nu_0 - \frac{\mu_B H_{ext}}{h} \qquad \nu_0 \qquad \nu + \frac{\mu_B H_{ext}}{h}$$

or
$$\frac{c}{\nu_0 - \mu_B H_{ext}/h} \qquad \frac{c}{\nu_0} \qquad \frac{c}{\nu_0 + \mu_B H_{ext}/h}$$

i.e., the splitting of spectral line in weak magnetic fields. The magnetic field splitting of lines of the triplet system of cadmium and the lines of the alkaline elements is much more complicated.

The other limitations of the Bohr model are:

(*i*) Bohr and even Bohr-Sommerfeld model fails to evaluate the intensities of spectral lines.

(*ii*) This model fails to calculate the energy and magnetic moment of the helium atom and encountered the same difficulties as in case of simple hydrogen atom.

(*iii*) This model cannot explain why the value of quantum number (now called principal quantum number) should have integral values.

(*iv*) Bohr model fails to explain the basis of the periodic system and molecular geometry.

(*v*) Bohr model is incapable of taking into account the interaction between the electrons and hence fails to predict the spectra of alkaline metals. For example, in case of alkali metals, the outermost electron (when present at a certain distance apart from the nucleus) does not experience its full charge. The charge is screened by the inner electrons and hence only fractional value of Ze, effective nuclear charge is experienced by the outermost electron.

(*vi*) This model cannot account for the two dimensional orbit when lines of force coming out of the positive charge should be uniform in all the directions, *i.e.*, it should have a spherical symmetry.

(*vii*) Since we cannot measure the quantities like radius of the orbit or velocity of the electron in any orbit by any experimental method, it casts doubt about the design of the model whether it gives correct description of these quantities or not.

(*viii*) Bohr's model fails to provide a method of calculating transition probabilities and intensities of spectral lines.

To conclude we can say that Bohr model is capable of giving satisfactory and semi-quantitative explanation only of an extremely limited number of experimental facts. This limitedness of Bohr's model (which is based on concepts of classical mechanics into which Bohr artificially introduced quantum laws alien to it) is rooted in the limitedness of classical concepts concerning microscopic particles; this led to the evolution of new concepts of dual character of photons and other fundamental particles.

1.3 MATTER WAVE OR de BROGLIE WAVES

The deflection of electrons by electrostatic and magnetic fields and accurate determination of ratio of their charges to mass ratio suggest very strongly that the electrons are particles. In 1924, de Broglie suggested that since light has much of character of particle, particles in particular electrons, by symmetry in nature, can be expected to exhibit wave like properties. The sand particles at the sea shore are evidence to the fact that matter waves, *i.e.*, water waves, possess energy. The photon, *i.e.*, corpuscular theory of light (formulated by Newton in 1672), which described such phenomenon as the photoelectric effect and Compton effect fails to explain the broad range of optical phenomenon such as interference and diffraction of light. These are better understood in terms of wave concepts (Huggens in 1678 formulated wave theory of light). Thus the wave theory concepts should by no means be rejected as a consequence of the success of the photon theory. Both the informations concerning the properties of light compel us to recognize that light possesses both corpuscular and wave properties. In certain phenomena photon properties predominate and in others wave properties predominate.

Let us examine the basic relations that connect the particle aspect of electromagnetic radiation with wave aspect. According to Planck, the energy of light particles or corpuscles and frequency of light (ν) are related to each other by the expression:

$$E = h\nu \qquad ...(1.46a)$$

The relativistic mechanics gives the total energy

$$E = E_0 + K \qquad ...(1.36)$$

In case of photon, $E = m_0 c^2$ and the total energy E = K, is entirely kinetic. However, according to de Broglie, frequency has no real place as momentum has in purely particle theory. The energy of photon, like any other portion of matter obeys equation:

$$E = mc^2 \qquad ...(1.36a)$$

From this it follows that mass of the photon is:

$$m = \frac{E}{c^2} = \frac{h\nu}{c^2}$$

The mass of the photon is the mass of the moving particle. The rest mass of the photon is nearly zero. The experimentalists claim it to be less than 0.6×10^{-20} MeV. Pressure of light and bent rays in the strong gravitational field are evidence for the mass of the photon. The energy expression for photon can also be expressed as:

$$E = mc^2 = pc \qquad ...(1.39)$$

where p is the momentum associated with photon. Equating equations (1.46a) and (1.39)

$$p = \frac{h\nu}{c} = \frac{h}{\lambda} \qquad ...(1.67)$$

This expression relates the wave character of photons, λ the wavelength and frequency ν to particle character, the momentum p. This direct relationship between corpuscular (momentum of photon) and the wave characteristics (frequency, wavelength) of light is an evidence of dual (wave-particle) character of light. But we would like to know the wavelength of a particle whose rest mass is different from zero. This is possible only if it is assumed that the relation between the momentum and wavelength is universal. According to de Broglie, if light wave can have a particle nature, then particles, such as electrons can have wave characteristics. Accordingly an electron can have a momentum-wave relation as

$$p = mv = \frac{h}{\lambda} \qquad ...(1.145)$$

where m is the relative mass. This expression shows that electron has a particle character. Now applying photon equation (1.67) to a particle equation (1.145), the wavelength of electron becomes:

$$\lambda = \frac{h}{mv} \qquad ...(1.146)$$

This is the famous de Broglie relation and holds good when the mass and the velocity of the particle are small.

Another interesting property of matter waves is that the velocity of particle (v) is the same as the group velocity (v_g) of the de Broglie waves (which represents a material point). The particle speed can be determined from $v = \dfrac{dE}{dp}$. The group velocity is defined from the wave theory as:

$$v_g = \frac{d\nu}{d\left(\frac{1}{\lambda}\right)} = -\lambda^2 \frac{d\nu}{d\lambda} \qquad ...(1.147)$$

The superposition of different waves of nearly the same frequency forms a sharply defined wave group (or wave packet) that represents material point. The wave particle or envelope formed from such a system of waves propagates with a velocity called group velocity (v_g), different from the velocity of either of the components velocities, called the phase velocities, which are approximately equal, *i.e.*, $v_{ph_1} = v_{ph_2} = v_{ph_3} = ... = v_{phn} = v_{ph}$ (Fig. 1.19). Using $\lambda = h/p$, ...(1.146)

$$d\lambda = -\frac{hdp}{p^2} \qquad ...(1.148)$$

We also know: $E^2 = E_0^2 + p^2c^2$ ...(1.149)

Thus $dE = \dfrac{pc^2}{E}dp$

Further $\nu = \dfrac{E}{h}$ and $d\nu = \dfrac{dE}{h} = \dfrac{pc^2}{Eh}dp$

Substituting $d\nu$ in equation (1.147):

$$v_g = -\lambda^2 \frac{pc^2}{Eh}dp \qquad ...(1.150)$$

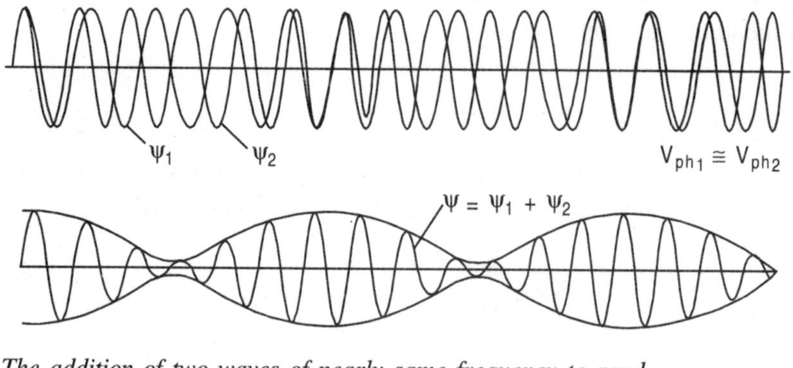

Figure 1.19: *The addition of two waves of nearly same frequency to produce wave groups (or beats)*

Combining equations (1.146), (1.148) and (1.150)

$$v_g = \frac{pc^2}{E}$$

or $$v_g = \frac{mvc^2}{mc^2} = v \quad ...(1.151)$$

Thus, the group velocity of particles (*i.e.*, wave group or wave packets) is equal to the velocity of a particle.

Let us now see whether or not this concept is valid for photons?

For a photon, $pc = E$

and group velocity of particle is:

$$v_g = \frac{pc^2}{E}$$

Comparing the two expressions, we get:

$$v_g = c$$

Thus, just like electrons, group velocity for photons (light wave packet) is equal to the velocity of the photon. The wave packet is sharply defined and its size Δx has been reduced considerably.

The diffraction of electrons from single crystals of nickel or diffraction by passing electrons through thin polycrystalline metal targets provides a concrete evidence to the wave nature of electron and hence de Broglie concept of matter waves. Note that in electron diffraction, electron rays are scattered by electrons and nuclei, whereas in *x*-ray diffraction, *x*-rays are scattered by electrons alone. On the other hand neutrons unlike *x*-rays, are not scattered by electrons but are deflected (scattered) from their paths in encounters with atomic nuclei. There are many substances whose atoms differ only slightly as to the number of electrons but differ radically as the properties of nuclei. In such cases *x*-rays cannot distinguish the atoms but neutron diffraction analysis is highly successful. This makes neutron diffraction analysis an important method of studying the structure of substances.

Let us now answer the question about the scope of this principle, nature of matter by considering the following examples.

Consider the simple particles, an electron. Its velocity in any of the orbit of hydrogen atom is of order of 10^6 ms^{-1}. Substituting the values of m_e and v in equation (1.146):

$$\lambda = \frac{6.626\times10^{-34}\text{ Js}}{(9.1\times10^{-31}\text{ kg})\,(10^6\text{m s}^{-1})} = 72.9 \times 10^{-10}\text{ m}$$

This value of λ is associated with X-rays and also nearly of the same value as the size of atoms. In other words, one can measure this wavelength and can associate the electron with dual character.

Now consider a particle of mass 100 g, moving with a velocity of 10 km s^{-1}. Substitute these values in equation (1.146) to calculate λ.

$$\lambda = \frac{6.626\times10^{-34}\ \text{Js}}{(0.1\ \text{kg})\ (10\ \text{km s}^{-1})} = 6.626 \times 10^{-34}\ \text{m}$$

In the latter case, the wavelength is extremely small and their behaviour can be explained on the basis of classical mechanics nicely. Equation (1.146) shows that wavelength associated with the matter depends on its mass and the velocity with which it is moving. Lighter and slow moving particles will have long wavelength and heavy and fast moving particles will have short wavelengths.

Example 1.10. Calculate the voltage required to accelerate an electron to have a velocity of 1.87 × 10^9 ms^{-1}. What will be the de Broglie's wavelength?

Solution. The equation required to calculate voltage is:

$$Ve = \frac{1}{2} m_e v^2$$

$$V = \frac{(1/2)\ (9.11\times10^{-31}\ \text{kg})\ (1.87\times10^{9}\ \text{m s}^{-1})^2}{1.6\times10^{-19}\ \text{C}} = 9.96 \times 10^{6}\ \text{V}$$

Equation (1.146) is used to calculate de Broglie wavelength

$$\lambda = \frac{6.626\times10^{-34}\ \text{Js}}{(9.11\times10^{-31}\ \text{kg})\ (1.87\times10^{9}\ \text{ms}^{-1})} = 0.389 \times 10^{-12}\ \text{m} = 0.389\ \text{pm}$$

1.3.1 HYDROGEN ATOM AND de BROGLIE HYPOTHESIS

Bohr's theory of hydrogen atom assumes that the angular momentum is quantized and its value is given by the relation:

$$p_\varphi = mvr = n\frac{h}{2\pi}, \quad n = 1, 2, 3, ... \qquad ...(1.80)$$

The above relation can also be obtained with the help of de Broglie's hypothesis. Since the matter can also have dual character, the electron moving in any orbit with velocity, v, will be associated with wave character. Let its wavelength be λ and can be calculated from:

$$\lambda = \frac{h}{m_e v} \qquad ...(1.146)$$

Now if we assume that in the stationary state orbits, the waves are also stationary, then their maxima and minima will not change with time. To fulfill this condition, it is necessary that in any orbit, circumference of the orbit should contain the integral multiples of waves. If this is not so, then after every one round of the orbit, the phase of wave will be different from the earlier one and the maxima and minima will change with time. If r is the radius of the circular orbit, then:

$$2\pi r = n\lambda, \qquad n = 1, 2, 3,... \qquad ...(1.152)$$

Combining equations (1.146) and (1.152)

$$2\pi r = \frac{nh}{m_e v}$$

$$m_e vr = p_\varphi = n\frac{h}{2\pi}$$

i.e., similar relation as assumed by Bohr.

1.4 HEISENBERG'S UNCERTAINTY PRINCIPLE

This principle was deduced by Heisenberg in 1927. This is perhaps the most general and fundamental principle of nature on which the conclusions of quantum mechanics diverge from those of classical mechanics. In classical mechanics it is assumed that one can simultaneously measure as many properties of a system as one imagines, *i.e.*, it allows to pinpoint the exact location of a particle at any given time. On the other hand Heisenberg principle imposes some fundamental limit to the simultaneous determination of certain pair of variables of particle called complimentary or conjugate pairs. The product of such pairs has the dimensions of action (*i.e.*, ML^2T^{-1}) and is associated with the same coordinate or degree of freedom. Examples of such pairs are: position and linear momentum, angular momentum and angle, energy and time. To conclude we say that *the two members of any conjugate pair of variables cannot be simultaneously measured accurately*. The uncertainty principle is a quantitative statement of the conclusion. It is not only a consequence of the wave-particle duality, *i.e.*, light simultaneously possess wave and corpuscular properties, but also dual behaviour of matter, *i.e.*, electron.

Let us suppose that a particle is moving with a velocity v, which is located somewhere with in wave packet that moves with velocity $v_g = v$ (Fig. 1.20). The group velocity of a wave packet can be written as:

$$v_g = \frac{dE}{dp} = \frac{dh\nu}{dp} = h\frac{d\nu}{dp} \quad \text{....(1.153)}$$

From Fig. 1.20 and equation (1.153), we get:

$$v_g = \frac{\Delta x}{\Delta t} = \frac{h\Delta\nu}{\Delta p}$$

or $$\Delta p\, \Delta x = h\Delta t\Delta\nu \quad \text{...(1.154)}$$

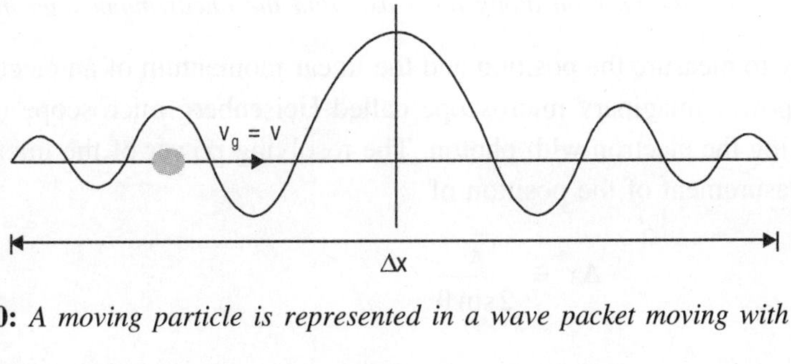

Figure 1.20: *A moving particle is represented in a wave packet moving with velocity* v_g

The minimum time required for the measurement of frequency of wave is equal to the time required for one complete wavelength to pass a reference point. Accordingly

$$\Delta t \geq \frac{1}{\Delta \nu} \quad \text{...(1.155)}$$

Comparing equations (1.154) and (1.155) we obtain the inequality.

$$\Delta p \; \Delta x \geq h \quad \text{...(1.156)}$$

which is the expression for Heisenberg's uncertainty principle. Due to small value of $h = 6.62 \times 10^{-34}$ Js, this relation is relevant in the microscopic world of atoms, protons, electrons, but not in the macroscopic world. The fundamental limit $\geq h$ implies not only that there is limit to the precision of a measurement but that the greater is the localization of the particle in position (*i.e.*, Δx is smaller), the greater will be the uncertainty in the measurement of its momentum and vice versa. This uncertainty of either position or momentum ($\geq h$) is evidently neither a matter of physical imperfection in the measuring apparatus not it is possible to improve on it by some other technique because it is the inherent or latent property of the system. The momentum in equation (1.156) is actually a component of momentum along the *x*-axis, so that it is more appropriate to write it as:

$$\Delta p_x . \; \Delta x \geq h \quad \text{...(1.157)}$$

There is no limitation in the precision of measurement of the components of momentum along the *y*- and *z*-directions (Δp_y and Δp_z) and the position in the *x*-direction.

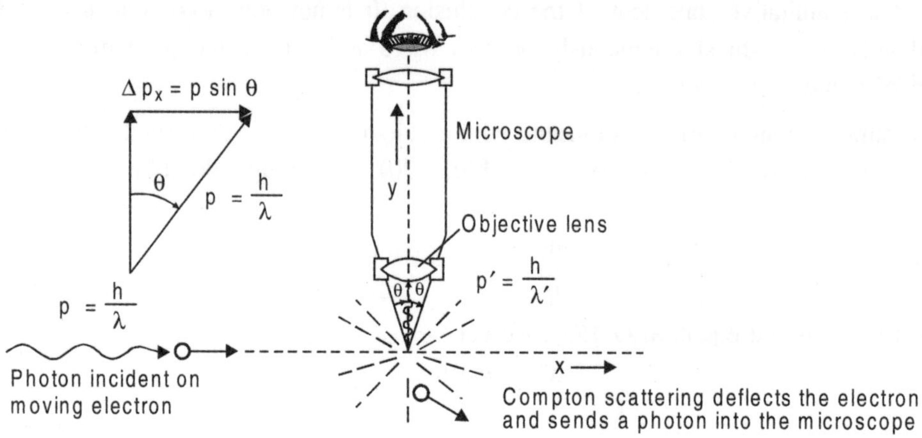

Figure 1.21: *Schematic diagram of Heisenberg microscope. Simultaneous measurements of the position along the x-axis and the linear momentum in the x-direction*

Let us now try to measure the position and the linear momentum of an electron with the aid of a high resolving power imaginary microscope called Heisenberg microscope (Fig. 1.21). This is done by illuminating the electron with photon. The resolving power of the microscope permits an uncertainty in measurement of the position of

$$\Delta x \cong \frac{\lambda}{2 \sin \theta} \quad \text{...(1.158)}$$

When the distance is less than Δx, the two points will be scanned as one. Hence Δx will represent the smallest uncertainty in the position of the traveling electron whose position we are trying to measure with the microscope. Equation (1.158) shows that smaller the wavelength λ of the illuminating light, the smaller the Δx, *i.e.*, the position of the electron can be fixed with more certainty. At the same time, the illuminating photon will interact with the electron through Compton Effect. The deflected electron will be seen provided the scattered photon enters the microscope within angle 2θ. Accordingly the uncertainty in the momentum of the photon in the x-direction will be:

$$\Delta p_x \geq p \sin\theta \qquad ...(1.159)$$

where $p = h/\lambda$ is the momentum of the incoming photon. According to law of conservation of momentum, equation (1.159) should also give the uncertainty in the scattered electron. Therefore for the scattered electron equations (1.158) and (1.159) show that:

$$\Delta p_x \Delta x = (p \sin\theta)\left(\frac{\lambda}{2\sin\theta}\right) = p.\ \lambda/2 \qquad ...(1.160)$$

or

$$\Delta p_x\ \Delta x \geq \frac{h}{\lambda}\frac{\lambda}{2} = \frac{h}{2}$$

which is the uncertainty principle. A more sophisticated approach will show that:

$$\Delta p_x\ \Delta x \geq \frac{h}{2\pi} \text{ or } \frac{h}{4\pi} \qquad ...(1.161)$$

Heisenberg's uncertainty principle can also be extended to other conjugate variable, *i.e.*, to energy and time. Knowing the relation between frequency and its time period we can write:

$$\Delta t \cdot \Delta v \geq 1 \qquad ...(1.162)$$

or

$$\Delta(E/h)\Delta t \geq 1$$

which becomes $\Delta E \cdot \Delta t \geq h$ ...(1.163)

and is called Heisenberg-Bohr uncertainty principle. Accordingly energy and time cannot be measured simultaneously with infinite precision. Again a more sophisticated derivation of equation (1.163) will give:

$$\Delta E \Delta t \geq \frac{h}{2\pi} \text{ or } \frac{h}{4\pi} \qquad ...(1.164)$$

1.4.1 DIFFRACTION PHENOMENON AND HEISENBERG PRINCIPLE

From Heisenberg's uncertainty principle one can make some important deductions and are described here. Based on corpuscular model of light and in conjugation with this law, it is possible to explain the diffraction phenomenon. Consider a long narrow slit of width b at a distance d from the source S as shown in Fig. 1.22. The distance between the source and slit can be varied at will and is kept large so that p_x of the photons approaching the slit has momentum only in the x-direction. The maximum value of p_x of photons reaching the slit will be:

$$p_x^{\max} \cong p \sin \theta = p \frac{b}{d} = \frac{h\nu}{c} \frac{b}{d} \qquad ...(1.165)$$

According to particle model of radiation, the number of particles (*i.e.*, intensity) will be maximum at P and will keep on decreasing on both sides of P and will be extremely small at points Q and Q′. The intensity will further decrease if we decrease the slit width. This is contrary to the experimental results. In reality the beam undergoes diffraction and intensity at the points such as Q and Q′ increases as the slit width is decreased. Thus the corpuscular model of radiation fails to explain the diffraction phenomenon. However corpuscular model in conjugation with natural law of uncertainty is capable of explaining the phenomenon of diffraction when a particle (photon or electron) passes through the slit we can say that the uncertainty in the position of particle is:

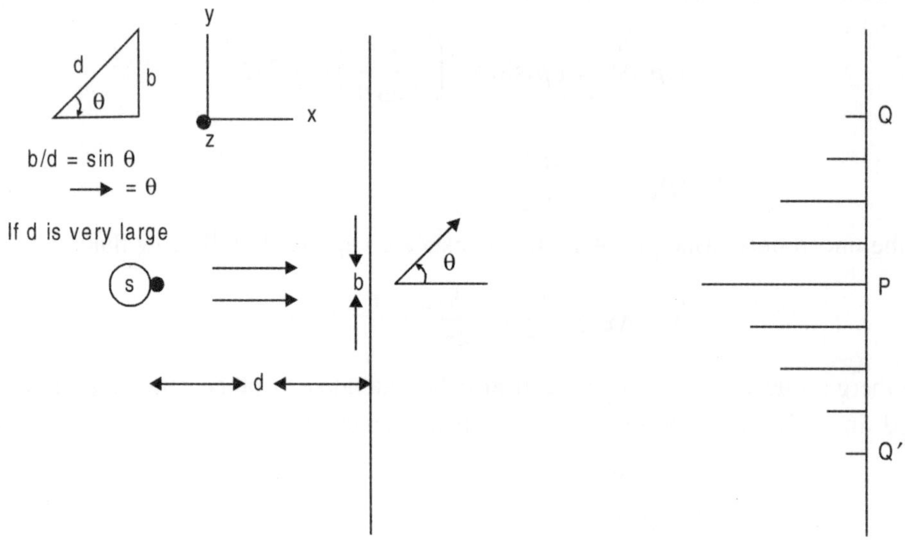

Figure 1.22: *Diffraction of particle (photon or electron) beam by a narrow slit of width b*

$$\Delta x \approx b \qquad ...(1.166)$$

According to uncertainty principle

$$\Delta p_x b \approx h \text{ or } \Delta p_x = \frac{h}{b} \qquad ...(1.167)$$

Thus we can say that when the photon passes through the slit, it imparts momentum $\approx h/b$ to the particle in the x-direction. Before the photon enters the slit, p_x (and hence Δp_x) can be made very small by increasing the distance d from the slit. By equation (1.165), as $d \to \infty$, $p_x \to 0$. Thus $\Delta p_x = 0$. In such a situation when photon passes through the slit, $\Delta p_x . \Delta x = 0$, which is not expected. Actually $\Delta p_x . \Delta x$ becomes zero before the photon enters the slit. In reality when the photon enters the slit it is localized with in distance b in the x-direction and hence $\Delta p_x \approx h/b$. Further before the photon enters the slit, $p_x \approx 0$, consequently

$$|p_x| \approx \Delta p_x \approx \frac{h}{b}$$

But $$p_x = p \sin\theta$$

Thus $p \sin\theta \approx \dfrac{h}{b}$ or $\sin\theta = \dfrac{h}{b}p$ ...(1.168)

Equation (1.168) reveals that the photon traveling at an angle of θ with the x-direction is inversely proportional to the slit width as expected on the basis of experimental results. This is indeed the diffraction phenomenon. Using de Broglie relation, momentum of photon is given by:

$$p = \frac{h\nu}{c} = \frac{h}{\lambda} \quad ...(1.145)$$

Consequently the equation (1.168) becomes

$$\sin\theta \approx \frac{\lambda}{b} \quad ...(1.169)$$

which conforms to the results based on the theory of diffraction. To conclude we can say that the wave-particle duality is a consequence of the uncertainty principle and vice versa.

1.4.2 BOHR'S HYDROGEN ATOM AND UNCERTAINTY PRINCIPLE

Based on Bohr's theory of hydrogen atom, it is possible to calculate the radii of various orbits in which the electron can move around the nucleus in different orbits, but there is no experimental verifications regarding the exactness of these quantities. Let us assume that we carry out such an experiment and the position of the electron in the first orbit can be determined by locating the position of photon, scattered by the electron, on the screen of a super microscope. Let λ and λ′ be the wavelengths of the photon, before and after the collision with the electron, moving in Bohr's first orbit. The shift in the wavelength of the photon can be calculated with the help of Compton Effect using equation (1.79).

$$(\lambda_s' - \lambda_0) = \frac{h}{m_0 c} = 2\sin^2\frac{\theta}{2} \quad ...(1.179)$$

The change in the energy of the electron after collision can be obtained as

$$\Delta E = h\Delta\nu = h(\nu_s' - \nu_0) = hc(\,1/\lambda_s' - 1/\lambda_0) = \frac{hc\,\Delta\lambda}{\lambda_s'\lambda_0} = \frac{hc\,\Delta\lambda}{\lambda_0^2}$$

$$= \frac{hc}{\lambda_0^2}\frac{h}{m_0 c} = \frac{h^2}{\lambda_0^2\, m_0} \quad ...(1.170)$$

if the change in the wavelength is not that large. Since we are only interested in locating the position of the electron in the first orbit, the energy gained by the electron or wavelength of photon used should be such that its corresponding energy should be less than that required for excitation to $n = 2$. This means that:

$$\Delta E \approx \frac{h^2}{\lambda_0^2 m} < \frac{e^2}{(4\pi\varepsilon_0)\,2a_0}\left(\frac{1}{1^2} - \frac{1}{2^2}\right) = \frac{3e^2}{(4\pi\varepsilon_0)\,8a_0} \quad ....(1.171)$$

where $a_0 = \dfrac{h^2(4\pi\varepsilon_0)}{4\pi^2 m_e e^2}$, *i.e.*, Bohr's unit of measuring radii.

$$\lambda_0^2 > \frac{8a_0 h^2 (4\pi\varepsilon_0)}{3m_e e^2}$$

or

$$\lambda_0 > \left[\frac{8a_0 h^2 (4\pi\varepsilon_0)}{3m_e e^2}\right]^{\frac{1}{2}} \quad ...(1.172)$$

Further from Bohr's unit for measuring radii,

$$\frac{h^2(4\pi\varepsilon_0)}{m_e e^2} = 4\pi^2 a_0$$

$$\lambda_0 > \left[\frac{8a_0}{3}\,.\,4\pi^2 a_0\right]^{\frac{1}{2}} = 4\left[\frac{2}{3}\right]^{\frac{1}{2}} \pi a_0 \quad ...(1.173)$$

After collision the momentum of the electron can change and we know that the magnitude of its component along the radius cannot be less that 0 and not more than $(2\Delta E\, m_e)^{1/2}$, where ΔE is the change in energy and m_e is the mass of the electron. That is, the value of the radial (or linear) momentum lies between 0 and $(2\Delta E m_e)^{1/2}$. The maximum uncertainty in the radial momentum will be:

$$\Delta p_r = (2m_e\, \Delta E)^{\frac{1}{2}} = \left(2m_e \frac{h^2}{\lambda_0^2 m_e}\right)^{\frac{1}{2}} = \sqrt{2}\,\frac{h}{\lambda_0} \quad ...(1.174)$$

Using Heisenberg uncertainty principle, one can calculate the uncertainty in the position of electron.

$$\Delta r = \frac{h}{\Delta p_x} \approx \frac{\lambda_0}{\sqrt{2}} \quad ...(1.175)$$

Combining equations (1.173) and (1.175)

$$\Delta r > 4\left(\frac{2}{3}\right)^{\frac{1}{2}}\left(\frac{1}{2}\right)^{\frac{1}{2}} \pi a_0 = 4\pi a_0 \left(\frac{1}{3}\right)^{\frac{1}{2}} = 7.26\, a_0 \quad ...(1.176)$$

Equation (1.176) clearly indicates that the error in measuring the position of the electron is much more than the actual values of a_0. Thus deriving an exact expression or talking of exact values of radii of Bohr's orbits does not mean anything and Bohr's theory also fails here.

Example 1.11. Using uncertainty principle determine: (*a*) the size of the hydrogen atom and (*b*) the total energy of the electron in an orbit.

Solution: (*a*) Proton being heavier than electron, only the motion of electron is considered. Let the electron be confined to a region of linear dimension $\approx a$. According to uncertainty principle:

$$p \approx \Delta p \approx \frac{h}{2\pi a}$$

Thus the kinetic energy of the electron is:

$$\text{K} = \frac{p^2}{2m_e} = \frac{h^2}{8\pi^2 a^2 m_e}$$

The total energy of the electron is given by:

$$\text{E} = \text{K} + \text{V} = \frac{h^2}{8\pi^2 a^2 m_e} - \frac{e^2}{(4\pi\varepsilon_0)a}$$

The system will have lowest energy provided:

$$\frac{d\text{E}}{da} = -\frac{h^2}{4\,\pi^2 a^3 m_e} + \frac{e^2}{(4\pi\varepsilon_0)a^2} = 0$$

The solution of this equation gives:

$$a = a_0 = \frac{h^2(4\pi\varepsilon_0)}{4\pi^2 e^2 m_e}$$

Substitute the values of respective parameters on the right hand side of the above equation to obtain the value of *a*.

$$a = a_0 = \frac{(6.62\times10^{-34}\ \text{J s})^2\ (1.113\times10^{-10}\ \text{C}^2\ \text{N}^{-1}\ \text{m}^{-2})}{4\times(3.14)^2\times(9.1091\times10^{-31}\ \text{kg})\ (1.602\times10^{-19}\ \text{C})^2}$$

$$= 0.529\times10^{-10}\left(\frac{\text{kg m}^2\ \text{s}^{-2}\ \text{m}^{-1}\ \text{s}^2}{\text{kg}}\right) = 0.0529\ \text{nm}$$

Thus the size of the hydrogen atom is a direct consequence of Heisenberg's uncertainty principle.

(*b*) In order to obtain the total energy of the system, replace '*a*' by $\frac{h^2(4\pi\varepsilon_0)}{4\pi^2 e^2 m_e}$ in the total energy equation.

$$\text{E} = \frac{h^2\ (4\pi^2 m_e e^2)^2}{(4\pi\varepsilon_0)^2\ 8\pi^2 m_e h^4} - \frac{e^2\times4\pi^2 m_e e^2}{(4\pi\varepsilon_0)\ (4\pi\varepsilon_0)h^2} = -\frac{2\pi^2 m_e e^4}{(4\pi\varepsilon_0)^2 h^2}$$

Substituting values of the respective parameters we get:

$$E = -\frac{(2)(3.142)^2(9.1091\times10^{-31}\,kg)(1.602\times10^{-19}\,C)^4}{(1.113\times10^{-10}C^2\,N^{-1}m^{-2})^2(6.626\times10^{-34}\,J\,s)^2} = -2.17\times10^{-19}\ kg\ m^2\ s^{-2}$$

$$= -2.17\times10^{-19}\ J = -13.6\ eV$$

Thus the ionization energy of hydrogen atom directly follows from the uncertainty principle.

Example 1.12. Using uncertainty principle estimate the energy of the ground state of two electron-atoms of nuclear charge *Ze*.

Solution. Let one of the two electrons is confined to a region of linear dimensions $\approx a_1$ and the second is localized in the region a_2. According to uncertainty principle:

$$p_1 \approx \Delta p_1 \approx \frac{h}{2\pi a_1}$$

and
$$p_2 \approx \Delta p_2 \approx \frac{h}{2\pi a_2}$$

Kinetic energy of the respective particle will be:

$$K_1 = \frac{p^2}{m_e} = \frac{h^2}{8\pi^2 m_e a_1^2} \quad \text{and} \quad K_2 = \frac{h^2}{8\pi^2 m_e a_2^2}$$

The total kinetic energy of the two electrons will be given by:

$$K = K_1 + K_2 = \frac{h^2}{8\pi^2 m_e}\left[\frac{1}{a_1^2} + \frac{1}{a_2^2}\right]$$

The total potential energy of the interaction with nucleus of charge Ze is given by:

$$V = V_1 + V_2 = -\frac{Ze^2}{(4\pi\varepsilon_0)\,a_1} - \frac{Ze^2}{(4\pi\varepsilon_0)\,a_2} = -\frac{Ze^2}{(4\pi\varepsilon_0)}\left[\frac{1}{a_1} + \frac{1}{a_2}\right]$$

and electrostatic force of repulsion between the two electrons will be $\approx \dfrac{e^2}{(4\pi\varepsilon_0)\,(a_1 + a_2)}$.

Thus the total energy of the system will be given by:

$$E(a_1, a_2) = K + V + \frac{e^2}{(4\pi\varepsilon_0)\,(a_1 + a_2)}$$

$$= \frac{h^2}{8\pi^2 m_e}\left[\frac{1}{a_1^2} + \frac{1}{a_2^2}\right] - \frac{Ze^2}{(4\pi\varepsilon_0)}\left[\frac{1}{a_1} + \frac{1}{a_2}\right] + \frac{e^2}{(4\pi\varepsilon_0)(a_1 + a_2)}$$

The energy of the system will be minimum provided:

$$\frac{\partial E}{\partial a_1} = -\frac{h^2}{4\pi^2 m_e\, a_1^3} + \frac{Ze^2}{(4\pi\varepsilon_0)a_1^2} - \frac{e^2}{(4\pi\varepsilon_0)(a_1 + a_2)^2} = 0$$

$$\frac{\partial E}{\partial a_2} = -\frac{h^2}{4\pi^2 m_e a_2^3} + \frac{Ze^2}{(4\pi\varepsilon_0)a_2^2} - \frac{e^2}{(4\pi\varepsilon_0)(a_1 + a_2)^2} = 0$$

Assuming $a_1 \approx a_2$, the solution of the above equation yields:

$$a_1 = a_2 = \frac{(4\pi\varepsilon_0)\, h^2}{4\pi^2 m_e e^2} \times \frac{1}{\left(Z - \frac{1}{4}\right)}$$

Substituting a_1 and a_2 in the total energy expression, we get:

$$E \approx -(Z - 1/4)^2 \; \frac{4\pi^2 m_e e^4}{(4\pi\varepsilon_0)^2 h^2} = -\,(Z - 1/4)^2\ (27.2\ \text{eV}) = -\,2\ (Z - 1/4)^2\ (13.6\ \text{eV})$$

Table 1.7: *Ground state energy (eV) of various two electron systems (in units of ionization energy of hydrogen atom, i.e., in units of 13.6 eV)*

System (Z)	Theoretical Value	Experimental Values
He (2)	–6.125	–5.807
Li^+ (3)	–15.125	–14.56
Be^{2+} (4)	–28.125	–27.31
B^{3+} (5)	–45.125	–44.06
C^{4+} (6)	–66.125	–64.80

The values of E for two electron systems computed from the above expression have been compared with experimental results in Table 1.7. The agreement between the theoretical and experimental data is good indeed.

Example 1.13. An electron with energy 100 eV is passed through a circular hole of radius 10^{-6} cm. What will be the uncertainty in the angle of emergence?

Solution. The momentum of electron with energy 100 eV is :

$$p = (2m_e E)^{1/2} = [(2 \times 0.9 \times 10^{-27}\ \text{g})\ (1.6 \times 10^{-10}\ \text{g cm}^2\ \text{s}^{-2})]^{1/2}$$

$$= (28.8 \times 10^{-38}\ \text{g}^2\ \text{cm}^2\ \text{s}^{-2})^{1/2}$$

$$= 5.36 \times 10^{-19}\ \text{g cm s}^{-1}.$$

According to uncertainty principle:

$$\Delta p_x\, \Delta x \approx \frac{h}{2\pi}$$

$$\Delta p_x = \frac{6.626 \times 10^{-27}\ \text{erg s}}{2 \times 3.14 \times 2 \times 10^{-6}\ \text{cm}} = 5.25 \times 10^{-22}\ \text{g cm s}^{-1}$$

The electron after passing through the hole will be diffracted at a certain angle. The uncertainty in the angle of emergence from the hole will be given by:

$$\Delta\theta \approx \frac{\Delta p_x}{p} = \frac{5.25\times10^{-22}\ \text{g cm s}^{-1}}{5.36\times10^{-19}\ \text{g cm s}^{-1}} = 9.7 \times 10^{-4}\ \text{radians}$$

or
$$\Delta\theta = \frac{(9.7\times10^{-4})(180)(60)}{3.14} = 3.3\ \text{minutes of arc.}$$

Example 1.14. (*a*) The dimensions of an atom are of the order of 10^{-8} cm and we would like to obtain an accuracy of 10^{-9} cm. Can we devise an experiment capable of establishing the site occupied by the electron at a given instant? (*b*) The thickness of a track of an electron moving with velocity of 1000 km s^{-1} in a cloud chamber is of the order of 10^{-2} cm. What is the uncertainty in the velocity of electron?

Solution. In principle such an experiment is possible. But according to Heisenberg uncertainty principle,

$$\Delta p_x\ \Delta x \geq \frac{h}{2\pi} = 1.055 \times 10^{-34}\ \text{kg m}^2\ \text{s}^{-1}$$

$$\Delta v_x\ \Delta x \geq \frac{1.055\times10^{-34}\ \text{kg m}^2\ \text{s}^{-1}}{9.1091\times10^{-31}\ \text{kg}}$$

$$\Delta v_x \geq \frac{1.055\times10^{-34}\ \text{kg m}^2\ \text{s}^{-1}}{10^{-11}\ \text{m}\times 9.1091\times10^{-31}\ \text{kg}} = 10^7\ \text{ms}^{-1} = 10^9\ \text{cm s}^{-1}$$

i.e., Δv_x exceeds 10^9 cm s^{-1}, which is absolutely meaningless. In other words, we cannot say anything about the velocity of electron.

Let us now think up an experiment that can actually be set up to make a more exact estimate of the velocity of atomic electron. But then we will have lost all the knowledge about the site where the electron is located. The inequality applied to an atomic electron shows that the corpuscular aspects does not operate in this case. The concept of trajectory of an electron becomes meaningless and nothing at all can be said about the paths of transition of an electron from one energy level to another.

(*b*) Proceeding as in part (*a*):

$$\Delta v \geq \frac{1.055\times10^{-34}\ \text{kg m}^2\ \text{s}^{-1}}{9.1091\times10^{-31}\ \text{kg} \times 10^{-4}\ \text{m}} \geq 1\ \text{ms}^{-1}$$

Example 1.15. The speed of a proton, a nuclear particle moving in the *x*-direction is measured to an accuracy of 10^{-6} ms^{-1}. Determine the limit of accuracy to which its position can be located: (*a*) along the *x*-axis, and (*b*) along the *y*-axis. (*c*) Compare the results so obtained with those of an electron.

Solution. (*a*) According to Heisenberg uncertainty principle:

$$\Delta p_x \Delta x \geq \frac{h}{2\pi} = \frac{6.626\times10^{-34}\ \text{J s}}{2\times3.14} = 1.055\times10^{-34}\ \text{kg m}^2\text{s}^{-1}$$

$$\Delta v_x = 10^{-6}\ \text{m s}^{-1},\ m_p = 1.672 \times 100^{-27}\ \text{kg}$$

Thus $$\Delta x \geq \frac{1.055\times10^{-34}\ \text{kg m}^2\ \text{s}^{-1}}{\Delta p_x} = \frac{1.055\times10^{-34}\ \text{kg m}^2\ \text{s}^{-1}}{m_p \times \Delta v_x}$$

$$= \frac{1.055 \times 10^{-34}\ \text{kg m}^2\ \text{s}^{-1}}{1.672\times10^{-27}\ \text{kg}\times10^{-6}\ \text{ms}^{-1}} = 0.063\,\text{m}$$

(*b*) The accuracy limit of determination of proton will be zero as Δx and Δv_x or Δp_y do not belong to the same coordinate.

(*c*) In case of an electron, $m_e = 9.1091 \times 10^{-31}$ kg, thus:

$$\Delta x \geq \frac{1.055\times10^{-34}\ \text{kg m}^2\text{s}^{-1}}{9.1091\times10^{-31}\ \text{kg}\times10^{-6}\ \text{ms}^{-1}} = 115\ \text{m}$$

Since $\frac{\Delta x(e)}{\Delta x(p)} = \frac{115\ \text{m}}{0.063\ \text{m}} = 1825$, so the accuracy of measurement of a proton is 1825 time more than that of an electron as expected. Mass of the proton is 1836 times more than that of an electron.

Example 1.16. The lifetime of an excited state of an atom is 10^{-8} s. What might be the minimum uncertainty in the determination of the energy of the excited state?

Solution. According to Bohr-Sommerfeld uncertainty principle:

$$\Delta E\,\Delta t \geq \frac{h}{2\pi} = 1.055 \times 10^{-34}\ \text{Js.}$$

$$\Delta E \geq \frac{1.055\times10^{-34}\ \text{Js}}{10^{-8}\text{s}} = 1.055 \times 10^{-26}\ \text{J} = 6.59 \times 10^{-8}\ \text{eV.}$$

This is called energy width of an excited state. Similarly from this principle we can determine the width of the excited state vibration, rotational and electronic states from their respective lifetimes in the excite states.

Example 1.17. The uncertainty in the position of an electron moving in a straight line is one nm. What might be the uncertainty in the measurement of: (*a*) its momentum, (*b*) its velocity, and (*c*) its kinetic energy?

Solution. Using Heisenberg uncertainty principle:

$$\Delta p_x\ \Delta x \geq \frac{h}{2\pi} = \frac{6.626\times10^{-34}\ \text{Js}}{2\times3.14} = 1.055 \times 10^{-34}\ \text{kg m}^2\ \text{s}^{-1}$$

Here $$m_e = 9.1091 \times 10^{-31}\ \text{kg},\ \Delta x = 1\ \text{nm} = 10^{-9}\ \text{m}$$

$$(a)\ \Delta p_x \geq \frac{1.055\times10^{-34}\ \text{kg m s}^{-1}}{10^{-9}\ \text{m}} = 1.055\times10^{-25}\ \text{kg m s}^{-1}$$

$$(b)\ \Delta v_x \geq \frac{1.055\times10^{-25}\ \text{kg ms}^{-1}}{9.1091\times10^{-31}\ \text{kg}} = 1.16\times10^{5}\ \text{m s}^{-1}$$

(c) Since kinetic energy $= \dfrac{p^2}{2m_e}$

$$\Delta E = \frac{(\Delta p)^2}{2m_e}$$

According to uncertainty principle in the measurement of kinetic energy is:

$$\Delta E \geq \frac{(1.055\times10^{-25}\ \text{kg ms}^{-1})^2}{2\times9.1091\times10^{-31}\ \text{kg}} = 6.10\times10^{-21}\ \text{kg m}^2\ \text{s}^{-2} = 6.10\times10^{-21}\ \text{J}$$

Example 1.18. A particle moving along x-direction has uncertainty in its position equals to its de Broglie wavelength. What might be the percentage uncertainty in its velocity?

Solution. de Broglie wavelength is defined as:

$$\lambda = \frac{h}{m_e v},\ \text{Thus}\ \Delta x = \lambda = \frac{h}{m_e v_x}$$

Using Heisenberg uncertainty principle:

$$\Delta p_x\, \Delta_x \geq \frac{h}{2\pi}$$

$$\Delta p_x \geq \frac{h}{2\pi\Delta x} = \frac{h}{2\pi}\frac{m_e v_x}{h} = \frac{m_e v_x}{2\pi}$$

Since $\Delta p_x = m_e \Delta v_x$

$$\therefore \quad \Delta v_x = \frac{\Delta p_x}{m_e} = \frac{m_e\, v_x}{2\pi\, m_e} = \frac{v_x}{2\pi}$$

The fractional uncertainty is:

$$\frac{\Delta v_x}{v_x} \geq \frac{1}{2\pi} = \frac{1}{2\times3.14} = 0.16 = 16\%$$

Thus all particles, say electron, proton, neutron etc. which follow de Broglie principle will have 16% uncertainty in their velocities if the uncertainty in their positions is equal to their de Broglie wavelengths, *i.e.*, the percentage uncertainty in velocity is independent of the mass of the particle if $\Delta x = \lambda$ (de Broglie).

Example 1.19. A pulsed laser produces a pulse of 10 ps duration. Determine the relative uncertainty in the measurement of the laser energy.

Solution. Using Bohr-Sommerfeld uncertainty principle:

$$\Delta E.\,\Delta t \geq \frac{h}{2\pi} = 1.055 \times 10^{-34} \text{ Js.}$$

Here $$\Delta t = 10 \text{ ps} = 10^{-12} \text{ s}$$

Thus $$\Delta E \geq \frac{1.055\times10^{-34} \text{ Js}}{10\times10^{-12}\text{s}} = 1.055 \times 10^{-23} \text{ J.}$$

Heisenberg Uncertainty Principle from Diffraction of Light (or of Electron)

Heisenberg uncertainty relations can be derived from the diffraction of light (or an electron) beam. Figure 1.23 shows the schematic diagram of the experiment. The electron beam directed along the axis y passes through the diaphragm D of width d, and gets to the screen. In its interaction with the diaphragm, an electron manifests itself as a wave, but it strikes the screen as hundred per cent particle. The de Broglie hypothesis predicts that the distribution of the electrons over the screen will be of the kind shown by the half circles. This means that the maximum intensity of the diffracted electrons will be found on the y-axis and the first minimum corresponds to angle φ given by the relation:

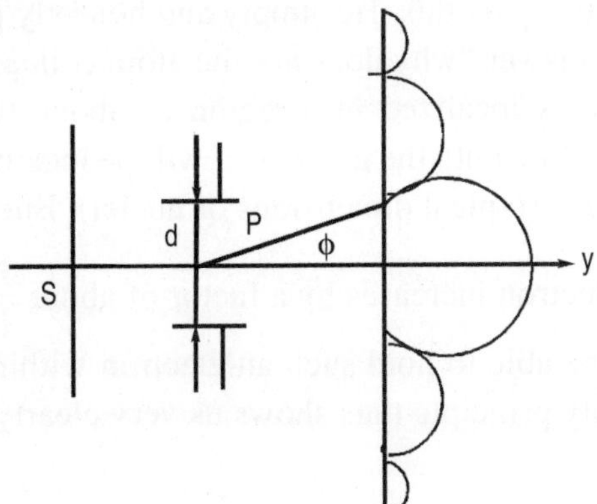

Figure 1.23: *Interaction of light or electron beam with small slit. S source of radiation, d is the width of the diaphragm and circles are the path of the photon or electron after passing through the diaphragm*

$$\sin \varphi = \frac{\lambda}{d} \text{ so much for the wave picture.} \quad ...(1.177)$$

Now let us treat the electrons as particles. At the instant when an electron passes through the diaphragm, its x-coordinate is only known with an error equal to the width of the diaphragm,

$$\Delta x = d \quad ...(1.178)$$

After the diaphragm, the electron changes the direction of its motion, which means that p_x, the projection of its momentum on the x-axis, will no longer be zero. If we consider only the first maximum; we have $0 \leq p_x \leq p \sin \varphi$.

But since $$\sin\varphi = \frac{\lambda}{d}, \text{ so that } \Delta p_x = p\frac{\lambda}{d} \quad ...(1.179)$$

Substituting the value of λ from de Broglie equation, *i.e.* $\lambda = \frac{h}{p}$, into equation (1.179) one gets:

$$\Delta p_x \approx \frac{h}{d} \quad \text{...(1.180)}$$

Or $$\Delta x\ \Delta p_x \approx h \quad \text{...(1.181)}$$

Quite similarly, we can write for the other two coordinates.

$$\Delta y\ \Delta p_y \approx h,\ \Delta z \approx h$$

These are Heisenberg's uncertainty relations after the name of Werner Heisenberg who first derived them rigorously.

Heisenberg Uncertainty Principle and Stability of the Atom

A very important question remained unanswered in the Bohr model, namely, why an electron, when rotating about the nucleus, does not radiate energy (which would lead to its fall on to the nucleus). Bohr has nothing to say on this. He simply and honestly postulated it. But let us try to use uncertainty principle to answer "why does not the atom collapse?" If an electron is moving in an orbit, we argue that it is localized in a region of about 10^{-10} m in size (such are the dimensions of the atom). If it falls onto the nucleus, it will be localized in a much smaller region, of about 10^{-14} to 10^{-15}m in size (typical dimensions of nuclei). But then the equation $\Delta x\ \Delta p_x \approx h$ shows that the velocity of electron increases by a factor of about $\frac{10^{-10}}{10^{-15}} = 10^5$. The electrostatic attractive forces would not be able to hold such an electron within the nucleus. It would leave it and go free. The uncertainty principle thus shows us very clearly that the electron simply can not fall.

Heisenberg Uncertainty Principle and Existence of Orbits

From classical mechanics, an electron in an atom has speed of the order of 10^6 m s^{-1}. Since the dimensions of the atom are something like 10^{-10} m, we can obtain from equation $\Delta x\ \Delta p_x \approx h$, the uncertainty in the electron velocity:

$$\Delta v_x \approx \frac{h}{m_e \Delta x} = \frac{6.626 \times 10^{-34}\ \text{Js}}{(9.1 \times 10^{-31}\ \text{kg})(10^{-10}\ \text{m})} = 7.28 \times 10^6\ \text{ms}^{-1}$$

that is to say, the error in determining the velocity is comparable with the velocity itself. This is in fact, the answer of our question. The motion of electron orbits turns out to be invalid. So the concept of probability was introduced in quantum mechanics. We may not speak of orbits. It is only meaningful to speak of the probability to find the electron at a given distance from the nucleus, *i.e.*, the concept of orbital came into being.

PROBLEMS

1. Equation (1.54) expresses the Planck's radiation distribution law in terms of frequency. Express the same equation in terms of wavelength.

2. Show that the expression for Rayleigh-Jeans law and Planck radiation distribution law has the units of energy per unit volume, *i.e.*, J m^{-3}.

3. Stefan-Boltzmann law gives the total energy radiated by black-body per unit area per unit time as $R = \frac{c}{4} E_v = \sigma T^4$, where E_v is the total radiation energy density and σ is the Stefan-Boltzmann constant. Using Planck's radiation distribution law, which gives energy density between ν and ν + *d*ν, show that the value of Stefan-Boltzmann constant $\sigma = \frac{2\pi^5 k_b^4}{15 h^3 c^2}$, where k_b is the Boltzmann constant. Calculate the value of σ using the values of constants. Compare it with experimentally known value 5.670 × 10^{-8} Jm^{-2} $K^{-4}s^{-1}$.

4. Using Planck's radiation distribution law, prove: (*a*) Wein's displacement law, *i.e.*, λ_{max} T = 2.90 × 10^{-3} m K and (*b*) Rayleigh-Jean's distribution law at lower frequencies.

5. For temperatures (*a*) 300 K, (*b*) 1000 K and (*c*) 10,000 K, calculate the value of λ_{max} in energy distribution function for a black body radiation.

6. The maximum wavelength observed in solar spectrum is at 500 nm. Calculate surface temperature of the Sun.

7. One of the hottest known stars (Sirius) possesses the maximum wavelength at 260 nm. Calculate its surface temperature.

8. Experimental value of Wein's displacement law is λ_{max} T = 2.90 × 10^{-3} mK and derived from Planck's distribution law is $\frac{hc}{4.965 k_b}$. Boltzmann constant, $k_b = \frac{R}{N_{av}}$ where R is gas constant (8.314 J K^{-1} mol^{-1}) and N_{av} is the Avogadro's number, calculate the value of Planck's constant and Avogadro's number.

9. Prove that the dimension of Planck's constant is that of angular momentum.

10. The work function of chromium metal is 4.40 eV. Calculate the threshold frequency of chromium metal.

11. When clean surface of silver is irradiated with 230 nm, the kinetic energy of the ejected electron is 1.28 × 10^{-19} J, whereas when it is irradiated with 260 nm, the kinetic energy of the ejected electron is 2.83 × 10^{-20} J. Calculate the value of Planck constant and work function of silver metal.

12. If the threshold frequency for lithium metal is 5.57 × 10^{14} Hz, calculate the stopping potential of lithium when irradiated with 350 nm.

13. Calculate the radius of 3rd, 5th and 7th Bohr orbits. Compare them with the radius of the first Bohr orbit radius for hydrogen atom. Do these values follow $r_n = n^2 r_1$, where r_1 is the radius of first Bohr orbit. (r_1 = 0.0529 nm)

14. Calculate the velocity of an electron in the 3rd, 5th and 7th Bohr orbits. Compare them with that of an electron in the first Bohr orbit. Do these values follow the $v_n = n\, v_1$, where v_1 is the velocity of an electron in the first Bohr orbit (v_1 = 2.18 × 10^6 m s^{-1}).

15. Calculate the energy of an electron in Bohr orbits when $n = 3$, 5 and 7.
16. Calculate the value of Rydberg's constant for He^+, Li^{2+} and B^{3+}. Compare these values for hydrogen atom neglecting the mass of the nucleus and taking the mass of the nucleus into account. Do these values follow any correlation?
17. Derive an expression for the energy of an electron in an atom $A^{(Z-1)+}$ starting from first principle for the Bohr model.
18. Using the expression derived in the problem 17, calculate the ionization energy of B^{4+}, C^{5+} and Be^{3+}.
19. Calculate the limiting values of the wavelengths: (*i*) Lyman, (*ii*) Balmer, (*iii*) Paschen and (*iv*) Bracket series. Label these to regions of the spectrum. Do any transitions belonging to these series lies at wavelength lower than the limiting values?
20. From Table 1.4 it is clear that Balmer series in the spectrum of hydrogen atom arises when the transition from higher energy levels ends up at $n = 2$. Indicate the values of n for which the Balmer series lie in the visible region (400 to 700 nm) of the spectrum.
21. Calculate the range of wavelength in which the (*i*) Lyman series, (*ii*) Balmer series, (*iii*) Paschen series, (*iv*) Bracket series and (*v*) Pfund series are observed.
22. If the value of Rydberg constant is 1,09,677.57 cm^{-1}, calculate the corresponding value of: (*i*) wavelength and (*ii*) the frequency of electromagnetic radiation. Convert R to electron volts and kJ mol^{-1} units.
23. A short lived species known as mesonic atom, consists of a proton and a negatively charged meson (μ, called muon). The charge of the muon is the same as that of an electron, but it is 207 times heavier than electron ($1\ \mu = 207\ m_e$). Assuming that Bohr's theory can be applied to the meson, calculate the (*i*) radius and (*ii*) the ground state energy of the first Bohr orbit. (*iii*) Also calculate the frequency of the transition associated with $n_1 = 1$ and $n_2 = 2$, as well as the ionization energy of the ground state.
24. A gold ball weighs 45 g. If it is traveling with 80 mph, calculate its de Broglie wavelength.
25. An electron follows a de Broglie wavelength of 12 pm. Calculate the accelerating voltage.
26. An α-particle ejected from radium has the mass of 6.68×10^{-27} kg and is moving with velocity of 1.5×10^7 m s^{-1}. Calculate the de Broglie wavelength.
27. Calculate the de Broglie wavelength of an electron moving in Bohr's 3rd orbit.
28. Using equation (1.80) *i.e.*, $mvr = \dfrac{nh}{2\pi}$, calculate the velocity of electron in $n = 3$.
29. Assume that if we can measure the velocity of gold ball in problem 24 to an accuracy of 5×10^{-8} m s^{-1} of its actual velocity. Calculate the uncertainty in distance (Δx) and comment over this value.
30. If the uncertainty in distance (Δx) of an electron in an atom is 100 pm, calculate the uncertainty in its velocity and comment.

CHAPTER

Quantum Mechanics

2

The failure of intrinsically contradictory and essentially semi-empirical theory of Bohr to solve the problem of atom and the evolution of new concepts on dual nature of micro-particles (namely electrons, protons, neutrons etc., alike photon, *e.g.* de Broglie theory of matter waves and Heisenberg Uncertainty principle) led to the birth of new mechanics or quantum mechanics. This is based on the wave like properties of electrons. This new mechanics teaches us how to think about the behaviour of micro-particles when neither corpuscular aspect nor their wave aspect is sufficient to interpret the phenomenon, like any physically observable quantities. To each of these physically observable quantities, a linear operator is related and the theory of linear operators is the mathematical apparatus of quantum mechanics. The two mathematical tools based on the theory of linear operators are due to Heisenberg in 1925 and Schrödinger in 1926. Heisenberg reduced the mathematical apparatus of new theory to matrix calculus, which operates with an assembly of coordinates of a system of oscillators representing each real atom, i.e. without using the operator concept. Heisenberg associated with each physical quantity a matrix called Heisenberg's matrix. On the other hand, Schrödinger reduced the mathematical apparatus of new theory to differential calculus. This theory introduces the operator concept on a wave function, also called matter field wave function that reflects the wave properties of a particle in the sense of de Broglie theory. The matrix calculus approach is difficult, but the two approaches are equivalent (*i.e.*, they arrive at the same results) although expressed in different mathematical languages. The differential calculus approach being simple and is generally adopted by chemists to solve the problems related to atoms and molecules.

Further we are eagerly awaiting the creation of a theory that would unite all the mechanics in a single whole unit and there is already a name for this as yet *uncreated*, it is called the relativistic quantum mechanics. A relativistic treatment of the quantum mechanics was pioneered by P. A. M. Dirac and led to a new picture of nature, including the concept of antimatter.

2.1 WAVE MECHANICS

Schrödinger took the clue from de Broglie's hypothesis that the particles can have wave motion and compared the motion of electrons with wave information occurring in a stretched strings or violin. Most of the mathematical ideas used were the principle of classical mechanics about waves and applied them to the electronic and molecular motions. The stationary states with which electrons or molecules might be associated are similar to the standing waves, set up in the above systems by applying certain boundary conditions. Later on this mathematical treatment was called Schrödinger wave equation and this equation relates the observable quantities to certain mathematical operators.

2.2 SCHRÖDINGER WAVE EQUATION

Let us consider that wave motion associated with the motion of electrons can be compared to the wave motion of the string waves associated with string tied at the ends (or to elastic vibration of a liquid sphere). For simplicity consider first the simplest case of wave motion along one axis, say along x-axis. If w is the amplitude at any point whose coordinate is x at any time t, the appropriate equation of wave motion is given by the expression:

$$\frac{\partial^2 w}{\partial x^2} = \frac{1}{u^2}\frac{\partial^2 w}{\partial t^2} \qquad ...(2.1)$$

where u is the velocity of propagation of waves. The above equation can be solved by separating the variables, *i.e.*, function $w(x, t)$ can be written as the product of two wave functions: $f(x)$ and $g(t)$, where $f(x)$ is a function of x only and $g(t)$ is only of time t.

$$w(x, t) = f(x)\, g(t) \qquad ...(2.2)$$

Now $$\frac{\partial^2 w(x,t)}{\partial t^2} = f(x)\frac{d^2 g(t)}{dt^2} \qquad ...(2.3)$$

and $$\frac{\partial^2 w(x,t)}{\partial x^2} = g(t)\frac{d^2 f(x)}{dx^2} \qquad ...(2.4)$$

Combining equations (2.1), (2.3) and (2.4), we get

$$g(t)\frac{d^2 f(x)}{dx^2} = \frac{1}{u^2} f(x)\frac{d^2 g(t)}{dt^2} \qquad ...(2.5)$$

The behaviour of standing waves, such as we are considering, with respect to time can be expressed by the equation (2.6).

$$g(t) = \text{A} \sin 2\pi\nu t \qquad ...(2.6)$$

where ν is the frequency of vibration and A is a constant known as the maximum amplitude. Differentiating (2.6) twice with respect to time one gets:

$$\frac{d^2 g(t)}{dt^2} = -4\pi^2\nu^2\text{A} \sin 2\pi\nu t = -4\pi^2\nu^2 g(t) \qquad ...(2.7)$$

Substituting $\frac{d^2 g(t)}{dt^2}$ in equation (2.5),

$$g(t)\ \frac{d^2 f(x)}{dx^2} = -\ \frac{4\pi^2 \nu^2}{u^2}\ f(x)\ g(t)$$

or

$$\frac{d^2 f(x)}{dx^2} = -\ \frac{4\pi^2}{\lambda^2}\ f(x) \qquad (\lambda = \frac{u}{\nu}) \qquad ...(2.8)$$

Equation (2.8) represents the behaviour of the standing waves in one direction and this treatment can be extended to the wave motion in three directions along x, y and z. Writing the wave function $\psi(x, y, z)$, in short form as ψ, the equation (2.8) can be written as:

$$\frac{\partial^2 \psi}{\partial x^2} + \frac{\partial^2 \psi}{\partial y^2} + \frac{\partial^2 \psi}{\partial z^2} = -\ \frac{4\pi^2}{\lambda^2}\ \psi(x, y, z) \qquad ...(2.9)$$

At this stage, Schrödinger made use of the assumption that the above equation can be applied to the particles, like electrons, protons, etc., which can have dual character. Using de Broglie's relation, $\lambda = \frac{h}{mv}$ where m is the mass and v is the velocity of the particle (it may be noted here that the actual velocity of the particle is different from the velocity of propagation u of the hypothetical wave motion), equation (2.9) becomes:

$$\frac{\partial^2 \psi}{\partial x^2} + \frac{\partial^2 \psi}{\partial y^2} + \frac{\partial^2 \psi}{\partial z^2} = -\ \frac{4\pi^2 m^2 v^2}{h^2}\ \psi(x, y, z)$$

$$= -\frac{8\pi^2 m}{h^2}\left(\frac{mv^2}{2}\right)\psi(x, y, z)$$

$$= -\ \frac{8\pi^2 m}{h^2}\ (\mathrm{K})\ \psi(x, y, z) \qquad ...(2.10)$$

Kinetic energy K can be expressed by the relation:

$$\mathrm{K} = \mathrm{E} - \mathrm{V} \qquad ...(2.11)$$

where E is the total energy and V is the potential energy, or function of coordinates. Combining equations (2.10) and (2.11) and rearranging, we get

$$-\ \frac{h^2}{8\pi^2 m}\left[\frac{\partial^2 \psi}{\partial x^2} + \frac{\partial^2 \psi}{\partial y^2} + \frac{\partial^2 \psi}{\partial z^2}\right] + \mathrm{V}\psi = \mathrm{E}\psi \qquad ...(2.12)$$

This expression, which is a second order differential equation, is generally called Schrödinger wave equation or wave equation of quantum mechanics. This equation represents the way in which the electron moves. It will be shown in latter sections that this equation is really helpful in solving many problems, concerning the structure of atoms and molecules. In one dimension wave function, time independent and time dependent Schrödinger wave equation can be written as:

$$-\frac{h^2}{8\pi^2 m}\frac{d^2\psi(x)}{dx^2} + V(x)\,\psi(x) = E\,\psi(x) \qquad ...(2.13)$$

$$-\frac{h^2}{8\pi^2 m}\frac{\partial^2\psi(x,t)}{\partial x^2} + V\psi(x,t) = \frac{ih^2}{4\pi^2}\frac{\partial\psi(x,t)}{\partial t} \qquad ...(2.14)$$

In equation (2.14), V is a function of both time and position. When the wave equation describes a stationary state in which ψ and V are not function of time, the equation (2.14) reduces to time independent form (2.13). The time independent equation is applicable only to conservative fields, i.e. those in which total energy is a constant of motion.

$$E = \frac{1}{2}mv^2 + V = \text{Constant} \qquad ...(2.15)$$

When kinetic energy is expressed in the form of momentum (p), the equation (2.15) reduces to

$$H = \frac{p^2}{2m} + V = E = \text{Constant}$$

As learnt earlier (Chapter 1), function H (p, x), in classical mechanics is called the Hamiltonian function of the system. Since only functions of position appear in time independent wave equation, this is called steady state form of Schrödinger wave equation.

2.3 PHYSICAL SIGNIFICANCE OF WAVE FUNCTION

Wave function is a solution of Schrödinger wave equation (very often contains the imaginary quantity iota, $\iota = \sqrt{-1}$) and its complete character originates from the concepts of wave nature of matter, as defined by de Broglie. Just as magnetic field or electrical field is a physical quantity, so is the square of the wave function. This function helps in describing the entire physical state of a particle or of a system of particles such as atom. Before attaching any physical meaning to this wave function, let us first accomplish a mutual correspondence between corpuscular and wave function concepts of matter. Based on corpuscular concept, the intensity of electrons beam is proportional to the number of electrons per cm^3. But in wave concept, since the wave intensity is proportional to the square of the amplitude, the intensity of the electron beam should be proportional to wave function ψ as its square or more precisely to the square of its modulus, $|\psi|^2 = \psi^*\psi$, where ψ^* is a complex conjugate of function ψ. Thus if $\psi = x + iy$, then $\psi^* = x - iy$ and $\psi^*\psi = (x + iy)(x - iy) = x^2 + y^2$, will always be a real quantity. On the other hand if ψ is a real quantity, then ψ and its complex conjugate ψ^* are the same.

In the light of above discussion, the wave function ψ must describe some thing about the location of electron at a given point, (say x) and time (say t). Thus in the former case the wave function $\psi(x)$ depends on position and it should be solution of the time independent Schrödinger wave equation (2.13). In the latter case wave function $\psi(x, t)$ depends on position and time, it should verify the time dependent Schrödinger wave equation (2.14). Thus the particle is more likely to be present at those places where the intensity is large. Max Born has related wave function to probability. According to Born, the probability of finding the particle within an element of length dx is:

$$P_x \propto \psi^* \psi \, dx \text{ or } |\psi|^2 \, dx \qquad ...(2.16)$$

The quantity $|\psi|^2$ is called the *probability density function*, where the quantity ψ itself is amplitude density. Since the maximum probability is always unity, so the probability of certainty of finding the particle in a complete space (between + ∞ to – ∞) will be equal to unity, *i.e.*,

$$\int_{-\infty}^{+\infty} \psi^*(x)\, \psi(x)\, dx = 1 \qquad ...(2.17)$$

This is called the normalization condition. The presence of electron at other places is always expressed in terms of a normalization constant, say A. Thus if ψ is not a normalized wave function, then

$$\int_{-\infty}^{+\infty} \psi^*(x)\, \psi(x)\, dx < 1 \qquad ...(2.18)$$

In order to achieve normalizing constant, we must look for a new wave function which should always be the solution of the wave equation. Let this function is $A\psi$, where A is a constant. Accordingly

$$\int_{-\infty}^{+\infty} A\psi^*(x)\, A\psi(x)\, dx = 1$$

or

$$A^2 \int_{-\infty}^{+\infty} \psi^*(x)\, \psi(x)\, dx = 1$$

or finally

$$\int_{-\infty}^{+\infty} \psi^*(x)\, \psi(x)\, dx = A^{-2} \qquad ...(2.19)$$

Here A is called the normalization constant and the new wave function $A\psi$ is the normalized wave function. The same treatment can be extended to the three dimensional wave function. The probability of certainty of finding the electron in volume element $dx\, dy\, dz$ or dV or $d\tau$ is:

$$\psi^*(x, y, z)\, \psi(x, y, z)\, d\tau$$

and total probability over the complete space is

$$\int_{-\infty}^{+\infty}\int_{-\infty}^{+\infty}\int_{-\infty}^{+\infty} \psi^*(x, y, z)\, \psi(x, y, z)\, d\tau \text{ (or } dx\, dy\, dz) = 1 \qquad ...(2.20)$$

If two wave functions, say, ψ_i and ψ_j are allowed for a given system, then the wave functions are said to be:

(*i*) Normalized provided they qualify the condition

$$\int_{-\infty}^{+\infty}\int_{-\infty}^{+\infty}\int_{-\infty}^{+\infty} \psi_j^*(x, y, z)\, \psi_j(x, y, z)\, d\tau \text{ (or } dx\, dy\, dz) = 1$$

$$\int_{-\infty}^{+\infty}\int_{-\infty}^{+\infty}\int_{-\infty}^{+\infty} \psi_i^* (x, y, z)\ \psi_i\ (x, y, z)\ d\tau \text{ (or } dx\ dy\ dz) = 1$$

and

(*ii*) Mutually orthogonal provided

$$\int_{-\infty}^{+\infty}\int_{-\infty}^{+\infty}\int_{-\infty}^{+\infty} \psi_i^* (x, y, z)\ \psi_j\ (x, y, z)\ d\tau \text{ (or } dx\ dy\ dz) = 0$$

or
$$\int_{-\infty}^{+\infty}\int_{-\infty}^{+\infty}\int_{-\infty}^{+\infty} \psi_j^* (x, y, z)\ \psi_i\ (x, y, z)\ d\tau \text{ (or } dx\ dy\ dz) = 0$$

If allowed wave functions qualify both the conditions, they are said to be ortho-normal wave functions. Both these relations can be summed up as:

$$\int_{-\infty}^{+\infty}\int_{-\infty}^{+\infty}\int_{-\infty}^{+\infty} \psi_i^* (x, y, z)\ \psi_j\ (x, y, z)\ d\tau \text{ (or } dx\ dy\ dz) = 1 \text{ if } i = j$$

$$= 0 \text{ if } i \neq j \qquad \text{...(2.21)}$$

Finally in order that the wave function is a solution of the Schrödinger wave equation, it must satisfy the following conditions:

(*i*) It must be consistent with the following relationships.

$$\lambda = \frac{h}{p};\ \nu = \frac{E}{h};\ H = \frac{p^2}{2m} + V$$

(*ii*) It must be linear in $\psi\ (x, t)$, *i.e.*, if $\psi_1\ (x, t)$, $\psi_2\ (x, t)$... $\psi_n\ (x, t)$ are solutions of Schrödinger wave equation then

$$\psi\ (x, t) = c_1\ \psi_1\ (x, t) + c_2\ \psi_2\ (x, t) + ... + c_n\ \psi_n\ (x, t) = \Sigma c_i\ \psi_i\ (x, t) \qquad \text{...(2.22)}$$

should also be the solution of Schrödinger wave equation. Here $c_1, c_2, ..., c_n$ are constants. This is known as the *principle of superposition.*

(*iii*) The first derivative of the function, *i.e.*, $\frac{\partial\psi(x,t)}{\partial x}$ should also be linear.

(*iv*) The function $\psi\ (x, t)$ as well as its first derivative should also be well behaved, *i.e.*, they must be:

(*a*) single valued. This restriction was imposed because we know that the probability density, $\psi^*\ \psi$ and probability $\psi^*\ \psi\ d\tau$ of finding the particle at any particular point will always have one and only one value. This is only possible if ψ is a single valued function.

(b) finite.

(c) continuous.

This means that if ψ is a function of θ, then $\psi(\theta) = \psi(\theta + 2n\pi)$ where $n = 1, 2, 3.....$

(v) When $x \to \pm\infty$, then $\psi(x, t) \to 0$, *i.e.*, $\lim_{x\to\pm\infty} \psi(x, t) \to 0$

Example 2.1: Determine the constant A_n for the wave function $\psi(x) = A_n \sin\left(\frac{2\pi nx}{L}\right)$, which is defined only within a region $0 \le x \le L$.

Solution: Applying the normalization condition, we get

$$\int_0^L \psi^*(x)\, \psi(x)\, dx = A_n \int_0^L \sin^2 \frac{2\pi nx}{L}\, dx = 1$$

or
$$A_n^2 \int_0^L \frac{1-\cos\frac{4\pi nx}{L}}{2}\, dx = 1$$

or
$$\frac{A_n^2}{2}\left[x - \frac{L}{4\pi n}\sin\frac{4\pi nx}{L}\right]_0^L = 1$$

or
$$\frac{A_n^2}{2}[L] = 1 \text{ or } A_n^2 = 2/L \text{ or } A_n = \sqrt{\frac{2}{L}}$$

Thus the normalized wave function is

$$\psi_n = \left(\frac{2}{L}\right)^{\frac{1}{2}} \sin\frac{2\pi nx}{L}$$

Example 2.2: Determine the constant A_n for a wave function $\psi(x, t) = A_n \sin \frac{n\pi L}{x} e^{-\frac{iEt\,2\pi}{h}}$ which is only defined in the region $0 \le x \le L$ at $t = 0$.

Solution: At $t = 0$ the wave function becomes

$$\psi(x, 0) = A_n \sin\frac{\pi nx}{L}$$

$|\psi(x, 0)|^2\, dx$ represents the probability of finding the particle between $x = 0$ and $x = L$ at $t = 0$. Applying the normalization condition, we get

$$\int_0^L \psi^*(x, 0)\, \psi(x, 0)\, dx = A_n^2 \int_0^L \sin^2\frac{\pi nx}{L}\, dx = 1$$

or
$$A_n^2 \int_0^L \frac{1-\cos\dfrac{2\pi nx}{L}}{2}\, dx = 1$$

or
$$\frac{A_n^2}{2}\left[x - \frac{L}{2\pi n}\sin\frac{2\pi nx}{L}\right]_0^L = 1$$

or
$$\frac{A_n^2}{2}[L] = 1 \quad \text{or} \quad A_n = \sqrt{\frac{2}{L}}$$

Thus at $t = 0$, both time dependent and time independent wave functions have the same solutions.

Example 2.3: Prove that $\psi(x, t) = A e^{2\pi i (p_x x - Et)/h}$ is a solution of the Schrödinger wave equation.

Solution: We can write the wave equation as:

$$\psi(x, t) = A e^{\frac{2\pi i p_x x}{h}} e^{-\frac{2\pi i E t}{h}} \qquad ...(a)$$

in which the variables x and t have been separated. Thus if the spatial part is written as

$$\psi(x) = A e^{\frac{2\pi i p_x x}{h}} \qquad ...(b)$$

then equation (a) can be written as:

$$\psi(x, t) = \psi(x)\, e^{-\frac{2\pi i E t}{h}} \qquad ...(c)$$

Now taking second order derivative of equation (c) with respect to position, we get

$$\frac{\partial^2 \psi(x,t)}{\partial x^2} = \frac{d^2\psi(x)}{dx^2} e^{-\frac{2\pi i E t}{h}} \qquad ...(d)$$

and taking the first derivative of equation (c) with respect to t, we obtain

$$\frac{\partial \psi(x,t)}{\partial x} = -\frac{2\pi i E}{h}\psi(x)\, e^{-\frac{2\pi i E t}{h}} \qquad ..(e)$$

Substitute (c), (d) and (e) in the time dependent Schrödinger wave equation (2.14)

$$-\frac{h^2}{8\pi^2 m}\frac{d^2\psi(x)}{dx^2} e^{-\frac{i 2\pi E t}{h}} + V\psi(x)\, e^{-\frac{i\, 2\pi E\, t}{h}} = \frac{ih}{2\pi}\frac{-2\pi i E}{h} e^{-\frac{i 2\pi E t}{h}} \psi(x)$$

Dividing it by $e^{-\frac{2\pi i E t}{h}}$, we get

$$-\frac{h^2}{8\pi^2 m}\frac{d^2\psi(x)}{dx^2} + V\psi(x) = E\psi(x)$$

which is a steady state Schrödinger wave equation. Hence $\psi\ (x, t)$ is a solution of Schrödinger wave equation.

Example 2.4: If ψ_i and ψ_j are the solutions of the Schrödinger wave equation, show that they are mutually orthogonal.

Solution: We write the one dimensional time independent Schrödinger wave equation in terms of both these functions.

$$\frac{d^2\psi_i}{dx^2} + \frac{8\pi^2 m}{h^2}(E_i - V_i)\,\psi_i = 0 \qquad \text{...}(a)$$

$$\frac{d^2\psi_j}{dx^2} + \frac{8\pi^2 m}{h^2}(E_j - V_j)\,\psi_j = 0 \qquad \text{...}(b)$$

Multiply equation (*a*) by ψ_j and equation (*b*) by ψ_i and then subtract. The resulting equation will be:

$$\psi_j\frac{d^2\psi_i}{dx^2} - \psi_i\frac{d^2\psi_j}{dx^2} + \frac{8\pi^2 m}{h^2}(E_i - E_j + V_j - V_i)\,\psi_i\,\psi_j = 0$$

or $$\frac{d}{dx}\left[\psi_j\frac{d\psi_i}{dx} - \psi_i\frac{d\psi_j}{dx}\right] + \frac{8\pi^2 m}{h^2}(E_i - E_j + V_j - V_i)\,\psi_i\psi_j = 0 \qquad \text{...}(c)$$

Now integrate the above equation (*c*) with respect to *x* over the whole space between the limits $+\infty$ and $-\infty$. Since ψ and V are zero at $+\infty$ and $-\infty$, the first term and the potential energy terms in equation (*c*) will be zero, we are left only with the integral:

$$\left[\psi_j\frac{d\psi_i}{dx} - \psi_i\frac{d\psi_j}{dx}\right]_{-\infty}^{+\infty} + \frac{8\pi^2 m}{h^2}(E_i - E_j + V_j - V_i)\int \psi_i\,\psi_j\,dx = 0$$

$$\frac{8\pi^2 m}{h^2}(E_i - E_j)\int_{-\infty}^{+\infty}\psi_i\psi_j\,dx = 0 \qquad \text{...}(d)$$

Since $E_i \neq E_j$, therefore

$$\int_{-\infty}^{+\infty}\psi_i\psi_j\,dx = 0 \qquad \text{...}(e)$$

Thus the wave functions ψ_i and ψ_j are mutually orthogonal. Further if $i = j$, then since $\int_{-\infty}^{+\infty} \psi_i \psi_j \, dx$ is necessarily a positive or finite, we must have $E_i = E_j$. But this does not mean that the corresponding wave functions will necessarily be orthogonal.

Example 2.5: Prove that the first derivative of the wave function is always continuous provided the potential energy V (x) is finite, whether or not it is continuous.

Solution: The one dimensional Schrödinger wave equation is:

$$\frac{d^2\psi}{dx^2} + \frac{8\pi^2 m}{h^2} [E - V(x)] \psi(x) = 0$$

Integrate it with in limits $x - \alpha$ and $x + \alpha$:

$$\int_{x-\alpha}^{x+\alpha} \frac{d^2\psi}{dx^2} dx = -\frac{8\pi^2 m}{h^2} \int_{x-\alpha}^{x+\alpha} [E - V(x)] \psi(x) \, dx$$

or

$$\left[\frac{d\psi}{dx}\right]_{x-\alpha}^{x+\alpha} = -\frac{8\pi^2 m}{h^2} \int_{x-\alpha}^{x+\alpha} [E - V(x)] \psi(x) \, dx$$

or

$$\psi'(x + \alpha) - \psi'(x - \alpha) = -\frac{8\pi^2 m}{h^2} \int_{x-\alpha}^{x+\alpha} [E - V(x)] \psi(x) \, dx$$

where $\psi' = \dfrac{d\psi}{dx}$, the first derivative of the wave function. Since V is assumed to be finite (it could be discontinuous). Because V $(x + \alpha)$ = V $(x - \alpha)$ as $\alpha \to 0$ and E is a constant quantity. Thus $\psi'\left(\dfrac{d\psi}{dx}\right)$ is continuous at any value of x, accordingly ψ has to be necessarily continuous everywhere.

Example 2.6: Normalize the wave function $\sin \dfrac{n\pi}{a} x$ within the range $-a$ to $+a$.

Solution: Applying the normalizing condition:

$$N^2 \int_{-a}^{+a} \sin^2 \frac{n\pi}{a} x dx = 1$$

or

$$\frac{N^2}{2} \int_{-a}^{+a} \left[1 - \cos \frac{2n\pi}{a} x\right] dx = 1$$

or $$\frac{N^2}{2}\left[x - \frac{a}{2n\pi}\sin\frac{2n\pi}{a}x\right]_{-a}^{+a} = 1$$

$$\frac{N^2}{2}[a-(-a)] = 1 \text{ or } N^2 = \frac{1}{a}$$

$$N = \left(\frac{1}{a}\right)^{\frac{1}{2}}$$

Normalized wave function is $\left(\frac{1}{a}\right)^{\frac{1}{2}} \sin\frac{n\pi}{a}x$.

2.4 ALGEBRAIC AND QUANTUM MECHANICAL OPERATORS

2.4.1 CONCEPT OF FUNCTIONS AND OPERATORS

The basic purpose of introducing the idea of operators is how to use this jargon of mathematics in quantum mechanics to calculate the values of observables. These are defined as physically observables and measurable physical quantities of physical systems. The values of observable, also called dynamical variables, are defined in classical mechanics. Many of the observables are not properties of the systems but are the characteristics of measuring system operating on the physical system. As has already been stated, there exists a linear operator for every observables and measurable physical quantity of a system. Thus the theory of linear operators is the mathematical apparatus for quantum mechanics. However before dealing with quantum mechanical operators, it is desirable to define a function (as operators act on functions), operators and properties of operators.

Function: A function is a mathematical rule that acts on a number to give a new number. Knowing the number x, we can find the number $y = f(x)$. Some of the examples of functions are:

Function, $f(x)$	Number x	$y = f(x)$
nx	2	$2n$ where n = 1, 2, 3,...
x^n	2	2^n, where n = 1, 2, 3,....
$x^{1/n}$	8	$8^{1/n}$, where n = 1, 2, 3,....
$\sqrt{x}$	4	2
$\sin x$	90°	1

Operator: An operator is a mathematical instruction which works on a function to give a new function and it is designated by putting a cap (^) on a particular sysmbol as a superscript ($\hat{O}$ is pronounced as O-cap or $\hat{L}$ is pronounced as el-cap) or writing in italics. However we will be using the former notation in the text. An operator $\hat{O}$ will map a given function $f(x)$ into a new function, $\psi(x) = \hat{O}[f(x)]$. For examples: (*i*) if $\hat{5}$ is a multiplication operator and $f(x)$ is a function $(5x^2 + \sin x)$, then $\hat{5}\ (5x^2 + \sin x) = 25x^2 + 5\sin x$. Similarly, (*ii*) if $\hat{D}$ is a differential operator, then $\hat{D}\ (5x^2 + \sin x) = \frac{d}{dx}(5x^2 + \sin x) = 10x + \cos x$, and (*iii*) if $\hat{I}$ is an integral operator and

$$\hat{I}\ (5x^2 + \sin x) = \int (5x^2 + \sin x)\, dx = \frac{5x^3}{3} - \cos x.$$

***Eigen Functions and Eigen Values*:** There is an important class of functions called as eigen functions. A function $f(x)$ is said to be an eigen function provided it qualifies the condition.

$$\hat{O}\ f(x) = O\ f(x) \qquad \text{...(2.23)}$$

Operator $\hat{O}$ acting on function $f(x)$ gives back the same function $f(x)$ multiplied by a constant (O). Similarly:

$$\hat{O}\ f(x, t) = \hat{\lambda}\ f(x, t) \qquad \text{...(2.24)}$$

(Operator $\hat{O}$) acting on function $f(x, t)$ = (Operator λ) acting on function $f(x, t)$. Here both $\hat{O}$ and λ are not scalar multipliers. The allowed solutions $f(x)$ are often called eigen functions of operator O and the corresponding O values are called eigen values of the operator O. The eigen values of the operator are the values of the corresponding physical quantities. The eigen values can be real or imaginary. A real physical quantity is defined by a Hermitian operator. If two eigen functions, say, $\alpha(x)$ and $\beta(x)$, of an operator O, belong to the same eigenvalue, they are said to be *degenerate*, *i.e.*,

$$\hat{O}\alpha(x) = O\alpha(x)$$

$$\hat{O}\beta(x) = O\beta(x)$$

The linear combination of such functions will also be an eigen function having the same eigen value,

$$\hat{O}\ [C_1\ \alpha(x) + C_2\ \beta(x)] = O\ [C_1\ \alpha(x) + C_2\ \beta(x)]$$

As we will see later that quantum mechanical functions are eigen functions and are also called wave functions. For example,

$$\frac{d}{dx}(e^{n\alpha x}) = n\alpha e^{n\alpha x} = n\alpha\ (e^{n\alpha x}).$$

Thus $e^{n\alpha x}$ is an eigen function of the operator $\frac{d}{dx}$ and $n\alpha$ is the eigen value of the operator $\frac{d}{dx}$. For $n = 1, 2, 3,\ldots$., the solutions of $e^{n\alpha x}$ are: $e^{\alpha x}$, $e^{2\alpha x}$, $e^{3\alpha x}$,..... and the respective eigen values are α, 2α, 3α..... of the operator d/dx.

The complete set of eigen values constitutes the spectrum of operator. An observable or a countable infinite set of O_0, O_1, O_2, is said to form a discrete spectrum. When the eigen values assume all values in certain interval, the spectrum is called continuous. In more complicated situations, the spectrum may exhibit both discrete and continuous parts. We must note here that the old quantum theory could formulate quantum conditions only for quantities that change abruptly and did not include the cases when the quantities change continuously.

If the eigen function f is a function of x, y, z, *i.e.*, $f = f(x, y, z)$, then:

$$\hat{L}\, f(x, y, z) = L\, f(x, y, z) \qquad ...(2.25)$$

where $\hat{L}$ is some operator and L is the eigen value. Now if we change x to $-x$, y to $-y$ and z to $-z$, and the above conditions hold for new function also, then we will have

$$\hat{L}\, f(-x, -y, -z) = L\, f(-x, -y, -z) \qquad ...(2.26)$$

Equations (2.25) and (2.26) reveal that both the functions are eigen functions of operator $\hat{L}$ with same eigen values. Thus $f(-x, -y, -z)$ is not different from $f(x, y, z)$, but it will be some multiple of $f(x, y, z)$ and vice versa, *i.e.*,

$$f(-x, -y, -z) = k\, f(x, y, z) \qquad ...(2.27)$$

$$f(x, y, z) = k\, f(-x, -y, -z) \qquad ...(2.28)$$

Combining equations (2.27) and (2.28), we obtain

$$f(x, y, z) = kf(-x, -y, -z) = k^2 f(x, y, z)$$

and

$$k^2 = 1 \text{ or } k = \pm 1.$$

Accordingly equation (2.27) reduces to:

$$f(-x, -y, -z) = \pm f(x, y, z)$$

Thus if $f(-x, -y, -z) = f(x, y, z)$, the function is said to have even parity and if $f(-x, -y, -z) = -f(x, y, z)$, it is said to have odd parity. For degenerate state, $f(-x)$ need not be a multiple of $f(x)$. Indeed if $f(-x)$ is not a multiple of $f(x)$, then Schrödinger wave equation, in terms of $f(x)$ and $f(-x)$, reveals that $f(-x)$ must be another independent solution corresponding to the same energy (same eigen value). Further any linear combination of $f(x)$ and $f(-x)$ could always be possible eigen function and we can always choose a linear combination $[f(x) + f(-x)]$ and $[f(x) - f(-x)]$ which are symmetric and anti-symmetric functions of x respectively. Thus even for degenerate states, we can always choose suitable linear combination which are either symmetric or anti-symmetric. This is called the principle of superposition.

2.4.2 PRINCIPLE OF SUPERPOSITION

Let us now verify the principle of superposition. Consider the Schrödinger wave equation in one dimension, *i.e.*,

$$-\frac{h^2}{8\pi^2 m}\frac{d^2\psi(x)}{dx^2} + V(x)\,\psi(x) = E\psi(x)$$

or
$$\frac{d^2\psi(x)}{dx^2} = -\frac{8\pi^2 m}{h^2}(E - V)\,\psi(x) = -\alpha^2\,\psi(x)$$

where $\alpha^2 = \frac{8\pi^2 m}{h^2}(E - V)$. According to principle of superposition if A cos αx and B sin αx, $Ce^{\pm\alpha x}$ are the solutions of the wave equation, then their linear combinations, *i.e.*, A cos αx + B sin αx, A cos αx + $Ce^{i\alpha x}$, B sin αx + $Ce^{\pm i\alpha x}$ and A cos αx + B sin αx + $Ce^{\pm i\alpha x}$ must also be the solutions of the wave equation. If $\psi(x) = C\,e^{i\alpha x}$ is a solution of the wave equation, then we can deduce the wave equation from this function. Taking the first derivative with respect to x:

$$\frac{d\psi(x)}{dx} = Ci\alpha\,e^{i\alpha x}$$

Taking the second derivative with respect to x:

$$\frac{d^2\psi(x)}{dx^2} = C\,(i\alpha)^2\,e^{i\alpha x} = -C\alpha^2\,e^{i\alpha x} = -\alpha^2\,\psi(x)$$

Hence $Ce^{i\alpha x}$ is a solution of the wave function. Similarly we can also prove that the function A cos αx + B sin αx + $Ce^{i\alpha x}$ is a solution of wave equation:

$$\psi(x) = A\cos\alpha x + B\sin\alpha x + Ce^{i\alpha x}$$

Differentiating this function twice with respect to x, we get

$$\frac{d\psi(x)}{dx} = -A\,\alpha\sin\alpha x + B\,\alpha\cos\alpha x + Ci\alpha\,e^{i\alpha x}$$

$$\frac{d^2\psi(x)}{dx^2} = -A\,\alpha^2\cos\alpha x - B\,\alpha^2\sin\alpha x + C\,(i\alpha)^2\,e^{i\alpha x}$$

$$\frac{d^2\psi(x)}{dx^2} = -\alpha^2\,[A\cos\alpha x + B\sin\alpha x + C\,e^{i\alpha x}] = -\alpha^2\,\psi(x)$$

Hence the result. These solutions at the moment may not appear to be useful, but we will see later that they are helpful in solving many problems provided certain constraints are imposed on these wave functions.

Example 2.7: Find out which one of the following functions are eigen functions to the operator $\hat{D}\,(d/dx)$. (*i*) kx^2, (*ii*) e^{-ikx}.

Solution: Using equation (2.23):

(*i*) $\frac{d}{dx}(kx^2) = 2kx$

It is not equal to constant multiplied by the same function, *i.e.*, kx^2. So kx^2 is not an eigen function to the operator D.

(*ii*) $$\frac{d}{dx}(e^{-ikx}) = -\,ike^{-ikx} = -\,ik\,(e^{-ikx})$$

It is an eigen function. All the exponential functions which are linearly dependent on variable are eigen functions to the first differential operator (D).

2.4.3 ALGEBRA OF OPERATORS

There exists a complete branch of mathematics, called operator algebra, but we shall discuss only those properties, which are useful in quantum mechanics. All the algebraic operators may or may not behave as quantum mechanical operators but the reverse is always true.

(*i*) **Addition of Operators:** If A and B are two operators, the sum of the two operators is defined as:

$$(\hat{A} + \hat{B})\, f(x) = \hat{A}\, f(x) + \hat{B}\, f(x) \qquad \text{...(2.29)}$$

For example, consider the function $5x^2 + \sin x$ and the two operators as $\hat{D}$ and $\hat{5}$. Thus:

$$(\hat{D} + \hat{5})\,(5x^2 + \sin x) = \left(\frac{d}{dx} + 5\right)(5x^2 + \sin x)$$

$$= \frac{d}{dx}(5x^2 + \sin x) + 5\,(5x^2 + \sin x)$$

$$= 10x + \cos x + 25x^2 + 5\sin x$$

$$= 25x^2 + 10x + 5\sin x + \cos x$$

(*ii*) **Product of Two Operators:** The product of two operators is defined as:

$$(\hat{A}\hat{B})\, f(x) = (\hat{A})\,[\hat{B}\, f(x)]$$

$$(\hat{A}^2)\, f(x) = (\hat{A}\,\hat{A})\, f(x) = \hat{A}\,[\hat{A}\, f(x)] \qquad \text{...(2.30)}$$

In this case, out of two operators, the one on right hand side operates first, followed by the next and then so on. For example, consider again the same function $(5x^2 + \sin x)$ and the operators are $\hat{x}$ and $\hat{D}$. The product can be written as:

$$(\hat{x}\hat{D})\,(5x^2 + \sin x) = \hat{x}\left[\frac{d}{dx}(5x^2 + \sin x)\right]$$

$$= x\,(10x + \cos x) = 10x^2 + x\cos x \qquad \text{...(2.31)}$$

Now the same operators can also be written as:

$$(\hat{D}\hat{x})\,(5x^2 + \sin x) = \hat{D}\,[\hat{x}\,(5x^2 + \sin x)] = \hat{D}\,[5x^3 + x\sin x]$$

$$= 15x^2 + \sin x + x\cos x \qquad \text{...(2.32)}$$

It is clear from equations (2.31) and (2.32) that the expressions obtained by writing two operators in different manner and then operating them on the function are not the same. Not always but this is quite common in operator algebra that the two operators $\hat{A}\hat{B}$ and $\hat{B}\hat{A}$ give different results when operated on a given function. If the two operators give the results $\hat{A}\hat{B} = \hat{B}\hat{A}$, we say that the operators $\hat{A}$ and $\hat{B}$ *commute* with each other. This relation generally holds if out of the two operators, at least one of them is a number. But other wise also, there are operators which commute with each other and will be discussed later. The difference between the two operators, defined as $\hat{A}\hat{B} - \hat{B}\hat{A}$, is called *commutator* and can be denoted as $[\hat{A}\hat{B}]$, *i.e.*,

$$[\hat{A}\hat{B}] = \hat{A}\hat{B} - \hat{B}\hat{A} \qquad ...(2.33)$$

The commutivity of operators expresses the possibility of measuring the corresponding physical quantities simultaneously. Or in other words, the non commutativity of operators makes a simultaneous measurement of the corresponding quantities impossible.

(iii) **Law of Associative Multiplication:** Associative law of multiplication is obeyed by the operators, *i.e.*,

$$\hat{A}\ (\hat{B}\hat{C}) = (\hat{A}\hat{B})\ \hat{C} \qquad ...(2.34)$$

The proof of the relation (2.34) can be obtained as follows.

Consider three operators, $\hat{5}, \hat{x}$ and $\hat{D}$. The function is again the same as discussed earlier $(5x^2 + \sin x)$. The left hand side can be written as:

$$\hat{A}\ (\hat{B}\hat{C})\ f(x) = \hat{5}\ (\hat{x}\ \hat{D})\ (5x^2 + \sin x) = \hat{5}\ (\hat{x})\ D\ (5x^2 + \sin x)$$

$$= \hat{5}\ (\hat{x})\ (10x + \cos x) = 5\ (10x^2 + x \cos x)$$

$$= 50x^2 + 5x \cos x$$

The right hand side can be written as:

$$(AB)\ (C)\ f(x) = (\hat{5}\ \hat{x})\ \hat{D}\ (5x^2 + \sin x).$$

$$= (\hat{5}\ \hat{x})\ (10x + \cos x) = 50x^2 + 5 \cos x$$

The expressions obtained by two methods are exactly same and thus proves the law of associative multiplication.

(iv) **Operators used in Quantum Mechanics are Linear:** An operator is defined as a linear if it obeys the following equations.

$$\hat{A}\ [f(x) + g(x)] = \hat{A} f(x) + \hat{A} g(x) \qquad ...(2.35)$$

$$\hat{A}\ [f(x)] = C\ [\hat{A} f(x)] \qquad ...(2.36)$$

where $f(x)$ and $g(x)$ are arbitrary functions and C is a constant. The operators, $\hat{5}$, $\hat{x}$, $\hat{D}$ or $\hat{D}^2$ are linear operators, whereas the square root operator is a non-linear operator. These linear operators act on functions of one or several variables. The variables can be either continuous, which is the case for the coordinates (position) of an object, or discontinuous, *i.e.*, assuming only discrete value which is the case for energy levels. Continuous variables can either take on any value or change within the certain domains. Discontinuous variables can take on both finite and infinite sequences of values. We will always assume that the values of the independent-variables of a function are real number, whereas the functions themselves, which the operators act on, can be complex. Typical operators that act on functions of continuous variables x are the multiplication of a function into x and the differentiation with respect to x.

$$\hat{O}\ [f(x)] = x f(x),\ \hat{O}\ [f(x)] = \frac{d}{dx} f(x)$$

In the first case, x plays a double role: it is the value of the independent variable of function $f(x)$ and it is the operator itself. The following identities can be derived for the linear operators.

$$(\hat{A}+\hat{B})\ \hat{C} = \hat{A}\hat{C} + \hat{B}\hat{C} \qquad ...(2.37)$$

$$\hat{A}\ (\hat{B}+\hat{C}) = \hat{A}\hat{B} + \hat{A}\hat{C} \qquad ...(2.38)$$

The equations (2.37) and (2.38) are actually derived using the sum and the product rules of the operators. For example:

$$\begin{aligned}(\hat{A} + \hat{B})\hat{C}\ f(x) &= (\hat{A}+\hat{B})\ [\hat{C}\ f(x)] \\ &= \hat{A}\hat{C}\ f(x) + \hat{B}\hat{C}\ f(x) \\ &= (\hat{A}\hat{C} + \hat{B}\hat{C})\ f(x)\end{aligned}$$

(*v*) **Operators used in Quantum Mechanics are Hermitian:** In mathematical sense, the operator $\hat{\alpha}$ is said to be Hermitian provided it satisfies the condition:

$$\int f_1^*(x)\ [\hat{\alpha} f_2(x)]\ dx = \int f_2(x)\ \hat{\alpha}^* f_1^*(x)\ dx \qquad ...(2.39)$$

where $f_1(x)$ and $f_2(x)$ belong to same class. If the operator α is expressed in matrix form and it is Hermitian if $\hat{\alpha} = \hat{\alpha}^t$, where α^t is the adjoint of α matrix. *Linear combination of Hermitian operators are also linear and the eigen values of Hermitian operators are always real.*

Let $\hat{\alpha}$ be a Hermitian operator of function $f_1(x)$ and let λ be its eigen value, *i.e.*

$$\hat{\alpha}\, f_1(x) = \lambda\, f_1(x) \qquad ...(2.40)$$

Taking the complex conjugate of each quantity in the equation (2.40)

$$\hat{\alpha}^*\, f_1^*(x) = \lambda^*\, f_1^*(x) \qquad ...(2.41)$$

Integrate after multiplying (2.40) by $f_1^*(x)$ and (2.41) by $f_1(x)$ respectively.

$$\int f_1^*(x)\,[\hat{\alpha}\, f_1(x)]\, dx = \lambda \int f_1^*(x)\, f_1(x)]\, dx \qquad ...(2.42)$$

and

$$\int f_1(x)\,[\hat{\alpha}^*\, f_1^*(x)]\, dx = \lambda^* \int f_1(x)\, f_1^*(x)]\, dx \qquad ...(2.43)$$

Since $\hat{\alpha}$ is Hermitian so

$$\int f_1^*(x)\,[\hat{\alpha}\, f_1(x)]\, dx = \int f_1(x)\,[\hat{\alpha}^*\, f_1^*(x)]\, dx$$

Thus,

$$\lambda \int f_1^*(x) f_1(x) dx = \lambda^* \int f_1(x)\, f_1^*(x)\, dx$$

This will be true provided $\lambda = \lambda^*$, *i.e.*, λ is a real quantity.

The other possible ways of writing a condition for Hermitian operators can be:

$$\int f_1^*(x)\,\hat{\alpha} f_2(x)\, dx = \left[\int f_2^*(x)\,\hat{\alpha} f_1(x)]\, dx\right]^*$$

or

$$\int f_1^*(x)\,[\hat{\alpha} f_2(x)]\, dx = \int [(\hat{\alpha} f_1(x)]^*\, f_2(x)\, dx$$

where f_1 and f_2 are two well behaved functions.

Example 2.8: Do the operators $\hat{x}$ and $\hat{D}$ commute with each other? If not, find the commutator.

Solution: The condition for two commuting operators is:

$$\hat{O}_1\hat{O}_2 f(x) = \hat{O}_2\hat{O}_1 f(x)$$

Let $f(x)$ be a function, then

$$\hat{O}_1\hat{O}_2\, f(x) = (x)\left(\frac{d}{dx}\right) f(x) = (x)\frac{df(x)}{dx} = x\,\frac{df(x)}{dx}$$

Similarly

$$\hat{O}_2\hat{O}_1\, f(x) = \left(\frac{d}{dx}\right)[x f(x)] = x\,\frac{df(x)}{dx} + f(x)$$

Since $(x)\left(\frac{d}{dx}\right) f(x) \neq \left(\frac{d}{dx}\right)[x f(x)]$, two operators do not commute. The commutator is defined as:

$$\hat{O}_1\hat{O}_2\, f(x) - \hat{O}_2\hat{O}_1\, f(x) = x\frac{df(x)}{dx} - f(x) - x\,\frac{df(x)}{dx} = -f(x)$$

2.4.4 QUANTUM MECHANICAL OPERATORS

In the last section, some characteristics of operators have been discussed. It has already been stated that our basic purpose of introducing the idea of operators was how to use this branch of mathematics in quantum mechanics to calculate the values of physically observables. An observable is a physically measurable quantity and its value is well defined in classical mechanics. As said earlier, two or more observables may not be measured simultaneously with infinite accuracy in quantum mechanics, but the value of the observable can always be obtained even if the scatter in measurement is large. So if we are in a position to transform the values of these physically observables, expressed in classical mechanics into the corresponding quantum mechanical operators, we shall be in a position to calculate the value of the observable (as we shall see later on) depending on whether a function is an eigen function or not respectively. Actually there are certain rules by which the values of these observables, expressed in classical mechanics can be transformed into their counterparts in quantum mechanics. To illustrate their use and method of converting the classical expression into quantum mechanical one, it can be seen from the following examples.

***Hamiltonian Operator*:** We know that one dimensional time independent Schrödinger wave equation is:

$$-\frac{h^2}{8\pi^2 m}\frac{d^2\psi(x)}{dx^2} + \text{V}(x)\,\psi(x) = \text{E}\psi(x) \qquad \text{...(2.13)}$$

where $\psi(x)$ is a wave function and defines the steady state of the system, V (x) is the potential energy and E is the total energy of the system. This equation is applicable only to conservative fields, *i.e.*, those in which total energy is a constant of the motion and defines the energy levels

$$\text{E} = \frac{1}{2}mv^2 + \text{V}(x)$$

of the system. When kinetic energy is expressed in terms of momentum, the equation is written as:

$$\text{H} = \frac{p^2}{2m} + \text{V} = \text{E} = \text{Constant} \qquad \text{...(2.44)}$$

In classical mechanics, the function H (p, x) is called the Hamiltonian of the system. The Schrödinger wave equation can be written into operator form. The inspection of the time independent equation (2.13) shows that it may be written as:

$$\left[-\frac{h^2}{8\pi^2 m}\frac{d^2}{dx^2} + \text{V}(x)\right]\psi(x) = \text{E}\psi(x) \qquad \text{...(2.45)}$$

The full expression in the brackets on the left hand side can now be defined as an operator

$$\hat{\text{H}} = -\frac{h^2}{8\pi^2 m}\frac{d^2}{dx^2} + \text{V}(x) \qquad \text{...(2.46)}$$

called Hamiltonian operator, because of its similarity with the Hamiltonian function of classical mechanics. Then, the time independent equation (2.44) can be written as:

$$\hat{H}\psi = E\psi \qquad ...(2.47)$$

The function ψ cannot be cancelled out of this equation, because $\hat{H}$ is not a scalar multiplier (but an operator), while E is the value of the total energy, a scalar quantity. Equation (2.47) is interpreted as:

(operator $\hat{H}$) acting on function, ψ = (total energy, E) multiplying the function, ψ

The three dimensional Schrödinger wave equation is written as:

$$-\frac{h^2}{8\pi^2 m}\left[\frac{\partial^2\psi}{\partial x^2}+\frac{\partial^2\psi}{\partial y^2}+\frac{\partial^2\psi}{\partial z^2}\right] + V\psi = E\psi \qquad ...(2.12)$$

where V = V (x, y, z) and $\psi = \psi\,(x, y, z)$. The three dimensional Hamiltonian operator is:

$$\hat{H} = -\frac{h^2}{8\pi^2 m}\left[\frac{\partial^2}{\partial x^2}+\frac{\partial^2}{\partial y^2}+\frac{\partial^2}{\partial z^2}\right] + V\,(x, y, z) \qquad ...(2.48)$$

or

$$\hat{H} = -\frac{h^2}{8\pi^2 m}\nabla^2 + V\,(x, y, z)$$

where

$$\hat{\nabla}^2 = \frac{\partial^2}{\partial x^2}+\frac{\partial^2}{\partial y^2}+\frac{\partial^2}{\partial z^2} \qquad ...(2.49)$$

is called Laplacian operator and will be discussed later.

Linear Momentum Operator: Let us now compare equations (2.44) and (2.45)

$$H = \frac{p_x^{\,2}}{2m} + V = E = \text{Constant}$$

$$\left[-\frac{h^2}{8\pi^2 m}\frac{d^2}{dx^2} + V\,(x)\right]\psi\,(x) = E\psi\,(x)$$

Both the equations are the same, provided that both sides of equation (2.45) are interpreted as operators and each is operating on the function ψ. Thus

$$\frac{p_x^{\,2}}{2m} + V = -\frac{h^2}{8\pi^2 m}\frac{d^2}{dx^2} + V$$

or $$\frac{p_x^2}{2m} = -\frac{h^2}{8\pi^2 m}\frac{d^2}{dx^2}$$

Finally $$\hat{p}_x = \frac{h}{2\pi i}\frac{d}{dx}\left(\text{or } -\frac{ih}{2\pi}\frac{d}{dx}\right) \quad ...(2.50)$$

is defined as the linear momentum operator. Similarly by coupling the three dimensional equation (2.12), we can represent the momentum operator $\hat{p}$ as:

$$\hat{p} = \frac{h}{2\pi i}\left[\frac{\partial}{\partial x}+\frac{\partial}{\partial y}+\frac{\partial}{\partial z}\right]$$

Accordingly we must use a particular derivative operator to represent that particular component of momentum p, *i.e.*

$$\hat{p}_x = \frac{h}{2\pi i}\frac{d}{dx} \quad ...(2.50\ (a))$$

The other components of the momentum operators are:

$$\hat{p}_y = \frac{h}{2\pi i}\frac{d}{dy}, \quad ...(2.50\ (b))$$

$$\hat{p}_z = \frac{h}{2\pi i}\frac{d}{dz} \quad ...(2.50\ (c))$$

***Laplacian Operator*:** Three dimensional Hamiltonian function for a particle of particular mass m and velocity v can be written as:

$$\text{H} = \frac{1}{2}mv^2 + \text{V}(x, y, z)$$

or $$\text{H} = \frac{1}{2}m(v_x^2 + v_y^2 + v_z^2) + \text{V}(x, y, z) \quad ...(2.51)$$

or $$\text{H} = \text{K} + \text{V}(x, y, z)$$

where v_x, v_y, and v_z are the components of velocity v in three different directions. There is no counterpart of K expressed in classical mechanics to change it into quantum mechanics, but there is an operator for linear momentum. So it is worth while to express equation (2.51) in terms of momentum, *i.e.*

$$\text{H} = \frac{1}{2m}[p_x^2 + p_y^2 + p_z^2] + \text{V}(x, y, z)$$

Now replacing p_x by the corresponding operator, *i.e.*, $\hat{p}_x = \frac{h}{2\pi i}\frac{d}{dx}$ and so on, one gets

$$\hat{H} = \frac{1}{2m}\left[\left(\frac{h}{2\pi i}\frac{d}{dx}\right)^2 + \left(\frac{h}{2\pi i}\frac{d}{dy}\right)^2 + \left(\frac{h}{2\pi i}\frac{d}{dz}\right)^2\right] + V\ (x,\ y,\ z)$$

$$= -\frac{h^2}{8\pi^2 m}\left[\frac{\partial^2}{\partial x^2} + \frac{\partial^2}{\partial y^2} + \frac{\partial^2}{\partial z^2}\right] + V\ (x,\ y,\ z)$$

$$= -\frac{h^2}{8\pi^2 m}\ \nabla^2 + V\ (x,\ y,\ z) \qquad \text{...(2.52)}$$

where $\nabla^2_{x,y,z} = \left[\frac{\partial^2}{\partial x^2} + \frac{\partial^2}{\partial y^2} + \frac{\partial^2}{\partial z^2}\right]$ and is called the Laplacian operator. As we will see later, that the Laplacian operator is generally more useful in its polar form in quantum mechanics than expressed in Cartesian coordinates. The polar form of this operator is:

$$\nabla^2_{r,\theta,\varphi} = \frac{1}{r^2}\frac{\partial}{\partial r}\left(r^2\frac{\partial}{\partial r}\right) + \frac{1}{r^2 \sin\theta}\frac{\partial}{\partial\theta}\left(\sin\theta\frac{\partial}{\partial\theta}\right) + \frac{1}{r^2\sin^2\theta}\frac{\partial^2}{\partial\varphi^2} \qquad \text{...(2.53)}$$

and it can be deduced from the polar coordinates r, θ, φ as follows. Recall from the relation between two coordinate systems, *i.e.*

$$z = r\cos\theta \qquad r = \left[x^2 + y^2 + z^2\right]^{\frac{1}{2}} \qquad \text{...(2.54)}$$

$$y = r\sin\theta\sin\varphi \qquad \theta = \cos^{-1}\left[\frac{z}{(x^2+y^2+z^2)^{\frac{1}{2}}}\right] \qquad \text{...(2.55)}$$

$$x = r\sin\theta\cos\varphi \qquad \varphi = \tan^{-1}\left(\frac{y}{x}\right) \qquad \text{...(2.56)}$$

Laplacian operator contains a second order differential equation and these terms can be transformed into spherical polar coordinates using chain rules.

$$\left(\frac{\partial}{\partial x}\right)_{y,z} = \left(\frac{\partial}{\partial r}\right)_{\theta,\varphi}\left(\frac{\partial r}{\partial x}\right)_{y,z} + \left(\frac{\partial}{\partial\theta}\right)_{r,\varphi}\left(\frac{\partial\theta}{\partial x}\right)_{y,z} + \left(\frac{\partial}{\partial\varphi}\right)_{r,\theta}\left(\frac{\partial\varphi}{\partial x}\right)_{y,z} \qquad \text{...(2.57)}$$

$$\left(\frac{\partial}{\partial y}\right)_{x,z} = \left(\frac{\partial}{\partial r}\right)_{\theta,\varphi}\left(\frac{\partial r}{\partial y}\right)_{x,z} + \left(\frac{\partial}{\partial\theta}\right)_{r,\varphi}\left(\frac{\partial\theta}{\partial y}\right)_{x,z} + \left(\frac{\partial}{\partial\varphi}\right)_{r,\theta}\left(\frac{\partial\varphi}{\partial y}\right)_{x,z} \qquad \text{...(2.58)}$$

$$\left(\frac{\partial}{\partial z}\right)_{x,y} = \left(\frac{\partial}{\partial r}\right)_{\theta,\varphi}\left(\frac{\partial r}{\partial z}\right)_{x,y} + \left(\frac{\partial}{\partial\theta}\right)_{r,\varphi}\left(\frac{\partial\theta}{\partial z}\right)_{x,y} + \left(\frac{\partial}{\partial\varphi}\right)_{r,\theta}\left(\frac{\partial\varphi}{\partial z}\right)_{x,y} \qquad \text{...(2.59)}$$

In all these expressions, each term on the right hand side is expressed by the product of two factors, for example, in the first term, the left hand factor is $\left(\frac{\partial}{\partial r}\right)_{\theta,\varphi}$, here the other variables θ

and φ are kept constant, whereas in the right hand factor $\left(\frac{\partial r}{\partial x}\right)_{y,z}$ the other two Cartesian coordinates are held constant. Each factor can be evaluated from the equations (2.54), (2.55) and (2.56).

$$\left(\frac{\partial r}{\partial x}\right)_{y,z} = \frac{x}{(x^2+y^2+z^2)^{1/2}} = \frac{r\sin\theta\cos\varphi}{r} = \sin\theta\cos\varphi \quad ...(2.60)$$

$$\left(\frac{\partial r}{\partial y}\right)_{x,z} = \frac{y}{(x^2+y^2+z^2)^{1/2}} = \frac{r\sin\theta\sin\varphi}{r} = \sin\theta\sin\varphi \quad ...(2.61)$$

$$\left(\frac{\partial r}{\partial z}\right)_{x,y} = \frac{z}{(x^2+y^2+z^2)^{1/2}} = \frac{r\cos\theta}{r} = \cos\theta \quad ...(2.62)$$

Similarly

$$\left(\frac{\partial \theta}{\partial x}\right)_{y,z} = \frac{\cos\theta\cos\varphi}{r}, \left(\frac{\partial \varphi}{\partial x}\right)_{y,z} = -\frac{\sin\varphi}{r\sin\theta} \quad ...(2.63)$$

$$\left(\frac{\partial \theta}{\partial y}\right)_{x,z} = \frac{\cos\theta\sin\varphi}{r}, \left(\frac{\partial \varphi}{\partial y}\right)_{x,z} = \frac{\cos\varphi}{r\sin\theta} \quad ...(2.64)$$

$$\left(\frac{\partial \theta}{\partial z}\right)_{x,y} = -\frac{\sin\theta}{r}, \left(\frac{\partial \varphi}{\partial z}\right)_{x,y} = 0 \quad ...(2.65)$$

Combining equations (2.57) – (2.65), we get

$$\left(\frac{\partial}{\partial x}\right)_{y,z} = \sin\theta\cos\varphi\left(\frac{\partial}{\partial r}\right)_{\theta,\varphi} + \frac{\cos\theta\cos\varphi}{r}\left(\frac{\partial}{\partial \theta}\right)_{r,\varphi} - \frac{\sin\varphi}{r\sin\theta}\left(\frac{\partial}{\partial \varphi}\right)_{r,\theta} \quad ...(2.66)$$

$$\left(\frac{\partial}{\partial y}\right)_{x,z} = \sin\theta\sin\varphi\left(\frac{\partial}{\partial r}\right)_{\theta,\varphi} + \frac{\cos\theta\sin\varphi}{r}\left(\frac{\partial}{\partial \theta}\right)_{r,\varphi} + \frac{\cos\varphi}{r\sin\theta}\left(\frac{\partial}{\partial \varphi}\right)_{r,\theta} \quad ...(2.67)$$

$$\left(\frac{\partial}{\partial z}\right)_{x,y} = \cos\theta\left(\frac{\partial}{\partial r}\right)_{\theta,\varphi} - \frac{\sin\theta}{r}\left(\frac{\partial}{\partial \theta}\right)_{r,\varphi} \quad ...(2.68)$$

Now

$$\left(\frac{\partial^2}{\partial x^2}\right)_{y,z} = \left[\left(\frac{\partial}{\partial x}\right)_{y,z}\left(\frac{\partial}{\partial z}\right)_{y,x}\right]$$

$$= \left[\sin\theta\cos\varphi\left(\frac{\partial}{\partial r}\right)_{\theta,\varphi} + \frac{\cos\theta\cos\varphi}{r}\left(\frac{\partial}{\partial \theta}\right)_{r,\varphi} - \frac{\sin\varphi}{r\sin\theta}\left(\frac{\partial}{\partial \varphi}\right)_{r,\theta}\right]$$

$$\left[\sin\theta\cos\varphi\left(\frac{\partial}{\partial r}\right)_{\theta,\varphi}+\frac{\cos\theta\cos\varphi}{r}\left(\frac{\partial}{\partial\theta}\right)_{r,\varphi}-\frac{\sin\varphi}{r\sin\theta}\left(\frac{\partial}{\partial\varphi}\right)_{r,\theta}\right]$$

$$= \sin\theta\cos\varphi\left(\frac{\partial}{\partial r}\right)_{\theta,\varphi}\left[\sin\theta\cos\varphi\left(\frac{\partial}{\partial r}\right)_{\theta,\varphi}+\frac{\cos\theta\cos\varphi}{r}\left(\frac{\partial}{\partial\theta}\right)_{r,\varphi}-\frac{\sin\varphi}{r\sin\theta}\left(\frac{\partial}{\partial\varphi}\right)_{r,\theta}\right]$$

$$+\frac{\cos\theta\cos\varphi}{r}\left(\frac{\partial}{\partial\theta}\right)_{r,\varphi}\left[\sin\theta\cos\varphi\left(\frac{\partial}{\partial r}\right)_{\theta,\varphi}+\frac{\cos\theta\cos\varphi}{r}\left(\frac{\partial}{\partial\theta}\right)_{r,\varphi}-\frac{\sin\varphi}{r\sin\theta}\left(\frac{\partial}{\partial\varphi}\right)_{r,\theta}\right]$$

$$-\frac{\sin\varphi}{r\sin\theta}\left(\frac{\partial}{\partial\varphi}\right)_{r,\theta}\left[\sin\theta\cos\varphi\left(\frac{\partial}{\partial r}\right)_{\theta,\varphi}+\frac{\cos\theta\cos\varphi}{r}\left(\frac{\partial}{\partial\theta}\right)_{r,\varphi}-\frac{\sin\varphi}{r\sin\theta}\left(\frac{\partial}{\partial\varphi}\right)_{r,\theta}\right]$$

$$= \left[\sin^2\theta\cos^2\varphi\frac{\partial^2}{\partial r^2}-\frac{\sin\theta\cos\theta\cos^2\varphi}{r^2}\frac{\partial}{\partial\theta}+\frac{\sin\theta\cos\theta\cos^2\varphi}{r}\frac{\partial^2}{\partial r\,\partial\theta}\right.$$

$$\left.+\frac{\sin\varphi\cos\varphi}{r^2}\frac{\partial}{\partial\varphi}-\frac{\sin\varphi\cos\varphi}{r}\frac{\partial^2}{\partial r\,\partial\varphi}\right]+$$

$$\left[\frac{\cos^2\theta\cos^2\varphi}{r}\frac{\partial}{\partial r}+\frac{\sin\theta\cos\theta\cos^2\varphi}{r}\frac{\partial^2}{\partial\theta\,\partial r}+\frac{\cos^2\theta\cos^2\varphi}{r^2}\frac{\partial^2}{\partial\theta^2}-\frac{\sin\theta\cos\theta\cos^2\varphi}{r^2}\frac{\partial}{\partial\theta}\right.$$

$$\left.+\frac{\cos^2\theta\sin\varphi\cos\varphi}{r^2\sin^2\theta}\frac{\partial}{\partial\varphi}-\frac{\cos\theta\sin\varphi\cos\varphi}{r^2\sin\theta}\frac{\partial^2}{\partial\theta\,\partial\varphi}\right]+\left[\frac{\sin^2\varphi}{r}\frac{\partial}{\partial r}-\frac{\sin\varphi\cos\varphi}{r}\frac{\partial^2}{\partial\varphi\,\partial r}\right.$$

$$\left.+\frac{\cos\theta\sin^2\varphi}{r^2\sin\theta}\frac{\partial}{\partial\theta}-\frac{\cos\theta\sin\varphi\cos\varphi}{r^2\sin\theta}\frac{\partial^2}{\partial\varphi\,\partial\theta}+\frac{\sin\varphi\cos\varphi}{r^2\sin^2\theta}\frac{\partial}{\partial\varphi}+\frac{\sin^2\varphi}{r^2\sin^2\theta}\frac{\partial^2}{\partial\varphi^2}\right] \quad ...(2.69)$$

Similarly

$$\frac{\partial^2}{\partial y^2} = \sin^2\theta\sin^2\varphi\frac{\partial^2}{\partial r^2}-\frac{\sin\theta\cos\theta\sin^2\varphi}{r^2}\frac{\partial}{\partial\theta}+\frac{\sin\theta\cos\theta\sin^2\varphi}{r}\frac{\partial^2}{\partial r\,\partial\theta}-$$

$$\frac{\sin\varphi\cos\varphi}{r^2}\frac{\partial}{\partial\varphi}+\frac{\sin\varphi\cos\varphi}{r}\frac{\partial^2}{\partial r\,\partial\varphi}+\frac{\cos^2\theta\sin^2\varphi}{r}\frac{\partial}{\partial r}+\frac{\sin\theta\cos\theta\sin^2\varphi}{r}\frac{\partial^2}{\partial\theta\,\partial r}$$

$$+\frac{\cos^2\theta\sin^2\varphi}{r^2}\frac{\partial^2}{\partial\theta^2}-\frac{\sin\theta\cos\theta\sin^2\varphi}{r^2}\frac{\partial}{\partial\theta}-\frac{\cos^2\theta\sin\varphi\cos\varphi}{r^2\sin\theta}\frac{\partial}{\partial\varphi}+\frac{\cos^2\varphi}{r}\frac{\partial}{\partial r}$$

$$+\frac{\cos\theta\sin\varphi\cos\varphi}{r^2\sin\theta}\frac{\partial^2}{\partial\theta\,\partial\varphi}+\frac{\sin\varphi\cos\varphi}{r}\frac{\partial^2}{\partial\varphi\,\partial r}+\frac{\cos\theta\cos^2\varphi}{r^2\sin\theta}\frac{\partial}{\partial\theta}$$

$$+\frac{\cos\theta\sin\varphi\cos\varphi}{r^2\sin\theta}\frac{\partial^2}{\partial\varphi\,\partial\theta}-\frac{\sin\varphi\cos\varphi}{r^2\sin^2\theta}\frac{\partial}{\partial\varphi}+\frac{\cos^2\varphi}{r^2\sin^2\theta}\frac{\partial^2}{\partial\varphi^2} \quad ...(2.70)$$

and

$$\frac{\partial^2}{\partial z^2} = \cos^2\theta \frac{\partial^2}{\partial r^2} + \frac{\sin\theta\cos\theta}{r^2}\frac{\partial}{\partial\theta} - \frac{\sin\theta\cos\theta}{r}\frac{\partial^2}{\partial r\,\partial\theta} + \frac{\sin^2\theta}{r}\frac{\partial}{\partial r}$$

$$- \frac{\sin\theta\cos\theta}{r}\frac{\partial^2}{\partial\theta\,\partial r} + \frac{\sin\theta\cos\theta}{r^2}\frac{\partial}{\partial\theta} + \frac{\sin^2\theta}{r^2}\frac{\partial^2}{\partial\theta^2} \qquad \text{...(2.70}a\text{)}$$

Thus

$$\frac{\partial^2}{\partial x^2} + \frac{\partial^2}{\partial y^2} + \frac{\partial^2}{\partial z^2} = [\sin^2\theta(\cos^2\varphi + \sin^2\varphi) + \cos^2\theta]\frac{\partial^2}{\partial r^2} + \left[-\frac{2\sin\theta\cos\theta}{r^2}(\cos^2\varphi + \sin^2\varphi)\right.$$

$$\left. + \frac{2\sin\theta\cos\theta}{r^2}\right]\frac{\partial}{\partial\theta} + \left[\frac{2\sin\theta\cos\theta}{r}(\cos^2\varphi + \sin^2\varphi) - \frac{2\sin\theta\cos\theta}{r}\right]\frac{\partial^2}{\partial r\,\partial\theta}$$

$$+\left[\frac{\cos\varphi\sin\varphi}{r^2} - \frac{\cos\varphi\sin\varphi}{r^2}\right]\frac{\partial}{\partial\varphi} + \left[\frac{\cos\varphi\sin\varphi}{r} - \frac{\cos\varphi\sin\varphi}{r}\right]\frac{\partial^2}{\partial r\,\partial\varphi}$$

$$+\left[\frac{\cos^2\theta}{r}(\cos^2\varphi + \sin^2\varphi) + \frac{\sin^2\theta}{r^2}\right]\frac{\partial}{\partial r} + \left[\frac{\cos^2\varphi}{r} + \frac{\sin^2\varphi}{r}\right]\frac{\partial}{\partial r}$$

$$+\left[\frac{\cos^2\theta}{r^2}(\cos^2\varphi + \sin^2\varphi) + \frac{\sin^2\theta}{r^2}\right]\frac{\partial^2}{\partial\theta^2} + \left[\frac{\sin\varphi\cos\varphi}{r^2} - \frac{\sin\varphi\cos\varphi}{r^2}\right.$$

$$\left. + \frac{\cos^2\theta\sin\varphi\cos\varphi}{r^2\sin^2\theta} - \frac{\cos^2\theta\sin\varphi\cos\varphi}{r^2\sin^2\theta} + \frac{\sin\varphi\cos\varphi}{r^2\sin^2\theta} - \frac{\sin\varphi\cos\varphi}{r^2\sin^2\theta}\right]\frac{\partial}{\partial\varphi}$$

$$+\left[\frac{2\sin\varphi\cos\varphi\cos\theta}{r^2\sin\theta} - \frac{2\sin\varphi\cos\varphi\cos\theta}{r^2\sin\theta}\right]\frac{\partial^2}{\partial\theta\,\partial\varphi} + \left[\frac{\sin\varphi\cos\varphi}{r} - \frac{\sin\varphi\cos\varphi}{r}\right]$$

$$\frac{\partial^2}{\partial\varphi\,\partial r} + \left[\frac{\sin^2\varphi + \cos^2\varphi}{r^2\sin^2\theta}\right]\frac{\partial^2}{\partial\varphi^2} + \left[\frac{\cos^2\varphi\cos\theta}{r^2\sin\theta} + \frac{\sin^2\varphi\cos\theta}{r^2\sin\theta}\right]\frac{\partial}{\partial\theta}$$

$$= \frac{\partial^2}{\partial r^2} + \frac{1}{r}\frac{\partial}{\partial r} + \frac{1}{r}\frac{\partial}{\partial r} + \frac{1}{r^2}\frac{\partial^2}{\partial\theta^2} + \frac{\cos\theta}{r^2\sin\theta}\frac{\partial}{\partial\theta} + \frac{1}{r^2\sin^2\theta}\frac{\partial^2}{\partial\varphi^2}$$

$$= \frac{\partial^2}{\partial r^2} + \frac{2}{r}\frac{\partial}{\partial r} + \frac{1}{r^2}\frac{\partial^2}{\partial\theta^2} + \frac{\cos\theta}{r^2\sin\theta}\frac{\partial}{\partial\theta} + \frac{1}{r^2\sin^2\theta}\frac{\partial^2}{\partial\varphi^2}$$

$$= \frac{1}{r^2}\frac{\partial}{\partial r}\left(r^2\frac{\partial}{\partial r}\right) + \frac{1}{r^2\sin\theta}\frac{\partial}{\partial\theta}\left(\sin\theta\frac{\partial}{\partial\theta}\right) + \frac{1}{r^2\sin^2\theta}\frac{\partial}{\partial\varphi^2}$$

or finally,

$$\frac{\partial^2}{\partial x^2} + \frac{\partial^2}{\partial y^2} + \frac{\partial^2}{\partial z^2} = \nabla^2_{x,y,z}$$

$$= \frac{1}{r^2}\frac{\partial}{\partial r}\left(r^2\frac{\partial}{\partial r}\right) + \frac{1}{r^2 \sin\theta}\frac{\partial}{\partial \theta}\left(\sin\theta\frac{\partial}{\partial \theta}\right) + \frac{1}{r^2 \sin^2\theta}\frac{\partial^2}{\partial \varphi^2} = \nabla^2_{r,\theta,\varphi} \quad ...(2.53)$$

***Energy Operator*:** Just like time independent Schrödinger wave equation, the time dependent equation may be converted to the Hamiltonian operator form.

$$-\frac{h^2}{8\pi^2 m}\frac{\partial^2 \psi}{\partial x^2} + V\psi = \frac{h}{2\pi i}\frac{\partial \psi}{\partial t} \text{ or } -\frac{hi}{2\pi}\frac{\partial \psi}{\partial t} \quad ...(2.71)$$

or $$\left(-\frac{h^2}{8\pi^2 m}\frac{\partial^2}{\partial x^2} + V\right)\psi = \frac{h}{2\pi i}\frac{\partial \psi}{\partial t} \quad ...(2.71a)$$

But $\hat{H} = -\frac{h^2}{8\pi^2 m}\frac{\partial^2}{\partial x^2} + V$ is the Hamiltonian operator, thus:

$$\hat{H}\psi = \frac{h}{2\pi i}\frac{\partial \psi}{\partial t} \quad ...(2.72)$$

A comparison of equation (2.72) with time independent Schrödinger wave equation into operator form, *i.e.*, $\hat{H}\psi = E\psi$ suggests that we can define an energy operator by the expression

$$\hat{E} = \frac{h}{2\pi i}\frac{\partial}{\partial t} \quad ...(2.73)$$

So in operator form the time dependent Schrödinger wave equation can be written as:

$$\hat{H}\psi(x, t) = \hat{E}\psi(x, t)$$

(operator $\hat{H}$) acting on function ψ = (operator $\hat{E}$) acting on function ψ

and both $\hat{H}$ and $\hat{E}$ are not simple scalar quantities.

***Angular Momentum Operators*:** This class of operators is related to angular momenta and are as important in quantum mechanics as the linear momentum, Hamiltonian, Laplacian, energy etc., operators are. The angular momentum M of a particle about the origin is related to its linear momentum p by the expression:

$$M = rxp \quad ...(2.74)$$

where r is the distance of the particle from the origin. If M_x, M_y and M_z are the components of M, *i.e.*, the angular momentum in the three directions x, y and z, then

$$\begin{aligned} M_x &= y\,p_z - z\,p_y \\ M_y &= z\,p_x - x\,p_z \\ M_z &= x\,p_y - y\,p_x \end{aligned} \quad ...(2.75)$$

The linear momentum operator is defined as $\frac{h}{2\pi i}\frac{\partial}{\partial x}$. Therefore the angular momentum corresponding to the component of angular momentum in the x-direction is:

$$\hat{\mathbf{M}}_x = \left(y\frac{h}{2\pi i}\frac{\partial}{\partial z} - z\,\frac{h}{2\pi i}\frac{\partial}{\partial y} \right) = \frac{h}{2\pi i}\left(y\frac{\partial}{\partial z} - z\frac{\partial}{\partial y} \right) \qquad ...(2.76a)$$

Similarly we can write for other two components, *i.e.* M_y and M_z as:

$$\hat{M}_y = \frac{h}{2\pi i}\left(z\frac{\partial}{\partial x} - x\frac{\partial}{\partial z} \right) \qquad ...(2.76b)$$

$$\hat{M}_z = \frac{h}{2\pi i}\left(x\frac{\partial}{\partial y} - y\frac{\partial}{\partial x} \right) \qquad ...(2.76c)$$

The total angular momentum:

$$M = iM_x + jM_y + kM_z \qquad ...(2.77)$$

M is not used as such but only its scalar product with some other vector or its square.

$$M^2 = M_x^2 + M_y^2 + M_z^2 \qquad ...(2.78)$$

is useful in quantum mechanics. The operator corresponding to M^2 is:

$$\hat{M}^2 = \left(\frac{h}{2\pi i}\frac{\partial}{\partial x}\right)^2 + \left(\frac{h}{2\pi i}\frac{\partial}{\partial y}\right)^2 + \left(\frac{h}{2\pi i}\frac{\partial}{\partial z}\right)^2 \qquad ...(2.79a)$$

or

$$\hat{M}^2 = -\frac{h^2}{4\pi^2}\left[\frac{\partial^2}{\partial x^2} + \frac{\partial^2}{\partial y^2} + \frac{\partial^2}{\partial z^2}\right]$$

or

$$\hat{M}^2 = -\frac{h^2}{4\pi^2}\nabla^2 \qquad ...(2.79b)$$

(**Note:** Value of $\hat{M}_x^2$ can also be obtained by using $\hat{M}_x\ \hat{M}_x = \left[\frac{h}{2\pi i}\left(y\frac{\partial}{\partial z} - z\frac{\partial}{\partial y} \right)\right]$ $\left[\frac{h}{2\pi i}\left(y\frac{\partial}{\partial z} - z\frac{\partial}{\partial y} \right)\right]$. Similarly for $\hat{M}_y\ \hat{M}_y$ and $\hat{M}_z\ \hat{M}_z$ and then adding them).

Just like Laplacian operator, the angular momentum operators in spherical polar coordinates are more useful in quantum mechanics. Here below we have worked out the transformation of angular momentum operator from Cartesian coordinates to polar coordinates. Using the relation given in equations (2.54 to 2.56) and writing right hand side of the equations, *i.e.*

$$r^2 = x^2 + y^2 + z^2 \qquad ...(2.54)$$

$$\cos\theta = \frac{z}{r} = \frac{z}{(x^2+y^2+z^2)^{\frac{1}{2}}} \qquad ...(2.55)$$

$$\tan\varphi = \frac{y}{x} \qquad ...(2.56)$$

Partial differentiation of first expression of this set with respect to x, y and z respectively results in the followings:

$$\left(\frac{\partial r}{\partial z}\right)_{x,y} = \frac{z}{r} = \cos\theta \qquad ...(2.80a)$$

and

$$\left(\frac{\partial r}{\partial y}\right)_{x,z} = \frac{y}{r} = \frac{r\sin\theta\cos\varphi}{r} = \sin\theta\sin\varphi \qquad ...(2.80b)$$

$$\left(\frac{\partial r}{\partial x}\right)_{y,z} = \frac{x}{r} = \frac{r\sin\theta\cos\varphi}{r} = \sin\theta\cos\varphi \qquad ...(2.80c)$$

Now partial derivatives of second expression in the set of equations (2.55) with respect to x, y and z respectively yield:

$$-\sin\theta\left(\frac{\partial\theta}{\partial x}\right)_{y,z} = (z)\left[-\frac{1}{2}(x^2+y^2+z^2)^{\frac{-3}{2}}.2x\right] = z\,x\,(x^2+y^2+z^2)^{\frac{-3}{2}} = -\frac{zx}{r^3}$$

or

$$\left(\frac{\partial\theta}{\partial x}\right)_{y,z} = \frac{(r\cos\theta)\,(r\sin\theta\cos\varphi)}{r^3\sin\theta} = \frac{\cos\theta\cos\varphi}{r} \qquad ...(2.81a)$$

Similarly we can arrive at:

$$\left(\frac{\partial\theta}{\partial y}\right)_{x,z} = \frac{\cos\theta\sin\varphi}{r} \qquad ...(2.81b)$$

$$\left(\frac{\partial\theta}{\partial z}\right)_{x,y} = -\frac{\sin\theta}{r} \qquad ...(2.81c)$$

Further from the partial differentiation of $\tan\varphi = y/x$ with respect to y and x respectively we arrive at the following results.

$$\tan\varphi = \frac{y}{x}$$

$$\sec^2\varphi\left(\frac{\partial\varphi}{\partial y}\right)_{x,z} = \frac{1}{x} \text{ or } \left(\frac{\partial\varphi}{\partial y}\right)_{x,z} = \frac{1}{x\sec^2\varphi} = \frac{\cos^2\varphi}{r\sin\theta\cos\varphi} = \frac{\cos\varphi}{r\sin\theta} \qquad ...(2.82a)$$

Similarly we get

$$\left(\frac{\partial\varphi}{\partial x}\right)_{y,z} = -\frac{\sin\varphi}{r\sin\theta} \text{ and } \left(\frac{\partial\varphi}{\partial z}\right)_{x,y} = 0 \qquad ...(2.82b)$$

Now replace the Cartesian coordinates and partial derivatives in the set of equations (2.76*a*) by their respective counterparts in polar coordinates and manipulate the resultant expressions to arrive at some meaningful expressions, *i.e.*,

$$\hat{M}_x = \frac{h}{2\pi i}\left(y\frac{\partial}{\partial z} - z\frac{\partial}{\partial y}\right)$$

$$= \frac{h}{2\pi i}\left[r\sin\theta\sin\varphi\left(\frac{\partial r}{\partial z}\frac{\partial}{\partial r} + \frac{\partial\theta}{\partial z}\frac{\partial}{\partial\theta} + \frac{\partial\varphi}{\partial z}\frac{\partial}{\partial\varphi}\right)\right.$$

$$\left. - r\cos\theta\left(\frac{\partial r}{\partial y}\frac{\partial}{\partial r} + \frac{\partial\theta}{\partial y}\frac{\partial}{\partial\theta} + \frac{\partial\varphi}{\partial y}\frac{\partial}{\partial\varphi}\right)\right]$$

$$= \frac{h}{2\pi i}\left[r\sin\theta\sin\varphi\left(\cos\theta\frac{\partial}{\partial r} - \frac{\sin\theta}{r}\frac{\partial}{\partial\theta} + 0\right)\right.$$

$$\left. - r\cos\theta\left(\sin\theta\sin\varphi\frac{\partial}{\partial r} + \frac{\cos\theta\sin\varphi}{r}\frac{\partial}{\partial\theta} + \frac{\cos\varphi}{r\sin\theta}\frac{\partial}{\partial\varphi}\right)\right]$$

$$= \frac{h}{2\pi i}\left[(r\sin\theta\sin\varphi\cos\theta - r\cos\theta\sin\theta\sin\varphi)\frac{\partial}{\partial r} + \right.$$

$$\left.(-\sin^2\theta\sin\varphi - \cos^2\theta\sin\varphi)\frac{\partial}{\partial\theta} + (0 - \cot\theta\cos\varphi)\frac{\partial}{\partial\varphi}\right]$$

$$= \frac{h}{2\pi i}\left[-\sin\varphi(\sin^2\theta + \cos^2\theta)\frac{\partial}{\partial\theta} - \cot\theta\cos\varphi\frac{\partial}{\partial\varphi}\right]$$

or finally

$$\hat{M}_x = -\frac{h}{2\pi i}\left[\sin\varphi\frac{\partial}{\partial\theta} + \cot\theta\cos\varphi\frac{\partial}{\partial\varphi}\right] \qquad ...(2.83a)$$

Similarly we can show that:

$$\hat{M}_y = \frac{h}{2\pi i}\left[\cos\varphi\frac{\partial}{\partial\theta} - \cot\theta\sin\varphi\frac{\partial}{\partial\varphi}\right] \qquad ...(2.83b)$$

$$\hat{M}_z = \frac{h}{2\pi i}\frac{\partial}{\partial\varphi} \qquad ...(2.83c)$$

Accordingly the operator $\hat{M}^2$ in terms of polar coordinates can be worked out as under:

$$\hat{M}^2 = \hat{M}_x^2 + \hat{M}_y^2 + \hat{M}_z^2$$

$$\hat{M}_x^2 = -\frac{h^2}{4\pi^2}\left(\sin\varphi\frac{\partial}{\partial\theta} + \cot\theta\cos\varphi\frac{\partial}{\partial\varphi}\right)\left(\sin\varphi\frac{\partial}{\partial\theta} + \cot\theta\cos\varphi\frac{\partial}{\partial\varphi}\right)$$

$$= -\frac{h^2}{4\pi^2}\left[\sin\varphi\frac{\partial}{\partial\theta}\left(\sin\varphi\frac{\partial}{\partial\theta} + \cot\theta\cos\varphi\frac{\partial}{\partial\varphi}\right) + \right.$$

$$\cot\theta\cos\varphi\frac{\partial}{\partial\varphi}\left(\sin\varphi\frac{\partial}{\partial\theta} + \cot\theta\cos\varphi\frac{\partial}{\partial\varphi}\right)$$

$$= -\frac{h^2}{4\pi^2}\left(\sin^2\varphi\frac{\partial^2}{\partial\theta^2} - \sin\varphi\cos\varphi\,\mathrm{cosec}^2\theta\frac{\partial}{\partial\varphi}\right.$$

$$+\cot\theta\cos\varphi\sin\varphi\frac{\partial^2}{\partial\theta\,\partial\varphi} + \cot\theta\cos^2\varphi\frac{\partial}{\partial\theta} + \sin\varphi\cos\varphi\cot\theta\frac{\partial^2}{\partial\varphi\partial\theta}$$

$$\left.-\cot^2\theta\cos\varphi\sin\varphi\frac{\partial}{\partial\varphi} + \cot^2\theta\cos^2\varphi\frac{\partial^2}{\partial\varphi^2}\right)$$

Similarly

$$\hat{M}_y^2 = -\frac{h^2}{4\pi^2}\left[\left(\cos\varphi\frac{\partial}{\partial\theta} - \cot\theta\sin\varphi\frac{\partial}{\partial\varphi}\right)\left(\cos\varphi\frac{\partial}{\partial\theta} - \cot\theta\sin\varphi\frac{\partial}{\partial\varphi}\right)\right.$$

$$= -\frac{h^2}{4\pi^2}\left[\cos\varphi\frac{\partial}{\partial\theta}\left(\cos\varphi\frac{\partial}{\partial\theta} - \cot\theta\sin\varphi\frac{\partial}{\partial\varphi}\right)\right.$$

$$\left.-\cot\theta\sin\varphi\frac{\partial}{\partial\varphi}\left(\cos\varphi\frac{\partial}{\partial\theta} - \cot\theta\sin\varphi\frac{\partial}{\partial\varphi}\right)\right]$$

$$= -\frac{h^2}{4\pi^2}\left[\cos^2\varphi\frac{\partial^2}{\partial\theta^2} + \sin\varphi\cos\varphi\,\mathrm{cosec}^2\theta\frac{\partial}{\partial\varphi} - \sin\varphi\cos\varphi\cot\theta\frac{\partial^2}{\partial\theta\partial\varphi}\right.$$

$$+\cot\theta\sin^2\varphi\frac{\partial}{\partial\theta} - \cot\theta\sin\varphi\cos\varphi\frac{\partial^2}{\partial\varphi\partial\theta}$$

$$\left.+\cot^2\theta\sin\varphi\cos\varphi\frac{\partial}{\partial\varphi} + \cot^2\theta\sin^2\varphi\frac{\partial^2}{\partial\varphi^2}\right]$$

$$\hat{M}_z^2 = -\frac{h^2}{4\pi^2}\frac{\partial^2}{\partial\varphi^2}$$

$$\hat{M}^2 = -\frac{h^2}{4\pi^2}\left[(\sin^2\varphi + \cos^2\varphi)\frac{\partial^2}{\partial\theta^2}\right.$$

$$+ (-\sin\varphi\,\cos\varphi\,\text{cosec}^2\theta + \sin\varphi\,\cos\varphi\,\text{cosec}^2\theta)\frac{\partial}{\partial\varphi}$$

$$+ (-\sin\varphi\,\cos\varphi\,\cot^2\theta + \sin\varphi\,\cos\varphi\,\cot^2\theta)\frac{\partial}{\partial\varphi}$$

$$+ (2\sin\varphi\,\cos\varphi\,\cot\theta - 2\sin\varphi\,\cos\varphi\,\cot\theta)\frac{\partial^2}{\partial\varphi\partial\theta}$$

$$\left. + \cot\theta\,(\cos^2\varphi + \sin^2\varphi)\frac{\partial}{\partial\theta} + \cot^2\theta\,(\sin^2\varphi + \cos^2\varphi)\frac{\partial^2}{\partial\varphi^2} + \frac{\partial^2}{\partial\varphi^2}\right]$$

$$= -\frac{h^2}{4\pi^2}\left[\frac{\partial^2}{\partial\theta^2} + \cot\theta\frac{\partial}{\partial\theta} + \cot^2\theta\frac{\partial^2}{\partial\varphi^2} + \frac{\partial^2}{\partial\varphi^2}\right]$$

$$= -\frac{h^2}{4\pi^2}\left[\frac{\partial^2}{\partial\theta^2} + \cot\theta\frac{\partial}{\partial\theta} + (\cot^2\theta + 1)\frac{\partial^2}{\partial\varphi^2}\right]$$

$$= -\frac{h^2}{4\pi^2}\left[\frac{\partial^2}{\partial\theta^2} + \frac{\cos\theta}{\sin\theta}\frac{\partial}{\partial\theta} + \frac{1}{\sin^2\theta}\frac{\partial^2}{\partial\varphi^2}\right]$$

Finally

$$\hat{M}^2 = -\frac{h^2}{4\pi^2}\left[\frac{1}{\sin\theta}\frac{\partial}{\partial\theta}\left(\sin\theta\frac{\partial}{\partial\theta}\right) + \frac{1}{\sin^2\theta}\frac{\partial^2}{\partial\varphi^2}\right] \qquad ...(2.84)$$

For ready reference the classical variables, *i.e.*, the observable physical properties of the system in classical mechanics and their corresponding operators in quantum mechanics are compiled in Table 2.1.

Table 2.1: *Quantum mechanical operators in term of Cartesian coordinates*

Observable	Quantum mechanical operator	Experssion for operator
Position	$\hat{r}$	$= \hat{x}\, i + \hat{y}\, j + \hat{z}\, k$
Linear Momentum	$\hat{p}$	$\hat{p} = \hat{p}_x i + \hat{p}_y j + \hat{p}_z k$ where $\hat{p}_x = \frac{h}{2\pi i}\frac{\partial}{\partial x}$ $\hat{p}_y = \frac{h}{2\pi i}\frac{\partial}{\partial y}$ $\hat{p}_z = \frac{h}{2\pi i}\frac{\partial}{\partial z}$
Time	$\hat{t}$	t
Energy	$\hat{E}$	$\frac{h}{2\pi i}\frac{\partial}{\partial t}$
Orbital angular momentum	$\hat{l}$	$\hat{l} = \hat{l}_x i + \hat{l}_y j + \hat{l}_z k$ where $\hat{l}_x = \frac{h}{2\pi i}\left(y\frac{\partial}{\partial z} - z\frac{\partial}{\partial y}\right)$ $\hat{l}_y = \frac{h}{2\pi i}\left(z\frac{\partial}{\partial x} - x\frac{\partial}{\partial z}\right)$ $\hat{l}_z = \frac{h}{2\pi i}\left(x\frac{\partial}{\partial y} - y\frac{\partial}{\partial x}\right)$
Spin angular momentum	$\hat{s}$	$\hat{s} = \hat{s}_x i + \hat{s}_y j + \hat{s}_z k$

Classical expression for various observables

Kinetic energy (K) = $\frac{p^2}{2m}$

Potential energy (V) proportional to r^{-1}

Dipole moment (μ) = er

Charge density ξ (R) = ξ (R – r)

Spin density ρ^{spin} (R) = $2s_z\,\rho$ (R – r)

Table 2.1 shows that the form of the operator, for a given physical quantity, depends on the choice of the variables for the functions on which the operator acts. The operators for the independent variable are always the multiplication into this variable.

Example 2.9: Using the following representation $\hat{p}_r = \dfrac{h}{2\pi i}\left(\dfrac{\partial}{\partial r} + \dfrac{1}{r}\right)$, prove that:

$$\frac{\hat{p}_r^2}{2m} = -\frac{h^2}{8\pi^2 m}\nabla^2 = \frac{p_r^2}{2m} + \frac{M^2}{2mr^2}$$

Solution:

$$\frac{\hat{p}_r^2}{2m}\psi = -\frac{h^2}{8\pi^2 m}\left[\left(\frac{\partial}{\partial r}+\frac{1}{r}\right)\left(\frac{\partial}{\partial r}+\frac{1}{r}\right)\right]\psi$$

$$= -\frac{h^2}{8\pi^2 m}\left[\left(\frac{\partial}{\partial r}+\frac{1}{r}\right)\left(\frac{\partial \psi}{\partial r}+\frac{\psi}{r}\right)\right]$$

$$= -\frac{h^2}{8\pi^2 m}\left[\left(\frac{\partial^2\psi}{\partial r^2} + \frac{\partial}{\partial r}\left(\frac{1}{r}\psi\right) + \frac{1}{r}\frac{\partial \psi}{\partial r} + \frac{1}{r^2}\psi\right]\right.$$

$$= -\frac{h^2}{8\pi^2 m}\left[\frac{\partial^2\psi}{\partial r^2} + \frac{1}{r}\frac{\partial \psi}{\partial r} - \frac{1}{r^2}\psi + \frac{1}{r}\frac{\partial \psi}{\partial r} + \frac{1}{r^2}\psi\right]$$

$$= -\frac{h^2}{8\pi^2 m}\left[\frac{\partial^2\psi}{\partial r^2} + \frac{2}{r}\frac{\partial \psi}{\partial r}\right] = -\frac{h^2}{8\pi^2 m}\left[\frac{1}{r^2}\frac{\partial}{\partial r}\left(r^2\frac{\partial \psi}{\partial r}\right)\right] \quad ...(a)$$

From equations (*a*) and (2.84)

$$\nabla^2\Psi = -\frac{h^2}{8\pi^2 m}\left[\frac{1}{r^2}\frac{\partial}{\partial r}\left(r^2\frac{\partial \Psi}{\partial r}\right)\right] + \frac{\hat{M}^2}{2mr^2}\Psi$$

$$\nabla^2 = \frac{p_r^2}{2m} + \frac{\hat{M}^2}{2mr^2}$$

From classical mechanics,

$$p^2 = p_r^2 + \frac{M^2}{r^2}$$

where p_r is the radial component of angular momentum. Thus,

$$\frac{p_r^2}{2m} = -\frac{h^2}{8\pi^2 m}\nabla^2 = \frac{p_r^2}{2m} + \frac{M^2}{2mr^2}$$

2.4.5 COMMUTATION RULES FOR ANGULAR MOMENTUM OPERATORS

The quantum mechanical operator may or may not commute but the algebraic operators do commute, *i.e.*, $\hat{A}\hat{B}\ f(x) = \hat{B}\hat{A}\ f(x)$. Let us now see whether or not the angular momentum operators $\hat{M}_x$, $\hat{M}_y$, $\hat{M}_z$ and $\hat{M}^2$ commute, *i.e.*,

$$\hat{M}_x \ \hat{M}_y \neq (\text{or} =) \ \hat{M}_y \ \hat{M}_x$$

$$\hat{M}_y \ \hat{M}_z \neq (\text{or} =) \ \hat{M}_z \ \hat{M}_y$$

$$\hat{M}_z \ \hat{M}_x \neq (\text{or} =) \ \hat{M}_x \ \hat{M}_z$$

$$\hat{M}^2 \ \hat{M}_x \neq (\text{or} =) \ \hat{M}_x \ \hat{M}^2$$

$$\hat{M}^2 \ \hat{M}_y \neq (\text{or} =) \ \hat{M}_y \ \hat{M}^2$$

$$\hat{M}^2 \ \hat{M}_z \neq (\text{or} =) \ \hat{M}_z \ \hat{M}^2$$

As examples, let us examine the following representative cases:

$$\hat{M}_x \ \hat{M}_y \neq (\text{or} =) \ \hat{M}_y \ \hat{M}_x$$

$$\hat{M}^2 \ \hat{M}_z \neq (\text{or} =) \ \hat{M}_z \ \hat{M}^2$$

From Table 2.1, we can write:

$$\hat{M}_x \ \hat{M}_y = \frac{h}{2\pi i}\left(y \frac{\partial}{\partial z} - z \frac{\partial}{\partial y}\right) \frac{h}{2\pi i}\left(z \frac{\partial}{\partial x} - x \frac{\partial}{\partial z}\right)$$

$$= -\frac{h^2}{4\pi^2}\left[y \frac{\partial}{\partial z}\left(z \frac{\partial}{\partial x} - x \frac{\partial}{\partial z}\right) - z \frac{\partial}{\partial y}\left(z \frac{\partial}{\partial x} - x \frac{\partial}{\partial z}\right)\right]$$

$$= -\frac{h^2}{4\pi^2}\left[y \frac{\partial}{\partial x} + yz \frac{\partial^2}{\partial z \, \partial x} - xy \frac{\partial^2}{\partial z^2} - z^2 \frac{\partial^2}{\partial y \, \partial x} + xz \frac{\partial^2}{\partial y \partial z}\right] \quad ...(2.85)$$

Similarly

$$\hat{M}_y \ \hat{M}_x = -\frac{h^2}{4\pi^2}\left[yz \frac{\partial^2}{\partial x \, \partial z} - z^2 \frac{\partial^2}{\partial x \, \partial y} - xy \frac{\partial^2}{\partial z^2} + x \frac{\partial}{\partial y} + xz \frac{\partial^2}{\partial y \, \partial z}\right] \quad ...(2.86)$$

Combining equations (2.85) and (2.86)

$$\hat{M}_x \ \hat{M}_y - \hat{M}_y \ \hat{M}_x = -\frac{h^2}{4\pi^2}\left[y \frac{\partial}{\partial x} - x \frac{\partial}{\partial y}\right] = \frac{h}{2\pi i} \frac{h}{2\pi i}$$

$$\left[y \frac{\partial}{\partial x} - x \frac{\partial}{\partial y}\right] = \frac{h}{2\pi i} \hat{M}_z$$

Similarly, $$\hat{M}_y \hat{M}_z - \hat{M}_z \hat{M}_y = \frac{h}{2\pi i} \hat{M}_x$$

And $$\hat{M}_z \hat{M}_x - \hat{M}_x \hat{M}_z = \frac{h}{2\pi i} \hat{M}_y \quad ...(2.87)$$

This set of rules is also called the commutation relation of the angular momentum operators. Let us examine the second case, *i.e.*,

$$\hat{M}^2 \hat{M}_z \neq (\text{or} =) \hat{M}_z \hat{M}^2$$

In this case for convenience, the polar form of the operators is used. Using equations (2.84) and (2.83), we can write:

$$\hat{M}^2 \hat{M}_z = -\frac{h^2}{4\pi^2}\left[\frac{1}{\sin\theta}\frac{\partial}{\partial\theta}\left(\sin\theta\frac{\partial}{\partial\theta}\right)+\frac{1}{\sin^2\theta}\frac{\partial^2}{\partial\varphi^2}\right]\left[\frac{h}{2\pi i}\frac{\partial}{\partial\varphi}\right]$$

$$= -\frac{h^3}{8\pi^3 i}\left[\frac{1}{\sin\theta}\frac{\partial}{\partial\theta}\left(\sin\theta\frac{\partial}{\partial\theta}\right)\left(\frac{\partial}{\partial\varphi}\right)+\frac{1}{\sin^2\theta}\frac{\partial^3}{\partial\varphi^3}\right]$$

Similarly $$\hat{M}_z \hat{M}^2 = -\frac{h^3}{8\pi^3 i}\left[\left(\frac{\partial}{\partial\varphi}\right)\frac{1}{\sin\theta}\frac{\partial}{\partial\theta}\left(\sin\theta\frac{\partial}{\partial\theta}\right)+\frac{1}{\sin^2\theta}\frac{\partial^3}{\partial\varphi^3}\right]$$

Thus, $\hat{M}^2 \hat{M}_z - \hat{M}_z \hat{M}^2 = 0$

Similarly, $\hat{M}^2 \hat{M}_y - \hat{M}_y \hat{M}^2 = 0$ and $\hat{M}^2 \hat{M}_x - \hat{M}_x \hat{M}^2 = 0$ ...(2.88)

Hence $\hat{M}^2$ commutes with, $\hat{M}_x$, $\hat{M}_y$ and $\hat{M}_x$. These commutation rules hold good for systems of particles as well as for a single particle. Also *equations (2.88) express the physical fact that the square of the angular momentum and any of the components of angular momentum can simultaneously have definite values.*

The commutation of $\hat{M}^2$ with each of the components of M can also be proved as follows:

$$\hat{M}_x^2 \hat{M}_z - \hat{M}_z \hat{M}_x^2 = \hat{M}_x (\hat{M}_x \hat{M}_z - \hat{M}_z \hat{M}_x) - (\hat{M}_z \hat{M}_x - \hat{M}_x \hat{M}_z)(\hat{M}_x)$$

$$= -\hat{M}_x (\hat{M}_z \hat{M}_x - \hat{M}_x \hat{M}_z) - (\hat{M}_z \hat{M}_x - \hat{M}_x \hat{M}_z)(\hat{M}_x)$$

$$= -\hat{M}_x\left(-\frac{h}{2\pi i}\right)\hat{M}_y - \left(-\frac{h}{2\pi i}\right)\hat{M}_y \hat{M}_x$$

$$= \frac{h}{2\pi i}(\hat{M}_x \hat{M}_y + \hat{M}_y \hat{M}_x)$$

Similarly

$$\hat{M}_y^2 \hat{M}_z - \hat{M}_z \hat{M}_y^2 = -\frac{h}{2\pi i}(\hat{M}_x \hat{M}_y + \hat{M}_y \hat{M}_x)$$

$$\hat{M}_z^2 \hat{M}_z - \hat{M}_z \hat{M}_z^2 = 0$$

Adding up the left hand sides and right hand sides:

$$(\hat{M}_x^2 + \hat{M}_y^2 + \hat{M}_z^2)\hat{M}_z - \hat{M}_z (\hat{M}_x^2 + \hat{M}_y^2 + \hat{M}_z^2) = 0$$

or

$$\hat{M}^2 \hat{M}_z - \hat{M}_z \hat{M}^2 = 0$$

Proceeding on the same lines we can also prove that:

$$\hat{M}^2 \hat{M}_x - \hat{M}_x \hat{M}^2 = 0$$

$$\hat{M}^2 \hat{M}_y - \hat{M}_y \hat{M}^2 = 0$$

2.4.6 HEISENBERG RULE FOR THE PRODUCT OF POSITION AND MOMENTUM OPERATORS

Let us consider a particle with momentum p and position q. Thus:

$$\hat{p} = \frac{h}{2\pi i}\frac{\partial}{\partial q}$$

$$\therefore \quad \hat{p}\,\hat{q}\,\psi = \frac{h}{2\pi i}\frac{\partial}{\partial q}(q\psi) = \frac{h}{2\pi i}\left[q\frac{\partial \psi}{\partial q} + \psi\right] \quad \text{...(2.89)}$$

and

$$\hat{q}\,\hat{p}\,\psi = q\frac{h}{2\pi i}\frac{\partial \psi}{\partial q} = \frac{h}{2\pi i}\,q\,\frac{\partial \psi}{\partial q} \quad \text{...(2.90)}$$

Commutation of equations (2.89) and (2.90) of position and momentum operators' yields:

$$(\hat{p}\hat{q} - \hat{q}\hat{p})\,\psi = \frac{h}{2\pi i}\,\psi \quad \text{...(2.91)}$$

This is Heisenberg rule for the commutation of position and momentum operators. Accordingly the angular momentum operators comprising of position and momentum operators of the type:

$$\hat{M}_x = \frac{h}{2\pi i}\left(y\frac{\partial}{\partial z} - z\frac{\partial}{\partial y}\right)$$

$$\hat{M}_y = \frac{h}{2\pi i}\left(z\frac{\partial}{\partial x} - x\frac{\partial}{\partial z}\right)$$

$$\hat{M}_z = \frac{h}{2\pi i}\left(x\frac{\partial}{\partial y} - y\frac{\partial}{\partial x}\right)$$

are non-commutation operators. But they are very much Hermitian operators.

Example 2.10: Prove that $\hat{M}_z$ commutes with Laplacian operator and with the potential energy operator (or with the Hamiltonian operator).

Solution: Let us consider the operator M_z acting on ψ

$$\hat{M}_z\,\psi = \frac{h}{2\pi i}\left(x\frac{\partial\psi}{\partial y} - y\frac{\partial\psi}{\partial x}\right)$$

Hence

$$(\nabla^2\hat{M}_z)\,\psi - (\hat{M}_z\nabla^2)\,\psi = \frac{h}{2\pi i}\left[\nabla^2\left(x\frac{\partial\psi}{\partial y} - y\frac{\partial\psi}{\partial x}\right) - \left(x\frac{\partial(\nabla^2\psi)}{\partial y} - y\frac{\partial(\nabla^2\psi)}{\partial x}\right)\right]$$

$$= \frac{h}{2\pi i}\left[2\frac{\partial^2\psi}{\partial x\,\partial y} - 2\frac{\partial^2\psi}{\partial y\,\partial x}\right) = 0$$

(Note: Proof of the above equation can be added if so desires, but we have left for the students to do so as an exercise)

We can prove in, a similar manner that $\hat{M}_z$ commutes with V (*r*) also, *i.e.*,

$$(\mathrm{V}\,\hat{M}_z)\,\psi - (\hat{M}_z\,\mathrm{V})\,\psi = \frac{h}{2\pi \mathrm{i}}\left[\mathrm{V}\left(x\frac{\partial\psi}{\partial y} - y\frac{\partial\psi}{\partial x}\right) - x\frac{\partial(\mathrm{V}\psi)}{\partial y} + y\frac{\partial(\mathrm{V}\psi)}{\partial x}\right]$$

$$= \frac{h}{2\pi i}\left(y\frac{\partial \mathrm{V}}{\partial x} - x\frac{\partial \mathrm{V}}{\partial y}\right)\psi = 0$$

Since if V = V (*r*), then $y\dfrac{\partial \mathrm{V}}{\partial x} - x\dfrac{\partial \mathrm{V}}{\partial y} = 0$

Thus, $\hat{M}_z$ commutes both with the Laplacian operator and with the potential energy, which implies that it commutes with the Hamiltonian operator. Accordingly we can write:

$$\hat{H}\,\hat{M}_x - \hat{M}_x\,\hat{H} = 0$$

$$\hat{H}\,\hat{M}_y - \hat{M}_y\,\hat{H} = 0 \qquad \text{...(2.92)}$$

$$\hat{H}\,\hat{M}_z - \hat{M}_z\,\hat{H} = 0$$

The set of equations (2.92) reveals that the components of angular momentum, *i.e.*, $\hat{M}_x$, $\hat{M}_y$ and $\hat{M}_z$ are constants of the motion in quantum mechanics. However as has been shown that these operators do not commute with each other, which implies that the physical quantities $\hat{M}_x$, $\hat{M}_y$ and $\hat{M}_z$ cannot have definite values simultaneously (with the exception of zero values), *i.e., the physical quantities* $\hat{M}_x$, $\hat{M}_y$ and $\hat{M}_z$ *can have definite values simultaneously provided their corresponding operators commute with each other*. Based on this problem we can make the following deductions:

Since we have proved that $\hat{M}^2$ can commute with any of its angular momentum components, $\hat{M}_x$, $\hat{M}_y$ and $\hat{M}_z$, this implies that $\hat{M}^2$ and $\hat{M}_x$, $\hat{M}^2$ and $\hat{M}_y$ or $\hat{M}^2$ and $\hat{M}_z$ can have simultaneously definite values.

On the other hand, since each of the components, $\hat{M}_x$, $\hat{M}_y$ and $\hat{M}_z$ commute with the Hamiltonian operator, their respective squares and the sum of their squares also commutes with it. Hence $\hat{M}^2$ will be a constant of motion:

$$\begin{aligned}\hat{H}\hat{M}_x^2 - \hat{M}_x^2\hat{H} &= 0\\ \hat{H}\hat{M}_y^2 - \hat{M}_y^2\hat{H} &= 0\\ \hat{H}\hat{M}_z^2 - \hat{M}_z^2\hat{H} &= 0\end{aligned} \qquad \text{...(2.92}a\text{)}$$

Adding up left and right sides, we get

$$\hat{H}\hat{M}^2 - \hat{M}^2\hat{H} = 0$$

We note that $\hat{H}$ is also a constant of motion. Thus we have three operators, $\hat{M}_z$ (say), $\hat{M}^2$ and $\hat{H}$ which commute with each other and are constants of motion. This suggests that we can choose function that satisfies the time independent Schrödinger wave equation and is also simultaneously eigen function of all the three operators, *i.e.*, it is a solution of the following equations:

$$\begin{aligned}\hat{H}\,\psi &= E\,\psi\\ \hat{M}^2\,\psi &= M\,\psi\\ \hat{M}_z\,\psi &= M_z\,\psi\end{aligned} \qquad \text{...(2.92}b\text{)}$$

2.4.7 HERMITICITY OF QUANTUM MECHANICAL OPERATORS

Let us now apply the theory of linear operators to accomplish the hermiticity of quantum mechanical operators as Hermitian operators are also linear. As an example let us examine the hermiticity of Hamiltonian operator. To recollect the memory, the Hamiltonian operator for a single particle in Cartesian coordinates is:

$$\hat{H} = -\frac{h^2}{8\pi^2 m}\left[\frac{\partial^2}{\partial x^2}+\frac{\partial^2}{\partial y^2}+\frac{\partial^2}{\partial z^2}\right]+V(x, y, z) \qquad ...(2.12)$$

The Hamiltonian function in classical mechanics form can be written as:

$$H = K + V = \frac{p^2}{2m} + V = \frac{1}{2m}\,[p_x^2 + p_y^2 + p_z^2] + V\,(x, y, z)$$

or

$$H = \frac{1}{2m}\left[\frac{1}{x}p_x^2\,x + \frac{1}{y}\,p_y^2\,y + \frac{1}{z}\,p_z^2\,z\right] + V\,(x, y, z) \qquad ...(2.93)$$

Now replace p's in the equation (2.93) by their linear operators, *i.e.*, $\hat{p}_x = \dfrac{h}{2\pi i}\,\dfrac{\partial}{\partial x}$,

$$\therefore \qquad \frac{1}{x}\,\frac{h}{2\pi i}\,\frac{\partial}{\partial x}(p_x.x) = \frac{1}{x}\frac{h}{2\pi i}\frac{\partial}{\partial x}\left(\frac{h}{2\pi i}\frac{\partial}{\partial x}.x\right)$$

$$= -\frac{h^2}{4\pi^2}\frac{1}{x}\left(x\frac{\partial^2}{\partial x^2}+\frac{\partial}{\partial x}\right)$$

or finally

$$\frac{1}{x}\,p_x^2\,x = -\frac{h^2}{4\pi^2}\left(\frac{\partial^2}{\partial x^2}+\frac{1}{x}\frac{\partial}{\partial x}\right)$$

Similarly

$$\frac{1}{y}\,p_y^2\,y = -\frac{h^2}{4\pi^2}\left(\frac{\partial^2}{\partial y^2}+\frac{1}{y}\,\frac{\partial}{\partial y}\right) \qquad ...(2.94)$$

$$\frac{1}{z}\,p_z^2\,z = -\frac{h^2}{4\pi^2}\left(\frac{\partial^2}{\partial z^2}+\frac{1}{z}\,\frac{\partial}{\partial z}\right)$$

Accordingly equation (2.93) reduces to:

$$\hat{H} = -\frac{h^2}{8\pi^2 m}\left[\left(\frac{\partial^2}{\partial x^2}+\frac{1}{x}\frac{\partial}{\partial x}\right)+\left(\frac{\partial^2}{\partial y^2}+\frac{1}{y}\frac{\partial}{\partial y}\right) + \left(\frac{\partial^2}{\partial z^2}+\frac{1}{z}\frac{\partial}{\partial z}\right)\right] + V\,(x, y, z) \qquad ..(2.95)$$

or
$$\hat{H} = -\frac{h^2}{8\pi^2 m}\left(\frac{\partial^2}{\partial x^2}+\frac{\partial^2}{\partial y^2}+\frac{\partial^2}{\partial z^2}\right) - \frac{h^2}{8\pi^2 m}\left(\frac{1}{x}\frac{\partial}{\partial x}+\frac{1}{y}\frac{\partial}{\partial y}+\frac{1}{z}\frac{\partial}{\partial z}\right) + V(x, y, z)$$

The condition for Hermiticity of operators is that $\hat{O} = \hat{O}^*$, where $\hat{O}^*$ is the Hermite conjugate of $\hat{O}$.

$$\therefore \quad \iiint_{-\infty}^{+\infty} \alpha^* \hat{H} \Psi \, dx\, dy\, dz = \iiint_{-\infty}^{+\infty} \Psi \hat{H}^* \alpha^* \, dx\, dy\, dz \qquad ...(2.96)$$

where α and ψ belongs to same class of quantum mechanical functions. Now consider the operator, $\frac{\partial^2}{\partial x^2}$. Integrating left hand side of equation (2.96)

$$\therefore \quad \iiint_{-\infty}^{+\infty} \alpha^* \frac{\partial^2 \Psi}{\partial x^2} dx\, dy\, dz = -\iint_{-\infty}^{+\infty} \alpha^* \frac{\partial \Psi}{\partial x} dy\, dz]_{-\infty}^{+\infty} - \iiint_{-\infty}^{+\infty} \frac{\partial \alpha^*}{\partial x}\frac{\partial \Psi}{\partial x} dx\, dy\, dz \qquad ...(2.97)$$

$\alpha^* = 0$ at infinity, $\frac{\partial \psi}{\partial x}$ is zero or finite.

$$\text{Hence} \quad \iiint_{-\infty}^{+\infty} \alpha^* \frac{\partial^2 \Psi}{\partial x^2} dx\, dy\, dz = -\iiint_{-\infty}^{+\infty} \frac{\partial \alpha^*}{\partial x}\frac{\partial \Psi}{\partial x} dx\, dy\, dz \qquad ...(2.98)$$

Now integrating the right hand side of equation (2.96), considering only $\frac{\partial^2}{\partial x^2}$ term,

$$\iiint_{-\infty}^{+\infty} \Psi \hat{H}^* \alpha^* \, dx\, dy\, dz = \iiint_{-\infty}^{+\infty} \Psi \frac{\partial^2 \alpha^*}{\partial x^2} dx\, dy\, dz$$

$$= \iint_{-\infty}^{+\infty} \Psi \frac{\partial \alpha^*}{\partial x} dy\, dz]_{-\infty}^{+\infty} - \iiint_{-\infty}^{+\infty} \frac{\partial \Psi}{\partial x}\frac{\partial \alpha^*}{\partial x} dx\, dy\, dz \qquad ...(2.99)$$

$\psi = 0$ at infinity and $\frac{\partial \alpha^*}{\partial x}$ is finite or zero. The first term on the right hand side of equation (2.99) is zero. Hence

$$\iiint_{-\infty}^{+\infty} \Psi \frac{\partial^2 \alpha^*}{\partial x^2} dx\, dy\, dz = -\iiint_{-\infty}^{+\infty} \frac{\partial \Psi}{\partial x}\frac{\partial \alpha^*}{\partial x} dx\, dy\, dz \qquad ...(2.100)$$

Comparing equations (2.98) and (2.100), we get

$$\therefore \qquad \iiint_{-\infty}^{+\infty} \alpha^* \frac{\partial^2 \psi}{\partial x^2}\, dx\, dy\, dz = \iiint_{-\infty}^{+\infty} \psi\, \hat{H}^* \alpha^*\, dx\, dy\, dz \qquad ...(2.101)$$

which is the required condition for Hermiticity of an operator. Thus the operator $\frac{\partial^2}{\partial x^2}$ is a Hermitian operator. Similarly we can prove that $\frac{\partial^2}{\partial y^2}$ and $\frac{\partial^2}{\partial z^2}$ are also Hermitian operators. Since $\hat{H}$ of equation (2.12) is the sum of Hermitian operators multiplied by some real constants, therefore it is Hermitian.

Let us now examine the operator $\frac{\partial}{\partial x}$ for its Hermiticity. The left hand side of equation (2.96) in terms of $\frac{\partial}{\partial x}$ is:

$$\iiint_{-\infty}^{+\infty} \alpha^* \frac{\partial \psi}{\partial x}\, dx\, dy\, dz = -\iint_{-\infty}^{+\infty} \alpha^* \psi\, dy\, dz]_{-\infty}^{+\infty} - \iiint_{-\infty}^{+\infty} \psi \frac{\partial \alpha^*}{\partial x}\, dx\, dy\, dz \qquad ...(2.102)$$

Since both $\psi = \alpha^* = 0$ at ∞, equation (2.102) reduces to

$$\iiint_{-\infty}^{+\infty} \alpha^* \frac{\partial \psi}{\partial x}\, dx\, dy\, dz = -\iiint_{-\infty}^{+\infty} \psi \frac{\partial \alpha^*}{\partial x}\, dx\, dy\, dz \qquad ...(2.103)$$

The right hand side of equation (2.96) in terms of $\frac{\partial}{\partial x}$ is:

$$\iiint_{-\infty}^{+\infty} \psi \left(\frac{\partial^*}{\partial x} \right) \alpha^*\, dx\, dy\, dz = -\iiint_{-\infty}^{+\infty} \alpha^* \frac{\partial \psi}{\partial x}\, dx\, dy\, dz \qquad ...(2.104)$$

A comparison of equations (2.103) and (2.104) shows that left hand side of both the equations is not equal, *i.e.,*

$$\int_{-\infty}^{+\infty}\iint \psi \frac{\partial \alpha^*}{\partial x}\, dx\, dy\, dz \neq \int_{-\infty}^{+\infty}\iint \alpha^* \frac{\partial \psi}{\partial x}\, dx\, dy\, dz$$

Consequently neither $\frac{\partial}{\partial x}$ nor $\frac{1}{x}\frac{\partial}{\partial x}$ are Hermitian operators. Similarly $\frac{1}{y}\frac{\partial}{\partial y}$ and $\frac{1}{z}\frac{\partial}{\partial z}$ will also not be Hermitian operators. The sum of $\frac{1}{x}\frac{\partial}{\partial x} + \frac{1}{y}\frac{\partial}{\partial y} + \frac{1}{z}\frac{\partial}{\partial z}$ will also be not an Hermitian operator. Thus the Hamiltonian operator written in equation (2.96) does not qualify as a quantum mechanical operator.

Example 2.11: Prove that $i \dfrac{\partial}{\partial x}$ or $- i \dfrac{\partial}{\partial x}$ is a Hermitian operator.

Solution: The condition for hermiticity of an operator is:

$$\iiint_{-\infty}^{+\infty} \alpha^* \hat{O}\, \psi \, dx\, dy\, dz = \iiint_{-\infty}^{+\infty} \psi\, \hat{O}^* \alpha^* dx\, dy\, dz \qquad ...(2.96)$$

provided $\hat{O} = \hat{O}^*$ where $\hat{O}^*$ is the Hermite conjugate of $\hat{O}$. In terms of $i \dfrac{\partial}{\partial x}$, the left hand side of equation (2.96) becomes:

$$\iiint_{-\infty}^{+\infty} \alpha^* i \frac{\partial}{\partial x} \psi \, dx\, dy\, dz = \iint_{-\infty}^{+\infty} i\alpha^* \psi \, dy\, dz]_{-\infty}^{+\infty} - \iiint_{-\infty}^{+\infty} \psi\, i \frac{\partial \alpha^*}{\partial x} dx\, dy\, dz$$

$$= - \iiint_{-\infty}^{+\infty} \psi\, i \frac{\partial \alpha^*}{\partial x} dx\, dy\, dz$$

$$= \mp \sqrt{-1} \iiint_{-\infty}^{+\infty} \psi \frac{\partial \alpha^*}{\partial x} dx\, dy\, dz \qquad (\because i = \pm \sqrt{-1})$$

after applying the condition that the functions are zero at infinity. Right hand side of equation (2.96) in terms of $i \dfrac{\partial}{\partial x}$ is

$$\iiint_{-\infty}^{+\infty} \psi\, i \frac{\partial \alpha^*}{\partial x} dx\, dy\, dz = \mp \sqrt{-1} \iiint_{-\infty}^{+\infty} \psi \frac{\partial \alpha^*}{\partial x} dx\, dy\, dz$$

Thus the left hand side and right hand side of Hermiticity condition in terms of operator $i \dfrac{\partial}{\partial x}$ are equal. Hence the operator $i \dfrac{\partial}{\partial x}$ is a Hermitian operator. Accordingly we can make some important deductions regarding the Hermiticity of an operator.

(*i*) Quantum mechanical operators and their total sum will always be Hermititain even if they are multiplied by some real constants or not.

(*ii*) If the inverse operator is Hermitian, the original operator is also Hermititian.

(*iii*) Non-Hermitian operators or their total sum will become Hermitian provided they are multiplied by some imaginary constant.

Example 2.12: Show that the angular momentum operator is an quantum mechanical operator.

Solution: We know that quantum mechanical operators are Hermitian. The condition for Hermiticity is defined by equation (2.96). Let us consider the angular momentum operator.

$$\hat{M}_x = \frac{h}{2\pi i}\left(y\frac{\partial}{\partial z} - z\frac{\partial}{\partial y}\right)$$

Replace the operator $\hat{O}$ of equation (2.96) by the operator, $\hat{M}_x$.

$$\iiint_{-\infty}^{+\infty} \alpha^* \hat{M}_x \Psi \, dx\, dy\, dz = \iiint_{-\infty}^{+\infty} \Psi \hat{M}_x^* \alpha^* \, dx\, dy\, dz \qquad ...(2.105)$$

Let us now examine the Hemiticity of $iy\frac{\partial}{\partial z}$. Left hand side of equation (2.96) in terms of $iy\frac{\partial}{\partial z}$ is:

$$\iiint_{-\infty}^{+\infty} \alpha^* iy\frac{\partial \Psi}{\partial z} dx\, dy\, dz = \iint_{-\infty}^{+\infty} i\alpha^*\Psi \, dy\, dz]_{-\infty}^{+\infty} - \iiint_{-\infty}^{+\infty} iy\, \Psi \frac{\partial \alpha^*}{\partial z} dx\, dy\, dz$$

$$= -\iiint_{-\infty}^{+\infty} iy\, \Psi \frac{\partial \alpha^*}{\partial z} dx\, dy\, dz$$

$$= \mp\sqrt{-1}\iiint_{-\infty}^{+\infty} y\, \Psi \frac{\partial \alpha^*}{\partial z} dx\, dy\, dz \qquad ...(2.106)$$

Similarly right hand side of equation (2.96) in terms of $iy\frac{\partial}{\partial z}$ is

$$\iiint_{-\infty}^{+\infty} \Psi\, iy\frac{\partial \alpha^*}{\partial z} dx\, dy\, dz = \pm\sqrt{-1}\iiint_{-\infty}^{+\infty} y\, \Psi \frac{\partial \alpha^*}{\partial z} dx\, dy\, dz \qquad ...(2.107)$$

A comparison of equations (2.106) and (2.107) reveals that

$$\iiint_{-\infty}^{+\infty} y\, \Psi \frac{\partial \alpha^*}{\partial z} dx\, dy\, dz = \iiint_{-\infty}^{+\infty} y\, \Psi \frac{\partial \alpha^*}{\partial z} dx\, dy\, dz$$

Hence the operator $iy\frac{\partial}{\partial z}$ is a Hermitian operator. Similarly we can also show that $iz\frac{\partial}{\partial y}$ is also a Hermitian operator. Thus the total sum of these operators, *i.e.*, $iy\frac{\partial}{\partial z} - iz\frac{\partial}{\partial y}$ multiplied by real constant (say $\frac{h}{2\pi}$) will also be Hermitian operator.

2.4.8 PROPERTIES OF ANGULAR MOMENTUM FROM COMMUTATION RULES

Let us deduce some typical properties of angular momenta from commutation rules.

(*i*) A set of eigen functions of commuting angular momentum operators is also the eigen functions of both the operators simultaneously.

Let us consider the commuting operators $\hat{M}^2$ and $\hat{M}_z$. If $Y_{l,m}(\theta, \varphi)$ is a set of eigen functions of these operators, then

$$\hat{M}^2 \, Y_{l,m} = k_l \, Y_{l,m} \qquad \text{...(2.108)}$$

$$\hat{M}_z \, Y_{l,m} = k_m \, Y_{l,m} \qquad \text{...(2.109)}$$

k_l and k_m are the eigen values associated with wave functions $Y_{l,m}$ for the respective operator $\hat{M}^2$ and $\hat{M}_z$ Equation (2.108) can be written as:

$$(\hat{M}_x^2 + \hat{M}_y^2 + \hat{M}_z^2) \, Y_{l,m} = k_l \, Y_{l,m} \qquad \text{...(2.110)}$$

Applying operator $\hat{M}_z^2$ to equation (2.109)

$$\hat{M}_z(\hat{M}_z Y_{l,m}) = k_m \, (\hat{M}_z Y_{l,m})$$

$$\hat{M}_z^2 \, Y_{l,m} = k_m^2 Y_{l,m} \qquad \text{...(2.111)}$$

Subtracting equation (2.111) from (2.110), we get

$$(\hat{M}_x^2 + \hat{M}_y^2) \, Y_{l,m} = (k_l - k_m^2) \, Y_{l,m} \qquad \text{...(2.112)}$$

Thus $Y_{l,m}$ is an eigen function of $(\hat{M}_x^2 + \hat{M}_y^2)$ as well as of $\hat{M}_z$ and $\hat{M}_z^2$.

(*ii*) The commutation rules aid in generating new set of angular momentum operators that help to obtain the form of set of eigen functions of the same class for the case of a single particle where the forms of operators are known.

$$\hat{M}_x \hat{M}_y - \hat{M}_y \hat{M}_x = \frac{ih}{2\pi} \hat{M}_z \qquad (a)$$

$$\hat{M}_y \hat{M}_z - \hat{M}_z \hat{M}_y = \frac{ih}{2\pi} \hat{M}_x \qquad (b) \qquad \text{...(2.87)}$$

$$\hat{M}_z \hat{M}_x - \hat{M}_x \hat{M}_z = \frac{ih}{2\pi} \hat{M}_y \qquad (c)$$

Multiply (2.87*b*) by i and add it to (2.87*c*); (the multiplication factor i is selected to get real eigen functions).

$$\hat{M}_z\ (\hat{M}_x - i\hat{M}_y) + (i\hat{M}_y - \hat{M}_x)\,\hat{M}_z = \frac{h}{2\pi}\ (i\hat{M}_y - \hat{M}_x)$$

or
$$\hat{M}_z\ (\hat{M}_x - i\hat{M}_y) = (\hat{M}_x - i\hat{M}_y)\ \hat{M}_z + \frac{h}{2\pi}\ (i\hat{M}_y - \hat{M}_x)$$

or finally
$$\hat{M}_z\ (\hat{M}_x - i\hat{M}_y) = (\hat{M}_x - i\hat{M}_y)\left(\hat{M}_z - \frac{h}{2\pi}\right) \qquad ...(2.113)$$

Further multiply (2.87*b*) by *i* and subtract it from (2.87*c*).

$$\hat{M}_z\ (\hat{M}_x + i\hat{M}_y) - (\hat{M}_x + i\hat{M}_y)\,\hat{M}_z = \frac{h}{2\pi}\ (\hat{M}_x + i\hat{M}_y)$$

$$\hat{M}_z\ (\hat{M}_x + i\hat{M}_y) = (\hat{M}_x + i\hat{M}_y)\ \hat{M}_z + \frac{h}{2\pi}\ (\hat{M}_x + i\hat{M}_y)$$

$$= (\hat{M}_x + i\hat{M}_y)\left(\hat{M}_z + \frac{h}{2\pi}\right) \qquad ...(2.114)$$

Before we procede further, let us define few terms.

$$\hat{M}_+ = (\hat{M}_x + i\hat{M}_y)$$

$$\hat{M}_- = (\hat{M}_x - i\hat{M}_y)$$

$\hat{M}_+$ is called as the *raising operators* and the $\hat{M}_-$ one is called *lowering operators*. These are also the examples of *ladder operators*. The terminology used for these operators will be clear in the end of this section.

Now operate both sides of equations (2.113) and (2.114) on the total angular wave function $Y_{l,\,m}\ (\theta,\ \varphi)$.

$$\hat{M}_z\ (\hat{M}_x \pm i\hat{M}_y)\,Y_{l,\,m}\ (\theta,\ \varphi) = (\hat{M}_x \pm i\hat{M}_y)\left(\hat{M}_z \pm \frac{h}{2\pi}\right) Y_{l,\,m}\ (\theta,\ \varphi)$$

$$= (\hat{M}_x \pm i\hat{M}_y)\left[\hat{M}_z\ Y_{l,m}(\theta,\ \varphi) \pm \frac{h}{2\pi}\ Y_{l,m}(\theta,\varphi)\right].$$

$$= (\hat{M}_x \pm i\hat{M}_y)\left[k_m\ Y_{l,m}(\theta,\ \varphi) \pm \frac{h}{2\pi}\ Y_{l,m}(\theta,\varphi)\right]$$

$$= \left(k_m \pm \frac{h}{2\pi}\right)\ (\hat{M}_x \pm i\hat{M}_y)\ Y_{l,\,m}\ (\theta,\ \varphi) \qquad ...(2.115)$$

Thus $(\hat{M}_x \pm i\hat{M}_y) Y_{l,m}(\theta, \varphi)$ seems to be an eigen function of $\hat{M}_z$ with eigen value of $(k_m \pm \frac{h}{2\pi})$ or zero. It may be also pointed out that the raising and lowering operators on the eigen functions with eigen value k_m, can generate a ladder of eigen values. The difference between the eigen values step by step is $\frac{h}{2\pi}$ $(\hbar)$, *i.e.*,

$$\ldots\ldots k_m - 2\frac{h}{2\pi},\; k_m - \frac{h}{2\pi},\; k_m,\; k_m + \frac{h}{2\pi},\; k_m + 2\frac{h}{2\pi},\ldots\ldots$$

Further, since $\hat{M}^2$ commutes with all the operators of equation (2.114), so

$$\hat{M}^2[(\hat{M}_x + i\hat{M}_y)\, Y_{l,m}(\theta, \varphi)] = k_l\, [\hat{M}_x + i\hat{M}_y] Y_{l,m}(\theta, \varphi)] \quad \ldots(2.116)$$

and $$\hat{M}^2\, [\hat{M}_x - i\hat{M}_y)\, Y_{l,m}(\theta, \varphi)] = k_l\, (\hat{M}_x - i\hat{M}_y)\, Y_{l,m}(\theta, \varphi) \quad \ldots(2.117)$$

i.e., $(\hat{M}_x + i\hat{M}_y)\, Y_{l,m}(\theta, \varphi)$ is an eigen function of $\hat{M}^2$ with eigen value of k_l or zero and $(\hat{M}_x - i\hat{M}_y)\, Y_{l,m}(\theta, \varphi)$ is also an eigen function of $\hat{M}^2$ with eigen value of k_l or zero. Further it can also be deduced that:

$$\hat{M}^2\, Y_{l,m}(\theta, \varphi) = l\,(l+1)\frac{h^2}{4\pi^2}\, Y_{l,m}(\theta, \varphi) \quad \ldots(2.118)$$

$$\hat{M}_z\, Y_{l,m}(\theta, \varphi) = m\,\frac{h}{2\pi}\, Y_{l,m}(\theta, \varphi) \quad \ldots(2.119)$$

The other important relations are:

$$(\hat{M}_x - i\hat{M}_y)\, Y_{l,m}(\theta, \varphi) = \frac{h}{2\pi}[l\,(l+1) - m\,(m-1)]^{1/2}\, Y_{l,m-1}(\theta, \varphi)$$

$$= \frac{h}{2\pi}[(l+m)\,(l-m+1)]^{1/2}\, Y_{l,m-1}(\theta, \varphi) \quad \ldots(2.120)$$

$$(\hat{M}_x + i\hat{M}_y)\, Y_{l,m}(\theta, \varphi) = \frac{h}{2\pi}[l\,(l+1) - m\,(m+1)]^{1/2}\, Y_{l,m+1}(\theta, \varphi)$$

$$= \frac{h}{2\pi}\, [(l-m)\,(l+m+1)]^{1/2}\, Y_{l,m+1}(\theta, \varphi) \quad \ldots(2.121)$$

It is also noted that results deduced entirely from commutation relations, will also hold good for any operators whose commutations relations are similar to those of angular momentum operators, for example, set of spin angular momentum operators $\hat{S}_x$, $\hat{S}_y$, $\hat{S}_z$ and $\hat{S}^2$ are analogous to the operators $\hat{M}_x$, $\hat{M}_y$, $\hat{M}_z$ and $\hat{M}^2$. The internal degrees of freedom or the corresponding intrinsic angular momentum of the electron *i.e.*, $\frac{h}{4\pi}$ is called the electron spin. The properties of electron's spin can be studied by using the commutation relations between the components of the conventional spin angular momentum. The eigen values of $\hat{S}^2$ is $[\frac{1}{2}(\frac{1}{2}+1)]\frac{h^2}{4\pi^2}$, and there are two eigen values of $\hat{S}_z$ *i.e.*, $\frac{1}{2}\frac{h}{2\pi}$ and $-\frac{1}{2}\frac{h}{2\pi}$. The spin angular momentum components can be spin eigen function for S = 1/2 is denoted by α and for S = –1/2 is denoted by β. The components of total angular momentum operators thus can be represented as:

$$(\hat{M}_x)_T = \hat{M}_x + \hat{S}_x, (\hat{M}_y)_T = \hat{M}_y + \hat{S}_y, (\hat{M}_z)_T = \hat{M}_z + \hat{S}_z \quad ...(2.122)$$

2.5 POSTULATES OF QUANTUM MECHANICS

In any game or any field of life, it is generally a custom to formulate certain rules so that a game can be played nicely and if any problem arises during the play, it can be solved nicely and amicably. In reality we are interested in the knowledge of real, observable quantities and these quantities can be measured or can be related to the experimental observations nicely. In quantum mechanics we are interested in understanding the structure of matter and other phenomenon, like, the interactions of radiation with matter etc. Before we start using these ideas of quantum mechanics to understand these quantities, it is necessary to formulate certain postulates. Further it may be made clear here that these postulates or rules are made in pure faith, so their proof may not be possible. As a consequence therefore, there is no unique set of postulates of quantum mechanics. Further there is a difference between the laws of thermodynamics and postulates of quantum mechanics. The former are results of the experiences of many years and can be tested or proven without any problem, whereas the latter are pure assumptions. These can be applied and the results obtained can be compared with the experimental results but can not be proven.

Before the postulates are stated, it is good to define two quantities, which are very common in quantum mechanics and have been mentioned early without proof.

(*a*) **Dynamical Variables:** Any property of the system in which we are interested, is called as dynamical variable. For example, the position *x*, the energy E, the angular momentum etc., are all dynamical variables.

(*b*) **Observables:** Any property of the system, which can be measured experimentally, is known as *observable*.

It may appear in beginning that the two quantities are same but it is not. There are certain quantities in which we are interested but cannot be measured experimentally. For example, it will be shown in Chapter 3 that one component of spin angular momentum vector can be measured experimentally, *i.e.*, the component along z-axis defines the angle θ between the spin angular momentum and the *z*-axis. Thus the spin angular momentum is a dynamical variable but the component along the *z*-axis can be both dynamical variable and observable.

These postulates are:

Postulate I: Any state of dynamical system of *n* particles can be described as fully as possible by a function $\psi\ (q_1, q_2, q_3, \ldots\ldots\ldots, q_{3n}, t)$, where $q_1, q_2, q_3, \ldots\ldots\ldots, q_{3n}$ are the coordinates of particles at any time *t*.

Postulate II: If a system is defined by a function ψ, then the probability of finding it between q_1 and $q_1 + dq_1$, q_2 and $q_2 + dq_2 \ldots\ldots\ldots\ldots\ldots q_{3n}$ and $q_{3n} + dq_{3n}$ at any time *t* is given by the relation:

$$\iiint_{-\infty}^{+\infty} \psi^* \psi \, d\tau = A^{-2} \qquad \ldots(1.123)$$

where $d\tau$ is the volume of the system, defined by $d\tau = dq_1\, dq_2 \ldots . dq_{3n}$ and ψ^* is a complex conjugate of the function ψ.

Postulate III: For every observable property of the system in classical mechanics, there exists a linear Hermitian Operator and physical properties of the observables can be calculated from mathematical properties of the associated operators.

Actually this postulate forms the bridge between the expression of the observable quantities of the system in classical mechanics and quantum mechanics. The rules to construct these operators have already been deduced and listed in Table 2.1.

Postulate IV: Let $\hat{O}$ be an operator, corresponding to physically observable quantity and the system can be represented by a function ψ. The precise value of this observable can only be calculated if the following equation is followed.

$$\hat{O}\psi = O\psi \qquad \ldots(2.124)$$

Here O is a constant and is the value of observable. The above equation imposes the restriction that ψ is an eigen function of the operator $\hat{O}$ and O is its eigen value. In other words the precise value of the observable can only be obtained or the values obtained for observable will be the same for each measurement if ψ is an eigen function to the operator $\hat{O}$. In real sense this problem reduces to finding ψ and O that satisfies the eigen value-eigen function equation (2.124).

Postulate V: If in postulate IV, ψ is not an eigen function to the operator $\hat{O}$, the measurements carried out every time will give a different value of the observable. In this case the average value of the observable is given by:

$$\langle O \rangle = \bar{O} = \frac{\int \psi^* \hat{O} \psi \, d\tau}{\int \psi^* \, \psi \, d\tau} \qquad \ldots(2.125)$$

Here $\langle O \rangle$ or $\bar{O}$ is the average or expectation value of the observable corresponding to the operator $\hat{O}$ and the above relation is called *mean value theorem.* If the wave function ψ is a normalized wave function then expression (2.125) reduces to (2.126).

$$\langle O \rangle = \bar{O} = \int \psi^* \hat{O}\psi \, d\tau \qquad ...(2.126)$$

The average value of the observable obtained by the expression is the average of the number of observations made to calculate the property corresponding to the operator $\hat{O}$ and it should not be related to the average in classical mechanics. *If ψ is an eigen function of operator, $\hat{O}$, the average value will be the same as the eigen value.* For example:

$$\langle O \rangle = \frac{\int \psi^* \hat{O}\, \psi \, d\tau}{\int \psi^* \psi \, d\tau} = \frac{\int \psi^* O \psi \, d\tau}{\int \psi^* \psi \, d\tau} = O\frac{\int \psi^* \psi \, d\tau}{\int \psi^* \psi \, d\tau} = O \qquad ...(2.127)$$

One postulate which we have not mentioned deliberately is regarding the time dependence part of the Schrödinger wave equation. It is simply because in the present book we are not going to deal any problem which is an explict function of time, *i.e.*, we will be dealing mainly with the steady state functions and their corresponding wave functions.

Example 2.13: If the operator for the energy along the x-axis is $-\dfrac{h^2}{8\pi^2 m}\dfrac{d^2}{dx^2}$ and the normalized wave function is $\sqrt{\dfrac{2}{a}}\,\sin\,\dfrac{n\pi}{a}\,x$, calculate the average value of the energy. x varies from 0 to a.

Solution: Use equation (2.124). Since the function is normalized, the denominator will be unity.

$$\langle E \rangle = \int_0^a \left(\frac{2}{a}\right)^{\frac{1}{2}} \sin\frac{n\pi x}{a} \left[-\frac{h^2}{8\pi^2 m}\frac{d^2}{dx^2}\right] \left(\frac{2}{a}\right)^{\frac{1}{2}} \sin\frac{n\pi x}{a}\, dx$$

$$= -\frac{2h^2}{8\pi^2 ma}\int_0^a \sin\frac{n\pi x}{a}\left[-\frac{n^2\pi^2}{a^2}\sin\frac{n\pi x}{a}\right]dx$$

$$= \frac{2h^2 n^2}{8ma^3}\int_0^a \sin^2\frac{n\pi x}{a}\, dx = \frac{2h^2 n^2}{8ma^3 2}\int_0^a \left[1-\cos\frac{2n\pi x}{a}\right] dx$$

$$= \frac{h^2 n^2}{8ma^3}\left[\int_0^a dx - \int_0^a \cos\frac{2n\pi x}{a}\, dx\right]$$

$$= \frac{h^2 n^2}{8ma^3}(a-0) = \frac{h^2 n^2}{8ma^2}$$

The value of average energy in the nth level is same as obtained in section 2.6.2 (equation 2.140). This is because the function is an eigen function of the Hamiltonian. In this case the average value is equal to the actual value.

2.6 APPLICATIONS OF SCHRÖDINGER EQUATION

2.6.1 FREE PARTICLE

The simplest case to consider is the motion of free particle. That is it is not bound to any thing and has no restrictions. Let the mass of the particle be m and moving with velocity v_x, along x-axis for simplicity. Since the potential energy in this case is zero, the total energy can be written as:

$$E = K + V = \frac{1}{2} mv_x^2 = \frac{p_x^2}{2m}$$

where p_x is the linear momentum along x-axis. As mentioned in section 2.4 the Hamiltonian operator can be:

$$\hat{H} = -\frac{h^2}{8\pi^2 m} \frac{d^2}{dx^2}$$

and the Schrödinger wave equation will be:

$$-\frac{h^2}{8\pi^2 m} \frac{d^2\psi}{dx^2} = E\,\psi \qquad ...(2.128)$$

or

$$\frac{d^2\psi}{dx^2} = -\frac{8\pi^2 mE}{h^2}\psi = -k^2\psi \qquad ...(2.129)$$

where

$$k^2 = \frac{8\pi^2 mE}{h^2} \qquad ...(2.130)$$

Expression (2.129) is a simple second order differential equation and its solution can be written as:

$$\Psi = A\cos kx + B\sin kx \qquad ...(2.131)$$

where A, B and k are constants, to be determined. It can be checked easily by differentiating twice ψ with respect to x that equation (2.131) is a solution of (2.129). In this case since there are no restrictions on the motion of the particle, A, B and k can have any arbitrary values. This means that unless we know the values of ψ at any point (this is called boundary conditions) the precise value of the function and the energy, E, cannot be evaluated from the Schrödinger wave equation. In other words, it is only the boundary conditions that specify the précised value of the observables. It can also be said that the motion of the particle under these conditions can be specified by the classical mechanical principles, *i.e.*, energy changes take place in a continuous fashion. In the next section, we shall impose certain conditions on the motion of the particle and thus will be able to find more precise solutions of the Schrödinger equation (2.131).

2.6.2 PARTICLE IN A ONE DIMENSIONAL BOX OF INFINITE WALLS

This is also a simplest problem but it is very good in the sense because: (*i*) it illustrates very nicely the principles of quantum mechanics, (*ii*) one does not have to suppose that energy levels are quantized but these automatically come up from the mathematical solutions of the Schrödinger wave equation, when certain boundary conditions are imposed.

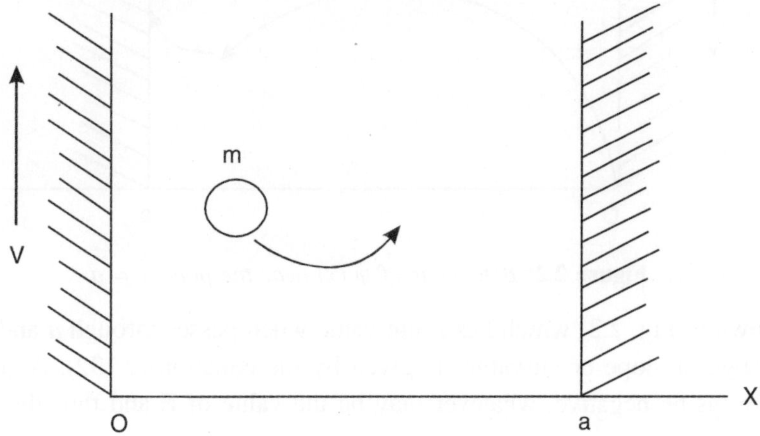

Figure 2.1: *Particle in one dimensional box. The potential energy is zero between x = 0 and x = a and infinity everywhere else. This is not the only choice of coordinates for this problem. Solutions with the origin at the middle of the box can also be solved*

Consider a particle of mass m, moving in a one dimensional box of length a, *i.e.*, particle is moving along the x-axis in the region $x = 0$ to $x = a$ as shown in Fig. 2.1. Consider further that potential energy within the box is constant and is infinite outside the box. For simplicity, it may be assumed that the potential energy is zero. Although it will not be proven, it can be shown that the wave function will not be affected and the total energy of different states will be displaced by a constant amount equal to the potential energy. In other words that particle is only allowed to move within the box[a]. One has to determine two things: (*a*) the energies of various states and (*b*) the wave function of the particle corresponding to each energy state so that other characteristics of the particle can be calculated. The most general form of the Schrödinger wave equation for this particle can be written as:

$$-\frac{h^2}{8\pi^2 m}\frac{d^2\psi}{dx^2} = (E - V)\,\psi \qquad ...(2.132)$$

Before this equation can be solved it is nice to find out the solution at various boundaries keeping in mind the postulates of quantum mechanics on the wave function ψ. Consider

a = just for information, the boundaries of box are not walls made up of any material but just hypothetical ones, that the particle does not leave the box.

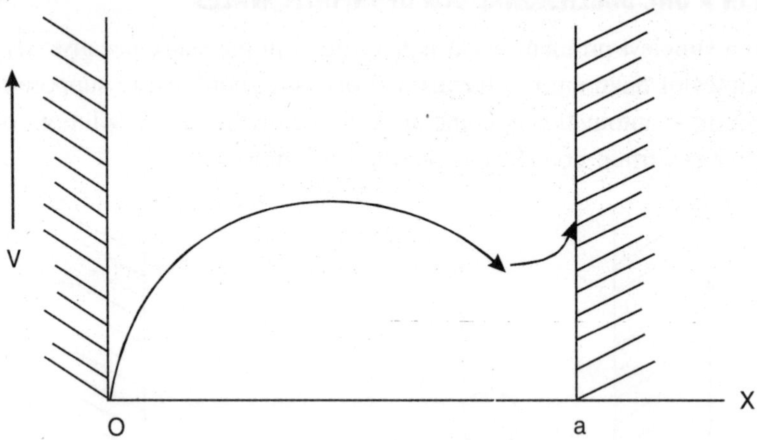

Figure 2.2: *Behaviour of $\psi(x)$ near the point $x = a$*

function A (shown in Fig. 2.2) which has finite value when passes through *a* and has finite slope. The rate of change of slope or curvature is given by the equation (2.132). At $x = 0$, V = ∞, so (E – V) will always be negative, whatever may be the value of E and thus the equation (2.132) will be:

$$\frac{h^2}{8\pi^2 m}\frac{d^2\psi}{dx^2} = \infty\ \psi \qquad ...(2.133)$$

This shows that the curvature increases very sharply and reaches infinity. The function will also behave like this, *i.e.*, will tend to ∞, but this is not acceptable in quantum mechanics. The only possible way to avoid this is that the function $\psi(x)$ itself becomes zero at $x = 0$. Similarly the function should also be zero at $x = a$. In other words the boundary conditions are:

$$\psi = 0,\ x = 0 \qquad ...(2.134a)$$

$$\psi = 0,\ x = a \qquad ...(2.134b)$$

Under the present conditions, equation (2.133) reduces to:

$$-\frac{h^2}{8\pi^2 m}\frac{d^2\psi}{dx^2} = \text{E}\psi$$

or
$$\frac{d^2\psi}{dx^2} = -\frac{8\pi^2 m\text{E}}{h^2}\ \text{E}\psi = -k^2\psi \qquad ...(2.135)$$

where
$$k^2 = \frac{8\pi^2 m\text{E}}{h^2} \qquad ...(2.135a)$$

As said in the last section, the general solution for this second order differential equation is:

$$\Psi = \text{A}\cos kx + \text{B}\sin kx \qquad ...(2.131)$$

where A, B and k are constants to be determined. These can be calculated by using boundary conditions.

At $x = 0$, $\psi = 0$, equation (2.131) becomes:

$$0 = \text{A} \sin k\, 0 + \text{B} \cos k\, 0 = \text{B}$$

i.e., B = 0, equation (2.131) reduces to:

$$\psi = \text{A} \sin kx \qquad \text{...(2.136)}$$

Applying second condition, $\psi = 0$ at $x = a$, equation (2.136) reduces to:

$$0 = \text{A} \sin ka$$

Accordingly either (*i*) A = 0 or (*ii*) $\sin ka = 0$. The first solution, A = 0, is not an acceptable because it will lead to $\psi = 0$, a trivial solution, as well as, it will also lead to the probability density function $\psi^* \psi$ and thereby probability to zero. This is against our assumption. The second solution leads to:

$\sin ka = 0 = \sin n\pi$, where $n = 0, 1, 2, 3, \ldots$

or $\quad ka = n\pi$ and thus $k = \dfrac{n\pi}{a}$ ...(2.137)

and thus equation (2.136) reduces to

$$\psi = \text{A} \sin \psi = \text{A} \sin \frac{n\pi}{a} x \qquad \text{...(2.138)}$$

The value of A can be determined by using the idea that total probability of finding the particle in the box is unity, *i.e.*,

$$1 = \int_0^a \psi^2\, dx = \int_0^a \text{A}^2 \sin^2 \frac{n\pi}{a} x\, dx$$

$$= \int_0^a \frac{\text{A}^2}{2} \left[1 - \cos \frac{2n\pi x}{a}\right] dx$$

$$= \text{A}^2 \int_0^a dx/2 - \frac{\text{A}^2}{2} \int_0^a \cos \frac{2n\pi x}{a}\, dx = \text{A}^2 \frac{a}{2}$$

or $\qquad \text{A} = \sqrt{\dfrac{2}{a}}$

Thus the normalized wave function is

$$\psi = \sqrt{\frac{2}{a}} \sin \frac{n\pi}{a} x \qquad \text{...(2.139)}$$

Energy levels: The energy of the particle can be found by substituting the value of k in equation (2.135*a*)

$$k^2 = \frac{n^2\pi^2}{a^2} = \frac{8\pi^2 m\text{E}}{h^2}$$

$$E = \frac{h^2}{8ma^2} n^2 \qquad n = 1, 2, 3,... \qquad ...(2.140)$$

It is clear from equation (2.140) that energy levels of the particle in one dimensional box of infinite walls are quantized and quantization of energy levels have come up from the acceptable mathematical solution of equation (2.135).

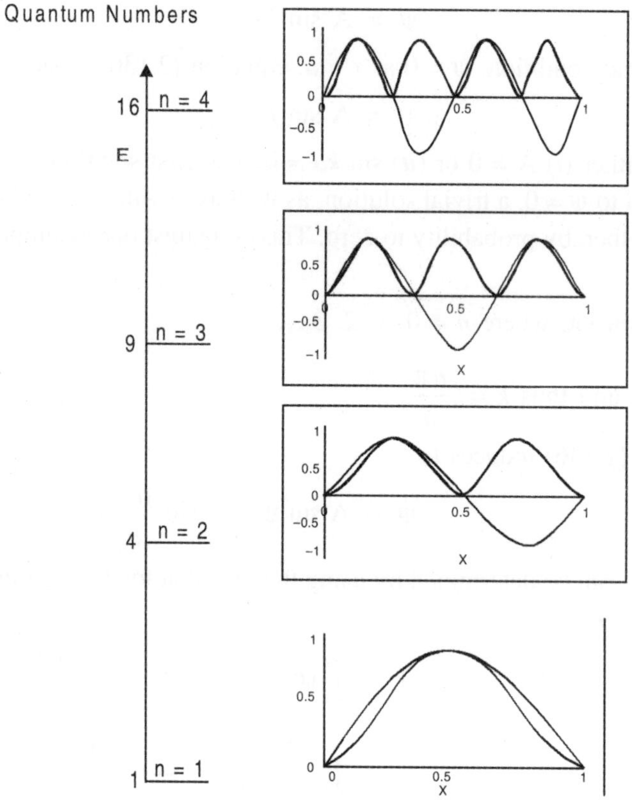

Figure 2.3: *Plot of energies, wave functions (thin lines) and square of the wave functions (thick lines) for different quantum numbers of particle in one dimensional box. Energy (E) is in units of* $\left(\frac{h^2}{8ma^2}\right)$.

To sum up, the acceptable wave functions and energies for the particle in a one dimensional box are:

$$\psi_n = \sqrt{\frac{2}{a}} \sin \frac{n\pi}{a} x, \; E_n = \frac{h^2}{8ma^2} n^2 \qquad n = 1, 2, 3,... \qquad .$$

ψ, ψ^2 and energy levels are plotted in Fig. (2.3). Each ψ is plotted at a height proportional to its energy. The solution of a particle in a one dimensional box reveals many features and also proves the assumptions involved in quantum mechanics. These can be viewed from the following points:

(1) Quantum numbers (n) only start from $n = 1$ and not from $n = 0$, although the latter is also a perfect solution of the equation (2.137). If $n = 0$, $E_n = 0$ and $\psi_n = 0$. The latter result tells that the probability of finding the particle in the box is zero, which is contradictory to our problem. The particle is present in the box. The $E_0 = 0$ states that the linear momentum is zero and thus can be measured precisely. $\Psi = 0$ also tells that the position of the particle can be measured accurately. This is against the Heisenberg's uncertainty principle.

(2) From Fig. 2.3 it is clear that for $n = 1$, the number of nodes is zero, for $n = 2$, it is one, for $n = 3$, it is two, and for $n = n$, the number of nodes will be $(n - 1)$. Further as the value of n increases (in other words as the number of nodes increases), the energy of the particle also increases. In other words, greater is the energy, greater are the number of nodes. *The node is a point where the wave function describing the behaviour of the particle is zero, as well as, the probability of finding the particle at the node is also zero.*

(3) From Fig. 2.3, the wave length of the particle is given by the relation:

$$\lambda = \frac{2a}{n} \qquad ...(2.141)$$

i.e., wave length will decrease with increase of quantum number or the energy of the particle. The same results can also be obtained with the help of de Broglie relation (1.14.6). It means that greater the energy, greater will be the linear momentum and smaller will be the value of wave length. It can be further observed that as $n \to \infty$, ψ will decrease very quickly and the number of waves present in the box will be very large. This will lead to the similar behaviour as observed in the classical mechanics (see also Bohr's correspondence principle section 1.2.5 or page 129).

(4) **Calculation of Average Value:** Postulate (5) can be applied to calculate the average value of linear momentum. The quantum mechanical operator for linear momentum is $\frac{h}{2\pi i}\frac{d}{dx}$. So using equation (2.126):

$$\langle p_x \rangle = \frac{\int_0^a \psi^* \hat{p}_x \psi \, dx}{\int_0^a \psi^* \psi \, dx} = \frac{\int_0^a \left(\frac{2}{a}\right)^{\frac{1}{2}} \sin\frac{n\pi x}{a} \left(\frac{h}{2\pi i}\frac{d}{dx}\right)\left(\frac{2}{a}\right)^{\frac{1}{2}} \sin\frac{n\pi x}{a}\, dx}{\int_0^a \left(\frac{2}{a}\right)^{\frac{1}{2}} \sin\frac{n\pi x}{a} \left(\frac{2}{a}\right)^{\frac{1}{2}} \sin\frac{n\pi x}{a}\, dx}$$

the denominator will be equal to unity since the function is normalized. So

$$\langle p_x \rangle = \frac{2}{a}\frac{h}{2\pi i}\frac{n\pi}{a}\int_0^a \sin\frac{n\pi x}{a} \cos\frac{n\pi x}{a}\, dx$$

$$= \frac{nh}{ia^2}\left[\frac{a}{2n\pi}\sin^2\frac{n\pi x}{a}\right]_0^a$$

$$= \frac{h}{2\pi i a}\left[\sin^2 \frac{n\pi}{a}a - \sin^2 \frac{n\pi 0}{a}\right] = 0$$

i.e., average value of the linear momentum is zero. From equation (2.140):

$$E = \frac{p^2}{2m} = \frac{h^2n^2}{8ma^2}$$

will lead to $p = \pm \frac{nh}{2a}$. The two results thus obtained leads to confusion, but can be understood as follows. The value of linear momentum calculated from the energy tells that it can either be $+\frac{nh}{2a}$ or $-\frac{nh}{2a}$. In other words, if a single measurement is made on p_x, its value will be either $\frac{nh}{2a}$ or $-\frac{nh}{2a}$ and one cannot predict before hand. So if large number of measurements is carried out, the average value will be zero. These results can be used to calculate the Heisenberg's uncertainty relation. The maximum uncertainty in the magnitude of p_x will be $\Delta p_x = \frac{nh}{2a} - \left(-\frac{nh}{2a}\right) = \frac{nh}{a}$. Similarly the maximum uncertainty in the position of finding the particle is $\Delta x = a/2 - (-a/2) = a$. So

$$\Delta p_x \, \Delta x = a \frac{nh}{a} = nh$$

The smallest value of $\Delta p_x \, \Delta x$ can be equal to h for $n = 1$. Though this is not the general form of Heisenberg uncertainty principle, it suggests that the product of uncertainty in position and linear momentum is related by Planck's constant.

Example 2.14: The wave function for the particle in one dimensional box of infinite walls in $n = 1$ state is $\psi_1 = \sqrt{\frac{2}{a}} \sin \frac{\pi}{a} x$. Calculate the probability of finding the particle in the region $0.02\ a$ and $0.03\ a$.

Solution: Use equation (2.16) and integrate it within limits $x = 0.02\ a$ and $x = 0.03\ a$. Since the function is real, $\psi = \psi^*$

$$\text{Probability (P)} = \int_{0.02a}^{0.03a} \left(\frac{2}{a}\right)^{\frac{1}{2}} \sin\frac{\pi x}{a}\left(\frac{2}{a}\right)^{\frac{1}{2}} \sin\frac{\pi x}{a}\,dx = 2/a \int_{0.02a}^{0.03a} \sin^2 \frac{\pi x}{a}\,dx$$

$$= 2/a \int_{0.02a}^{0.03a} \frac{1 - \cos\frac{2\pi x}{a}}{2}\,dx = \frac{1}{a}\int_{0.02a}^{0.03a} dx - \frac{1}{a}\int_{0.02a}^{0.03a} \cos \frac{2\pi x}{a}\,dx$$

$$= \frac{1}{a}[X]_{0.02a}^{0.03a} - \frac{1}{a}\left[\frac{a}{2\pi}\sin\frac{2\pi x}{a}\right]_{0.02a}^{0.03a}$$

$$= \frac{1}{a}[0.03a - 0.02a] - \frac{1}{2\pi}\left[\sin\frac{2\pi}{a}(0.03a) - \sin\frac{2\pi}{a}(0.02a)\right]$$

$$= 0.01 - \frac{7}{44}[0.1874 - 0.1253] = 0.01 - 9.88 \times 10^{-3} = 1.20 \times 10^{-4}$$

(5) Bohr's Correspondence Principle: It states that the results obtained by using quantum mechanical and classical mechanics are identical, when (*a*) the quantum numbers describing the systems are very large, (*b*) energy of the particle is large, and (*c*) micro-particle is moving over large distance. This statement can be nicely illustrated by the following example.

Example 2.15: (*a*) Calculate the difference between E_2 and E_1 for electron, moving over a distance of 0.5 nm and (*b*) for oxygen molecule moving over a distance of 20 cm.

Solution: (*a*) Substituting the values of respective terms and $n_1 = 1$ and $n_2 = 2$ in equation (2.140).

$$\Delta E = E_2 - E_1 = \frac{h^2}{8ma^2}(2^2 - 1^2) = \frac{3h^2}{8ma^2}$$

$$= \frac{(3)\,(6.62\times10^{-34}\text{ J s})^2}{(8)\,(9.11\times10^{-31}\text{ kg})\,(0.5\times10^{-9}\text{ m})^2} = 7.21 \times 10^{-19}\text{ J}$$

$$\Delta E = h\nu = h\frac{c}{\lambda} \text{ or } \lambda = \frac{hc}{\Delta E}$$

or $$\lambda = \frac{(6.62\times10^{-34}\text{ J s})\,(3.0\times10^{8}\text{ ms}^{-1})}{7.21\times10^{-19}\text{ J}} = 275.5\text{ nm}$$

$$(b)\ \Delta E = \frac{(3)\,(6.62\times10^{-34}\text{ J s})^2(6.02\times10^{23}\text{mol}^{-1})}{(8)\,(32\times10^{-3}\text{ kg})\,(0.2\text{ m})^2} = 7.72 \times 10^{-44}\text{ J}$$

From the above example (*a*) it is clear that wavelength required for the promotion of an electron from one energy state to another state falls in the U.V. region and this is the region where molecules or atoms absorb. Thus very simple model, which does not include the specific form of potential energy, gives the idea about the quantization of states and predicts results which are quite agreeable with the experimental results. Whereas the results of example (*b*) indicate that difference between the respective energy states is very small. According to the classical mechanics, the magnitude of the translational energy at room temperature, calculated with the help of equipartition principle of energy is $\frac{1}{2}k_b T$ (or $\frac{1}{2} \times 1.38 \times 10^{-23}\text{ J K}^{-1} \times 298\text{K} =$

2.0×10^{-21} J) per degree of freedom. On comparing the results of example (*b*) with this value, one arrives at the conclusion that though the translational energy levels are quantized, but the gap between the successive energy levels in case of oxygen molecule is so small that it can be considered as if the energy is changing in a continuous manner. In conclusion one can arrive at the following remarks that *the results predicted by quantum mechanics at large energy values, for heavier particles (masses) and over the distance more than molecular dimensions approaches the results predicted by classical mechanics.*

(6) The eigen function (wave function) corresponding to each eigen value (energy of the state) of the same operator (Hamiltonian) forms orthonormal set *i.e.*,

$$\int \psi_m^* \, \psi_n \, d\tau = 0 \quad \text{for } m \neq n$$

$$\int \psi_m^* \, \psi_n \, d\tau = 1 \quad \text{for } n = m$$

The second part is already verified. The first part can be verified by taking a general example:

$$\int \psi_m^* \, \psi_n \, d\tau = \frac{2}{a}\int_0^a \sin\frac{n\pi}{a}x \, \sin\frac{m\pi}{a}x \, dx = \frac{2}{a}\int_0^a \frac{\left[\cos\dfrac{(n-m)\pi x}{a} - \cos\dfrac{(n+m)\pi x}{a}\right]}{2} dx$$

$$= \frac{1}{a}\int_0^a \cos(n-m)\frac{\pi}{a}x \, dx - \frac{1}{a}\int_0^a \cos(n+m)\frac{\pi}{a}x \, dx$$

$$= \frac{1}{a}\left[\frac{a}{(n-m)\pi} \sin\frac{(n-m)\pi}{a}x\right]_0^a - \frac{1}{a}\left[\frac{a}{(n+m)\pi} \sin\frac{(n+m)\pi}{a}x\right]_0^a$$

$$= \frac{1}{(n-m)\pi}\left[\sin(n-m)\pi - \sin 0\right] - \frac{1}{(n+m)\pi}\left[\sin(n+m)\pi - \sin 0\right]$$

$$= 0$$

(7) Besides being very simple and hypothetical model it still can give very nicely the structural aspects of the molecules qualitatively. For example, consider the conjugate dienes. If one assumes that the π electrons of the molecules are only localized between the double bonds, the energy difference between the states will be very large, as can be seen from the example (2.15*a*). Whereas if the π electrons are localized over the complete length of the molecule, the energy spacing between states will decrease and the lowest energy transition will shift towards the longer wave length. It will be seen in Chapter 4 that π electrons in the conjugated dienes are not localized between the respective double bond but these are completely delocalized over the whole chain length of the molecule. The wave length or frequency corresponding to the respective transition will be discussed in chapter 4.

Example 2.16: A free electron model can be applied to the linear conjugated chain molecules. Show that the longest wave length transition is depicted by the relation.

$$\bar{\nu} = \frac{h}{8\, m_e\, c\,(\text{N}+1)\, l^2}$$

where N is the number of carbon atoms, l is the bond length between two carbon atoms equivalent to benzene molecule, c is the velocity of light, m_e is the mass of electron and h is the Planck's constant.

Solution: The equation depicting energy levels is:

$$\text{E}_n = \frac{h^2}{8 m_e a^2} n^2 \qquad \text{...(2.142)}$$

Since each orbital can have two electrons, the highest occupied orbital will be $\frac{\text{N}}{2}$. Lowest unoccupied orbital will be $\frac{\text{N}}{2} + 1$. Substituting the values of these quantum numbers in equation (2.140), we can define the gap between two states as:

$$\bar{\nu} = \frac{\text{E}_{\frac{\text{N}}{2}+1} - \text{E}_{\frac{\text{N}}{2}}}{hc}$$

$$= \frac{h^2}{8m_e\, a^2 h\, c}\left[\left(\frac{\text{N}}{2}+1\right)^2 - \left(\frac{\text{N}}{2}\right)^2\right] = \frac{h\,(\text{N}+1)}{8m_e\, ca^2}$$

'a' length of the chain can be approximated to:

'a' = number of actual number of bonds between N carbon atoms + 2 = (N – 1 + 2) l

where l is the bond length in benzene molecule.

$$\bar{\nu} = \frac{h\,(\text{N}+1)}{8m_e c\,(\text{N}+1)^2\, l^2} = \frac{h}{8m_e c\,(\text{N}+1) l^2}$$

Example 2.17: The radius of nucleus is 10^{-14} m. Applying free electron model, show that the electron cannot be found in the nucleus.

Solution: The minimum kinetic energy can be obtained as:

$$\text{E} = \text{K} = \frac{h^2\,(1)^2}{8m_e a^2} = \frac{(6.62\times10^{-34}\ \text{J s})^2}{(8)\,(9.11\times10^{-31}\ \text{kg})\,(10^{-14}\ \text{m})^2}$$

$$= 6.01 \times 10^{-10}\ \text{J}$$

and potential energy (V) is:

$$V = -\frac{e^2}{(4\pi\varepsilon_0)\, r} = -\frac{(1.6\times10^{-19}\ \text{C})^2}{(1.113\times10^{-10}\ \text{C}^2\ \text{N}^{-1}\ \text{m}^{-2})\ (10^{-14}\ \text{m})}$$

$$= -\ 2.3 \times 10^{-14}\ \text{J}$$

Since kinetic energy is much larger than the potential energy, the electron will be repelled out of the nucleus and thus cannot stay in the nucleus.

2.6.3 NON-INTERACTING PARTICLES: (THEOREM REGARDING SEPARATION OF VARIABLES)

Consider two particles, q_1 and q_2, which do not interact with each other. Let the particle q_1 be represented by the coordinates (x_1, y_1, z_1) and the particle q_2 by (x_2, y_2, z_2), $\hat{H}_1$ and $\hat{H}_2$ be the Hamiltonian operators for the particle one and two respectively. Since there is no interaction between the two particles, q_1 and q_2, the Hamiltonian for the complete system can be written as:

$$\hat{H} = \hat{H}_1 + \hat{H}_2$$

and the Schrödinger wave equation will be:

$$\hat{H}\ \psi = (\hat{H}_1 + \hat{H}_2)\, f(q_1, q_2) = E\, f(q_1\ q_2) \qquad ...(2.143)$$

Let us try to solve the above equation by separation of variables like:

$$f(q_1, q_2) = Q(q_1)\, Q(q_2) \qquad ...(2.144)$$

Equation (2.142) becomes

$$(\hat{H}_1 + \hat{H}_2)\, Q(q_1)\, Q(q_2) = E\, Q(q_1)\, Q(q_2)$$

$$\hat{H}_1\ Q(q_1)\, Q(q_2) + \hat{H}_2\, Q(q_1)\, Q(q_2) = E\, Q(q_1)\, Q(q_2) \qquad ...(2.145)$$

Since Hamiltonian operator $\hat{H}_1$ only involves the coordinates and linear momentum of the particle 1, the function representing the behaviour of particle 2, will be just acting like a constant term for $\hat{H}_1$. *In other words it also means that the given operator will operate only on the functions of those particles, which are represented by that operator.* Thus

$$\hat{H}_1\ Q(q_1)\, Q(q_2) = Q(q_2)\ \hat{H}_1\ Q(q_1) \qquad ...(2.146)$$

Combining equations (2.144) and (2.145)

$$Q(q_2)\ \hat{H}_1\ Q(q_1) + Q(q_1)\ \hat{H}_2\ Q(q_2) = E\, Q(q_1)\, Q(q_2)$$

Divide this equation by $Q(q_1)\, Q(q_2)$, we get

$$\frac{\hat{H}_1\ Q(q_1)}{Q(q_1)} + \frac{\hat{H}_2\ Q(q_2)}{Q(q_2)} = E \qquad ...(2.147)$$

The left hand side of equation (2.146) contains two terms and each term is representing the behaviour of individual particle, *i.e.*, if the coordinates of the particle 1 are varied, only the first term will be affected, whereas term 2 will be behaving like a constant. Now rewrite equation (2.146) as:

$$\frac{\hat{H}_1 Q(q_1)}{Q(q_1)} = E - \frac{\hat{H}_2 Q(q_2)}{Q(q_2)} \qquad ...(2.148)$$

Since E is a constant and the second term also acts as a constant as the coordinates of only particle 1 are varied. In other words the left side of the equation (2.147) will be equal to a constant even if the coordinates of particle 1 are varied. Let it be equal to E_1, *i.e.*,

$$\frac{\hat{H}_1 Q(q_1)}{Q(q_1)} = E - \frac{\hat{H}_2 Q(q_2)}{Q(q_2)} = E_1 \qquad ...(2.149)$$

Thus $$\hat{H}_1 Q(q_1) = E_1 Q(q_1) \qquad ...(2.150)$$

$$\hat{H}_2 Q(q_2) = (E - E_1) Q(q_2) = E_2 Q(q_2) \qquad ...(2.151)$$

where $$E_1 + E_2 + E \qquad ...(2.152)$$

It means that, when our system consists of two non-interacting particles, we can reduce the two particles problem to separate one particle problem by solving the following equations.

$$\hat{H}_1 Q(q_1) = E_1 Q(q_1) \text{ and } \hat{H}_2 Q(q_2) = E_2 Q(q_2) \qquad ...(2.153)$$

and total energy will be the sum of the individual energies of each particle given by equation (2.151).

The above results can be generalized to any number of non-interacting particles. For example, consider n such non-interacting particles such that

$$\hat{H} = \hat{H}_1 + \hat{H}_2 + \ldots\ldots\ldots\ldots\ldots\ldots + \hat{H}_n \qquad ...(2.154)$$

$$\psi(q_1, q_2, q_3, \ldots\ldots, q_n) = \psi_1(q_1)\,\psi_2(q_2) \ldots\ldots\ldots\ldots \psi_n(q_n) \qquad ...(2.155)$$

$$E = E_1 + E_2 + \ldots\ldots\ldots\ldots + E_n \qquad ...(2.156)$$

and $$\hat{H}_1 \psi = E_1 \psi_1;\ \hat{H}_2 \psi_2 = E_2 \psi_2;\ \ldots\ldots\ \hat{H}_n \psi_n(q_n) = E_n \psi_n \qquad ...(2.157)$$

Similarly, the above problem can be applied to one particle system whose Hamiltonian is the sum of separate terms for each coordinate:

$$\hat{H} = \hat{H}_x(p_x, x) + \hat{H}_y(p_y, y) + \hat{H}_z(p_z, z) \qquad ...(2.158)$$

In this case, the wave function $\psi(x, y, z)$ can be written as:

$$\psi(x, y, z) = X(x)\, Y(y)\, Z(z) \qquad ...(2.159)$$

$$E = E_x + E_y + E_z \qquad ...(2.160)$$

and $\hat{H}_x X(x) = E_x X(x)$; $\hat{H}_y Y(y) = E_y Y(y)$; $\hat{H}_z Z(z) = E_z Z(z)$ ...(2.161)

2.6.4 PARTICLE IN A THREE DIMENSIONAL BOX

This problem is similar to the problem of a particle in one dimensional box, but it illustrates two problems very clearly, (*i*) *separation of variables* and (*ii*) *degeneracy*. Consider a particle in a three dimensional box of length *a*, breadth *b*, and height *c*. Also assume that potential energy inside the box is constant and can be taken as zero and infinite outside the box. The boundary conditions can be derived as discussed in case of particle in one dimensional box, *i.e.*, $\psi = 0$, when $x = y = z = 0$ and $\psi = 0$ when $x = a$, $y = b$ and $z = c$ respectively. The Hamiltonian can be written as

$$H = \frac{1}{2}mv_x^2 + \frac{1}{2}mv_y^2 + \frac{1}{2}mv_z^2 = \frac{1}{2}mv^2$$

where *m* is the mass of the particle, v_x, v_y and v_z are the components of the velocity *v* along the *x*, *y* and *z* axes, respectively. The Hamiltonian can be transformed to:

$$H = \frac{1}{2m}[p_x^2 + p_y^2 + p_z^2] = \frac{1}{2m}p^2$$

The corresponding Hamiltonian operator will be

$$\hat{H} = -\frac{h^2}{8\pi^2 m}\left[\frac{\partial^2}{\partial x^2} + \frac{\partial^2}{\partial y^2} + \frac{\partial^2}{\partial z^2}\right]$$

and the Schrödinger wave equation becomes

$$-\frac{h^2}{8\pi^2 m}\left[\frac{\partial^2\psi}{\partial x^2} + \frac{\partial^2\psi}{\partial y^2} + \frac{\partial^2\psi}{\partial z^2}\right] + V\psi = E\psi \qquad ...(2.12)$$

Since the total Hamiltonian represents the sum of the terms which only depends on the respective coordinates, the total wave function ψ can be represented as:

$$\Psi(x, y, z) = X(x)\,Y(y)\,Z(z) \qquad ...(2.161)$$

Combining the equations (2.158) and (2.163) and then dividing by X (*x*) Y (*y*) Z (*z*), the equation (2.12) reduces to:

$$-\frac{h^2}{8\pi^2 m}\left[\frac{d^2X(x)}{X(x)\,dx^2} + \frac{d^2Y(y)}{Y(y)\,dy^2} + \frac{d^2Z(z)}{Z(z)\,dz^2}\right] = E \qquad ...(2.163)$$

Write equation (2.162) as:

$$-\frac{h^2}{8\pi^2 m}\frac{d^2X(x)}{X(x)\,dx^2} = E + \frac{h^2}{8\pi^2 m}\frac{d^2Y(y)}{Y(y)\,dy^2} + \frac{h^2}{8\pi^2 m}\frac{d^2Z(z)}{Z(z)\,dz^2} \qquad ...(2.164)$$

If the coordinates x is varied, each term in the right hand side of equation (2.165) will not change as these are independent of x. In other words, the left hand side of the equation (2.165) will be a constant even if the x-coordinate is changed. Equating the right hand side of (2.165) equal to E_x, we get

$$-\frac{h^2}{8\pi^2 m}\frac{d^2X(x)}{X(x)\,dx^2} = E_x \qquad ...(2.165)$$

Since x, y and z are symmetrical to each other, equation (2.166) can be further written as:

$$-\frac{h^2}{8\pi^2 m}\frac{d^2Y(y)}{Y(y)\,dy^2} = E_y \qquad ...(2.166)$$

$$-\frac{h^2}{8\pi^2 m}\frac{d^2Z(z)}{Z(z)\,dz^2} = E_z \qquad ...(2.167)$$

The above equation can be rearranged to give the following forms:

$$\frac{d^2X(x)}{dx^2} = -\frac{8\pi^2 m E_x}{h^2} X(x) \qquad ...(2.168)$$

$$\frac{d^2Y(y)}{dy^2} = -\frac{8\pi^2 m E_y}{h^2} Y(y) \qquad ...(2.169)$$

$$\frac{d^2Z(z)}{dz^2} = -\frac{8\pi^2 m E_z}{h^2} Z(z) \qquad ...(2.170)$$

and

$$E = E_x + E_y + E_z \qquad ...(2.171)$$

Equations (2.169) to (2.171) are similar to the Schrödinger wave equation (2.138), for the particle in one dimensional box. Thus the solutions for each second order differential equations are:

$$X(x) = \left(\frac{2}{a}\right)^{1/2} \sin\frac{n_x\pi}{a}x, \quad E_x = \frac{h^2}{8ma^2} n_x^2 \qquad ...(2.172)$$

$$Y(y) = \left(\frac{2}{b}\right)^{1/2} \sin\frac{n_y\pi}{b}y, \quad E_y = \frac{h^2}{8mb^2} n_y^2 \qquad ...(2.173)$$

$$Z(z) = \left(\frac{2}{c}\right)^{1/2} \sin\frac{n_z\pi}{c} z, \quad E_z = \frac{h^2}{8mc^2} n_z^2 \qquad \text{...(2.174)}$$

where
$$n_x = 1, 2, 3, \ldots\ldots$$
$$n_y = 1, 2, 3, \ldots\ldots\ldots$$
$$n_z = 1, 2, 3, \ldots\ldots\ldots$$

Also

$$\psi(x, y, z) = \left(\frac{8}{abc}\right)^{1/2} \sin\frac{n_x\pi}{a} x \; \sin\frac{n_y\pi}{b} y \; \sin\frac{n_z\pi}{c} z \qquad \text{...(2.175)}$$

$$E = \frac{h^2}{8m}\left[\frac{n_x^2}{a^2} + \frac{n_y^2}{b^2} + \frac{n_z^2}{c^2}\right] \qquad \text{...(2.176)}$$

Since individual function is normalized, the total function will also be normalized, *i.e.*,

$$\int_0^a\int_0^b\int_0^c \psi^*(x,y,z)\,\psi(x,y,z)\,d\tau = \int_0^a X^*(x)\,X(x)\,dx \int_0^b Y^*(y)\,Y(y)\,dy \int_0^c Z^*(z)\,Z(z)\,dz$$

Cubic Box: The other interesting feature of this problem can be understood if it is considered that all the three sides of the box are equal, *i.e.*, the box is a cubical one. Equation (2.177) can be written as:

$$E = \frac{h^2}{8ma^2}\left[n_x^2 + n_y^2 + n_z^2\right] \qquad \text{...(2.177)}$$

and the energy of any state can be calculated by substituting the value of quantum numbers of the state. Table 2.2 gives some of the allowed energies of a particle confined to a cube of infinite high walls. The energy of the lowest state (ground state) is calculated by making the substitution $n_x = n_y = n_z = 1$ and the value is $\frac{3h^2}{8ma^2}$. The other useful and important observation is that some of the states with different quantum numbers have the same energy. For example, the states with quantum numbers (2, 1, 1), (1, 2, 1) and (1, 1, 2) have the same energy as $\frac{6h^2}{8ma^2}$. Whereas, the wave functions, corresponding to each set of quantum numbers ($\psi_{2,1,1}$, $\psi_{1,2,1}$, $\psi_{1,1,2}$) are three different independent wave functions and therefore represent different states of the system. These states are called degenerate states and are defined as: "*Those states or energy levels are called degenerate states if these have different functions but have the same energy*". *The number of different and independent states having the same energy is called degeneracy*, for example in the above case the degeneracy is three.

Table 2.2: *The energies of the first few states of a particle in a box in terms of* $\frac{h^2}{8ma^2}$

n_x	n_y	n_z	E	n_x	n_y	n_z	E
1	1	1	3	2	2	2	12
2	1	1	6	1	2	3	14
1	2	1	6	1	3	2	14
1	1	2	6	2	1	3	14
2	2	1	9	2	3	1	14
1	2	2	9	3	1	2	14
2	1	2	9	3	2	1	14
3	1	1	11				
1	3	1	11				
1	1	3	11				

These ideas of separation of variables or degeneracy are very simple but are very useful in understanding the structure of atoms and molecules, which involve many electrons. In latter case, the total wave functions of multi-electron system can be approximated to the product of wave functions of a single electron and thus the total energy will be sum of the energy of the individual electron. Although, it will be shown later that this approximation is too simple to get accurate results, this is a first step to understand the structure of multielectron atoms and molecules.

Degeneracy: In this section we shall like to prove a theorem regarding degeneracy. If there are *n* different independent wave functions which have the same energy (E) to the Hamiltonian operator ($\hat{H}$), any linear combination of these functions is also an eigen function of the Hamiltonian operator with the same eigen value E. Let there be *n* independent wave functions, $\psi_1, \psi_2, \psi_3, \ldots\ldots, \psi_n$ and E be the energy corresponding to the Hamiltonian operator $\hat{H}$.

$$\hat{H}\psi_1 = E\,\psi_1;\ \hat{H}\psi_2 = E\,\psi_2;\ldots\ldots\ldots\ldots;\ \hat{H}\,\psi_n = E\,\psi_n$$

Let the linear combination be:

$$\psi = C_1\,\psi_1 + C_2\,\psi_2 + C_3\,\psi_3 + \cdots\cdots\cdots + C_n\,\psi_n$$

We have to show that $\hat{H}\,\psi = E\,\psi$ or

$$\hat{H}\,(C_1\,\psi_1 + C_2\,\psi_2 + C_3\,\psi_3 + \cdots\cdots + C_n\,\psi_n) = E\,(C_1\,\psi_1 + C_2\,\psi_2 + C_3\,\psi_3 + \cdots + C_n\,\psi_n)$$

Since we know that Hamiltonian is a linear operator, the left hand side can be written as:

$$\hat{H}\,(C_1\,\psi_1 + C_2\,\psi_2 + C_3\,\psi_3 + \cdots\cdots + C_n\,\psi_n) = \hat{H}\,(C_1\,\psi_1) + \hat{H}\,(C_2\,\psi_2) + \cdots + \hat{H}\,(C_n\,\psi_n)$$

$$= C_1\,(\hat{H}\psi_1) + C_2\,(\hat{H}\psi_2) + \cdots\cdots + C_n\,(\hat{H}\psi_n)$$

$$= E\,(C_1\,\psi_1) + E\,(C_2\,\psi_2) + \cdots\cdots + E\,(C_n\,\psi_n)$$

the desired result.

2.6.5 PARTICLE ON A CIRCULAR RING

Consider a particle of mass m, moving on the periphery of a ring, with a radius r. Since the distance of the particle from the centre remains fixed and there is no other particle, the potential energy acting on the particle will be constant and can be taken as zero. Thus the total energy of the particle will be only the kinetic energy and can be written as:

$$H = E = \frac{1}{2} m \left(v_x^2 + v_y^2\right) \qquad ...(2.178)$$

where v_x and v_y are the components of the velocity along the x-and y-axis respectively. Since the particle is having a circular motion, it is nice to express the behaviour of the particle in terms of circular coordinates. Let at any particular time, the position of the particle be represented by (x, y) in terms of Cartesian coordinates and (r, θ) in terms of Circular coordinates. The relation between the two systems is:

$$x = r \cos \theta,\ y = r \sin \theta \qquad ...(2.179)$$

as shown in Fig. 2.4. The corresponding components of velocities along x-and y-axis can be obtained by differentiating x and y with respect to time (t).

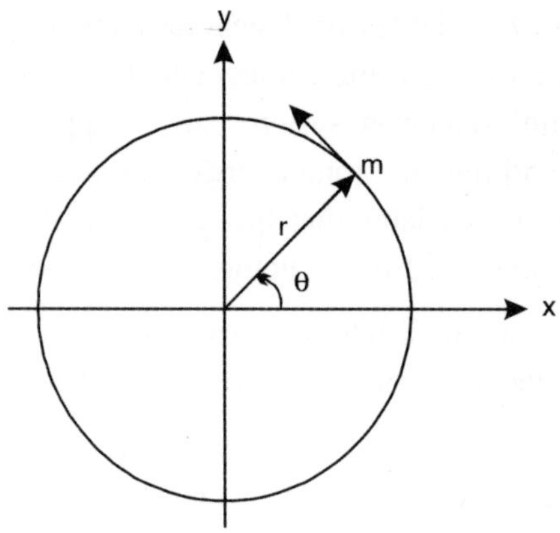

Figure 2.4: *Particle of mass m moving on circular ring of radius r*

$$v_x = \frac{dx}{dt} = -r \sin \theta \frac{d\theta}{dt}; \quad v_y = \frac{dy}{dt} = r \cos \theta \frac{d\theta}{dt} \qquad ...(2.180)$$

Substituting the values of v_x and v_y in (2.177)

$$H = \frac{1}{2} m \left[r^2 \sin^2 \theta \left(\frac{d\theta}{dt}\right)^2 + r^2 \cos^2 \theta \left(\frac{d\theta}{dt}\right)^2 \right]$$

$$= \frac{1}{2} m r^2 \left(\frac{d\theta}{dt}\right)^2 = \frac{1}{2I} (I w)^2 \qquad ...(2.181)$$

where w is the angular velocity and I is the moment of inertia, counter part of the mass in the linear motion. Substituting the value of angular momentum (I w) from Table 2.1, the Hamiltonian operator can be written as:

$$\hat{H} = -\frac{h^2}{8\pi^2 I}\frac{d^2}{d\theta^2} \qquad ...(2.182)$$

and the Schrödinger wave equation is:

$$-\frac{h^2}{8\pi^2 I}\frac{d^2\psi}{d\theta^2} = E\psi \qquad ...(2.183)$$

or

$$\frac{d^2\psi}{d\theta^2} = -\frac{8\pi^2 IE}{h^2}\psi = -m^2\psi \qquad ...(2.184)$$

where

$$m^2 = \frac{8\pi^2 IE}{h^2} \qquad ...(2.185)$$

Equation (2.183) is similar to the expression (2.135) and its solution can also be written as:

$$\psi = A\,e^{im\theta} \qquad ...(2.186)$$

where A is the normalization constant and can be obtained as:

$$1 = \int_0^{2\pi} \psi^* \psi\, d\theta = \int_0^{2\pi} A\,e^{-im\theta} A\,e^{im\theta}\, d\theta = A^2\int_0^{2\pi} d\theta = A^2\, 2\pi$$

$$A = \left(\frac{1}{2\pi}\right)^{\frac{1}{2}}$$

The normalized wave function is:

$$\psi = \left(\frac{1}{2\pi}\right)^{\frac{1}{2}} e^{im\theta} \qquad ...(2.187)$$

The boundary condition of this system so long it remains on the circle is that after complete one revolution, value of the function is the same at particular point, *i.e.*, $\psi(\theta) = \psi(2\pi + \theta)$. Substituting this value in (2.188)

$$\psi = A\,e^{im\theta} = A\,e^{im(2\pi + \theta)} = A\,e^{im\theta}\,.e^{2\pi mi}$$

or

$$e^{2\pi mi} = 1 = \cos 2m\pi + i \sin 2m\pi \qquad ...(2.188)$$

The relation (2.190) is satisfied only if m can have the values 0, ±1, ±2, ±3...

The energy levels of these states can be expressed as:

$$E = \frac{h^2}{8\pi^2 I} m^2, \quad m = 0, \pm1, \pm2, \pm3, \pm4, \ldots \quad .. \qquad ...(2.189)$$

and the wave function

$$\psi = \left(\frac{1}{2\pi}\right)^{\frac{1}{2}} e^{im\theta} \quad \text{...(2.190)}$$

Two points are worth noting.

(*i*) Except $m = 0$, all the other states are doubly degenerate.

(*ii*) The value of quantum number $m = 0$ is also an allowed solution. This is perfectly valid. Although when $m = 0$, the total energy will be zero and thus the momentum of the particle is precisely defined. This ($m = 0$) does not make function zero but states that ψ is independent of θ. In other words, the particle can exist in all possible directions or the uncertainty in the position of the particle is infinity. This satisfies the Heisenberg's uncertainty principle.

Example 2.18: Assume that a circle of radius 0.139 nm touches the six corners of benzene ring. Calculate the wave length of the lowest energy transition, assuming that benzene electrons move on the circular ring.

Solution: Since there are six π electrons, the highest occupied orbital will corresponds to when $m = 1$ and lowest unoccupied to $m = 2$.

$$E_1 = \frac{h^2}{8\pi^2 I}, \; E_2 = \frac{4h^2}{8\pi^2 I},$$

$$\nu = \frac{\Delta E}{h} = \frac{h}{8\pi^2 I}(4 - 1) = \frac{3h}{8\pi^2 I}$$

$$= \frac{(3)\,(6.62 \times 10^{-34}\ \text{J s})}{(8)\,(22/7)^2\,(9.11 \times 10^{-31}\ \text{kg})\,(1.39 \times 10^{-10}\ \text{m})^2}$$

$$= 1.428 \times 10^{15}\ \text{s}^{-1}.$$

$$\lambda = \frac{c}{\nu} = \frac{3 \times 10^8\ \text{m s}^{-1}}{1.428 \times 10^{15}\ \text{s}^{-1}} = 2.1 \times 10^{-7}\text{m} = 210\ \text{nm}$$

Experimentally observed long wave length transition is at 257 nm.

2.6.6 POWER SERIES METHOD

The cases considered so far in using Schrödinger wave equation to get the useful information involve the potential energy V (x), to be constant and is taken as equal to zero. This has made the second order differential equation (Schrödinger wave equation) very simple and can be solved exactly. But this is not the end of the story. There are many situations in real systems, where the potential energy is also a function of position of the system and thus has a variable value. Under such situations, the Schrödinger wave equation though becomes complicated, but can also be solved. The accurate values of the eigen functions and the corresponding eigen values can be calculated. The most common procedure is to use a power series method, *i.e.*, function which can be expressed as:

$$f(x) = a_0 + a_1 x + a_2 x^2 + a_3 x^3 \ldots\ldots\ldots\ldots\ldots\ldots \quad ...(2.191)$$

$$= \sum_{n=0}^{\infty} a_n x^n$$

where a_n are the coefficients to be determined. In this section, we shall take a known case, which has already been solved in section (2.6.2) and will show that the results obtained by both the methods are same.

Consider the case of a particle in one dimensional box and the second order differential equation is:

$$f''(x) + m^2 f(x) = 0 \quad ...(2.135)$$

where m^2 is a real, positive number and is equal to $\frac{8\pi^2 mE}{h^2}$. Also $m = 1, 2, 3, \ldots\ldots$ The solution used in the section (2.6.2) is:

$$f(x) = \text{A} \cos mx + \text{B} \sin mx \quad ...(2.136)$$

where A and B are constants. This kind of solution is quite common where the above differential equation (2.135) has constant coefficients and the roots of the auxiliary equations are pure imaginary. The equation (2.135) can also be solved with the help of power series method.

Let us assume that solution can be obtained in the form of Taylor series about $x = 0$ and the power series is:

$$f(x) = \sum_{n=0}^{\infty} a_n x^n = a_0 + a_1 x + a_2 x^2 + a_3 x^3 \ldots\ldots\ldots \quad ...(2.192)$$

Differentiating equation (2.189) with respect to x:

$$f'(x) = a_1 + 2 a_2 x + 3 a_3 x^2 + \ldots\ldots\ldots\ldots\ldots + n a_n x^{n-1} = \sum_{n=1}^{\infty} n a_n x^{n-1} \quad ...(2.193)$$

In this equation we have assumed that term by term differentiation is valid for the series. (This is not always true for infinite series). Similarly $f''(x)$ can be obtained as:

$$f''(x) = 2 a_2 + 32 a_3 x + 43 a_4 x^2 + \ldots + n(n-1) a^n x^{n-2} = \sum_{n=2}^{\infty} n(n-1) a_n x^{n-2} \quad ...(2.194)$$

Combining equations (2.138), (2.135) and (2.192):

$$\sum_{n=2}^{\infty} n(n-1) a_n x^{n-2} + m^2 \sum_{n=0}^{\infty} a_n x^n = 0 \quad ...(2.195)$$

The two terms in equation (2.192) can be combined to form one term provided (*i*) the summation limits in each sum is the same and (*ii*) the power of x is the same in each term. This can be done by substituting $n = k + 2$ in the first term, *i.e.*,

$$\sum_{n=2}^{\infty} n(n-1)a_n x^{n-2} = \sum_{k=0}^{\infty} (k+2)(k+1)a_{k+2} x^k \qquad ...(2.196)$$

$$= \sum_{n=0}^{\infty} (n+2)(n+1)a_{n+2} x^n \qquad ...(2.197)$$

Equations (2.194) and (2.195) are the same because the variables used in these equations are dummy variables, *i.e.*, it is immaterial which letter is used to denote the variable. This can be seen from the equation (2.196) and (2.197).

$$\sum_{n=0}^{4} a_n x^n = a_0 + a_1 x + a_2 x^2 + a_3 x^3 + a^4 x^4 \qquad ...(2.198)$$

$$\sum_{m=0}^{4} a_m x^m = a_0 + a_1 x + a_2 x^2 + a_3 x^3 + a_4 x^4 \qquad ...(2.199)$$

Substituting (2.195) in equation (2.193), we get

$$\sum_{n=0}^{\infty} (n+2)(n+1)a_{n+2} x^n + m^2 \sum_{n=0}^{\infty} a_n x^n = 0$$

or $$\sum_{n=0}^{\infty} [(n+2)(n+1)a_{n+2} + m^2 a_n]x^n = 0 \qquad ...(2.200)$$

If equation (2.198) is valid for all values of x, the coefficients of each term of x should be equal to zero, *i.e.*,

$$(n+2)(n+1)a_{n+2} + a_n m^2 = 0$$

$$a_{n+2} = -\frac{m^2}{(n+2)(n+1)} a_n \qquad ...(2.201)$$

Equation (2.199) is called *recursion relation.* This equation gives the relation between the coefficients of the power series. If we know the value of a_0, the other coefficients a_2, a_4, a_6........ can be calculated in terms of a_0 from equation (2.199). Similarly a_3, a_5, a_7.......can be calculated in terms of a_1. Expression (2.199) helps only in calculating either even coefficients or odd coefficients. Since the above procedure does not impose any restriction on the values of a_0 and a_1, the coefficients a_0 and a_1 can have any arbitrary value. Let a_0 = A and $a_1 = m$ B. Using equation (2.199), one can calculate the values of other constants. For example,

$$a_0 = \text{A},\ a_2 = -\frac{m^2\text{A}}{2\cdot 1} = (-1)^1 \frac{m^2\ \text{A}}{2!}$$

$$a_4 = -\frac{m^2}{4\cdot 3} a_2 = (-1)^2 \frac{m^4\ \text{A}}{4\cdot 3\cdot 2\cdot 1} = (-1)^2 \frac{m^4\ \text{A}}{4!}$$

$$a_6 = -\frac{m^2}{6\cdot 5} a_4 = (-1)^3 \frac{m^6\ \text{A}}{6\cdot 5\cdot 4\cdot 3\cdot 2\cdot 1} = (-1)^3 \frac{m^6\ \text{A}}{6!}$$

$$a_{2l} = (-1)^l \frac{m^{2l}}{2l!} \text{A},\ l = 0, 1, 2, 3, \ldots\ldots\ldots \qquad \ldots(2.202)$$

Similarly

$$a_3 = (-1)\frac{m^2}{3{\cdot}2{\cdot}1}\, a_1 = (-1)^1 \frac{m^3\,\text{B}}{3!}$$

$$a_5 = -\frac{m^2}{5{\cdot}4}\, a_3 = (-1)^2 \frac{m^5\,\text{B}}{5!}$$

$$a_{2l+1} = (-1)^l \frac{m^{2l+1}\,\text{B}}{(2l+1)!},\ l = 0, 1, 2, 3, \ldots\ldots\ldots\ldots \qquad \ldots(2.203)$$

Equation (2.135) can be written as:

$$f(x) = \sum_{n=0}^{\infty} a_n x^n = \sum_{n=0,2,4,6\ldots}^{\infty} a_n x^n + \sum_{n=1,3,5,\ldots}^{\infty} a_n x^n$$

$$= \text{A} \sum_{l=0}^{\infty} (-1)^l \frac{m^{2l}\, x^{2l}}{(2l)!} + \text{B} \sum_{l=0}^{\infty} (-1)^l \frac{m^{2l+1}\, x^{2l+1}}{(2l+1)!} \qquad \ldots(2.204)$$

The term in equation (2.202) represents the Taylor series for cos mx and sin mx. Thus:

$$f(x) = \text{A} \cos mx + \text{B} \sin mx \qquad \ldots(2.131)$$

i.e., solution obtained by power series is the same as that obtained by earlier method.

2.6.7 SIMPLE (OR LINEAR) HARMONIC OSCILLATOR

This problem is little different than discussed earlier in the sense that in this case the total energy is the sum of kinetic energy and potential energy, *i.e.*, potential energy is also a variable function. This problem like previous one, explains many new features which can be used in the coming sections. For example, (*i*) introduction of power series method in solving Schrödinger wave equation, (*ii*) this problem is similar to vibrational motion of molecules. Thus calculation of wave function and the corresponding energy levels helps in understanding the vibrational spectrum of molecules, quantum theory of radiation etc. and (*iii*) it allows us to understand the features of quantum system along with its transition to classical domain.

Simple harmonic motion is defined as the motion where the restoring force is proportional to the displacement from its mean position. From Newton's law of motion

$$f = m\frac{d^2x}{dt^2} = -kx \qquad \ldots(2.205)$$

where m is the mass of the particle, x is the displacement from the mean position and k is the constant, known as force constant. The general solution for the second order differential equation is:

$$x = A \sin (2 \pi \nu_0 t + \delta) \quad ...(2.206)$$

where A is the amplitude and ν_0 is the classical frequency defined by:

$$\nu_0 = \frac{1}{2\pi}\left(\frac{k}{m}\right)^{1/2} \Rightarrow k = 4\pi^2 m \nu_0^2 \quad ...(2.207)$$

and δ is the phase angle. In order to calculate the energies of states and their corresponding wave functions, the first step is to express total energy, E, in the form of Hamiltonian, which is a function of H (p_x, x), coordinate and corresponding momentum. The potential energy term for harmonic oscillator can be calculated by the following expression and has been shown in Fig. 2.5.

$$V(x) = \int_0^x -f\, dx = \int_0^x k\, x\, dx = \frac{1}{2} k x^2 = \frac{1}{2} 4 \pi^2 \nu_0^2 m x^2 = 2 \pi^2 \nu_0^2 m x^2 \quad ...(2.208)$$

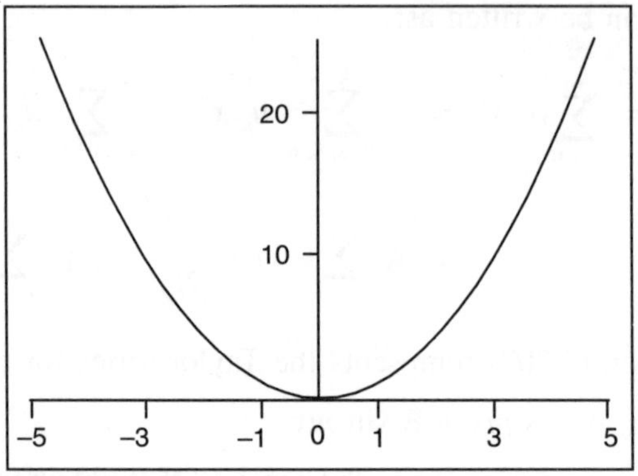

Figure 2.5: *Diagram showing the potential energy for the harmonic oscillator*

The total energy can be written as:

$$H = E = \frac{1}{2} m v_x^2 + V(x) = \frac{p_x^2}{2m} + 2 \pi^2 \nu_0^2 m x^2$$

The corresponding Hamiltonian operator

$$\hat{H} = -\frac{h^2}{8 \pi^2 m}\frac{d^2}{dx^2} + 2 \pi^2 \nu_0^2 m x^2 \quad ...(2.209)$$

and the Schrödinger wave equation is:

$$-\frac{h^2}{8 \pi^2 m}\frac{d^2\psi}{dx^2} + 2 \pi^2 \nu_0^2 m x^2 \psi = E \psi$$

$$\text{or} \quad \frac{d^2\psi}{dx^2} + \left(\frac{8 \pi^2 m E}{h^2} - \frac{16 \pi^4 \nu_0^2 m^2 x^2}{h^2}\right)\psi = 0 \quad ...(2.210)$$

Let $\alpha = \dfrac{8\pi^2 mE}{h^2}$, $\beta = \dfrac{4\pi^2 \nu_0 m}{h}$

So equation (2.207) can be written as:

$$\frac{d^2\psi}{dx^2} + (\alpha - \beta^2 x^2)\,\psi = 0 \qquad \text{...(2.211)}$$

Before we attempt to solve the equation (2.208), we must keep in mind the restrictions imposed on the wave function, *i.e.*, it should be single valued, finite, continuous and integrable. Further, we also know from our experience with the particle in a box problem that boundary conditions are necessary to the introduction of quantization of energy. In case of particle in a box problem, the potential energy increases abruptly to infinity at the walls or outside the box, whereas in this case the potential energy increases slowly to infinity at large displacement. This means that ψ will reduce to zero slowly at $x = -\infty$ and $x = \infty$. When potential energy is very large, *i.e.*, $\beta >> \alpha$, then α can be neglected. Equation (2.208) reduces to:

$$\frac{d^2\psi}{dx^2} = \beta^2 x^2\,\psi \qquad \text{...(2.212)}$$

and the solution of this equation will be:

$$\psi = e^{\pm\frac{\beta x^2}{2}} \qquad \text{...(2.213)}$$

Before we proceed further, let us try to verify whether these are proper solutions or not.

$$\frac{d\psi}{dx} = \pm\,\beta x e^{\pm\frac{\beta x^2}{2}}$$

$$\frac{d^2\psi}{dx^2} = \pm\,\beta\, e^{\pm\frac{\beta x^2}{2}} \pm \beta^2 x^2 e^{\pm\frac{\beta x^2}{2}} \qquad \text{...(2.214)}$$

At large value of x, $x^2 \gg 1$, equation (2.211) reduces to:

$$\frac{d^2\psi}{dx^2} = \pm\,\beta^2 x^2 e^{\pm\frac{\beta x^2}{2}}$$

and this is similar to (2.209). Out of two solutions, the one $e^{\frac{\beta x^2}{2}}$ is not acceptable because it tends to infinity as x increases and thus can be neglected. Thus $\psi = e^{-\frac{\beta x^2}{2}}$ is the solution which is only applicable at $x = \infty$. The accurate solution, which is applicable over the complete region, can be obtained by multiplying the asymptotic solution with a power series function, *i.e.*,

$$\psi = e^{-\frac{\beta x^2}{2}} f(x) = e^{-\frac{\beta x^2}{2}} \Sigma a_n x^n \qquad \text{...(2.212)}$$

$$\frac{d\psi}{dx} = -\beta x e^{-\frac{\beta x^2}{2}} f(x) + e^{-\frac{\beta x^2}{2}} f'(x)$$

$$\frac{d^2\psi}{dx^2} = -\beta e^{-\frac{\beta x^2}{2}} f(x) + \beta^2 x^2 e^{-\frac{\beta x^2}{2}} f(x) - \beta x e^{-\frac{\beta x^2}{2}} f'(x) - \beta x e^{-\frac{\beta x^2}{2}} f'(x) + e^{-\frac{\beta x^2}{2}} f''(x)$$

$$= [-\beta f(x) + \beta^2 x^2 f(x) - 2\beta x f'(x) + f''(x)]\, e^{-\frac{\beta x^2}{2}} \quad ...(2.215)$$

Combining equations (2.208) and (2.213), we get

$$f''(x) - 2\beta x\ f'(x) + [\alpha - \beta] f(x) = 0 \quad ...(2.216)$$

Now $f(x)$ can be expanded as power series as:

$$f(x) = \sum_{n=0}^{\infty} a_n x^n = a_0 + a_1 x + a_2 x^2 + a_3 x^3 \ldots \ \ldots \quad ...(2.217)$$

$$f'(x) = a_1 + 2a_2 x + 3a_3 x^2 + \ldots\ldots + n a_n x^{n-1} = \sum_{n=1}^{\infty} n a_n x^{n-1}$$

$$= \sum_{n=0}^{\infty} n a_n x^{n-1} \quad ...(2.218)$$

$$f''(x) = 2a_2 + 3\cdot 2 a_3 x + 4\cdot 3 a_4 x^2 + \ldots + n(n-1) a^n x^{n-2} = \sum_{n=2}^{\infty} n(n-1) a_n x^{n-2}$$

$$= \sum_{n=0}^{\infty} (n+2)(n+1) a_{n+2} x^n \quad ...(2.219)$$

Substitute the values of (2.215), (2.216) and (2.217) in equation (2.214),

$$\sum_{n=0}^{\infty} (n+2)(n+1) a_{n+2} x^n - 2\beta \sum_{n=0}^{\infty} n a_n x^n + (\alpha - \beta) \sum_{n=0}^{\infty} a_n x^n = 0$$

$$\sum_{n=0}^{\infty} [(n+2)(n+1) a_{n+2} - (2n\beta - \alpha + \beta) a_n]\, x^n = 0 \quad ...(2.220)$$

For non-trivial solution of equation (2.218), the coefficients of each term of equation (2.221) will be zero, *i.e.*,

$$a_{n+2} = \frac{2n\beta - \alpha + \beta}{(n+2)(n+1)} a_n = \frac{(2n+1)\beta - \alpha}{(n+2)(n+1)} a_n \quad ...(2.222)$$

Equation (2.219) is known as *two term recursion relation*. This formula helps in calculating $a_2, a_4, a_6, \ldots$ (even coefficients) in terms of a_0 and $a_3, a_5, a_7 \ldots$ (odd coefficients) in terms of

a_1. Both a_0 and a_1 are arbitrary coefficients, similar to that used in the section (2.6.6). If we set a_1 equal to zero, the power series will only contain the even coefficients and thus the even powers of x, multiplied by the exponential factor.

$$\psi = e^{-\beta x^2/2} \sum_{0,\,2,\,4...}^{\infty} a_n\, x^n = e^{-\beta x^2/2} \sum_{0,1,2,...}^{\infty} a_{2l}\, x^{2l} \qquad ...(2.223)$$

If we set a_0 equal to zero, we get another independent solution:

$$\psi = e^{-\beta x^2/2} \sum_{1,\,3,\,5,...}^{\infty} a_n\, x^n = e^{-\beta x^2/2} \sum_{0,\,1,\,2,...}^{\infty} a_{2l+1}\, x^{2l+1} \qquad ...(2.224)$$

The general solution of Schrödinger equation is:

$$\psi = \mathrm{A}\, e^{-\beta x^2/2} \sum_{0,\,1,\,2,...}^{\infty} a_{2l+1}\, x^{2l+1} + \mathrm{B}\, e^{-\beta x^2/2} \sum_{0,\,1,\,2,...}^{\infty} a_{2l}\, x^{2l} \qquad ...(2.225)$$

where A and B are arbitrary constants. This equation is similar to the equation (2.201).

Next step is to see whether after applying the boundary conditions, equation (2.222) still leads to a well behaved solution. To note how the two series behave at large value of x, we have to examine how the coefficients of these series behave for large value of x. The ratio of the coefficients of x^{2l+2} and x^{2l} in the second series is obtained by substituting $n = 2l$ in (2.219):

$$\frac{a_{2l+2}}{a_{2l}} = \frac{2\,(2l)\,\beta - \alpha + \beta}{(2l+2)\,(2l+1)} = \frac{4l\,\beta + \beta - \alpha}{(2l+2)\,(2l+1)} \qquad ...(2.226)$$

For large value of l, it reduces to:

$$\frac{a_{2l+2}}{a_{2l}} = \frac{(4\,\beta l)}{(2l)\,(2l)} = \frac{\beta}{l}$$

Similarly setting $n = 2l + 1$, in equation (2.219), the ratio a_{2l+3} / a_{2l+1} also leads to β / l for large value of l, *i.e.*, for large value of l, the ratio of the coefficient of successive terms of x leads to same value (β / l). Now consider the expansion of power series for the function $e^{\beta x^2}$.

$$e^{\beta x^2} = 1 + \beta\, x^2 + \frac{(\beta\, x^2)^2}{2!} + \frac{(\beta\, x^2)^3}{3!} + \ldots\ldots\ldots\ldots + \frac{(\beta\, x^2)^l}{l!} + \frac{(\beta\, x^2)^{l+1}}{(l+1)!}$$

The ratio of the coefficients of x^{2l+1} and x^{2l} is given as:

$$\frac{\beta^{l+1}}{(l+1)!}\,\frac{l!}{\beta^l} = \frac{\beta}{l+1} \approx \frac{\beta}{l} \text{ for large value of } l$$

This indicates that ratio of the successive coefficients in each of the infinite series in solution (2.222) is the same as in the power series for $e^{\beta x^2}$. Thus it can be concluded that each term in equation (2.222) leads to $e^{\beta x^2}$.

Replacing $e^{\beta x^2}$ for the power series in each term of (2.222), the equation will reduce to $\psi = e^{\frac{\beta x^2}{2}}$, which is not an acceptable solution in quantum mechanics. This is because the value of ψ tends to infinity as x increases. If it is some how possible to break the series after certain terms ψ will tend to zero for large values of x. This is only possible if for some value of $n = v$, the numerator in the recursion formula becomes zero. This will led to a_{v+2}, a_{v+4},.... all equal to zero and one of the series in (2.222) will have a finite number of terms. Setting the coefficient a_v equal to zero in (2.219):

$$\alpha - \beta - 2\,\beta\, v = 0$$

$$\alpha = 2\,\beta\,(v + 1/2) \qquad \text{...(2.227)}$$

$$\frac{8\pi^2 mE}{h^2} = \frac{8\pi^2 \mathrm{v}_0 m}{h}(\mathrm{v} + 1/2) \qquad \text{...(2.228)}$$

$$E = h\,\mathrm{v}\mathrm{v}_0\,(v + 1/2),\ v = 0, 1, 2, 3, \ldots\ldots\ldots \qquad \text{...(2.229)}$$

and the recursion formula becomes:

$$a_{n+2} = \frac{2\,\beta\,(n - v)}{(n+1)\,(n+2)}\,a_n \qquad \text{...(2.230)}$$

By quantizing the energy according to equation (2.226), we have made one of the series to break off after finite number of terms. To get rid of the second series in (2.222), we can set the arbitrary constant, that multiplies the power series and $e^{-\frac{\beta x^2}{2}}$ equal to zero. This will leave us with the wave function, ψ, which is a product of $e^{-\frac{\beta x^2}{2}}$ and finite series containing either even or odd power of x, depending on whether v is even or is odd. Equation (2.222) leads to only non-negative integral values of v and also the energy levels which are equally spaced. *Similar to particle in a box problem, it is the boundary conditions that make the energy quantized.* So, one can visualize the importance of boundary conditions in quantum mechanics.

***Energy levels*:** There is one difference in the energy levels, given by the relation (2.226) and that is given by old quantum theory, *i.e.*, even at $v = 0$, the energy of the oscillator will be $1/2\ h\,\mathrm{v}_0$, known *as zero point energy or ground state energy* and will be non-zero even at absolute zero. The meaning of zero point energy can be understood nicely from the Heisenberg's uncertainty principle. If energy of the lowest state of harmonic oscillator is zero, kinetic and potential energy will also be zero, because none of them is having negative values. Zero kinetic energy will lead to zero linear momentum or Δp_x will be zero. Zero potential energy means that the harmonic oscillator is at its mean position Δx is zero. But we know that both can not have zero values or cannot be determined simultaneously accurately. This leads to the fact that particle must have zero point energy. Other characteristics of energy levels will be discussed in Chapter 5.

***Wave functions*:** Let us try to find out the wave functions for some of the energy levels with smaller values of v. For $v = 0$ all terms of odd power of x will be equal to zero and similarly from equation (2.212), other coefficients are also zero, *i.e.*,

$$\psi_0 = a_0 e^{-\frac{\beta x^2}{2}} \qquad ...(2.231)$$

The normalization constant a_0 can be calculated from the relation:

$$1 = \int_{-\infty}^{+\infty} \psi_0^2 \, d\tau = a_0^2 \int_{-\infty}^{+\infty} e^{-\beta x^2} \, dx = 2 \, a_0^2 \int_0^{+\infty} e^{-\beta x^2} \, dx$$

where $$\int_0^{+\infty} e^{-\beta x^2} dx = \frac{1}{2}\left(\frac{\pi}{\beta}\right)^{\frac{1}{2}} \qquad ...(2.232)$$

Hence $$a_0 = \left(\frac{\beta}{\pi}\right)^{\frac{1}{4}} \qquad ...(2.233)$$

and $$\psi_0 = \left(\frac{\beta}{\pi}\right)^{\frac{1}{4}} e^{-\frac{\beta x^2}{2}} \qquad ...(2.234)$$

if it is chosen that the phase of the normalization constant is zero. The wave function is a Gaussian function and is shown in Fig. (2.6*a*). For $v = 1$, there is no even powers of x in the polynomial factor and $a_3 = a_5 = \ldots\ldots.= 0$.

Hence $$\psi_1 = a_1 \, x e^{-\frac{\beta x^2}{2}} \qquad ...(2.235)$$

After normalization, we have

$$\psi_1 = \left(\frac{4\,\beta^3}{\pi}\right)^{\frac{1}{4}} x e^{-\frac{\beta x^2}{2}} \qquad ...(2.236)$$

and the function is plotted in Fig. (2.6*b*).

Similarly for $v = 2$,

$$\psi_2 = (a_0 + a_2 \, x^2) \, e^{-\frac{\beta x^2}{2}} \qquad ...(2.237)$$

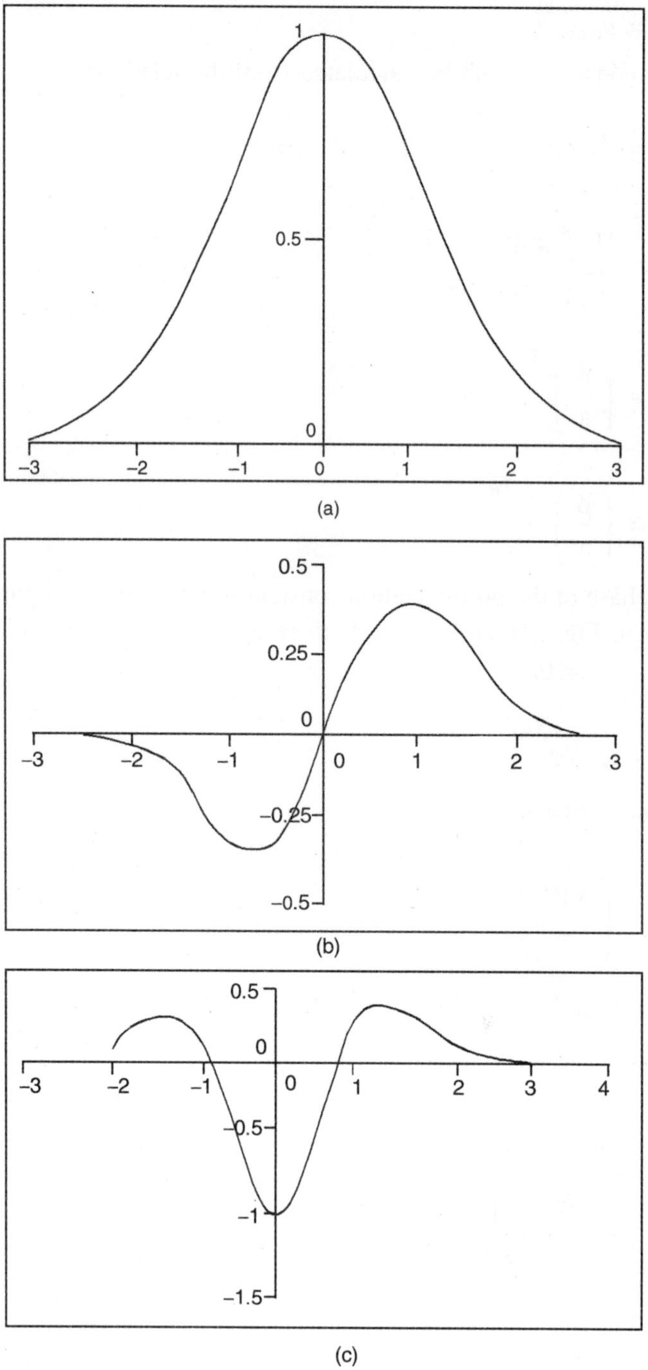

Figure 2.6: *Plot of wave functions of the harmonic oscillator (a) v = 0, (b) v = 1 and (c) v = 2*

From equation (2.227)

$$a_2 = \frac{-4\beta a_0}{2} = -2\beta a_0$$

and $$\psi_2 = a_0 (1 - 2\beta x^2) e^{\frac{\beta x^2}{2}} \qquad ...(2.238)$$

The normalized wave function is:

$$\psi_2 = \left(\frac{\beta}{4\pi}\right)^{\frac{1}{4}} (2\beta x^2 - 1) e^{-\frac{\beta x^2}{2}}$$

and the plot is shown in Fig. (2.6*c*).

For larger values of *v*, one can find out the wave function from the polynomial known as Hermite polynomial, *i.e.*,

$$\psi_v = \left(\frac{\beta}{4\pi}\right)^{\frac{1}{4}} \left(\frac{1}{2^v v!}\right)^{1/2} e^{-\frac{\beta x^2}{2}} \mathrm{H}_v (\beta^{\frac{1}{2}} x) \qquad ...(2.239)$$

where the Hermite polynomial is defined as:

$$\mathrm{H}_v (q) = (-1)^v e^{q^2} \frac{d^v (e^{-q^2})}{dq^v}, \; q = \beta^{\frac{1}{2}} x \qquad ...(2.240)$$

where $v = 0, 1, 2, 3,$........ The first five members of the series are given below.

$$\mathrm{H}_0 (q) = (-1)^0 e^{q^2} \frac{d^0 (e^{-q^2})}{dq^0} = 1$$

$$\mathrm{H}_1 (q) = (-1)^1 e^{q^2} \frac{d^1 (e^{-q^2})}{dq^1} = -e^{q^2} (-2q) e^{-q^2} = 2q$$

$$\mathrm{H}_2 (q) = (-1)^2 e^{q^2} \frac{d^2 (e^{-q^2})}{dq^2} = e^{q^2} \frac{d}{dq}[-2q e^{-q^2}] = e^{q^2} [-2e^{-q^2} + 4q^2 e^{-q^2}] = (4q^2 - 2)$$

$$\mathrm{H}_3 (q) = (-1)^3 e^{q^2} \frac{d^3 (e^{-q^2})}{dq^3} = 8q^3 - 12q$$

$$\mathrm{H}_4 (q) = (-1)^4 e^{q^2} \frac{d^4 (e^{-q^2})}{dq^4} = 16q^4 - 48q^2 + 12$$

Table 2.3: The first few Hermite Polynomials

$H_0(q) = 1$	$H_3(q) = 8q^3 - 12q$
$H_1(q) = 2q$	$H_4(q) = 16q^4 - 48q^2 + 12$
$H_2 = 4q^2 - 2$	$H_5(q) = 32q^5 - 160q^3 + 120q$

In Fig. 2.6, we have learnt that the wavefucntions for the harmonic oscillator can extend up to $-\infty$ to $+\infty$. This means that the probability of finding the oscillator can extend up to ∞ on both sides of the mean postion. Although this observation is very interesting from quantum mechanics point of view, but from classical mechanics point of view this is not feasable. This kind of behaviour is a characteristic of quantum mechanical systems and is responsible for the phenomenon called quantum mechanical tunneling. In these kinds of phenomena, the particle can tunnel into the regions which are forbidden based on classical mechanics. This has been discussed in details in next chapter. Here we will only determine the lmits of amplitudes between which the oscillator can oscillate.

Consider the ground state of harmonic oscillator. The probability that the displacement of a oscillator lies between x and $x + dx$ is given by:

$$P(x)\ dx = \Psi_0^2(x)dx = \left(\frac{\beta}{\pi}\right)^{\frac{1}{2}} e^{-\beta x^2} dx$$

The total energy in this state $= \dfrac{1}{2} h\,\omega_e = \dfrac{1}{2}\dfrac{h}{2\pi}\sqrt{\dfrac{k}{\mu}}$

In classical mechanics, the oscillator only possesses potential energy at the extreme positions of oscillations and this depicts the maximum amplitude an oscillator can achieve. Thus the maximum potential energy will be $\frac{1}{2}kA^2$. Equationg the two energies, we get

$$\frac{1}{2}k\,A^2 = \frac{1}{2}\frac{h}{2\pi}\sqrt{\frac{k}{\mu}}$$

or

$$A^2 = \frac{h}{2\pi}\frac{1}{\sqrt{k\mu}} \Rightarrow A = \left[\frac{h}{2\pi\sqrt{k\mu}}\right]^{\frac{1}{2}} \quad ...(2.241)$$

Equation (2.238) gives the well defined value of the amplitude of classical harmonic oscillator. Although we have mentioned above that quantum mechanics does not give well defined value of amplitude, but the amplitude can be correlated by calculating $\langle x^2 \rangle$ as a measure of square of the amplitude and in particular we can equate root mean square value of x $\left(\langle x^2 \rangle^{\frac{1}{2}}\right)$ as a measure of amplitude., *i.e.*,

$$A_{rms} = r_{rms} = \langle x^2 \rangle^{\frac{1}{2}}$$

The root mean square amplitude can be related to the experimentally observable quantities like frequency of absorption as follows.

$$A_{rms} = r_{rms} = \langle x^2 \rangle^{\frac{1}{2}} = \int_{-\infty}^{+\infty} \left(\frac{\beta}{\pi}\right)^{\frac{1}{2}} e^{-\beta x^2} x^2 dx$$

$$= \left[\left(\frac{\beta}{\pi}\right)^{\frac{1}{2}} \left\{\frac{1}{2\beta}\left(\frac{\pi}{\beta}\right)^{\frac{1}{2}}\right\}\right]^{\frac{1}{2}} = \left[\frac{1}{2\beta}\right]^{\frac{1}{2}}$$

$$= \left[\frac{1}{4\pi(\sqrt{\mu k}}\right]^{\frac{1}{2}} \qquad ...(2.242)$$

Using equation $\omega_e = \dfrac{h}{2\pi}\sqrt{\dfrac{k}{\mu}}$ and eliminating the force constant (k), we get

$$A_{rms} = \left[\frac{h}{8\pi^2 c\bar{\nu}_{obs}\mu}\right]^{\frac{1}{2}} \qquad ...(2.243)$$

Thus, knowing the vibrational frequency one can calculate the maximum amplitude of the oscillation

Example 2.19: Find the wave function for the Harmonic oscillator for $v = 5$, using Hermite polynomial.

Solution: Using equation (2.237) the Hermite polynomial for $v = 5$, is

$$H_5\ (q) = (-1)^5 e^{q^2} \frac{d^5(e^{-q^2})}{dq^5} = -e^{q^2} \frac{d^4}{dq^4}[-2qe^{-q^2}] = -e^{q^2} \frac{d^3}{dq^3}[-2e^{-q^2} + 4q^2e^{-q^2}]$$

$$= -e^{q^2} \frac{d^2}{dq^2}[4qe^{-q^2} + 8qe^{-q^2} - 8q^3e^{-q^2}] = -e^{q^2} \frac{d^2}{dq^2}[12qe^{-q^2} - 8q^3e^{-q^2}]$$

$$= -e^{q^2} \frac{d}{dq}[12e^{-q^2} - (24q^2 + 24q^2)e^{-q^2} + 16q^4e^{-q^2}]$$

$$= -e^{q^2} \frac{d}{dq}[12e^{-q^2} - 48q^2e^{-q^2} + 16q^4e^{-q^2}]$$

$$= -e^{q^2}[-24qe^{-q^2} - 96qe^{-q^2} + 96q^3e^{-q^2} + 64q^3e^{-q^2} - 32q^5e^{-q^2}]$$

$$= 32q^5 - 160q^3 + 120q = 16[2q^5 - 10q^3 + 15/2q]$$

Substituting the values of $H_5(q)$ in equation (2.236), the normalized wave function for $v = 5$ is:

$$\psi_5 = \left(\frac{\beta}{\pi}\right)^{\frac{1}{4}} \left(\frac{1}{2^5\, 5!}\right)^{\frac{1}{2}} e^{-\frac{\beta x^2}{2}} (32q^5 - 160q^3 + 120q)$$

$$= 16\left(\frac{\beta}{\pi}\right)^{\frac{1}{4}} \left(\frac{1}{2^5\, 5!}\right)^{\frac{1}{2}} e^{-\frac{\beta x^2}{2}} (2q^5 - 10q^3 + \frac{15}{2}q)$$

Example 2.20: The Hermite polynomials for $v = 1$ and $v = 2$ are $H_1(q) = 2q$ and $H_2(q) = 4q^2 - 2$ respectively. Show that the wave functions of a Harmonic oscillator in the stationary states 1 and 2 are orthogonal. The total wave function for states 1 and 2 for harmonic oscillator are:

$$\psi_1(q) = 2\,N_1\, q\, e^{-\frac{q^2}{2}} \text{ and } \psi_2(q) = N_2\, e^{-\frac{q^2}{2}} \left(4q^2 - 2\right), \text{ where } q = (\beta)^{1/2} x$$

Solution: Wave functions are orthogonal if $\int_{-\infty}^{+\infty} \psi_1\, \psi_2\, dq = 0$. Substituting the values of the wave functions and x:

$$\int_{-\infty}^{+\infty} \psi_1\, \psi_2\, dx = \frac{2\,N_1 N_2}{\beta^{1/2}} \int_{-\infty}^{+\infty} q e^{-q^2/2}\, (4q^2 - 2)\, e^{-q^2/2}\, dq \qquad (dq = \sqrt{\beta}\, dx)$$

$$= \frac{2\,N_1\, N_2}{\beta^{1/2}} \int_{-\infty}^{+\infty} 4q^3 e^{-q^2} dq - \frac{4\,N_1\, N_2}{\beta^{1/2}} \int_{-\infty}^{+\infty} q e^{-q^2} dq$$

Both the integrals are odd functions so the values of both the integrals are zero. Thus ψ_1 and ψ_2 are orthogonal functions.

Example 2.21: The fundamental vibrational frequency and the equilibrium bond distance of CO molecule are 2170.2 cm^{-1} and 112.8 pm respectively. Calculate the root mean square displacement in the ground state of CO molecule and compare it with the equilibriun bond distance value.

Solution: The reduce mass of CO is:

$$\mu = \frac{(12)\,(16)\,(10^{-3}\text{kg g}^{-1})}{(12+16)\,(6.02\times 10^{23}\text{mol}^{-1})} = 1.0998 \times 10^{-26}\text{ kg}$$

The root mean square displacement can be calculated by using equation (2.240), *i.e.*,

$$A_{rms} = \left[\frac{(6.626\times 10^{-34}\text{ Js})}{(8)\left(\frac{22}{7}\right)^2 (3\times 10^{10}\text{cm s}^{-1})\,(2.1702\times 10^3\text{cm}^{-1})\,(1.0998\times 10^{-26}\text{kg})}\right]^{\frac{1}{2}} = 3.42\text{ pm}$$

Comparing with the equilibrium bond distance [$\frac{A_{rms}}{r_e} = 0.0303$ (or 3.03%)], the root mean square displacement is only 3.03%.

Behaviour of the wave function can be discussed from the following points of view:

(*i*) Figure 2.6*a* gives the probability of finding the particle in the lowest energy state, obtained by classical mechanics as well as by quantum mechanics. The results obtained by two methods do not agree with each other. In classical mechanics, the particle is confined to certain boundaries, keeps on oscillating between the end points and the probability is maximum at the boundaries. Whereas in quantum mechanics ψ_0^2 is maximum at the origin, falls off rapidly and extends over region which are beyond the boundaries, *i.e.*, classically forbidden zones. This means that particle can go to the regions where the total energy is less than the potential energy. In other words, this will lead to the negative value of kinetic energy. The result seems to be unusual in classical mechanics, has no paradox in quantum mechanical language and can be explained with the help of Heisenberg's uncertainty principle. To verify that the particle is in the region which is classically forbidden, we must make a measurement of its position. This measurement changes the state of system, *i.e.*, interaction of the oscillator with the measuring apparatus transfers sufficient energy to the oscillator that it can be driven to the classically forbidden region. An accurate measurement of position x introduces large uncertainty in the linear momentum and hence in kinetic energy.

(*ii*) It will be noticed that for $v = 0, 2, 4$, the curves are symmetrical with respect to change from positive to negative value of x, *i.e.*, the function is called an even function. The condition for even function is $f(-x) = f(x)$. Whereas for $n = 1, 3$, the curves are antisymmetrical with respect to such a change in the sign in x, *i.e.*, it is an odd function. $f(-x) = -f(x)$.

(*iii*) Since H_v is a polynomial of degree v, ψ_v will have v zero or v nodes. In other words the probability of finding the particle at these points is zero and these are nodes beside the point at infinity. Further as the energy (*i.e.*, as the value of v increases) ψ_v^2 matches closely with the classical results, means probability of finding the particle increases at the boundaries and decreases at the middle. This proves the Bohr's correspondence principle (*i.e.*,the results obtained with the help of classical mechanics and with the help of quantum mechanics at higher energies are the same).

(*iv*) All the results obtained in case of particle in a box problem, *i.e.*, the wave functions are normalized, orthogonal for two different quantum numbers and calculation of average values of the observables like average value of distance, energy, linear momentum etc., can be calculated in case of harmonic oscillator also. Only the mathematical part will be complicated.

(*v*) Lastly couple of useful relations about Hermite polynomial are given below:

$$q\,H_v(q) = 1/2\,H_{v+1}(q) + v\,H_{v-1}(q) \quad ...(2.244)$$

$$\frac{d}{dq}[H_v(q)] = 2\,v\,H_{v-1}(q) \quad ...(2.245)$$

2.6.8 RIGID ROTATOR

Rigid rotator is a two body problem in which the particles of masses m_1 and m_2, separated by a fixed distance r, rotates around an axis through the centre of gravity, as shown in Fig. 2.7. This problem is very helpful in finding out the energy levels for the rotational motion and in understanding the theory of rotational spectrum. The total Hamiltonian can be written as:

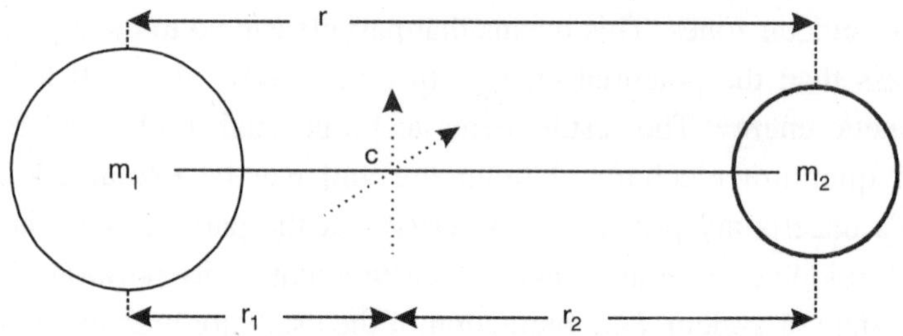

Figure 2.7: *A rigid diatomic molecule with masses m_1 and m_2 joined by thin rod of length $r = r_1 + r_2$. The centre of mass is at C*

$$\hat{H} = \frac{1}{2}m_1(v_x^2 + v_y^2 + v_z^2) + \frac{1}{2}m_2(v_x^2 + v_y^2 + v_z^2) \quad ...(2.246)$$

This is exactly similar to the problem discussed in section 1.1.2 and the two particles body system can be transformed into two one particle systems. The Schrödinger wave equations are:

$$-\frac{h^2}{8\pi^2 M}\left[\frac{\partial^2 F}{\partial X^2} + \frac{\partial^2 F}{\partial Y^2} + \frac{\partial^2 F}{\partial Z^2}\right] = E_1\,F \quad ...(2.247)$$

$$-\frac{h^2}{8\pi^2 \mu}\left[\frac{\partial^2 G}{\partial x^2} + \frac{\partial^2 G}{\partial y^2} + \frac{\partial^2 G}{\partial z^2}\right] + V(x, y, z) = E_2\,G \quad ...(2.248)$$

The first equation describes the behaviour of particle of mass $M = m_1 + m_2$, moving freely and we are not interested in the solution of this problem as it has been already discussed in section 2.6.2. The second equation gives the behaviour of a particle of mass $\mu = \dfrac{m_1\,m_2}{m_1 + m_2}$, moving under the influence of potential energy V. Since this is a rotational motion, it is nice to represent the Hamiltonian in spherical polar coordinates. As stated in section 2.4.4 (equation 2.53), the equation (2.245) can be written in spherical polar coordinates as:

$$-\frac{h^2}{8\pi^2\mu}\left[\frac{1}{r^2}\frac{\partial}{\partial r}\left(r^2\frac{\partial G}{\partial r}\right)+\frac{1}{r^2\sin\theta}\frac{\partial}{\partial\theta}\left(\sin\theta\frac{\partial G}{\partial\theta}\right)+\frac{1}{r^2\sin^2\theta}\frac{\partial^2 G}{\partial\varphi^2}\right]+VG$$

$$= E_2G \qquad \text{...(2.249)}$$

In the rigid rotator problem, since r the distance between the two particles remains fixed, the potential energy will be constant and can be taken as zero. Consequently equation (2.244) reduces to:

$$-\frac{h^2}{8\pi^2\mu}\left[\frac{1}{r^2\sin\theta}\frac{\partial}{\partial\theta}\left(\sin\theta\frac{\partial G}{\partial\theta}\right)+\frac{1}{r^2\sin^2\theta}\frac{\partial^2 G}{\partial\varphi^2}\right]= E_2\, G \qquad \text{...(2.250)}$$

or

$$\frac{1}{r^2\sin\theta}\frac{\partial}{\partial\theta}\left(\sin\theta\frac{\partial G}{\partial\theta}\right)+\frac{1}{r^2\sin^2\theta}\frac{\partial^2 G}{\partial\varphi^2}=-\frac{8\pi^2\mu}{h^2}E_2\, G$$

...(2.251)

Multiply equation (2.248) by $r^2 \sin^2\theta$ and rearrange:

$$\sin\theta\frac{\partial}{\partial\theta}\left(\sin\theta\frac{\partial G}{\partial\theta}\right)+\frac{\partial^2 G}{\partial\varphi^2}=-\frac{8\pi^2\mu r^2\sin^2\theta}{h^2}E_2\, G$$

or

$$\sin\theta\frac{\partial}{\partial\theta}\left(\sin\theta\frac{\partial G}{\partial\theta}\right)+\frac{8\pi^2\mu r^2\sin^2\theta}{h^2}E_2 G=-\frac{\partial^2 G}{\partial\varphi^2} \qquad \text{...(2.252)}$$

Left hand side of equation (2.247) depends only on the variable θ, whereas the right hand side on the variable φ. If θ is kept fixed, the left hand side of equation (2.247) will remain constant even if we vary φ. Using the principle of separation (section 2.6.3),

$$G(\theta, \varphi) = \Theta(\theta)\,\Phi(\varphi) \qquad \text{...(2.253)}$$

Combining equations (2.248) and (2.247) we get

$$\frac{\sin\theta}{\Theta(\theta)}\frac{d}{d\theta}\left(\sin\theta\frac{\partial\Theta}{\partial\theta}\right)+\frac{8\pi^2\mu r^2\sin^2\theta}{h^2}E_2=-\frac{1}{\Phi(\varphi)}\frac{\partial^2\Phi}{\partial\varphi^2}=m^2$$

Thus

$$\frac{\partial^2\Phi}{\partial\varphi^2}=-m^2\,\Phi(\varphi) \qquad \text{...(2.254)}$$

$$\frac{1}{\sin\theta}\frac{d}{d\theta}\left(\sin\theta\frac{\partial\Theta}{\partial\theta}\right)+\left(\frac{8\pi^2\mu r^2E_2}{h^2}-\frac{m^2}{\sin^2\theta}\right)\Theta=0 \qquad \text{...(2.255)}$$

Solution of Φ (φ) part: Equation (2.249) is a second order linear differential equation and is similar to equation (2.135). The solution of this equation can also be written as:

$$\Phi\ (\varphi) = Ae^{im\varphi} \qquad ...(2.256)$$

where A and m are constants. The only boundary condition is that after full one rotation of the particle, it is at the same position, *i.e.*, $\Phi\ (\varphi + 2\pi) = \Phi\ (\varphi)$. Substituting this in equation (2.251), we get

$$A\ e^{im\varphi} = A\ e^{im(\varphi + 2\pi)} = A\ e^{im2\pi}\ e^{im\varphi}$$

or
$$e^{im2\pi} = 1$$

or
$$\cos 2m\pi + i \sin 2m\pi = 1 \qquad ...(2.257)$$

Equation (2.252) is satisfied only if $m = 0, \pm 1, \pm 2, ...$ No other values of m can satisfy this equation. The value of A can be calculated by normalizing equation (2.251), *i.e.*,

$$1 = \int_0^{2\pi} \Phi^*\ \Phi\ d\varphi = \int_0^{2\pi} Ae^{-im\varphi}\ Ae^{im\phi}\ d\varphi = A^2 \int_0^{2\pi} d\varphi = A^2 2\pi$$

or
$$A = \left(\frac{1}{2\pi}\right)^{\frac{1}{2}}$$

or
$$\Phi\ (\varphi) = \left(\frac{1}{2\pi}\right)^{\frac{1}{2}} e^{\pm im\varphi} \qquad ...(2.258)$$

It is clear from the above solution that except for $m = 0$, all values are doubly degenerate, *i.e.*, the solution is satisfied for $+ m$ and $- m$ values. As stated in section (2.6.4), linear combination of the degenerate wave functions is also an eigen function of the Hamiltonian, with the same eigen value. Thus, the other solutions can be:

$$\Phi\ (\varphi) = \left(\frac{1}{2\pi}\right)^{\frac{1}{2}} \text{ for } m = 0$$

$$\Phi\ (\varphi) = \left(\frac{1}{2\pi}\right)^{\frac{1}{2}} \cos |m|\varphi$$

$$= \left(\frac{1}{2\pi}\right)^{\frac{1}{2}} \sin |m|\varphi \qquad ...(2.259)$$

for
$$|m| = 1, 2, 3,... \quad ..$$

Although both of the solutions are equally probable, the latter solutions are more useful as these are real functions and these will be used more often in hydrogen atom problem.

Example 2.22: Obtain the real form of the wave function $\Phi\ (\varphi) = \left(\frac{1}{2\pi}\right)^{\frac{1}{2}} \cos |m|\varphi$ and $\left(\frac{1}{2\pi}\right)^{\frac{1}{2}} \sin |m|\varphi$ from $\Phi\ (\varphi) = e^{\pm im\varphi}$.

Solution: Since $\Phi(\varphi) = e^{+im\varphi}$ and $\Phi(\varphi) = e^{-im\varphi}$ are solutions of the differential equation (2.251), their linear combination will also be the solutions of equation (2.251). Taking plus combination:

$$e^{+im\varphi} + e^{-im\varphi} = 2 \cos |m|\varphi$$

and taking negative combination and dividing by i :

$$\frac{e^{+im\varphi} - e^{-im\varphi}}{i} = 2 \sin |m|\varphi$$

Solution of Θ (θ) Equation

Equation (2.252) can be written as:

$$\frac{1}{\sin\theta}\frac{d}{d\theta}\left(\sin\theta \frac{\partial\Theta}{\partial\theta}\right) - \frac{m^2}{\sin^2\theta}\Theta + \alpha\,\Theta = 0 \qquad ...(2.252a)$$

where $\alpha = \dfrac{8\pi^2\mu\, r^2 E_2}{h^2}$

This equation can be solved nicely if the following substitution is made.

$$z = \cos\theta,\ 1 - z^2 = \sin^2\theta \text{ and } \Theta(\theta) = Z(z)$$

$$\frac{d\Theta}{d\theta} = \frac{dZ}{dz}\frac{dz}{d\theta} = -\sin\theta\frac{dZ}{dz} \qquad ...(2.260)$$

Equation (2.252a) with the help of (2.257) can be written as:

$$\frac{d}{dz}\left[(1 - z^2)\frac{dZ}{dz}\right] + \left(\alpha - \frac{m^2}{1 - z^2}\right)Z = 0$$

$$(1 - z^2)\frac{d^2Z}{dz^2} - 2z\frac{dZ}{dz} + \left(\alpha - \frac{m^2}{1 - z^2}\right)Z = 0 \qquad ...(2.261)$$

Since θ varies from 0 to π, z will vary from 1 to – 1. Both these points are the regular points. As usual we will first find the boundary conditions. The only difficulty in this case is that some of the terms in equation (2.256) become zero and some become infinite on substituting the limits of the variable z. To solve this problem try the following kind of power series.

$$Z = (1 - z^2)^s\, G(z) \qquad ...(2.262)$$

where s is a constant to be determined and G (z) is a power series.

$$Z' = (1 - z^2)^s G'(Z) - 2sz(1 - z^2)^{s-1} G(z) \qquad ...(2.263)$$

$$Z'' = (1 - z^2)^s G''(Z) - 2sz(1 - z^2)^{s-1}G'(Z) - 2sz(1 - z^2)^{s-1}G'(Z) - 2s(1 - z^2)^{s-1} G(z) + 4s(s - 1)z^2(1 - z^2)^{s-2} G(z) \qquad ...(2.264)$$

$$= (1 - z^2)^s G''(Z) - 4sz(1 - z^2)^{s-1} G'(Z) + [4s(s - 1)z^2(1 - z^2)^{s-2} - 2s(1 - z^2)^{s-1}]\, G(z)$$

Substituting the values of Z, Z′ and Z″ in the following form of (2.256)

$$(1 - z^2)^2 Z'' - 2z(1 - z^2) Z' + [\alpha(1 - z^2) - m^2] Z = 0 \quad ...(2.258a)$$

and after rearranging and dividing by $(1 - z^2)^s$, the following equation is obtained.

$$(1 - z^2)^2 G''(z) - 2z(2s + 1)(1 - z^2) G'(z) + [4s(s - 1) z^2 - 2s(1 - z^2) + 4sz^2 + \alpha(1 - z^2) - m^2] G(z) = 0 \quad ...(2.265)$$

Substituting the value of $z = \pm 1$ in equation (2.261), we gets

$$[4s(s - 1) + 4s - m^2] G(z) = 0$$

or

$$(4s^2 - m^2) G(z) = 0 \quad ...(2.267)$$

For any non-trivial solution of (2.262) the coefficients must be equal to zero, *i.e.*

$4s^2 - m^2 = 0$, which leads to $s = \pm\frac{m}{2}$

s can have both $\frac{m}{2}$ and $-\frac{m}{2}$ values. So we use the absolute value of $s = |\frac{m}{2}|$, in equation (2.259).

$$Z = (1 - z^2)^{\left|\frac{m}{2}\right|} G(z) \quad ...(2.267)$$

Removing the absolute sign for convenience sake, one can write equation (2.263) as:

$$Z = (1 - z^2)^{\frac{m}{2}} G(z) \quad ...(2.268)$$

After differentiating Z (z) twice with respect to z, substituting the values of $Z''(z)$, $Z'(z)$ and Z (z) in equation (2.260), collecting the similar terms and dividing by $(1 - z^2)^{\left|\frac{m}{2}\right|}$ will lead to the following form of equation.

$$(1 - z^2) G''(z) - 2(m + 1) z G'(z) + [\alpha - m(m + 1)] G(z) = 0 \quad ...(2.269)$$

Now using the power series method

$$G(z) = a_0 + a_1 z + a_2 z^2 + a_3 z^3 + ... = \sum_0^\infty a_n z^n \quad ...(2.270)$$

$$G'(z) = a_1 + 2a_2 z + 3a_3 z^2 + ... = \sum_1^\infty n a_n z^{n-1} = \sum_0^\infty n a_n z^{n-1} \quad ...(2.271)$$

$$G''(z) = 2a_2 + 3.2a_3 z + 4.3a_4 z^2 .... = \sum_2^\infty n(n - 1) a_n z^{n-2}$$

$$= \sum_0^\infty (n + 2)(n + 1) a_{n+2} z^n \quad ...(2.272)$$

Substituting the values of G (z), $G'(z)$ and $G''(z)$ in equation (2.264), we get

$$(1 - z^2)\sum_0^{\infty} (n + 2)(n + 1) a_{n+2} z^n - 2(m + 1) z \sum_0^{\infty} n a_n z^{n-1}$$

$$+ [\alpha - m(m + 1)] \sum_0^{\infty} a_n z^n = 0$$

$$\sum_0^{\infty} (n + 2)(n + 1) a_{n+2} z^n - \sum_0^{\infty} (n + 2)(n + 1) a_{n+2} z^{n+2} - 2(m + 1) \sum_0^{\infty} n a_n z^n$$

$$+ [\alpha - m(m + 1) \sum_0^{\infty} a_n z^n = 0$$

$$\sum_0^{\infty} (n + 2)(n + 1) a_{n+2} z^n - \sum_0^{\infty} n(n - 1) a_n z^n - 2(m + 1) \sum_0^{\infty} n a_n z^n$$

$$+ [\alpha - m(m + 1) \sum_0^{\infty} a_n z^n = 0$$

$$\sum_0^{\infty} [(n+2)(n+1) a_{n+2} - n(n-1) a_n - 2n(m+1) a_n + \alpha a_n - m(m+1) a_n] z^n = 0 \quad ...(2.273)$$

To have a non-trivial solution of (2.268), the coefficients of each term must be equal to zero, *i.e.*,

$$(n + 2)(n + 1) a_{n+2} = [n(n - 1) + 2n(m + 1) - \alpha + m(m + 1)] a_n$$

$$a_{n+2} = \frac{n(n - 1) + 2n(m + 1) + m(m + 1) - \alpha}{(n + 2)(n + 1)} a_n$$

$$= \frac{(n + m)(n + m + 1) - \alpha}{(n+2)(n+1)} a_n \quad ...(2.274)$$

Similar to the harmonic oscillator problem, the solution of the Schrödinger wave equation for rigid rotator gives the wave function which consists of either the even powers of z (if the odd coefficients are taken as zero) or the odd powers of z (if the even coefficients are taken as zero). These coefficients can be determined from the recursion formula (2.269). It can be shown that power series obtained with the help of recursion formula does not give a well behaved eigen functions. Thus as has been carried out in case of harmonic oscillator, we must break one of the series after certain number of terms, say $n = k$, for given value of α. The other series can be eliminated by setting either $a_0 = 0$ or $a_1 = 0$, depending on whether k is odd or k is even. This can be done by setting the numerator of (2.270) equal to zero, *i.e.*,

$$\alpha = (k + m)(k + m + 1), \quad k = 0, 1, 2, 3,... \quad ...(2.275)$$

Since m takes on the values of 0, 1, 2, 3... , the quantity $(k + m)$ will have the values of 0, 1, 2, 3,... . We, therefore, define another quantum number J, as

$$J = k + m, \quad J = 0, 1, 2, 3... \qquad ...(2.276)$$

From (2.271) it is clear that the values of m can be either equal to J or less than J, *i.e.*,

$$m \leq J \qquad ...(2.277)$$

Thus the possible values of m can be:

$$m = J, J-1, J-2, ...1, 0, -1, ...-J+1, -J \qquad ...(2.278)$$

Substituting the value of α in equation (2.270)

$$\frac{8\pi^2 \mu r^2 E_J}{h^2} = J(J+1)$$

or

$$E_J = \frac{h^2}{8\pi^2 \mu r^2} J(J+1) \qquad ...(2.279)$$

Rigid rotator problem has revealed the following information:

(*i*) The rotational energy levels are also quantized.

(*ii*) Except for J = 0, the other rotational states are degenerate. The degeneracy of the states is 2 J + 1.

(*iii*) Calculation of rotational wave functions are deferred till hydrogen atom problem.

***Rigid Rotator in a Plane*:** This is the special case of a motion of particle in space. Let us consider the particle in the x y-plane. In such a situation, $\theta = \pi/2$. Consequently equation (2.249) reduces to:

$$\frac{d^2G}{d\varphi^2} = -\frac{8\pi^2 I E}{h^2} G \qquad ...(2.280)$$

which is analogous to the azimuthal equation.

$$\frac{d^2\Phi}{d\varphi^2} = -m^2 \Phi \qquad ...(2.251)$$

Thus the eigen functions of the equation in a plane are $\psi = \Phi_{\pm}(\varphi)$, with the energy eigen values $E = \frac{h^2}{8\pi^2 I} m^2$. These eigen functions of energy are also the eigen functions of $\hat{M}_z$ with eigen values $\pm m \frac{h}{2\pi}$, *i.e.*,

$$\hat{M}_z \psi = \pm m \frac{h}{2\pi} \psi$$

1. PROBLEMS

1. Carry out $\frac{d^2}{dx^2} f(x)$ on the following functions.

   (i) $f(x) = x^3 + 4x^2 + 2x + 8 = 0$ (ii) $f(x) = x^2 + \cos x$
   (iii) $f(x) = C_1\, e^{+\alpha\, x} + C_2\, e^{-\alpha . x}$ (iv) $f(x) = 4x^3$
   (v) $f(x) = A \sin kx + B \cos kx$

2. Calculate $y = \hat{O} f(x)$, where $\hat{O}$ is the operator and function $f(x)$ are given below.

	$\hat{O}$	$f(x)$
(i)	$\int_0^{2\pi} d\varphi$	$C_1 \sin\varphi + C_2 \cos\varphi$
(ii)	$\int_0^{\pi} d\theta$	$A \cos 2\theta$
(iii)	$\frac{\partial^2}{\partial x^2}+\frac{\partial^2}{\partial y^2}+\frac{\partial^2}{\partial z^2}$	$x^4\, y^3\, z^2$

3. Find out whether the following operators are linear or non-linear. (i) √, (square root), (ii) reciprocal, (iii) ln or log, (iv) multiplication by 0 (zero), (v) $\hat{x}^2$, (vi) $\hat{D}$, (vii) $\hat{D}^2$, (viii) cos (ix) $(\ )^2$, (x) $\hat{x}^2\hat{D}^2$, (xi) exp. Calculate the function $f(x)$ at $x = 5$. $f(x) = x^2 + 5x + 6$.

4. Find out whether the following functions are eigen functions of momentum operator $\left(\frac{h}{2\pi i}\frac{\partial}{\partial x}\right)$.

   (i) $f(x) = e^{-\alpha x}$, (ii) $f(x) = e^{-\alpha x^2}$, (iii) $f(x) = A \sin kx + B \cos kx$, (iv) $f(x) = Nxe^{-\alpha x}$.

5. For each function, $f(x)$, an operator is given. Find out whether these functions are eigen functions or not.

   $\hat{O} = 4\frac{d^2}{dx^2}+3\frac{d}{dx}+1$, $f(x) = e^{i\alpha x}$ and $f(x) = A \sin kx$.

   $\hat{O} = \frac{\partial^2}{\partial z^2}$ $f(x) = y^2 \cos kz$ and $f(x) = y^2\, e^{-\alpha x}$

6. Consider a function $f(x) = Ae^{kx}$, where $k = a + ib$. For which values of constants a and b the function is a well behaved function, where x varies between $-\infty$ and $+\infty$

7. Laplacian operator is given as: $\nabla^2 = \frac{\partial^2}{\partial x^2}+\frac{\partial^2}{\partial y^2}+\frac{\partial^2}{\partial z^2}$. Show that $(e^{i\alpha x})\,(e^{i\beta y})\,(e^{i\gamma z})$ is an eigen function of this operator.

8. In an ordinary algebra (A + B) (A – B) = $A^2 - B^2$. In operator algebra if $\hat{A}$ and $\hat{B}$ are operators, calculate $(\hat{A} + \hat{B})(\hat{A} - \hat{B})$. Also find out under what condition the results obtained by operator algebra and ordinary algebra are same.

9. The given relation $\langle \hat{B} + \hat{C} \rangle \neq \langle \hat{B} \rangle + \langle \hat{C} \rangle$. *i.e.*, average value of the sum of the operators is not equal to the sum of the average values of the individual operators.

10. In ordinary algebra, numbers follow the commutative law, whereas in operator algebra it may or may not be. Check that: $(\hat{D} + \hat{x})^2 \neq \hat{D}^2 + 2x\hat{D} + \hat{x}^2$.

11. The operators are given as: (*i*) $\frac{d}{dx} - x$, (*ii*) $x + iy$, (*iii*) $\frac{d}{dx} + x^2$ (*iv*) $\frac{d^2}{dx^2} + x$. Calculate $\hat{O}^2$ and $\hat{O}^3$. [Hint: Before starting operations, do include the function $f(x)$].

12. Calculate the commutator $[\hat{A}\hat{B}]$, where $\hat{A}$ and $\hat{B}$ are:

$\hat{A}$	$\hat{B}$
$\frac{d^2}{dx^2} - x$	$\frac{d}{dx} + x^2$
$\frac{d^2}{dx^2} + x$	$\frac{d}{dx} - x^2$
$\sqrt{}$	$\sqrt{}$
x	p_x
$\frac{d}{dx} - x$	$\frac{d}{dx} + x$
$\frac{d}{dx}$	x^2

13. Show that $\hat{D}\hat{x}$ and $\hat{1} + \hat{x}\hat{D}$ commute with each other.

14. Calculate the commutator between $[z^3 \hat{D}]$.

15. Find out whether $\hat{p}_x^2$ and $(\hat{p}_x)^2$ are equal or not for any arbitrary function.

16. Show that $(\hat{L}_x^2 + \hat{L}_y^2) Y_l^{|m_l|}(\theta, \varphi) = [l(l+1) - m_l^2] \frac{h^2}{4\pi^2} Y_l^{|m_l|}(\theta, \varphi)$

17. Prove that $\hat{L}^2$ commute with $\hat{L}_x, \hat{L}_y$ and $\hat{L}_z$, *i.e.*,$[\hat{L}^2, \hat{L}_x] = [\hat{L}^2, \hat{L}_y] = [\hat{L}^2, \hat{L}_z] = 0$.

18. Although taking general case, it is little difficult to show that $\langle L_x \rangle = \langle L_y \rangle = 0$, but taking few spherical Harmonics, it can be shown that $\langle L_x \rangle = \langle L_y \rangle = 0$

19. Average values of L_x and L_y ($\langle L_x \rangle$ and $\langle L_y \rangle$) are zero. Explain whether these values satisfy Heisenberg Uncertainty Principle.

20. Starting from Cartesian coordinates, show that the values of $\hat{L}_x$ and $\hat{L}_y$ in spherical polar coordinates are:

$$\hat{L}_x = \frac{h}{2\pi i}\left[-\sin\varphi\frac{\partial}{\partial\theta} - \cot\theta\,\cos\varphi\frac{\partial}{\partial\varphi}\right]$$

$$\hat{L}_y = \frac{h}{2\pi i}\left[\cos\varphi\frac{\partial}{\partial\theta} - \cot\theta\,\sin\varphi\frac{\partial}{\partial\varphi}\right]$$

21. Prove that: $[\hat{L}_x, \hat{L}_y] = \frac{ih}{2\pi}\hat{L}_z$, $[\hat{L}_y, \hat{L}_z] = \frac{ih}{2\pi}\hat{L}_x$ $[\hat{L}_z, \hat{L}_x] = \frac{ih}{2\pi}\hat{L}_y$.

22. Why cannot the ground state energy of a particle in a one dimensional box be zero, whereas that of particle in a circular ring can be zero? Explain briefly. Is the Heisenberg Uncertainty Principle valid in both cases?

23. If Ψ is an eigen function to Hamiltonian operator then AΨ is also an eigen function to Hamiltonian operator. A is a constant.

24. Calculate the probability that particle in a one dimensional box of length a is found to be between:

(*i*) 0 to $\frac{a}{4}$, (*ii*) $\frac{a}{4}$ to $\frac{3a}{4}$, and (*iii*) $\frac{5a}{8}$ and $\frac{7a}{8}$.

25. Applying de Broglie particle wave equation to a particle in one dimensional box of length a, calculate the energy of the particle. Note that it agrees with the value obtained by the expression.

26. The average value of any quantity odd or even is given as:

$$\langle x\rangle = \sum_{j=1}^{n} x_j\, P_j = \sum_{j=1}^{n} x_j\, f(x_j)$$

where P_j is the probability of occurrence for the event x and $f(x_j)$ is the probability of realizing the number x. Suppose there are four events. Their values and P_j are given below. Calculate the average value of x.

x	2	3	4	5
P_j	0.15	0.25	0.4	0.2

27. Similarly the average value of the square of the event is defined as:

$$\langle x^2\rangle = \sum_{j=1}^{n} x_j^2\, P_j = \sum_{j=1}^{n} x_j^2\, f(x_j)$$

This is also called second moment of distribution $f(x_j)$ and is similar to moment of inertia. Calculate $\langle x^2\rangle$ using the data of problem 26. Evaluate whether $[\langle x\rangle]^2$ equal to $\langle x^2\rangle$ or not. The answer is not and is a general result, *i.e.*, $[\langle x\rangle]^2 \neq \langle x^2\rangle$.

28. The result of problems 26 and 27 have shown that $[\langle x\rangle]^2 \neq \langle x^2\rangle$. Derive the general expression

$\sigma_x^2 = \langle [x-\langle x\rangle]^2\rangle = \sum_{j=1}^{n} (x_j - \langle x\rangle)^2 f(x_j)$. σ_x^2 is known as the second central moment or the variance (*i.e.*, spread of distribution about its mean, physically better quantity than $\left(\langle x^2\rangle\right)$, where x_j is the value of the jth event. (Hint: Expand $\Sigma\ \left(x_j-\langle x\rangle\right)^2\ f_j$ and apply equations given in problems 26 and 27)

Note: The definitions of $< x >$, $< x^2 >$ and σ_x^2 given in problems 26, 27 and 28 are for events, which have discrete values. The similar definitions for events varying in a continuous manner are:

$$\langle x\rangle = \int_{-\infty}^{+\infty} x\,f(x)\,dx$$

$$\langle x^2\rangle = \int_{-\infty}^{+\infty} x^2\,f(x)\,dx$$

and

$$\sigma_x^2 = \int_{-\infty}^{+\infty} (x-\langle x\rangle)^2\,f(x)\,dx$$

29. General wave function for a particle in one dimensional box of length a is $= \sqrt{\frac{2}{a}}\ \sin\ \frac{n\pi}{a}x$. Calculate the values of $\langle x\rangle$, $\langle x^2\rangle$ and σ^2_x. (Hint: The probability density distribution function is $\frac{2}{a}\ \sin^2\frac{n\pi}{a}x$.

30. Calculate the values of $\langle p_x\rangle, \langle p_x^2\rangle$ and $\sigma^2_{p_x}$ for a particle in one dimensional box. Use the probability density function given in problem 29.

31. If σ_x and σ_{p_x} represent the spread of position and momentum from its mean values respectively (also known as the uncertainty in measurements of the position and momentum) of the particle in one dimensional box, show that $\sigma_x\,\sigma_{p_x}$ gives the value of Heisenberg's Uncertainty Principle.

32. Show that the wave functions $\psi_2 = \sqrt{\frac{2}{a}}\ \sin\frac{2\pi}{a}x$ and $\psi_4 = \sqrt{\frac{2}{a}}\ \sin\frac{4\pi}{a}x$ are orthogonal.

33. Show that the wave functions $\psi_n = \sqrt{\frac{2}{a}}\ \sin\frac{n\pi}{a}x$ and $\psi_m = \sqrt{\frac{2}{a}}\ \sin\frac{m\pi}{a}x$ are orthogonal if $m \neq n$ and are normalized if $n = m$.

34. π-electrons in butadiene can be considered as free electrons moving over the framework of butadiene molecule. If this molecule is assumed to be of length two C = C bonds (0.135 nm each), one C – C bond (0.154 nm) and the radius of carbon atom on both sides (0.077 nm each side of the molecule), calculate the longest wavelength transition. Compare this value with the experimentally known value of $4.61 \times 10^4\ cm^{-1}$ and comment.

35. Use the similar procedure as described in problem 34 to hexatriene and octatetraene to calculate the longest wavelength transition. Compare these values with those determined experimentally (268 nm and 304 nm).

36. β-Carotene consists of 11 double bonds. Using the similar method as explained in problem 34, calculate the value of lowest energy transition and compare it with experimental value (480 nm). (Note: Departure of calculated λ_{max} from the experimental values increases as the number of conjugated double bonds increases)

37. Size of nucleus is 10^{-14} nm approximately. Consider the hydrogen atom (consists of one proton, *i.e.*, one unit of positive charge) and one electron (one unit of negative charge). Will the presence of electron on the periphery of nucleus lead to stable model of hydrogen atom?

38. Similarly the results of problem 37 can be extended to an atom of atomic number Z (*i.e.*, it consists of $+Ze$ as the nuclear charge and presence of Z electrons to balance the nuclear charge to keep the neutral atom). Show that electrons cannot be present in the nucleus.

39. Particle in one dimensional box (section 2.6.2) can also be solved by assuming that particle is moving within the boundaries $-\frac{a}{2}$ to $+\frac{a}{2}$, *i.e.*, the origin is in the middle of the box. Calculate (*a*) energy and (*b*) wave functions of the particle. (*c*) Show that $\sigma_x \sigma_{px} > h/4\pi$. Do you think that the experimentally observable properties depend on the choice of origin of the *x*-axis.

40. Using the results of problem 39, show that as the value of n increases and becomes very large the average value and mean square deviation value agrees with the classical value, *i.e.*, it is uniform throughout the box.

41. In three dimensional box of length a, breadth b and height c, the average value of position of a particle in this box is at a point $\left(\frac{a}{2}, \frac{b}{2}, \frac{c}{2}\right)$.

42. Similar to a particle in one dimensional box, the average value of momentum in two dimensional box is also zero and one can extend it to a three dimensional box. Explain this briefly.

43. Consider two regions of a three dimensional box as (0 to $\frac{a}{4}$, 0 to $\frac{b}{4}$ and 0 to $\frac{c}{4}$) and ($\frac{3a}{4}$ to a, $\frac{3b}{4}$ to b and $\frac{3c}{4}$ to c). Calculate the probability of finding the particle in these regions and comment.

44. Porphyrine is a square molecule of side 0.5 nm and consists of 18π electrons. Considering these electrons to be moving in a two dimensional box, calculate the value of the longest wavelength transition in nm.

45. Consider a cubic box of length 0.02 nm. Calculate the number of quantum states present in the energy region of $\frac{5h^2}{8ma^2}$ to $\frac{24h^2}{8ma^2}$ for an electron.

46. Calculate the probability of a particle in a three dimensional region given as ($\frac{5a}{8}$ to $\frac{7a}{8}$, $\frac{2a}{8}$ to $\frac{5a}{8}$, $\frac{3a}{8}$ to $\frac{5a}{8}$).

47. (*a*) Consider a particle of mass m, confined to two dimensional box of length a and breadth b. Set up Schrödinger wave equation and show that:

$$\psi_{n_x n_y}(x, y) = \sqrt{\frac{4}{ab}} \sin\frac{n_x \pi}{a} x \sin\frac{n_y \pi}{b} y$$

and the corresponding allowed energy levels are:

$$E_{n_x, n_y} = \frac{h^2}{8m}\left[\frac{n_x^2}{a^2} + \frac{n_y^2}{b^2}\right], \text{ where } n_x = n_y = 1, 2, 3\ldots$$

(*b*) Calculate the probability that the particle in the ground state and the first state is found within region $\frac{a}{3}$ to $\frac{2a}{3}$, $\frac{b}{3}$ to $\frac{2b}{3}$.

48. Consider naphthalene molecule, which contains 10-π electrons. Assume that all the carbon atoms lie on a circle of radius 0.01397 nm. Calculate the value of lowest energy transition and compare it with the experimental value of ~ 320 nm.

49. Consider an electron moving on a ring of radius 0.5 nm. Calculate its probability if it is lying between an angle of $\frac{\pi}{3}$ to $\frac{2\pi}{3}$.

50. Consider the following molecules containing four fused rings

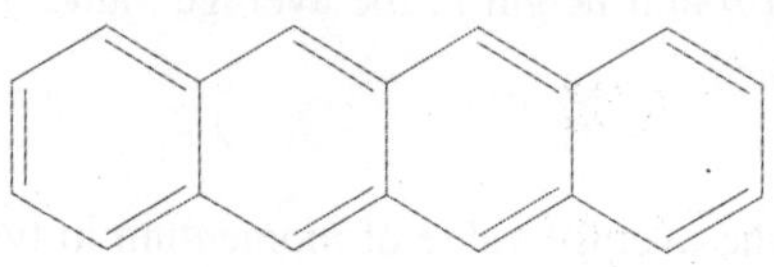

(a)

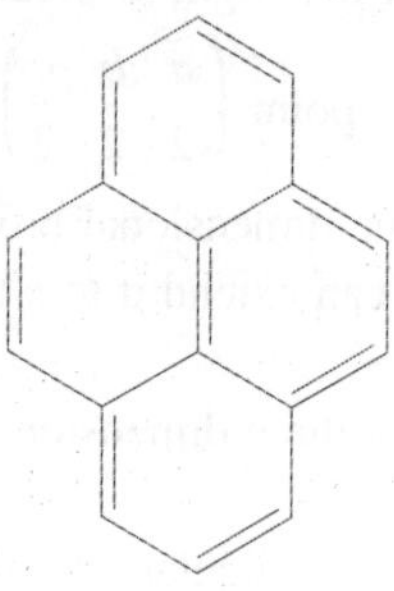

(b)

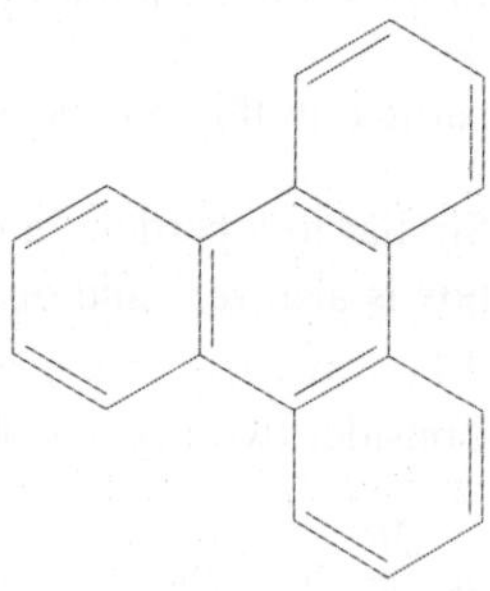

(c)

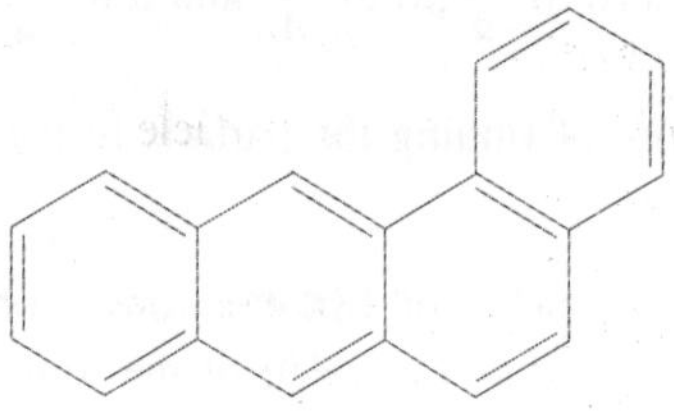

(d)

Assuming that all the carbon atoms lie on a circle of radius for (*a*) 0.242 nm, (*b*) 0.34925 nm, (*c*) 0.354 nm and (*d*) 0.242 nm respectively, calculate the wavelength of the lowest energy transition in each case and compare with the experimental data. Comment if deviation is observed.

51. Calculate the average value of the position $(\langle x \rangle)$ of a particle of mass *m* in the second excited state moving in a one dimensional box of length *a*.

52. Calculate $\Psi_2(x)$ and $\Psi_3(x)$ using equations (2.236) and (2.237). Compare them with those given in Table 2.3 and show that these are orthogonal to each other. Also prove that these are normalized,

53. Calculate $\Psi_5(x)$ using equation (2.236) and plot this function of $\pm x$. Also plot $\psi_5^2(x)$ as function of $\pm x^2$ and compare with that obtained for $\psi_3^2(x)$. Explain why the amplitude square in case of $\psi_5^2(x)$ is smaller in the middle as compared to that of $\psi_3^2(x)$. In general as the value of *v* increases, the amplitude square of the wings is larger than that in the middle.

54. Calculate the value of $\langle x \rangle$ for $\Psi_0(x)$, $\Psi_1(x)$ and $\Psi_2(x)$. Compare these values and comment.

55. Calculate the value of $\langle p_x \rangle$ for $\Psi_0(x)$, $\Psi_1(x)$ and $\Psi_2(x)$.

56. Calculate the value of $\langle x^2 \rangle$ and $\langle p_x^2 \rangle$ as well as σ_x^2 and $\sigma_{p_x}^2$ for the ground state of Harmonic oscillator. Show that $[\langle x^2 \rangle]^{\frac{1}{2}}$ $[\sigma_x^2]^{\frac{1}{2}}$ follows Heisenberg Uncertainty Principle.

57. State the virial theorem and verify for $v = 1$ state of harmonic oscillator. Show that kinetic energy is equal to the potential energy.

58. Show that the Hemite polynomial function is odd when v is odd and even when v is even, whereas square of the Hermit polynomial functions is even whether v is odd or even.

59. Show that the product of two odd or even Hermite polynomial functions will always be even, whereas that of odd and even Hermite polynomial functions is odd.

60. Show that the first derivative of an odd Hermite wave function is even and that of even function is odd.

61. Calculate $\langle x^2 \rangle$ and $\langle p_x^2 \rangle$ for $v = 1$ quantum state. Comparing these values with those for $v = 0$ quantum state. Can we extend these results to make a general case for vth quantum state.

62. Show that $e^{-\alpha x^2}$ is not a asymptotic solution of equation (2.208).

63. Using Hermite polynomials (2.236) verify that the Hermite functions given in Table 2.3.

64. Assuming that $\psi(x) = \text{H}(x)\, e^{-\frac{\alpha x^2}{2}}$ is a complete solution of $\frac{d^2\psi}{dx^2} + (\lambda - \alpha^2 x^2)\psi = 0$, show that the Hermite's differential equation is:

$$\frac{d^2\psi}{dx^2} - 2x\frac{d\psi}{dx} + (\frac{\lambda}{\alpha} - 1)\, \text{H}(x) = 0$$

65. Show that $\text{H}_4(q) = 16q^4 - 48q^2 + 12$ is a solution of the Hermite differential equation.

66. Derive equation (2.238) using equations (2.236) and (2.237).

67. Show that by equating the coefficients of each term in equation (2.218), one can get $(n + 1)\,(n + 2)\, a_{n+2} + [(2n + 1)\,\beta - \alpha]\, a_n = 0$.

CHAPTER

3

Atomic Structure

In the last chapter, we have discussed the simple mathematical formulations of quantum mechanics and have applied these ideas to some simple systems to illustrate their applicability and usefulness. It has been shown earlier that the results predicted by quantum mechanics agree nicely with the experimental ones, for example the results of vibrational motion predicted by wave mechanics and similarly many other problems can be solved by the use of quantum mechanics.

In this chapter, the postulates of quantum mechanics will first be applied to hydrogen atom (one electron and one proton system) not only because its theoretical solution is easy to carry out but it forms the basis for discussion of more complex atoms. For example, it was the spectra of hydrogen atom, which was empirically expressed by the simple relation, called Balmer series or Balmer relation and later on by a more general relation called Rydberg-Ritz relation. Bohr also laid foundation of atomic structure by starting with the mathematical solution of hydrogen atom. Due to these reasons, the ideas of wave mechanics will be applied to hydrogen atom and it will be shown that the wave mechanical treatment of more complex atoms can often be expressed in terms of functions, which are closely related to that of hydrogen atom or similar ions. Moreover, one often finds that quantitative or semi-quantitative wave mechanical treatment to explain many features in chemistry, involves the functions similar to the functions used in hydrogen atom.

3.1 HYDROGEN ATOM (ONE ELECTRON SYSTEM)

Hydrogen atom is the simplest atom in the periodic table. Thus it is easy to handle and solve it using Schrödinger wave equation. The problem of hydrogen like atoms, *i.e.,* an atom or ion consisting of a nucleus and one electron, for example He^+ (Z = 2), Li^{2+} (Z = 3), is a problem of motion of an electron in Coulomb field of the nucleus. A heavy nucleus of charge Ze is assumed to be at rest at the origin of the rectangular coordinate system and electron of mass m_e revolving around it at a distance r under the electrostatic field of attraction of the system (Fig. 3.1). The potential energy (V) of the atom is:

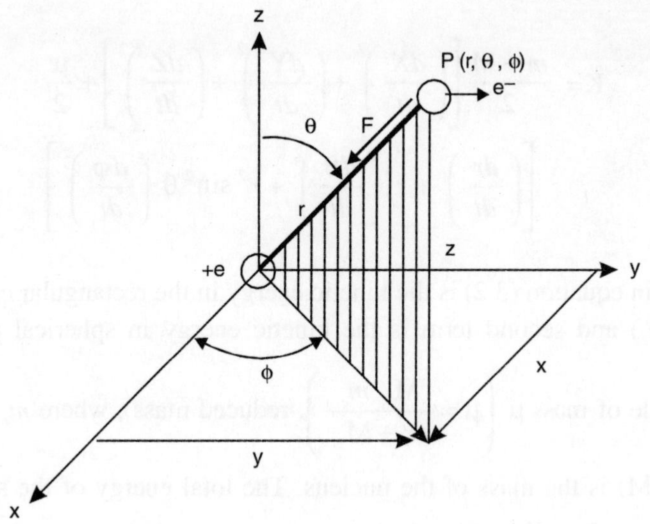

Figure 3.1: *Schematic representation of hydrogen atom. The electron of charge –e at P revolves about the massive proton nucleus of charge +e at the origin O. The force F is the electrostatic attraction of the electron to proton*

$$V = -\frac{Ze^2}{4\pi\varepsilon_0 r} \quad ...(3.1)$$

From quantum mechanical point of view, the electron is represented by a wave system bounded by the potential well of Coulomb field. The wave equation must be represented by three-dimensional space. Since hydrogen like atoms (ions) are in steady state, the time independent Schrödinger wave equation is used to solve the problem. The geometrical symmetry of the problem suggests that it is convenient to handle this problem in spherical polar coordinates rather than in Cartesian coordinates. The rectangular coordinates are related to the polar coordinates by:

$$x = r \sin\theta \cos\varphi$$

$$y = r \sin\theta \sin\theta$$

$$z = r \cos\theta$$

where θ varies from 0 to π, while φ varies from 0 to 2π. The Laplacian operator in Cartesian coordinates $\nabla^2_{x,y,z}$ has already been transformed into its counter part in spherical polar coordinates (2.53) *i.e.,* $\nabla^2_{r,\theta,\varphi}$ is:

$$\frac{\partial^2}{\partial x^2}+\frac{\partial^2}{\partial y^2}+\frac{\partial^2}{\partial z^2} = \frac{1}{r^2}\frac{\partial}{\partial r}\left(r^2\frac{\partial}{\partial r}\right)+\frac{1}{r^2\sin\theta}\frac{\partial}{\partial\theta}\left(\sin\theta\frac{\partial}{\partial\theta}\right)+\frac{1}{r^2\sin^2\theta}\frac{\partial^2}{\partial\varphi^2} \quad ...(2.53)$$

Also recall from section (1.1.2), the total kinetic energy of hydrogen like systems is:

$$K = \frac{m+M}{2}\left[\left(\frac{dX}{dt}\right)^2 + \left(\frac{dY}{dt}\right)^2 + \left(\frac{dZ}{dt}\right)^2\right] + \frac{\mu}{2}$$

$$\left[\left(\frac{dr}{dt}\right)^2 + r^2\left(\frac{d\theta}{dt}\right)^2 + r^2 \sin^2\theta \left(\frac{d\varphi}{dt}\right)^2\right] \qquad ...(3.2)$$

The first term in equation (3.2) is the kinetic energy in the rectangular coordinates of a particle of mass ($M_N + m_e$) and second term is the kinetic energy in spherical polar coordinates of a hypothetical particle of mass μ $\left(\mu = \frac{M_N m_e}{m_e + M_N}\right.$, reduced mass), where m_e (m) is the mass of the electron and M_N (M) is the mass of the nucleus. The total energy of the system is:

$$E = K + V(x, y, z)$$

and the corresponding Hamiltonian is:

$$H = \frac{p^2}{2m} + V(x, y, z) \qquad ...(3.3)$$

The momentum p is actually a composite of radial momentum p_r and Cartesian momentum p_c, *i.e.*, $p = p_r + p_c$. Thus equation (3.3) can be expressed as:

$$H = \frac{p_c^2(x, y, z)}{2(M_N + m_e)} + \frac{p_{r\,(r,\theta,\varphi)}^2}{2\mu} + V(x, y, z) \qquad ...(3.4)$$

Accordingly the Hamiltonian operator is:

$$\hat{H} = -\frac{h^2}{8\pi^2(M_N + m_e)}\left[\frac{\partial^2}{\partial X^2} + \frac{\partial^2}{\partial Y^2} + \frac{\partial^2}{\partial Z^2}\right] - \frac{h^2}{8\pi^2\mu}\left[\frac{1}{r^2}\frac{\partial}{\partial r}\left(r^2\frac{\partial}{\partial r}\right)\right.$$

$$\left. + \frac{1}{r^2 \sin\theta}\frac{\partial}{\partial\theta}\left(\sin\theta\frac{\partial}{\partial\theta}\right) + \frac{1}{r^2\sin^2\theta}\frac{\partial^2}{\partial^2\varphi}\right] - \frac{Ze^2}{4\pi\varepsilon_0 r} \qquad ...(3.5)$$

The Schrödinger wave equation for hydrogen like atoms is:

$$-\frac{h^2}{8\pi^2(M_N + m_e)}\left[\frac{\partial^2}{\partial X^2} + \frac{\partial^2}{\partial Y^2} + \frac{\partial^2}{\partial Z^2}\right]\psi' - \frac{h^2}{8\pi^2\mu}\left[\frac{1}{r^2}\frac{\partial}{\partial r}\left(r^2\frac{\partial}{\partial r}\right)\right.$$

$$\left. + \frac{1}{r^2 \sin\theta}\frac{\partial}{\partial\theta}\left(\sin\theta\frac{\partial}{\partial\theta}\right) + \frac{1}{r^2\sin^2\theta}\frac{\partial^2}{\partial\varphi^2}\right]\psi' - \frac{Ze^2}{4\pi\varepsilon_0\, r}\psi' = E'\psi' \qquad ...(3.6)$$

or $$\frac{h^2}{8\pi^2(M_N+m_e)}\left[\frac{\partial^2}{\partial X^2}+\frac{\partial^2}{\partial Y^2}+\frac{\partial^2}{\partial Z^2}\right]\psi'+\frac{h^2}{8\pi^2\mu}\left[\frac{1}{r^2}\frac{\partial}{\partial r}\left(r^2\frac{\partial}{\partial r}\right).\right.$$

$$\left.+\frac{1}{r^2\sin\theta}\frac{\partial}{\partial\theta}\left(\sin\theta\frac{\partial}{\partial\theta}\right)+\frac{1}{r^2\sin^2\theta}\frac{\partial^2}{\partial\varphi^2}\right]\psi'+\left(E'+\frac{Ze^2}{4\pi\varepsilon_0 r}\right)\psi' = 0 \quad ...(3.7)$$

and wave function ψ' is considered to depend on Cartesian coordinates (X, Y, Z) and polar coordinates (r, θ, φ), *i.e.*,

$$\psi' = \psi'(X, Y, Z, r, \theta, \varphi) \quad ...(3.8)$$

Assume that the equation (3.7) has solutions that may be expressed as product of function of Cartesian and polar coordinates, *i.e.*,

$$\psi' = F(X, Y, Z)\,\psi(r, \theta, \varphi) \quad ...(3.9)$$

This is because the first term of equation (3.7) depends on the Cartesian coordinates (X, Y, Z) and the second term only on polar coordinates (r, θ, φ). Using the expression for functions as expressed in equation (3.9), we can write Schrödinger wave equation (3.7) as:

$$-\frac{h^2}{8\pi^2(M_N+m_e)}\left[\frac{\partial^2}{\partial X^2}+\frac{\partial^2}{\partial Y^2}+\frac{\partial^2}{\partial Z^2}\right]F(X, Y, Z) = E''F(X, Y, Z)$$

or $$\nabla^2_{X,Y,Z}F(X, Y, Z)+\frac{8\pi^2(M_N+m_e)}{h^2}(E'')F(X,Y,Z) = 0 \quad ...(3.10)$$

where E″ is a constant, and

$$-\frac{h^2}{8\pi^2\mu}\left[\frac{1}{r^2}\frac{\partial}{\partial r}\left(r^2\frac{\partial}{\partial r}\right)+\frac{1}{r^2\sin\theta}\frac{\partial}{\partial\theta}\left(\sin\theta\frac{\partial}{\partial\theta}\right)+\frac{1}{r^2\sin^2\theta}\frac{\partial^2}{\partial\varphi^2}\right]\psi(r, \theta, \varphi)$$

$$-\frac{Ze^2}{4\pi\varepsilon_0 r}\psi(r,\theta,\varphi) = E\psi(r,\theta,\varphi)$$

or $$\left[\frac{1}{r^2}\frac{\partial}{\partial r}\left(r^2\frac{\partial}{\partial r}\right)+\frac{1}{r^2\sin\theta}\frac{\partial}{\partial\theta}\left(\sin\theta\frac{\partial}{\partial\theta}\right)+\frac{1}{r^2\sin^2\theta}\frac{\partial^2}{\partial\varphi^2}\right]\psi(r, \theta, \varphi)$$

$$+\frac{8\pi^2\mu}{h^2}\left[\frac{Ze^2}{4\pi\varepsilon_0 r}+E\right]\psi(r, \theta, \varphi) = 0 \quad ...(3.11)$$

where $$E = E' - E''$$

The rectangular coordinate part of the equation (3.10) does not contain potential energy term but only the variable X, Y, and Z and total energy is only the kinetic energy. This is identical to the wave equation for a free particle (2.6.2) and hence represents the translational energy of the atom as a whole. However the polar coordinate part, *i.e.*, equation (3.11) relates the motion of electron bound to the nucleus and is of interest to us. It also allows the separation of variables r, θ, φ. Introducing three independent functions $R(r)$, $\Theta(\theta)$ $\Phi(\varphi)$:

$$\psi = R(r)\ \Theta(\theta)\ \Phi(\varphi) \qquad ...(3.12)$$

We transform this equation to the form:

$$\left[\frac{1}{r^2}\frac{\partial}{\partial r}\left(r^2\frac{\partial}{\partial r}\right)\right] R(r)\,\Theta(\theta)\,\Phi(\varphi) + \left[\frac{1}{r^2\sin\theta}\frac{\partial}{\partial\theta}\left(\sin\theta\frac{\partial}{\partial\theta}\right) + \frac{1}{r^2\sin^2\theta}\frac{\partial^2}{\partial\varphi^2}\right]$$

$$R(r)\,\Theta(\theta)\,\Phi(\varphi) + \frac{8\pi^2\mu}{h^2}\left[\frac{Ze^2}{4\pi\varepsilon_0 r} + E\right] R(r)\,\Theta(\theta)\,\Phi(\varphi) = 0 \qquad ...(3.13)$$

Dividing equation (3.13) by $R(r)\,\Theta(\theta)\,\Phi(\varphi)$ and multiplying by r^2, we obtain

$$\frac{1}{R(r)}\frac{\partial}{\partial r}\left(r^2\frac{dR(r)}{dr}\right) + \frac{1}{\Theta(\theta)\sin\theta}\frac{d}{d\theta}\left(\sin\theta\frac{d\Theta(\theta)}{d\theta}\right) + \frac{1}{\Phi(\varphi)\sin^2\theta}\frac{d^2\Phi}{d\varphi^2}$$

$$+\frac{8\pi^2\mu r^2}{h^2}\left(\frac{Ze^2}{4\pi\varepsilon_0 r} + E\right) = 0 \qquad ...(3.14)$$

Multiply equation (3.14) by $\sin^2\theta$ and rearrange it as follows:

$$\frac{\sin^2\theta}{R(r)}\frac{\partial}{\partial r}\left(r^2\frac{dR(r)}{dr}\right) + \frac{\sin\theta}{\Theta(\theta)}\frac{d}{d\theta}\left(\sin\theta\frac{d\Theta(\theta)}{d\theta}\right) + \frac{8\pi^2\mu r^2\sin^2\theta}{h^2}\left(\frac{Ze^2}{4\pi\varepsilon_0 r} + E\right)$$

$$= -\frac{1}{\Phi(\varphi)}\frac{d^2\Phi}{d\varphi^2} \qquad ...(3.14a)$$

It is clear from equation (3.14*a*) that the right hand side is a function of φ only. If θ and r are kept constant, the left hand side of the equation (3.14*a*) will also be constant. In other words, even if we vary φ, the right hand side of the equation (3.14*a*) will always be a constant quantity and let it be equal to α^2. Thus, we get

$$\frac{1}{\Phi(\varphi)}\frac{d^2\Phi}{d\varphi^2} = -\alpha^2 \qquad ...(3.15)$$

and

$$\frac{\sin^2\theta}{R(r)}\frac{\partial}{\partial r}\left(r^2\frac{dR(r)}{dr}\right) + \frac{\sin\theta}{\Theta(\theta)}\frac{d}{d\theta}\left(\sin\theta\frac{d\Theta(\theta)}{d\theta}\right)$$

$$+\frac{8\pi^2\mu r^2\sin^2\theta}{h^2}\left(\frac{Ze^2}{4\pi\varepsilon_0 r} + E\right) = \alpha^2 \qquad ...(3.16)$$

Divide equation (3.16) by $\sin^2\theta$ and rearrange the resultant equation as follows and equate it to λ:

$$\frac{1}{R(r)}\frac{\partial}{\partial r}\left(r^2\frac{dR(r)}{dr}\right)+\frac{8\pi^2\mu\, r^2}{h^2}\left(\frac{Ze^2}{4\pi\varepsilon_0 r}+E\right)$$

$$= -\frac{1}{\Theta(\theta)\sin\theta}\frac{d}{d\theta}\left(\sin\theta\frac{d\Theta(\theta)}{d\theta}\right)+\frac{\alpha^2}{\sin^2\theta}=\lambda \qquad ...(3.16a)$$

Right hand side of equation (3.16*a*) is only a function of θ, whereas the left hand side is of function of r only. Thus the equation (3.16*a*) has been equated to a constant λ. This we may write as:

$$-\frac{1}{\Theta(\theta)\sin\theta}\frac{d}{d\theta}\left(\sin\theta\frac{d\Theta(\theta)}{d\theta}\right)+\frac{\alpha^2}{\sin^2\theta}=\lambda \qquad ...(3.17)$$

and

$$\frac{1}{R(r)}\frac{d}{dr}\left(r^2\frac{dR(r)}{dr}\right)+\frac{8\pi^2\mu r^2}{h^2}\left(\frac{Ze^2}{4\pi\varepsilon_0 r}+E\right)=\lambda \qquad ...(3.18)$$

This can be written as:

$$\frac{d}{dr}\left(r^2\frac{dR(r)}{dr}\right)+\frac{8\pi^2\mu r^2}{h^2}\left(\frac{Ze^2}{4\pi\varepsilon_0 r}+E\right)R(r)=\lambda R(r)$$

or

$$r^2\frac{d^2R(r)}{dr^2}+2r\frac{dR(r)}{dr}+\frac{8\pi^2\mu r^2}{h^2}\left(\frac{Ze^2}{4\pi\varepsilon_0 r}+E\right)R(r)=\lambda R(r)$$

Divide by r^2 and rearrange:

$$\frac{d^2R(r)}{dr^2}+\frac{2}{r}\frac{dR(r)}{dr}+\left(\frac{8\pi^2\mu E}{h^2}+\frac{8\pi^2\mu Ze^2}{4\pi\varepsilon_0 rh^2}-\frac{\lambda}{r^2}\right)R(r)=0 \qquad ...(3.18a)$$

where the constants α and λ are to be determined. The Schrödinger wave equation (3.14) in terms of three variables r, θ, φ for hydrogen like atoms has now been subdivided into three parts, each one being dependent only on one of the variables, and in the process two constants have been introduced into the problem. The three sub-equations can also be rewritten as:

Azimuthal part is

$$\frac{d^2\Phi}{d\varphi^2}=-\alpha^2\,\Phi(\varphi) \qquad ...(3.19)$$

Polar part is

$$\frac{1}{\sin\theta}\frac{d}{d\theta}\left(\sin\theta\,\frac{d\Theta(\theta)}{d\theta}\right)+\left[\lambda-\frac{\alpha^2}{\sin^2\theta}\right]\Theta\,(\theta)=0 \qquad ...(3.20)$$

Radial part is:

$$\frac{d^2R(r)}{dr^2}+\frac{2}{r}\frac{dR(r)}{dr}+\left[A^2+\frac{2\,B}{r}-\frac{\lambda}{r^2}\right]R\,(r)=0 \qquad ...(3.21)$$

where $A^2 = \dfrac{8\pi^2\mu E}{h^2}$ and for brevity $B = \dfrac{4\pi^2\mu Ze^2}{4\pi\varepsilon_0\, h^2}$

3.1.1 SOLUTION OF AZIMUTHAL EQUATION

This equation describes the behaviour of the wave function as the electron rotates about *z*-axis. This form of equation (3.19) is similar to that of the motion of the particle moving in a ring, which had among the solutions, the two real and one complex periodic functions.

$$\begin{aligned}\Phi(\varphi) &= C\,\sin\,\alpha\varphi\\ \Phi(\varphi) &= C\,\cos\,\alpha\varphi\\ \Phi(\varphi) &= C\,e^{i\alpha\varphi}\end{aligned} \qquad ...(3.22)$$

This set of functions should also be the solution of the azimuthal equation (3.19). This can be proved as follows, *i.e.,* differentiating twice any of the functions in set (3.22) with respect to φ, we obtain:

$$\frac{d\Phi(\varphi)}{d\varphi} = C\,i\,\alpha\,e^{i\alpha\varphi}$$

$$\frac{d^2\Phi(\varphi)}{d\varphi^2} = C\,(i\,\alpha)^2 e^{i\alpha\varphi} = -\,\alpha^2\,C\,e^{i\alpha\varphi} = -\,\alpha^2\,\Phi(\varphi)$$

Hence it gives the desired equation. The function Φ (φ) must belong to the class of well behaved functions, *i.e.*, it should be a single valued function. On physical grounds, as discussed earlier (2.6.5, p. 138), when the electron completes one circle about the *z*-axis, the solution Φ (φ) must have the same value when angle φ is reached again, *i.e.*, φ + 2απ. This condition will be satisfied by the set of functions provided α is an integral multiple of 2π. Since φ is in radians, α must be an integral value. Further this includes both zero and rotation in opposite directions. The only allowed values of α, are:

$$\alpha = 0, \pm 1, \pm 2, \pm 3, \pm 4..... \qquad ...(3.23)$$

In this case, the condition of integral values of α is followed provided:

$$e^{i\alpha\varphi} = e^{i\alpha(\varphi + 2\pi)},$$

i.e., $\Phi(\varphi) = \Phi(\varphi + 2\pi\alpha)$

The characteristics of α are those of a quantum number and particular restriction on it suggests that it is an analogue of angular momentum quantum number m_φ of the Bohr-Sommerfeld model. As we will see later on that α is called magnetic quantum number and is denoted by m_l. The value of α/m_l as its square will give the eigen values proper to various eigen functions from set of equations (3.22) which are in turn the permitted solutions of azimuthal equation. Condition for normalization is:

$$\int_0^{2\pi} \Phi^*(\varphi)\, \Phi(\varphi)\, d\varphi = 1 \text{ or } \int_0^{2\pi} C^2\, e^{\pm i m_l \varphi}\, e^{\mp i m_l \varphi}\, d\varphi = 1$$

$$\int_0^{2\pi} C^2\, e^0\, d\varphi = 1 \text{ or } C^2\, [\varphi]_0^{2\pi} = 1$$

Finally $$C = \left(\frac{1}{2\pi}\right)^{\frac{1}{2}}$$

The normalized azimuthal wave function is:

$$\Phi(\varphi) = \left(\frac{1}{2\pi}\right)^{\frac{1}{2}} e^{i m_l \varphi} \qquad ...(3.24)$$

where $m_l = 0, \pm 1, \pm 2, \pm 3, \pm 4...$.

Normalized azimuthal wave functions, $\Phi(\varphi)$ for hydrogen like atoms are listed in Table 3.1.

Table 3.1: *Normalized azimuthal wave functions* $\Phi_{ml}(\varphi)$ *or hydrogen like atoms* $(\Phi_{m_l} = (1/2\pi)^{1/2}\, e^{i m_l \varphi})$

α/m_l	$\Phi_{m_l}(\varphi)$	
0	$\Phi_0(\varphi) = \frac{1}{\sqrt{2\pi}}$	
+1	$\Phi_{+1}(\varphi) = \frac{1}{\sqrt{2\pi}}\, e^{i\varphi}$	$\Phi_{+1}(\varphi) + \Phi_{-1}(\varphi) = \frac{1}{\sqrt{2\pi}}\cos\varphi$
−1	$\Phi_{-1}(\varphi) = \frac{1}{\sqrt{2\pi}}\, e^{-i\varphi}$	$\Phi_{+1}(\varphi) - \Phi_{-1}(\varphi) = \frac{1}{\sqrt{2\pi}}\sin\varphi$
+2	$\Phi_{+2}(\varphi) = \frac{1}{\sqrt{2\pi}}\, e^{2i\varphi}$	$\Phi_{+2}(\varphi) + \Phi_{-2}(\varphi) = \frac{1}{\sqrt{2\pi}}\cos 2\varphi$
−2	$\Phi_{-2}(\varphi) = \frac{1}{\sqrt{2\pi}}\, e^{-2i\varphi}$	$\Phi_{+2}(\varphi) - \Phi_{-2}(\varphi) = \frac{1}{\sqrt{2\pi}}\sin 2\varphi$

$\sin x = \frac{e^{ix} - e^{-ix}}{2i}$, $\cos x = \frac{e^{ix} + e^{-ix}}{2}$, $e^{2in\pi} = 1$, The complex circular functions are all periodic. Since the azimuthal probability per unit volume P_φ is $\Phi^*_{ml}(\varphi)\,\Phi_{ml}(\varphi) = (C e^{iml\varphi})(C^* e^{-im_l\varphi}) = C\,C^*$ = constant, the azimuthal probability density is independent of φ, the azimuthal angle.

Example 3.1. (*a*) Show that the superposition of the wave functions C cos $\alpha\varphi$ and C sin $\alpha\varphi$ is also a solution of the azimuthal wave function and (*b*) determine the normalization constant of the superposition wave function.

Solution: The superposition wave function is:

$$\Phi(\varphi) = C\cos\alpha\varphi + C\sin\alpha\varphi$$

In order that this wave function is a solution of the azimuthal equation, the second derivative of $\Phi(\varphi)$ should give rise to an expression that must conform to the azimuthal equation, *i.e.*,

$$\frac{d\Phi(\varphi)}{d\varphi} = -C\,\alpha\sin\alpha\varphi + C\,\alpha\cos\alpha\varphi$$

$$\frac{d^2\Phi(\varphi)}{d\varphi^2} = -C\alpha^2\cos\alpha\varphi - C\alpha^2\sin\alpha\varphi = \alpha^2[C\cos\alpha\varphi + C\sin\alpha\varphi]$$

$$= -\alpha^2\Phi(\varphi)$$

or $$\frac{d^2\Phi(\varphi)}{d\varphi^2} + \alpha^2\Phi(\varphi) = 0$$

which is the expected azimuthal equation.

(*b*) The condition for normalization of azimuthal wave function is:

$$\int_0^{2\pi} \Phi^*(\phi)\,\Phi(\varphi)\,d\varphi = 1$$

$\therefore$ $$\int_0^{2\pi} [(C\sin\alpha\varphi + C\cos\alpha\varphi)^2\,d\varphi] = 1$$

$$\int_0^{2\pi} [(C^2\sin^2\alpha\varphi + C^2\cos^2\alpha\varphi + 2C^2\sin\alpha\varphi\cos\alpha\varphi)\,d\varphi] = 1$$

$$C^2\int_0^{2\pi} [(1 + 2\sin\alpha\varphi\cos\alpha\varphi)\,d\varphi] = 1$$

Integrating and applying limits

$$C^2[\varphi]_0^{2\pi} + 0 = 1 \text{ or } C^2\,2\pi = 1$$

or $$C = \left(\frac{1}{2\pi}\right)^{\frac{1}{2}}$$

Accordingly the normalized superposition azimuthal function is:

$$\Phi\ (\varphi) = \left(\frac{1}{2}\right)^{\frac{1}{2}}\ [\cos\alpha\varphi + \sin\alpha\varphi]$$

Example 3.2: Form the real functions from the functions defined by the equation $\Phi(\varphi) = \left(\frac{1}{2\pi}\right)^{\frac{1}{2}} e^{im_l\varphi}$ for $m_l = \pm 2$ and normalize if these are not.

Solution: Substitute the values of m_l in the above equation.

$$\Phi_2(\varphi) = \left(\frac{1}{2\pi}\right)^{\frac{1}{2}} e^{i2\varphi}$$

$$\Phi_{-2}\ (\varphi) = \left(\frac{1}{2\pi}\right)^{\frac{1}{2}} e^{-2i\varphi}$$

Taking linear combinations, we get

$$\Phi_+(\varphi) = \left(\frac{1}{2\pi}\right)^{\frac{1}{2}} [e^{i2\varphi} + e^{-2i\varphi}]$$

$$= \left(\frac{1}{2\pi}\right)^{\frac{1}{2}} [\cos 2\varphi + i\sin 2\varphi + \cos 2\varphi - i\sin 2\varphi]$$

$$= \frac{2}{\sqrt{2\pi}}\cos 2\varphi = \left(\frac{1}{2\pi}\right)^{\frac{1}{2}}\cos 2\varphi$$

Similarly

$$\Phi_i\ (\varphi) = \left(\frac{1}{2\pi}\right)^{\frac{1}{2}} \frac{1}{i}[e^{i2\varphi} - e^{-2i\varphi}]$$

$$= \left(\frac{1}{2\pi}\right)^{\frac{1}{2}} [\cos 2\varphi + i\sin 2\varphi - \cos 2\varphi + i\sin 2\varphi]$$

$$= \frac{2}{\sqrt{2\pi}}\sin 2\varphi = \left[\frac{2}{\pi}\right]^{\frac{1}{2}}\sin 2\varphi$$

These wave functions may be normalized or may not be. Using normalizing relation:

$$\int_0^{2\pi} \Phi_+^*(\phi)\,\Phi_+(\varphi)\,d\varphi = \int_0^{2\pi} C\left(\frac{2}{\pi}\right)^{1/2}\cos 2\varphi.\, C\left(\frac{2}{\pi}\right)^{1/2}\cos 2\varphi\, d\varphi$$

where C is the normalization constant.

$$= C^2 \int_0^{2\pi} \frac{2}{\pi} \cos^2 2\varphi \, d\varphi = C^2 \frac{2}{\pi} \int_0^{2\pi} \frac{1 + \cos 4\varphi}{2} d\varphi$$

$$= \frac{C^2}{\pi} [\varphi]_0^{2\pi} + \frac{C^2}{\pi} \left[\frac{\sin 4\varphi}{4} \right]_0^{2\pi}$$

$$= \frac{C^2}{\pi} 2\pi + 0 = 2C^2$$

So $2C^2 = 1$ and that gives $C = \frac{1}{\sqrt{2}}$, taking positive value. So the normalized wave function is:

$$\Phi_+ = \left(\frac{1}{\pi}\right)^{\frac{1}{2}} \cos 2\varphi$$

Similarly it can also be shown that the normalized wave function for Φ_- is

$$\Phi_- = \left(\frac{1}{\pi}\right)^{\frac{1}{2}} \sin 2\varphi$$

3.1.2 SOLUTION OF POLAR EQUATION (*i.e.*, Θ PART)

The polar equation (3.20) describes the behaviour of the wave function in polar angle θ about the origin and can be written as:

$$\frac{1}{\sin\theta} \frac{d}{d\theta} \left(\sin\theta \frac{d\Theta(\theta)}{d\theta} \right) + \left[\lambda - \frac{m_e^2}{\sin^2\theta} \right] \Theta(\theta) = 0 \qquad \text{...(3.25)}$$

In this equation (3.25), α^2 has been replaced by m_l^2, the value obtained from the solution of azimuthal equation (3.19). This equation is an analogue of Θ (θ) equation (2.258) of rigid rotator. The value of λ, obtained as J(J + 1) there, can be written as $l(l + 1)$ where l stands for orbital angular momentum quantum number. The solution of equation (3.25) becomes easy if it is cast into the Associated Legendre polynomial form. The transformation from Θ to Z form is carried out by making the following substitutions:

$z = \cos\theta$, $1 - z^2 = \sin^2\theta$, $\Theta(\theta) = Z(z)$ and $\frac{d\Theta}{d\theta} = \frac{dZ}{dz}\frac{dz}{d\theta} = -\sin\theta \frac{dZ}{dz}$ into Θ (θ) equation (3.25). Since θ varies from 0 to π, z will vary from +1 to –1. Accordingly

$$\frac{d}{dz}\left[(1-z^2)\frac{dZ}{dz}\right] + \left[l(l+1) - \frac{m_l^2}{1 - z^2}\right] Z = 0$$

or $$(1-z^2)\frac{d^2Z}{dz^2}-2z\frac{dZ}{dz}+\left[\ell(\ell+1)-\frac{m_\ell^2}{1-z^2}\right]Z = 0 \qquad ...(3.26)$$

This is a Legendre polynomial of degree l and order m_l. Since m_l appears in this equation as m_l^2, so $\Theta_{l,m_l} = \Theta_{l,-m_l}$. The normalized wave functions are therefore expressed as:

$$\Theta_{lm_l}(\theta) = \left[\frac{(2l+1)\,(l-|m_l|)!}{2\,(l+|m_l|)!}\right]^{\frac{1}{2}} P_l^{|m_l|}(\cos\theta) \qquad ...(3.27)$$

where m_l stands for the absolute value of m_l, which is an integer and $P_l^{m_l}(\cos\theta)$ is:

$$P_l^{|m_l|}(\cos\theta) = \frac{(-1)^l}{2^l\, l!}(1-\cos^2\theta)^{|m_l|/2}\frac{d^{l+|m_l|}}{d\cos\theta^{l+|m_l|}}(\cos^2\theta-1)^l \qquad ...(3.28)$$

Equation (3.27) will have an acceptable solution provided $l - m_l$ is also an integer. This is only acceptable if l is also an integer. It has already been shown in section (2.6.8) that the values of J/l are only integers and the values of $m_j/m_l \leq J/l$.

The limits on values of m_l can be looked from different angle now. Since m_l can take on only positive, negative and zero values, from the known properties of Legendre polynomials as applied to hydrogen like systems (Fig. 3.1), where the polar angle may take on values from 0 to π, the solutions given by equation (3.27) [*i.e.*, $P_l^{|m_l|}(\cos\theta)$] exists only when l is an integer and when it is equal to or greater than the absolute value of m_l. This imposes a condition on the second quantum number, *i.e.*, the orbital angular momentum quantum number l. A summary of the situation up to this point is that:

$$l = 0, 1, 2, 3\ldots$$

$$m_l = 0, \pm 1, \pm 2, \pm 3\ldots \quad ..\pm l \qquad ...(3.29)$$

In general, for any given value of l, there are $(2l + 1)$ possible values of m_l. Thus a state corresponding to a given value of l will have $(2l + 1)$ degenerate states in m_l. As an example, for $l = 2$, the possible number of degenerate states in $m_l = 2 \times 2 + 1 = 5$, corresponds to –2, –1, 0, +1, +2. When magnetic filed is applied to the hydrogen atom, the eigen values are separated, *i.e.*, the degeneracy of the states in m_l gets removed and hence m_l takes on the status of magnetic quantum number.

The polar probability density function, *i.e.*, θ dependent probability density function is given by:

$$P_\theta = \Theta^*_{l,m_l}\,\Theta_{l\,m_l} \qquad ...(3.30)$$

Since $\Theta_{l\,m_l}$ depends on l, m_l, as well as on θ, effect of P_θ is different for various states as determined by the quantum numbers l and m_l (Table 3.2). Because $\Theta^*_{l,m_l}\,\Theta_{l m_l}$ is independent of φ, a plot of $P_\theta = \Theta^*_{l,m_l}\,\Theta_{l\,m_l}$ versus θ for various states, as determined by quantum numbers l and m_l, in each case exhibits a symmetry of revolution with respect to the z-axis. In view of the above discussion, the normalized polar wave functions, as deduced from equations (3.27) and (3.28) have been compiled in Table 3.2.The i.ıformation compiled in Table 3.2 reveals that for:

(1) $l = 0$, $m_l = 0$ state, the plot of P_θ versus θ in polar coordinates is a sphere of radius 1/2. As shown in Fig. 3.2*a*, the straight line from the origin to the curve has a length equal to $\Theta^*_{l,m_l}\,\Theta_{l\,m_l}$ and represents the probability that the electron is in the direction of that line.

(2) $l = 1$, $m_l = 0$ state, the plot of P_θ against θ is shown in, Fig. 3.2*b*. The plot shows two lobes with a maximum at $\theta = 0$ and minimum at $\theta = \pm\frac{\pi}{2}$. These lobes are symmetrical about z-axis.

(3) $l = 1$, $m_l = \pm 1$ states, the plot between P_θ versus θ again shows two lobes but symmetrical about y-axis and x-axis respectively, Fig. 3.2 *c*. The probability is maximum along y-axis when $\theta = \pm\frac{\pi}{2}$ and zero in the vertical direction when $\theta = 0$ or π.

(4) $l = 2$, $m_l = 0$, the plot between P_θ and θ shows two big lobes symmetrical about z-axis and two small lobes symmetrical to y-axis, Fig. 3.2*d*. The ratio of maximum probability when $\theta = 0$ or π and when $\theta = \pm\frac{\pi}{2}$

$$= \frac{\frac{5}{8}(3\cos^2 0 - 1)^2}{\frac{5}{8}(3\cos^2 \pi/2 - 1)^2} = \frac{2.5}{0.625} = 4$$

The probability is zero at the origin, *i.e.*, at the point of intersection of two types of lobes.

(5) $l = 2$, $m_l = \pm 1$ state, the plot of P_θ against θ shows four lobes, two symmetrical about the yz-line and two about the yz-line Fig. 3.2*e*. The probability is maximum when $\theta = 45°$, 135°, 225° and 315° respectively and zero when $\theta = \pm\frac{\pi}{2}$ and 0 and π.

(6) $l = 2$, $m_l = \pm 2$ state, the plot between P_θ and θ will consist of two lobes symmetrical about y-axis, Fig. 3.2 *f*. The probability will be maximum when $\theta = 0$ and π and minimum

when $\theta = \frac{\pi}{2}$

Table 3.2: *The normalized polar wave functions and polar probability density function for hydrogen like atoms*

State*	l,	m_l	$\Theta_{l\,m_l}$	$P_\theta = \Theta^*_{l,m_l}\,\Theta_{\ell\,m\ell}$
s	0	0	$\Theta_{0,\,0} = \left(\frac{1}{2}\right)^{\frac{1}{2}}$	$= \frac{1}{2}$
p	1	0	$\Theta_{1,\,0} = \left(\frac{3}{2}\right)^{\frac{1}{2}} \cos\theta$	$= \frac{3}{2}\cos^2\theta$
	1	±1	$\Theta_{1,\,\pm 1} = \left(\frac{3}{4}\right)^{\frac{1}{2}} \sin\theta$	$= \frac{3}{4}\sin^2\theta$
d	2	0	$\Theta_{1,\,0} = \left(\frac{5}{8}\right)^{\frac{1}{2}} (3\cos^2\theta - 1)$	$= \frac{5}{8}(3\cos^2\theta - 1)^2$
d	2	± 1	$\Theta_{2,\,\pm 1} = \left(\frac{15}{4}\right)^{\frac{1}{2}} \sin\theta\cos\theta$	$= \frac{15}{4}\sin^2\theta\cos^2\theta$
d	2	± 2	$\Theta_{2,\,\pm 2} = \left(\frac{15}{16}\right)^{\frac{1}{2}} \sin^2\theta$	$= \frac{15}{6}\sin^4\theta$

The number of states is taken over from the nomenclature of optical spectroscopy, because, it was utilized by the earlier workers in atomic spectra. The nomenclature is based on the values of *l*. In atomic spectroscopy, the letters, *s*, *p*, *d* and *f* stands for the sharp, principle, diffuse and fundamental spectroscopic series, whereas the other letters (like *g*, *h*, etc. ...) are the continuation of the alphabets after *f*.

The Bohr's theory does not provide any evidence to the location of electrons in the positions shown in Fig. 3.2. The probability density distribution for hydrogen like atoms in three-dimensional space will be discussed by inducting radial probability distribution and deferred till the solution of radial equation.The normalized spherical harmonics, *i.e.*, the product of two wave functions, dependent on angles θ and φ, $Y_{l\,m_l}(\theta\;\varphi)$ can be obtained by multiplying Θ^*_{l,m_l} by

$\Phi_{m_l}(\varphi) = \left(\frac{1}{2\pi}\right) e^{im_l\varphi}$, *i.e.*,

$$Y_{l\,m_l}(\theta\;\varphi) = \Theta^*_{l,m_l}\;\Phi_{m_l}(\varphi) \qquad ...(3.31)$$

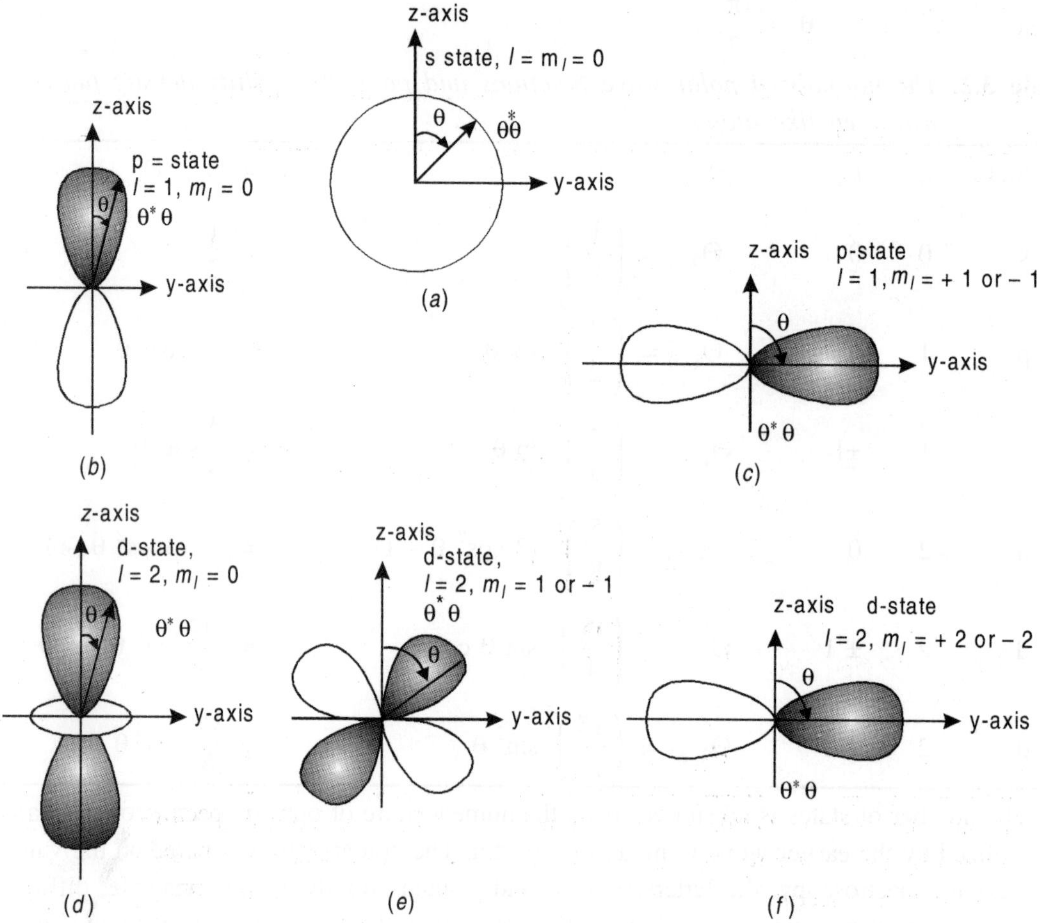

Figure 3.2: *Probability density $P_\theta = \Theta^*_{l,m_l}\ \Theta_{l\,m_l}$ versus θ. Since P_θ is independent of φ there is a symmetry of revolution about z-axis*

This total wave function is a solution of the Schrödinger wave equation for rigid rotator in space. It may be mentioned here that spherical harmonics, $Y_{l\,m_l}(\theta\ \varphi)$ form orthonormal set with respect to $\sin\theta\ d\theta\ \theta\varphi$ and not just $d\theta\ d\varphi$. The quantity $\sin\theta\ d\theta\ d\varphi$ possesses the following physical meaning. We know that in spherical polar coordinates, the small volume element is represented by $r^2\ \sin\theta\ dr\ d\theta\ d\varphi$. If we consider r as constant, as assumed in case of rigid rotator, and set equal to unity for convenience sake, then the spherical coordinate volume element becomes a surface element, $dA = \sin\theta\ d\theta\ d\varphi$. If we integrate this quantity, it becomes 4π, *i.e.,* surface area of unit radius. Thus $\sin\theta\ d\theta\ d\varphi$ is a surface area element on the surface of sphere of unit radius. These wave functions satisfy simultaneously the equations

$$\hat{M}^2\ Y_{l\,m_l}(\theta\ \varphi)\ =\ l\,(l+1)\ \frac{h^2}{4\pi^2}\ Y_{l\,m_l}(\theta\ \varphi) \qquad ...(3.32)$$

$$\hat{M}_z\, Y_{l\, m_l}(\theta\ \varphi) = \pm \frac{h}{2\pi}\, Y_{l\, m_l}(\theta\ \varphi) \qquad ...(3.33)$$

where the terms and symbols have their usual meanings. Thus, these functions are:

(*i*) Eigen functions of the Hamiltonian operator with eigen values for the energy $E = \frac{h^2}{8\pi^2 \mu r^2}\, l\,(l + 1)$ for hydrogen like atoms and $E = \frac{h^2}{8\pi^2 \mu r^2}\, J(J + 1)$ for rigid rotator.

(*ii*) Eigen functions of the total angular momentum operator with eigen values

$$\mathbf{M^2} = l(l+1)\ \frac{h^2}{4\pi^2}$$

and

(*iii*) Eigen functions of the *z*-component of the angular momentum with eigen values,

$$\mathbf{M_z} = \pm \frac{h}{2\pi} m_l$$

Example 3.3: Calculate the normalized wave function Θ for $l = 2$, $m_l = 2$.

Solution: The normalized wave function Θ (θ) can be obtained by substituting the values in the equation (3.27), *i.e.*,

$$\Theta\ (\theta) = \left[\frac{(4\ +\ 1)\ (2\ -\ 2)!}{2\ (4)!}\right]^{\frac{1}{2}} P_2^2\ (\cos\theta) = \left[\frac{5}{48}\right]^{\frac{1}{2}} P_2^2\ (\cos\theta)$$

Now using equation (3.28)

$$P_2^2\ (\cos\theta) = \frac{(-1)^2}{2^2\ 2!}(1-\cos^2\theta)^{\frac{2}{2}} \frac{d^4}{d(\cos\theta)^4}[(\cos^2\theta - 1)^2]$$

$$= \frac{1}{8}(\sin^2\theta)\frac{d^4}{d(\cos\theta)^4}[\cos^4\theta - 2\cos^2\theta + 1] = 3\sin^2\theta$$

Thus $$\Theta\ (\theta) = \left[\frac{5}{48}\right]^{\frac{1}{2}} 3\sin^2\theta = \left[\frac{15}{16}\right]^{\frac{1}{2}} \sin^2\theta$$

Note: To make differentiation easy, substitute cos θ = *z*.

Example 3.4: Show that the wave function obtained in example 3.3 is normalized.

Solution: The wave function is normalized if:

$$\int_0^{\pi} \Theta^*(\theta)\ \Theta(\theta)\ \sin\theta\ d\theta = 1$$

$$i.e., \int_0^\pi \left(\frac{15}{16}\right)^{\frac{1}{2}} \sin^2\theta \left(\frac{15}{16}\right)^{\frac{1}{2}} \sin^2\theta \sin\theta \, d\theta = \frac{15}{16}\int_0^\pi \sin^5\theta \, d\theta$$

Using the relation: $\int \sin^n x \, dx = -\frac{\sin^{n-1} x \cos x}{n} + \frac{n-1}{n}\int \sin^{n-2} x \, dx$, we get

$$\frac{15}{16}\int_0^\pi \sin^5\theta \, d\theta = \frac{15}{16}\left[-\frac{\sin^4\theta\cos\theta}{5}\right]_0^\pi + \frac{15}{16}\times\frac{4}{5}\int_0^\pi \sin^3\theta \, d\theta = \frac{3}{4}\int_0^\pi \sin^3\theta \, d\theta$$

$$= \frac{3}{4}\left[-\frac{\sin^2\theta\cos\theta}{3}\right]_0^\pi + \frac{3}{4}\times\frac{2}{3}\int_0^\pi \sin\theta \, d\theta = \frac{1}{2}\int_0^\pi \sin\theta \, d\theta$$

$$= \frac{1}{2}[-\cos\theta]_0^\pi = -\frac{1}{2}(-1-1) = 1$$

i.e., wave function given in example 3.3 is normalized.

Certain useful relations for the Legendre polynomials are:

(*i*) Orthonormal relations for the Legendre polynomials are:

0 when $l \neq l'$

$$\int_{-l}^{+l} P_{l'}^{|m_l|}(z)\, P_l^{|m_l|}(z)\, dz = \frac{2}{2l+1}\times\frac{(l|m_l|)!}{(l-|m_l|)!} \qquad ...(3.33a)$$

(*ii*) Recursion relation is:

$$z\, P_l^{|m_l|}(z) = \left(\frac{l+|m_l|}{2l+1}\right) P_{l-1}^{|m_l|}(z) + \left(\frac{l-|m_l|+1}{2l+1}\right) P_{l+1}^{|m_l|}(z) \qquad ...(3.33b)$$

3.1.3 SOLUTION OF RADIAL EQUATION

Now turning to the solution of the equation for radial function (3.18), this equation can be written as:

$$-\frac{h^2}{8\pi^2\mu r^2}\left[\frac{1}{R(r)}\frac{d}{dr}\left(r^2\frac{dR(r)}{dr}\right)\right] + \left[\left(\frac{h^2 l(l+1)}{8\pi^2\mu\, r^2} + V(r)\right) - E\right] R(r) = 0 \qquad ...(3.18)$$

We have learnt in the last section that the angular momentum is quantized and conserved in a central field, just as it is conserved classically in a central filed. Equation (3.18) has the direct physical interpretation that the total energy E is the sum of a radial kinetic energy, an angular kinetic energy, and the potential energy. Further equation (3.18) can be written as:

$$\frac{d^2R}{dr^2} + \frac{2}{r}\frac{dR}{dr} + \left[\frac{8\pi^2\mu E}{h^2} + \frac{8\pi^2\mu Ze^2}{4\pi\varepsilon_0 r h^2} - \frac{l(l+1)}{r^2}\right] R = 0 \qquad ...(3.34)$$

Substituting $a_0 = \frac{4\pi\varepsilon_0 h^2}{4\pi^2 \mu e^2}$, *i.e.*, the value of Bohr's radius in equation (3.34), we get

$$\frac{d^2R}{dr^2} + \frac{2}{r}\frac{dR}{dr} + \left[\frac{4\pi\varepsilon_0 2E}{e^2 a_0} + \frac{2Z}{a_0 r} - \frac{l\,(l+1)}{r^2}\right] R = 0 \qquad ...(3.35)$$

Similar to earlier cases, this equation can also be solved by the power series method, but the solution of this equation leads to recursion relation, which contains three terms rather than two terms. As discussed in case of harmonic oscillator, let us try to examine the behaviour of the solution at large value of *r*. For large value of *r*, equation (3.35) reduces to:

$$\frac{d^2R}{dr^2} + \left[\frac{4\pi\varepsilon_0\ 2E}{e^2 a_0}\right] R = 0$$

or

$$\frac{d^2R}{dr^2} = -\frac{4\pi\varepsilon_0\ 2E}{e^2 a_0} R \qquad ...(3.36)$$

$$R\ (r) = A\frac{d^2R}{dr^2} = Ae^{\pm\left(-\frac{4\pi\varepsilon_0\ 2E}{e^2 a_0}\right)^{\frac{1}{2}} r} \qquad ...(3.37)$$

is a solution of the above equation for large value of *r*, where A is a normalizing constant. This solution is called an asymptotic solution. Now the value of E can be negative or positive. Since the potential energy is having a negative value, the positive value of E means that the magnitude of kinetic energy (which is a positive quantity) is greater than the potential energy. In other words, this means that the electron is not bound to the nucleus and thus having a motion which is similar to the behaviour of a free particle. This means that the functions representing this behaviour will be imaginary functions and all possible values of energies are allowed. Further, the values of energies obtained from the behaviour of free particles are continuous. The functions corresponding to these energies are called continuum eigen functions. Similar to free particle, the eigen functions are not normalized in usual sense, but behaviour of the angular part of the wave functions can still be expressed by the spherical harmonics.

The negative value of E means that the value of potential energy is always greater than that of kinetic energy. In other words, the electron is bound to the nucleus and thus this problem can be called as *bound state of hydrogen like atoms*. Under these circumstances the terms in the parenthesis are positive and can be equated to C, *i.e.*, $C = \left(-\frac{4\pi\varepsilon_0 2E}{e^2 a_0}\right)^{\frac{1}{2}}$. Further the positive value of the exponent will lead to positive solution and this will lead to infinity as the value of *r* increases to infinity. This kind of function is not an allowed function and thus can be neglected. The acceptable or general solution over the complete range of *r* can be written as:

$$R\ (r) = e^{-Cr}\ S\ (r) \qquad ...(3.38)$$

where S (r) is a polynomial.

$$R'(r) = -Ce^{-Cr} S(r) + e^{-Cr}$$

$$R''(r) = C^2e^{-Cr} S(r) - Ce^{-Cr} S'(r) - Ce^{-Cr} S'(r) + e^{-Cr}$$

$$= [S''(r) - 2CS'(r) + C^2 S(r)] e^{-Cr}$$

Substituting the values of R(r), R$'(r)$, and R$''(r)$ in equation (3.35) and multiplying the resulting equation by $r^2 e^{Cr}$ one gets:

$$r^2 S''(r) + r(2 - 2Cr) S'(r) + \left[\left(\frac{2Z}{a_0} - 2C\right)r - l(l+1)\right] S = 0 \quad ...(3.39)$$

Since r varies from 0 to ∞, the equation (3.39) should be valid for $r = 0$. Substituting the value of $r = 0$ in equation (3.39), one gets:

$$l(l+1) S(0) = 0$$

or
$$l(l+1) b_0 = 0$$

where b_0 is the first coefficient of the power series S (r). For all non-zero values of l, this relation leads to S (0) = b_0, = 0 which is not a good solution as it will lead to solution with vanishing terms. In order to avoid this, the polynomial series should be multiplied by r^s, where s is a constant, to be determined. Thus S (r) can be written as:

$$S(r) = r^s K(r) \quad ...(3.40)$$

where
$$K(r) = \sum_{j=0}^{\infty} b_j r^j \quad ...(3.41)$$

Now

$$S'(r) = s r^{s-1} K(r) + r^s K'(r)$$

$$S''(r) = s(s-1) r^{s-2} K(r) + s r^{s-1} K'(r) + s r^{s-1} K'(r) + r^s K''(r)$$

or

$$S''(r) = r^s K''(r) + 2s r^{s-1} K'(r) + s(s-1) r^{s-2} K(r)$$

Substituting the values of S(r), S$'(r)$, and S$''(r)$ in equation (3.39) and dividing by r^s, one gets:

$$r^2 K''(r) + (2rs + 2r - 2Cr^2) K'(r) + \left[s(s-1) + 2s - 2Crs - l(l+1) + \left(\frac{2Z}{a_0} - 2C\right)r\right] K(r) = 0 \quad ...(3.42)$$

The equation (3.42) must satisfy the required condition at $r = 0$,

i.e.,
$$[s(s-1) + 2s - l(l+1)] K(0) = 0$$

or
$$[s(s+1) - l(l+1)] b_0 = 0$$

Since b_0 cannot be equal to zero, so the coefficient

$s(s + 1) - l(l + 1) = 0$

The quadratic equation in s gives rise to $s = l$ and $s = -(l + 1)$. Out of these two solutions the solution $s = -(l + 1)$ is not considered as this will lead to an unacceptable function. So equation (3.40) reduces to:

$$S(r) = r^l K(r) = r^l \sum_{j=0}^{\infty} b_j r^j \qquad ...(3.43)$$

Combining equations (3.39), (3.40), $S'(r)$ and $S''(r)$, we get

$$r K''(r) + (2l + 2 - 2C r) K'(r) + \left(\frac{2Z}{a_0} - 2C - 2C l\right) K(r) = 0 \qquad ...(3.44)$$

Now $$K(r) = \sum_{j=0}^{\infty} b_j r^j$$

$$K'(r) = \sum_{j=1}^{\infty} j b_j r^{j-1} = \sum_{j=0}^{\infty} (j+1) b_{j+1} r^j$$

$$K''(r) = \sum_{j=2}^{\infty} j(j-1) b_j r^{j-2} = \sum_{j=0}^{\infty} (j+1) j b_{j+1} r^{j-1}$$

Substituting the values of $K'(r)$ and $K''(r)$ in equation (3.44), we get

$$\sum_{j=0}^{\infty} j(j+1) b_{j+1} r^j + (2l + 2) \sum_{j=0}^{\infty} (j + 1) b_{j+1} r^j - 2C \sum_{j=0}^{\infty} j b_j r^j$$

$$+ \left(\frac{2Z}{a_0} - 2C - 2C l\right) \sum_{j=0}^{\infty} b_j r^j = 0$$

or

$$\sum_{j=0}^{\infty} \left[j(j+1) b_{j+1} + (2l + 2)(j + 1) b_{j+1} - 2C j b_j \right.$$

$$\left. + \left(\frac{2Z}{a_0} - 2C - 2C l\right) b_j \right] r^j = 0$$

Setting the coefficients of $r^j = 0$, one gets the recursion relation:

$$b_{J+1} = \frac{2C j - \frac{2Z}{a_0} + 2C + 2Cl}{j(j+1) + 2(j+1)(l+1)} b_j = \frac{2C(j+1+l) - \frac{2Z}{a_0}}{(j+1)(j+2l+2)} b_j \qquad ...(3.45)$$

As usual, let us examine the behaviour of the series expressed by (3.41) for large value of r. At large value of r, the behaviour of the series is determined by the term with the large value of

j. Thus let us examine the behaviour of the ratio of the coefficients b_{j+1}/b_j, *i.e.*, neglecting 1, l and $\frac{2Z}{a_0}$ terms in equation (3.45)

$$\frac{b_{j+1}}{b_j} \cong \frac{2Cj}{j(j+2)} = \frac{2C}{j} \qquad ...(3.46)$$

Now let us examine the behaviour of the series for e^{2Cr}.

$$e^{2Cr} = 1 + 2Cr + \frac{1}{2!}(2Cr)^2 + ... + \frac{(2C)^j r^j}{j!} + \frac{(2C)^{j+1} r^{j+1}}{(j+1)!} + ... \qquad ...(3.47)$$

$$\frac{(2C)^{j+1}}{(j+1)!}\frac{j!}{(2C)^j} = \frac{2C}{j+1} \cong \frac{2C}{j} \text{ for large value of } j. \qquad ...(3.48)$$

This shows that infinite power series (3.41) behaves like e^{2Cr} for large values of r. In other words the radial function for large value of r can be written as:

$$R(r) \sim e^{-Cr}\, r^l\, e^{2Cr} = r^l e^{Cr} \qquad ...(3.49)$$

This indicates that this function will become infinite at large values of r and thus is not an acceptable function. In order to avoid this situation (called as *infinite catastrophe, similar to earlier problem*), the series must terminate after certain value of $j = k$, the coefficient b_k reduces to zero. This will make the coefficients b_{k+1}, b_{k+2}, b_{k+3}....... equal to zero too. Substitute $j = k$ in numerator of equation (3.45) and equate it to zero.

$$2C(k + 1 + l) - \frac{2Z}{a_0} = 0$$

or

$$\frac{2Z}{a_0} = 2C(k + 1 + l), \text{ where } k = 0, 1, 2, ... \qquad ...(3.50)$$

and k and l are integers. We can define a new integer

$$n = k + l + 1, \quad n = 1, 2, 3......... \qquad ...(3.51)$$

Since the minimum value of k and l starts from zero, the minimum value of n will starts from one. It is quite clear from equation (3.51) that the value of $l \le (n - k - 1)$ and thus puts a limit on the value of l, *i.e.*, it goes from 0 to $n - 1$, *i.e.*, $l = 0, 1, 2, 3..........(n - 1)$. Equation (3.50) can be written as:

$$C = \frac{Z}{n\,a_0} = \left(-\frac{4\pi\varepsilon_0 2E}{a_0 e^2}\right)^{1/2}$$

or

$$\frac{Z^2}{n^2 a_0^2} = -\frac{4\pi\varepsilon_0 2E}{a_0 e^2}$$

or $$E = E_n = -\frac{e^2Z^2}{4\pi\varepsilon_0\, 2a_0}\frac{1}{n^2} = -\frac{2\pi^2 e^4\mu Z^2}{(4\pi\varepsilon_0)^2h^2}\frac{1}{n^2} \qquad ...(3.52)$$

or $$E\,(cm^{-1}) = E_n\,(cm^{-1}) = -\frac{2\,\pi^2 e^4\mu\, Z^2 10^{-2}}{(4\pi\varepsilon_0)^2h^3c}\frac{1}{n^2} \qquad ...(3.52a)$$

after substituting the value of a_0. Equation (3.52) represents the energy levels of the bound state of hydrogen like atoms. These are shown in Fig. 3.3. The expression so obtained coincides exactly with the expression for the energy of hydrogen like atoms (1.116) in Bohr theory, which is characterized by the principal quantum number n.

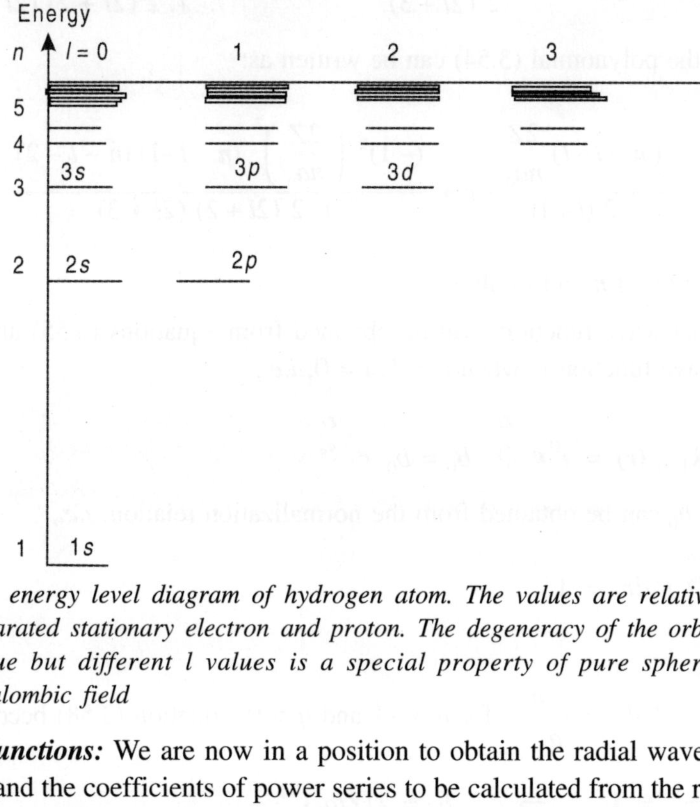

Figure 3.3: *The energy level diagram of hydrogen atom. The values are relative to an infinitely separated stationary electron and proton. The degeneracy of the orbitals with same n value but different l values is a special property of pure spherical symmetrical Coulombic field*

Radial Wave Functions: We are now in a position to obtain the radial wave functions using the equation (3.41) and the coefficients of power series to be calculated from the recursion relation (3.45). Using equation (3.50), we can rewrite the recursion relation (3.45) as:

$$b_{j+1} = \frac{(j + l + 1 - n)\ \dfrac{2Z}{na_0}}{(j+1)\,(j+2l+2)}b_j \qquad ...(3.53)$$

The polynomial (3.41) with the help of (3.51) becomes as:

$$K(r) = \sum_{j=0}^{n-l-1} b_j r^j \qquad ...(3.54)$$

and radial wave function becomes:

$$R_{n,\,l}\,(r) = r^l\, e^{-\frac{Zr}{na_0}} \sum_{j=0}^{n-l-1} b_j\, r^j \qquad ...(3.55)$$

With help of recursion relation (3.53), the coefficients b_j can be calculated and are:

$$b_1 = -\frac{(n-l-1)\dfrac{2Z}{na_0}}{2(l+1)}\, b_0$$

$$b_2 = -\frac{(n-l-2)\dfrac{2Z}{na_0}}{2\,(2l+3)}\, b_1 = \frac{(-1)^2 (n-l-1)\,(n-l-2)\left(\dfrac{2Z}{na_0}\right)^2}{1.\,2\,(2l+2)\,(2l+3)}\, b_0$$

Accordingly the polynomial (3.54) can be written as:

$$K\,(r) = b_0\left[1 - \frac{(n-l-l)\dfrac{2Z}{na_0}}{2\,(l+1)}\, r + \frac{(-1)^2\left(\dfrac{2Z}{na_0}\right)^2 (n-l-1)\,(n-l-2)}{1.\,2\,(2l+2)\,(2l+3)}\, r^2 + \cdot\right] \qquad ...(3.56)$$

This is a function of n and l only.

Different radial wave functions can be obtained from equations (3.55) and (3.56). Thus, the simplest radial wave function is when $n = 1$, $l = 0$, *i.e.*,

$$R_{1,\,0}\,(r) = r^0 e^{-\frac{Zr}{a_0}}\, b_0 = b_0\, e^{-\frac{Zr}{a_0}} \qquad ...(3.57)$$

The constant b_0 can be obtained from the normalization relation, *i.e.*,

$$b_0^2 \int_0^\infty e^{-\frac{2Zr}{a_0}}\, r^2 dr = 1 \qquad ...(3.58)$$

Using $\int_0^\infty x^n\, e^{-qx}\, dx = \frac{n}{q^{n+1}}$ for $n > -1$ and $q > 0$, equation (3.58) becomes:

$$b_0^2\, \frac{2}{\left(\dfrac{2Z}{a_0}\right)^3} = 1 \quad \Rightarrow \quad b_0 = 2\,(Z/a_0)^{3/2} \qquad ...(3.59)$$

and equation (3.57) reduces to:

$$R_{1,\,0}\,(r) = 2(Z/a_0)^{3/2}\, e^{-\frac{Zr}{a_0}} \qquad ...(3.60)$$

Similarly for $n = 2$, l can be 0 or 1. Thus, for $n = 2$, $l = 0$, equation (3.55) becomes:

$$R_{2,\,0}\,(r) = r^0\, e^{-\frac{Zr}{a_0}}\,(b_0 + b_1 r) = b_0\left[1 - \frac{2-0-1}{2}\left(\frac{Z}{a_0}\right) r\right] e^{-\frac{Zr}{a_0}}$$

$$= b_0 \left[1 - \frac{Z}{2a_0} r\right] e^{-\frac{Zr}{a_0}} \quad ...(3.61)$$

For $n = 2$, $l = 1$, equation (3.55) becomes:

$$R_{2,1}(r) = b_0\, r\, e^{-\frac{Zr}{2a_0}} \quad ...(3.62)$$

The constants of integration can be obtained using normalization relations.

The radial wave functions for higher values of n and corresponding l values can be obtained by this method, but becomes cumbersome. These radial wave functions can be obtained from the generalized Laguerre polynomials, *i.e.*,

$$R_{n,l}(r) = C\left(\frac{2Zr}{na_0}\right)^l e^{-\frac{Zr}{a_0}} L_{n+1}^{2l+1}\left(\frac{2Zr}{na_0}\right) \quad ...(3.63)$$

or

$$R_{n,l}(r) = e^{-nr}\; L_{n,l}(r)$$

The term $L_{n+1}^{2l+1}(x)$ is a derivative of the Laguerre polynomials and is defined as:

$$L_{n+1}^{2l+1}(x) = \frac{d^{2l+1}}{dx^{2l+1}}\left[e^x \frac{d^{n+l}}{dx^{n+l}}(x^{n+l}\, e^{-x})\right] \quad ...(3.64)$$

where $x = \dfrac{2Zr}{na_0}$ and C is the normalization constant obtained from the condition:

$\int_0^\infty R^2 r^2 dr = 1$. Substituting the values of the respective parameters in the preceding equation and solving for C, the normalization constant, one gets:

$$\int_0^\infty R^2 r^2 dr = 1 = C^2 \int_0^\infty x^{2l} e^{-x}\left[L_{n+l}^{2l+1}(x)\right]^2 x^2 dx \left(\frac{na_0}{2Z}\right)^3$$

because

$$r^2 = \left(\frac{na_0}{2Z}\right)^2 x^2 \text{ and } dr = \frac{na_0}{2Z} dx$$

or

$$\left(\frac{na_0}{2Z}\right)^3 C^2 \int_0^\infty x^{2l+2} e^{-x}\left[L_{n+l}^{2l+1}(x)\right]^2 dx = 1 \quad ...(3.65)$$

The integral is analogus to the associated Laguerre functions. Accordingly the value of the integral is given by:

$$\int_0^\infty x^{2l+2} e^{-x}\left[L_{n+l}^{2l+1}(x)\right]^2 dx = \frac{2n[(n+l)!]^3}{(n-l-1)!} \quad ...(3.65a)$$

$$\therefore \qquad C = \left(\frac{2Z}{na_0}\right)^{\frac{3}{2}} \left[\frac{(n-l-1)!}{2n\,[(n+l)!]^3}\right]^{\frac{1}{2}} \qquad ...(3.66)$$

The generalized form of normalized radial wave function for various logical combinations of n and l now can readily be deduced from equation (3.67), obtained by combining equations (3.63), (3.64) and (3.66).

$$R_{n,\,l}(r) = -\left(\frac{2Z}{na_0}\right)^{\frac{3}{2}} \left[\frac{(n-l-1)!}{2n[(n+l)!]^3}\right]^{\frac{1}{2}} \left(\frac{2Zr}{na_0}\right)^{l} e^{-\frac{Zr}{na_0}}\, L_{n+l}^{2l+1}\left(\frac{2Zr}{na_0}\right) \qquad ...(3.67)$$

Here negative sign is introduced to make the function positive. These radial wave functions are real and independent of the angular coordinate's θ and φ and magnetic quantum number m_l. The functions $R_{n,\,l}(r)$, are the eigen functions of the Hamiltonian operator and thus possess the property of being orthogonal and normalized, *i.e.,* orthonormal set.

$$\int_0^{\infty} R_{n,\,l}(r) R_{n',\,l}(r)\, r^2\, dr = \delta_{n\,n'} \qquad ...(3.68)$$

These functions, however, do not make up a complete set because the Hamiltonian

Table 3.3: *The normalized hydrogen like radial wave functions*

n	l	State	Radial wave function $R_{n,\,l}$
1	0	1*s*	$R_{1,\,0} = 2\left(\frac{Z}{a_0}\right)^{\frac{3}{2}} e^{-\sigma}$
2	0	2*s*	$R_{2,\,0} = \frac{1}{2\sqrt{2}}\left(\frac{Z}{a_0}\right)^{\frac{3}{2}} (2-\sigma)\, e^{-\frac{\sigma}{2}}$
2	1	2*p*	$R_{2,\,1} = \frac{1}{2\sqrt{6}}\left(\frac{Z}{a_0}\right)^{\frac{3}{2}} \sigma\, e^{-\frac{\sigma}{2}}$
3	0	3*s*	$R_{3,\,0} = \frac{2}{81\sqrt{3}}\left(\frac{Z}{a_0}\right)^{\frac{3}{2}} (27 - 18\,\sigma + 2\sigma^2)\, e^{-\frac{\sigma}{3}}$
3	1	3*p*	$R_{3,\,1} = \frac{4}{81\sqrt{6}}\left(\frac{Z}{a_0}\right)^{\frac{3}{2}} (6\,\sigma - \sigma^2)\, e^{-\frac{\sigma}{3}}$
3	2	3*d*	$R_{3,\,2} = \frac{4}{81\sqrt{3}}\left(\frac{Z}{a_0}\right)^{\frac{3}{2}} \sigma^2 e^{-\frac{\sigma}{3}}$
where		$\sigma = \frac{Z}{a_0} r$	

operator has a continuous spectrum in addition to the discrete spectrum. The normalized radial wave functions for hydrogen like atoms have been compiled in Table 3.3.

Example 3.5: Determine the radial wave function for $n = 3$, $l = 0$, *i.e.*, 3s orbital.

Solution: First determine $L_{n+l}^{2l+1}(x)$, *i.e.*,

$$L_{n+l}^{2l+1}(x) = L_3^1(x) = \frac{d^{2l+1}}{dx^{2l+1}}\left[e^x \frac{d^{n+l}}{dx^{n+l}}(x^{n+l}e^{-x})\right] = \frac{d}{dx}\left[e^x \frac{d^3}{dx^3}(x^3e^{-x})\right]$$

$$= \frac{d}{dx}\left[e^x \frac{d^2}{dx^2}(3x^2e^{-x} - x^3e^{-x})\right]$$

$$= \frac{d}{dx}\left[e^x \frac{d}{dx}(6xe^{-x} - 3x^2e^{-x} - 3x^2e^{-x} + x^3e^{-x})\right]$$

$$= \frac{d}{dx}\left[e^x \frac{d}{dx}(6xe^{-x} - 6x^2e^{-x} + x^3e^{-x})\right]$$

$$= \frac{d}{dx}[e^x(6e^{-x} - 6xe^{-x} - 12xe^{-x} + 6x^2e^{-x} + 3x^2e^{-x} - x^3e^{-x})]$$

$$= \frac{d}{dx}(6 - 18x + 9x^2 - x^3) = -3(x^2 - 6x + 6)$$

$$R_{3,0}(r) = -\left(\frac{2Z}{3a_0}\right)^{3/2}\left[\frac{(3-0-1)!}{2\times 3\,[(3n+0)!]^3}\right]^{1/2} e^{-x/2}\, x^0\, [-3(x^2 - 6x + 6)]$$

$$= 3\left(\frac{8}{27}\right)^{1/2}\left(\frac{Z}{a_0}\right)^{3/2}\left[\frac{2}{6\times 216}\right]^{1/2} e^{-x/2}\,(x^2 - 6x + 6)$$

$$= \frac{1}{9\sqrt{3}}\left(\frac{Z}{a_0}\right)^{3/2} e^{-x/2}\,(x^2 - 6x + 6)$$

where $$x = \left(\frac{2Z}{na_0}\right)r.$$

Radial Probability: It is informative to analyze the distribution of probability density as a function of the distance of electron from the nucleus. The radial part of the wave function is given by equation (3.67) and the *probability density*, also called *probability per unit volume*, for finding the electron at a given point along the radius is:

$$P_r = R^*_{n,l}\, R_{n,l}\,; \qquad R_{n,l} = e^{-n\,r}\, r^l\, L_{n,l}(r)$$

The element of volume depends only on the radius of the spherical shell enclosed by two concentric spheres of radii r and $r + dr$ respectively. This volume element is:

$$d\tau_r = 4\pi r^2 \, dr \qquad ...(3.69)$$

The radial probability of finding the electron in volume element of $d\tau_r$ is:

$$P_r = R^*_{n,l} R_{n,l} \, d\tau_r = 4\pi r^2 R^2_{n,l} \, dr \qquad ...(3.70)$$

or
$$P_r = e^{-2nr} r^{2(l+1)} L^2_{n,l}(r) \; 4\pi dr\text{: } L_{n\,l} \text{ is a polynomial given by:}$$

$$L_{n\,l} = b_0 + b_1 r + b_2 r^2 + \cdot \quad + b_{n-l-1} r^{n-l-1}$$

where b's are constants. Thus the radial probability is the probability of finding the electron between two concentric spheres of radii r and $r + dr$. The charge density at this radius is just the electric charge multiplied by this probability.

The plot of radial wave functions ($R_{n,l}$) and radial probability (P_r) against the radial distance measured in units of Bohr radius, *i.e.*, $\frac{r}{a_0}$, where $a_0 = 0.053$ nm, for some low quantum states of hydrogen like atoms have been shown in Figs. 3.4 and 3.6 respectively. The nature of curves in Figs. (3.4) and (3.6) reveals that there exists no correlation, as expected, between the positions of maxima (or minima) and the integral multiple of a_0. The only common characteristics of all the curves are that the values of the functions $R_{n,\,l}$ or probability P_r decreases to zero asymptotically for large value of r.

For the 1s state ($n = 1$, $l = 0$), equation (3.70) becomes:

$$P_r = R^*_{1,0} \; R_{1,0} \, d\tau_r = R^2_{1,0} \, d\tau_r = e^{-2r} r^2 b_0^2 \; 4\pi dr$$

Note that when $r = 0$, $P_r = 0$, and the maximum probability is obtained when $r = a_0 = 0.053$ nm, which is in agreement with radius of orbit for the ground state as computed by Bohr theory. As r approaches $5a_0$, P_r becomes very small and approaches zero.

The 2s state ($n = 2$, $l = 0$) has a probability:

$$P_r = e^{-4r} r^2 (b_0 + b_1 r)^2 \; 4\pi dr$$

and $P_r = 0$ for $r = 0$. Now there are two maxima, one at $r = a_0$ and the other at $r \cong 5a_0$, with the second being larger than the first one.

The probability for the 3s state ($n = 3$, $l = 0$) is

$$P_r = e^{-6r} r^2 (b_0 + b_1 r + b_2 r^2)^2 \; 4\pi dr$$

and the three maxima are now found to occur around a_0, $4a_0$ and $12a_0$, with the latter being the largest.

The electrons found in the 2p state have the maximum radial probability at about $4a_0$ and those in the 3p state have maxima at about $2a_0$ and $12a_0$. Note that in each case latter maximum is more pronounced than the others. The probability of locating the electrons at one of these distances along the radius is greater here, and we have the mechanism for locating the electrons in certain shells about the nucleus. Unlike Bohr Theory, however, there are other positions where there is a definite probability of finding the electron. In each case, the probability of finding the electron at $r = 0$ is zero. The constants b_0, b_1, b_2... .can be found using normalization condition as said earlier (3.68).

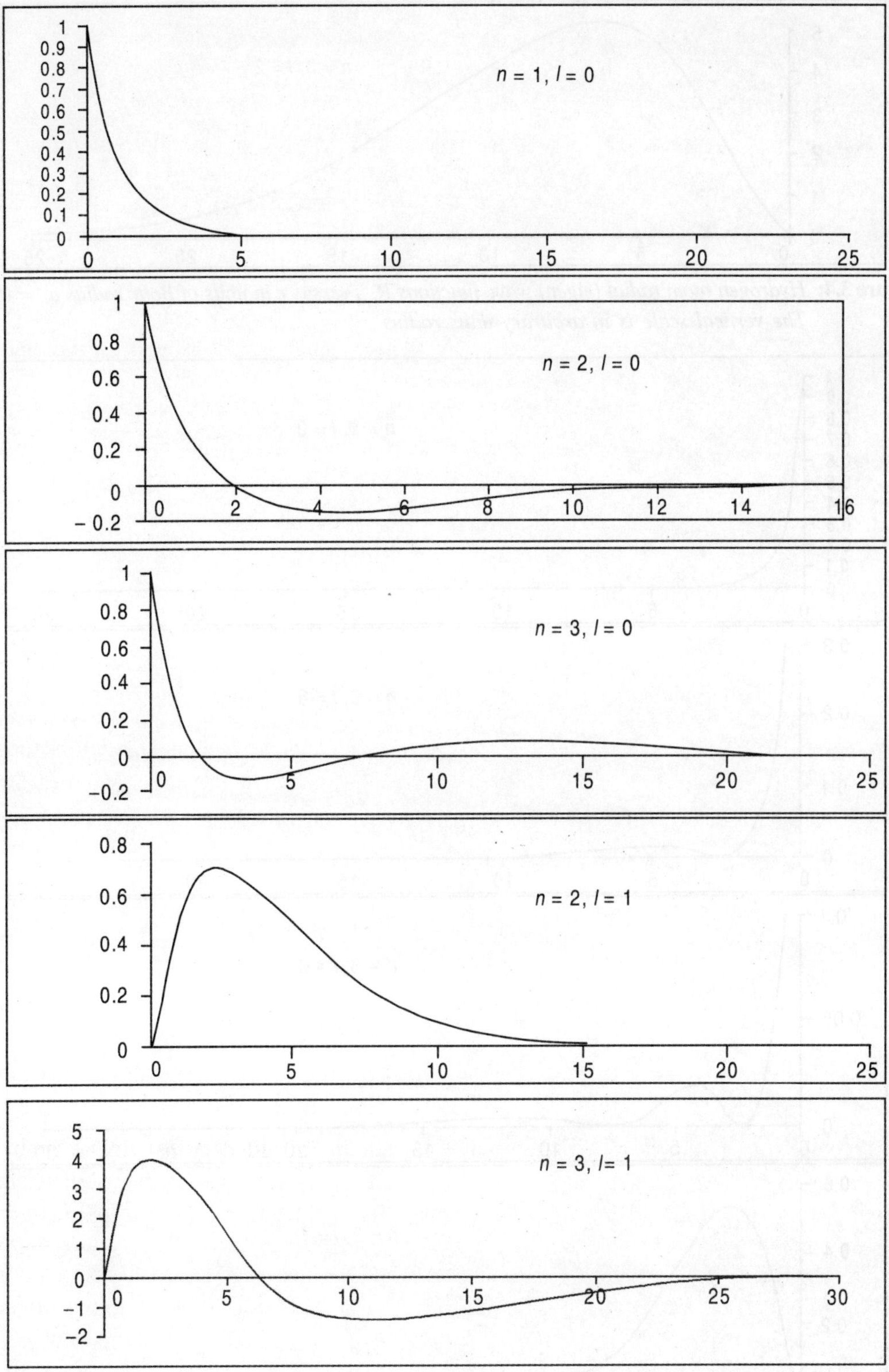
n = 1, l = 0
n = 2, l = 0
n = 3, l = 0
n = 2, l = 1
n = 3, l = 1

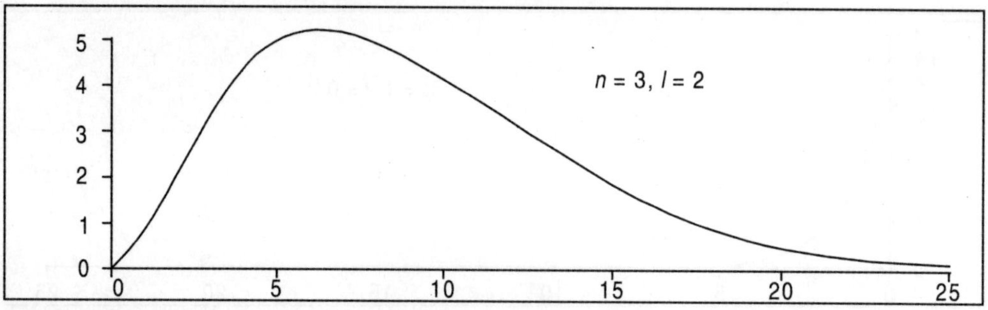

Figure 3.4: *Hydrogen atom radial (eigen) wave functions $R_{n,\,l}$ versus r in units of Bohr radius a_0 = 0.053 nm. The vertical scale is in arbitrary units radius*

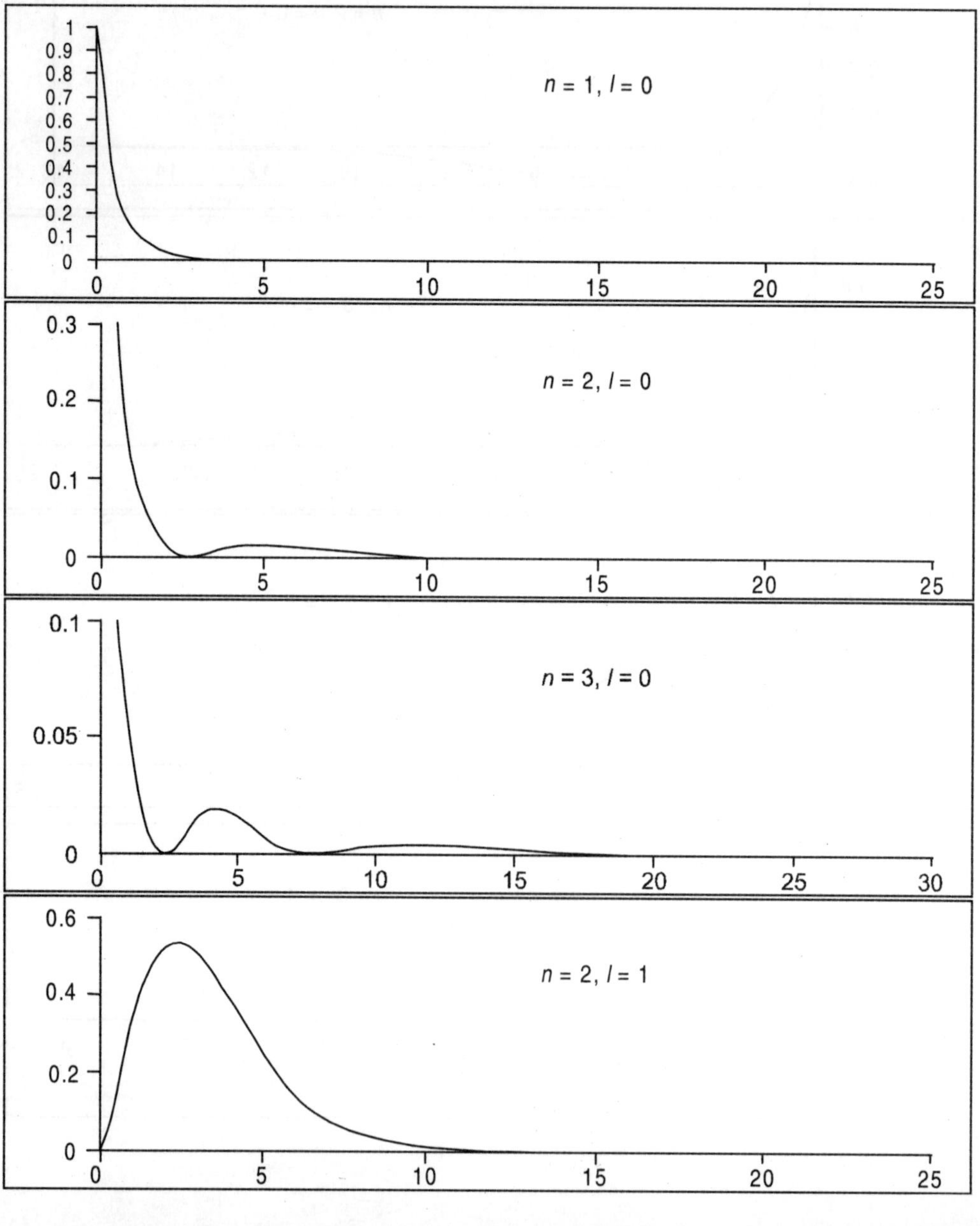

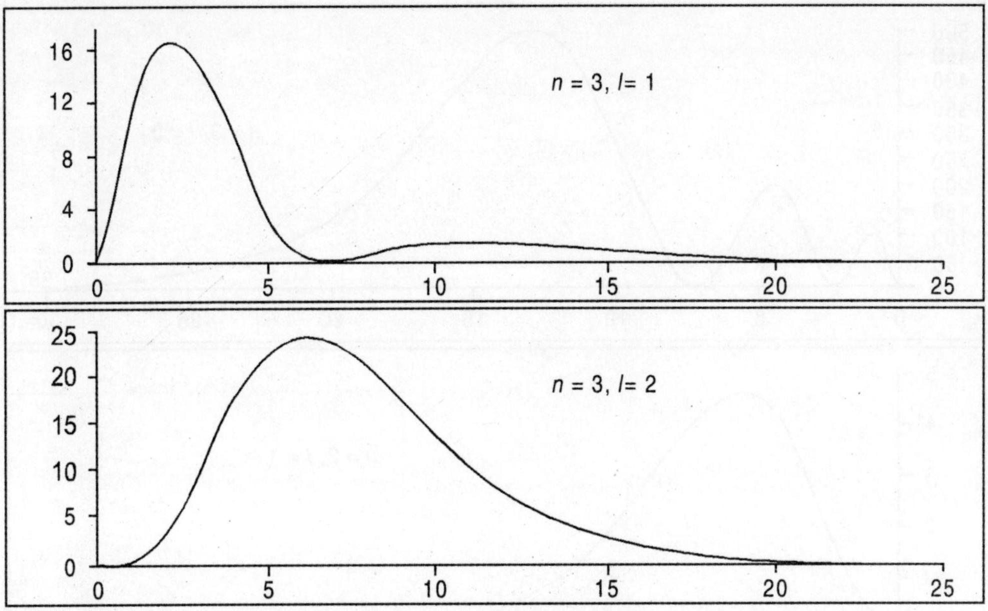

Figure 3.5: *Hydrogen atom Radial Probability Density functions $P_r = R^*_{n,l} R_{n,l}$ versus distance r in terms of Bohr radius a_0 = 0.053 nm. Vertical scale is in arbitrary units*

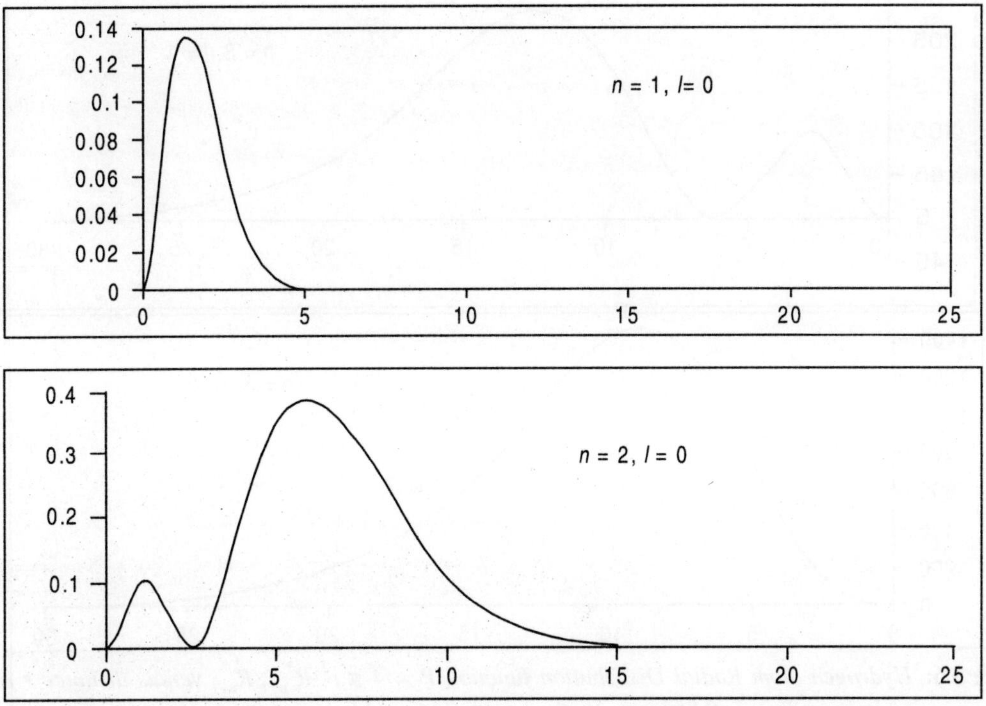

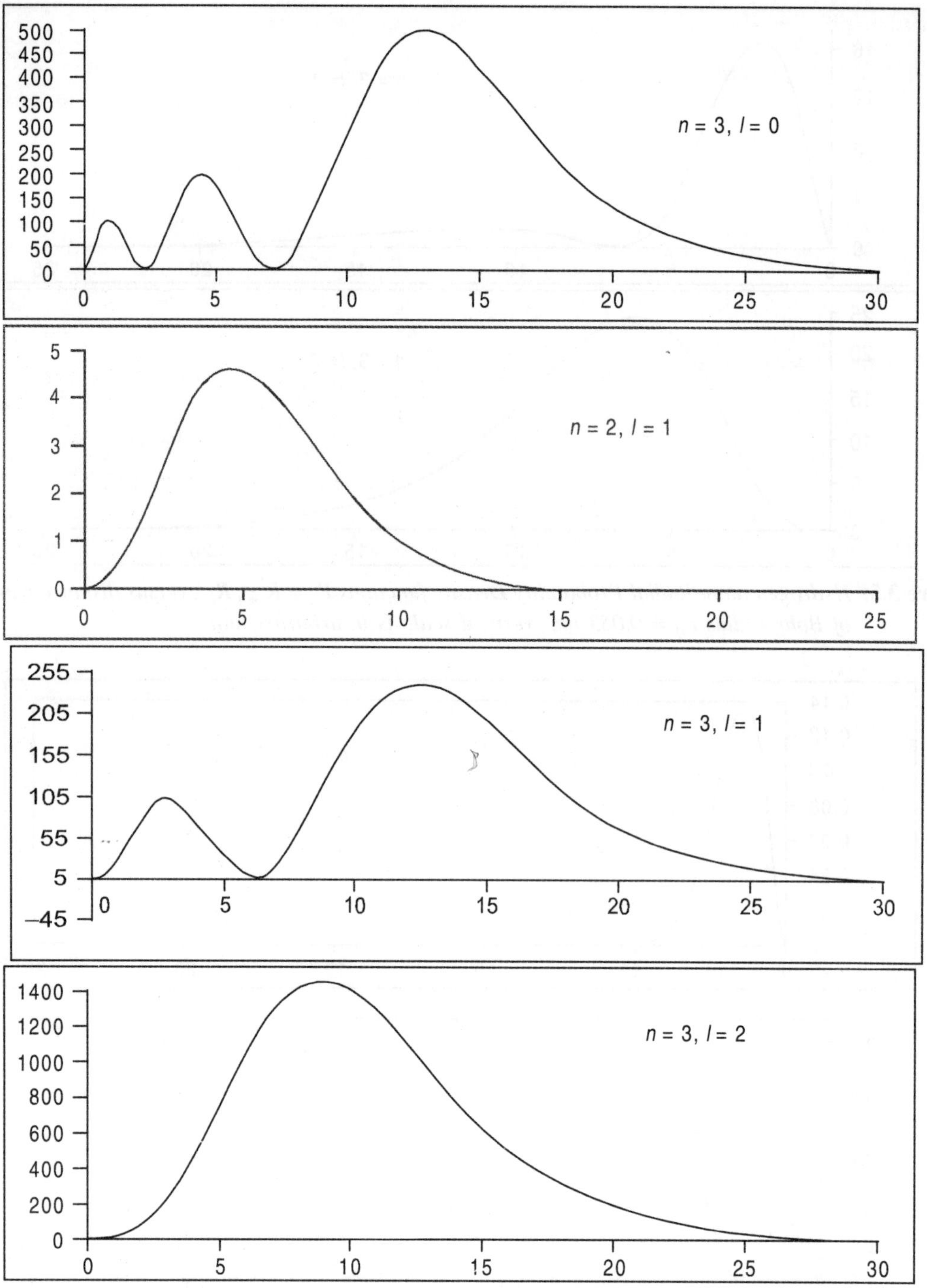

Figure 3.6: *Hydrogen atom Radial Distribution function $P = 4\,\pi\, r^2\, R^*_{n,\,l}\, R_{n,\,l}$ versus distance r in units of Bohr radius a_0 0.053 nm. Vertical scale is in arbitrary units*

$$\frac{1}{4\pi}\int_{r=0}^{r=\infty} R^*_{n,\,l}\, R_{n,\,1}\, d\tau_r = \frac{1}{4\pi}\int_{r=0}^{r=\infty} \mathrm{R}^2_{n,\,l}\, d\tau_r = 1$$

3.1.4 QUANTUM NUMBERS AND ENERGY FOR HYDROGEN LIKE ATOMS

From the solution of Schrödinger wave equation (H ψ = E ψ), it is clear that there are three quantum numbers, which can describe the wave function and other characteristics of the hydrogen like atoms. It is thus analogous to the solution of a particle in a three dimensional box problem. But still the solution of Schrödinger wave equation for a moving electron in a central Coulombic field of hydrogen and hydrogen like atoms in spherical polar coordinates gives rise to many new and different results. For example, the energies of different quantum states are given by the equation (3.52), *i.e.*,

$$E_n = -\frac{2\pi^2 \mu e^4 Z^2}{(4\pi\varepsilon_0)^2 h^2}\frac{1}{n^2},\quad n = 1, 2, 3\ldots\ldots \qquad \ldots(3.52)$$

$$= -\frac{R_H\, hc\, Z^2}{n^2} = -\frac{Z^2}{n^2}E_H \qquad \ldots(3.71)$$

where $R_H = \frac{2\,\pi\mu e^4}{(4\pi\varepsilon_0)^2 ch^3}$ and $E_H = R_H\, hc$. R_H is a Rydberg constant for hydrogen atom and E_H is the energy of the hydrogen atom in its lowest state. The expression for calculating the values of energy levels is similar to that derived with the help of Bohr's quantum theory. Unlike the particle in a three dimensional box, the energy of any atomic state of hydrogen atom or hydrogen like atoms depends only on one quantum number (n) as the square of its inverse, although the wave function is dependent on all the three quantum numbers, *i.e.*, n, l, m_l. The total wave function for electron in central Coulombic field of hydrogen like atoms for the bound state is described by the equation:

$$\Psi_{n,l,m_l}(r, \theta, \varphi) = R_{n,l}(r)\, Y_{l,m_l}(\theta, \varphi) \qquad \ldots(3.72)$$

or

$$\Psi_{n,l,m_l}(r, \theta, \varphi) = R_{n,1}(r)\, \Theta_{l,m_l}(\theta)\, \Phi_{m_l}(\varphi) \qquad \ldots(3.73)$$

The normalization condition for the radial wave function for the discrete spectrum is given by:

$$\int_0^\infty |R_{n,l}(r)|^2 r^2\, dr = 1 \qquad \ldots(3.74)$$

and the normalization condition for the total wave function of the bound states is given by:

$$\int_0^\infty R_{n,l}^2(r)\, r^2 dr \int_0^\pi \Theta_{lm_l}^2(\theta)\sin\theta\, d\theta \int_0^{2\pi} \Phi_{m_l}^2(\varphi) d\varphi \qquad \ldots(3.75)$$

Functions (3.72) and (3.73) are simultaneous eigen functions of the Hamiltonian operator $\hat{H}$, the square of the angular momentum $\hat{M}^2$ and the component of angular momentum $\hat{M}_z$ along the

z-axis. Due to this very reason in a state described either by (3.72) or (3.73), these three quantities have definite values simultaneously.

Quantity (Operator)	**Eigen value**
$\hat{H}$	$E_{n,\,l}$ or E
$\hat{M}^2$	$l(l+1)\dfrac{h^2}{4\pi^2}$
$\hat{M}_z$	$m_l\dfrac{h}{2\pi}$

Thus a state is characterized by three quantum numbers (n, l, m_l). The label n is called the *principal quantum number* (since it determines the total energy) and is usually defined as the sum of:

$$n = n_r + l + 1 \qquad \text{...(3.76)}$$

where n_r determines the number of zeros (nodes) of $R_{n,\,l}(r)$. The label n_r is called the *radial quantum number* and l the *azimuthal quantum number*. Such definition of n is possible because when an eigen function of differential operator of type in a radial equation (3.51) belongs to a discrete spectrum, it is characterized by the number of its nodes. Since n_r is non-negative, the principal quantum number is greater than the azimuthal quantum number at least by one. Thus, the principal quantum number determines the total numbers of nodes (*i.e.*, $n - 1$), including radial and angular nodes.

For a given value of n, the azimuthal quantum number l can take on values:

$$l = 0, 1, 2, 3 \ldots \qquad (n - 1)$$

and also implies the following useful information:

(1) This quantum number specifies the number of angular nodes. By equation (3.76), the total number of radial nodes is: $n - l - 1$. Since the total number of nodes including radial and angular are: $n - 1$, the total number of angular nodes = $n - 1 - n + l + 1 = l$

(2) Second important information it provides is about the total orbital angular momentum ($\mathbf{M}^2$). The radial form of Schrödinger equation when written as:

$$\frac{1}{r^2}\frac{d}{dr}\left(r^2\frac{dR}{dr}\right)+\frac{8\pi^2\mu}{h^2}\left[E-V(r)-\frac{h^2}{8\pi^2\mu r^2}\,l(l+1)\right]R = 0 \qquad \text{...(3.18)}$$

yields

$$\frac{h^2}{4\pi^2}\,l(l+1)\,r^2(2\mu E_k^*) = r^2p^2 \qquad \text{...(3.77)}$$

where E_k^* is the kinetic energy and p is the linear momentum. The quantity linear momentum multiplied by radius is just angular momentum $\mathbf{M}$ and can now be expressed as:

$$\mathbf{M}^2 = \frac{h^2}{4\pi^2} l(l+1)$$

or $$\mathbf{M} = \frac{h}{2\pi}\sqrt{l(l+1)}\ , \quad l = 0, 1, 2,..............(n-1) \qquad ...(3.78)$$

For large values of l, equation (3.78) reduces to:

$$\mathbf{M} \cong \frac{h}{2\pi} l \qquad ...(3.79)$$

which is similar to Bohr's postulate

$$\mathbf{M} = \frac{h}{2\pi} n \qquad ...(1.80)$$

The only difference is that the value of $n = 0$ is not allowed in Bohr's treatment, while $l = 0$ is permitted in Schrödinger wave equation.

Further just as the energy (E_n) increases with increase in n or the number of nodes, the orbital angular momentum also increases with increase of l or angular nodes. Similar to the Bohr's value of orbital angular momentum, the value given by quantum mechanics is also quantized, but this value is not an integral multiple of $\frac{h}{2\pi}$ except for large value of l (as proved above). This difference for small value of l can be explained as follows: Bohr's model for hydrogen atom is planar and thus the value of the orbital angular momentum, p_φ, is a scalar quantity (*i.e.,* some number). On the other hand, quantum mechanical picture of hydrogen atom is rather three-dimensional one. If the electron is confined to *xy*-plan, the *z*-component of the position as well as that of velocity will be zero. In other words, the uncertainties in the positions and momentum under these conditions are zero. This violates the basic principle of uncertainty. Now with three-dimensional model, the components of angular momentum along *x*, *y* and *z*-axis can be $\mathbf{M}_x$, $\mathbf{M}_y$ and $\mathbf{M}_z$ respectively. With suitable choice of axis, the component $\mathbf{M}_z$ and p_φ can be related to angular momentum of planar model of Bohr's model. The radial equation (3.71) reveals that the energy levels $E_{n,\,l}$ do not depend on m_l as expected, since $m_l\ \frac{h}{2\pi}$ is the value of angular momentum component along *z*-axis, as direction of the axis is well defined. But if the magnetic filed is applied to the system (say in the *z* direction), the energy levels will depend on m_l as well. For this reason m_l is called *magnetic quantum number*. The value of *z*-component of orbital angular momentum is given by the expression:

$$\mathbf{M}_z = m_l \frac{h}{2\pi}\ , \quad m_l = 0.\ \pm 1, \pm 2, \pm 3....................\pm l \qquad ...(3.80)$$

The sign of m_l points out whether the orbital angular momentum is along the positive or negative value of *z*-axis. There is one sharp distinction from Bohr's model that the angular momentum defined by Bohr is a function of magnetic quantum number m_l and is not of principal quantum number n. Further $\mathbf{M}_z$ is the only one of the components of the orbital angular momentum. Thus the absolute value of $\mathbf{M}_z$ will be equal to or less than $\frac{h}{2\pi}\sqrt{l(l+1)}$. Accordingly the total energy of hydrogen and hydrogen like atoms in a magnetic field is:

$$E_{n,l,m_l} = -\frac{2\pi^2\mu Z^2 e^4}{(4\pi\varepsilon_0)^2 h^2}\frac{1}{n^2} + \mu_B m_l H_{ext} \quad ...(3.81)$$

The contribution of the first term is due to Coulomb interaction between electron and nucleus and of the second term is due to interaction between external magnetic field (H_{ext}) and magnetic moment (μ_B) of the circulating electron. Recall that for a given n and a given l, there are $2l + 1$values possible for m_l (0, ±1, ±2, ±3,..........±l). Each permitted value of m_l produces a different value for E_{n,l,m_l}, thus splitting the original energy level into $2l + 1$ levels.

(3) As will be seen later, the magnetic quantum number describes the orientation of the atomic orbitals in the three dimensional space.

3.1.5 TOTAL NUMBER OF ENERGY STATES

Lastly it can be seen from the above results that except for $n = 1$, there can be many different values of l, m_l, for the same value of n, *i.e.*,

$$n = 1, 2, 3, \ldots$$

$$l = 0, 1, 2, \ldots, n-1 \qquad n \text{ values}$$

$$m_l = 0, \pm 1, \pm 2, \pm 3, \ldots, \pm l \qquad 2l+1 \text{ values}$$

Every allowed combination of the three quantum numbers, n, l, m_l defines a state of the atom and every state is characterized by a corresponding eigen function, Ψ_{n,l,m_l}. One of the solutions of the Schrödinger wave equation for the hydrogen and hydrogen like atoms is:

$$\Psi_{n,l,m_l} = R_{n,l}(r)\,\Theta_{l,m_l}(\theta)\,\Phi_{m_l}(\varphi)$$

in which

$$R_{n,l}(r) = e^{-nr} r^l L_{n,l}(r)$$

$$\Theta_{l,m_l}(\theta) = P_{l,m_l}(\cos\theta)$$

$$\Phi_{m_l}(\varphi) = A\,e^{im_l\varphi}$$

When the atom is not under the influence of external magnetic field and contains one electron, the eigen values of the energy are:

$$E_n = -\frac{Z^2}{n^2}E_H \quad ...(3.82)$$

and the energy is independent of l and m_l. In such a situation, there is not a one to one correspondence between the eigen values and eigen functions. The Schrödinger wave equation will have two or more solutions for one value of the energy and the solutions in this case are said to be degenerate, *i.e., the states having the same energy but characterized by different eigen functions, are said to be degenerate states.* On the other hand, when there is only one independent solution for a given eigen value of energy, the solution is non-degenerate and the state is said to be non-degenerate. The degeneracy can be removed by applying some external field e.g., magnetic or electric field. The total number of degenerate states for a given value of n and l can be calculated as follows:

For a given value of n, the azimuthal quantum number l can have the following values.

$$l = 0, 1, 2, \ldots\ldots\ldots (n - 1)$$

For every value of l, the magnetic quantum number m_l has the following $(2l + 1)$ values:

$$m_l = 0, \pm 1, \pm 2, \pm 3, \ldots\ldots\ldots\ldots\ldots \pm l$$

Thus the total number of eigen functions or possible degenerate states is:

$$g = \sum_{l=0}^{l=n-1} (2l+1) = 1 + 3 + 5 + \ldots\ldots\ldots\ldots\ldots (2n - 1) \qquad \ldots(3.83)$$

This is an arithmetic progression where sum is given by:

$$S = \left(\frac{\alpha + \beta}{2}\right) n$$

where α is the value of the first term, β is the value of the last term and n is the total number of terms.

$$g = \left[\frac{1 + (2n - 1)}{2}\right](n) = n^2 \qquad \ldots(3.84)$$

It must be kept in mind that n^2 degeneracy for hydrogen or hydrogen like atoms is only because of the spherically symmetrical nature of potential energy field. If this spherical symmetry is removed, the energy E_n will not only be the function of n, but also of other quantum numbers, as will be clear from the nature of complex atoms.

It has already been stated that there is an accepted spectroscopic notation for energy state, *i.e.,* terms. Terms with the same value of n but different values of l are denoted by Latin letters, *s, p, d, f, g, h*........ For instance,

Term	State	Spectroscopic notation	Term	State	Spectroscopic notation
$n = 1, l = 0$	s	1s	$n = 3, l = 0$	s	3s
$n = 2, l = 0$	s	2s	$n = 3, l = 1$	p	3p
$n = 2, l = 1$	p	2p	$n = 3, l = 2$	d	3d

The origin of the letters *s, p, d, f* is historic and has to do with the designation of the observed spectral lines for sodium-atom. The letters *s, p, d, f,* stand for sharp, principal, diffuse and fundamental. Thus, according to Schrödinger equation (3.82), no spectral term is degenerate, whereas in reality all terms except, '*s*' terms (corresponding to $l = 0$) are doublets, *i.e.,* consists of two adjacent terms (fine structure of spectral lines). The fine structure of the spectral lines in the atomic spectra of hydrogen or hydrogen like atoms can be interpreted in terms of spin quantum number, introduced by Dirac in his theory called Dirac relativistic quantum theory of electron. The concept of spin quantum number will be discussed in detail later on in section 3.7.

Example 3.6: Calculate the degeneracy of the energy states with $n = 3$.

Solution: Using equation (3.84)

$$g = 3^2 = 9$$

i.e., for $n = 3$, $l = 0, 1, 2$, *i.e.,* there is one '*s*' state, three '*p*' states and five '*d*' quantum states. All nine quantum states for hydrogen like atoms have the same energy for $n = 3$.

Example 3.7: Construct a table of all possible states and eigen functions for $n = 4$.

Solution: For $n = 4$, the permitted values of l and m_l are:

$l\ (= n - 1)$	$m_l\ (= 2l + 1)$
0	0
1	0, ±1
2	0, ±1, ±2
3	0, ±1, ±2, ±3

The possible number of states $g = 4^2 = 16$ and eigen functions for the possible states for $n = 4$ are:

State n	l	m_l	Eigen functions	Number of states	Sates n	l	m_l	Eigen functions	Number of states
4	0	0	$\Psi_{4,0,0}$	1	4	2	+2	$\Psi_{4,2,+2}$	
4	1	−1	$\Psi_{4,1,-1}$		4	3	−3	$\Psi_{4,3,-3}$	
4	1	0	$\Psi_{4,1,0}$	3	4	3	−2	$\Psi_{4,3,-2}$	
4	1	+1	$\Psi_{4,1,+1}$		4	3	−1	$\Psi_{4,3,-1}$	
4	2	−2	$\Psi_{4,2,-2}$		4	3	0	$\Psi_{4,3,0}$	7
4	2	−1	$\Psi_{4,2,-1}$		4	3	+1	$\Psi_{4,3,+1}$	
4	2	0	$\Psi_{4,2,0}$	5	4	3	+2	$\Psi_{4,3,+2}$	
4	2	+1	$\Psi_{4,2,+1}$		4	3	+3	$\Psi_{4,3,+3}$	

3.1.6 WAVE FUNCTIONS FOR HYDROGEN LIKE ATOMS

Having discussed the quantum numbers and methods to calculate the wave functions for individual variable, let us now summarize and analyze what we have learnt about the total wave function for hydrogen and hydrogen like atoms, with positive charge equal to Ze at the nucleus? We have seen that the Schrödinger wave equation for hydrogen like atoms.

$$\hat{H}\,\psi = E\,\psi$$

has the general solution

$$\Psi_{n,l,m_l}(r,\theta,\varphi) = R_{n,l}(r)\,\Theta_{l,m_l}(\theta)\,\Phi_{m_l}(\varphi)$$

$$= [e^{-n\,r}\,r^{l}\,L_{n,\,l}(r)]\,[P_{l,m_l}(\cos\theta)]\,A\,e^{im_l\varphi}$$

where $L_{n,\,l}(r)$ are called the associated Laguerre polynomial, $P_{l,m_l}(\cos\theta)$ are called the Legendre polynomial, the $L_{n,\,l}$ polynomials are given by:

$$L_{n,\,l} = b_0 + b_1 r + b_2 r^2 + \ldots + b_{n-l-1}\,r^{n-l-1} \qquad \ldots(3.85)$$

The middle term of the expression clearly reveals that each function is dependent on a certain set of quantum numbers and these quantum numbers define that particular function. The quantum numbers can take on following ranges of values.

$$n = 1, 2, 3, \ldots$$

$$l = 0, 1, 2, \ldots .n-1 \qquad n \text{ values}$$

$$m_l = 0, \pm 1, \pm 2, \pm 3, \ldots ..\pm l \qquad 2l+1 \text{ values}$$

For example, the function $\Phi_{m_l}(\varphi)$ can be defined precisely if we know the quantum number m_l, the precise determination of function $\Theta_{l,m_l}(\theta)$ needs two quantum numbers, (l, m_l) and $R_{n,\,l}(r)$ requires the knowledge of quantum numbers n, l. The functions $\Phi_{m_l}(\varphi)$, $\Theta_{l,m_l}(\theta)$, and $R_{n,\,l}(r)$ can be calculated with the aid of equations (3.24), (3.27) and (3.63) respectively and are listed in Table 3.4, for $n = 1, 2, 3$ with respective values of l and m_l. Like any wave function in quantum mechanics, these wave functions also possess certain characteristics. For example:

(*i*) Every allowed contribution of three quantum numbers defines a state of atom, and for every state there is a corresponding eigen function $\Psi_{n,l,m_l}(r,\theta,\varphi)$. The function $\Psi_{n,l,m_l}(r,\theta,\varphi)$ is in general complex. If $\Psi^*_{n,l,m_l}(r,\theta,\varphi)$ is its complex conjugate, then the real quantity

$$P = \Psi^*_{n,l,m_l}(r,\theta,\varphi)\;\Psi_{n,l,m_l}(r,\theta,\varphi) \qquad \ldots(3.86)$$

represents the *probability density function (probability per unit volume)* of finding the electron at a given point in space from the nucleus of the hydrogen like atoms. The beauty of these wave functions is that these functions form a orthonormal set even if one of the three quantum numbers is different (*i.e., it is not necessary that all the three or two of the three quantum numbers have to be different to have a orthonormal set*). The integral (3.86) vanishes except for $n = n'$, $l = l'$ and $m_l = m_{l'}$. In other words, the integral in Φ will be zero if $m_l \neq m_{l'}$, if $m_l = m_l$ but $l \neq l'$ the integral in Θ will be zero and lastly even if $m_l = m_{l'}$, $l = l'$, the integral in R (r) will be zero if $n \neq n'$.

The last Column in Table 3.4 gives the complete wave function, where the Φ part has been expressed in real functions because its description in real form is quite convenient and gives easy relations of angular part to the Cartesian coordinates. The wave functions expressed in Cartesian coordinates (*e.g., x, y, z, xy, xz,* etc.) helps in explaining the directional properties of the chemical bonds and also this type of description is very common in chemistry. As usual the total probability of finding the electron in a bound state is unity,

$$\int_0^\infty \int_0^\pi \int_0^{2\pi} \Psi_{n,l,m_l} \Psi_{n,l,m_l} d\tau = \int_0^\infty R_{n,l}(r) R_{n,l}(r)\, r^2 dr$$

$$\int_0^\pi \Theta_{l\,m_l}(\theta)\, \Theta_{l\,m_l}(\theta) \sin\theta\, d\theta \int_0^{2\pi} \Phi_{m_l}(\varphi)\, \Phi_{m_l}(\varphi)\, d\varphi = 1 \qquad ...(3.87)$$

That is the individual wave functions and hence the total wave function is normalized. The charge density distribution for the atom can be obtained by multiplying the probability density by the electronic charge.

(*ii*) The orthonormal relationship between the respective wave functions is observed, *i.e.,*

$$\int_0^\infty \int_0^\pi \int_0^{2\pi} \Psi_{n,l,m_l} \Psi_{n,l,m_l} d\tau = \int_0^\infty R_{n,l}(r) R_{n',l'}(r) r^2 dr$$

$$\int_0^\pi \Theta_{l,m_l}(\theta)\, \Theta_{l',m'_l}(\theta) \sin\theta\, d\theta \int_0^{2\pi} \Phi_{m_l}(\varphi)\, \Phi_{m'_l}(\varphi) d\varphi \qquad ...(3.88)$$

Table 3.4: *Real Hydrogen-like Wave functions.* $\sigma = (Z\,r/a_0)$

n	l	m_l	Wave function
1	0	0	$\Psi_{1s} = \dfrac{1}{\sqrt{\pi}}\left(\dfrac{Z}{a_0}\right)^{3/2} e^{-\sigma}$
2	0	0	$\Psi_{2s} = \dfrac{1}{4\sqrt{2\pi}}\left(\dfrac{Z}{a_0}\right)^{3/2} (2-\sigma)\, e^{-\sigma/2}$
2	1	0	$\Psi_{2p_z} = \dfrac{1}{4\sqrt{2\pi}}\left(\dfrac{Z}{a_0}\right)^{3/2} \sigma\, e^{-\sigma/2} \cos\theta$
2	1	± 1	$\Psi_{2p_x} = \dfrac{1}{4\sqrt{2\pi}}\left(\dfrac{Z}{a_0}\right)^{3/2} \sigma\, e^{-\sigma/2} \sin\theta\cos\varphi$
2	1	± 1	$\Psi_{2p_y} = \dfrac{1}{4\sqrt{2\pi}}\left(\dfrac{Z}{a_0}\right)^{3/2} \sigma\, e^{-\sigma/2} \sin\theta\sin\varphi$

3	0	0	$\Psi_{3s} = \frac{1}{81\sqrt{3\pi}}\left(\frac{Z}{a_0}\right)^{3/2}(27-18\sigma+2\sigma^2)\, e^{-\sigma/3}$
3	1	0	$\Psi_{3p_z} = \frac{\sqrt{2}}{81\sqrt{\pi}}\left(\frac{Z}{a_0}\right)^{3/2}(6-\sigma)\,\sigma\, e^{-\sigma/3}\cos\theta$
3	1	± 1	$\Psi_{3p_x} = \frac{\sqrt{2}}{81\sqrt{\pi}}\left(\frac{Z}{a_0}\right)^{3/2}(6-\sigma)\,\sigma\, e^{-\sigma/3}\sin\theta\cos\varphi$
3	1	± 1	$\Psi_{3p_y} = \frac{\sqrt{2}}{81\sqrt{\pi}}\left(\frac{Z}{a_0}\right)^{3/2}(6-\sigma)\,\sigma\, e^{-\sigma/3}\sin\theta\sin\varphi$
3	2	0	$\Psi_{3d_{z^2}} = \frac{1}{81\sqrt{6\pi}}\left(\frac{Z}{a_0}\right)^{3/2}\sigma^2 e^{-\sigma/3}(3\cos^2\theta-1)$
3	2	± 1	$\Psi_{3d_{zx}} = \frac{\sqrt{2}}{81\sqrt{\pi}}\left(\frac{Z}{a_0}\right)^{3/2}\sigma^2 e^{-\sigma/3}\sin\theta\cos\theta\cos\varphi$
3	2	± 1	$\Psi_{3d_{yz}} = \frac{\sqrt{2}}{81\sqrt{\pi}}\left(\frac{Z}{a_0}\right)^{3/2}\sigma^2 e^{-\sigma/3}\sin\theta\cos\theta\sin\varphi$
3	2	± 2	$\Psi_{3d_{x^2-y^2}} = \frac{1}{81\sqrt{2\pi}}\left(\frac{Z}{a_0}\right)^{3/2}\sigma^2 e^{-\sigma/3}\sin^2\theta\cos 2\varphi$
3	2	± 2	$\Psi_{3d_{xy}} = \frac{1}{81\sqrt{2\pi}}\left(\frac{Z}{a_0}\right)^{3/2}\sigma^2 e^{-\sigma/3}\sin^2\theta\sin 2\varphi$

Example 3.8: Show that the respective wave functions $\Psi_{1,0,0} = \frac{1}{\sqrt{\pi}}\left(\frac{Z}{a_0}\right)^{\frac{3}{2}} e^{-\frac{Zr}{a_0}}$ and

$$\psi_{2,0,0} = \frac{1}{4\sqrt{2\pi}}\left(\frac{Z}{a_0}\right)^{\frac{3}{2}}\left(2-\frac{Zr}{a_0}\right)e^{-\frac{Zr}{a_0}}$$ are orthogonal.

Solution: Substitute the respective terms in equation (3.88).

$$\int \Psi_{1,0,0}\Psi_{2,0,0}\,d\tau = \int_0^\infty \frac{1}{\sqrt{\pi}}\left(\frac{Z}{a_0}\right)^{\frac{3}{2}} e^{-\frac{Zr}{a_0}} \frac{1}{4\sqrt{2}\,\pi}\left(\frac{Z}{a_0}\right)^{\frac{3}{2}}\left(2-\frac{Zr}{a_0}\right)e^{-\frac{Zr}{a_0}} r^2 dr$$

$$\int_0^\pi \sin\theta\, d\theta \int_0^{2\pi} d\varphi$$

$$= \frac{1}{4\pi\sqrt{2}}\left(\frac{Z}{a_0}\right)^3 \int_0^\infty \left(2-\frac{Zr}{a_0}\right)e^{-\frac{2Zr}{a_0}} r^2 dr\, 4\pi$$

$$= \frac{2}{\sqrt{2}}\left(\frac{Z}{a_0}\right)^3 \int_0^\infty e^{-\frac{2Zr}{a_0}} r^2 dr - \frac{1}{\sqrt{2}}\left(\frac{Z}{a_0}\right)^4 \int_0^\infty e^{-\frac{2Zr}{a_0}} r^3\, dr$$

$$= \frac{1}{\sqrt{2}}\left(\frac{Z}{a_0}\right)^3 \left[2\frac{2}{\left(\frac{3Z}{2a_0}\right)^3}\right] - \frac{1}{\sqrt{2}}\left(\frac{Z}{a_0}\right)^4 \left[\frac{Z}{a_0}\frac{3}{(3Z/2a_0)^4}\right]$$

$$= \frac{1}{\sqrt{2}}\left[\frac{Z^3}{a_0^3}\frac{32\, a_0^3}{27\, Z^3}\right] - \frac{1}{\sqrt{2}}\left[\frac{Z^4}{a_0^4}\frac{96\, a_0^4}{81\, Z^4}\right] = \frac{1}{\sqrt{2}}\left[\frac{32}{27} - \frac{32}{27}\right] = 0$$

The example has clearly revealed that wave function $\Psi_{1,0,0}$ and $\psi_{2,0,0}$ are orthogonal to each other even though l and m_l are the same for both the wave functions.

Example 3.9: Check the orthogonality of: (*a*) the angular parts of $3d_{z^2}$ and $3d_{x^2-y^2}$, and (*b*) the angular parts of $3d_{z^2-y^2}$ and $3d_{z^2-x^2}$ states.

Solution: The normalized angular wave functions for $3d_{z^2}$ and $3d_{x^2-y^2}$, are:

$$3d_{z^2} = \left(\frac{5}{16\pi}\right)^{\frac{1}{2}} (3\cos^2\theta - 1),$$

$$3d_{x^2-y^2} = \left(\frac{5}{16\pi}\right)^{\frac{1}{2}} \sin^2\theta\,(\cos^2\varphi - \sin^2\varphi) = \left(\frac{5}{16\pi}\right)^{\frac{1}{2}} \sin^2\theta\,\cos 2\varphi$$

These functions are orthogonal if:

$$\int_0^\pi \int_0^{2\pi} 3d_{z^2}\, 3d_{x^2-y^2}\, \sin\theta\, d\theta\, d\varphi = 0$$

$$= \frac{5}{16\pi}\int_0^\pi (3\cos^2\theta - 1)\,\sin^2\theta\,\sin\theta\, d\theta \int_0^{2\pi} \cos 2\varphi\, d\varphi$$

$$= \frac{5}{16\pi}\left[\int_0^\pi 3\cos^2\theta\,\sin^3\theta\, d\theta - \int_0^\pi \sin^3\theta\, d\theta\right] \int_0^{2\pi} \cos 2\varphi\, d\varphi$$

$$= \frac{5}{16\pi}\left[3\,\frac{4}{15} - \frac{4}{3}\right]\left[\frac{\sin 2\varphi}{2}\right]_0^{2\pi} = 0 \quad \text{as} \quad \left[\frac{\sin 2\varphi}{2}\right]_0^{2\pi} = 0$$

(*b*) The normalized angular parts of $3d_{z^2-y^2}$ and $3d_{z^2-x^2}$ are:

$$3d_{z^2-x^2} = \left(\frac{5}{16\pi}\right)^{\frac{1}{2}} (\cos^2\theta - \sin^2\theta\,\cos^2\varphi)$$

and $$3d_{z^2-y^2} = \left(\frac{5}{16\pi}\right)^{\frac{1}{2}} (\cos^2\theta - \sin^2\theta \sin^2\varphi)$$

Now

$$\int_0^{\pi}\int_0^{2\pi} 3d_{z^2-x^2}\, 3d_{z^2-y^2} \sin\theta \, d\theta \, d\varphi$$

$$= \frac{15}{16\pi}\int_0^{\pi}\int_0^{2\pi} (\cos^2\theta - \sin^2\theta \cos^2\varphi)(\cos^2\theta - \sin^2\theta \sin^2\varphi) \sin\theta \, d\theta \, d\varphi$$

$$= \frac{15}{16\pi}\int_0^{\pi}\int_0^{2\pi} (\cos^4\theta - \sin^2\theta \cos^2\theta \sin^2\varphi - \sin^2\theta \cos^2\theta \cos^2\varphi + \sin^4\theta$$

$$\cos^2\varphi \sin^2\varphi) \sin\theta \, d\theta \, d\varphi$$

$$= \frac{15}{16\pi}\int_0^{\pi} \cos^4\theta \sin\theta \, d\theta \int_0^{2\pi} d\varphi - \frac{15}{16\pi}\int_0^{\pi}\int_0^{2\pi} \sin^2\theta \cos^2\theta \, (\sin^2\varphi + \cos^2\varphi)$$

$$\sin\theta \, d\theta \, d\varphi + \frac{15}{16\pi}\int_0^{\pi} \sin^5\theta \, d\theta \int_0^{2\pi} \sin^2\varphi \cos^2\varphi \, d\varphi$$

Using the following definite integrals and solving one integral at a time:

$$\int \cos^m x \sin^n x \, dx = \frac{\cos^{m-1} ax \sin^{n+1} ax}{(m+n)a} + \frac{m-1}{(m+n)a}\int \cos^{m-2} ax \sin^n ax \, dx$$

$$\int \sin^n x \, dx = -\frac{\sin^{n-1} ax \cos ax}{na} + \frac{n-1}{na}\int \sin^{n-2} ax \, dx$$

$$\int (\sin ax)(\cos^m ax) \, dx = -\frac{\cos^{m+1} ax}{(m+1)a}$$

$$\int (\sin^m ax)(\cos ax) dx = -\frac{\sin^{m+1} ax}{(m+1)a}$$

i.e.,

$$\int_0^{\pi} \cos^4\theta \sin\theta \, d\theta = \left[-\frac{\cos^5\theta}{5}\right]_0^{\pi} = -\frac{1}{5}[\cos^5\pi - \cos^5 0] = \frac{2}{5}$$

$$\int_0^{\pi} \sin^3\theta \cos^2\theta \, d\theta = \left[\frac{\cos\theta \sin^4\theta}{3+2}\right]_0^{\pi} + \frac{1}{5}\int_0^{\pi} \sin^3\theta \, d\theta = \frac{1}{5}\int_0^{\pi} \sin^3\theta \, d\theta$$

$$= \frac{1}{5}\left[-\frac{\cos\theta\,\sin^2\theta}{3}\right]_0^\pi + \frac{1}{5}\,\frac{2}{3}\int_0^\pi \sin\theta\,d\theta = \frac{2}{15}\int_0^\pi \sin\theta\,d\theta$$

$$= -\frac{2}{15}\,[\cos\theta]_0^\pi = -\frac{2}{15}[\cos\pi - \cos 0] = \frac{4}{15}$$

$$\int_0^\pi \sin^5\theta\,d\theta = \left[-\frac{\sin^4\theta\,\cos\theta}{5}\right]_0^\pi + \frac{4}{5}\int_0^\pi \sin^3\theta\,d\theta = \frac{4}{5}\int_0^\pi \sin^3\theta\,d\theta$$

$$= \frac{4}{5}\left[-\frac{\sin^2\theta\,\cos\theta}{3}\right]_0^\pi + \frac{4}{5}\,\frac{2}{3}\int_0^\pi \sin\theta\,d\theta = \frac{8}{15}\int_0^\pi \sin\theta\,d\theta$$

$$= \frac{8}{15}[-\cos\theta]_0^\pi = \frac{16}{15}$$

$$\int_0^{2\pi} \cos^2\varphi\,\sin^2\varphi\,d\varphi = \left[\frac{\sin^3\varphi\,\cos\varphi}{4}\right]_0^{2\pi} + \frac{1}{4}\int_0^{2\pi} \sin^2\varphi\,d\varphi = \frac{1}{4}\int_0^{2\pi} \sin^2\varphi\,d\varphi$$

$$= \frac{1}{4}\left[-\frac{\sin\varphi\,\cos\varphi}{2}\right]_0^{2\pi} + \frac{1}{4}\cdot\frac{1}{2}\int_0^{2\pi} d\varphi = \frac{\pi}{4}$$

Substituting the values of the respective integrals in the main equation, we get

$$= \frac{15}{16\pi}\left[\frac{2}{5}\,2\pi - \frac{4}{15}\,2\pi + \frac{16}{15}\,\frac{\pi}{4}\right] = \frac{15}{16}\left[\frac{4}{5} - \frac{8}{15} + \frac{4}{15}\right] = 1/2$$

Thus the angular parts of $3d_{z^2-y^2}$ and $3d_{z^2-x^2}$ states are *not* orthogonal to each other. *But note that the linear combinations of* $3d_{z^2-y^2}$ *and* $3d_{z^2-x^2}$ *are orthogonal. (See equations* 3.97 *and* 3.98*).*

3.1.7 GEOMETRICAL DESCRIPTION OF HYDROGEN LIKE ATOMIC ORBITALS

Geometric description of Bohr's orbits is quite simple because these are defined as the fixed path of an electron moving in a circle. The radius of these orbits can be calculated with the help of equation (1.85). Whereas in quantum mechanics, the behaviour of electron is described in three-dimensional model and its description is more complex. Before we describe this behaviour, let us define the term *orbital* or to be more specific *atomic orbital*, which we have used quite often. *Spatial part of one electron wave function or a wave function, which describes the behaviour of one electron in hydrogen like atoms, is called an atomic orbital.* This definition is only valid for one electron system and does not hold for multi-electron systems although we use it very regularly and commonly. In this section we shall concentrate ourselves with one-electron atoms.

There are two fundamental ways of depicting the orbital. First method deals with plotting the graphs of the functions and the second method deals with drawing contour surfaces of constant probability density. There is a lot of confusion between the two methods. Many books plot the angular part of the wave functions and called these as the orbital contour surfaces, which is wrong. We shall try to keep the distinction between the two methods very clear.

Graphical Representation of Orbitals: The wave function of any orbital is a function of three variables, *i.e.*, r, θ, and φ. To plot the variation of ψ as a function of three variables, we need a four dimensional space. This prevents us from plotting these functions to any meaningful representation of ψ and also it is not easy to grasp and follow the graphical representations. To understand and to have better feeling, we draw the graphs of individual functions, *i.e.*, the plotting of R (r) gives information about the behaviour of an electron as a function of its distance form the nucleus and is called radial function. This plot gives no information about the angular behavior of ψ. Similarly the plot of $\Theta(\theta)\,\Phi(\varphi)$ by keeping r fixed gives the angular description of ψ and is called angular plots, also defined as $Y(\theta, \varphi)$.

Let us define couple of terms used quite often in the coming section. The probability of finding the electron in the regions of space where r is varying from r to $r + dr$, from θ to $\theta + d\theta$ and from φ to $\varphi + d\varphi$ is:

$$\psi^{*}\psi d\tau = R^{2}_{n,l}(r)\Theta^{2}_{l,m_l}\Phi^{2}_{m_l}(\varphi)\,r^{2}\sin\theta dr d\theta d\varphi \qquad ...(3.89)$$

Now let us define the probability of finding electron having its coordinate r to $r + dr$ with no restriction on the values of θ and φ. In other words, we want to calculate the probability of finding the electron in a thin spherical shell centered at the origin with the inner radius equal to r and outer radius $r + dr$. This can be obtained by adding the probabilities calculated from equation (3.89) for all possible values of θ and φ. Thus the probability of finding the electron between r to $r + dr$ is:

$$R^{2}_{n,\,l}(r)\,r^{2}\,dr\int_{0}^{\pi}\int_{0}^{2\pi}\Theta^{2}_{l,m_l}(\theta)\,\Phi^{2}_{m_l}(\varphi)\sin\theta\,d\theta\,d\varphi = 4\pi R^{2}_{n,\,l}(r)\,r^{2}dr \qquad ...(3.90)$$

because the angular functions are normalized, *i.e.*,

$$\int_{0}^{\pi}\int_{0}^{2\pi}\Theta^{2}_{l,m_l}(\theta)\,\Phi^{2}_{m_l}(\varphi)\sin\theta\,d\theta\,d\varphi = 1$$

The function $R^{2}_{n,\,l}(r)\,r^{2}$, which determines the probability of finding the electron at a distance r from the nucleus, is called the *radial distribution functions*. Thus the total probability density is determined by the product of radial probability function and angular probability P_θ function as it affects the radial probability P_r. Thus the quantity $\Psi^{*}\Psi\, d\tau$ is the probability that the electron is located with in the volume element $d\tau$. Thus one can divide the complete space into small volume elements and calculate the average or some representative value of $\Psi^{*}\Psi$ within each volume element and then represent the value $\Psi^{*}\Psi$ by the density dot in a picture. Figure 3.7 depicts such plot for number of orbitals. In the p-state when $m_l = 0$, Figs. (3.6) and (3.7*b*), the two lobes are symmetrical about the z-axis and the most probable position of the electron is indicated by the density points in Fig. (3.7*b*). The Bohr Theory makes no provision for locating the electron in the positions indicated by Fig. (3.2), Fig. (3.6) and Fig. (3.7). But there is some correspondence for the cases $l = n - 1$. Figure (3.2) indicates that the most probable location of the electron corresponds closely to the radius determined by the Bohr Theory.

Example 3.10: Calculate the values of the quantum numbers n, l, m_l for the hydrogen like wave function, which has the expression:

$$\Psi_{n,l,m_l} = N \left(\frac{Zr}{a_0}\right) e^{-\frac{Zr}{3a_0}} \sin^2\theta \cos 2\varphi$$

Solution: From Table 3.4, it is clear that the radial wave function is simple power of r when l is having its maximum value; otherwise it is a series or sum of terms. In the former case the value of $l = n - 1$. In the present case, $l = 2 = n - 1$, so $n = 3$. Similarly the angular wave function involving θ is a simple power of sin θ if m_l is having a maximum value. So $l = m_l = 2$ and is supplemented by the angular function involving φ part. So $n = 3$, $l = 2$, $m_l = 2$.

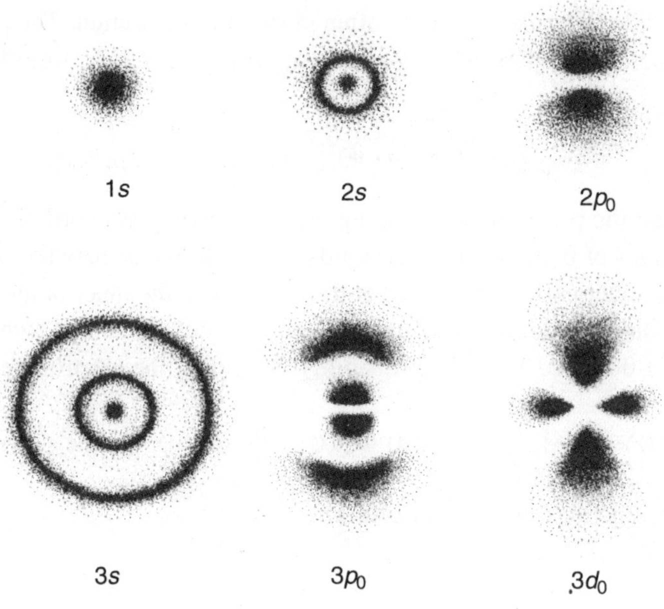

Figure 3.7: *The Radial density distribution functions in three dimensions of* Ψ^2_{n,l,m_l} *is represented by the density points. The figures are obtained by combining Figs. 3.2 and 3.6*

Radial Wave Function: The ground state wave function for hydrogen like atoms is:

$$\psi_{1s} = \psi_{1,0,0} = \left(\frac{Z^3}{\pi a_0^3}\right)^{\frac{1}{2}} e^{-\frac{Zr}{a_0}}$$

where Z is the atomic charge. The function is independent of θ and φ and is called 1s orbital, *i.e.*, ψ_{1s}. ψ and ψ^2 as a function of r is plotted in Figs. (3.4) and (3.5) respectively. It is clear form the figures that ψ and ψ^2 are not zero at $r = 0$. The ψ^2 plot indicates that the probability density of finding the electron along any distance r from the nucleus is the maximum at the nucleus. Physically this is also correct as is visualized from the potential energy function which has a maximum value at $r = 0$. But we are more interested in the probability of finding the electron at a particular distance r and not along any fixed value of θ and φ. This is given by the *radial distribution function* $r^2R^2(r)$ or $4\pi r^2\psi^2$. This plot as a function of r is shown in Fig. 3.6. Since radial distribution function contains r, the function is zero at $r = 0$, increases to a maximum value (because at smaller r, r^2 will be a dominant term) and then decreases to zero at $r = \infty$ (because of the presence of exponential term). The behaviour of this plot is quite different from Bohr's model.

(*i*) Unlike Bohr's model the electron is not localized over an orbit of certain fixed distance from the nucleus, but completely smeared out over the whole space. The only common thing is that the electron is most likely to be located at a distance (known as most probable distance) equal to that given by Bohr's model for $n = 1$. The most probable distance can be calculated by differentiating the radial distribution function with respect to r and equating to zero, *i.e.*,

$$\frac{d}{dr}\left[4\pi r^2 \frac{Z^3}{\pi a_0^3} e^{-\frac{2Zr}{a_0}}\right] = 0$$

$$4\frac{Z^3}{a_0^3}\left[2re^{-\frac{2Zr}{a_0}} - \frac{2Zr^2}{a_0} e^{-\frac{2Zr}{a_0}}\right] = 0$$

or
$$2r\left(1 - \frac{Zr}{a_0}\right) = 0$$

or
$$r = 0 \text{ or } r = \frac{a_0}{Z} \qquad \text{...(3.91)}$$

The value of $r = 0$ can be ignored as it gives absurd results. For hydrogen atom, $r = a_0$. This indicates that the most probable value of r for $1s$ orbital, decreases with increase of Z, *i.e.*, the electron is confined over a smaller distance from the nucleus. This is because at shorter distance the exponential term under weighs the r^2 term.

(*ii*) Size of Bohr's orbits are very well defined and can be calculated numerically. On the other hand the size of the orbital is very vague, because the electron is scattered over a complete space. Few models have been suggested to define the size of the orbital. First one is the value of r where the probability of finding the electron is maximum, *i.e.*, most probable distance ($r = a_0$). But if one uses this value and calculates the total probability, it is only ~ 33%. The other model defines the size of the orbital as the average value of r, which is calculated as:

$$\langle r \rangle = \int_0^{\infty} \psi^* \, r \, \psi \, d\tau = \frac{Z^3}{\pi a_0^3} \int_0^{\infty} e^{-\frac{Zr}{a_0}} r e^{-\frac{Zr}{a_0}} r^2 \sin\theta \, dr \, d\theta \, d\varphi \qquad ...(3.92)$$

$$= \frac{4Z^3}{a_0^3} \int_0^{\infty} r^3 e^{-\frac{2Zr}{a_0}} dr = \frac{4Z^3}{a_0^3} \left[\frac{3}{\left(\frac{2Z}{a_0}\right)^4} \right] = \frac{4Z^3}{a_0^3} (6) \frac{a_0^4}{16Z^3} = \frac{3}{2} a_0$$

The average value of r in general for any orbital is given by:

$$\langle r \rangle_{n,l} = \int_0^{\infty} \psi^*_{n,l} \, r \, \psi_{n,l} \, d\tau = \frac{n^2 a_0}{Z} \left[1 + \frac{1}{2} \left\{ 1 - \frac{l\,(l+1)}{n^2} \right\} \right] \qquad ...(3.93)$$

For hydrogen atom $\langle r \rangle = 1.5 \, a_0$.

(*iii*) Other useful method of describing the size of the orbital is to calculate the value of r for a fixed value of the total probability. For example, the value of r, which specifies that the electron will be found 90% in this sphere, can be obtained as:

$$\int_0^{r} \frac{Z^3}{\pi a_0^3} \psi_{1,0,0}^2 \, d\tau = \frac{4Z^3}{a_0^3} \int_0^{r} r^2 e^{-\frac{2Zr}{a_0}} dr = 0.9 \qquad ...(3.94)$$

Integrating the above expression and solving for r, one gets $r = 2.6 \, a_0$, which is larger than r_{mp} and $\langle r \rangle$. All these calculations have shown that the useful size of hydrogen varies from a_0 to $2.6 \, a_0$ or 0.53×10^{-10} m to 1.4×10^{-10} m. In other words the dimensions of the atoms are of the order of angstrom (10^{-8} cm) or 10^{-10} m. The experimental determination of the size of atoms also gives this kind of order of magnitude and thus consistent with the above calculations.

For $n = 2$, there can be four different quantum states, $\psi_{2,0,0}$, $\psi_{2,1,-1}$, $\psi_{2,1,0}$, and $\psi_{2,1,+1}$. The state $\psi_{2,0,0}$ can be called as ψ_{2s} orbital and the other three as p orbitals ($2p_0$, $2p_{-1}$ and $2p_{+1}$). It has been shown in section (3.1.1) that ψ_{2p-1} and ψ_{2p+1} (Table 3.1) functions containing imaginary parts, can be combined to form the real functions. These can be denoted as ψ_{2pz} ψ_{2px} and ψ_{2py}. Since the radial functions depend on the values of n and l, it has a different value for ψ_{2s}, but has the same value for all the three p orbitals. Further the exponential part is also different from that of ψ_{1s}. Similarly for $n = 3$, there can be nine different states. On the similar grounds as explained above, the radial functions for three p orbitals and five d orbitals are similar to each other.

The wave functions, probability density functions and radial distribution functions for $2s$, $3s$, $4s$ etc. are plotted in Figs. 3.4, 3.5 and 3.6 respectively. One thing common in these plots and plot for $1s$ is that the radial functions and probability density functions are not zero at $r = 0$, whereas the radial distribution functions are zero at $r = 0$. Unlike $1s$ orbital, the radial functions for higher s orbitals become zero at other points, besides when $r = \infty$.

This behaviour is similar to that observed for particle in a box and simple harmonic oscillator. Thus these points are called *nodes*. Larger the value of n, the greater will be the number of nodes and consequently one can correlate the presence of nodes with the energy of the states. The number of radial nodes observed can be equated to $n - l - 1$, as mentioned earlier, *i.e.,* the number of nodes present in $2s$ and $3s$ orbitals is equal to one and two respectively. Other interesting point in the radial distribution function for higher s orbitals is that there are more than one maximum in the plot. This indicates that the electron in s orbitals can go closer to the nucleus. In other words, the probability of finding the electron is quite large near the nucleus. This point is quite helpful in explaining many different characteristics of s orbitals from p, d, f,....... orbitals; for example, shielding of the nucleus by different orbitals, penetration of the electrons closer to the nucleus in different orbitals etc.

The radial functions, probability density functions and radial distribution functions decrease very slowly to zero value for larger value of n. This is because the exponential terms become smaller. In the physical sense, the presence of electron is scattered over a larger distance and charge cloud is more smeared out or diffused. This also tells that the size of the atom will increase with increase of n and similar results are also noticed from experimental observations.

The wave functions, probability density functions and radial distribution functions for $2p$, $3p$, and $3d$ are plotted in Figs. 3.4, 3.5 and 3.6 respectively. Because of the presence of r term, the radial functions for all the orbitals other than s are zero at $r = 0$. Other features of these plots are similar to those discussed for ns orbitals, *i.e.,* (*i*) number of radial nodes is given by the relation $(n - l - 1)$. This means that there is no node in $2p$ and $3d$ orbitals, but one node in $3p$ orbital. (*ii*) As the value of n increases, the size of p or d orbitals increases or the electron is scattered over a larger distance from the centre of the nucleus.

Example 3.11 (*a*): Prove that for 1s hydrogen atom, the most probable distance of the electron is 0.053 nm, (*ii*) also calculate the average value of distance of electron as well as its square from the nucleus.

Solution:

(*a*) For $1s$ state of hydrogen atom, $n = 1$, $l = 0$. Radial wave function is (from Table 3.4):

$$R_{1,0} = 2\left(\frac{1}{a_0}\right)^{\frac{3}{2}} e^{-\frac{r}{a_0}}$$

Thus, the radial distribution function R (r) for $1s$-state is:

$$R(r) = 4\pi r^2 (R_{1,0})^2 = 4\pi r^2 \left[2\left(\frac{1}{a_0}\right)^{\frac{3}{2}} e^{-\frac{r}{a_0}}\right]^2 = 16\pi \frac{1}{a_0^3} e^{-\frac{2r}{a_0}} r^2$$

The condition for the most probable distance is given by $dR_r/dr = 0$

Thus $$\frac{d}{dr}\left[16\pi\,\frac{1}{a_0^3}\,r^2 e^{-\frac{2r}{a_0}}\right] = 0$$

$$16\pi\frac{1}{a_0^3}\left[2re^{-\frac{2r}{a_0}} - \frac{2r^2}{a_0}\,e^{-\frac{2r}{a_0}}\right] = 0$$

$$16\pi\,\frac{1}{a_0^3}\,e^{-\frac{2r}{a_0}}\left[2r - \frac{2r^2}{a_0}\right] = 0$$

i.e., either $e^{-\frac{2r}{a_0}} = 0,\ r = 0$ or $r = a_0$

But $e^{-\frac{2r}{a_0}} \neq 0$, so $r = 0$ can be discarded as it does not give good results. The other values of $r = a_0 = 0.053$ nm, is the most probable smallest distance of electron from the nucleus of hydrogen atom. This distance is also termed as Bohr's radius of hydrogen atom.

(*b*) The average distance of electron from the nucleus is given by the integral:

$$\langle r\rangle = \int_0^\infty R_{1,\,0}\ r\ R_{1,\,0}\ r^2 dr = \frac{4}{a_0^3}\int_0^\infty e^{-\frac{2r}{a_0}}\,r^3 dr$$

Since $$\int_0^\infty e^{-\alpha r}\,r^n dr = \left(\frac{1}{\alpha}\right)^{n+1} n\,!$$

So the integral reduces to: $$\frac{4}{a_0^3}\left[\left(\frac{a_0}{2}\right)^4\cdot 3\right] = \frac{4}{a_0^3}\frac{a_0^4}{16}6 = 1.5a_0$$

The average value of r^2 is:

$$\langle r^2\rangle = \int_0^\infty R_{1,0}\ r^2 R_{1,0}\ r^2 dr = \frac{4}{a_0^3}\int_0^\infty e^{-\frac{2r}{a_0}}\,r^4 dr = \frac{4}{a_0^3}\left[\left(\frac{a_0}{2}\right)^5\cdot 4!\right]$$

$$= \frac{4}{a_0^3}\,\frac{a_0^5}{32}\cdot 24 = 3a_0^2.$$

Example 3.12: The standard deviation of the distance of electron from the nucleus (Δr) in the ground state of hydrogen is defined as: $\Delta r = \left[\langle r^2\rangle - (\langle r\rangle)^2\right]^{\frac{1}{2}}$. Calculate its value.

Solution: Using the results of example 3.11, $\langle r^2 \rangle = 3\,a_0^2$ and $\langle r \rangle = 1.5\,a_0$.

$$\Delta r = [3a_0^2 - (1.5\,a_0)^2]^{\frac{1}{2}}\ [3\,a_0^2 - 2.25\,a_0^2]^{\frac{1}{2}} = 0.867\,a_0$$

Example 3.13: (*i*) Calculate the most probable distance of 2s orbital of hydrogen atom. (*ii*) Calculate the radial probability density at these values of r_{mp} and comment on the results.

Solution:

(*i*) Taking the value of ψ_{2s} wave function from Table 3.4, the radial distribution density function is given as:

$$R(r) = \left[\frac{1}{4\sqrt{2\pi}}\left(\frac{1}{a_0}\right)^{\frac{3}{2}}\right]^2 \left(2 - \frac{r}{a_0}\right)^2 r^2 e^{-\frac{r}{a_0}} . 4\pi$$

$$= \frac{1}{8a_0^3}\left(4 - \frac{4r}{a_0} + \frac{r^2}{a_0^2}\right) r^2 e^{-\frac{r}{a_0}}$$

$$= \frac{1}{8a_0^5}\left(4a_0^2 r^2 - 4a_0 r^3 + r^4\right) e^{-\frac{r}{a_0}}$$

Differentiating with respect to *r* and equating the derivative to zero, we get

$$\frac{dR(r)}{dr} = \frac{1}{8a_0^5}\left(8a_0^2\,r - 12a_0 r^2 + 4r^3\right) e^{-\frac{r}{a_0}}$$

$$- \frac{1}{8a_0^5}\left(4a_0^2\,r^2 - 4a_0\,r^3 + r^4\right) \frac{1}{a_0} e^{-\frac{r}{a_0}}$$

or
$$= \frac{r}{8a_0^6}\left(8a_0^3 - 12a_0^2\,r + 4a_0\,r^2 - 4a_0^2\,r + 4a_0 r^2 - r^3\right) e^{-\frac{r}{a_0}} = 0$$

i.e., $e^{-\frac{r}{a_0}} \neq 0$ and $r = 0$ will not give the good results, so:

$$r^3 - 8a_0\,r^2 + 16a_0^2\,r - 8\,a_0^3 = 0$$

$$\left(r^2 - 8a_0^3\right) - 8a_0 r\,(r - 2a_0) = 0$$

$$(r - 2a_0)\left(r^2 + 2a_0 r + 4a_0^2 - 8a_0 r\right) = 0$$

$$(\mathrm{r} - 2a_0)\left(r^2 + 2a_0 r + 4a_0^2 - 8a_0 r\right) = 0$$

The above equation has three roots. One of these is $r = 2a_0$, and the other two can be calculated from:

$$r^2 - 6a_0 r + 4a_0^2 = 0$$

i.e., $$r = \frac{6a_0 \pm (36a_0^2 - 16a_0^2)^{\frac{1}{2}}}{2} = \frac{6a_0 \pm 2\sqrt{5}\,a_0}{2}$$

or $$r = (3 + \sqrt{5})\, a_0 \text{ and } r = (3 - \sqrt{5})\, a_0.$$

R (r) at $r = 2a_0$ is:

$$= \frac{1}{8a_0^5}\left[\left(4a_0^2\,(2a_0)^2 - 4a_0\,(2a_0)^3 + (2a_0)^4\right)\right] e^{-\frac{2a_0}{a_0}}$$

$$= \frac{1}{8a_0^5}\left(16a_0^4 - 32a_0^4 + 16a_0^4\right) e^{-2} = 0$$

R (r) at $r = (3 + \sqrt{5})\, a_0$ comes out to be $0.191\ a_0$

and at $r = (3 - \sqrt{5})\, a_0$, the value is $0.052\ a_0$

The above results indicate that at $r = 2a_0$, the probability distribution function is zero. This means that we have a minimum in the radial distribution function between the two maxima. This can be understood as follows: The wave function ψ_{2s} has a node at $r = 2a_0$ at which the probability density function and the probability distribution function are zero, *i.e.*, in going from ψ_{2s} to ψ_{2s}^2 the position of the node does not change, with the difference that ψ_{2s} changes sign at node, whereas ψ_{2s}^2 remains positive through out. Even the radial distribution, functions obtained by multiplying the probability density function by $4\pi r^2$ does not change the nodal point. The only difference is that the function is large at large value of r relative to its value at small value of r. Thus the first maximum is at $r = (3 - \sqrt{5})\, a_0$ and the second maximum is at $r = (3 + \sqrt{5})\, a_0$. The plots have been shown in Figs. 3.4, 3.5 and 3.6 and results agree with these values.

Example 3.14: Verify that the probability density functions of all the orbitals with $l \geq 1$ are symmetrical about z-axis.

Solution: The wave functions of all the orbitals with $l \geq 1$ contain the Φ part as:

$$\Psi_{n,l,m_l} = \text{N}\ \text{R}_{n,\,l}\,(r)\ \Theta_{l,m_l}\,(\theta)\ e^{im_l\varphi}$$

The probability density function is defined as:

$$\text{P}_{n,l,m_l} = \Psi^*_{n,l,m_l}\ \Psi_{n,l,m_l} = \text{N}^2\ \text{R}^2_{n,\,l}\,(r)\ \Theta^2_{l,m_l}\,(\theta)\ e^{im_l\varphi}\ e^{-im_l\varphi}$$

$$= \text{N}^2\ \text{R}^2_{n,\,l}\,(r)\ \Theta^2_{l,m_l}\,(\theta)$$

Since P_{n,l,m_l} is independent of φ, all the probability density functions are symmetrical about the z-axis.

Example 3.15: The radial part of the wave function of $2p$ ($n = 2$, $l = 1$) orbital of hydrogen atom is: $R_{2,1}(r) = \left(\frac{Z}{a_0}\right)^{5/2} \frac{1}{2\sqrt{6}} re^{-\frac{Zr}{2a_0}}$. Plot $R_{2,1}(r)$ and $4\pi^2 r^2 R^2_{2,1}(r)$ versus r.

Solution: For hydrogen atom Z = 1

$$R_{2,1}(r) = \left(\frac{1}{a_0}\right)^{\frac{5}{2}} \frac{1}{\sqrt{24}} re^{-\frac{r}{2a_0}} \text{ and } 4\pi r^2 R^2_{2,1}(r) = \frac{4}{a_0^5}\frac{1}{24} r^4 e^{-\frac{r}{a_0}} \pi.$$

Values of $R_{2,1}(r)$ and $4\pi r^2 R^2_{2,1}(r)$ as a function of r, are given in Table 3.5. The radial wave function and radial density distribution function are plotted in Figs. 3.4 and 3.6 as a function of distance r respectively. The only difference is that the vertical scale is arbitrary and does not correspond to with the number given in Table 3.5.

Table 3.5: *Values of $R_{2,1}(r)$ and $4\pi r^2 R^2_{2,1}$ at different values of r*

Sr. No.	$r(m)$	$R_{2,1}(r)$ $(10^{-14} m^{-3/2})$	$4\pi r^2 R^2_{2,1}$ $(10^{10} m^{-1})$	Sr. No	$r(m)$	$R_{2,1}(r)$ $(10^{-14} m^{-3/2})$	$4\pi r^2 R^2_{2,1}$ $(10^{10} m^{-1})$
1	0.5	3.13	–	9	4.5	0.64	1.16
2	1.0	3.90	2.1	10	5.0	0.44	0.69
3	1.5	3.64	4.16	11	5.5	0.31	0.39
4	2.0	3.03	5.11	12	6.0	0.21	0.22
5	2.5	2.36	4.85	13	7.0	0.09	0.06
6	3.0	1.77	3.91	14	8.0	0.04	0.02
7	3.5	1.28	2.81	15	9.0	0.02	0.004
8	4.0	0.92	1.87	16	10.0	0.001	0.0009

Angular Wave Functions

As we have observed in case of hydrogen atom and will observe later for poly-electronic atoms that the potential energy can be expressed as spherical symmetric for any isolated atoms. Thus the angular part of the atomic wave functions must be given by the spherical harmonics of Table 3.4, *i.e.,* the angular part of atomic wave functions can be shown by the *polar plots* of functions $\Theta\,\Phi$ in one or more planes passing through the origin. Polar plot of any angular function $f(\theta)$ is drawn on a *polar graph paper*. Polar graph paper consists of a central point called origin. Straight lines originate from central point radiating outward at every 5° or 10°. Complete procedure to plot polar function $f(\theta)$ is shown in example 3.16 and plot is drawn in Fig. 3.8.

Example 3.16: The angular wave function for hydrogen atom for l = 2, m_l = ±2 is $\Psi_{3d_{x^2-y^2}} = N\,R(r)\sin^2\theta\cos 2\varphi$. Prepare the plot for this orbital in xy-plane.

Solution: For fixed value of r and $\theta = \pi/2$, the angular wave function in xy-plane will vary as shown in Table 3.6 (Consider $N\,R(r)\sin^2\theta = 1$ in arbitrary units). Plot of cos 2φ is shown in Fig. 3.8.

Table 3.6: *Values of* $\Psi_{3d_{x^2-y^2}}$ *for different values of* φ. $(N\ R\ (r)\ sin^2\theta = 1)$

Φ	Cos 2φ	Φ	cos 2φ	Φ	cos 2φ	Φ	cos 2φ
0	1.0	90	–1.0	180	1.0	270	–1.0
10	0.94	100	– 0.94	190	0.94	280	– 0.94
20	0.76	110	– 0.766	200	0.766	290	– 0.766
30	0.5	120	– 0.5	210	0.5	300	– 0.5
40	0.174	130	– 0.174	220	0.174	310	– 0.174
50	– 0.174	140	0.174	230	– 0.174	320	0.174
60	– 0.5	150	0.5	240	– 0.5	330	0.6
70	– 0.766	160	0.766	250	– 0.766	340	0.766
80	– 0.94	170	0.94	260	– 0.94	350	0.94

The polar plots of different orbitals are shown in Figs. 3.9. In such a graph, the magnitude of $\Theta\ \Phi$ function, at any particular values of angle θ and φ is given by the radial distance from the origin. The graphs shown in Fig. 3.9 are cross-section of the complete three dimensional graphs and thus show the behaviour of $\Theta\ \Phi$ in only a single plane, as a function of either θ or φ alone. However the planes selected in Fig. 3.9 are those that most clearly display the shapes of the individual wave functions. Further, since

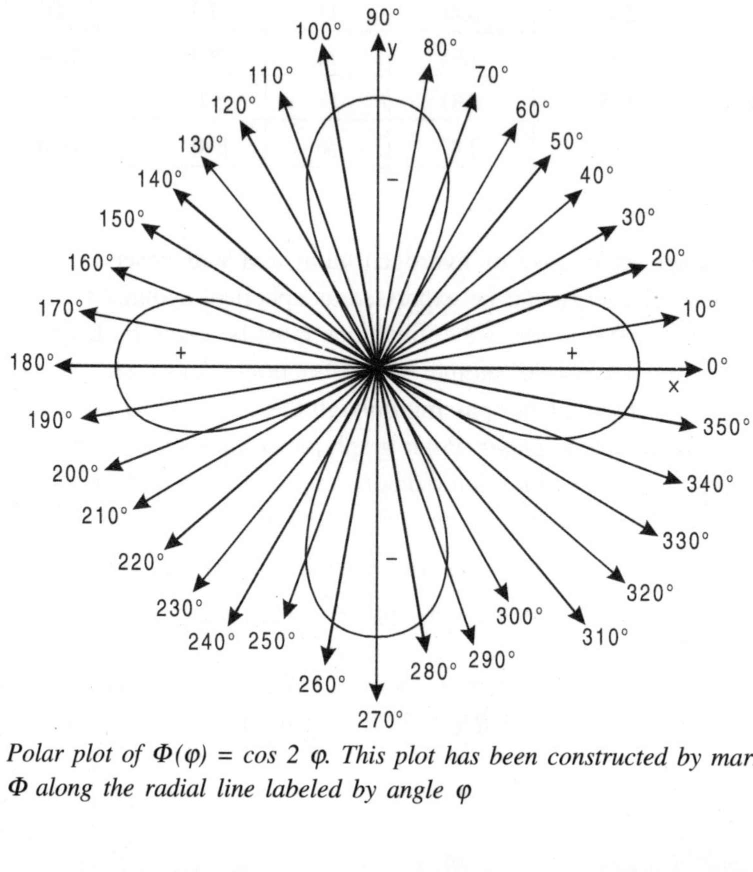

Figure 3.8: *Polar plot of* $\Phi(\varphi) = \cos 2\ \varphi$. *This plot has been constructed by marking a distance* Φ *along the radial line labeled by angle* φ

$\Theta_{l,m_l}(\theta)\,\Psi_{m_l}(\varphi)$ is independent of n; we can expect orbitals of same l but different n to be similar in shape.

The simplest angular wave function is of s orbital, *i.e.*, $l = 0$, $m_l = 0$. Since this function is totally independent of θ and φ and is a function of r only, this plot can be drawn in any direction or plane. Since its value at a given distance r from the nucleus will be the same, the contour diagram obtained by joining the points of the same magnitude will generate circles in a given plane. The joining of these circles will generate a sphere, *i.e.*, it will be spherically symmetric about the nucleus. It possesses no nodal plane and agrees with the fact that the number of nodal planes in angular wave functions is given by the value of azimuthal quantum number (l).

The next wave function is of p orbitals for which $l = 1$. Thus it will have three orbitals with $m_l = +1, 0, -1$. For $m_l = 0$, the angular function is cos θ and it corresponds to z orbital. For fixed value of r, the function will be proportional to cos θ and its plot as a function of θ in the range of 0 to π will generate two half circles along the z-axis touching tangentially to each other at the origin (Fig. 3.9 *a*). As we know that this function is independent of φ part, the graph can be drawn in any direction to the z-axis, in other words the plot is symmetrical about z-axis and thus will generate two circular lobes touching tangentially to each other at the origin (Fig. 3.9*b*). The upper lobe corresponds to positive values of cos θ and the lower lobe corresponds to negative values of cos θ. Since as said earlier, p_z orbital is independent of φ, the orbital is symmetrical about z-axis and the polar plot in xz-plane is similar to that in the yz-plane (see example 3.16 for $d_{x^2-y^2}$). It may be mentioned here that such a polar plot does not reflect the shape of an orbital and it only shows its direction or orientation. Fig.3.9*b* shows that this function is directed towards z-axis and hence is called p_z orbital. As mentioned earlier this function is independent of angle φ, thus, it is symmetrical about z-axis and complete three dimensional plot of this function can be obtained by rotating this orbital around z-axis. Similarly the angular plots of p_x and p_y functions are similar to p_z with the difference that the p_x is symmetric around x-axis (Fig. 3.9*c*) and p_y is symmetric around y-axis (Fig. 3.9*d*). These plots have been obtained by taking $\theta = 90$. It is also evident from the plots that all the p-orbital functions have only one nodal plane; xy-plane in case of p_z, xz-plane in case of p_y and yz-plane in case of p_x orbital.

For $l = 2$, there are five d orbitals, *i.e.*, $m_l = 0$; $m_l = \pm 1$, d_{xz}, d_{yz}; $m_l = \pm 2$, $d_{x^2-y^2}$, d_{xy} (Figs. 3.9 *e* to *i*.). The angular plot of d_{z^2}, displays large positive lobes along the z-axis and small negative lobe in the xy-plane. Since d_{z^2} is independent of φ, it will be symmetrical about the z-axis and the polar plot in xz- and yz-planes will be identical. The angular plot of other four d-orbitals, are similar, *i.e.*, it consists of four lobes with alternating positive and negative signs. These orbitals are d_{xz}, d_{yz}; d_{xy}, $d_{x^2-y^2}$ and vary with θ and φ in the same manner as do the functions xz, yz, xy, and $x^2 - y^2$ respectively. It is clear from the plots that all the d-functions except d_{z^2} have identical plots but possess different orientations. For $l = 2$, there will be two nodal

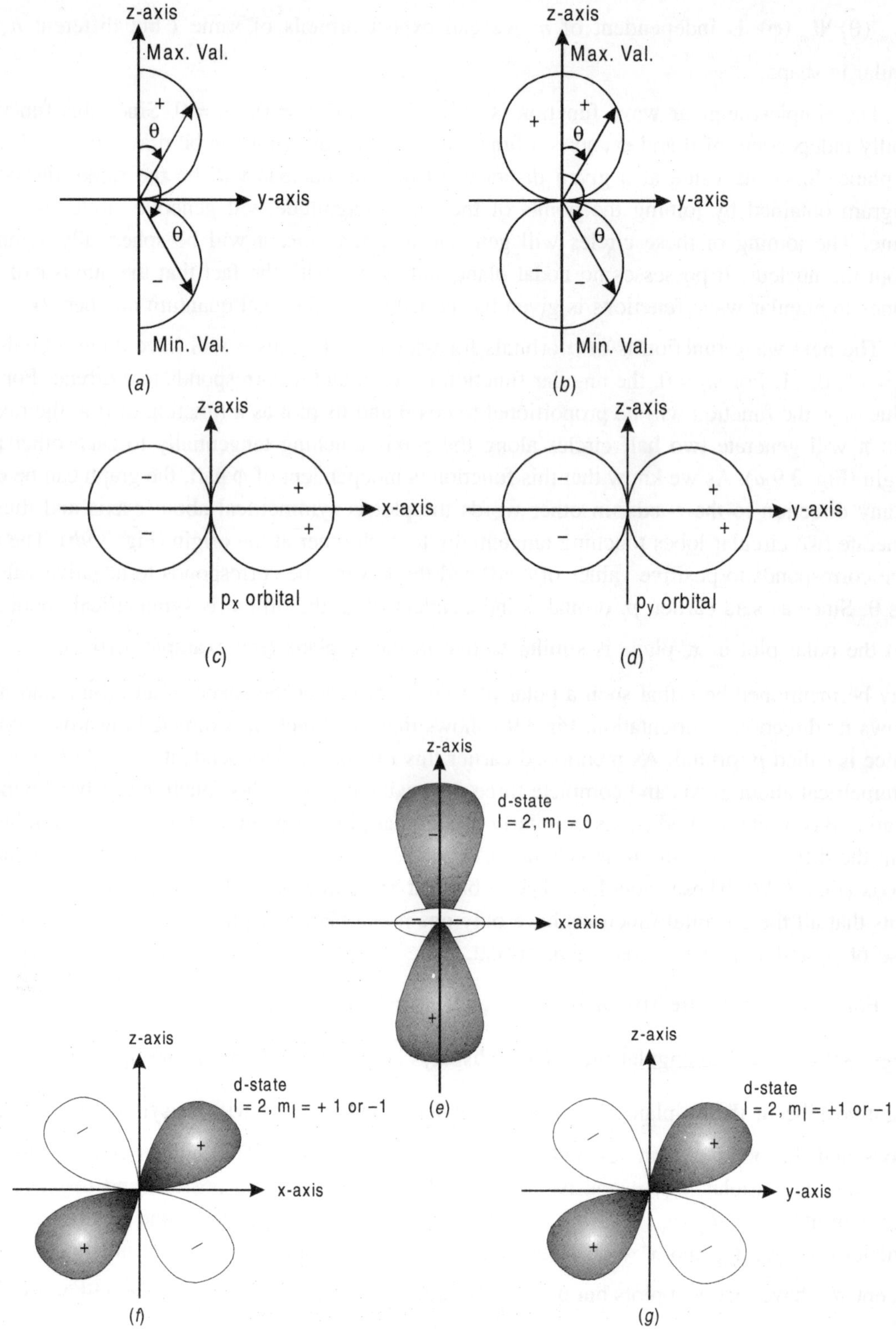
z-axis
Max. Val.
y-axis
Min. Val.
(a)
z-axis
Max. Val.
y-axis
Min. Val.
(b)
z-axis
x-axis
p_x orbital
(c)
z-axis
y-axis
p_y orbital
(d)
z-axis
d-state
l = 2, m_l = 0
x-axis
(e)
z-axis
d-state
l = 2, m_l = + 1 or −1
x-axis
(f)
z-axis
d-state
l = 2, m_l = +1 or −1
y-axis
(g)

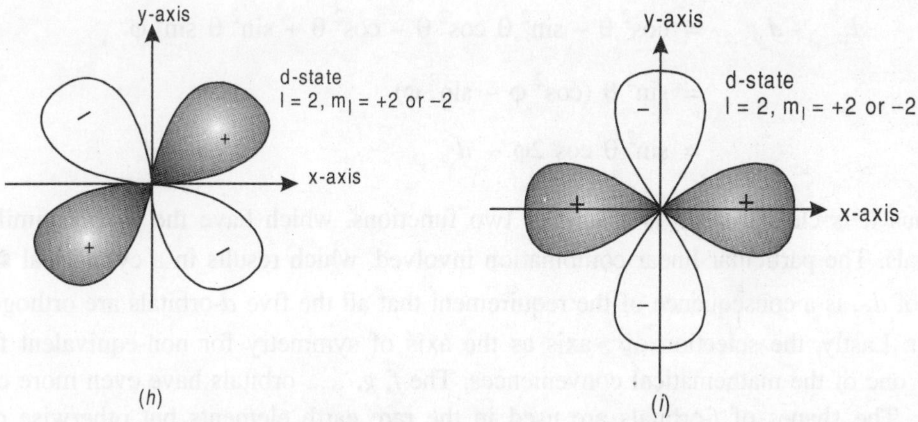

Figure 3.9: *(a) Plot of angular wave function of p_z orbital versus θ. θ has been varied only between 0 to π. (b) Graph obtained by full rotation of the curve obtained in (a) around z-axis as explained in the text. (c) Plot of p_x orbital in the xz-plan, taking θ = 90°. (d) Plot of p_y orbital in the yz-plane taking θ = 90°. (e) Plot of d_{z^2} in x-z plane. (f) Plot of d_{xz} orbital in xz-plane, (f) Plot of d_{yz} orbital in yz-plane. (h) Plot of d_{xy} in xy-plane and (i) Plot of d_{x2-y2} orbital in xy-plane*

surfaces in the angular plots of the *d*-functions. Excepth for d_{z^2}, the nodal surfaces are planes intersecting each other at right angles, whereas for d_{z^2} the nodal surfaces are right circular cones.

One thing is very evident from the above plots that the angular plot of d_{z^2} is not apparent from its function, compared to the other four *d*-orbitals. To make it clear, let us consider two alternative functions, which correspond to $d_{z^2-x^2}$ and $d_{z^2-y^2}$. Their angular functions will be given by:

$$d_{z^2-x^2} \cong \cos^2\theta - \sin^2\theta\cos^2\varphi \quad ...(3.95)$$

$$d_{z^2-y^2} \cong \cos^2\theta - \sin^2\theta\sin^2\varphi \quad ...(3.96)$$

The angular plots of these functions will be similar to other four *d*-orbitals, but the directions will be different. For example, in case of $d_{z^2-x^2}$, the positive lobes will be along the *z*-axis and negative lobes will be along the *x*-axis. Similarly in $d_{z^2-y^2}$, the positive lobes will be along the *z*-axis and negative lobes along the *y*-axis. Now if we take the linear combinations of these functions, *i.e.,*

$$\begin{aligned} d_{z^2-x^2} + d_{z^2-y^2} &= \cos^2\theta - \sin^2\theta\cos^2\varphi + \cos^2\theta - \sin^2\theta\sin^2\varphi \\ &= 2\cos^2\theta - \sin^2\theta(\sin^2\varphi + \cos^2\varphi) \\ &= 2\cos^2\theta - \sin^2\theta = 3\cos^2\theta - 1 \sim 3z^2 - r^2 = d_{z^2} \quad ...(3.97) \end{aligned}$$

and $$d_{z^2-x^2} - d_{z^2-y^2} = \cos^2\theta - \sin^2\theta\cos^2\theta - \cos^2\theta + \sin^2\theta\sin^2\varphi$$

$$= \sin^2\theta\,(\cos^2\varphi - \sin^2\varphi)$$

$$= \sin^2\theta\cos 2\varphi \sim d_{x^2-y^2} \qquad ...(3.98)$$

Thus it is clear that d_{z^2} is a sum of two functions, which have the shapes similar to other d-orbitals. The particular linear combination involved, which results in a cylindrical symmetrical shape of d_{z^2}, is a consequence of the requirement that all the five d-orbitals are orthogonal to one another. Lastly, the selection of z-axis as the axis of symmetry for non-equivalent functions is simply one of the mathematical conveniences. The f, g, ….. orbitals have even more complicated shapes. The shapes of f-orbitals are used in the rare earth elements but otherwise one seldom needs to consider these.

The distribution of the electrons is governed not by the wave function itself but by its square. Therefore the squares of these functions are plotted in Figs. 3.10. It is clear that the squares of p- and d-orbitals are even more strongly directional than the functions themselves, since squaring weighs more heavily those angles for which Y (θ, φ) was already large.

In the above discussion, we have plotted the radial and angular functions separately, simply to understand their characteristics. If we want to obtain the electron distribution in the above discussion, we have plotted the radial and angular functions separately, simply to understand their characteristics. If we want to obtain the electron distribution in an atom, we must multiply the $|Y(\theta, \varphi)|^2$ of Fig. 3.10 by the square of the radial wave function R(r).

Now let us try to analyze what we have learnt? We have seen that in case of p-and d-orbitals, the wave functions describing these orbitals show a strong directional character. But in addition to wave functions describing these orbitals, some other solutions of the Schrödinger wave equation are also possible. For example, consider p_x and p_y orbitals with corresponding wave functions $\psi(p_x)$ and $\psi(p_y)$. These are the solutions of Schrödinger wave equation and it can also be shown that their linear combination is also a solution of Schrödinger wave equation. This can be done by, substituting

$$\psi(p_x + p_y) = \psi(p_x) + \psi(p_y) \qquad ...(3.99)$$

$$\nabla^2\psi(p_x + p_y) + \frac{8\pi^2\mu}{h^2}(E - V)\,\psi(p_x + p_y) = 0 \qquad ...(3.100)$$

or $$\nabla^2\psi(p_x) + \nabla^2\psi(p_y) + \frac{8\pi^2\mu}{h^2}(E - V)\,\psi(p_x) + \frac{8\pi^2\mu}{h^2}(E - V)\,\psi(p_y) = 0$$

$$\left[\nabla^2\psi(p_x) + \frac{8\pi^2\mu}{h^2}(E - V)\,\psi(p_x)\right] + \left[\nabla^2\psi(p_y) + \frac{8\pi^2\mu}{h^2}(E - V)\,\psi(p_y)\right] = 0 \qquad ...(3.101)$$

Since $\psi(p_x)$ and $\psi(p_y)$ are individually also solutions of the Schrödinger wave equation, that is,

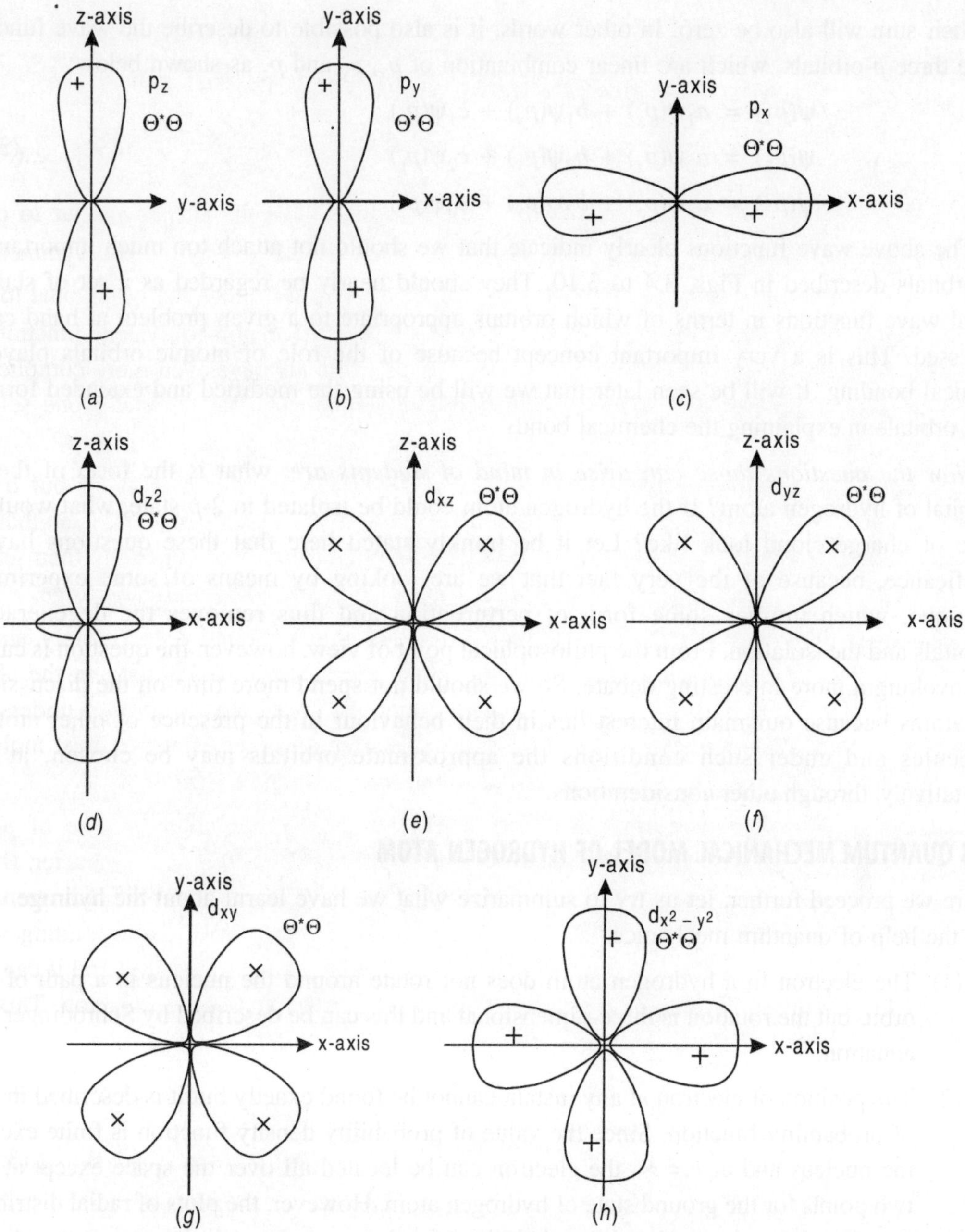

Figure 3.10: Polar plot of $\Theta^*\ \Theta$ versus angle θ. (*a*) p_z orbital, (*b*) p_y orbital, (*c*) p_x orbital, (*d*) d_{z^2} orbital, (*e*) d_{xz} orbital, (*f*) d_{yz} orbital, (*g*) d_{xy} orbital and (*h*) $d_{x^2-y^2}$ orbital in the planes mentioned in the figures

$$\nabla^2\psi\ (p_x)\ +\ \frac{8\pi^2\mu}{h^2}\ (\text{E}\ -\ \text{V})\ \psi(p_x)\ =\ 0 \qquad \text{...(3.102)}$$

$$\nabla^2\psi(p_y)\ +\ \frac{8\pi^2\mu}{h^2}\ (\text{E}\ -\ \text{V})\ \ \psi(p_Y)\ =\ 0$$

and their sum will also be zero. In other words, it is also possible to describe the wave functions of the three p-orbitals, which are linear combination of p_x, p_y and p_z as shown below:

$$\begin{aligned}\psi(p_1) &= a_1\psi(p_x) + b_1\psi(p_y) + c_1\psi(p_z)\\ \psi(p_2) &= a_2\psi(p_x) + b_2\psi(p_y) + c_2\psi(p_z)\\ \psi(p_3) &= a_3\psi(p_x) + b_3\psi(p_y) + c_3\psi(p_z)\end{aligned} \qquad ...(3.103)$$

The above wave functions clearly indicate that we should not attach too much importance to the orbitals described in Figs. 3.4 to 3.10. They should nearly be regarded as a set of standard orbital wave functions in terms of which orbitals appropriate to a given problem at hand can be expressed. This is a very important concept because of the role of atomic orbitals played in chemical bonding. It will be seen later that we will be using the modified and extended forms of these orbitals in explaining the chemical bonds.

Now the questions those can arise in mind of students are: what is the form of the true p-orbital of hydrogen atom? If the hydrogen atom could be isolated in 2-p state, what would the shape of charge cloud look like? Let it be frankly stated here that these questions have no significance, because of the very fact that we are looking by means of some experimental technique, which implies some form of perturbation and thus removes the degeneracy of p-orbitals and the isolation. From the philosophical point of view, however, the question is capable of provoking a more interesting debate. So we should not spend more time on the discussion of free atoms because our main interest lies in their behaviour in the presence of other atoms in molecules and under such conditions the approximate orbitals may be chosen, at least qualitatively, through other considerations.

3.1.8 QUANTUM MECHANICAL MODEL OF HYDROGEN ATOM

Before we proceed further, let us try to summarize what we have learnt about the hydrogen atom with the help of quantum mechanics.

(1) The electron in a hydrogen atom does not rotate around the nucleus in a path of fixed orbit, but the rotation is three-dimensional and this can be described by Schrödinger wave equation.

(2) The position of electron at any instant cannot be found exactly but it is described in terms of probability function. Since the value of probability density function is finite except at the nucleus and at $r = \infty$, the electron can be located all over the space except at these two points for the ground state of hydrogen atom. However, the plots of radial distribution functions have shown that the probability is having a maximum value at a certain distance from the nucleus. Interestingly, the value of r where the probability of finding the electron for the ground state of hydrogen atom is at the same distance from the nucleus as the radius (a_0) of the first Bohr's orbit, which is $h^2/4\pi^2 me^2$, *i.e.*, 0.0529 nm and the maximum probability is equal in all possible directions so long the distance of electron from the nucleus is same (spherical symmetric). In other orbitals, when $l \neq 0$ the maximum probability lies along the preferred direction.

(3) To have any meaningful physical representation of the orbital and thus to have any quantum mechanical analogue of classical orbit, the orbital can be visualized as a three-dimensional surface around the nucleus, where the electron can be found for most of the time (say for example 90%). In other words, the total probability of finding the electron with in this surface is 90%.

The other point to be mentioned here is that though the solution of Schrödinger wave equation gives the wave functions in terms of spherical harmonics, we, the chemists prefer the solutions as given in Figs. 3.9 and 3.10, obtained by taking linear combination of the complex wave functions, as given in Table 3.4. The reason is simple that spherical harmonics represent the wave functions in imaginary quantities and the Figs. 3.9 and 3.10 represent in real form. These real wave functions have also directional properties which is very useful to chemists in building the molecular orbital wave functions out of atomic orbital wave functions. If the atomic orbitals have proper directional properties, it is possible to use the chemical intuition to decide which, are the more important atomic orbitals to use to describe molecular orbitals (Chapter 4).

3.2. VIRIAL THEOREM

There is a powerful theorem in classical mechanics that deals with the motion of a system of particles in a finite region of space. *The theorem states that the time energy value of kinetic energy is related to the virial, which is a certain linear function of the derivatives, with respect to the rectangular coordinates, of potential energy* V. The coefficients of the derivatives are proportional to the corresponding coordinates. For one mass point, the virial is:

$$V_{vir} = x\frac{\partial V}{\partial x} + y\frac{\partial V}{\partial y} + z\frac{\partial V}{\partial z} \qquad ...(3.104)$$

where V is the potential energy. Recall that the equations of motions are:

$$m\ddot{x} = -\frac{\partial V}{\partial x},\ m\ddot{y} = -\frac{\partial V}{\partial y},\ m\ddot{z} = -\frac{\partial V}{\partial z} \qquad ...(3.105)$$

Thus

$$V_{vir} = -m(x\ddot{x} + y\ddot{y} + z\ddot{z}) \qquad ...(3.106)$$

and

$$\frac{d}{dt}(x^2 + y^2 + z^2) = 2(x\dot{x} + y\dot{y} + z\dot{z}) \qquad ...(3.106a)$$

$$\frac{d^2}{dt^2}(x^2 + y^2 + z^2) = 2(\dot{x}^2 + \dot{y}^2 + \dot{z}^2) + 2(x\ddot{x} + y\ddot{y} + z\ddot{z})$$

or $$\frac{d^2}{dt^2}(x^2 + y^2 + z^2) = \frac{4K}{m} - \frac{2}{m}V_{vir} \qquad ...(3.107)$$

where $$K = \frac{1}{2}m(\dot{x}^2 + \dot{y}^2 + \dot{z}^2)$$

It is easy to see that the time average value of the left hand side of equation (3.107) vanishes since it is equal to the difference of the values of equation (3.106*a*) at two distant moments of time, divided by the time interval. Thus:

$$2\langle K\rangle = \langle V_{vir}\rangle \quad ...(3.108)$$

This is the essence of the virial theorem of classical mechanics for the case of one mass point. The theorem can be generalized for a system of mass points. If the potential energy is a homogeneous function of degree *n* in the Cartesian coordinates, the definition of the virial theorem (3.104) gives

$$V_{vir} = nV \quad ...(3.109)$$

and hence $$2\langle K\rangle = n\langle V\rangle \quad ...(3.110)$$

Thus the theorem states that if the potential energy of a particle can be expressed in the form of $V = x^n$, *its average potential energy is related to its average kinetic energy* (K) *by the expression* (3.110) *and is know as virial theorem.* In case of hydrogen atom, the potential energy is given by the expression $(-e^2 r^{-1})$, *i.e.,* $n = -1$. Thus $\langle K\rangle$ is related to $\langle V\rangle$ by the expression $\langle K\rangle = -\frac{1}{2}\langle V\rangle$.

The virial theorem is valid in case of Bohr's model for hydrogen atom and can be shown as follow. The kinetic energy of the electron in hydrogen atom is; $K = \frac{1}{2}m_e v^2$, where m_e is the mass of the electron and $V = -e^2 r^{-1}$. According to equation (1.83), neglecting the permittivity factor,

$$m_e v^2 = -e^2 r^{-1} \quad ...(1.83)$$

$$K = \frac{1}{2}m_e v^2 = \frac{1}{2}(-e^2 r^{-1}) = -\frac{1}{2}V$$

Thus $$2K = -V \quad ...(3.111)$$

The above expression can also be proved quantum mechanically. The expression for average values of each quantity can be calculated with the help of equation (2.123). Thus

$$\langle V\rangle = \int \psi^*_{n,l,m_l}(r,\theta,\varphi)\left[-\frac{Ze^2}{r}\right]\psi_{n,l,m_l}(r,\theta,\varphi)\, r^2\sin\theta\, dr\, d\theta\, d\varphi \quad ...(3.112)$$

For simplicity, consider the ground state wave function of hydrogen atom (*i.e.,* 1*s*) Equation (3.112) can be written as:

$$\langle V\rangle = -Ze^2\frac{Z^3}{\pi a_0^3}\int_0^\infty e^{-\frac{2Zr}{a_0}}\, r\, dr\int_0^\pi \sin\theta\, d\theta\int_0^{2\pi} d\varphi = -\frac{4Z^4e^2}{a_0^3}\int_0^\infty e^{-\frac{2Zr}{a_0}}\, rdr$$

$$= -\frac{4Z^4e^2}{a_0^3}\left[\frac{1}{\left(\frac{2Z}{a_0}\right)^2}\right] = -\frac{Z^2e^2}{a_0} \quad ...(3.113)$$

Similarly the average kinetic energy is obtained by:

$$\langle K \rangle = \left\langle \frac{p^2}{2m_e} \right\rangle = \int_0^\infty \left(\frac{Z^3}{\pi a_0^3}\right)^{\frac{1}{2}} e^{-\frac{Zr}{a_0}} \left[-\frac{h^2}{8\pi^2 m_e}\nabla^2\right] \left(\frac{Z^3}{\pi\, a_0^3}\right)^{\frac{1}{2}} e^{-\frac{Zr}{a_0}}\, d\tau \qquad ...(3.114)$$

Since the ground state wave function for hydrogen and hydrogen like atoms is only a function of r, the θ and φ parts of Laplacian operator will not operate on the given function. So

$$\langle K \rangle = -\frac{h^2}{8\pi^2 m_e}\frac{Z^3}{\pi a_0^3}\int_0^\infty e^{-\frac{Zr}{a_0}}\left[\frac{1}{r^2}\frac{d}{dr}\left(r^2\frac{d}{dr}e^{-\frac{Zr}{a_0}}\right)\right] r^2\, dr \int_0^\pi \sin\theta\, d\theta \int_0^{2\pi} d\varphi$$

$$= -\frac{h^2}{8\pi^2 m_e}\frac{Z^3}{\pi a_0^3}(4\pi)\left(-\frac{Z}{a_0}\right)\int_0^\infty e^{-\frac{Zr}{a_0}}\left[\frac{1}{r^2}\frac{d}{dr}\left(r^2\, e^{-\frac{Zr}{a_0}}\right)\right] r^2 dr$$

$$= \frac{h^2}{2\pi^2 m_e}\frac{Z^4}{a_0^4}\int_0^\infty e^{-\frac{Zr}{a_0}}\left[2r\, e^{-\frac{Zr}{a_0}} - \frac{Z}{a_0}r^2 e^{-\frac{Zr}{a_0}}\right] dr$$

$$= \frac{h^2}{2\pi^2 m_e}\frac{Z^4}{a_0^4}\left[2\int_0^\infty e^{-\frac{2Zr}{a_0}}\, r\, dr - \frac{Z}{a_0}\int_0^\infty e^{-\frac{2Zr}{a_0}}\, dr\right]$$

$$= \frac{h^2}{2\pi^2 m_e}\frac{Z^4}{a_0^4}\left[2\left(\frac{1}{\left(\frac{2Z}{a_0}\right)^2}\right) - \left(\frac{Z}{a_0}\right)\left(\frac{2}{\left(\frac{2Z}{a_0}\right)^3}\right)\right]$$

$$= \frac{h^2}{2\pi^2 m_e}\frac{Z^4}{a_0^4}\left[\frac{a_0^2}{2Z^2} - \frac{a_0^2}{4Z^2}\right] = \frac{h^2}{2\pi^2 m_e}\frac{Z^4}{a_0^4}\frac{a_0^2}{4\, Z^2}$$

$$= \frac{h^2}{8\pi^2 m_e}\frac{Z^2}{a_0^2} = \frac{h^2 Z^2}{8\pi^2 m_e a_0}\frac{4\pi^2 m_e e^2}{h^2}\left[a_0 = \frac{h^2}{4\pi^2 m_e e^2}\right]$$

$$= \frac{e^2 Z^2}{2a_0}. \qquad ...(3.115)$$

Thus the above treatment has shown that virial theorem is applicable to the ground state of hydrogen atom. Similarly for any given state, with quantum numbers n, l, m_l, we can obtain the following relations:

$$\langle V \rangle_{n,\, l,\, m_l} = 2E_n = -\frac{e^2 Z^2}{a_0 n^2} \qquad ...(3.116)$$

$$\langle \mathrm{K} \rangle_{n,\, l,\, m_l} = -\,\mathrm{E}_n = \frac{e^2 Z^2}{2a_0 n^2} \qquad \text{...(3.117)}$$

3.3 QUANTUM MECHANICAL TUNNELING

Results of quantum mechanics have explained many facets of chemical reactions, whereas there are no counterparts in classical mechanics. One of these phenomena is known as quantum mechanical tunneling. The description of wave functions, probability density functions or radial distribution functions, both in case of harmonic oscillator, as well as, in case of hydrogen atom have indicated that the particle can be found all over the space, starting from $r = 0$ to $r = \infty$. These results are very different from classical mechanics. In classical mechanics, when the kinetic energy of the particle is zero, particle turns back under the influence of potential energy. That is the particle can be located with in the boundaries created by the potential energy barriers and cannot cross out of these boundaries. These boundaries in case of hydrogen atom can be obtained from the relation:

$$\mathrm{V}(r) = -\frac{e^2}{r} = \mathrm{E} = -\frac{e^2}{2a_0} \qquad \text{...(3.118)}$$

i.e., the classical mechanics limits of the boundaries are when $r = 2a_0$. But it has been seen that the electron can be found out even beyond these boundaries or one can say that the electron has tunneled into or through the potential energy barriers. The consequences of this effect are very strange. For example, if the particle is found beyond the classical limits, it means that the potential energy of the particle is more than the kinetic energy. To keep the total energy constant, it means that the kinetic energy will have negative value or the velocity of the particle will be imaginary. The above kind of explanation is very absurd and difficult to accept in classical mechanics. In quantum mechanics one does not have to give this kind of explanation and this kind of behaviour can be explained as follows. When one talks of the position of the electron or particle in general in any region, this means that the measurements have been carried out. This knowledge about the position will introduce the uncertainty in the kinetic energy. It can be shown that the measurement of the position of the particle in the classically forbidden region will introduce an error in the measurement of kinetic energy to such an extent that the negative value of kinetic energy will be taken care. This means that the measurement about the position of the particle and the consequences to be followed can be explained nicely with the help of quantum mechanics. Similar kind of explanation has been used to explain a model for radioactive decay of nuclei and to explain the very high values of the rate constants, observed for certain chemical reactions, *e.g.,* electron transfer reactions, proton transfer reactions etc.,

Now at this stage it is worth while to discuss the tunneling effect in case of thin potential wall which acts as potential barrier, since tunneling in such potential barriers plays an important role in understanding the various processes in the evolution of altogether new branch in biological sciences called *quantum genetics*. This is being described in brief after the quantum mechanical analysis of thin potential wall as potential barrier.

3.3.1 THIN POTENTIAL WALL AS POTENTIAL BARRIER

Let us consider the thin potential wall of the type shown in Fig. 3.11. Classically for a particle of kinetic energy E_K incident on the barrier, if $E_K < V$, the particle gets reflected by the barrier and if $E_K > V$, the particle gets transmitted through the barrier. But we will see that quantum mechanically, even for $E_K < V$, there may be a finite probability of the particle tunneling through the barrier. Further for $E_K > V$, there may be a finite probability for reflection. Let us consider $E_K < V$. This situation is represented by putting the first jump at $x = 0$, the origin, and the second one at $x = A$. This divides the axis into three regions:

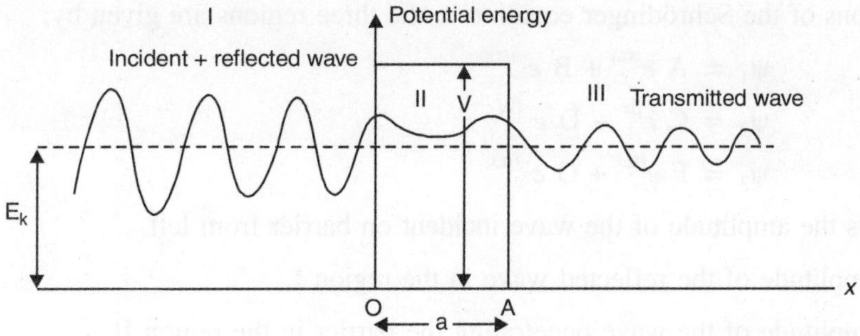

Fig. 3.11: *A beam of particles of kinetic energy E_K incident on a potential barrier $V > E_K$ with a width OA = a*

I, $x < 0$, P. E. = 0

II, $0 < x < a$, P. E. = V and

III, $x > 0$, P. E. = 0

If ψ_1, ψ_2 and ψ_3 are the respective wave functions (which are continuous, single valued and finite every where along the x-axis) in the regions I, II and III as indicated in Fig. 3.11, the corresponding Schrödinger equations are:

$$\text{I} \quad -\frac{h^2}{8\pi^2 m}\frac{d^2\psi_1}{dx^2} = E_K\psi_1 \qquad \text{since } V_I = 0$$

$$\text{II} \quad -\frac{h^2}{8\pi^2 m}\frac{d^2\psi_2}{dx^2} + V\,\psi_2 = E_K\psi_2 \qquad \text{where } V_{II} = V \qquad ...(3.119)$$

$$\text{III} \quad -\frac{h^2}{8\pi^2 m}\frac{d^2\psi_3}{dx^2} = E_K\psi_3 \qquad \text{since } V_{III} = 0$$

The rearrangements of these equations yield:

$$\text{I} \quad \frac{d^2\psi_1}{dx^2} + \alpha^2\,\psi_1 = 0$$

$$\text{II} \quad \frac{d^2\psi_2}{dx^2} - \beta^2 \psi_2 = 0 \qquad ...(3.120)$$

$$\text{III} \quad \frac{d^2\psi_3}{dx^2} + \alpha^2 \psi_3 = 0$$

where $\alpha^2 = \frac{8\pi^2 m E_K}{h^2}$ and $\beta^2 = \frac{8\pi^2 m}{h^2}(V - E_K)$

The solutions of the Schrödinger equation in the three regions are given by:

$$\text{I} \qquad \psi_1 = A\,e^{i\alpha x} + B\,e^{-i\alpha x} \qquad ...(3.121)$$

$$\text{II} \qquad \psi_2 = C\,e^{\beta x} + D\,e^{-\beta x}$$

$$\text{III} \qquad \psi_3 = F\,e^{i\alpha x} + G\,e^{-i\alpha x}$$

Where A is the amplitude of the wave incident on barrier from left,

B is the amplitude of the reflected wave in the region I,

C is the amplitude of the wave penetrating the barrier in the region II,

D is the amplitude of the reflected wave from the surface at A, in the region II,

F is the amplitude of the transmitted wave in region III, and

G is the amplitude of a (non-existing) reflected wave in the region III.

We set G = 0, since there cannot be a wave propagating in the x-direction in the region $x > a$.

Continuity of ψ_1 and $\frac{d\psi_1}{dx}$ at $x = 0$, as well as, of ψ_2 and $\frac{d\psi_2}{dx}$ at $x = 0$ results in:

$$A + B = C + D \qquad (i.e., \Psi_1 = \Psi_2 \text{ at } x = a) \qquad ...(3.122)$$

and
$$\frac{d\psi_1}{dx} = i\alpha A e^{i\alpha x} - i\alpha B e^{-i\alpha x}$$

$$\frac{d\psi_2}{dx} = \beta C e^{\beta x} - \beta D e^{-\beta x}$$

Since
$$\frac{d\psi_1}{dx} = \frac{d\psi_2}{dx} \text{ at } x = 0$$

$$\therefore \quad i\alpha (A - B) = \beta(C - D)$$

or
$$\frac{i\alpha}{\beta}(A - B) = C - D \qquad ...(3.123)$$

Thus, from equations (3.122) and (3.123)

$$C = \frac{1}{2}\left(1 + \frac{i\alpha}{\beta}\right) A + \frac{1}{2}\left(1 - \frac{i\alpha}{\beta}\right) B \qquad ...(3.124)$$

and

$$D = \frac{1}{2}\left(1 - \frac{i\alpha}{\beta}\right) A + \frac{1}{2}\left(1 + \frac{i\alpha}{\beta}\right) B \qquad ...(3.125)$$

Thus,

$$\frac{C}{D} = \frac{\left(1 + \frac{i\alpha}{\beta}\right) A + \left(1 - \frac{i\alpha}{\beta}\right) B}{\left(1 - \frac{i\alpha}{\beta}\right) A + \left(1 + \frac{i\alpha}{\beta}\right) B} \qquad ...(3.126)$$

Continuity conditions at $x = a$, yields:

$$Ce^{\beta a} + De^{-\beta a} = Fe^{i\alpha a} \qquad ...(3.127)$$

and

$$\frac{d\psi_2}{dx} = \frac{d\psi_3}{dx}$$

i.e.,

$$\beta Ce^{\beta a} - \beta De^{-\beta a} = i\alpha Fe^{i\alpha a} \qquad ...(3.128)$$

From equations (3.127) and (3.128), we get

$$Ce^{\beta a} + De^{-\beta a} = Fe^{i\alpha a}$$

$$Ce^{\beta a} - De^{-\beta a} = \frac{i\alpha}{\beta} Fe^{i\alpha a}$$

Thus,

$$C = \frac{1}{2}\left(1 + \frac{i\alpha}{\beta}\right) Fe^{-\beta a} e^{i\alpha a} \qquad ...(3.129)$$

$$D = \frac{1}{2}\left(1 - \frac{i\alpha}{\beta}\right) Fe^{\beta a} e^{i\alpha a} \qquad ...(3.130)$$

and

$$\frac{C}{D} = \frac{\left(1 + \frac{i\alpha}{\beta}\right)}{\left(1 - \frac{i\alpha}{\beta}\right)} e^{-2\beta a} = \frac{\left(1 + \frac{i\alpha}{\beta}\right) A + \left(1 - \frac{i\alpha}{\beta}\right) B}{\left(1 - \frac{i\alpha}{\beta}\right) A + \left(1 + \frac{i\alpha}{\beta}\right) B} \qquad ...(3.131)$$

After equating with equation (3.126):

or

$$\frac{e^{-\beta a}}{e^{\beta a}} = \frac{\left(1 - \frac{i^2\alpha^2}{\beta^2}\right) A + \left(1 - \frac{i\alpha}{\beta}\right)^2 B}{\left(1 - \frac{i^2\alpha^2}{\beta^2}\right) A + \left(1 + \frac{i\alpha}{\beta}\right)^2 B}$$

or
$$\frac{e^{\beta a}}{e^{-\beta a}} = \frac{\left(1+\frac{\alpha^2}{\beta^2}\right)A + \left(1+\frac{i\alpha}{\beta}\right)^2 B}{\left(1+\frac{\alpha^2}{\beta^2}\right)A + \left(1-\frac{i\alpha}{\beta}\right)^2 B}$$

$$\frac{e^{\beta a} - e^{-\beta a}}{e^{\beta a} + e^{-\beta a}} = \frac{\left[\left(1+\frac{i\alpha}{\beta}\right)^2 - \left(1-\frac{i\alpha}{\beta}\right)^2\right]B}{2\left(1+\frac{\alpha^2}{\beta^2}\right)A + \left[\left(1+\frac{i\alpha}{\beta}\right)^2 + \left(1-\frac{i\alpha}{\beta}\right)^2\right]B}$$

$$\frac{\sinh \beta a}{\cosh \beta a} = \frac{\left(\frac{4i\alpha}{\beta}\right)B}{2\left(1+\frac{\alpha^2}{\beta^2}\right)A + 2\left(1-\frac{\alpha^2}{\beta^2}\right)B}$$

$$\frac{(\alpha^2+\beta^2)\sinh \beta a}{2i\alpha\beta \cosh \beta a} = \frac{B}{A + B\left(\frac{\beta^2-\alpha^2}{\beta^2+\alpha^2}\right)}$$

or
$$A + B\left(\frac{\beta^2-\alpha^2}{\beta^2+\alpha^2}\right) = \frac{B\, 2i\alpha\beta \cosh \beta a}{(\alpha^2+\beta^2)\sinh \beta a}$$

$$\frac{A}{B} + \frac{\beta^2-\alpha^2}{\beta^2+\alpha^2} = \frac{2i\alpha\beta \cosh \beta a}{(\alpha^2+\beta^2)\sinh \beta a}$$

$$\frac{A}{B} = \frac{2i\alpha\beta \cosh \beta a}{(\alpha^2+\beta^2)\sinh \beta a} - \frac{\beta^2-\alpha^2}{\beta^2+\alpha^2}$$

$$= \frac{2i\alpha\beta \cosh \beta a - (\beta^2-\alpha^2)\sinh \beta a}{(\alpha^2+\beta^2)\sinh \beta a}$$

or finally

$$\frac{B}{A} = \frac{(\alpha^2+\beta^2)\sinh \beta a}{(\alpha^2-\beta^2)\sinh \beta a + 2i\alpha\beta \cosh \beta a} \quad ...(3.132)$$

Further in order to obtain F/A, we proceed as follows:

$$A + B = C + D$$

$$A - B = \frac{\beta}{i\alpha}(C - D)$$

Substituting the values of C and D in the preceding equation and simplifying it we obtain:

$$2A = \frac{Fe^{i\alpha a}}{i\alpha\beta}[2i\alpha\beta\,(e^{\beta a} + e^{-\beta a}) + (\alpha^2 - \beta^2)\,(e^{\beta a} - e^{-\beta a})]$$

$$= \frac{F}{i\alpha\beta e^{-i\alpha a}}[4i\,\alpha\beta\cos h\beta a + 2(\alpha^2 - \beta^2)\sin h\beta a]$$

$$\frac{F}{A} = \frac{2\,i\,\alpha\,\beta\,e^{-\alpha\,a}}{(\alpha^2 - \beta^2)\sin h\,\beta a + 2\,i\,\alpha\beta\,\cos h\beta a} \qquad ...(3.133)$$

Since the probability density associated with a wave function is proportional to the square of the amplitude of that function, we can define the barrier transmission coefficient as:

$$T = \frac{|F|^2}{|A|^2} \qquad ...(3.134)$$

And reflection coefficient for the barrier surface at $x = 0$ as:

$$R = \frac{|B|^2}{|A|^2} \qquad ...(3.135)$$

If the barrier is high compared to the total energy if the particle or is thick compared to the wavelength of the wavefunction, then the transmission coefficient becomes.

$$T \approx \frac{16E_k}{V}\left(1 - \frac{E_k}{V}\right)e^{-(2a/\hbar)\sqrt{2m(V-E_k)}} \qquad ...(3.134a)$$

where 'a' is the physical thickness of the barrier.

We have reached a remarkable conclusion that if a particle with energy E_k is incident on a thin energy barrier of height greater than E_k, there is a finite probability of the particle penetrating the barrier. This phenomenon, called tunnel effect, is as a result in the quantum mechanics that is not allowed in the classical treatment as state earlier also.

Thus the probability of transmission that a 1.0 eV electron will penetrate a potential barrier of 4.0 eV when the barrier width is 2.0 Å will be

$$T \approx \frac{16(1.0\,\text{eV})}{4.0\,\text{eV}}\left(1 - \frac{0.1\,\text{eV}}{4.0\,\text{eV}}\right) \times \exp - \frac{2\times2\times10^{-10}\,\text{m}}{1.05\times10^{-34}\,\text{Jsec}}\sqrt{\frac{2(9.1\times10^{-31}\,\text{hg})\,(4-1)}{(1.6\times10^{-19}\,\text{J})}}$$

$$\approx 0.084$$

Accordingly only about light 1.0 eV electrons out of every hundred penetrate the barrier.

It is calculated on the basis of tunnel effect that if an α particle impinges on the inside of the nuclear force barrier that holds the nucleus together, stands about one chance in 10^{38} of penetrating the barrier and escaping from the nucleus. This escape, however, constitutes radioactive α-decay.

The nucleus has a diameter of about 10^{-14} m, and the α-particle moves within it with a velocity of about 10^7 m/sec, so that it makes about 10^{21} collisions/sec with the barrier. Thus, there are

$$\frac{10^{38}\text{ collisions}}{10^{21}\text{ collisions/sec}} = 10^{17}\text{ sec/penetration}$$

This is about 3 × 10^9 year required for an α particle to make a probable escape, and it provides some understanding of the long radioactive half life of uranium which is over a billion years. The quantum mechanical tunnel effect in α decay exhibits extreme difference in radioactive lifetimes (ranging from billons of years to milliseconds) for rather small variations in the height of potential barrier. In case of polonium, the proble escape time for an α particle is of the order of

$$\frac{10^{17}\text{ collisions}}{10^{21}\text{ collisions/sec}} \approx 10^{-4}\text{ sec.}$$

Furthermore the quantum mechanical treatment of the proton tunneling in case of H-bonded complex A—H···B through the double will potential barrier of the type shown in Fig 3.11(*a*) gives a proton transfer frequency in the range from 5 × 10^{13}/sec to 5 × 10^{11}/sec. This range is in agreement with the range of proton transfer frequencies found experimentally by means of modern relaxation and NMR spectroscopic techniques.

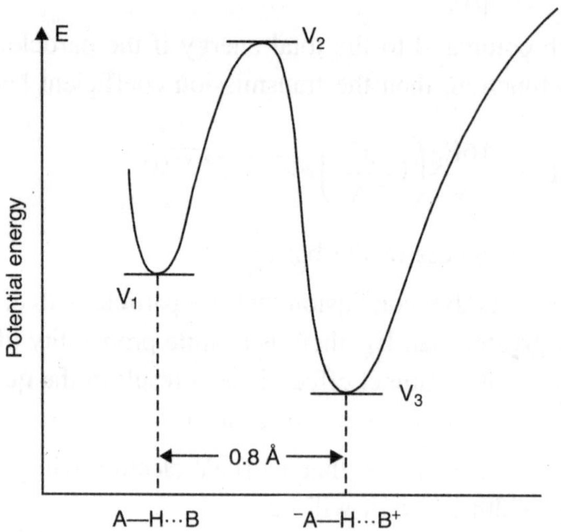

Figure 3.11(*a*): *A typical double wall potential. The distance between two minima is 0.8 Å, a value appropriate for an O—H···OH bond*

Finally we would like to say little on quantum genetics which mainly is based on proton tunneling. It deals with some aspects on the biological problems of heredity, mutations, aging and tumors in view of the quantum theory of DNA molecule *i.e.,* spontaneous proton tunneling in DNA, electronic structure of DNA, induced proton tunneling in DNA.

Further discussion on this topic is beyond the scope of this book.

3.4 HAMILTONIAN AND ITS SYMMETRY

The wave function that describes a stationary state of a many electron system must be an eigen function of the appropriate Hamiltonian. The Hamiltonian for a many electron atom in the absence of magnetic field is expressed as:

$$\hat{H} = -\frac{h^2}{8\pi^2 m}\sum_{k=1}^{n}\nabla_k^2 + \sum_{k=1}^{n} V(x_k, y_k, z_k) + \sum_{k,\,k>l}^{n}\frac{e^2}{[(x_k - x_l)^2 + (y_k - y_l)^2 + (z_k - z_l)^2]^{\frac{1}{2}}} \quad ...(3.136)$$

where ∇_k^2 is the Laplacian operator that acts on the kth electron, $V(x_k, y_k, z_k)$ is the potential energy of the external field, external in relation to the n-electron. For instance, the potential energy of the external field produced by the nucleus of the atom or by the nuclei in case of a molecule and the double sum is the potential energy of interaction between the electrons. If the system of electrons is in an external magnetic field, the extra terms that depend on spin needs to be introduced into the Hamiltonian expression.

The energy levels and stationary states of the system are determined from the equation:

$$\hat{H}\psi = E\psi \quad ...(2.47)$$

where $\hat{H}$ is given by equation (3.136). We know that though $\hat{H}$ does not contain spin variables, the energy levels E depend on spin angular momentum (s), since the symmetry properties of Schrödinger wave function ψ depends on s.

For an atom, $\hat{H}$ is spherically symmetric, *i.e.*, it is independent of rotations of coordinates axis in space. Thus we can assume, have also learnt in chapter 2, that Schrödinger wave function (a function of coordinates) is simultaneously the eigen function of the operator of the square of the orbital angular momentum (the quantum number l) and that of the operator of one of the components of the angular momentum (the quantum number m_l). If in addition, this eigen function possesses symmetry properties, corresponding to definite value of s, then it can be used to build a function that contains spins also, *i.e.*,

$$\psi = \psi(r_1, s_1;\ r_2, s_2; \ldots \quad ..r_n, s_n) \quad ...(3.137)$$

This function qualifies the condition of Pauli Exclusion Principle and is simultaneous eigen function of the following five operators: the Hamiltonian, the square of the angular momentum, the square of the spin angular momentum, the square of the total (spin and orbital) angular momentum and the component of the total angular momentum along the coordinate axis. The function is built using the so, called vector model.

The Hamiltonian of diatomic molecule has only axial symmetry but not spherical symmetry. Axial symmetry can also be used to introduce quantum numbers, such as rotational quantum number J and to partially determine the wave functions.

If we use spherical or axial symmetry that a system possesses, we can introduce certain quantum numbers and thus classify the energy levels. But symmetry alone cannot be fixed a criterion to determine the levels themselves and the stationary states. An exact solution of equation $\hat{H}\psi = E\psi$ presents insurmountable mathematical difficulties, except in case of one electron systems. For this reason, the development of approximate methods for solving the $\hat{H}\psi = E\psi$ equation is most important.

3.5 APPROXIMATE METHODS

Three most important approximate methods are: (*i*) the perturbation method, (*ii*) the variational method and (*iii*) the JWKB (after Jeffrys, Wentzel, Krammers and Brillouin) method. In the variational method, a trial wave function is selected to gather information about the ground state of the system, but the method becomes cumbersome when applied to higher excited states. On the other hand the JWKB method gives information about the states of the problem provided: (1) the potential energy variation of the system varies smoothly and, (2) the Schrödinger wave equation is separable to one dimensional equation. Here we will describe only the first two methods. The third method is beyond the scope of this book, but the readers interested in this method can consult any standard textbook on quantum mechanics.

3.5.1 THE PERTURBATION METHOD

The first step to solve any quantum mechanical problem is to simplify it and then solve it exactly. The second step is to calculate those corrections to the simplified problem that enable us to estimate the influence of the small terms omitted in the first step. The general method for calculating such corrections is called the perturbation method.

Let us assume that there exists an operator, say Hamiltonian that can be represented as a sum of two terms.

$$\hat{H} = \hat{H}^0 + \lambda\hat{H}' \qquad \text{...(3.138)}$$

The first term, $\hat{H}^0$ is the Hamiltonian of unperturbed problem or the zeroth order Hamiltonian and the second term $\lambda\hat{H}'$ is the correction or perturbation. In general the value of parameter λ is very small and lies between 0 and 1. It is a measure of the extent of perturbation from the unperturbed system. When $\lambda \to 0$, $\hat{H} \to \hat{H}^0$. This is termed as zeroth order perturbation. When $\lambda \to 1$, $\lambda\,\hat{H}' \longrightarrow \hat{H}'$ and $\hat{H}' = \hat{H}^0 + \hat{H}'$. This is the case of first order perturbation and in most of the applications λ is assumed to be unity. Before we proceed further it may be mentioned here that perturbation is different for degenerate states and non-degenerate states. In this section we will only study the effect of perturbation on the non-degenerate system. In case the system consists of both degenerate and non-degenerate energy levels, the treatment in this section will only be applicable to non-degenerate levels.

Let us now assume that eigen functions ψ_i^0 and eigen values E_i^0 of $\hat{H}^0$ are exactly known and we have found the solution to:

$$\hat{H}^0 \psi_i^0 = E_i^0 \psi_i^0 \qquad ...(3.139)$$

which represents the unperturbed system. Now our next aim is to find the approximate expressions for the eigen functions and eigen values of the Hamiltonian $\hat{H}$, *i.e.,* solve the equation

$$\hat{H} \psi_i = E_i \psi_i \qquad ...(3.140)$$

$$\left(\hat{H}_i^{(0)} + \lambda \hat{H}'\right) \psi_i = E_i \psi_i \qquad ...(3.141)$$

where the subscript i denotes the state of the system. This method can be used to solve many problems in quantum mechanics. For example, the effect of magnetic field on the system can also be studied by this method. In order to solve the posed problem, we proceed as follows. Since we have assumed that $\hat{H}$ is a function of parameter λ, the wave function, ψ_i and E_i will also be the functions of λ. In other words, as a first step we expand both ψ_i and E_i as a Taylor series with the assumption that, $\lambda \hat{H}' \ll \hat{H}$, *i.e.,* perturbation is very small as compared to the unperturbed value.

$$\psi_i = \psi_i^{(0)} + \lambda \psi_i^{(1)} + \lambda^2 \psi_i^{(2)} + \ldots\ldots\ldots\ldots \qquad ...(3.142)$$

$$E_i = E_i^{(0)} + \lambda E_i^{(1)} + \lambda^2 E_i^{(2)} + \ldots\ldots\ldots \qquad ...(3.143)$$

As $\lambda \to 0$, $\psi_i = \psi_i^{(0)}$ and $E_i = E_i^{(0)}$, and thus gives the values of wave functions and energy levels of the unperturbed system. The second and third terms give the first order and second order corrections to $\psi_i^{(0)}$ and $E_i^{(0)}$. Further higher order terms can be added in equations (3.142), (3.143) and (3.141). Now substitute the values of ψ_i and E_i from equations (3.142) and (3.143) in equation (3.141), *i.e.,*

$$(\hat{H}_i^{(0)} + \lambda \hat{H}')(\psi_i^{(0)} + \lambda \psi_i^{(1)} + \lambda^2 \psi_i^{(2)} + \ldots)$$

$$= (\psi_i^{(0)} + \lambda \psi_i^{(1)} + \lambda^2 \psi_i^{(2)} + \ldots)(E_i^{(0)} + \lambda E_i^{(1)} + \lambda^2 E_i^{(2)} + \ldots) \qquad ...(3.144)$$

After expanding the terms and then rearranging the terms in equation (3.144) according to the increasing power of λ, we get

$$(\hat{H}^{(0)}\psi_i^{(0)} - E_i^{(0)}\psi_i^{(0)}) + (\hat{H}^0\psi_i^{(1)} + \hat{H}'\psi_i^{(0)} - E_i^0\psi_i^{(1)} - E_i^1\psi_i^{(0)})\lambda$$

$$+ (\hat{H}^{(0)}\psi_i^{(2)} + \hat{H}'\psi_i^{(1)} - E_i^{(0)}\psi_i^{(2)} - E_i^{(1)}\psi_i^{(1)} - E_i^{(2)}\psi_i^{(0)})\lambda^2 = 0 \qquad ...(3.145)$$

Equation (3.145) will be satisfied for each value of λ and this is only possible if the coefficients of each power of λ is zero individually, *i.e.,*

$$\text{for } \lambda^0;\ \hat{H}^{(0)}\psi_i^{(0)} - E_i^{(0)}\psi_i^{(0)} = 0 \qquad ...(3.146)$$

$$\text{for } \lambda^1;\ \hat{H}^0\psi_i^{(1)} + \hat{H}'\psi_i^{(0)} - E_i^0\psi_i^{(1)} - E_i^1\psi_i^{(0)} = 0 \qquad ...(3.147)$$

for λ^2 $\hat{H}^{(0)}\psi_i^{(2)} + \hat{H}'\psi_i^{(1)} - E_i^{(0)}\psi_i^{(2)} - E_i^{(1)}\psi_i^{(1)} - E_i^{(2)}\psi_i^{(0)} = 0$ ...(3.148)

and similar terms for λ^3, λ^4,..........

Equation (3.146) represents the Schrödinger wave equation for the unperturbed system and its solution is similar to that we have solved earlier, *i.e.,* for simple harmonic oscillator and hydrogen atom. In other words, this equation (3.146) can be solved accurately. Equations (3.147) and (3.148) give the first order and second order corrections to the unperturbed wave function ψ_i^0 and energy E_i^0. These correction terms can be calculated as follows.

Multiply equation (3.147) by ψ_i^{*0} from the left and integrate it. This can be written as:

$$\int \psi_i^{*0}\hat{H}^0\psi_i^{(1)}d\tau - E_i^0\int \psi_i^{*0}\ \psi_i^{(1)}d\tau = -\int \psi_i^{*0}\hat{H}'\psi_i^0 d\tau + E_i^{(1)}\int \psi_i^{*0}\psi_i^0 d\tau \qquad ...(3.149)$$

Since $\hat{H}^0$ is a Hermitian operator, then,

$$\int \psi_i^{*0}\hat{H}^0\psi_i^{(1)}d\tau = \int \psi_i^{*(1)}\hat{H}^0\psi_i^0 d\tau = \int \psi_i^{*(1)}E_i^0\psi_i^0 d\tau = E_i^0\int \psi_i^{*(1)}\psi_i^0 d\tau$$

Thus equation (3.149) reduces to:

$$\int \psi_i^{*0}\hat{H}'\psi_i^0 d\tau - E_i^{(1)}\int \psi_i^{*0}\psi_i^0 d\tau = 0$$

or

$$E_i^{(1)} = \int \psi_i^{*0}\hat{H}'\psi_i^0 d\tau \qquad ...(3.150)$$

because ψ_i^0 forms an orthonormal set, *i.e.,* $\int \psi_i^{*0}\psi_i^0 d\tau = 1$. The integral on the right hand side of equation (3.150) can also be written as H'_{ii}, *i.e.,*

$$E_i^{(1)} = H'_{ii} \qquad ...(3.151)$$

Physical meaning of equation (3.151) suggests that the first order correction term $E_i^{(1)}$ to the unperturbed energy E_i^0 is the average value of the perturbation operator, $\hat{H}'$, obtained by using the unperturbed wave function ψ_i^0 of the unperturbed system.

Next question is to solve the correction to the unperturbed wave function, ψ_i^0, *i.e.,* $\psi_i^{(1)}$. This is done with the help of well known theorem in quantum mechanics, *i.e.,* any well behaved function can be expressed as a linear combination of a set of eigen functions. Hence:

$$\psi_i^{(1)} = a_{1i}\ \psi_1^0 + a_{2i}\ \psi_2^{(0)} + a_{3i}\ \psi_3^0 + \ .... = \sum_j a_{ji}\ \psi_j^0 \qquad ...(3.152)$$

Using equation (3.152), equation (3.147) can be written as:

$$(\hat{H}^0 - E_i^0)\sum_j a_{ji}\psi_j^0 = -(\hat{H}' - E_i^{(1)})\psi_i^0 \qquad ...(3.153)$$

Multiply equation (3.153) from the left hand side by ψ_k^0 and integrate them. Thus:

$$\int \psi_k^0\, \hat{H}^0 \sum_j a_{ji} \psi_j^0 d\tau - E_i^0 \int \psi_k^0 \sum_j a_{ji} \psi_j^0 d\tau = - \int \psi_k^0\, \hat{H}^i\, \psi_i^0 d\tau + E_i^0 \int \psi_k^0\, \psi_i^0 d\tau \qquad ...(3.154)$$

or $$\sum_j a_{ji} \left[\int \psi_k^0\, \hat{H}^0\, \psi_j^0 d\tau - E_i^0 \int \psi_k^0\, \psi_j^0 d\tau \right] = - \int \psi_k^0\, \hat{H}^i\, \psi_i^0 d\tau + E_i^0 \int \psi_k^0\, \psi_i^0 d\tau$$

Using the following information, *i.e.,*

(*i*) $$\hat{H}^0\, \psi_j^0 = E_j^0\, \psi_j^0$$

and (*ii*) the wave functions ψ_k^0, ψ_i^0 and Ψ_j^0 are orthonormal, $k = i$, equation (3.154) reduces to equation (3.150). For $k \neq i$, all the terms in the summation will be equal to zero except those with $k = j$. Thus equation (3.154) reduces to:

$$a_{ki} \left[\int \psi_k^0\, \hat{H}^0\, \psi_k^0\, d\tau - E_i^0 \int \psi_k^0\, \psi_k^0\, d\tau \right] = - \int \psi_k^0\, \hat{H}'\, \psi_i^0\, d\tau$$

or $$a_{ki}\, [E_k^0 - E_i^0] = - \int \psi_k^0\, \hat{H}'\, \psi_i^0\, d\tau \qquad ...(3.155)$$

Therefore

$$a_{ki} = - \frac{\int \psi_j^0\, \hat{H}'\, \psi_i^0 d\tau}{E_k^0 - E_i^0} = \frac{\int \psi_j^0\, \hat{H}'\, \psi_i^0 d\tau}{E_i^0 - E_k^0} \qquad ...(3.156)$$

Thus the first order correction to the wave function ψ_i^0 can be written as:

$$\psi_i^{(1)} = \sum_{j \neq i} \left[\frac{\int \psi_j^0\, \hat{H}'\, \psi_i^0\, d\tau}{E_i^0 - E_j^0} \right] \psi_j^0 \qquad ...(3.157)$$

In other words, it may be stated that first order correction terms to the unperturbed wave function is obtained with help of unperturbed wave functions and the unperturbed eigen values. The above treatment can be better understood if we solve some examples.

Example 3.17: Consider a particle in a box with infinite walls and a magnetic field is applied and its strength varies as F = *kx*, where *k* is a constant. Calculate the first order correction to the energy levels.

Solution: First we have to formulate this problem. In this case the magnetic field is acting as a perturbation and is a part of the potential energy. Thus the total Hamiltonian can be written as:

$$\hat{H} = - \frac{h^2}{8\pi^2 m} \frac{d^2}{dx^2} + kx \qquad ...(2.158)$$

In this case the unperturbed Hamiltonian operator is:

$$\hat{H}^0 = -\frac{h^2}{8\pi^2 m}\frac{d^2}{dx^2}$$

and perturbation operator is

$$\hat{H}' = kx.$$

Earlier results have shown that the unperturbed wave function

$$\psi^0 = \sqrt{\frac{2}{a}}\sin\frac{n\pi x}{a}$$

and the corresponding unperturbed energy levels are given as:

$$E^0 = \frac{n^2h^2}{8ma^2}$$

The first order correction to energy E^0 can be obtained using the equation (3.150) and the unperturbed Hamiltonian operator as:

$$E_i^{(1)} = \int \psi_i^0\,\hat{H}'\,\psi_i^0 d\tau = \int_0^a \left(\frac{2}{a}\right)^{\frac{1}{2}} \sin\frac{n\pi x}{a}\,(kx)\left(\frac{2}{a}\right)^{\frac{1}{2}}\sin\frac{n\pi x}{a}dx = \frac{2k}{a}\int_0^a x\,\sin^2\frac{n\pi x}{a}\,dx$$

$$= \frac{2k}{a}\int_0^a x\left(\frac{1-\cos\frac{2n\pi x}{a}}{2}\right)dx = \frac{2k}{2a}\left[\int_0^a x\,dx - \int_0^a x\cos\frac{2n\pi x}{a}\,dx\right]$$

$$= \frac{k}{a}\left(\frac{a^2}{2}-0\right) - \frac{k}{a}\left[\frac{a}{2\pi n}\left(x\sin\frac{2\pi nx}{a}\right)_0^a - \int_0^a \frac{a}{2\pi n}\sin\frac{2\pi nx}{a}\,dx\right]$$

$$= \frac{ka}{2} - \frac{a}{2\pi n}\int_0^a \sin\frac{2\pi nx}{a}\,dx = \frac{ka}{2} + \frac{k}{a}\frac{a^2}{4\pi^2 n^2}\left[\cos\frac{2\pi nx}{a}\right]_0^a$$

$$= \frac{ka}{2}$$

Thus the $$E = \frac{n^2h^2}{8ma^2} + \frac{ka}{2}. \quad \text{...(3.159)}$$

Example 3.18: The potential energy expression for the anharmonic oscillator is given as: $V = \frac{1}{2}kx^2 + \frac{1}{6}\chi x^3 + \frac{1}{24}\delta x^4$. Calculate the first order correction to the ground state energy term of this oscillator.

Solution: In the anharmonic oscillator, the total Hamiltonian can be expressed as:

$$\hat{H} = -\frac{h^2}{8\pi^2\mu}\frac{d^2}{dx^2} + \frac{1}{2}kx^2 + \frac{1}{6}\chi x^3 + \frac{1}{24}\delta x_4$$

Thus unperturbed Hamiltonian operator is: $\hat{H}^0 = -\frac{h^2}{8\pi^2\mu}\frac{d^2}{dx^2} + \frac{1}{2}kx^2$

The first order perturbation to Hamiltonian is: $\hat{H}' = \frac{1}{6}\chi x^3 + \frac{1}{24}\delta x_4$

The ground state unperturbed wave function of the harmonic oscillator is:

$$\psi^0(x) = \left(\frac{\alpha}{\pi}\right)^{\frac{1}{4}} e^{-\frac{\alpha x^2}{2}}$$

where $\alpha = \left(\frac{4\pi^2 k\mu}{h^2}\right)^{\frac{1}{2}}$. The first order correction to the ground state energy is given by equation (3.150), *i.e.,*

$$E_0^1 = \int_{-\infty}^{\infty}\left(\frac{\alpha}{\pi}\right)^{\frac{1}{4}} e^{-\frac{\alpha x^2}{2}}\left(\frac{1}{6}\chi x^3 + \frac{1}{24}\delta x^4\right)\left(\frac{\alpha}{\pi}\right)^{\frac{1}{4}} e^{-\frac{\alpha x^2}{2}} dx$$

$$= \left(\frac{\alpha}{\pi}\right)^{\frac{1}{2}}\int_{-\infty}^{\infty}\left(\frac{1}{6}\chi x^3 + \frac{1}{24}\delta x^4\right) e^{-\alpha x^2} dx \qquad \text{...(3.160)}$$

The first integral of equation (3.160) will be zero as the integrand is an odd. The second integral is:

$$E_0^1 = \left(\frac{\alpha}{\pi}\right)^{\frac{1}{2}}\frac{\delta}{24}\int_{-\infty}^{\infty} x^4 e^{-\alpha x^2} dx = \left(\frac{\alpha}{\pi}\right)^{\frac{1}{2}}\frac{\delta}{12}\int_{0}^{\infty} x^4 e^{-\alpha x^2} dx$$

This integral can be found out using the standard tables and is:

$$= \frac{3\sqrt{\pi}}{8\alpha^{\frac{5}{2}}}$$

and thus $$E_0^1 = \left(\frac{\alpha}{\pi}\right)^{\frac{1}{2}}\frac{\delta}{12} = \frac{3\sqrt{\pi}}{8\alpha^{\frac{5}{2}}} = \frac{\delta}{32\alpha^2} = \frac{h^2\delta}{128\,k\mu}$$

The total ground state energy will be:

$$E = \frac{1}{2}h\omega_e + \frac{h^2\delta}{128\pi^2 k\mu} \qquad ...(3.161)$$

The perturbation theory will be applied to calculate the energy levels of the helium atom in the next section.

3.5.2 ATOM (ION) WITH TWO ELECTRONS

The Schrödinger wave equation for helium atom or for ion with two electrons (Li^+, Be^{2+}, B^{3+}, C^{4+}.....) has the form:

$$\frac{\partial^2\psi}{\partial x_1^2} + \frac{\partial^2\psi}{\partial y_1^2} + \frac{\partial^2\psi}{\partial z_1^2} + \frac{\partial^2\psi}{\partial x_2^2} + \frac{\partial^2\psi}{\partial y_2^2} + \frac{\partial^2\psi}{\partial z_2^2} +$$

$$\frac{8\pi^2 m_e}{h^2}\left[E + \frac{Ze^2}{r_1} + \frac{Ze^2}{r_2} - \frac{e^2}{r_{12}}\right]\psi = 0 \qquad ...(3.162)$$

where E is the total energy of the atom, r_1 and r_2 are the distances of electrons 1 and 2 from the nucleus respectively. Disregarding the zero order approximation, the energy of the mutual repulsion of the electrons, $\frac{e^2}{r_{12}}$, we obtain a solution to the equation (3.162) in the form of a product of two functions Ψ_{n_1} (x_1, y_1, z_1) and $\Psi_{n_2}(x_2, y_2, z_2)$:

$$\psi(x_1, y_1, z_1; x_2, y_2, z_2) = \Psi_{n_1}(x_1, y_1, z_1)\,\Psi_{n_1}(x_2, y_2, z_2) \qquad ...(3.163)$$

Each of which is a solution of one of the equations of the form:

$$\frac{\partial^2\psi_k}{\partial x_k^2} + \frac{\partial^2\psi_k}{\partial y_k^2} + \frac{\partial^2\psi_k}{\partial z_k^2} + \frac{8\pi^2 m_e}{h^2}\left[E_k + \frac{Ze^2}{r_k}\right]\psi_k = 0 \qquad ...(3.164)$$

obtained by separation of initial wave equation. Thus, to a zero order approximation, the problem of two electron atoms is to reduce the problem of an atom of one electron. The solution of this equation will be similar to that of hydrogen atom. Accordingly the energy of an atom in an appropriate quantum state will be:

$$E_n = -\frac{2\pi^2 m_e e^4 Z^2}{h^2 n^2} \qquad ...(3.165)$$

Thus the total energy of two-electron atom (ion) to a zero order approximation is equal to the sum of the energies of the individual electrons.

$$E = E_{n_1} + E_{n_2} \qquad ...(3.166)$$

Based on the first order approximation, let us solve the problem of the ground state of the helium atom and the corresponding ions. According to perturbation theory (the theory of mutual repulsion of electrons) is expressed by the integral:

$$\langle E \rangle = \frac{\int |\psi|^2 \frac{e^2}{r_{12}} \, d\tau}{\int |\psi|^2 \, d\tau} \qquad ...(3.167)$$

which on normalization of the wave function ψ, reduces to the integral.

$$\langle E \rangle = \int |\psi|^2 \frac{e_2}{r_{12}} \, d\tau \qquad ...(3.167a)$$

The integral (3.167*a*) is taken over all the positions of electrons. The function ψ is the product of the wave functions for the ground state of hydrogen like atom.

$$\psi_1 = \frac{1}{\sqrt{\pi}} \left(\frac{Z}{a_0}\right)^{\frac{3}{2}} e^{-\frac{\rho}{2}} \quad \text{(unperturbed motion)}$$

i.e.,

$$\psi = \psi_{r_1} \psi_{r_2} = \frac{1}{\pi} \left(\frac{Z}{a_0}\right)^3 e^{-\frac{\rho_1}{2}} e^{-\frac{\rho_2}{2}} \qquad ...(3.168)$$

where $\rho_k = \frac{2Z}{a_0} r_k$. The small element of volume ($d\tau$) in polar coordinates can be written as:

$$d\tau = \left(\frac{a_0}{2Z}\right)^6 \rho_1^2 \, d\rho_1 \sin\theta_1 \, d\theta_1 \, d\varphi_1 \, \rho_2^2 \, d\rho_2 \sin\theta_2 \, d\theta_2 \, d\varphi_2 \qquad ...(3.169)$$

Substituting the values of $d\tau$ together with the wave function into energy of perturbation $\langle E \rangle$ equation (3.167*a*), we get

$$\langle E \rangle = \frac{Ze^2}{2^5 \pi^2 a_0} \int_0^\infty \int_0^\pi \int_0^{2\pi} \int_0^\infty \int_0^\pi \int_0^{2\pi} \frac{e^{-(\rho_1 + \rho_2)}}{\rho_{12}} \rho_1^2 \, d\rho_1$$

$$\sin\theta_1 \, d\theta_1 \, d\varphi_1 \, \rho_2^2 \, d\rho_2 \sin\theta_2 \, d\theta_2 \, d\varphi_2 \qquad ...(3.170)$$

where $\rho_{12} = \frac{2Z}{a_0} r_{12}$. Evaluation of the integral in this expression yields $20\pi^2$. Thus

$$\langle E \rangle = \left(\frac{Ze^2}{2^5 \pi^2 a_0}\right)(20\pi^2) = \frac{5}{4}\left(\frac{Ze^2}{2a_0}\right)$$

$$= \frac{5}{4}\left(\frac{2\pi^2 Z e^4 m_e}{h^2}\right) = -\frac{5}{4} ZE_H$$

where E_H is the energy of the normal hydrogen atom. Thus to a first approximation, the total energy of two electron system, we have

$$E = -\frac{2\pi^2 m_e e^4 Z^2}{h^2 n^2} - \frac{2\pi^2 m_e e^4 Z^2}{h^2 n^2} + \frac{5}{4}\left(\frac{2\pi^2 Z e^4 m_e}{h^2}\right)$$

$$= -\left(2Z^2 - \frac{5}{4}Z\right)E_H \qquad ...(3.171)$$

The energy values of total ionization of a helium atom and of ions with two electrons calculated from the formula are compared with respective experimental values in Table 3.7.

Data of Table 3.7 show that in the first order perturbation, theory gives satisfactory agreement with the experiment. On the other hand, the Hyllerass variational method, which is being discussed in next section, gives total agreement of calculated values of the first ionization potential of helium atom and of ions with two electrons and the measured values of this quantity.

Table 3.7: *A comparison of theoretical and measured ionization energies values of helium atom of ions with two electrons (eV)*

Elements	Ionization energy			% Deviation
	$-E_0$	$-E_1$	$-E_{expt}$	
He	108.76	74.78	78.98	5.3
Li^+	244.71	193.73	198.04	2.2
Be^{2+}	435.04	367.06	371.51	1.2
B^{3+}	679.79	594.78	599.43	0.8
C^{4+}	978.84	876.88	881.83	0.6

% Deviation = $(E_1 - E_{expt}) \times 100/E_1$; E_0 is from zero perturbation, E_1 is from first order perturbation and E_{expt} is the measured value of ionization energy.

3.5.3 EXCHANGE DEGENERACY

Up till now we have considered cases when to a zero order approximation the problem reduces to the solution of wave equations for hydrogen like atoms in the ground state, *i.e.,* cases of the ground state of atoms with two electrons.

Let us now solve the general problem when any states of hydrogen like atoms are possible, so that the wave function of the unperturbed problem (zero order perturbation) must be taken in the form given in the preceding section. If all the coordinates of the first electron (x_1, y_1, z_1), are denoted by number 1 and the coordinates of the second electron by 2, then the wave function ψ can be written in the form:

$$\psi = \Psi_{n_1}(1)\,\Psi_{n_2}(2) \qquad ...(3.172)$$

However, this solution is not the only one that corresponds to the given values of the principal quantum numbers n_1 and n_2 and consequently, to the given value of energy of atom but also in the case when the two electrons are indistinguishable, *i.e.,* when the first electron is in state n_2 and the second electron in n_1.

$$\psi = \Psi_{n_1}(2)\,\Psi_{n_2}(1) \qquad ...(3.173)$$

We have thus two states of the systems, *i.e.,* $\psi = \Psi_{n_1}(1)\,\Psi_{n_2}(2)$ and $\psi = \Psi_{n_1}(2)\,\Psi_{n_2}(1)$ with the same energy (a doubly degenerate system). This degeneracy, which is due to the indistinguishability of electrons and their interchangeability that follows there from, is called *exchange degeneracy.*

Due to exchange degeneracy, a more general solution of the problem is given by the linear combinations of both the preceding particular solutions, *i.e.,*

$$\psi_s = \psi = \Psi_{n_1}(1)\,\Psi_{n_2}(2) + \Psi_{n_1}(2)\,\Psi_{n_2}(1) \qquad ...(3.174)$$

$$\psi_a = \psi = \Psi_{n_1}(1)\,\Psi_{n_2}(2) - \Psi_{n_1}(2)\,\Psi_{n_2}(1) \qquad ...(3.175)$$

The first of these solutions, ψ_s, which does not change through an interchange of electrons (interchange of number 1 and 2), is *symmetrical* while the second ψ_a, which changes its sign if the electrons are interchanged, is called *anti-symmetrical.*

When we take into consideration the perturbation, *i.e.,* the mutual repulsion of electrons $\dfrac{e^2}{r_{12}}$, exchange degeneracy disappears and to each function, these now correspond to different energy values E_s and E_a into which the initial doubly degenerate levels split. Combining equations (3.174) and (3.175) and taking of the square of modulus of the combined functions, we get

$$|\psi|^2 = \psi^*_{n_1}(1)\,\psi_{n_1}(1)\,\psi^*_{n_2}(2)\,\psi_{n_2}(2) + \psi^*_{n_1}(2)\,\psi_{n_1}(2)\,\psi^*_{n_2}(1)\,\psi_{n_2}(1)$$

$$\pm\, 2\,\psi^*_{n_1}(1)\,\psi_{n_1}(2)\;\;\psi^*_{n_2}(2)\,\psi_{n_2}(1) \qquad ...(3.176)$$

Substituting these values of $|\psi|^2$ into the mean energy of perturbation (3.156) and noting that by virtue of symmetry of the perturbing term $\dfrac{e^2}{r_{12}}$, relative to both electrons (1 and 2) the first two terms (3.176) yields the same results, we get

$$\langle E \rangle = \frac{\iint \left[|\psi_{n_1}(1)|^2 |\psi_{n_2}(2)|^2 \pm \psi_{n_1}(1)\,\psi^*_{n_1}(2)\,\psi_{n_2}(1)\,\psi^*_{n_2}(2) \right] \dfrac{e^2}{r_{12}}\, d\tau_1\, d\tau_2}{\iint \left[|\psi_{n_1}(1)|^2 |\psi_{n_2}(2)|^2 \pm \psi_{n_1}(1)\,\psi^*_{n_1}(2)\,\psi_{n_2}(1)\,\psi^*_{n_2}(2) \right] d\tau_1\, d\tau_2} \qquad ...(3.177)$$

Equation (3.177) contains four integrals: two in the numerator and two in the denominator. Let us first consider the integrals in the numerator. Let the first integral be donated by C.

$$C = \int |\psi_{n_1}(1)|^2 \frac{e^2}{r_{12}} \int |\psi_{n_2}(2)|^2\, d\tau_1\, d\tau_2 \qquad ...(3.178)$$

Since quantity $e\,|\psi_{n_1}(1)|^2$ represents the charge density of first electron while the quantity $e\,|\psi_{n_2}(2)|^2$ is the charge density of the second electron, the integral C expresses the energy of

Coulombic interaction of the electrons provided the electrons are distinguishable. For this reason, the integral C is called Coulombic integral (usually it is divided by e^2).

The second integral in the numerator of equation (3.177) is designated by:

$$A = \int \psi_{n_1}(1)\,\psi^*_{n_2}(2)\,\frac{e^2}{r_{12}}\,\psi_{n_1}(2)\,\psi^*_{n_2}(1)\,d\tau_1\,d\tau_2 \quad ...(3.179)$$

This integral does not have such a simple physical meaning. It arises from exchange degeneracy caused by the indistinguishability of electrons, cannot be interpreted on the basis of classical theory or mechanics. This integral is called an exchange integral (also when it is divided by e^2) and arises only because of quantum mechanics.

As has been pointed out that the concept of Coulombic (C) and exchange (A) integrals is based on the indistinguishability of the electrons and thus contradicts experimental results. Hence, it follows that division of atomic energy into Coulombic (C) and exchange (A) energies, is arbitrary and is the result of an inexact interpretation of the physical meaning of the integral C. An exact physical meaning is retained only in the sum of integrals C ± A, which according to equation (3.177) is equal to the mean value of Coulomb potential energy of mutual electron repulsion, since the denominator of expression (3.177) is equal to unity on the condition that functions ψ_{n1} (1) and ψ_{n2} (2) are normalized. Accordingly the first term in the denominator of this expression:

$$\int |\psi_{n_1}(1)|^2\,|\psi_{n_2}(2)|^2\,d\tau_1\,d\tau_2 = 1, \text{ since } \int |\psi_{n_1}(1)|^2 d\tau_1 = \int |\psi_{n_2}(2)|^2 d\tau_2 = 1$$

Further these functions are also orthogonal, *i.e.*, $\int \Psi_{n_1}(1)\Psi_{n_2}(1)\,d\tau_1 = \int \Psi_{n_1}(2)\Psi_{n_2}(2)\,d\tau_2 = 0.$

The second term in the denominator: $\int \Psi_{n_1}(1)\Psi^*_{n_1}(2)\,d\tau_1 \int \Psi_{n_2}(1)\Psi^*_{n_2}(2)\,d\tau_2 = 0$

Thus, for the perturbation energy:

$$\langle E \rangle = C \pm A \quad ...(3.180)$$

Due to indistinguishably of electrons, the quantity $|\Psi(1, 2)|^2$, (3.176), which determines the electron probability or density of electron cloud must be symmetric with respect to transposition of electrons, *i.e.*, it must not change sign in the process: $|\Psi(1, 2)|^2 = |\Psi(2, 1)|^2$. This leads to suggest that the wave function itself may be both symmetric and antisymmetric: $|\Psi(1, 2)| = \pm\,|\Psi(2, 1)|$; this property is indeed satisfied by the wave function ψ_s and ψ_a.

3.6 THE VARIATION METHOD

The other approximate method is the variation method. This method is quite useful in calculating the energies of certain states of the system, especially in computing the energy of the ground state, in which we are generally interested. The main draw back of this method is that it gives the upper limit to the energy of the interested state. Further it also does not give any indication how far this value of energy is from the true one. But with all these limitations, this method is still very useful. It will be shown first that the value of the energy of any state calculated by this method is greater than the true value. This will be followed by the procedure to apply this method to any problem (section 3.6.2). Let the wave function in which we are interested be φ and it is normalized. The choice of this function is purely arbitrary, but the correct selection of φ helps in reducing the labour and also gives a value of E quite close to the true value of E_0. In practice several parameters are put into the trial variable function φ and then vary these parameters, so as to minimize the variation integral (3.181) or (3.182). Successful use of variation method depends on our intuition to make shrewd choice or guess for the trial function. This just needs experience. If $\hat{H}$ is the Hamiltonian operator, the average value of E can be calculated from:

$$\langle E \rangle = \int \varphi^* \hat{H} \varphi d\tau \qquad ...(3.181)$$

if φ^* is normalized and

$$\langle E \rangle = \frac{\int \varphi^* \hat{H} \varphi \, d\tau}{\int \varphi^* \varphi \, d\tau} \qquad ...(3.182)$$

if φ is not normalized. If φ, known as the variation function, is close to true function ψ_0, then

$$\langle E \rangle = \int \psi^* \hat{H} \psi \, d\tau = E_0 \qquad ...(3.183)$$

because $\hat{H}\ \psi_0 = E_0\ \psi_0$. If φ is not equal to ψ_0, then φ can be expressed in terms of complete set of orthonormal functions, $\psi_0, \psi_1, \psi_2 \ldots\ldots \Psi_n$, *i.e.*,

$$\varphi = \sum_n C_n \psi_n \qquad ...(3.184)$$

with

$$\sum_n C_n^* C_n = 1, \qquad ...(3.185)$$

where C_n are the arbitrary variables to be determined and there values are such that one can obtain the minimum value of E, given by the equation (3.181). Further φ satisfies the same boundary conditions as ψ_i because φ is very well behaved function. Substituting the value of φ in equation (3.181), we get

$$\langle E \rangle = \int \sum_n C_n^* \psi_n^* \hat{H} \sum_n C_n \psi_n \, d\tau = \sum_n C_n^* C_n \int \psi_n^* \hat{H} \psi_n \, d\tau$$

$$= \sum_n C_n^* C_n E_n \qquad ...(3.186)$$

Subtracting E_0, the lowest value of the energy from equation (3.186), we obtain

$$\langle E \rangle - E_0 = \sum_n C_n^* C_n (E_n - E_0) \qquad ...(3.187)$$

Since E_n is always either greater than or equal to E_0 for all values of n and further the coefficients $\sum_n C_n^* C_n$ of course are either positive or zero, the right hand side of equation (3.187) will be either positive or zero. This proves the above stated fact that the value of E will be always greater than the true value, *i.e.,*

$$\langle E \rangle \geq E_0 \qquad ...(3.188)$$

If the function φ is not normalized, even then this theorem is applicable, with the only difference that E will be given by the expression:

$$\langle E \rangle = \frac{\int \varphi^* \hat{H} \varphi \, d\tau}{\int \varphi^* \varphi \, d\tau} \qquad ...(3.182)$$

Linear Variation Function: A special kind of variation function, widely used in the study of molecules, is the linear variation function. A linear variation function involving a linear combination of n linearly independent functions $\psi_1, \psi_2, \psi_3 \ldots\ldots \psi_n$ is:

$$\varphi_i = C_{i1} \psi_1 + C_{i2} \psi_2 + \ldots \quad + C_{in} \psi_n \qquad ...(3.189)$$

where φ_i is a trial function and coefficients are parameters to be determined by minimizing the variation integral (3.182). The functions ψ_i must satisfy the boundary conditions of the given problem. In this study, we shall restrict ourselves to real φ so that the C_i's and ψ_i's are also real.

Equation (3.182) with the help of equation (3.189) can be written as:

$$\langle E \rangle = \frac{\int \varphi^* \hat{H} \varphi \, d\tau}{\int \varphi^* \varphi \, d\tau} = \frac{\sum_{i=1}^{n} \sum_{j=1}^{n} C_i C_j H_{ij}}{\sum_{i=1}^{n} \sum_{j=1}^{n} C_i C_j S_{ij}} \qquad ...(3.190)$$

Let us define few terms before we apply the variation theorem:

$$\int \varphi_i^* \varphi_j \, d\tau = \int \sum_{i=1}^{n} C_i \psi_i \sum_{j=1}^{n} C_j \psi_j \, d\tau = \sum_{i=1}^{n} \sum_{j=1}^{n} C_i C_j \int \psi_i \psi_j \, d\tau \qquad ...(3.191)$$

Now $\int \psi_i \psi_j \, d\tau = S_{ij}$ ...(3.192)

is the overlap integral. Thus integral in equation (3.191) reduces to

$$\sum_{i=1}^{n} \sum_{j=1}^{n} C_i C_j S_{ij} \qquad ...(3.193)$$

One thing must be kept in mind that the integral S_{ij} is not necessarily equal to δ_{ij}. Since there is no reason to suppose that the function ψ_i are mutually orthogonal, these are not necessarily the eigen functions of any operator. Now the numerator of equation (3.182) can be written as:

$$\int \varphi_i^* \hat{H} \varphi_i \, d\tau = \int \sum_{i=1}^{n} C_i \psi_i \, \hat{H} \sum_{j=1}^{n} C_j \psi_j \, d\tau$$

$$= \sum_{i=1}^{n} \sum_{j=1}^{n} C_i C_j \int \psi_i H \psi_j \, d\tau \qquad ...(3.194)$$

Using abbreviation $\int \psi_i \ \hat{H} \ \psi_j = H_{ij}$ ...(3.195)

Equation (3.190) can be written as (3.196).

$$\langle E \rangle \sum_{i=1}^{n}\sum_{j=1}^{n} C_i C_j S_{ij} = \sum_{i=1}^{n}\sum_{j=1}^{n} C_i C_j H_{ij} \qquad ...(3.196)$$

Since the variation integral is a function of n independent variables $C_1, C_2 \ldots .C_n$, a necessary condition for a minimum in E is that its partial derivatives with respect to each of the variable must be zero, *i.e.*,

$$\frac{\partial}{\partial C_i}\langle E \rangle \sum_{i=1}^{n}\sum_{j=1}^{n} C_i C_j S_{ij} + \langle E \rangle \frac{\partial}{\partial C_i} \sum_{i=1}^{n}\sum_{j=1}^{n} C_i C_j S_{ij}$$

$$= \frac{\partial}{\partial C_i} \sum_{i=1}^{n}\sum_{j=1}^{n} C_i C_j H_{ij} \qquad ...(3.197)$$

Now let us solve the individual term. $\frac{\partial}{\partial C_i}\langle E \rangle = 0$, because the energy must be minimum for a particular proper set of coefficients. Thus left hand side of equation (3.197) reduces to:

$$\langle E \rangle \frac{\partial}{\partial C_i} \sum_{i=1}^{n}\sum_{j=1}^{n} C_i C_j S_{ij} = \sum_{i=1}^{n}\sum_{j=1}^{n}\left(C_j \frac{\partial C_i}{\partial C_i} + C_i \frac{\partial}{\partial Ci} C_j \right) S_{ij} \qquad ...(3.198)$$

As C_j are independent variables, therefore $\frac{\partial C_j}{\partial C_i} = 1$ for $i = j$ and zero for $i \neq j$ ...(3.199)

$$i.e., \quad \frac{\partial}{\partial C_i} \sum_{i=1}^{n}\sum_{j=1}^{n} C_i C_j S_{ij} = \sum_{i=1}^{n}\sum_{j=1}^{n} (C_j + C_i) S_{ij} = \sum_{i=1}^{n} C_i S_{ij} + \sum_{j=1}^{n} C_j S_{ji} \qquad ...(3.200)$$

Since ψ_i are real functions, $\int \psi_i \psi_j \, d\tau = \int \psi_j \psi_i \, d\tau$ (*i.e.*, $S_{ij} = S_{ji}$. Combining this with equation (3.200), we get

$$\frac{\partial}{\partial C_i} \sum_{i=1}^{n}\sum_{j=1}^{n} C_i C_j S_{ij} = 2 \sum_{i=1}^{n} C_i S_{ij} \qquad ...(3.201)$$

It can also be shown that $H_{ij} = H_{ji}$ as follows. The Hamiltonian operator is the sum of two terms, *i.e.*, term containing double differential (kinetic energy term) and a term containing $1/r$ (potential energy term). Since the latter term is only a multiplication term, it will lead to the same results even if we write as:

$$\int \psi_i \frac{1}{r} \psi_j \, d\tau = \int \psi_j \frac{1}{r} \psi_i \, d\tau$$

The first term is again a sum of $\left(\frac{\partial^2}{\partial x^2}+\frac{\partial^2}{\partial y^2}+\frac{\partial^2}{\partial z^2}\right)$ three terms. Considering only the partial derivative with respect to x only, contribution to H_{ij} by x can be written as,

$$H_{ij} = \int \psi_i \left(\frac{\partial^2}{\partial x^2}\right) \psi_j \, dx = \int \psi_i \left(\frac{\partial^2 \psi_j}{\partial x^2}\right) dx \qquad ...(3.202)$$

Integrating by parts will lead to the following expression.

$$H_{ij} = \left[\psi_i \frac{\partial \psi_j}{\partial x}\right] - \int \frac{\partial \psi_i}{\partial x}\frac{\partial \psi_j}{\partial x} dx$$

$$= \left[\psi_i \frac{\partial \psi_j}{\partial x}\right] - \left[\psi_j \frac{\partial \psi_i}{\partial x}\right] + \int \frac{\partial^2 \psi_i}{\partial x^2} \psi_j \, dx \qquad ...(3.203)$$

For acceptable wave functions ψ_i and ψ_j must vanish at infinity. Thus, equation (3.203) reduces to (3.204) as the integration limits are between $+\infty$ to $-\infty$:

$$H_{ij} = \int \psi_i \left(\frac{\partial^2}{\partial x^2}\right) \psi_j \, dx = \int \frac{\partial^2 \psi_i}{\partial x^2} \psi_j \, dx = H_{ji} \qquad ...(3.204)$$

Replacing S_{ij} and H_{ij} in equation (3.201) it can be shown that:

$$\frac{\partial}{\partial C_i} \sum_{i=1}^{n}\sum_{j=1}^{n} C_i C_j H_{ij} = 2 \sum_{i=1}^{n} C_i H_{ij} \qquad ...(3.205)$$

Combining equations (3.197), (3.201) and (3.205), we get

$$2\,E \sum_{i=1}^{n} C_i S_{ij} = 2 \sum_{i=1}^{n} C_i H_{ij} \qquad ...(3.206)$$

or

$$\sum_{i=1}^{n} (H_{ij} - E\,S_{ij})\, C_i = 0 \qquad ...(3.207)$$

The summation (3.207) is a set of n simultaneous linear homogeneous equations in n unknown coefficients (C_1, C_2.........C_n) and can also be expressed as:

$(H_{11} - ES_{11})\,C_1 + (H_{12} - ES_{12})\,C_2 + \ldots\ldots\ldots + (H_{1n} - ES_{1n})\,C_n = 0$

$(H_{21} - ES_{21})\,C_1 + (H_{22} - ES_{22})\,C_2 + \ldots\ldots\ldots + (H_{2n} - ES_{2n})\,C_n = 0$

....

....

$(H_{n1} - ES_{n1})\,C_1 + (H_{n2} - ES_{n2})\,C_2 + \ldots\ldots\ldots + (H_{nn} - ES_{nn})\,C_n = 0$

To have a non-trivial solution for the above set of simultaneous equations, the determinant formed from the coefficients must be zero, *i.e.*,

$$\begin{vmatrix} H_{11}-ES_{11} & H_{12}-ES_{12} \ldots\ldots\ldots\ldots\ldots H_{1n}-ES_{1n} \\ H_{21}-ES_{21} & H_{22}-ES_{22} \ldots\ldots\ldots\ldots\ldots H_{2n}-ES_{2n} \\ \ldots\ldots\ldots\ldots\ldots\ldots\ldots\ldots\ldots\ldots & \\ \ldots\ldots\ldots\ldots\ldots\ldots\ldots\ldots\ldots\ldots & \\ H_{n1}-ES_{n1} & H_{n2}-ES_{n2} \ldots\ldots\ldots\ldots\ldots H_{nn}-ES_{nn} \end{vmatrix} = 0 \qquad \ldots(3.208)$$

The determinant so formed is called *secular determinant* and the set of simultaneous equations is called the set of *secular equations*. The expansion of this determinant will give an *n*th order equation in the unknown E with *n* roots. It can be shown that these roots are real. Then the roots so obtained give the approximate values of the energies of the various states. These values are upper limits to the energies of the respective states. The approximate ground state wave functions can be obtained by substituting the value of E_0 in the secular equations. The coefficients C_2, C_3,............C_n can be obtained in terms of C_1. The value of C_1 can be calculated using normalization condition.

The problem in this method, before the advent of electronic computers, was the solution of *n*th order equation. But with the advent of fast computers, solution of 100th degree equation is also possible. The solution is further facilitated if some of the wave functions are orthogonal, because it will lead to some δ_{ij} to be equal to zero and this condition will make some of the non-diagonal elements or terms of secular determinant zero. Lastly the above procedure or equations are also valid even if some of the functions are not real.

3.6.1 GROUND STATE ENERGY OF THE HYDROGEN LIKE ATOMS BY THE VARIATION METHOD

Let the trial wave function for hydrogen atom for its ground state is:

$$\psi = e^{-ar} \qquad \ldots(3.209)$$

where *a* is the variation parameter. We will calculate

$$\langle E \rangle = \frac{\int \varphi^* \, \hat{H} \, \varphi \, d\tau}{\int \varphi^* \, \varphi \, d\tau} \qquad \ldots(3.210)$$

and set $\dfrac{\partial \langle E \rangle}{\partial a} = 0$. This will give us the value of *a*, for which E is minimum. The problem will be solved provided we are able to evaluate the two integrals $\int \psi^* \, H \, \psi \, d\tau$ and $\int \psi^* \, \psi \, d\tau$. The Hamiltonian operator of the hydrogen like atoms is:

$$\hat{H} = -\frac{h^2}{8\pi^2 m_e}\nabla^2 - \frac{Ze^2}{r}$$

Further since ψ depends only on *r*, the radial part of $\hat{H}$ needs to be considered, *i.e.*,

$$\hat{H}_{\text{rad}} = -\frac{h^2}{8\pi^2 m_e}\frac{1}{r^2}\frac{\partial}{\partial r}\left(r^2\frac{\partial}{\partial r}\right) - \frac{Ze^2}{r}$$

$$\nabla^2\psi = \left[\frac{1}{r^2}\frac{\partial}{\partial r}\left(r^2\frac{\partial}{\partial r}\right)\right]e^{-ar} = \frac{1}{r^2}\frac{\partial}{\partial r}[r^2(-ae^{-ar})]$$

$$= -\frac{a}{r^2}(-ar^2 e^{-ar} + 2re^{-ar}) = \left(a^2 - \frac{2a}{r}\right)e^{-ar}$$

The limits of this integration are determined by considering the electron to be in a spherical shell at a distance r to $r + dr$ from the nucleus as r goes from zero to infinity. The volume of spherical shell at a distance r from the nucleus is given by $4\pi r^2 dr$. Substituting the value of $\nabla^2\psi$ state in equation (3.210) and applying the integral limits, we get

$$\langle E_a \rangle = \frac{\int_0^\infty e^{-ar}\left[-\frac{h^2}{8\pi^2 m_e}\left(a^2 - 2\frac{a}{r}\right)e^{-ar} - \frac{Ze^2}{r}e^{-ar}\right]4\pi r^2 dr}{\int_0^\infty e^{-2ar}4\pi r^2 dr}$$

$$= \frac{-\frac{h^2a^2}{8\pi^2 m_e}\int_0^\infty r^2 e^{-2ar}dr + \frac{2h^2 a}{8\pi^2 m_e}\int_0^\infty re^{-2ar}\,dr - Ze^2\int_0^\infty re^{-2ar}\,dr}{\int_0^\infty r^2 e^{-2ar}dr}$$

Since

$$\int_0^\infty x^n e^{-ax}\,dx = \frac{\Gamma(n+1)}{a^{n+1}},\ n > -1,\ a > 0$$

$$\frac{n!}{a^{n+1}},\ (a > 0,\ n \text{ is positive integer}).$$

where $\Gamma(n)$ is finite if $n > 0$, $\Gamma(n+1) = n\Gamma(n)$. $\Gamma(n) = (n + 1)!$ If n = integer > 0.

$$\therefore \langle E_a \rangle = \frac{-\frac{h^2a^2}{8\pi^2 m_e}\left[\frac{2}{(2a)^{2+1}}\right] + \frac{2h^2 a}{8\pi^2 m_e}\left[\frac{1}{(2a)^{1+1}}\right] - Ze^2\left[\frac{1}{(2a)^{1+1}}\right]}{\frac{2!}{[2a]^{2+1}}}$$

$$= \frac{-\frac{h^2a^2}{8\pi^2 m_e}\left[\frac{2}{8a^3}\right] + \frac{2h^2 a}{8\pi^2 m_e}\left[\frac{1}{4a^2}\right] - \frac{Ze^2}{4a^2}}{\frac{2}{8a^3}}$$

$$= -\frac{h^2a^2}{8\pi^2 m_e} + \frac{2h^2a^2}{8\pi^2 m_e} - Ze^2 a = \frac{h^2a^2}{8\pi^2 m_e} - Ze^2 a$$

Now taking the first derivative with respect to 'a' and equating to zero, we get

$$\frac{\partial \langle E_a \rangle}{\partial a} = \frac{2h^2 a}{8\pi^2 m_e} - Ze^2 = 0$$

$$a = \frac{4\pi^2 m_e Ze^2}{h^2} \quad \text{...(3.211)}$$

and $$E_{\min}(a) = -\frac{2\pi^2 m_e Z^2 e^4}{h^2} = -13.6\, Z^2\, (eV)$$

which conforms to the ionization energy data of some one electron systems listed in Table 3.7.

The trial wave functions $\psi_1 = e^{-\frac{ar}{2}}$ and $\psi_2 = e^{-2ar}$ yield:

$$a = \frac{2\pi^2 m_e Ze^2}{h^2}, \qquad E_{\min}(a) = -\frac{\pi^2 m_e Z^2 e^4}{h^2} = -6.8\ eV$$

$$a = \frac{4\pi^2 m_e Ze^2}{h^2} \qquad E_{\min}(a) = -\frac{8\pi^2 m_e Z^2 e^4}{h^2} = -27.2\ eV$$

Though the trial functions ψ, ψ_1, and ψ_2 are close to each other for small and large values of r but the trial wave function $\psi = e^{-ar}$ is considered to be the best since it gives the expected energy data for one electron systems.

It is to be kept in mind that the best form of wave functions depends on the particular problem under consideration. If one is interested in large r behaviour of the function for calculating a physical quantity for the system, the function that gives better energy eigen values may not be an appropriate wave function. As an example the students are advised to use the following trial wave functions.

$$\psi_1 = A_1\, e^{-\alpha \frac{r}{a_0}} \qquad E^1_{\min}(\alpha) \text{ where } \alpha = \left(\frac{4\pi^2 m_e e^2}{h^2}\right) a_0 = 1$$

$$\psi_2 = A_2\, e^{-\alpha \left(\frac{r}{a_0}\right)^2} \qquad E^2_{\min}(\alpha) \text{ where } \alpha = \frac{8}{9}\pi$$

$$\psi_3 = A_3 \frac{r}{a_0}\, e^{-\alpha \frac{r}{a_0}} \qquad E^3_{\min}(\alpha) \text{ where } \alpha = \frac{3}{2}$$

and plot the curves between ψ's and r/a_0 (= 0, 0.5, 1.0, 1.5, 2.0, 2.5, 3.0, 3.5, 4.0). It will be found that ψ_3 is closer to the wave functions ψ_1 and ψ_2 for large values of r but energy corresponding to ψ_2 wave function will be minimum and close to the expected value.

3.6.2 HELIUM ATOM

The next atom after hydrogen is the helium atom. It contains two electrons, two protons (two units of positive charge) and two neutrons (*i.e.,* four units of mass). The helium atom can be represented pictorially in Fig. 3.12. Although two electrons cannot be distinguished from each other, these have been labeled as 1 and 2 for convenience sake.

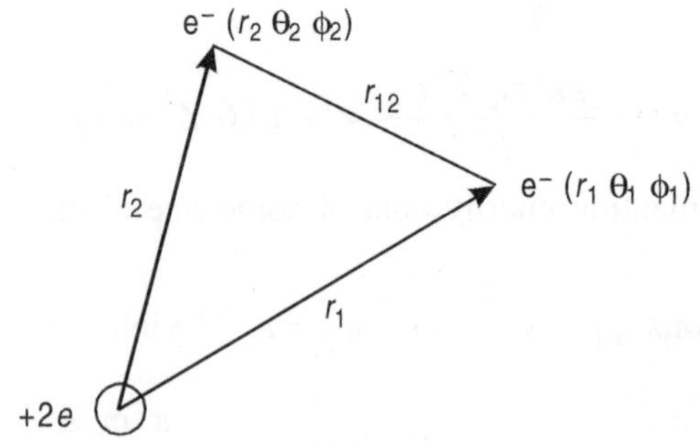

Figure 3.12: *Schematic diagram of helium atom depicting the polar coordinates*

Since we are only interested in the relative motion of electrons with respect to nucleus in helium or helium like atoms, the kinetic energy of nucleus can be neglected. Thus the total Hamiltonian operator for the helium atom can be written as:

$$\hat{H} = -\frac{h^2}{8\pi^2 m_e}[\nabla^2(1) + \nabla^2(2)] - \frac{Ze^2}{r_1} - \frac{Ze^2}{r_2} + \frac{e^2}{r_{12}} \quad ...(3.212)$$

The terms in the square brackets represent the kinetic energy operators for the electrons 1 and 2 respectively. The rest of the three terms represent the potential energy terms, the third and fourth terms are attractive potential energy terms between the electron and nucleus, the fifth term is the repulsive potential energy term between the two electrons. The corresponding Schrödinger wave equation for helium atom will be:

$$\hat{H}\,\psi = E\,\psi \quad ...(3.213)$$

Unlike hydrogen like atoms, equation (3.213) cannot be solved accurately, *i.e.,* there is no analytical solution which when operated by Hamiltonian operator, gives back the same function multiplied by a constant number, known as total energy. If one compares the equation (3.213) with the similar one for hydrogen atom, it is clear that it is not possible to separate out the variables because of the presence of $\frac{e^2}{r_{12}}$ term. This means that one cannot write the equations depending on only one variable, as has been done in case of particle in a three dimensional box or for hydrogen atom. The repulsive potential energy is called *electron repulsion or electron correlation term* and calculation of its actual magnitude poses the most difficult problem for the quantum chemists. Its magnitude can be calculated by using certain approximations. Most crude approximation is to neglect the term $\frac{e^2}{r_{12}}$. Physically this means that each electron is attracted by the nucleus of two units of positive charge and their behaviour is not affected by the presence of each other, *i.e.*, they behave as independent particles. If we neglect this term, the Hamiltonian operator (3.212) can be written as:

$$\hat{H} = \left[-\frac{h^2}{8\pi^2 m_e}\nabla^2(1) - \frac{Ze^2}{r_1}\right] + \left[-\frac{h^2}{8\pi^2 m_e}\nabla^2(2) - \frac{Ze2}{r_2}\right]$$

$$= \hat{H}_0(1) + \hat{H}_0(2) \qquad ...(3.214)$$

where $\hat{H}_0(1)$ and $\hat{H}_0(2)$ are the Hamiltonian operators corresponding to the electrons 1 and 2 respectively. This means that the total energy is a sum of the energies of electrons 1 and 2. Under such circumstances, the total wave function ψ can be written as the product of one-electron wave functions, similar to hydrogen like atoms, with two units of positive charge on the nucleus, *i.e.*,

$$\psi(1, 2) = \psi_1 \psi_2 \qquad ...(3.215)$$

and the total energy

$$E = E_1 + E_2 \qquad ...(3.216)$$

where ψ_1 and ψ_2 are hydrogen like wave functions with Z = 2. For the ground state of helium atom, ψ_1 and ψ_2 will be 1*s* orbitals and E_1 and E_2 are the lowest energy states, *i.e.*, 1*s* orbital. Substituting (3.215) in (3.213) and separating the variables as has been done in the previous chapter the following equations can be written:

$$\hat{H}_0(1)\,\Psi_1 = \left[-\frac{h^2}{8\pi^2 m_e}\nabla^2(1) - \frac{Ze^2}{r_1}\right]\psi_1 = E_1\Psi_1 \qquad ...(3.217)$$

$$\hat{H}_0(2)\,\Psi_2 = \left[-\frac{h^2}{8\pi^2 m_e}\nabla^2(2) - \frac{Ze^2}{r_2}\right]\psi_2 = E_2\Psi_2 \qquad ...(3.218)$$

Equations (3.217) and (3.218) are similar to that of hydrogen like atoms with Z = 2 and can be solved accurately. The 1*s* functions and corresponding energies for helium atom can be written as:

$$\psi_1 = \left(\frac{2^3}{\pi a_0^3}\right)^{\frac{1}{2}} e^{-\frac{2r_1}{a_0}}, \qquad E_1 = 2^2 E_H = 4E_H \qquad ...(3.219)$$

$$\psi_2 = \left(\frac{2^3}{\pi a_0^3}\right)^{\frac{1}{2}} e^{-\frac{2r_2}{a_0}}, \qquad E_2 = 2^2 E_H = 4E_H \qquad ...(3.220)$$

Thus $E = E_1 + E_2 = 4E_H + 4\,E_H = 8E_H = 8\,(-13.6\ eV) = -108.8\ eV$ ...(3.220*a*)

Although the value of E so obtained by this method is larger than the experimentally determined value of –79.0 *e*V, the results are on the right track. The only problem is that the energy predicted for the lowest state is very low as compared to the experimental value. This could be due to the neglect of electron repulsion term, which is having a positive value.

The above treatment has clearly indicated that the electron correlation term $\frac{e^2}{r_{12}}$ in the potential energy term plays an important role, but the problem arises how this term can be included and what is its physical significance? This is accomplished as follows. Let us first understand the physical significance. Introduce the first electron into helium nucleus. This will be similar to the hydrogen like atoms with two units of positive charge and wave function representing the behaviour of this electron in ground state will be 1*s* hydrogen atom wave function with Z = 2. Since 1*s* orbital is spherical symmetrical, its negative charge cloud will be surrounding the nucleus uniformly. In other words, the charge cloud will be shielding the nucleus equally in all the directions. If the electron 2 is present at infinity, ($r_2 = \infty$) to start with, it will experience the nucleus which is completely shielded by the charge cloud of the electron 1, *i.e.*, Z = 1. But on approaching near the nucleus, it will experience attractive interaction from the nucleus, as well as, the repulsive interaction from the charge of electron 1. In other words it will be attracted by the nucleus less than it would have been in one electron He^+ ion. One can also express this by saying that in presence of electron 1, the electron 2 is not experiencing the full two units of positive charge of the nucleus. Now if r_2 keeps on decreasing and approaches zero *i.e.*, close to nucleus it will experience the unshielded nuclear charge equal to +2, as the charge cloud of the electron 1, which is shielding the nucleus or radial distribution function due to electron 1 near the nucleus will be very small. Thus, the actual value of charge on the nucleus (effective nuclear charge Z_{eff}) will be a function of distance of the electron 2 from the nucleus and its value varies between one and two. This clearly indicates that the wave functions represented by the equations (3.219) and (3.220) should be modified by introducing effective nuclear charge (Z_{eff}) in place of the actual nuclear charge (Z). The normalized new wave functions can be written as:

$$\psi_1 = \left(\frac{Z_{eff}^3}{\pi a_0^3}\right)^{\frac{1}{2}} e^{-\frac{Z_{eff}\, r_1}{a_0}} \qquad \text{...(3.221)}$$

$$\psi_2 = \left(\frac{Z_{eff}^3}{\pi a_0^3}\right)^{\frac{1}{2}} e^{-\frac{Z_{eff}\, r_2}{a_0}} \qquad \text{...(3.222)}$$

Same Z_{eff} value is used for both the electrons because in the ground state both the electrons will be having equivalent distribution. The only problem now arises is how to find the values of Z_{eff}. This can be evaluated by using variation method, *i.e.*, finding out the value of Z_{eff} for which energy is minimum. The average value of energy is given by the expression:

$$\langle E \rangle = \int \psi^* \hat{H} \psi \, d\tau = \int \psi^*{}_1 \, \psi^*{}_2 \, \hat{H} \psi_1 \, \psi_2 \, d\tau_1 \, d\tau_2$$

because both the functions ψ_1 and ψ_2 are normalized. Substituting the values of ψ_1 and ψ_2 and $\hat{H}$, we get

$$\langle E\rangle = \int \frac{Z_{eff}^3}{\pi a_0^3} e^{-\frac{Z_{eff}\, r_1}{a_0}} e^{-\frac{Z_{eff}\, r_2}{a_0}} \left[-\frac{h^2}{8\pi^2 m_e}\nabla^2(1) - \frac{h^2}{8\pi^2 m_e}\nabla^2(2) - \frac{Ze^2}{r_1} - \frac{Ze^2}{r_2} + \frac{e^2}{r_{12}} \right] \frac{Z_{eff}^3}{\pi a_0^3} e^{-\frac{Z_{eff}\, r_1}{a_0}} e^{-\frac{Z_{eff}\, r_2}{a_0}} \, d\tau_1\, d\tau_2 \qquad ...(3.223)$$

The expression (3.223) can be simplified with the help of following expressions taking E_H as negative:

$$-\frac{h^2}{8\pi^2 m_e}\nabla^2(1)\,\psi_1 = \frac{Z_{eff}e^2}{r_1}\,\psi_1 - Z_{eff}^2 E_H \psi_1 \qquad ...(3.224)$$

$$-\frac{h^2}{8\pi^2 m_e}\nabla^2(2)\,\psi_2 = \frac{Z_{eff}e^2}{r_2}\,\psi_2 - Z_{eff}^2\, E_H \psi_2 \qquad ...(3.225)$$

Writing (3.223) as follows:

$$\begin{aligned}\langle E\rangle = & \int \psi_1^* \psi_2^* \left[-\frac{h^2}{8\pi^2 m_e}\nabla^2(1)\right] \psi_1 \psi_2 \, d\tau_1\, d\tau_2 \\ & + \int \psi_1^* \psi_2^* \left[-\frac{h^2}{8\pi^2 m_e}\nabla^2(2)\right] \psi_1 \psi_2 \, d\tau_1\, d\tau_2 \\ & - Ze^2 \int \psi_1^* \psi_2^* \left(\frac{1}{r_1} + \frac{1}{r_2}\right) \psi_1 \psi_2 \, d\tau_1\, d\tau_2 \\ & + e^2 \int \psi_1^* \psi_2^* \frac{1}{r_{12}} \psi_1 \psi_2 \, d\tau_1\, d\tau_2\end{aligned}$$

and using equations (3.224) and (3.225) we can write the above equation as:

$$\begin{aligned}\langle E\rangle = & \int \psi_1^* \psi_2^* \left[-Z_{eff}^2\, E_H + \frac{Z_{eff}\, e^2}{r_1}\right] \psi_1 \psi_2 \, d\tau_1\, d\tau_2 \\ & + \int \psi_1^* \psi_2^* \left[-Z_{eff}^2\, E_H + \frac{Z_{eff}\, e^2}{r_2}\right] \psi_1 \psi_2 \, d\tau_1\, d\tau_2 \\ & - Ze^2 \int \psi_1^* \psi_2^* \left(\frac{1}{r_1} + \frac{1}{r_2}\right) \psi_1 \psi_2 \, d\tau_1\, d\tau_2 \\ & + e^2 \int \psi_1^* \psi_2^* \frac{1}{r_{12}} \psi_1 \psi_2 \, d\tau_1\, d\tau_2\end{aligned}$$

$$= -2\,Z_{eff}^2\,E_H \int \psi_1^* \psi_1\, d\tau_1 \int \psi_2^* \psi_2\, d\tau_2 + (Z_{eff} - Z)\, e^2$$

$$\int \psi_1^* \frac{1}{r_1} \psi_1\, d\tau_1 \int \psi_2^* \psi_2\, d\tau_2 + (Z_{eff} - Z)\, e^2$$

$$\int \psi_2^* \frac{1}{r_2} \psi_2\, d\tau_2 \int \psi_1^* \psi_1\, d\tau_1 + e^2 \int \psi_1^* \psi_2^* \frac{1}{r_{12}}\, \psi_1 \psi_2\, d\tau_1\, d\tau_2$$

$$= -2Z_{eff}^2 E_H + (Z_{eff} - Z)\, e^2 \left[\int \psi_1^* \frac{1}{r_1} \psi_1\, d\tau_1 + \int \psi_2^* \frac{1}{r_2} \psi_2\, d\tau_2\right]$$

$$+ e^2 \int \psi_1^* \psi_2^* \frac{1}{r_{12}}\, \psi_1 \psi_2\, d\tau_1\, d\tau_2 \qquad \text{...(3.226)}$$

Since both the functions ψ_1 and ψ_2 are of $1s$ orbital, the second and third terms of equation (3.226) are similar and will lead to the same value. Substituting the values of ψ_1 in the second term of equation (3.226), we get

$$(Z_{eff} - Z)\, e^2 \left[\int \psi_1^* \frac{1}{r_1} \psi_1\, d\tau_1 = \frac{Z_{eff}^3\, e^2}{\pi a_0^3} \int_0^\infty e^{-\frac{2Z_{eff}\, r_1}{a_0}}\, r_1\, dr_1 \int_0^\pi \sin\theta\, d\theta \int_0^{2\pi} d\varphi\right.$$

$$= \frac{4Z_{eff}^3\, e^2}{a_0^3} \int_0^\infty e^{-\frac{2Z_{eff}\, r_1}{a_0}}\, r_1\, dr = \frac{4\, Z_{eff}^3\, e^2}{a_0^3} \left[\frac{1}{\left(\frac{2\, Z_{eff}}{a_0}\right)^2}\right]$$

$$= \frac{4Z_{eff}^3\, e^2}{a_0^3} \left[\frac{a_0^2}{4\, Z_{eff}^2}\right] = \frac{Z_{eff}\, e^2}{a_0} = 2\, Z_{eff}\, E_H \qquad \text{...(3.227)}$$

The value of integral

$$e^2 \int \psi_1^* \psi_2^* \frac{1}{r_{12}}\, \psi_1 \psi_2\, d\tau_1\, d\tau_2 = \frac{5}{4} Z_{eff} E_H \qquad \text{...(3.228)}$$

has been solved in example 3.17. Combining equations (3.226), (3.227) and (3.228)

$$\langle E \rangle = -2Z_{eff}^2 E_H + 4Z_e (Z_{eff} - Z) E_H + \frac{5}{4}\, Z_{eff} E_H \qquad \text{...(3.229)}$$

Differentiating $\langle E \rangle$ with respect to Z_{eff} and equating the resulting equation to 0, we get:

$$\frac{d\langle E \rangle}{dZ_{eff}} = -\,4Z_{eff}\, E_H + 8Z_{eff}\, E_H - 4\, ZE_H + \frac{5}{4} E_H = 0$$

or
$$Z_{eff} = \frac{1}{4}\left(4Z - \frac{5}{4}\right) = \frac{1}{4}\left(8 - \frac{5}{4}\right) = \frac{27}{16} \qquad ...(3.230)$$

where Z = 2. Substituting the value of Z_{eff} in (3.229), we get

$$\langle E \rangle = -2\left(\frac{27}{16}\right)^2 (13.6\ eV) = -77.45\ eV \qquad ...(3.231)$$

Using the effective value of charge on nucleus, the energy of the lowest state is only out by a factor of 1.9%. In accord with the variation theorem, the true ground state energy is less than that of calculated by the variational theorem. With this kind of approximation, *i.e.*, considering the independent electron model and a simple exponential wave function (with the form given by equations 3.221 and 3.222), the agreement between the experimental and calculated values is quite good. The values of ionization energies of ions with two electrons, *viz.*, Li^+, Be^{2+}, B^{3+}, C^{4+}, can also be calculated in a similar fashion and compared with respective experimental values, *i.e.*, 24.47, 75.28, 153.10, 258.1 ±0.2 and 389.9 ± 0.4 eV. The energy of H^- and H_2^+ can also be calculated by this method. The other important deduction made to make is that each electron is screened from the nucleus of charge plus two by $\frac{5}{16}$. The latter number is a measure of the screening constant effect that the average distribution of one electron has upon the nucleus charge experienced by the other electron.

The next question is how to improve the results obtained by variation method. The answer to this question is that when we write trial function as the product of separate function of each electron, we are making approximation. Since $\frac{e^2}{r_{12}}$ term is present in the Hamiltonian operator, the Schrödinger equation for helium atom cannot be separated out and the true ground state wave function cannot be written as the product of separate functions for each electron. This means that we have to modify the trial wave functions as expressed in (3.221) and (3.222). Hylleras and others have modified the variational wave functions by introducing the interelectronic distance r_{12} more explicitly. The function containing as many as 1078 terms have been employed to calculate the energy of helium atom and better accuracy has been achieved. But at this stage one has to compromise with better accuracy in energy and function but we may loose a physical picture, which one gets from simple trial functions. As a chemist, physical picture about the structure of atoms or molecules is more important than getting an extremely accurate value of energy and functions. This kind of situation will arise many times in quantum mechanics and according to Mulliken, *the more accurate the calculations become, the more concepts tend to vanish into thin air.*

Example 3.19: Evaluate the integral $e^2 \int \psi_1^* \psi_2^* \frac{1}{r_{12}} \psi_1 \psi_2 \, d\tau_1 \, d\tau_2$.

Solution: Substituting the values of ψ_1 and ψ_2, from equations (3.221) and (3.222), we get

$$e^2 \int \psi_1^* \psi_2^* \frac{1}{r_{12}} \psi_1 \psi_2 \, d\tau_1 \, d\tau_2$$

$$= e^2 \int \frac{Z_{eff}^3}{\pi a_0^3} e^{-\frac{Z_{eff} r_1}{a_0}} e^{-\frac{Z_{eff} r_2}{a_0}} \frac{1}{r_{12}} \frac{Z_{eff}^3}{\pi a_0^3} e^{-\frac{Z_{eff} r_1}{a_0}} e^{-\frac{Z_{eff} r_2}{a_0}} d\tau_1 \, d\tau_2$$

$$= e^2 \left(\frac{Z_{eff}^3}{\pi a_0^3} \right)^2 \int e^{-\frac{2Z_{eff} r_1}{a_0}} e^{-\frac{Z_{eff} r_2}{a_0}} \frac{1}{r_{12}} d\tau_1 \, d\tau_2$$

The integral represents the mutual electrostatic repulsive energy of two spherical symmetrical distribution of electronic charge with density $e^{-\frac{2Z_{eff} r_1}{a_0}}$ and $e^{-\frac{2Z_{eff} r_2}{a_0}}$ respectively. It can be calculated by evaluating the potential due to the first distribution by integrating over $d\tau_1$ and then evaluating the energy of second distribution in the field of first one.

The potential of spherical shell of radius r to $r + dr$ and total charge $4\pi r^2 e^{-\frac{2Z_{eff} r_1}{a_0}}$ at a point p is:

$$4\pi r^2 e^{-\frac{2Z_{eff} r_1}{a_0}} dr_1 \frac{1}{r_1} \qquad \text{if } p < r_1$$

and

$$4\pi r^2 e^{-\frac{2Z_{eff} r_1}{a_0}} dr_1 \frac{1}{p} \qquad \text{if } p > r_1$$

i.e., potential is constant with in the shell and has the same value outside the shell if the entire charge were located at the origin. The potential of complete distribution is:

$$\Phi(p) = \frac{4\pi}{p} \int_0^p r_1^2 e^{-\frac{2Z_{eff} r_1}{a_0}} dr + 4\pi \int_p^\infty r_1 e^{-\frac{Z_{eff} r_1}{a_0}} dr_1$$

Integrating by parts and using the limits one gets:

$$\Phi(p) = -\frac{4\pi a_0^2}{4Z_{eff}^2} e^{-\frac{2Z_{eff} p}{a_0}} - \frac{8\pi a_0^3}{4Z_{eff}^3 p} e^{-\frac{2Z_{eff} p}{a_0}} + \frac{8\pi a_0^3}{8Z_{eff}^3 p}$$

The integral then has the value

$$= \frac{e^2 Z_{eff}^6}{\pi^2 a_0^6} \int \Phi(r_2) e^{-\frac{2Z_{eff} r_2}{a_0}} d\tau_2$$

$$= \frac{e^2 Z_{eff}^6}{\pi^2 a_0^6} \int \left[-\frac{4\pi a_0^2}{4Z_{eff}^2} e^{-\frac{2Z_{eff} r_2}{a_0}} - \frac{8\pi a_0^3}{8Z_{eff}^3 r_2} e^{-\frac{2Z_{eff} r_2}{a_0}} \right.$$

$$\left. + \frac{8\pi a_0^3}{8Z_{eff}^3 r_2} e^{-\frac{2Z_{eff} r_2}{a_0}} \right] 4\pi r_2^2 e^{-\frac{2Z_{eff} r_2}{a_0}} dr_2$$

$$= \frac{e^2 Z_{\text{eff}}^6}{\pi^2 a_0^6}\left[\frac{4\pi^2 a_0^2}{Z_{\text{eff}}^2}\int_0^{\infty} r_2^2\, e^{-\frac{4Z_{\text{eff}} r_2}{a_0}}\, dr_2 - \frac{4\pi^2 a_0^3}{Z_{\text{eff}}^3}\int_0^{\infty} r_2\, e^{-\frac{4Z_{\text{eff}} r_2}{a_0}}\, dr_2 + \frac{4\pi^2 a_0^3}{Z_{\text{eff}}^3}\int_0^{\infty} r_2\, e^{-\frac{4Z_{\text{eff}} r_2}{a_0}}\, dr_2\right]$$

Integrating by parts and applying the limits:

$$\frac{e^2 Z_{\text{eff}}^6}{\pi^2 a_0^6}\left[\frac{\pi^2 a_0^5}{Z_{\text{eff}}^5} - \frac{\pi^2 a_0^5}{4Z_{\text{eff}}^5} - \frac{\pi^2 a_0^5}{8Z_{\text{eff}}^5}\right] = \frac{5e^2 Z_{\text{eff}}}{8\, a_0} = \frac{5}{4}\,\frac{e^2 Z_{\text{eff}}}{2a_0} = \frac{5}{4}\, Z_{\text{eff}}\, E_H.$$

3.7 ELECTRON SPIN OR SPIN ANGULAR MOMENTUM

If the atomic absorption lines are viewed under high-resolution spectrometer, it can be seen that any of the atomic lines are not single lines, but a mixture of two or three lines. This is known as *fine structure* in atomic absorption spectroscopy. Doublet (two lines, one at 589 nm and the other at 589.6 nm) nature of sodium spectrum in this region is well known example. Similar structure is also observed in other region of sodium spectrum. This clearly indicates that many close lying states are present in the atoms.

In 1925, Uhlenbeck and Gouldsmit, just before the advent of quantum mechanics predicted that besides having orbital angular momentum about the nucleus, the electrons are also having another intrinsic property, *i.e.,* the electrons possess another kind of angular momentum, called *spin angular momentum* or *simply spin*. This vector quantity or motion of electron can be compared with earth's motion around its own axis, as well as, around the Sun. If one considers the electron as a sphere of charge spinning around one of its diameters, one can explain this spin angular momentum, arising out of the motion of the electron around its own axis. The difficulty arises if one wants to picturize this motion or one wants to find a model in the classical mechanics. To be very frank, there is no counter part of this behaviour in classical mechanics and further we cannot expect to explain the behaviour of the microscopic particles based on models taken from our experience in macroscopic world. Similar behaviour has also been observed for other fundamental particles.

It was only 1928, when Dirac developed the relativistic quantum mechanics; the spin quantum number came out naturally. Dirac's treatment of relativistic quantum mechanics also predicted the existence of *positron*, *i.e.,* particle, which has a mass of electron but with a one unit of positive charge. This can also be called an *anti-electron*. The existence of positron was discovered in only 1932.

Since we are studying only non-relativistic quantum mechanics in this book, the idea of induction of spin can only be through an additional hypothesis. As we have seen earlier that for every physically observable quantity, we first write an expression for that observable in classical mechanics. Then with the help of quantum mechanical operator we can write the desired expression in quantum mechanics. For example the kinetic energy of any state can be written in the form of linear momentum and then can be transformed into the desired quantum mechanical language with the help of proper operator. But as said earlier there is no counter part of spin of the electron in classical mechanics, so we cannot use this method to construct the operators for the spin. But behaviour of the spin of the electron about its own axis can be compared with that of orbital angular momentum. Analogous to the orbital angular momentum operators, $\hat{L}$, $\hat{L}_x$, $\hat{L}_y$ and $\hat{L}_z$, the similar operators for the spin has been introduced as $\hat{S}^2$, $\hat{S}_x$, $\hat{S}_y$ and $\hat{S}_z$. $\hat{S}^2$ is for the square of the magnitude of total spin angular momentum of the particle and $\hat{S}_z$ is the operator for the z-component of the particle's spin angular momentum. The properties and results, similar to orbital angular momentum are observed from the spin operators.

For example, the total spin angular momentum is quantized according to:

$$\mathbf{S} = \sqrt{s(s+1)}\,\frac{h}{2\pi} \qquad \text{...(3.232)}$$

In which the spin angular momentum quantum number has only the single unique value of 1/2. The only possible value of **S** is:

$$\mathbf{S} = \sqrt{\frac{1}{2}\left(\frac{1}{2}+1\right)}\,\frac{h}{2\pi} = \frac{1}{2}\sqrt{3}\,\frac{h}{2\pi} \qquad \text{...(3.233)}$$

The number '*s*' is called spin quantum number, but it is not used with other quantum numbers since it has only the value of $\frac{1}{2}$ and does not really distinguish one state from another. The only value of the spin angular momentum can have $\mathbf{S} = \frac{1}{2}\sqrt{3}\,\frac{h}{2\pi}$ and for this reason it is as much a fundamental characteristic of the electron as the mass or the charge. Thus the magnetic field produced by the nucleus ($\mathbf{H}_p$) as it circles the electron interacts with the magnetic dipole moment μ_s from the spinning electron (Fig. 3.13*a*).

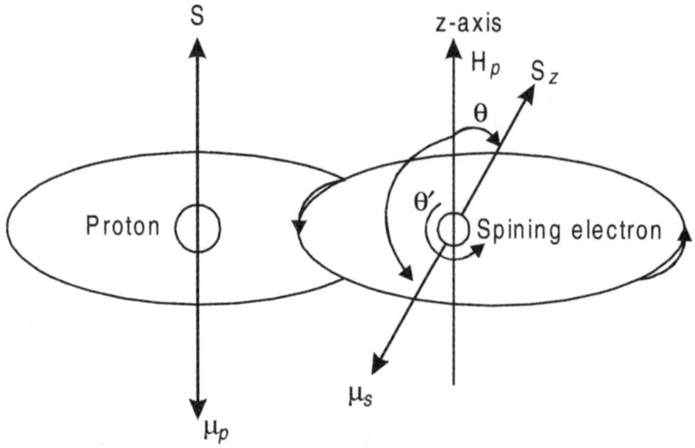

Figure 3.13a: *Motion of electron in the orbital around the proton and around its own axis. μ_p is the magnetic moment produced by the orbital motion of the electron around nucleus*

This results in the additional energy term.

$$\Delta E_s = -\mu_s \mathbf{H}_p = -\mu_s \mathbf{H}_p \cos\theta' = \mu_s \mathbf{H}_p \cos\theta \qquad ...(2.234)$$

which must be added to the energy from the electrostatic interaction of the electron and proton. This interaction is called *spin-orbit interaction* and is the source of additional energy levels, which produces the fine structure of the spectrum. Thus in the presence of applied magnetic field, the total energy of the atom is:

$$E = -\frac{2\pi^2\mu Z^2 e^4}{h^2}\frac{1}{n^2} + \Delta E_s \qquad ...(3.235a)$$

$$= -\frac{2\pi^2\mu Z^2 e^4}{h^2}\frac{1}{n^2} + \mu_s \mathbf{H}_p \cos\theta \qquad ...(3.235b)$$

Again similar to the Zeeman effect, if the direction along the magnetic field ($\mathbf{H}_p$) is chosen as the z-direction (Fig. 3.13*b*), the component of s in the z-direction is quantized and is given by:

$$s_z = s\cos\theta = m_s \frac{h}{2\pi} \qquad ...(3.236)$$

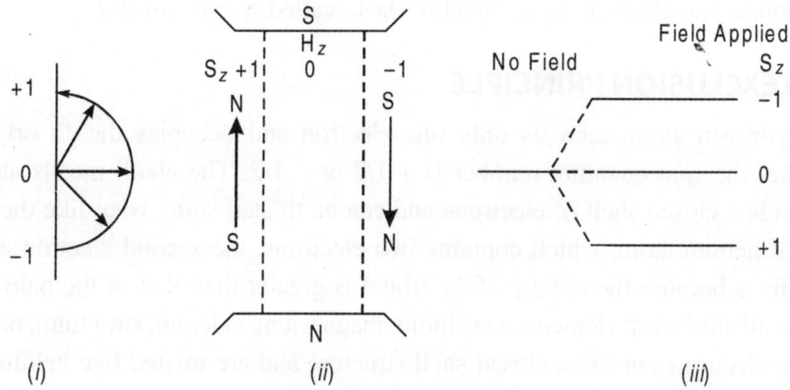

Figure 3.13b: *(i) Depicts the three orientations of the spin of electrons with spin quantum number s = 1, (ii) The direction of the resulting magnetic dipole moment μ_z in the presence of external magnetic field H_z, (iii) the three energy levels allowed to the electrons*

where unlike the value of l or m_l, $m_s = \pm\frac{1}{2}$ and is called the *magnetic spin quantum number*. $m_s = \frac{1}{2}$ is designated as spin up (↑) or α and $m_s = -\frac{1}{2}$ as spin down (↓) or β.

Since s is always $\frac{1}{2}$, the state of the atom can be completely represented by the quantum numbers n, l, m_l, m_s. For every l, there are $2l + 1$values of m_l and hence for every value of s, there are $2s + 1$ values of m_s. The list of complete set of quantum numbers have been given in Table 3.8.

Further, the wave functions discussed for the hydrogen atom or hydrogen like atoms so far are incomplete, because these wave functions describe only the motion of the electron

Table 3.8: *List of complete set of quantum numbers*

Quantum numbers	Values	Number of possible values
Principal, n	1, 2, 3,.....	Any number
Orbital, l	$0, 1, 2, \ldots\ldots\ldots\ldots (n-1)$	n
Magnetic m_l	$0, \pm1, \pm2, \pm3, \ldots\ldots\ldots \pm l$	$2l + 1$
Spin, s	$\frac{1}{2}$	1
Magnetic spin, m_s	$\pm\frac{1}{2}$	$2s + 1$

around the nucleus (*i.e.,* space part) and do not include the motion of the electron around its own axis, *i.e.,* spin part. Thus the total wave function of the electron will be the product of space part and spin part of the wave functions, *i.e.,*

$$\psi = \psi_{\text{space}} \, \psi_{\text{spin}} \qquad \ldots(3.237)$$

The complete one-electron wave function ψ is called a *spin orbital.*

3.8 PAULI EXCLUSION PRINCIPLE

In case of hydrogen atom, there is only one electron and occupies the 1s orbital. It does not matter, whether the spin quantum number is + 1/2 or – 1/2. The alkali metals also have a single electron out side a closed shell of electrons and can be treated some what like the hydrogen atom. But in case of helium atom, which contains two electrons, the second electron also occupies the 1s orbital. This is because the energy of 2s orbital is greater than that of the pairing energy of the 1s orbital. The alkaline earth elements, beryllium, magnesium, calcium, strontium, barium and radium have also two electrons outside a closed shell structure and are treated like helium atom. Now the question arises whether the second electron is having the same spin as that of the first electron or different from that of the first one, as the other three quantum numbers of the two electrons are the same (1, 0, 0). Pauli solved this problem by enunciating a principle known as *Pauli Exclusion Principle.* This states that "*no two electrons can in a given atom may have the same set of quantum numbers n, l, m_l, m_s*. So it clarifies that helium or in general, an orbital can have maximum of two electrons provided their spins are anti-parallel. In the classical theory, the most stable state of electrons is when they are at their lowest energy state, the ground state. Thus, from hydrogen atom on up to heaviest elements, all of the electrons should be in the first orbit, with the orbit being smaller for the heavier elements. As a consequence, each individual atom should differ in size, *i.e.,* lead < carbon < hydrogen. In other words we can say that if the Pauli Exclusion Principle does not hold good, the atoms should vary greatly in size, but in reality this is not so. All of the atoms are approximately are of the same size.

Later on Pauli's Exclusion Principle was given more general definition, because of the identical and indistinguishable nature of the electrons and other fundamental particles. In classical mechanics, the motion of each particle, even if these are indistinguishable can be described with the help of Newton's law of motion. But this is not possible in quantum mechanics because of the natural law of uncertainty. That is in case of microscopic particles, each particle can be distinguished from each other provided they have different masses or charges or spins etc.,

Thus the wave function ψ (spatial) is insufficient for a complete characterization of the appropriate quantum number of the atom, since this requires that the electron spin also be taken into account. Introducing of spin coordinates of the electrons in addition to the coordinates x, y, z and assuming that the variables in the total wave function which includes the spin coordinates as well are separable, we can represent the total wave function, known as spin orbital:

$$\psi_{\text{total}} = \psi_{\text{spatial}}\, \psi_s \qquad \text{...(3.237}a\text{)}$$

where ψ_s is the function of spin coordinates (spin function). Like the function ψ, the spin function ψ_s also possesses the properties of symmetry. If we represents the magnetic spin quantum number ($m_s = +1/2$) by α and $m_s = -1/2$, by β, we can represent the general solution of the spin part of the wave equation containing two electrons (say 1 and 2) in the form of following four functions.

$$\psi_s = \alpha\,(1)\,\alpha\,(2)$$

$$\psi_s = \beta\,(1)\,\beta\,(2)$$

$$\psi_s = \alpha\,(1)\,\beta\,(2) + \alpha\,(2)\,\beta\,(1)$$

$$\psi_{(a)} = \alpha\,(1)\,\beta\,(2) - \alpha\,(2)\,\beta\,(1)$$

The first three functions are symmetric say ψ_s (s) with respect to transposition of electrons the latter is anti-symmetric ψ_s (a). Thus, for the first three functions total value of $m_s = ms\,(1) + m_s\,(2) = 0, \pm 1$, and that for the last one $m_s = 0$. Hence it may be concluded that the first three functions refer to triply degenerate state of an atom characterized by the total spin S = 1, the latter to a singlet state with spin S = 0. Thus according to the symmetry of the functions ψ and ψ_s, the total wave function ψ_{total} (ψ_t) may be symmetric or anti-symmetric. The functions $\psi_t\,(s) = \psi\,(s)\,\psi_s\,(s)$ and $\psi_t\,(s) = \psi\,(a)\,\psi_s\,(a)$ are symmetric, and functions: $\psi_t\,(a) = \psi\,(s)\,\psi_s\,(a)$ and $\psi_t\,(a) = \psi\,(a)\,\psi_s\,(s)$ are anti-symmetric. Assuming two electrons with quantum numbers identical, transposition of these electrons will yield a state identical with the initial state, and consequently $\psi_{1,\,2} = \psi_{2,\,1}$. For the anti-symmetric function we have: $\psi_{1,\,2} = -\,\psi_{2,\,1}$. Hence, the condition for satisfying the Pauli Exclusion Principle, is the condition of anti-symmetry of the total wave function. Such is the quantum mechanical formulation of the Pauli Exclusion Principle. Thus according to *quantum mechanical picture of Pauli Principle when the coordinates of any two identical particles are changed, the sign of total wave function must also a change*. However, all particles do not obey the Pauli Exclusion Principle. It has been found that particles with spin $s = \frac{1}{2}$ obey the Pauli Exclusion Principle and follow Fermi statistics. These are known as *fermions* (have half integral spins) and include neutrinos and atoms containing odd number of particles.

Other particles those do not follow the Pauli Exclusion Principle obey laws of *Bose-Einstein statistics*. These are called *Bosons* (having integral spins) and include photons, α-particles and atoms composed of an even number of protons and neutrons,

Later on Pauli, also showed that relativistic quantum mechanics indicates that particles with half-integral spins require anti-symmetric wave function, whereas the particles with integral spins require symmetric wave functions. Experimental results have also proven these facts.

The above stated principle can be explained with an example. Consider the helium atom. It has two electrons and both are present in the 1*s* orbital in the ground state. This means that the space part of the wave function for each electron in the ground state is same (1*s*). Thus, the spin part of each electron in the total wave function has to be different. The space part of the wave function is 1*s* (1) 1*s* (2), as mentioned by equation (3.172). The spin part can be written as:

$$\alpha\,(1)\,\alpha\,(2) \quad ...(3.238)$$

$$\beta\,(1)\,\beta\,(2) \quad ...(3.239)$$

$$\alpha\,(1)\,\beta\,(2) \quad ...(3.240)$$

$$\alpha\,(2)\,\beta\,(1) \quad ...(3.241)$$

The first two arrangements lead to total spin as 1 and –1 respectively. The last two arrangements results into zero spin. The last two arrangements cannot be used because these spin functions state that the electrons can be distinguished, which in reality we cannot, as stated earlier. This can be achieved by writing the spin wave functions as:

$$\psi_+ = \frac{1}{\sqrt{2}}\,[\alpha\,(1)\,\beta\,(2) + \alpha\,(2)\,\beta\,(1)] \quad ...(3.242)$$

$$\psi_- = \frac{1}{\sqrt{2}}\,[\alpha\,(1)\,\beta\,(2) - \alpha\,(2)\,\beta\,(1)] \quad ...(3.243)$$

Out of these four combinations of the spins, only one leads to the anti-symmetric wave function, *i.e.*,

$$\psi_t\,(a) = 1s\,(1)\,1s\,(2)\,[\alpha\,(1)\,\beta\,(2) - \alpha\,(2)\,\beta\,(1)] \quad ...(3.244)$$

The interchange of electrons in the above expression gives:

$$\psi = 1s\,(2)\,1s\,(1)\,[\alpha\,(2)\,\beta\,(1) - \alpha\,(1)\,\beta\,(2)] = -\,\psi$$

Thus ψ is anti-symmetric with respect to the relabeling of the pair of electrons. The one line expression can also be written in the form of determinant as

$$\psi = \frac{1}{\sqrt{2}}\begin{vmatrix} 1s\,(1)\,\alpha\,(1) & 1s\,(1)\,1\beta\,(1) \\ 1s\,(2)\,\alpha\,(2) & 1s\,(2)\,1\beta\,(2) \end{vmatrix} \quad ...(3.245)$$

The wave function written in this form is known as a *Slater determinant*. The property of anti-symmetry of the wave function is very nicely displayed in this form also, *i.e.,* if the labels of the electrons 1 and 2 are interchanged, this will effectively change the rows or columns of the determinant. Now one of the properties of determinant is that if a successive rows or columns of the determinant are interchanged, there is a change in sign.

The earlier definition of Pauli Exclusion Principle can also be explained using Slater's determinant. In case of helium atom, since both the electrons are in 1*s* orbital ($n = 1$, $l = 0$, $m_l = 0$) and if both the electrons have the same spin [α (1) = β (2)], the above determinant will become:

$$\psi = \frac{1}{\sqrt{2}} \begin{vmatrix} 1s\,(1)\,\alpha\,(1) & 1s\,(1)\,1\alpha\,(1) \\ 1s\,(2)\,\alpha\,(2) & 1s\,(2)\,1\alpha\,(2) \end{vmatrix} \qquad ...(3.246)$$

This will be identically zero, because all determinant are zero if any two rows or two columns are same. In other words the wave function for such a configuration is zero, which is not true. This means that Slator's determinant is a better way of writing the wave function of a system and for *n* electrons system it takes the following form:

$$\psi\,(1, 2, ... \; .2N) = \frac{1}{\sqrt{(2N)!}} \begin{vmatrix} \varphi_1\alpha(1) & \varphi_1\beta(1) & ..... & \varphi_N\alpha(1) & \varphi_N\beta(1) \\ \varphi_1\alpha(2) & \varphi_1\beta(2) & ..... & \varphi_N\alpha(2) & \varphi_N\beta(2) \\ ... & ... & ..... & ... & ... \\ ... & ... & ..... & ... & ... \\ \varphi_1\alpha(2N) & \varphi_1\beta(2N) & ..... & \varphi_N\alpha(2N) & \varphi_N\beta(2N) \end{vmatrix} \qquad ...(3.247)$$

where $\frac{1}{\sqrt{(2N)!}}$ is the normalization constant. The electrons in a given column have the same spin whereas the elements in a given row involve the same electron.

Based on Pauli Exclusion Principle and exchange degeneracy let us examine the energy levels of helium atom. According to Pauli Exclusion Principle for a two electron atom (*e.g.*, helium atom), the two wave functions are possible.

$$\psi_t\,(a) = \psi\,(s)\,\psi_s\,(a)$$

$$\psi_t\,(a) = \psi\,(a)\,\psi_s\,(s)$$

If the two electrons are equivalent they will have the same value of n, l, m_l. In addition Pauli Exclusion Principle excludes the state characterized by the function $\psi_t\,(a) = \psi\,(a)\,\psi_s\,(s)$, since in this case $m_s\,(1) = m_s\,(2)$, which contradicts this principle. Thus, for the ground state of helium atom $n_1 = n_2$, $l_1 = l_2$ and $m_l\,(1) = m_l\,(2)$. Since the orbital angular momenta of both the electrons and the total spin angular momentum of both the electrons are zero, hence the statistical weight is unity.

On the other hand in case of two nonequivalent electrons, $n_1 = 1$, $n_2 = 2$, both anti-symmetric states are possible:

$$\psi_t\,(a) = \psi\,(s)\,\psi_s\,(a)$$

$$\psi_t\,(a) = \psi\,(a)\,\psi_s\,(s)$$

The first one is a singlet (S = 0)) and has statistical weight 1. The second is a triplet (S = 1) and has statistical weight 3. The perturbation energy corresponding to a singlet state is $\langle E \rangle = C + A$ and for triplet state it is $\langle E \rangle = C - A$. Thus the triplet state will be

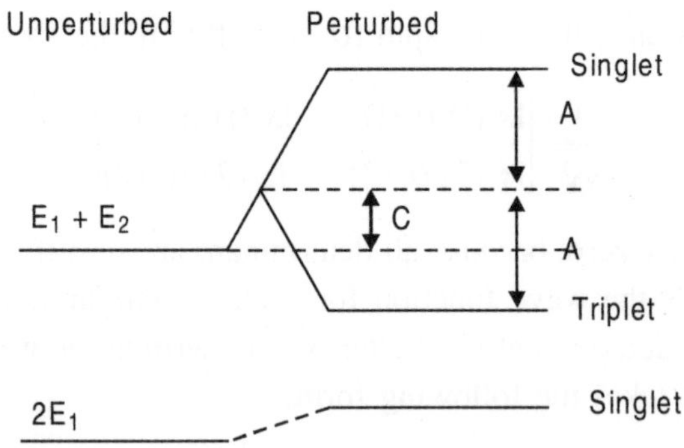

Figure 3.14: *Energy level diagram of few lower states of helium atom*

situated below the singlet state. Figure 3.14 indicates that the ground level of helium atom must be a singlet and first excited level a triplet. This conclusion not only conforms to the experimental data on the atom but also in the case of many electron atoms or ions with two outer electrons for instance, Mg, Zn. Cd, Hg, or Al^+, Tl^+ etc., All the fore going atoms or ions are diamagnetic in the ground state, thus indicating that these states are singlet. Accordingly the magnetic moment of the helium atom in the ground state is zero, unlike the magnetic moment in the triplet state, of which measurements yield $2\mu_B$. It may be added that the energy levels of the helium atom, calculated to a first approximation, coincides with the observed energy values to within 20 to 30%. Further note that one of the great achievements of two principles of quantum theory of atom (quantization of electron states and Palui Exclusion Principle) together with a third rule, according to which the ground state of every atom is the state of least energy, provides physical proof of the periodic system of elements, *i.e.,* complete explanation of the properties of elements by means of the periodic table) and is being discussed in section 3.12.

3.9 MULTI-ELECTRON ATOMS

We have seen that the method of perturbation theory, like the variation method of Hylleraas, although solved the problem of an atom with a small number of electrons, such as the helium atom and also certain other problems, but encountered great difficulties when applied to many electron systems where it is very hard to account for the perturbation energy. For this reason, other methods of calculations came into being for solving such systems. The statistical methods have proved partially fruitful. One such statistical method, which is based on Schrödinger wave equation, is due to Hartree. In this method, the electric field of all the electrons is replaced by a certain average time constant field, which Hartree called a *self-consistent field.* Later on this method was modified by, Fock and Slator. Now it is commonly known as *Hartree-Fock self-consistent field* (HFSCF) method.

Before we start discussing Hartree-Fock SCF method, it will be advisable to learn about the Hartree's natural system of units. In problems connected with atoms, the natural system of units (atomic units) introduced by Hartree is used. In this system the electron charge, the electron mass and Planck's constant divided by 2π are put equal to unity. The atomic unit of length will be equal to 0.529 Å (0.0529 nm), the atomic unit of speed is the 1/137th part of the speed of light, and the atomic unit of energy will be twice the ground state energy of hydrogen atom, *i.e.,* 27.21 eV (1 hartree). The units are deduced as follows: The standard units for charge, mass and Planck's constant in Hartree system are:

$$e = 4.8 \times 10^{-10} \text{ esu, } 1 \text{ esu} = g^{\frac{1}{2}} \text{ cm}^{\frac{3}{2}} \text{ s}^{-1}$$

$$\text{mass} = 9.11 \times 10^{-28} \text{ g}$$

$$\frac{h}{2\pi} = \frac{1}{2\pi} \times 6.62x10^{-27} \text{ ergs}, 1 \text{ erg} = \text{g cm}^2 \text{ s}^{-2} \text{ and}$$

$$1 \text{ erg} = 6.2421 \times 10^{11} \text{ eV}$$

Thus in natural system of units, the unit of length is:

$$a_0 = \left(\frac{h}{2\pi}\right)^2 \frac{1}{m_e e^2}$$

$$= \frac{(6.626 \times 10^{-27} \ g \text{ cm}^2 \text{ s}^{-2} \text{ s})^2}{(4)\ (3.14)^2 (9.11 \times 10^{-28} \text{g})\ (4.8 \times 10^{-10} \text{ g}^{\frac{1}{2}} \text{ cm}^{\frac{3}{2}} \text{ s}^{-1})^2}$$

$$= 0.529 \times 10^{-8} \text{ cm} = 0.529 \text{ Å}$$

and unit of energy is:

$$E_0 = \left(\frac{2\pi}{h}\right)^2 m_e e^2 = \frac{e^2}{a_0} = \frac{(4.8 \times 10^{-10} \text{ g}^{\frac{1}{2}} \text{ cm}^{\frac{3}{2}} \text{ s}^{-1})^2}{0.529 \times 10^{-8} \text{ cm}}$$

$$= 4.335 \times 10^{-11} \text{g cm}^2 \text{ s}^{-2}$$

$$= 4.335 \times 10^{-11} \text{ erg} = 4.335 \times 10^{-11} \text{ erg} \times 6.2321 \times 10^{11} \text{ eV erg}^{-1}$$

$$= 27.0595 \text{ eV, called a hartree.}$$

The unit of velocity will be $\left(\frac{2\pi}{h}\right) e^2 = \frac{(2)\ (3.14)\ (4.810^{-10} \text{ g}^{\frac{1}{2}} \text{ cm}^{\frac{3}{2}} \text{ s}^{-1})^2}{6.626 \times 10^{-27} \text{ g cm}^2 \text{ s}^{-2} \text{ s}}$

$$= 2.183 \times 10^8 \text{ cm s}^{-1} = \frac{1}{137^{\text{th}}} \text{ the part of speed of light.}$$

To start with let us write the complete Hamiltonian of n electron atom.

$$\hat{H} = -\frac{h^2}{8\pi^2 m_e}\sum_{i=1}^{n}\nabla_i^2 - \sum_{i=1}^{n}\frac{Ze^2}{r_i} + \sum_{j>i}^{n}\sum_{i=1}^{n}\frac{e^2}{r_{ij}} \quad ...(3.248)$$

The above equation in Hartree form can be written as;

$$\hat{H} = -\frac{1}{2}\sum_{i=1}^{n}\nabla_i^2 - \sum_{i=1}^{n}\frac{Z}{r_i} + \sum_{j>i}^{n}\sum_{i=1}^{n}\frac{1}{r_{ij}} \quad ...(3.248a)$$

The first term in equation (3.248) contains the kinetic energy operators for n electrons, the second term is the attractive potential energy terms between the nucleus of Z units of positive charge (Z = n for neutral atom) and each of the electron and third term contains the repulsive energy terms between the various electrons. The restriction $j > i$ in the summation avoids counting the same inter-electronic repulsive terms twice and the terms like $\frac{e^2}{r_{ii}}$. The Schrödinger wave equation ($\hat{H}\ \psi = E\ \psi$) for n electron system cannot be solved accurately due to the presence of terms like $\frac{e^2}{r_{ii}}$. The crude approximation we have learnt so far is to neglect the inter-electronic repulsive terms completely. Then the Hamiltonian can be expressed as the sum of one-electrons only. This will lead to n Schrödinger wave equations, similar to hydrogen like equations with Z units of positive charge on the nucleus. The total wave functions (zero order wave function) so obtained would be a product of n hydrogen like (one electron) orbitals (wave functions).

$$\psi^0 = \psi_1 (r_1, \theta_1, \varphi_1)\ \psi_2 (r_2, \theta_2, \varphi_2)........\ \psi_n (r_n, \theta_n, \varphi_n) \quad ...(3.249)$$

where the hydrogen like orbitals are:

$$\psi (r, \theta, \varphi) = R_{n,\,l,} (r)\ \Theta_{l,\,m_l} (\theta)\ \Phi_{m_l} (\varphi) \quad ...(3.250)$$

The other thing we have learnt is the Pauli's Exclusion Principle, *i.e.,* two electrons placed in one orbital with spin opposite to each other and thus the ground state electronic wave function can be written for any atom. We also know that the kind of wave function expressed by the equation (3.250) explains the results qualitatively and cannot be used for quantitative purposes. This is because this model considers that: (*i*) each electron in different orbitals experiences the full Z units of positive charge of the nucleus and (*ii*) the inter-repulsion potential energy terms are neglected. This treatment was modified, by including the effective nuclear charge in the one-electron wave function, *i.e.,* thus, minimizing the variational integral to calculate the energy of the ground state. Before we discuss the actual method, let us assume one other approximation, *i.e.,* the hydrogen like wave functions can be written as the product of radial part and angular part (*i.e.,* spherical harmonics).

Harteree method involves the followings steps. It basically tries to calculate the effective potential energy in which one of the electrons under consideration moves in the electrostatic field of the other electrons. Let us consider the electron 1, which is moving in the electrostatic field of (n – 1) electrons, which is spread over the complete space and varies continuously. This involves the calculation of inter-electronic repulsion terms like $V_{1,2}$, $V_{1,3}$ ....$V_{1,n}$. According to Coulomb's law, the inter-electronic repulsion energy between the two electrons is given by:

$$V_{1,2} = \frac{q_1 q_2}{r_{12}} \qquad ...(3.251)$$

where $q_1 = e$ and q_2 is the average value of the charge cloud because of electron 2. Since the charge density function is given by $f_2^2(r_2, \theta_2, \varphi_2)$, the charge due to electron 2, present over the small volume $d\hat{\tau}_2$ will be $e\ f_2^2(r_2, \theta_2, \varphi_2)\ d\tau_2$. Thus equation (3.251) becomes:

$$V_{1,2} = e^2 \int \frac{f_2^2(r_2, \theta_2, \varphi_2)}{r_{12}}\, d\tau_2 \qquad ...(3.252)$$

The inter-electronic repulsion potential energy terms between electron 1 and other electrons can be calculated similarly. Thus

$$V_{1,2} + V_{1,3} + V_{1,4} + \ldots\ldots\ldots\ldots + V_{1,n} = e^2 \sum_{j=2}^{n} \int \frac{f_j^2(r_j, \theta_j, \varphi_j)}{r_{ij}}\, d\tau_j \qquad ...(3.253)$$

Thus the total potential energy of electron 1 will be:

$$V(r_1, \theta_1, \varphi_1) = e^2 \sum_{j=2}^{n} \int \frac{f_j^2(r_j, \theta_j, \varphi_j)}{r_{ij}}\, d\tau_j - \frac{Ze^2}{r_1} \qquad ...(3.254)$$

Now let us make one more approximation, *i.e.*, the potential energy is spherically symmetric (the effective potential acting on any electron in an atom is a function of r only). This can be calculated by averaging $V(r_1, \theta_1, \varphi_1)$ over the angles to arrive at a potential energy, which depends only on r_1, *i.e.*,

$$V(r_1) = \frac{\int_0^{\pi}\int_0^{2\pi} V(r_1, \theta_1, \varphi_1)\sin\theta\, d\theta\, d\varphi}{\int_0^{\pi}\int_0^{2\pi} \sin\theta\, d\theta\, d\varphi} \qquad ...(3.254a)$$

Thus the Schrödinger wave equation for electron 1 will be:

$$\left[-\frac{h^2}{8\pi^2 m_e}\nabla^2(1) + V(r_1)\right] g_1(1) = \varepsilon_1\, g_1(1) \qquad ...(3.255)$$

where $g_1(r_1)$ is the improved wave function for electron 1 and ε_1 is the energy of the corresponding electron at this stage. Since this equation is similar to hydrogen like atoms, the results obtained will be similar to the characteristics of hydrogen like atoms, *i.e.*, the angular part of the function $g_1(1)$ is given by spherical harmonics, given by the quantum number l_1 and m_{l1}, whereas the radial part in $g_1(1)$ is given by the quantum number n_1.

Now consider the electron 2, moving in the charge cloud of rest of the electrons, calculated with the help of modified function of electron 1 and the old functions of other electrons. With the help of $V(r_2)$, obtained in the similar manner as for electron 1, we can write the Schrödinger wave equation for electron 2. This will give the improved function for electron 2 as g_2 (2). We continue this process until we have a set of improved functions for all the n electrons. We then go back to electron 1 and repeat the process as described above with improved set of wave functions (*i.e.*, $g_1, g_2 \ldots\ldots\ldots g_n$). We continue to calculate the improved orbitals until we get no further change from the one iteration to the next one. The final set of orbitals gives the Hartree self-consistent field wave functions.

Now let us calculate the total energy of n electrons in an atom. Before we add $\varepsilon_1 + \varepsilon_2 + \varepsilon_3 + \ldots\ldots + \varepsilon_n$ to obtain the total energy, we should be very careful and let us examine how we have calculated ε_i. If we analyze ε_1, it contains inter-electronic repulsion terms $V_{1,2}$, $V_{1,3}$, $V_{1,4} \ldots\ldots V_{1,n}$. Similarly ε_2 also includes $V_{1,2}$, $V_{2,3} \ldots\ldots\ldots V_{2,n}$ and so on. That is the inter-electronic repulsion terms have been included twice. So

$$E = \sum_{i=1}^{n} \varepsilon_i - \sum_{j>i} \sum_{i=1}^{n} J_{ij} \qquad \ldots(3.256)$$

where $J_{i,j}$ is the repulsive energy term between the electrons i and j. Further modifications were carried out by Fock and Slater and these are:

(*i*) Though care has been taken to arrange the electrons according to the Pauli's Exclusion Principle, but the spin of the electrons have not been defined explicitly. To improve the variation functions, one can use the wave functions described by Slater's determinant. This is called Hartree-Fock method.

(*ii*) It is not necessary to use the hydrogen like wave functions for the trial variation function. Roothan proposed the Hartree-Fock orbitals as the linear combinations of a complete set of known functions, called basis functions. The most commonly used set of basis functions is the set of generalized Slater's orbitals, which have the normalized form as:

$$\Psi_{n,l,m_l}(r,\theta,\varphi) = \frac{[2Z_e/a_0]^{\frac{n^*+1}{2}}}{[(2n^*)!]^{\frac{1}{2}}} r^{n^*-1} e^{-\frac{Z_e r}{a_0 n^*}} Y_l^{|m_l|}(\theta, \varphi) \qquad \ldots(3.257)$$

where $Y_l^{|m_l|}(\theta, \varphi)$ is the spherical harmonic, for $n = 1, 2, 3$, write $n^* = n$, but for $n = 4, 5, 6\ldots\ldots$, write $n^* = 3.7, 4.0, 4.2\ldots$ respectively. Z_e is the effective nuclear charge experienced by the particular electron in the desired atomic orbital, *i.e.*, total positive charge on the nucleus minus the shielding effect of the electrons present closer to the nucleus than the particular electron under consideration. Slater has given certain rules to calculate the shielding constant (σ) of the nucleus. The effective nuclear charge is defined as:

$$Z_{eff} = Z - \sigma \qquad \ldots(3.258)$$

where Z is the actual atomic number. These rules are:

(*a*) Divide atomic orbitals into the following groups.

(1*s*); (2*s*, 2*p*); (3*s*, 3*p*); (3*d*); (4*s*, 4*p*); (4*d*); (4*f*); (5*s*, 5*p*); (5*d*)

(*b*) The contribution to the shielding constant σ, from different electrons present in the atom, is obtained from the following set of instructions.

(*i*) From an orbital in a group, orbital out side the one (*i.e.*, coming after) considered is zero, *i.e.*, consider an orbital 3*s*, the contribution to the shielding constant from 3*d*, 4*s*, 4*p* etc., will be zero.

(*ii*) 0.35 from each other electron in the group considered, except that 0.3 is used in the 1*s* group.

(*iii*) For an *s* and *p* orbital, 0.85 from each electron with *n* one less than the orbital considered and 1.0 from each electron still further in.

(*iv*) For a *d* orbital, 1.0 for each electron inside the group.

Furthermore, the energy level of a multi-electron atom or ion can now be written as:

$$E_n = R\frac{Z - \sigma}{n^2} \qquad ...(3.258a)$$

According to this formula, the quantity $\left(\frac{E_n}{R}\right)^{\frac{1}{2}}$ for various atoms and ions with one outer electron must be a linear function of nuclear charge Z.

$$\left(\frac{E_n}{R}\right)^{\frac{1}{2}} = \frac{(Z - \sigma)^{\frac{1}{2}}}{n} \qquad ...(3.258b)$$

Now considering the shielding constant σ for the combining levels (*i.e.*, involved in the given transition) say *n* and *m* to be approximately the same, we obtain from equation (3.258*b*)

$$\left(\frac{\nu_{n,m}}{R}\right)^{\frac{1}{2}} = \frac{\left(\sqrt{m^2 - n^2}\right)(Z - \sigma)^{\frac{1}{2}}}{mn} \qquad ...(3.258c)$$

Thus, the linear function of Z is satisfied also by the square root of the frequency of similar lines (for instance, of lines that are the first members in the series etc.), the relationship (3.258*c*) expresses the so called Moseley law, first established for *x*-rays spectra. Formulae (3.258*b*) and (3.258*c*) have been shown to be valid by numerous experimental findings. The regularity expressed by (3.258*b*) for various energy levels (terms) of the sodium atom and the ions Mg^+, Al^{2+}, Si^{3+} etc. is shown in Fig. 3.15. From the intercept of the lines one can determine the values of shielding constant σ.

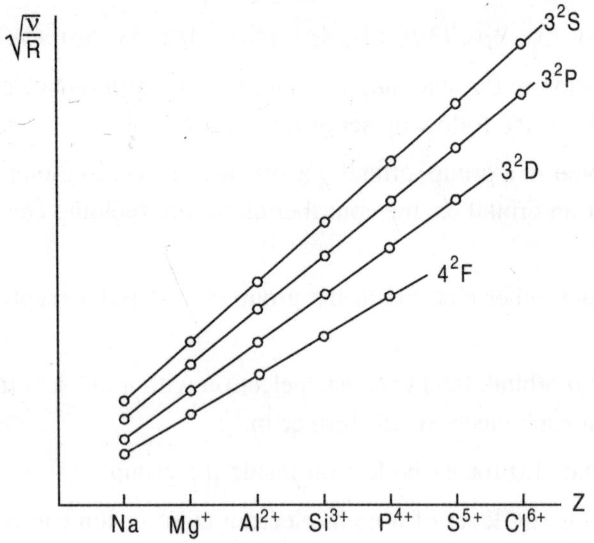

Figure 3.15: $\left(\frac{\nu}{R}\right)^{\frac{1}{2}}$ *versus Z for the lower terms of Na, Mg^+, Al^{2+}*

Example 3.20: Calculate the effective nuclear charge for: (*a*) N, (*b*) Na, (*c*) S, (*d*) Zn.

Solution:

(*a*) For N, Z = 7

$$Z_{eff}(1s) = 7 - 0.3 = 6.7.$$

$$Z_{eff}(2s) = 7 - (2 \text{ electrons from } 1s) - (4 \text{ electrons from } 2s \text{ and } 2p)$$
$$= 7 - 2 \times 0.85 - 4 \times 0.35 = 7 - 1.7 - 1.4 = 3.90$$

$$Z_{eff}(2p) = 7 - 2 \times 0.85 - 4 \times 0.35 = 3.90$$

(*b*) For Na, Z = 11

$$Z_{eff}(3s) = 11 - (2 \text{ electrons from } 1s) - (8 \text{ electrons from } 2s \text{ and } 2p)$$
$$= 11 - 2 \times 1.0 - 8 \times 0.85 = 11 - 2.0 - 6.8 = 2.2$$

(*c*) For S, Z = 16

$$Z_{eff}(3s) = 16 - (2 \text{ electrons from } 1s) - (8 \text{ electrons from } 2s \text{ and } 2p)$$
$$- (5 \text{ electrons from } 2s \text{ and } 2p)$$
$$= 16 - 2 \times 1 - 8 \times 0.85 - 5 \times 0.35 = 16 - 2 - 6.8 - 1.75 = 5.45$$

(*d*) For Zn, Z = 30

$$Z_{eff}(3s) = 30 - (2 \text{ electrons from } 1s) - (8 \text{ electrons from } 2s \text{ and } 2p)$$
$$- (8 \text{ electrons from } 3s \text{ and } 3p) - (10 \text{ electrons from } 3d)$$
$$- (\text{one electron from } 4s)$$

$$= 30 - 2 \times 1 - 8 \times 1 - 8 \times 0.85 - 10 \times 0.85 - 1 \times 0.35$$

$$= 30 - 2 - 6.8 - 8.5 - 0.35 = 4.35$$

$$Z_{eff}(3d) = 30 - (2 \text{ electrons from } 1s) - (8 \text{ electrons from } 2s \text{ and } 2p)$$

$$- (8 \text{ electrons from } 3s \text{ and } 3p) - (9 \text{ electrons from } 3d)$$

$$= 30 - 2 \times 1 - 8 \times 1 - 8 \times 1.0 - 9 \times 0.35 = 30 - 2 - 8 - 8 - 3.15 = 8.85.$$

3.10 SCREENING EFFECT OF THE INNER SHELL ELECTRON ORBITAL

Lithium atom has three electrons. According to Pauli Exclusion Principle, only two electrons can be placed into 1*s* orbital with opposite spin. Third electron can enter into either 2*s* orbital or one of the 2*p* orbitals. In case of hydrogen atom, all the four orbitals with $n = 2$ (2*s*, $2p_x$, $2p_y$ and $2p_z$) have the same energy (degenerate), but this is not the case with multi-electron atoms. This is due to the partially shielding of the nucleus by different inner orbitals.As seen in the previous

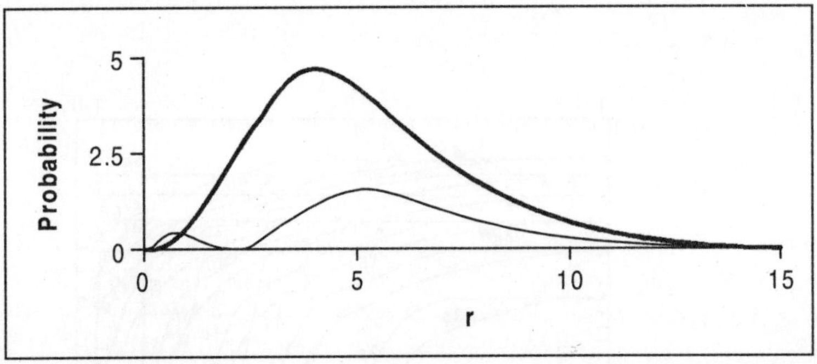

Figure 3.16: *Probability versus r (in terms of a_0) plot of 2s and 2p on the same plot. Thin line is 2s and thick line 2p*

section, the extent of shielding for 2*s* and 2*p* electron is different. This is because the radial distribution function for the two kinds of orbitals is not the same. It is clear from Fig. 3.16 that 2*s* orbital has penetrated more into the charge cloud of 1*s* orbital than 2*p* orbital. This is because, besides the major maximum in the radial distribution, 2*s* orbital has a minor maximum inside the charge cloud of 1*s* orbital. The radial distribution function for 2*p* orbital is small in this region. Due to this reason, 2*s* electron will be more close to the nucleus and will experience more effective positive charge of the nucleus than 2*p* electrons. The net effect of penetration of the 2*s* orbital close to the nucleus and shielding of the nucleus is removal of the degeneracy of 2*s* and 2*p* orbitals; 2*s* being lower in energy than those of 2*p* orbitals. Similar principle can be applied in calculating the energies of the orbitals with higher values of *n*, *i.e.*, order for energies of 3*s*, 3*p* and 3*d* orbitals are $3s < 3p < 3d$ and similar trend is followed for the orbitals with $n > 3$. The splitting of these levels, which are degenerate in hydrogen atom, arises from the inter-electron repulsions. In the limit $Z \rightarrow \infty$, orbitals with the same value of *n* are again degenerate, because the inter-electronic repulsions become insignificant in comparison to electron-nucleus attraction. This is clear from Fig. 3.17, where the energies of the orbitals are plotted as a function of the nuclear charge.

The position or relative energies of certain orbitals change with changing Z. For example, the energy of 4*s* orbital is higher than that of 3*d* orbital for hydrogen atom, but for Z in the region 7 to 21, the energy of 4*s* orbital is lower than that of 3*d*. For large value of Z, the energy of 3*d* orbital is again lower than that of 4*s*. This is because the 4*s* orbital will be stabilized more than 3*d* orbital. But once the electron starts occupying the 3*d* orbital, the energy of 3*d* orbital decreases very sharply when Z > 21. The electrons in 3*d* orbitals do not shield each other very well, so there is sudden drop in 3*d* orbital energy. Fig. 3.17 shows that the separation between *ns* and *np* orbitals is not very much large than that between *np* and *nd* orbitals. This gives rise to the familiar $ns^2\ np^6$ stable octet.

The situation becomes more complex with increase of *n* and the number of levels grows with the growth of *n*. The complexity for higher values of *n* is because of the smaller gap between the energy levels as compared to that for the lower values of *n*; consequently various levels are not well defined. Thus, the true or exact electron configuration for the heavier atoms can be obtained from atomic spectrum or careful calculations or both. The atomic energy levels have been arranged as:

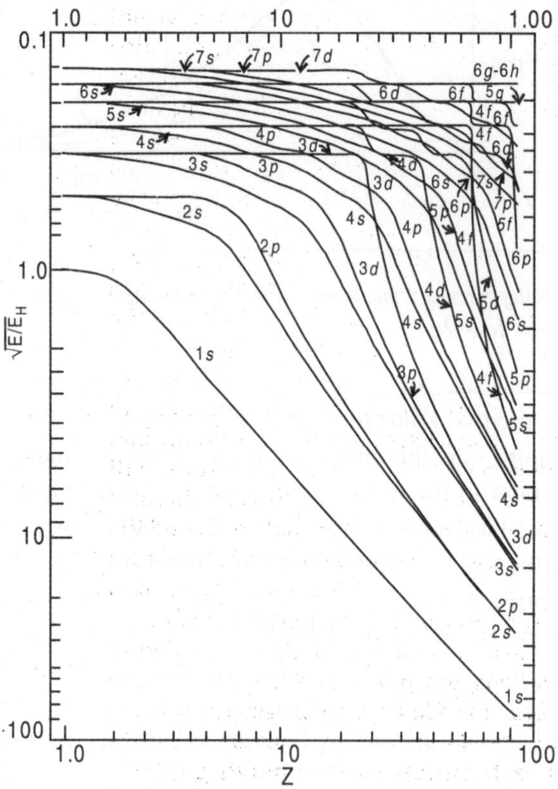

Figure 3.17: *Plot of Atomic Orbital Energy versus Atomic Number*

1*s* < 2*s* < 2*p* < 3*s* < 3*p* < 4*s* < 3*d* < 4*p* < 5*s* < 4*d* < 5*p* < 6*s* < 4*f* < 5*d* < 6*p* < 7*s*

Students do counter difficulty in remembering the above order of energy levels. The most common method, commonly called the *uncle Wiggling path*, is shown in Fig.3.18. We should follow the path as indicated by arrows in Fig. 3.18. Recently a better method has also been suggested and procedure is as follows. (*i*). Draw a grid of 7 × 7 squares. (*ii*) Label the diagonal with known *s* orbital as shown in Fig. 3.19. (*iii*). Fill in each row with all possible orbitals allowed for that energy level (or for that *n*). (*iv*). Read the orbital sequence by following the columns down beginning at the left. If all the orbitals in each column are written before advancing to the next column, the proper filling order is determined.

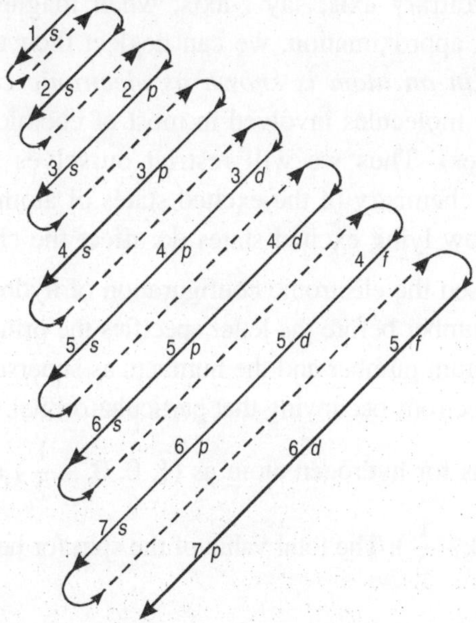

Figure 3.18: *Uncle Wiggly path for filling order of orbitals*

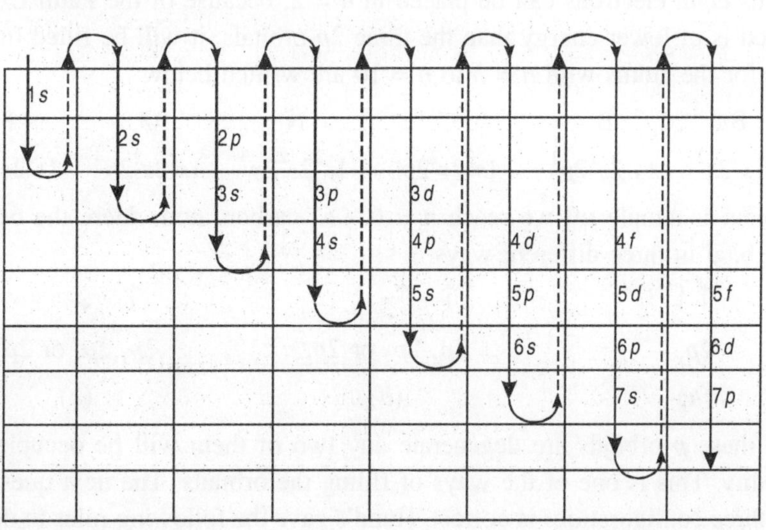

Figure 3.19: *The orbital sequence by following the columns down beginning at the left*

3.11 AUF-BAU PRINCIPLE

After discussing the energies of the atomic orbitals and the Pauli Exclusion Principle, we are in a position to determine the structure of atoms and thus to discuss the characteristics of the periodic table. It is generally observed that for most purposes it is enough to specify the principal quantum number (n) and the azimuthal quantum number (l), *i.e.*, the energy of atomic orbital. The chemical reactivity is mostly determined by these quantum numbers and the values of m_l and m_s change the energies of quantum states slightly, preferably in the presence of magnetic or electric field (for example, m_l only specifies the difference in the orientation of electron's angular momentum with respect to some chosen arbitrary axis, say z-axis, when magnetic field is applied along this direction). Thus to the first approximation, we can neglect their effect. *A given assignment of n and l to all the electrons in an atom is known as electronic configuration.* Further we have also observed that atoms or molecules involved in most of chemical reactions, are in their lowest energy states (ground states). Thus we will restrict ourselves to the ground state electronic configuration, although the chemistry of the excited states of atom are receiving a lot of attention and sometimes especially low lying excited states do affect the chemical properties.

We have already discussed the electronic configuration of hydrogen and helium atoms (*i.e.*, $1s^1$ and $1s^2$ respectively). The number before the letter specifies the principal quantum number, the letter specifies the azimuthal quantum number and the numbers as superscript at the right hand side of the latter gives the number of electrons occupying that particular orbital. Thus in short hand notation, one can write the wave functions for hydrogen atom as $(1, 0, 0, +\frac{1}{2})$ or $(1, 0, 0, -\frac{1}{2})$ and for helium atom as $(1, 0, 0, +\frac{1}{2}; 1, 0, 0, -\frac{1}{2})$. The total value of the spin for helium is $\frac{1}{2} - \frac{1}{2} = 0$ and the two electrons are paired.

For $n = 2$, the possible values of l can be 0 and 1. For $l = 0$, $m_l = 0$ and for $l = 1$, m_l can be +1, 0, –1. Thus eight electrons can be placed in $n = 2$, because of the Pauli Exclusion Principle. Since $2s$ orbital is of lower energy than the three $2p$ orbitals, it will be filled first. The electronic configuration for the atoms with $n = 3$ to $n = 10$ are written below:

Li	Be	B	C	N	O	F	Ne
$1s^22s^1$	$1s^22s^2$	$1s^22s^22p^1$	$1s^22s^22p^2$	$1s^22s^22p^3$	$1s^22s^22p^4$	$1s^22s^22p^5$	$1s^22s^22p^6$

The problem is simple till we reach $n = 6$, *i.e.*, carbon atom. Here the 6th electron can be placed in p orbital in three different ways.

↑↓	↑ ↑	↑ ↓
$2p_x$	$2p_x$ $2p_y$ or $2p_z$	$2p_x$ $2p_y$ or $2p_z$
(*a*)	(*b*)	(*c*)

Since all these p orbitals are degenerate any two of them will be occupied and these have equal probability. This is one of the ways of filling the orbitals. The next question is to find out which of the three configurations is correct. Hund's gave the following rules to decide the filling of the orbitals, which are degenerate. These are:

(*i*) Orbitals, with equal energies are always singly occupied by the electrons, till all of these are filled one each.

(*ii*) Electrons occupying the orbitals singly of the same energy have parallel spins.

Thus, the above rules clearly tell that the arrangement (*b*) is favourable over (*a*) and (c). The former effect can be qualitatively understood very easily from the fact that *repulsive potential from the two similar charges will be minimum if these charges are kept as far apart as possible*. The latter effect has its origin from the magnetic property associated with the motion of the negatively charged electron around its own axis passing through its centre. As we know that spin quantum number is quantized and can have only two possible values, *i.e.,* ± 1/2. In other words the spin axis may be aligned either parallel to or opposed to some arbitrary directions, *i.e.,* it is possible to speak of the electron (or magnetic field produced by the motion of the electron) as aligned parallel to or opposed to an applied magnetic field. It may be pointed out here that we shall be most commonly concerned with the relative spins of the two electrons when these are close to each other and we can mention that the spins are correlated. Without going into rigorous calculations, it may be stated here that *spin correlation energy is positive when the electrons are paired and negative when spins are the same*. Due to this reason electrons with the same spin tend to avoid one another and do not occupy the same region of space, *i.e., paired electrons, with opposed spins, exhibit greater interaction than electrons with like spins which tend to avoid each other*. This statement is very important because it means that in systems of two or more electrons, the spin-spin interactions greatly influence the spatial distribution of the electrons.

For $n = 3$, the l can be 0, 1, 2 and for each value of l, we have m_l as 0; 1, 0, –1 and 2, 1, 0, –1, –2. Since $\varepsilon_{3s} < \varepsilon_{3p} < \varepsilon_{3d}$, the orbitals are filled in the sequence of 3*s* and then 3*p*, *i.e.,* atoms from $n = 11$ (Na) to $n = 18$ (Ar). Further the atoms with $n \geq 11$, will also have orbitals $n = 1$ and $n = 2$ filled. This is often described by saying that all these atoms have neon like core of 10 electrons, corresponding to the configurations $1s^2 2s^2 2p^6$. Thus, the electronic configuration for Na and onward can be written as Ne$3s^1$ and so on.

After argon, *i.e.,* $n = 18$, there is a reversal in the sequence of atomic orbitals. As explained in section 3.10, the energy of 4*s* orbital is lower than that of 3*d* up to Z = 20 and thus in the elements after argon, add electrons to 4*s* rather than to 3*d*. Thus potassium (K) with atomic number 19 has the ground state electronic configuration $1s^2 2s^2 2p^6 3s^2 3p^6 4s^1$ (Ar $4s^1$), similar to the alkali metals (Li, Na). Ca with atomic number 20, Ar $4s^2$, is similar to alkaline earth elements like Mg (Ne $3s^2$). After $n = 20$, since 4*s* orbital is completely filled, the filling of 3*d* orbital begins with the next element scandium (Sc). The five 3*d* orbitals will contain 10 electrons, thus the fourth period, which corresponds to filling of 4*s*, 3*d* and 4*p* orbitals contain 18 elements, ending with the element of Krypton (Kr), in which 4*p* is completely filled. The next element starts with the filling of 5*s* orbital, *i.e.,* rubidium (Rb), similar to K. The fifth period, similar to 4th period, also contains 18 elements (filling of 5*s*, 4*d*, 5*p*). The sixth period starts with filling of 6*s* 5*d* ≈ 4*f*, 6*p* orbitals and contains 32 elements. The seventh period, similar to the sixth period, also consists of 32 elements, filling 7*s*, 6*d* ≈ 5*f*, 7*p* orbitals, although 17 of them have been discovered by late 1960.

The ground state electronic configuration of all the elements, as determined from experimental observations is given in Table 3.9. In most cases, the ground state electronic configurations agree with the order of energy levels, specified earlier.

3.12 PERIODIC TABLE AND PERIODICITY IN PROPERTIES

In the last few sections of this chapter, we have learnt about the structure of atoms, *i.e.,* to calculate the energy levels and the corresponding wave functions (atomic orbitals) and to arrange the electrons in different atomic orbitals, what we call as the electronic configuration. Now next question before us is to see whether the atomic properties and their reactions with other or similar atoms can be related to the electronic configuration or not. In the following section we shall try to examine the connection between the electronic configuration and the periodic variation in the properties of the elements.

3.12.1 MODERN PERIODIC TABLE

In the original periodic table, the elements were arranged according to the mass of the respective elements. Because of this kind of arrangements, problems arose in three pairs of elements, for example, argon and potassium; nickel and cobalt; tellurium and iodine and thus these pairs of elements were arranged arbitrarily. Later on, when the atomic structure was understood, it was found that the elements in the periodic table can be arranged nicely with increase in the atomic

Table 3.9: *Some atomic properties of the elements*

Atomic number	Element	Configuration	Ground state term symbol	First ionization energy, (eV)	Covalent radius, (pm)	Atomic radius (pm)	Electron affinity kJ mol^{-1}
1	H	$1s$	$^2S_{1/2}$	13.595	32	79	72.8
2	He	$1s^2$	$^1S_0$	24.580	93	49	–21
3	Li	$[He]2s^1$	$^2S_{1/2}$	5.390	123	205	59.8
4	Be	$[He]2s^2$	$^1S_0$	9.320	90	140	–18
5	B	$[He]2s^12p^1$	$^2P_{1/2}$	8.296	82	117	23
6	C	$[He]2s^12p^2$	$^3P_0$	11.264	77	91	122.5
7	N	$[He]2s^12p^3$	$^4S_{3/2}$	14.54	73	75	–7
8	O	$[He]2s^12p^4$	$^3P_2$	13.614	73	65	141
9	F	$[He]2s^12p^5$	$^2P_{3/2}$	17.420	72	57	322
10	Ne	$[He]2s^12p^6$	$^1S_0$	21.559	71	51	–29
11	Na	$[Ne]3s^1$	$^2S_{1/2}$	5.138	154	233	52.9
12	Mg	$[Ne]3s^2$	$^1S_0$	7.644	136	172	–21
13	Al	$[Ne]3s^23p^1$	$^2P_{1/2}$	5.984	118	182	44
14	Si	$[Ne]3s^23p^2$	$^3P_0$	8.149	111	146	133.6

15	P	[Ne]$3s^2 3p^3$	$^{4}S_{3/2}$	11.000	106	123	71.7
16	S	[Ne]$3s^2 3p^4$	$^{3}P_{2}$	10.357	102	109	200.4
17	Cl	[Ne]$3s^2 3p^5$	$^{2}P_{3/2}$	13.010	99	97	348.7
18	Ar	[Ne]$3s^2 3p^6$	$^{1}S_{0}$	15.755	98	88	–35
19	K	[Ar]$4s^1$	$^{2}S_{1/2}$	4.339	203	277	48.3
20	Ca	[Ar]$4s^2$	$^{1}S_{0}$	6.111	174	223	–188
21	Sc	[Ar]$4s^2 3d^1$	$^{2}D_{3/2}$	6.560	144	209	
22	TI	[Ar]$4s^2 3d^2$	$^{3}F_{2}$	6.830	132	200	
23	V	[Ar]$4s^2 3d^3$	$^{4}F_{3/2}$	6.740	122	192	
24	Cr	[Ar]$4s^1 3d^5$	$^{7}S_{3}$	6.736	118	185	
25	Mn	[Ar]$4s^2 3d^5$	$^{6}S_{5/2}$	7.432	117	179	
26	Fe	[Ar]$4s^2 3d^6$	$^{5}D_{4}$	7.900	117	172	
27	Co	[Ar]$4s^2 3d^7$	$^{4}F_{9/2}$	7.860	116	167	
28	Ni	[Ar]$4s^2 3d^8$	$^{3}F_{4}$	7.633	115	162	
29	Cu	[Ar]$4s^1 3d^{10}$	$^{2}S_{1/2}$	7.724	117	157	
30	Zn	[Ar]$4s^2 3d^{10}$	$^{1}S_{0}$	9.391	125	153	
31	Ga	[Ar]$4s^2 3d^{10} 4p^1$	$^{2}P_{1/2}$	6.000	126	181	36
32	Ge	[Ar]$4s^2 3d^{10} 4p^2$	$^{3}P_{0}$	7.880	122	152	116
33	As	[Ar]$4s^2 3d^{10} 4p^3$	$^{4}S_{3/2}$	9.810	120	133	77
34	Se	[Ar]$4s^2 3d^{10} 4p^4$	$^{3}P_{2}$	9.750	116	122	195
35	Br	[Ar]$4s^2 3d^{10} 4p^5$	$^{2}P_{3/2}$	11.840	114	112	324.5
36	Kr	[Ar]$4s^2 3d^{10} 4p^6$	$^{1}S_{0}$	13.996	112	103	–39
37	Rb	[Kr]$5s^1$	$^{2}S_{1/2}$	4.176	216	298	46.9
38	Sr	[Kr]$5s^2$	$^{1}S_{0}$	5.692	191	245	–146
39	Y	[Kr]$5s^2 4d^1$	$^{2}D_{3/2}$	6.500	162	227	
40	Zr	[Kr]$5s^2 4d^2$	$^{3}F_{2}$	6.950	145	216	
41	Nb	[Kr]$5s^2 4d^3$	$^{4}F_{3/2}$	6.770	134	208	
42	Mo	[Kr]$5s^1 4d^5$	$^{7}S_{3}$	7.100	130	201	
43	Tc	[Kr]$5s^2 4d^5$	$^{6}S_{5/2}$	7.280	127	195	
44	Ru	[Kr]$5s^2 4d^6$	$^{5}D_{4}$	7.364	125	189	
45	Rh	[Kr]$5s^2 4d^7$	$^{4}F_{9/2}$	7.460	125	183	
46	Pd	[Kr]$5s^2 4d^8$	$^{3}F_{4}$	8.330	128	179	
47	Ag	[Kr]$5s^1 4d^{10}$	$^{2}S_{1/2}$	7.574	134	175	

48	Cd	[Kr]$5s^24d^{10}$	$^1S_0$	8.991	148	171	
49	In	[Kr]$4d^65s^25p^1$	$^2P_{1/2}$	5.785	144	205	34
50	Sn	[Kr]$4d^65s^25p^2$	$^1P_0$	7.342	141	172	121
51	Sb	[Kr]$4d^65s^25p^3$	$^4S_{3/2}$	8.639	140	153	101
52	Te	[Kr]$4d^65s^25p^4$	$^3P_2$	9.010	136	142	190.2
53	I	[Kr]$4d^65s^25p^5$	$^2P_{3/2}$	10.454	133	132	295.3
54	Xe	[Kr]$4d^65s^25p^6$	$^1S_0$	12.127	131	124	–41
55	Cs	[Xe]$6s^1$	$^2S_{1/2}$	3.893	235	324	45.5
56	Ba	[Xe]$6s^2$	$^1S_0$	5.210	198	278	–46
57	La	[Xe]$5d^16s^2$	$^2D_{3/2}$	5.610	169	274	
58	Ce	[Xe]$4f^15d^16s^2$	$^3H_4$	6.910	165	270	
59	Pr	[Xe]$4f^36s^2$	$^4I_{9/2}$	5.760	165	267	
60	Nd	[Xe]$4f^46s^2$	$^5I_4$	6.310	164	264	
61	Pm	[Xe]$4f^56s^2$	$^6H_{5/2}$	5.554	163	262	
62	Sm	[Xe]$4f^66s^2$	$^7F_0$	5.600	162	259	
63	Eu	[Xe]$4f^76s^2$	$^8S_{7/2}$	5.670	185	256	
64	Gd	[Xe]$4f^75d^16s^2$	$^9D_2$	6.160	161	254	
65	Tb	[Xe]$4f^96s^2$	$^6H_{15/2}$	6.740	159	251	
66	Dy	[Xe]$4f^{10}6s^2$	$^5I_8$	6.820	159	249	
67	Ho	[Xe]$4f^{11}6s^2$	$^4I_{15/2}$	6.016	158	247	
68	Er	[Xe]$4f^{12}6s^2$	$^3H_6$	6.080	157	245	
69	Tm	[Xe]$4f^{13}6s^2$	$^2F_{5/2}$	5.810	156	242	
70	Yb	[Xe]$4f^{14}6s^2$	$^1S_0$	6.200	174	240	
71	Lu	[Xe]$4f^{14}5d^16s^2$	$^2D_{1/2}$	5.000	156	225	
72	Hf	[Xe]$5d^26s^2$	$^3F_4$	6.650	144	216	
73	Ta	[Xe]$5d^36s^2$	$^4F_{3/2}$	7.880	134	209	
74	W	[Xe]$5d^46s^2$	$^5D_0$	7.980	130	202	
75	Re	[Xe]$5d^56s^2$	$^6S_{5/2}$	7.870	128	197	
76	Os	[Xe]$5d^66s^2$	$^5D_4$	8.700	126	192	
77	Ir	[Xe]$5d^76s^2$	$^4F_{9/2}$	9.000	127	187	
78	Pt	[Xe]$5d^86s^2$	$^3D_3$	9.000	130	183	
79	Au	[Xe]$5d^{10}6s^1$	$^2S_{1/2}$	9.220	134	179	
80	Hg	[Xe]$5d^{10}6s^2$	$^1S_0$	10.430	149	176	

81	Tl	$[Xe]5d^6 6s^2 6p^1$	$^2P_{1/2}$	6.106	148	208	30
82	Pb	$[Xe]5d^6 6s^2 6p^2$	$^3P_0$	7.415	147	181	35.2
83	Bi	$[Xe]5d^6 6s^2 6p^3$	$^4S_{3/2}$	7.287	146	163	101
84	Po	$[Xe]5d^6 6s^2 6p^4$	$^3P_2$	8.430	146	153	186
85	At	$[Xe]5d^6 6s^2 6p^5$	$^2P_{3/2}$	_	145	143	270
86	Rn	$[Xe]5d^6 6s^2 6p^6$	$^1S_0$	10.740	_	134	–41
87	Fr	$[Rn]6s^1$	$^2S_{1/2}$	–	–	–	
88	Ra	$[Rn]6s^2$	$^1S_0$	5.277		–	
89	Ac	$[Rn]6d^1 6s^2$	$^2D_{1/2}$	5.170		0	
90	Th	$[Rn]6d^2 7s^2$	$^3F_{3/2}$	6.950	165	_	
91	Pa	$[Rn]5f^2 6d^1 7s^2$	$^4K_{13/2}$	5.890	_	_	
92	U	$[Rn]5f^3 6d^1 7s^2$	$^5L_7$	6.100	142	0	
93	Np	$[Rn]5f^2 6d^1 7s^2$	$^6L_{13/2}$	6.190	–	–	
94	Pu	$[Rn]5f^6 7s^2$	$^7F_0$	5.100	–	–	
95	Am	$[Rn]5f^7 7s^2$	$^8S_{7/2}$	6.000	–	–	
96	Cm	$[Rn]5f^7 6d^1 7s^2$	$^9D_6$	6.02	–	–	
97	Bk	$[Rn]5f^9 7s^2$	$^6H_{15/2}$	–			
98	Cf	$[Rn]5f^{10} 7s^2$	$^5I_8$	–			
99	Es	$[Rn]5f^{11} 7s^2$	$^4H_{5/2}$	–			
100	Fm	$[Rn]5f^{12} 7s^2$	$^3H_6$	–			
101	Md	$[Rn]5f^{13} 7s^2$	$^2F_{7/2}$	–			
102	No	$[Rn]5f^{14} 7s^2$	$^1S_0$	–			
103	Lw	$[Rn]5f^{14} 6d^1 7s^2$	$^2D_{3/2}$	–			

number, rather than with increase in the atomic mass and thus the above mentioned anomalies were removed automatically. *The modern periodic law states that the properties of the elements are a periodic function of the atomic number.*

Based on the above mentioned procedure of arranging the elements, the modern table (also known as long form) consists of eighteen columns and seven rows, as shown in Fig. 3.20. Actually, this has been reduced, by putting fourteen elements containing 4*f* sub-shell, starting with lanthanum (known as lanthanides) and 5*f* sub-shell (known as actinides) and arranging them separately at the bottom of the periodic table. *The elements belonging to the same column are known as groups.* The numbering of the groups corresponds to the number of electrons present in the outer most shell, which is also called *valence shell*. This is also known as the valence number or one of the common valence numbers of the group of elements.

The electronic configuration for the rare gases is ns^2np^6 (ns^2, for helium) is commonly designated as [ ]. Thus, except for hydrogen, the electronic configuration of the elements which are in the first column or first group, viz., Li, Na, K, Rb, Cs, Fr is [ ] ns^1. The first group elements are known as *alkali metals*. The second group of elements (Be, Mg, Ca, Sr, Ba, and Ra), called *alkaline earth metals,* have [ ] ns^2 electronic configuration. To have a break but continuity in the periodicity, the groups 13 to 17 have the valence electronic configuration as ns^2np^x where x can have the value 1 to 5. All these elements, which nearly form half of the periodic table, are known as *main group elements*. These elements have mainly s and p electrons in their outer most shell or in their valence shell. Most of the remaining elements are known as *transition elements* and are located in the middle of the periodic table. The group number starts from 3 and ends at 12, *i.e.,* 10 groups. In the fourth period, starting from Ca (Z = 20) and ending at Ga (Z = 31), there are ten elements, involving the filling of $3d$ orbitals. With two exceptions all the elements have electronic configuration as $[Ar]4s^23d^x$ where x can have the value from 1 to 10.

Similarly between Sr (Z = 38) and In (Z = 49), there are other 10 elements which involve the filling of $4d$ orbitals, i.e. $[Kr]5s^24d^1$ to $[Ar]4d^{10}$. Lastly starting with La (Z = 57) we have 10 other elements ending at Hg (Z = 80), with electronic configuration $[Xe]6s^25d^1$ to $[Xe]6s^25d^{10}$. Other notations commonly used for these elements are: $3d$ elements are called as the *first transition series*, $4d$ elements, the *second transition series* and $5d$ elements the *third transition series* respectively. All these elements form a block of elements whose neutral atoms or ions involve the incomplete d sub-shell. Although we are not going to discuss the chemistry of these elements, but the electronic configuration of these elements lead to many characteristic physical and chemical properties.

The elements Zn, Cd and Hg are unique in the sense that these elements do not possess oxidation state higher than +2, similar to alkaline earth elements, but are different in the sense that instead of having a tightly bound rare gas structure for their divalent cations, these divalent ions have easily polarizable nd^{10} sub-shell.

Lastly, in between La and Hf and after Ac, we have two sets of fourteen elements each, which involve the filling of $4f$ and $5f$ elements. The former is called *lanthanides* and the latter is called *actinides* series. These are also called as *f block elements*. Similar to transition elements, their properties are also similar to each other. But these are less studied, because of their poor abundance in nature and also most of the $5f$ block elements are man made. Besides the notation used above regarding the periodic table, there is also another kind of notation, for example I*a*, I*b*; II*a*, II*b* etc., This notation is somewhat arbitrary and slowly loosing its significance.

In the coming section few atomic properties, which are quite common and useful, will be discussed.

Periodic Table of the Elements

1 H 1.008																	2 He 4.003
3 Li 6.94	4 Be 9.01											5 B 10.81	6 C 12.011	7 N 14.01	8 O 16.00	9 F 19.00	10 Ne 20.18
11 Na 22.99	12 Mg 24.31											13 Al 26.98	14 Si 28.09	15 P 30.97	16 S 32.06	17 Cl 35.45	18 Ar 39.95
19 K 39.10	20 Ca 40.08	21 Sc 44.96	22 Ti 47.90	23 V 50.94	24 Cr 52.00	25 Mn 54.94	26 Fe 55.85	27 Co 58.93	28 Ni 58.71	29 Cu 63.55	30 Zn 65.37	31 Ga 69.72	32 Ge 72.59	33 As 74.92	34 Se 78.96	35 Br 79.90	36 Kr 83.80
37 Rb 85.47	38 Sr 87.62	39 Y 88.91	40 Zr 91.22	41 Nb 92.91	42 Mo 95.94	43 Tc 98.91	44 Ru 101.07	45 Rh 102.91	46 Pd 106.4	47 Ag 107.87	48 Cd 112.40	49 In 114.82	50 Sn 118.69	51 Sb 121.75	52 Te 127.60	53 I 126.90	54 Xe 131.30
55 Cs 132.91	56 Ba 137.34	57 La 138.91	72 Hf 178.49	73 Ta 180.95	74 W 183.85	75 Re 186.2	76 Os 190.2	77 Ir 192.2	78 Pt 195.09	79 Au 196.97	80 Hg 200.59	81 Tl 204.37	82 Pb 207.19	83 Bi 208.98	84 Po (209)	85 At (210)	86 Rn (222)
87 Fr (223)	88 Ra 226.03	89 Ac (227)	104 Unq* (261)	105 Unp* (262)	106 Unh* (263)	107 Uns* (262)	108 Uno* (265)	109 Una* (266)									

Lanthanides	58 Ce 140.12	59 Pr 140.91	60 Nd 144.24	61 Pm (145)	62 Sm 150.35	63 Eu 151.96	64 Gd 157.25	65 Tb 158.93	66 Dy 162.50	67 Ho 164.93	68 Er 167.26	69 Tm 168.93	70 Yb 173.04	71 Lu 174.97
Actinides	90 Th 232.04	91 Pa (231)	92 U 238.03	93 Np (237)	94 Pu (244)	95 Am (243)	96 Cm (247)	97 Bk (249)	98 Cf (249)	99 Es (254)	100 Fm (257)	101 Md (258)	102 No (259)	103 Lr (260)

*Symbol (and name) provisional.

Numbers in parentheses: available radioactive isotope of longest half-life.

Figure 3.20: *Peliodic table of the elements*

3.12.2 IONIZATION ENERGY

Ionization energy is the energy required to remove the electron completely from an atom, molecule or ion present in the gaseous state. Thus an atom can have as many ionization energies as the number of electrons. The first and second ionization energies are defined as follows:

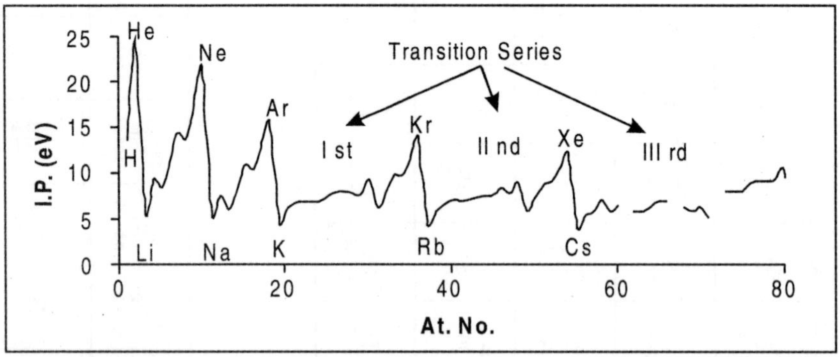

Figure 3.21: *Ionization energies of various atoms as a function of atomic number up to 80*

$$A\ (g) \rightarrow A^{+}\ (g) + e^{-} \quad ...(3.259)$$

$$A^{+}\ (g) \rightarrow A^{2+}\ (g) + e^{-} \quad ...(3.260)$$

Higher ionization energies can also be defined in a similar manner. The first ionization energies of various atoms are compiled in Table 3.9 and ionization energies as a function of atomic numbers are plotted in Fig. 3.21. Before we discuss the main features of the Fig. 3.21, it may be stated that ionization energy of an electron can be equated to the following expression.

$$\text{I.E.} \cong \frac{Z_{eff}}{n^2}\left(\frac{e}{2a_0}\right) \quad ...(3.261)$$

As usual, Z_{eff} is the effective nuclear charge expressed by the respective electron, n is the principal quantum number, e is the electronic charge and a_0 the Bohr radius. The main features of the Fig. 3.21 are:

(1) The maximum ionization energies in any period are due to rare gases and the minimum ionization energies are due to alkali metals. This is quite obvious because the rare gases represent the closed shell configuration and thus are very stable, resist to disruption either to get ionized or to form chemical bonds. On the other hand, electronic configuration of alkali metals is ns^1 and the odd electron in the outer most shell is well shielded from the nucleus by the closed shell of rare gases. In other words, the electron can be easily removed.

(2) In a given period the ionization energies keep on increasing with increase of atomic number, whereas in a given group the ionization energy decreases with increase of atomic number. This is because electrons with the same principal quantum number are nearly at the same average distance from the nucleus and thus cannot shield the nucleus effectively. With the increase of atomic number, one unit of positive charge is added to

the nucleus. But because of poor shielding, the incoming electron only partly screens the nucleus and an increase in the ionization energy is observed according to equation (3.261), through the series s, s^2, s^2p, s^2p^2... ..s^2p^6.

(3) Although the ionization energy increases with increase of atomic number in a given period, the increase is not monotonic. Some discrepancies are observed. For example, consider second period. Ionization energy of Be ([He]$2s^2$) is greater than that of Li ([He]$2s^1$), but the ionization energy of B ([He] $2s^2 2p^1$) is less than that of Be. This is because the energy of $2p$ orbital is higher than that of $2s$ orbital and $2p$ orbital is less penetrating than $2s$, thus is more shielded. In other words, electron in $2p$ can be easily removed. There is a further increase in the ionization energies of the elements carbon ([He]$2s^2 2p^2$), nitrogen ([He]$2s^2 2p^3$), but decrease is observed in case of oxygen ([He]$2s^2 2p^4$). The reason being that nitrogen is half filled, whereas the addition of electron will be in the orbitals, which are already singly occupied and thus repulsion will be observed. This will allow the electron to be removed easily.

(4) Variation in the ionization energies for other elements can be explained in the similar manner, except that their ionization energies are less. This is due to the presence of more core electrons, which are shielding the nucleus. The other point worth mentioning is the ionization energies of the fourth, fifth, sixth periods etc., The elements present in group 1, 2 and 13 to 18 follow the similar trend as explained in case of elements belonging to second period, but the ionization energies of $3d$, $4d$ and $5d$ transition series behave differently. As in the case of electronic configurations, the proximity of the $(n-1)$ d and ns orbital energies possess interesting but subtle questions. The fact that the order of levels may be different in the ion and in the neutral atom, plus there is a gradual change of the role of d-orbital from an outer to an inner orbital, accounts for the observed variation. In these cases, it is better to think of the ionization energy as the energy of the neutral atom, but a complete discussion is beyond the scope of this book.

3.12.3 ELECTRON AFFINITY

The electron affinity (EA) of neutral atom is defined as the energy released when an electron with zero translational energy, is captured by a neutral atom. The final energy of the electron is

$$A + e^- \rightarrow A^- \quad \text{...(3.262)}$$

negative. The EA of the atoms are compiled in Table 3.9. The values of EA measured for the atoms are much less than the values of ionization energies known. This is because of the experimental difficulties encountered in measuring the energy of A^-. Before we discuss the trends observed in EA, it may be mentioned here that by normal convention, the EA is the negative of the energy or enthalpy released in the equation (3.262), *i.e.,* EA = – ΔH. This unfortunate practice has been accepted so much that if it is changed now, it will create a lot deal of confusion. It may be better understood by defining the process as the ionization energy of the negative ion, *i.e.,*

$$A^- \rightarrow A + e^- \quad \text{...(3.263)}$$

The trends observed in EA are similar to those observed in case of ionization energies, *i.e.*, (*i*) EA increase if one moves from the left to right in a period. This is because the EA, like ionization energy, also depends on electronic configuration. It has a maximum value for elements in a given period those need minimum number of electrons to attain a closed shell configuration. For example, halogens, F has EA ≈ 3.4 eV, whereas that of Li is 0.6 eV. Oxygen atom, which is short of two electrons, has EA ≈ 1.4 eV. Although O^{2-} has a closed shell structure, it appears to be unstable in the free state, so the second EA of oxygen is zero. (*ii*) EA decreases with increase of atomic number in a given group, for example, EA of F is 3.4 eV and that of Iodine is 3.1 eV.

An interesting departure from the above trend is that EA of second period elements are lower than the third period members of the same group, *i.e.*, EA (Li) < EA (Na); EA (O) < EA (S); and EA (F) < EA (Cl). There seems to be no adequate explanation for this phenomenon at the present time, although in case of fluorine, it is often said that due to smaller size of fluorine atom than chlorine atom, the addition of extra electron increases the repulsion, which is more than that noticed in case of chlorine atom.

3.12.4 ATOMIC AND IONIC RADII

While discussing the nature of the wave functions for atoms and ions in section 3.2.5, we have already learnt that atoms and ions do not have specific size. Thus, their values depend on the method used to determine the distance between the atoms or ions in a molecule, which are formed under the influence of interatomic interactions. Although, the Tables have been prepared for covalent radii (measured generally from the interatomic distance between homonuclear diatomic molecules), ionic radii (measured between the centers of two adjacent ions to be equal to the sum of ionic radii), the values compiled in Table 3.9, are the averages of the bonds in covalent, metallic and ionic crystals and in molecules. Atomic radii for rare gases are the exceptions and these are the Van der Waal's radii in pure condensed phases. The orbital radii, which are also compiled in Table 3.9, are the radii of the principal maximum of the charge density of the outermost atomic orbital. These radii, had been calculated by J. T. Weber and D. T. Chromer (J. Chem. Phys., **42,** 4116 1965). It is clear from the values of atomic radii in Table 3.9 that the trends observed are similar to that noticed in ionization energies, that is, atomic radius decreases with increase of atomic number in a given period, whereas it increases with increase of atomic number in a given group. This trend can be explained with the help of equation (3.264), which is nearly similar to that used to calculate Bohr's radii of different orbitals:

$$r \approx \frac{n^2}{Z_{\text{eff}}^2} a_0, \quad a_0 = \frac{h^2}{4\pi^2 m_e} \qquad \text{...(3.264)}$$

where each term has its usual meanings. The above mentioned, results can be easily explained with equation (3.264). We know that the principal quantum number (n) remains the same for all the elements in a given row, whereas the effective nuclear charge increases with increase in the atomic number and thus a decrease in the atomic radius. On the other hand, in a given column (group) of the periodic table, effective nuclear charge nearly remains the same with increase of atomic number but the principal quantum number does change by a value of one and this increases the atomic radii.

Deviations in the values of atomic radii, determined experimentally and predicted by equation (3.264) can be expected. For example, the atomic radii of rare gases are larger than those, which can be predicted by equation (3.264). This is because these are Van der Waal's radii. Similarly the radius of sodium in sodium chloride (which is ionic in nature, 95 pm) is much smaller than that of sodium in diatomic sodium (which is covalent, 154 pm), whereas atomic radius of metallic sodium (233 pm) is greater than the two values. In the end it may be mentioned that one should be careful in comparing the atomic radii reported in different tables, because all these value are based on the nature of chemical bonding in the molecules.

3.13 ANGULAR MOMENTUM

3.13.1 FOR ONE ELECTRON

In the previous sections of this chapter the emphasis has been laid on the calculation of energies of atomic orbitals and their corresponding wave functions, leading to write the electronic configurations of atoms. One point is obvious from these electronic configurations for many atoms that these are not unique. What we mean is that for a given electronic configuration, there are number of ways by which m_l and m_s values can be assigned. For example, consider the electronic configuration of carbon atom in its ground state, $1s^2 2s^2 2p^2$. Two electrons in the *p* orbitals can be present in any of the three orbitals (p_x, p_y, p_z) with different spin values and consistent with Pauli Exclusion Principle. These different states can have different values of energies and thus more correct designation of states is required. In other words the understanding of motion of electron (angular momentum) associated with m_l and m_s are necessary in describing, specially, the spectral characteristics of atoms. In the following section, we will discuss first the physical meaning of these angular momenta for one electron system, followed by the similar discussion for multi-electron atoms.

Before we proceed further, let us define certain terms.

(*i*) ***Orbital Angular Momentum:*** An electron moving in its orbital around the nucleus possesses an orbital angular momentum. As the wave function of an electron is an eigen function to the Hamiltonian operator and gives the energy of the state or orbital, it is also an eigen function to the square of the orbital angular momentum operator $(\hat{L}^2)^{\frac{1}{2}}$ and eigen value of this operator $(\hat{L}^2)$ is:

$$\mathbf{l} = \sqrt{l\,(l+1)}\;\frac{h}{2\pi} \qquad ...(3.265)$$

This is also quantized and is represented in the units of $\frac{h}{2\pi}$, where *h* is a Planck's constant. **l** (bold face) and *l* (italics) denote the orbital angular momentum and azimuthal quantum number. In other words, azimuthal quantum number is a measure of the orbital angular momentum. Further **l** and *l* are so closely related to each other that these are often mistaken and are used quite often for each other. For example, we often say that orbital angular momentum is 3 for azimuthal quantum number 3, whereas it means that the value is $\sqrt{3} \times 4.\dfrac{h}{2\pi}$, *i.e.*, $\sqrt{12}.\dfrac{h}{2\pi}$. So one should be careful whether we are talking about the azimuthal quantum number or orbital angular momentum.

Orbital angular momentum is a vector quantity. So its magnitude and direction must be specified precisely. Using mathematical notation, vector can be represented by pointing an arrow and length gives the magnitude of the vectors. Figure 3.22 gives the description of *l*. The other important characteristic of orbital angular momentum is its direction. Unlike classical mechanics, it cannot have all possible directions and it is always specified, often known as *space quantization*. There are many ways to do so. One of the ways is to choose the direction externally, *i.e.*, the direction of magnetic field or electrical field applied

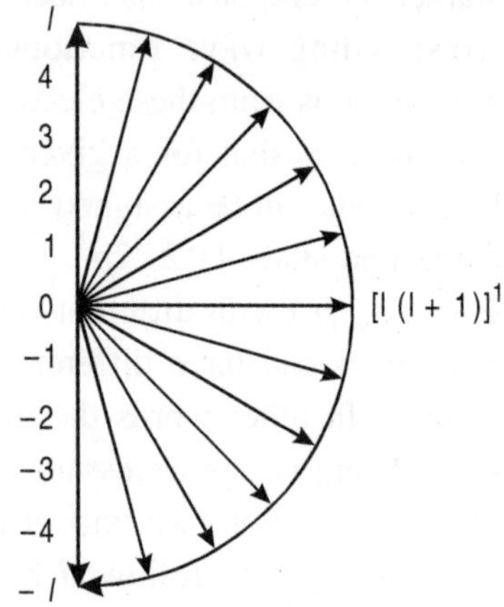

Figure 3.22: *The allowed directions of the angular momentum vector for an electron with l value, also known as space quantization*

externally. The other method is internally, perhaps by specifying the direction of the orbital angular momentum of any electron. Once this is done, angular momentum will have such a direction that its component along this direction will be an integral multiple of $\dfrac{h}{2\pi}$. It cannot possess any arbitrary direction. In general this direction coincides with the *z*-axis, but it can be specified by another axis. The component of the orbital angular momentum is $\boldsymbol{m_l}$ and the corresponding quantum number m_l, magnetic quantum number as mentioned earlier. Since there will be $2l + 1$ magnetic quantum numbers for each value of l, the values of the components of orbital angular momentum will also be $2l + 1$. Figure 3.23 shows the various components of $l = 1$ and $l = 2$ respectively. The positive and negative values only depict the clockwise and anti-clockwise motion of electron in an orbital around the nucleus. Thus the value of m_l, can be depicted by the relation:

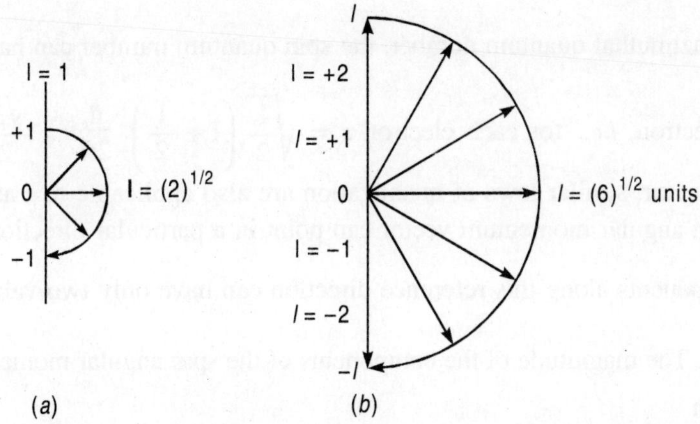

Figure 3.23: *The allowed components of the orbital angular momentum of an electron for (a) l = 1 and (b) l = 2*

$$\boldsymbol{m}_l = m_l \; h/2\pi \qquad ...(3.266)$$

where $\quad m_l = l, l-1, l-2, \ldots \quad .0, \ldots \quad -(l-1), -l$

It may be mentioned here again that wave function of an electron is an eigen function to the z-component of the orbital angular momentum.

Lastly, the orbital energy of an electron does not depend on the direction of the orbital angular momentum, but only on its magnitude. Thus, for each value of l, there will be $2l + 1$ degenerate quantum states or orbitals. The degeneracy of $2l + 1$ quantum states can be removed if an external magnetic field is applied and thus, energy of each quantum state corresponding to each value of m_l will be different (Zeeman Effect).

(*ii*) ***Electron Spin Angular Momentum:*** Besides moving in an orbital around the nucleus, the electron also spins around its own axis and both these motions take place simultaneously. This is known as spin angular momentum quantum number s and the corresponding angular momentum $\boldsymbol{s}$ is quantized in the units of $\frac{h}{2\pi}$. It can be depicted as:

$$\boldsymbol{s} = \sqrt{s\,(s+1)}\,\frac{h}{2\pi} \qquad ...(3.267)$$

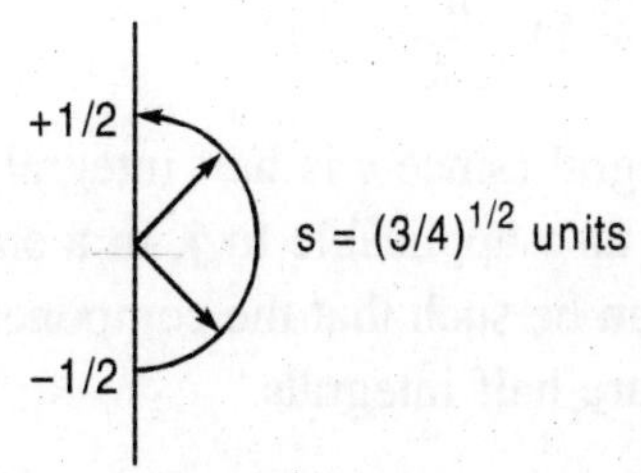

Figure 3.24: *The allowed directions of the electronic spin angular momentum vector*

Unlike azimuthal quantum number, the spin quantum number can have a value of $\frac{1}{2}$ for each electron, *i.e.*, for each electron $s = \sqrt{\frac{1}{2}\left(1+\frac{1}{2}\right)}\frac{h}{2\pi} = \frac{\sqrt{3}}{2}\cdot\frac{h}{2\pi}$. Since s is also a vector, similar laws of quantization are also applicable to s as are applicable to l, *i.e.*, spin angular momentum vector can point in a particular direction in such a way that its components along this reference direction can have only two values, *i.e.*, either $+\frac{1}{2}$ or $-\frac{1}{2}$. The magnitude of the components of the spin angular momentum is given by the equation.

$$m_s = \frac{1}{2}\frac{h}{2\pi} \quad \text{or} \quad m_s = -\frac{1}{2}\frac{h}{2\pi} \qquad ...(3.268)$$

It may also be specified here that similar to z-component of orbital angular momentum, the wave function of the electron is also an eigen function of the z-component of the spin angular momentum and it is either $\frac{1}{2}$ multiple of $\frac{h}{2\pi}$ or $-\frac{1}{2}$ multiple of $\frac{h}{2\pi}$. The two $(2s + 1)$ directions are shown in Figure 3.24.

(*iii*) ***Total Angular Momentum:*** As the name suggests, it is the resultant of the orbital and spin angular momenta of the electron. So we have to devise procedure to calculate the value of total angular momentum. It is defined as:

$$\mathbf{j} = \mathbf{l} + s \qquad ...(3.269)$$

Here $\mathbf{j}$ is the total angular momentum. Since $\mathbf{l}$ and s are vectors, $\mathbf{j}$ is also a vector and is a vector sum of $\mathbf{l}$ and s. Further the spin and orbital angular momenta has certain directions, so they can only be added in certain manner. They must be added in such a manner that various j values differ from one another by integral numbers and also j must be positive. The possible values of j are:

$$j = (l + s),\ (l + s - 1),\ (l + s - 2),\ldots \qquad |l - s| \qquad ...(3.270)$$

Similar to $\mathbf{l}$ and s, $\mathbf{j}$ can be related to the total angular momentum quantum number j through

$$\mathbf{j} = \sqrt{j(j + 1)}\ \frac{h}{2\pi} \qquad ...(3.271)$$

where $\mathbf{j}$ can be half integral (since s is half integral for one electron atom). The laws of quantum mechanics are also applicable to $\mathbf{j}$, in a similar manner as these are to $\mathbf{l}$ and s, *i.e.*, the direction of $\mathbf{j}$ can be such that the components of total angular momentum along the particular direction are half integrals.

$$m_j = \pm j,\ \pm (j - 1),\ \pm (j - 2), \quad + \frac{1}{2} \qquad ...(3.272)$$

$$\text{and} \qquad \boldsymbol{m}_j = m_j \frac{h}{2\pi} \qquad \text{...(3.273)}$$

i.e., for each value of **j**, there can be 2**j** + 1components. Two methods are quite common to calculate the value of ***j***. The first method is called the method of vector summation method and is applicable to one electron system, while the second method known as the summation of *z*-components is applicable to multi-electron atoms.

(1) ***Vector summation:*** This method is based on the law of *parallelogram of forces*, *i.e.*, as **l** and *s* are vectors, these can be represented in magnitude and direction by two sides of parallelogram (may be in the form of arrows). After completing the parallelogram, the resultant of the two vectors is given by the diagonal of the parallelogram. Exactly similar method can be used for this vector summation. But there is one important difference.

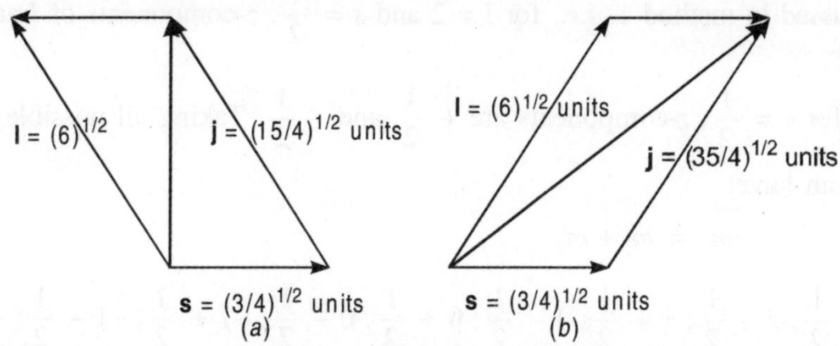

Figure 3.25: *The two energy states having different total angular momentum which can arise as a result of the vector addition of* $\boldsymbol{l} = (6)^{1/2}$ *units and* $s = (3/4)^{1/2}$ *units*

The laws of quantum mechanics put a restriction on the angle between **l** and *s*. The angle between **l** and *s* should be such that the values of **j** obtained agree with those given by the equation (3.271) with half-integral **j**. Thus, the value of **j** can be: $\frac{\sqrt{3}}{2}\frac{h}{2\pi}$, $\frac{\sqrt{15}}{2}\frac{h}{2\pi}, \frac{\sqrt{35}}{2}\frac{h}{2\pi}$,.......... for $j = \frac{1}{2}, \frac{3}{2}, \frac{5}{2}$ and so on. The above procedure has been illustrated in Fig. 3.25, for $l = 2\left(l = \sqrt{6}\frac{h}{2\pi}\right)$ and $s = \frac{1}{2}\left(s = \frac{\sqrt{3}}{2}\frac{h}{2\pi}\right)$. In Fig. 3.25*a*, the summation gives $\mathbf{j} = \frac{\sqrt{15}}{2}\frac{h}{2\pi}$, which corresponds to $j = \frac{3}{2}$, while Fig. 3.25*b*, yields $\mathbf{j} = \frac{\sqrt{35}}{2}\frac{h}{2\pi}$ for $j = \frac{5}{2}$. One point can be noted from the above procedure is that the similar values of *j* can be obtained if the summation of quantum numbers *l* and *s* are used to calculate the quantum number *j*, *i.e.*, using equation (3.270), the possible values of *j* can be:

$$j = 2 + \frac{1}{2} = \frac{5}{2} : j = 2 - \frac{1}{2} = \frac{3}{2}.$$

Note that $j = \frac{1}{2}$ is not a permitted value of the total angular momentum quantum number, since the smallest value of j is $l - s = 2 - \frac{1}{2} = \frac{3}{2}$. Although this approach is simpler, and useful only for one electron system, cannot be extended to multi-electron atoms.

(2) ***Summation of z-components:*** This method is also based on the mathematical principle that *the summation of the components of any two vectors along any direction (say z-axis) yield the component of the resultant vector.* Consider the same example as discussed in method 1, *i.e.*, for $l = 2$ and $s = \frac{1}{2}$. z-components of l are: 2, 1, 0, –1, –2 and for $s = \frac{1}{2}$, z-components are $+\frac{1}{2}$ and $-\frac{1}{2}$. Taking all possible sums of the two, we can have:

$$m_j = m_l + m_s \qquad ...(3.274)$$

i.e., $$2 + \frac{1}{2}, 2 - \frac{1}{2}; 1 + \frac{1}{2}, 1 - \frac{1}{2}; 0 + \frac{1}{2}, 0 - \frac{1}{2}; -1 + \frac{1}{2}, -1 - \frac{1}{2}; -2 + \frac{1}{2}, -2 - \frac{1}{2}$$

i.e., $$\frac{5}{2}, \frac{3}{2}, \frac{3}{2}, \frac{1}{2}, \frac{1}{2}, -\frac{1}{2}, -\frac{1}{2}, -\frac{3}{2}, -\frac{3}{2}, -\frac{5}{2}.$$

In the above list of ten values of m_j, the maximum value of m_j is $\frac{5}{2}$ and this must correspond to $j = \frac{5}{2}$. Further, the components of $j = \frac{5}{2}$ are $\frac{5}{2}, \frac{3}{2}, \frac{1}{2}, -\frac{1}{2}, -\frac{3}{2}, -\frac{5}{2}$. Removing these components from the list of ten, we are left with $\frac{3}{2}, \frac{1}{2}, -\frac{1}{2}, -\frac{3}{2}$. These values of m_j conforms to $j = \frac{3}{2}$. In other words, we can say that all these ten values are accounted for if we conclude that $j = \frac{5}{2}$ and $j = \frac{3}{2}$ are formed from $l = 2$ and $s = \frac{1}{2}$. The results obtained are consistent with those obtained by vector addition method or given by equation (3.270).

The above results can be looked from different angle. For example, $\mathbf{j} = \frac{\sqrt{35}}{2}\frac{h}{2\pi}$ can be obtained from $\mathbf{l} = \sqrt{6}\frac{h}{2\pi}$ and $s = \frac{\sqrt{3}}{2}\frac{h}{2\pi}$, when two vectors point in the same direction, *i.e.,* they reinforce each other, whereas the value $\mathbf{j} = \frac{\sqrt{15}}{2}\frac{h}{2\pi}$ is obtained when **l** and s oppose each other or they are aligned in opposite direction. In other words two values of total angular momentum are obtained, depending on the direction of two vectors. This will lead to the formation of two energy states. Since the azimuthal quantum number for both states is the same ($l = 2$), these two states will be denoted by the same symbol (D, see section 3.1.5). The distinction between the two states will be made by writing the j values as subscript at the right hand side of letter D, *i.e.,* $D_{\frac{5}{2}}$ or $D_{\frac{3}{2}}$ (the capital letter is used for the resultant of the angular momentum of all the electrons in an atom and a small letter for the state of a single electron). In case of hydrogen atom this distinction is trivial since it contains only one electron. The two states generated in this way are called doublet and is denoted by writing $^2D_{\frac{5}{2}}$ and $^2D_{\frac{3}{2}}$. These are also called as *state symbol* or *term symbol* (section 3.14) and are read as *doublet five half* or *doublet three half.*

Similar treatment can be applied to different values of l (electron configuration) for hydrogen atom, *i.e.,* the two ^{2}F states arising from the configuration f^1 are $^2F_{\frac{7}{2}}$ and $^2F_{\frac{5}{2}}$.

It is evident that every term ^{2}L of hydrogen atom except ^{2}S is really two energy levels, which will be at slightly different energies. The number of levels is $2s + 1$, except in case where $l = 0$. In this case the contribution to the total angular momentum from orbital angular momentum is zero and the resultant (total) angular momentum would only be $s = \frac{\sqrt{3}}{2}\frac{h}{2\pi}$ for $s = \frac{1}{2}$. Since the negative value of the quantum number is not possible, so $s = -\frac{1}{2}$ is not allowed. This means that it is only the z-component of the spin angular momentum, which can have positive or negative value. Thus the possible term symbol will be $^2S_{\frac{1}{2}}$.

Lastly it may be mentioned that if it is assumed that orbital and spin motions of the electron did not interact or influence each other, the two J levels would have the same energy. The coupling or interaction between orbital and spin angular momenta is very small for hydrogen atom, amounting to fraction of wave number, which is a very small portion of the energy of the term. For example, the ^{2}S – ^{2}P term separation is 83,000 cm^{-1} but the separation between $^2P_{1/2}$ – $^2P_{3/2}$ levels, is less than 0.1 cm^{-1}.

3.13.2 MULTI-ELECTRON ATOMS

Before we discuss the procedures to calculate the resultant of different momenta from the individual electrons, there are two possible ways to carry out.

(*i*) For light atoms with Z < 40, the total orbital angular momentum are nearly separately conserved and so L and S are almost good quantum numbers. This means that the resultant orbital angular momentum and spin angular momentum are calculated separately and then are combined vectorially to calculate J, *i.e.*,

$$L = \Sigma l_i,\ S = \Sigma s_i\ , \qquad J = L + S \qquad ...(3.275)$$

This procedure is known as *spin-orbit (s–o) coupling* or *Russell-Saunders(R-S) coupling*.

(*ii*) For heavy atoms with Z > 40, the total angular momentum is calculated by summing the orbital and spin angular momentum for each electron individually and then the *j* values for each electron is summed up to calculate the resultant J values, *i.e.*,

$$j_i = l_i + s_i,\ J = \Sigma j_i \qquad ...(3.276)$$

This is known as *j-j coupling*. In the following sections, former method will be discussed.

Resultant Orbital Angular Momentum: The resultant orbital angular momentum, designated as **L**, can be calculated by three different methods.

(*i*) This method is similar to that discussed for the summation of orbital and spin angular momentum for a single electron. Let l_1 and l_2.....be the orbital angular momentum of different electrons. These are combined in such a manner that the resultant **L** value can be given by the equation (3.277).

$$\mathbf{L} = \sqrt{L\,(L + 1)}\ \frac{h}{2\pi},\quad L = 0, 1, 2, 3............. \qquad ...(3.277)$$

where L stands for the total orbital angular momentum quantum number. This clearly suggests that the value of **L** can be only 0, $\sqrt{2}\ \frac{h}{2\pi}$, $\sqrt{6}\ \frac{h}{2\pi}$, $\sqrt{12}\ \frac{h}{2\pi}$ ...

(*ii*) As mentioned earlier, each angular momentum is closely related to the respective angular momentum quantum number. Thus, the total orbital angular momentum quantum number can be obtained from l_1 and l_2 as:

$$L = l_1 + l_2,\ l_1 + l_2 - 1,\ l_1 + l_2 - 2, ...|l_1 - l_2| \qquad ...(3.278)$$

where modulus |.....| indicates that only positive value of L is possible and the maximum possible value of L will be either $2l_1 + 1$ or $2l_2 + 1$, which ever is smaller. This method is the simplest and is most useful when both the electrons are present in orbitals with different *n* and *l* value, also known as *non-equivalent electrons*.

(*iii*) In this method, different values of the component of orbital angular momentum are summed, by taking various combinations. From these sets of data, various allowed values of L are picked up. This method is very useful for electrons present in the orbitals with same n and l values, known as *equivalent electrons*.

Resultant Spin Angular Momentum: Resultant spin angular momentum is also calculated by using three methods described for the angular momentum. On the same lines, **S** stands for spin angular momentum and S for spin angular momentum quantum number. Briefly these methods can be mentioned as:

(*i*) Vectorial addition of spin angular momentum of each electron can lead to:

$$\mathbf{S} = \sqrt{S(S+1)}\,\frac{h}{2\pi} \qquad \text{...(3.279)}$$

Since each electron can have $s = \frac{1}{2}$, the S values will be integer or zero for even electrons or half integers or $\frac{1}{2}$ for odd electrons.

(*ii*) Addition of individual spin angular momentum quantum numbers. For N-electrons:

$$S = \Sigma s_i,\ \Sigma s_i - 1,\ \Sigma s_i - 2 \ldots\ldots\ldots\ldots \qquad \text{...(3.279}a\text{)}$$

$$= N,\ N-1,\ N-2 \ldots\ldots\ldots\ldots 0, \qquad \text{where N is even.}$$

$$= \frac{N}{2},\ \frac{N}{2} - 1,\ \frac{N}{2} - 2 \ldots\ldots \qquad \text{where N is odd.}$$

This is similar to that obtained to calculate orbital angular momentum quantum number. This method is the simplest.

(*iii*) Addition of s_z of each electron gives rise to S_z.

Total Angular Momentum: The procedures to calculate the total angular momentum (J) from L and S are exactly the same as mentioned earlier for one electron. A notable point to be mentioned is that quantum number J in the equation (3.280)

$$\mathbf{J} = \sqrt{J(J+1)}\,\frac{h}{2\pi} \qquad \text{...(3.280)}$$

should be integral if S is integral and half integral if S is the half integral. Three quantum numbers J, L and S are related to each other as:

$$J = L + S,\ L + S - 1,\ L + S - 2 \ldots\ldots\ldots\ldots\ldots\ldots |L - S| \qquad \text{...(3.281)}$$

where the smallest value of J will be the positive value of L – S.

For example if L = 1, S = 1, the maximum and minimum values of J = 2 and 0 (1 – 1) respectively. Thus possible values of J are 2, 1, 0. Similarly for L = 2 and S = $\frac{3}{2}$, the maximum value of J can be $2 + \frac{3}{2} = \frac{7}{2}$ and the minimum value of j will be $2 - \frac{3}{2} = \frac{1}{2}$. Thus, possible values of J can be $\frac{7}{2}, \frac{5}{2}\left(\frac{7}{2}-1\right), \frac{3}{2}\left(\frac{7}{2}-2\right), \frac{1}{2}$. It can be seen from equation (3.281) that when L > S, the number of possible J values for a given value of L is 2S + 1. On the other hand if L < S, the number of possible J values for a given value of L is 2L + 1. In particular, for S terms (L = 0, 2L + 1 = 1), the examples given above belong to the first category and the number of J values obtained satisfies the 2S + 1 rule (*i.e.*, S = 1, 2S + 1 = 3, three J values for ^{3}P system and four J values for S = $\frac{3}{2}$ and L = 2). Now consider ^{4}P system. Here S = $\frac{1}{2}(4-1) = \frac{3}{2}$ and L = 1, also L < S. In this case number of J values will be 2 × 1 + 1 = 3(2L + 1). This can also be obtained from equation (3.281). Maximum value of J = $1 + \frac{3}{2} = \frac{5}{2}$ and minimum value $|L - S| = \left|1 - \frac{3}{2}\right| = \frac{1}{2}$. Thus, the third value is $1 + \frac{3}{2} - 1 = \frac{3}{2}$.

Example 3.21: The electronic configuration of some state of an atom is shown as p^1d^1, *i.e.*, one electron is present in *p* orbital and one in *d* orbital. Calculate the resultant orbital angular momentum.

Solution: Using method (*i*), for p-electron, $l_1 = 1$ and d-electron $l_2 = 2$. This will give us $\mathbf{l_1} = \sqrt{2}\,\frac{h}{2\pi}$ and $\mathbf{l_2} = \sqrt{6}\,\frac{h}{2\pi}$. As shown in Fig. 3.26, there can be only three and three possible ways by which we sum up these orbital angular momentums to get the values of L conforming to equation (3.275). These are $\sqrt{2}\,\frac{h}{2\pi}$ $\sqrt{6}\,\frac{h}{2\pi}$ and $\sqrt{12}\,\frac{h}{2\pi}$ corresponding to the total orbital angular momentum quantum number L = 1, 2 and 3 respectively. Using method (*ii*) and equation (3.275), the maximum value of L = 1 + 2 = 3 and the minimum value of L = |1 – 2| = 1. Thus,

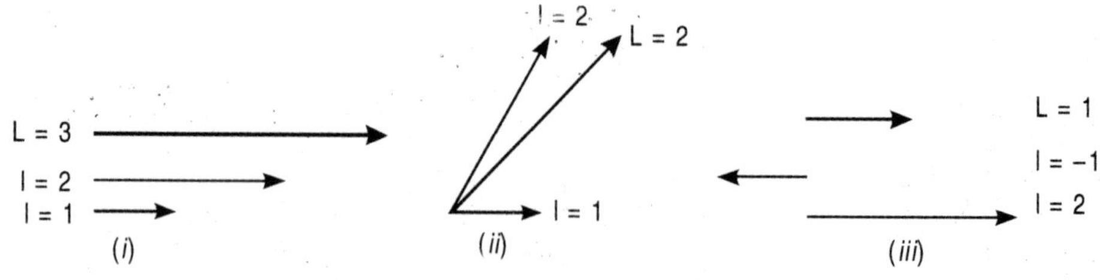

Figure 3.26: *Different possible vector combinations of orbital angular momentum of two electrons,* $l_1 = 1,\ l_2 = 2$

the three possible values of L are 3, 2, 1 and these corresponds to $\mathbf{L} = \sqrt{12}\,\frac{h}{2\pi}, \sqrt{6}\,\frac{h}{2\pi}, \sqrt{2}\,\frac{h}{2\pi}$, same as obtained above.

3.14 STATE SYMBOL OR TERM SYMBOLS

In the previous sections so far we have given importance to calculate (*i*) the total orbital angular momentum (**L**), (*ii*) total spin angular momentum (**S**) and (*iii*) total angular momentum (**J**) using equations (3.275) by adding the respective angular momentum of each electron except the quantity **J**, which has been determined by adding **L** and **S** vectorially for multi-electron atoms (Further the contributions to these angular momenta are only from the orbitals which are partially filled or valence electrons). Methods used to calculate **J** was based on Russell-Sanders coupling (*i.e.*, $Z < 40$) as most of the times a chemist is involved with the chemistry of these elements. We have also learnt that these states arising from different **J** values differ in energy unless these are degenerate and thus can lead to affect the spectra of an atom. Before we learn the effect of these quantities on the atomic spectra, it will be nice if we learn to depict the states by some symbols, known as *spectroscopic state* or *state symbols*, also know as *term symbols*. It is depicted as:

$$^{2S+1}L_J$$

where the left hand side superscript stands for spin multiplicity, right had subscript stands for total angular momentum and L stands for total orbital angular momentum. By analogy with *s*, *p*, *d*, *f*....., symbols which are used to describe the orbital angular momentum (l = 0, 1, 2, 3......), letters, as described below, are also used to describe the values of L.

for	L = 0	1	2	3	4	5........
symbol	S	P	D	F	G	H........

Although we have defined the term symbol by three angular momentum quantum numbers, it may be mentioned here that the L and S parts of the term symbol are enough to find the complete term, symbol, because the value of J can be easily obtained from equation (3.281).

In general, as said earlier, spin multiplicity (2S + 1) is associated with number of possible J values or number of components only when L > S. This is not true when L < S. For example, we have learnt earlier that all the term symbols belonging to S has only one component or one J value even though 2S + 1 value, say 3, *i.e.*, ^{3}S. Thus it may be made clear here that in reality the 2S + 1 does not reflect the number of components or J values, but the magnitude of its additional spin angular momentum vectors. For example, consider terms like ^{3}S, ^{3}P, ^{3}D....... In all these cases, the total spin angular momentum S = 1. This value of S gives three components or three values of J for ^{3}P and ^{3}D, but only one component for ^{3}S term as L = 0. In spite of this fact, the term ^{3}S behaves like ^{3}P, ^{3}D..... terms since the value of S for all terms is equal to one.

General method used to calculate the spectroscopic states or terms symbols is as follows:

(*a*) Write all possible combinations of different wave functions (orbital angular momentum and spin angular momentum) which are allowed by Pauli Exclusion Principle.

(*b*) Calculate the values of M_L and M_S for each combination, with the help of equations (3.279*a*).

(*c*) Calculate the term symbols for each combinations having largest values of M_L and M_S, *i.e.*, $S = M_S$ and $L = M_L$.

(*d*) To this highest value term assign all the other sub-states, which correspond to the rest of the *z*-component of these respective angular momenta.

(*e*) Repeat the step (*c*) and (*d*) for the next highest M_L and M_S combinations in such a manner that all the combinations obtained in (*a*) are taken care, without being getting repeated.

Although the above rules are applicable to all systems, they can be further simplified by dividing the electrons into two classes.

Non-equivalent Electrons: As mentioned earlier, non-equivalent electrons are those electrons which belong to different orbitals, *i.e.,* at least with different *n* or *l*. For example electrons occupying *p, d*.... have different either *n* or *l* values. In this case Pauli Exclusion Principle is always satisfied and all the possible combinations are allowed.

If there are more than two electrons, the term symbols can be obtained as follows: (*a*) Calculate the M_L and M_S values by combining two electrons. (*b*), Then combine each M_L and M_S values of the third electron and so on. This procedure can be understood by studying the example 3.22.

Example 3.22: Calculate the spectroscopic term symbols for the electronic configuration $2p^1 3p^1 3d^1$ system.

Solution: In this case the values of either *n* or *l* are different for each electron and thus falls into the category of non-equivalent electrons. Further, since *n* or *l* values are different for each electron, Pauli Exclusion Principle will be obeyed and thus each combination will be allowed.

In the above system, first take $2p^1 3p^1$. Here $l_1 = 1$ and $l_2 = 2$. Thus:

$$L = 1 + 1 = 2$$
$$= 1 + 1 - 1 = 1$$
$$= |1 - 1| = 0$$

$$S = \frac{1}{2} + \frac{1}{2} = 1, \qquad 2S + 1 = 2 \times 1 + 1 = 3$$

$$= \frac{1}{2} - \frac{1}{2} = 0, \qquad = 2 \times 0 + 1 = 1$$

The term symbols due to coupling of $2p^1 3p^1$ electrons can be ^3D, ^3P, ^3S and ^1D, ^1P, ^1S. Now each of six states can couple with $3d^1$ electron to give the resultant L and S. For example, $^3D3d^1$ gives

$$L = 2 + 2 = 4$$
$$= 2 + 2 - 1 = 3$$
$$= 2 + 2 - 2 = 2$$
$$= 2 + 2 - 3 = 1$$
$$= |2 - 2| = 0$$

$$S = 1 + \frac{1}{2} = \frac{3}{2}, \qquad 2S + 1 = \frac{3}{2} \times 2 + 1 = 4$$

$$= \left|1 - \frac{1}{2}\right| = \frac{1}{2}, \qquad = \frac{1}{2} \times 2 + 1 = 2$$

The term symbols for the coupling of $^3D3d^1$ configuration are:

^4G, ^4F, ^4D, ^4P, ^4S and ^2G, ^2F, ^2D, ^2P, ^2S.

For $^3P3d^1$, the term symbols are: ^4F, ^4D, ^4P and ^2F, ^2D, ^2P

For $^3S3d^1$, the term symbols are: ^4D, and ^2D

Similarly for combinations:

$^1D3d^1$, the terms are: ^2G, ^2F, ^2D, ^2P, ^2S

$^1P3d^1$ the term are; ^2F, ^2D, ^2P

and $^1P3d^1$, the term is: ^2D

It can be seen that some of the terms symbols occur more than once in different combinations and they can be combined to give:

^4G, ^4F (2), ^4D (3), ^4P (2), ^4S and ^2G (2), ^2F (4), ^2D (6), ^2P (4), ^2S (2)

The number in the parenthesis tells how many times the particular state is formed by different combination of L and S.

Now once L and S values have been obtained, we can calculate the J values. Take one of the state say ^4G.

Here L = 4, S = $\frac{3}{2}$, the maximum value of J = $4 + \frac{3}{2} = \frac{11}{2}$ and the minimum value of J is $4 - \frac{3}{2} = \frac{5}{2}$. Thus J values can be $\frac{11}{2}, \frac{9}{2}, \frac{7}{2}, \frac{5}{2}$ and term symbols are:

$^4G_{11/2}$, $^4G_{9/2}$, $^4G_{7/2}$ and $^4G_{5/2}$.

Similarly for ^4F, the term symbols are: $^4F_{9/2}$, $^4F_{7/2}$, $^4F_{5/2}$ and $^4F_{3/2}$

For ^4D, these are; $^4D_{7/2}$, $^4D_{5/2}$, $^4D_{3/2}$ and $^4D_{1/2}$

For ^4P, the maximum value of J = $1 + \frac{3}{2} = \frac{5}{2}$ and the minimum value is $\left|1 - \frac{3}{2}\right| = 1/2$.

Thus the term symbols are; $^4P_{5/2}$, $^4P_{3/2}$, $^4P_{1/2}$ and for ^4S, it is only $^4S_{3/2}$.

On the other hand for;

^2G; $^2G_{9/2}$, $^2G_{7/2}$

^2F; $^2F_{7/2}$, $^2F_{5/2}$

^2D; $^2D_{5/2}$, $^2D_{3/2}$

^2P; $^2P_{3/2}$, $^2P_{1/2}$

^2S; $^2S_{1/2}$

Example 3.23: The term symbol for the state is: (*a*) $^3G_5$, (*b*) $^2D_{5/2}$. Calculate the values of orbital, spin and total angular momentum for each state.

Solution: For $^3G_5$, 2S + 1= 3 gives S = 1

G stands for L = 4 and J = 5. So the total orbital angular momentum, total spin angular momentum and total angular momentum are 4, 1 and 5 respectively.

(*b*) Similarly for $^2D_{5/2}$, L = 2, J = $\frac{5}{2}$ and 2S + 1 = 2 gives S = 1/2.

Equivalent Electrons: Equivalent electrons are those electrons, which belong to the same orbital, *i.e.*, same n and l values, for example, np^2, nd^3, $3d^5$, etc., Due to this electronic configuration, some of the states, which are obtained when the electrons are non-equivalent, vanish because of Pauli Exclusion Principle. The procedure is exactly same as described above to calculate the term symbols for non-equivalent electrons. The total number of microstates (TM) obtained is given by the expression (3.282).

$$TM = \frac{(4l + 1)!}{(4l + 2 - N)!\,(N!)} \qquad ...(3.282)$$

where N is the number of electrons and l is the azimuthal quantum number. Three methods will be used to calculate the term symbols by giving an example.

Method I: This method will be discussed by taking an electronic configuration as $2p^13p^1$ and $2p^2$, *i.e.*, non-equivalent electrons and equivalent electrons.

Example 3.24: Derive the term symbols for the electronic configurations: (*a*) $2p^13p^1$ and (*b*) $2p^2$.

Solution:

(*a*) In case of $2p^13p^1$, the Pauli Exclusion Principle will be always observed for any combination of l and s because one of the quantum numbers (n) is different. Thus, the possible values of L = 1 + 1 = 2, L = 1 + 1 – 1 = 1 and L = 1 – 1 = 0.

$$S = \frac{1}{2} + \frac{1}{2} = 1,\ S = \frac{1}{2} - \frac{1}{2} = 0$$

So the states without J values can be: ^3D, ^3P, ^3S, and ^1D, ^1P, ^1S.

and with J values are: $^3D_3$, $^3D_2$, $^3D_1$; $^3P_2$, $^3P_1$, $^3P_0$; $^3S_1$ and $^1D_2$, $^1P_1$, $^1S_0$

(*b*) In this case $n_1 = n_2$, $l_1 = l_2 = 1$. Since n and l values of both the electrons are same, so they are called equivalent electrons. One has to be careful while calculating the total orbital angular, total spin angular and total angular momentum as some restrictions will be imposed while coupling them.

(*i*) In this $n_1 = n_2$, $l_1 = l_2$. Now if m_l (1) = m_l (2), then Pauli Exclusion Principle requires that m_s (1) ≠ m_s (2), *i.e.*, one of the two quantum numbers m_l or m_s will have to be different.

(*ii*) Indistinguishability of electrons: we have learnt earlier that the electrons are indistinguishable and if the energies of two electrons are exchanged, there is no experimental procedure by which we can make sure that such an exchange has taken place. In other words if all the values of four quantum numbers (n, l, m_l, m_s) are exchanged for two electrons, we cannot distinguish the final arrangement from the initial arrangement. Our main aim is to make distinction of different states while calculating the total angular momentum and then term symbol. We must try to avoid those states, which have been obtained by merely exchanging all the four quantum numbers n, l, m_l, and m_s.

Making use of these restrictions, let us try to combine the various angular momenta to calculate the resultant one.

(*a*) Knowing that $n_1 = n_2$ and $l_1 = l_2$, if we assume that $m_l\,(1) = m_l\,(2)$, then the values of m_s can be either $m_s\,(1) = \frac{1}{2}$, $m_s\,(2) = -\frac{1}{2}$ or $m_s\,(1) = -\frac{1}{2}$, $m_s\,(2) = \frac{1}{2}$. We know that these two situations have arisen because we cannot distinguish the two electrons. Thus the two arrangements are similar and we can choose only one.

(*b*) Now suppose $m_s\,(1) = m_s\,(2)$. Knowing that $l = 1$ can have three values of m_l, *i.e.*, 1, 0, –1 for each electron. The following combinations can be made.

$$m_s\,(1) = m_s\,(2);\ m_l\,(1) = 1,\ m_l\,(2) = 0 \text{ or } -1$$
$$m_l\,(1) = 0,\ m_l\,(2) = 1 \text{ or } -1$$
$$m_l\,(1) = -1,\ m_l\,(2) = 1 \text{ or } 0$$

Each combination, *i.e.*, (1, 0) or (0, 1); (0, –1) or (–1, 0) and (1, –1) or (–1, 1) are similar to each other and thus we will choose only three out of six, *i.e.*, $m_l\,(1) = 1$, $m_l\,(2) = 0$; $m_l\,(1) = 1$, $m_l\,(2) = -1$; $m_l\,(1) = 0$, $m_l\,(2) = -1$.

(*c*) Lastly for $m_s\,(1) \neq m_s\,(2)$ and $m_l\,(1) \neq m_l\,(2)$, the interchange of positions can lead to one pair (e.g. $m_s\,(1) = \frac{1}{2}$, $m_s\,(2) = -\frac{1}{2} \rightarrow m_s\,(1) = -\frac{1}{2}$, $m_s\,(2) = \frac{1}{2}$), which can be distinguished from each other. Here the electron is already having different values of m_l and thus the distinction, can be made by changing the spin value. As said earlier, the indistinguishable state can be obtained if all the four numbers n, l, m_l and m_s are exchanged.

Keeping the above restrictions in mind, different microstates which can be obtained by fifteen combinations are compiled in Table 3.10 by using steps *a* and *b* of the method described above. Column 1 gives the serial number. Columns 2 to 5 give the different values of $m_l\,(1)$, $m_l\,(2)$, $m_s\,(1)$ and $m_s\,(2)$ respectively.

Columns six and seven give the total M_L and M_S values, obtained by addition of respective values of these angular momenta. Using step (*c*) of the rules, start with the largest value of $M_L = 2$ and $M_S = 0$, the term symbol is ^1D. Using step (*e*), the microstates associated with ^1D are (2. 0), (1, 0), (0, 0), (–1, 0) and (–2, 0), *i.e.*, sr. no.s 1, 3, 7, 12 and 15 respectively. This combination of L = 2 and S = 0 can have only one value of J, *i.e.*, L + S = |L – S| = 2 + 0 = 2. So the term symbol is $^1D_2$.

Now to assign the next state, choose the next largest value of M_L = +1 and M_S = + 1 and will give ^3P term symbol. Since M_S = S = 1, there can be three components along the *z*-axis, *i.e.*, +1, 0, –1 and these can combine with each value of M_L (1, 0, –1), the possible states are:

^3P state, $M_S = +1$, $M_L = 1, 0, -1$ (2, 6, 11)

$M_S = 0$, $M_L = 1, 0, -1$ (4, 8, 13)

$M_S = -1$ $M_L = 1, 0, -1$ (5, 10, 14)

The possible values of J can be 2, 1, 0 and term symbols are $^3P_2$, $^3P_1$, $^3P_0$. The only microstate left is (0, 0), *i.e.*, (9) and is due to $M_L = 0$, $M_S = 0$, $^1S_0$.

The above example has shown that the spectroscopic states or term symbols arising from the equivalent electrons are much less than that arising from the same number of non-equivalent electrons. This is because in the latter case $n_1 \neq n_2$, thus Pauli Exclusion Principle and indistinguishability principle do not impose any restriction on the values of m_l and m_s.

Method II: The term symbols for the equivalent electrons if they are two in number, *i.e.*, np^2, nd^2 etc., can be obtained as follows.

(*a*) Make a table of m_l values of one of the electrons as row and the m_l values of the second

Table 3.10: *Microstates of two equivalent p electrons* ($n_1 = n_2 = 2$, $l_1 = l_2 = 1$)

Sr. No.	m_l (1)	m_l (2)	m_s (1)	m_s (2)	M_L [m_l (1) + m_l (2)]	M_S [m_s (1) + m_s (2)]
1	+1	+1	$+\frac{1}{2}$	$-\frac{1}{2}$	+2	0
2	+1	0	$+\frac{1}{2}$	$+\frac{1}{2}$	+1	+1
3	+1	0	$+\frac{1}{2}$	$-\frac{1}{2}$	+1	0
4	+1	0	$-\frac{1}{2}$	$+\frac{1}{2}$	+1	0
5	+1	0	$-\frac{1}{2}$	$-\frac{1}{2}$	+1	– 1
6	+1	– 1	$+\frac{1}{2}$	$+\frac{1}{2}$	0	+1
7	+1	– 1	$+\frac{1}{2}$	$-\frac{1}{2}$	0	0
8	+1	– 1	$-\frac{1}{2}$	$+\frac{1}{2}$	0	0

9	+1	− 1	$-\frac{1}{2}$	$-\frac{1}{2}$	0	− 1
10	0	0	$+\frac{1}{2}$	$-\frac{1}{2}$	0	0
11	0	− 1	$+\frac{1}{2}$	$+\frac{1}{2}$	− 1	+ 1
12	0	− 1	$+\frac{1}{2}$	$-\frac{1}{2}$	− 1	0
13	0	− 1	$-\frac{1}{2}$	$+\frac{1}{2}$	− 1	0
14	0	− 1	$-\frac{1}{2}$	$-\frac{1}{2}$	− 1	−1
15	− 1	− 1	$+\frac{1}{2}$	$-\frac{1}{2}$	− 2	0

electron as columns as shown in example 3.25.

(*b*) Calculate each element of the square table by the addition of two m_l values of the elements.

(*c*) Elements lying on the diagonal of the table will be singlet and occurring off diagonal as triplet.

(*d*) Make use only the M_L values from one side of the diagonal as those on the other side are merely correspond to a mirror image.

Example 3.25: Derive the term symbols for the electronic configuration np^2 using method II and compare with those obtained in example 3.24*a*.

Solution: m_l values for each electron in np^2 are 1. 0. –1 and table can be constructed as:

$m_l(1)$ / $m_l(2)$	+1	0	−1
−1	0	−1	−2
0	+1	0	−1
+1	+2	+1	0

The absolute values of the diagonal elements are 2 and 0. Thus states D and S are singlet (rule *c*), *i.e.*, ^1D and ^1S. If you draw the dotted line as shown in the table, the elements enclosed in it, *i.e.*, M_L = +2, +1, 0, –1, –2 and 0 form the microstates for L = 2 and L = 0.

The off-diagonal element L = 1, will be triplet state, *i.e.*, ^3P and the elements $M_L = -1, 0 -1$ form the microstates. The total angular momentum quantum number (J) can be obtained as done in example 3.24. Thus, the microstates are: $^1D_2$, $^1S_0$, $^3P_2$, $^3P_1$, $^3P_0$ and are similar to ones obtained in example 3.24*a*.

Method III: The above method (II) is very useful if the number of electrons is two. If the number of electrons is more than two, the following procedure is very helpful. This involves the following steps:

(*a*) Calculate the M_S values for the possible arrangements obtained by grouping the electrons, taking care of Pauli Exclusion Principle. The possible arrangements for each M_S value and the total number of possible arrangements for unpaired electrons from 0 to 7 (as deduced from the empirical formula 2^n, where n is the number of unpaired electrons) are given in Table 3.11.

Table 3.11: *Possible M_S values for grouping of arrangements with 0 to 7 unpaired electrons*

No. unpaired electrons					M_S values				Total no. of arrangements
0						0			1
1					$\frac{1}{2}$	$-\frac{1}{2}$			2
2					1	0 (2)	−1		4
3				$+\frac{3}{2}$	$+\frac{1}{2}$ (3)	$-\frac{1}{2}$ (3)	$-\frac{3}{2}$		8
4			+2	+1 (4)	0 (6)	−1 (4)	−2		16
5		$+\frac{5}{2}$	$+\frac{3}{2}$ (5)	$+\frac{1}{2}$ (10)	$-\frac{1}{2}$ (10)	$-\frac{3}{2}$ (5)	$-\frac{5}{2}$		32
6	+3	+2 (6)	+1 (15)	0 (20)	−1 (15)	−2 (6)	− 3		64
7	$+\frac{7}{2}$	$+\frac{5}{2}$ (7)	$+\frac{3}{2}$ (21)	$+\frac{1}{2}$ (35)	$-\frac{1}{2}$ (35)	$-\frac{3}{2}$ (21)	$-\frac{5}{2}$ (7)	$\frac{7}{2}$	128

Note: Number in the parenthesis gives the possible arrangements by which the possible M_S values can be obtained. Following rules are used to construct the above table.

(*i*) Total number of arrangements = 2^n, where n is the number of unpaired electrons.

(*ii*) Number of M_s values are one more than the number of electrons.

(*iii*) Maximum and minimum values of M_s is $\pm\frac{1}{2}n$.

(*iv*) Allowed M_s values are $\Delta M_s = \pm 1$.

(*iv*) Weight of each ΔM_s value is determined from the coefficients of the bionomial expression.

$$(1 + x)^n = 1 + nx + \frac{n(n+1)}{2!} x^2 + \frac{n(n+1)\ (n+2)}{3!} + \ldots$$

(*b*) For each value of M_S, construct a table to calculate the M_L values for different possible arrangements of electrons with Pauli Exclusion Principle in mind.

(*c*) Finally construct the major array that contains all the arrangements. This can be done, by realizing that each M_L value that occurs in arrangements with one unpaired electron has two possible M_S values. Whereas each M_L value that occurs in arrangements with three unpaired electrons has eight M_S values and so on. Thus, the tolal number of arrangements (*i.e.,* M_s values) for each M_L value that occurs in arrangements with *n* unpaired electrons can be deduced from the empirical formula 2^n. It should be kept in mind that total number of microstates in the major array should be the same as that obtained by equation (3.282).

(*d*) The term symbols and the number of times the particular state is occurring can be obtained by solving this major array table as shown in example 3.24.

Before we move to solve an example, discussion on this topic is not complete before we take care of the following points:

(1) ***Closed Shell Electronic Configuration:*** In this configuration all the orbitals for a given value of *l* are completely full, for example, ns^2, np^6, nd^{10},... Considering ns^2 configuration, L = 0 + 0 = 0 and S = $\frac{1}{2} - \frac{1}{2}$ = 0. Thus the term symbol is $^1S_0$ as J = 0 + 0 = 0. In case of np^6, the three *np* orbitals, the quantum numbers are (n, 1, 1, $\pm \frac{1}{2}$), (n, 1, 0, $\pm \frac{1}{2}$) and (n, 1, –1, $\pm \frac{1}{2}$). This will lead to M_S = S = 0, M_L = L = 0 and J = 0 + 0 = 0. The term symbol for np^6 will also be $^1S_0$, similar to ns^2. In other words, the completely filled sub-shells will always give L = S = 0 and thus we can ignore the electrons in the completely filled sub-shells when considering the electronic configurations of other electrons. For example in case of nitrogen atom, the electrons present in $1s^2 2s^2$ can be neglected and only the electrons present in *p* orbitals are considered.

(2) Without going into details, it may be mentioned here that atoms having electronic configurations as np^x and np^{6-x}; nd^x and nd^{10-x} and so on will have the similar term symbols. For example the term symbols for carbon ($2p^2$) and oxygen ($2p^4$) are similar. The only difference comes while arranging the components of total angular momentum or J values of the terms (Hunds rules).

(3) It is noteworthy that the term symbol gives information only about the total orbital angular momentum, total spin angular momentum and total angular momentum of the whole atom. It tells us nothing about the states of electrons, as well as, about the number of electrons contributing to these momenta. For example, consider a term symbol ^2P. The left hand side superscript only tells us that S = $\frac{1}{2}$ can be obtained from the odd number of electrons and one of them is unpaired. The actual arrangements of electrons could be

(↑), (↑↑↓), (↑↓↑↓↑) etc. and all of them give rise to $S = \frac{1}{2}$. Similarly the value of L = 1 can be obtained if there is one electron in *p* orbital or one electron in *p* orbital and two electrons in *s* orbital and so on many other arrangements of electrons in an atom.

From spectroscopic point of view it is not that important to know the above points. Only thing, which is really important is the energy of the state of the atom and thus can calculate the transition frequency. If one is really interested to know the states of electron, one can do so by including the

Table 3.12: *Term symbols of non-equivalent electrons for different n*

Electronic configuration	Terms
s s	^1S, ^3S
s p	^1P, ^3P
s d	^1D, ^3D
p p	^1S, ^1P, ^1D, ^3S, ^3P, ^3D
p d	^1P, ^1D, ^1F, ^3P, ^3D, ^3F
d d	^1S, ^1P, ^1D, ^1F, ^1G, ^3S, ^3P, ^3D, ^3F, ^3G
s s s	^2S (2), ^4S
s s p	^2P (2), ^4P
s p p	^2S (2), ^2P (2), ^2D (2), ^4S, ^4P, ^4D
s p d	^2P (2), ^2D (2), ^2F (2), ^4P, ^4D, ^4F
p p p	^2S (2), ^2P (6), ^2D (4), ^2F (2), ^4S, ^4P (3), ^4D (2), ^4F
p p d	^2S (2), ^2P (4), ^2D (6), ^2F (4), ^2G (2), ^4S , ^4P (2), ^4D (3), ^4F (2), ^4G
p d f	^2S (2), ^2P (4), ^2D (6), ^2F (6), ^2G (6), ^2H (4), ^2I (2), ^4S, ^4P (2), ^4D (3), ^4F (3), ^4G (3), ^4H (2), ^4I

Table 3.13: *Term symbols for equivalent electrons (n and l are the same)*

Electronic configuration	Terms
s^2	^1S
p^2 or p^4	^1S, ^1D, ^3P
p^3	^2P, ^2D, ^4S
p^1 or p^5	^2P
p^6	^1S
d^2	^1S, ^1D, ^1G, ^3P, ^3F
d^3	^2P, ^2D (2), ^2F, ^2G, ^2H, ^4P, ^4F
d^4	^1S (2), ^1D (2), ^1F, ^1G (2), ^1I, ^3P (2), ^3D, ^3F (2), ^3G, ^3H, ^5D
d^5	^2S, ^2P, ^2F (2), ^2G (2), ^2H, ^2I, ^4P, ^4D, ^4F, ^4G, ^6S

term symbol. For example, ^2P state symbol can be obtained from the electron configuration $1s^2 2p^1$ or $1s^1 2p^1 3s^1$ etc.,

Few examples have been solved for different electronic configurations to get the term symbols. But it may be mentioned here that many other electronic configurations do occur for the ground, as well as, other states of atoms in which two or more than two electrons contribute to the total energy. Similar procedure can be applied to calculate the term symbols. Tables 3.12 and 3.13 give the term symbols for the non-equivalent and equivalent electrons present in the electronic configuration of atoms.

Example 3.26: Derive the term symbols for the electronic configuration of p^3.

Solution: There are three electrons and the possible M_S values, according to step 1, can be $\frac{3}{2}$ and $\frac{1}{2}$.

The total M_L values can be calculated for each M_S value as in Table 3.11. a (step *b*).

m_l	+1	0	−1	M_L $(\Sigma m_l(i))$	m_l	+1	0	−1	M_L $(\Sigma m_l(i))$
	X	/		2			X	/	−1
	X		/	1		/		X	−1
	/	X		1			/	X	−2

Cross (X) means, two electrons with opposite spins and dash (/) means only one electron in any particular orbital.

There will be only one group for $M_S = \frac{3}{2}$, because there are three *p* orbitals with $m_l = 1$, 0, −1 and total $M_L = 0$.

Now the major array can be constructed as (step *c*), *i.e.*, by adding up all the arrangements for each M_L and M_S values from the two sub-tables. For example, the M_L values of 2, 1, 0, −1, −2 for $M_S = \frac{1}{2}$ are obtained by 1, 2, 2, 1 ways respectively, whereas for $M_S = \frac{3}{2}$, M_L value is zero, but M_S values of $\frac{3}{2}, \frac{1}{2}, -\frac{1}{2}, -\frac{3}{2}$ can be obtained by 1, 3, 3, 1 ways (Table 3.12).

M_L	M_S			
	$\frac{3}{2}$	$\frac{1}{2}$	$-\frac{1}{2}$	$-\frac{3}{2}$
2		1	1	
1		2	2	
0	1	3	3	1
−1		2	2	

For calculation of term symbols, remove each term one by one from the major array as shown below.

M_L	M_S			
	3/2	1/2	−1/2	−3/2
2		1	1	
1		2	2	
0	1	3	3	1
−1		2	2	
−2		1	1	

$\xrightarrow{^2D}$ L = 2, S = $\frac{1}{2}$, sub-array removed

M_L	M_S			
	3/2	1/2	−1/2	−3/2
1		1	1	
0	1	2	2	1
−1		1	1	

$\xrightarrow{^2P}$ L = 1, S = ½, sub-array removed

M_L	M_S			
	3/2	1/2	−1/2	−3/2
0	1	1	1	1

$\xrightarrow{^4S}$ L = 0, S = 3/2, sub-array removed

The possible values of J for each combination are: $^4S_{3/2}$, $^2D_{5/2}$, $^2D_{3/2}$, $^2P_{3/2}$, $^2P_{1/2}$.

Hund's rules: In the last few sections we have learnt that different angular momenta interact with each other and these interactions lead to states with different energies. Larger the interaction, larger is the energy difference. Hund's rules are used to arrange the different spectroscopic terms or states according to increasing energy levels. These are:

(*a*) The state, which has the highest spin multiplicity, has the lowest energy.

(*b*) If the two states have the same spin multiplicity, the state with greater L value will be lower in energy than that with lower L value.

(*c*) If the orbital is less than half filled, the state with smallest J value will be lower in energy and the energies of states increases with increase in J values. Whereas if the orbital is more than half filled, the state with largest J value will be lower in energy and energies of states increases with decrease in J values.

Example 3.27: Using Hund's rules arrange the term symbols obtained from np^2 electronic configuration in order of increasing energy.

Solution: The term symbols obtained from np^2 electronic configuration are: ^1D, ^1S and ^3P. Using Hund's rules (*a*), ^3P state will be of lowest energy as it has maximum spin multiplicity. Further since *p* orbital are less than half filled, the terms including J will be $^3P_0$, $^3P_1$, $^3P_2$ (rule *c*).

Now both ^1D and ^1S are singlet states, state ^1D will be of lower energy than ^1S (rule 2). Depending on the interactions, the different terms can be arranged in increasing order of energy as shown in Fig. 3.27. It is clear from Fig. 3.27 that the total number of microstates is 15, agreeing with those calculated using equation (3.282).

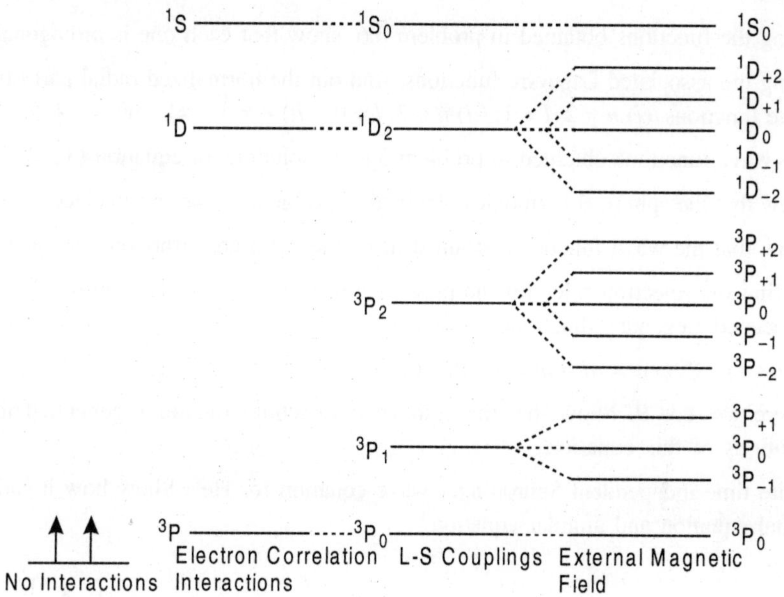

Figure 3.27: *Term splitting in the np^2 system, indicating the effect of various kinds of interactions*

PROBLEMS

3.1 Consider that an electron is moving in a plane under the influence of circular central force field. (*a*) Write the Schrödinger wave equation. (*b*) Write the Laplacian operator in Cartesian coordinates. (*c*) Convert this Laplacian operator to the plane polar coordinates. (*d*) Using the method of separation of variables, write two equations each depending on one variable. (*e*) Solve angular as well as the radial equations.

3.2 Legendre polynomials are the solutions to equation (3.27) when $m = 0$. Show that the first few Legendre polynomials as given in Table 3.2 are the solutions of equation (3.27). (Few Legendre polynomials are: $P_0(x) = 1$; $P_1(x) = x$, $P_2(x) = \frac{1}{2}(3x^2 - 1)$; $P_3(x) = \frac{1}{2}(5x^3 - 3x)$; $P_5(x) = \frac{1}{8}(35x^4 - 30x^2 + 3)$.

3.3 Show that the following combinations of Associated Legendre polynomial are orthogonal. (*i*) P_2^0 and P_3^0, (*ii*) P_3^1 and P_3^2, (*iii*) P_3^1 and P_3^0.

3.4 Find the normalization constants of the associated Legendre polynomial functions as: (*i*) P_3^2, (*ii*) P_3^1 and (*iii*) P_4^2.

3.5 Show that the associated Legendre polynomial functions, obtained in problem 3.3 are normalized.

3.6 Using associated Legendre polynomial function, find out the normalized spherical harmonic functions for (*i*) $l = 2, m_l = 1$; (*ii*) $l = 4, m_l = -3$; (*iii*) $l = 4, m_l = -2$; (*iv*) $l = 3, m_l = 0$; (*v*) $l = 3, m_l = 2$.

3.7 Using the functions obtained in problem 3.6, show that each one is orthogonal to each other.

3.8 Using the associated Laguerre functions, find out the normalized radial part of the hydrogen like wave functions. (*i*) $n = 2, l = 1$; (*ii*) $n = 3, l = 0$, (*iii*) $n = 3, l = 1$, (*iv*) $n = 4, l = 2$; (*v*) $n = 4, l = 3$.

3.9 The wave functions obtained in problem 3.8 are solutions of equation (3.35).

3.10 Show that the spherical harmonics obtained in problem 3.6 are normalized.

3.11 Show that the wave functions obtained in problem 3.8 are orthogonal to each other.

3.12 To find the selection rules for the pure rotational transitions, the following recursion formula of associated Legendre functions is used.

$$(2l + 1)\, x\, P_l^m(x) = (l - m + 1)\; P_{l-1}^m(x) + (l + m)\; P_{l-1}^m(x)$$

where $x = \cos\theta$. Show that the associated Legendre functions generated in problem 3.6 are solutions of this equation.

3.13 Write time independent Schrödinger wave equation for He^+. Show how it can be separated into radial equation and angular equation?

3.14 The ground state wave function (1*s*) for the He^+ ion is given as $\psi_{1s} = \left[\frac{8}{\pi a_0^3}\right]^{\frac{1}{2}} e^{-\frac{2r}{a_0}}$. Calculate the most probable distance from the nucleus for the electron occupying 1*s* orbital in He^+. Also calculate the average value as well $\langle r^2 \rangle^{\frac{1}{2}}$ and compare it with the most probable distance.

3.15 Show that $\psi_{3.0,0}$ and $\psi_{2,1,0}$ are normalized wave functions for hydrogen like atom.

3.16 Prove that hydrogen like wave functions $\psi_{2,1,1}$ and $\psi_{2,1,0}$ are orthogonal to each other.

3.17 Calculate the most probable value of r for an electron in hydrogen like wave function for (*i*) 2*s*, (*ii*) 2*p*, (*iii*) 3*p*, (*iv*) 3*d*, (*v*) 4*s*.

3.18 Calculate the average value of r $(\langle r \rangle)$ for an electron in hydrogen like wave function for (*i*) 2*s*, (*ii*) 2*p*, (*iii*) 3*p*, (*iv*) 3*d*, (*v*) 4*s*. Compare these values with those obtained from general relations.

Note: The general result of definite integral is;

$$(-1)^n \int_0^\infty r^n\, e^{-\beta r} dr = (-1)^n \frac{n!}{\beta^{n+1}}$$

3.19 Calculate that the probability of an electron following hydrogen like wave functions, e.g. $1s$ over $r = a_0$. Comment over this value by comparing it with Bohr's model.

3.20 Repeat the problem 3.19 when $r = 2a_0$, $5a_0$ and $10a_0$.

3.21 Calculate the value of radius of sphere that tells the probability of $1s$ electron for hydrogen atom is: (*i*) 60%, (*ii*) 80% and (*iii*) 95%.

3.22 Calculate the average value of potential energy for $1s$ hydrogen and show that it is twice the total energy of hydrogen atom in the ground state.

3.23 Repeat the problems 3.21 and 3.22 for $2s$ electron of hydrogen atom.

3.24 Calculate the average value of r for $1s$ orbital of He^+, Li^{2+}, B^{3+} using hydrogen atom like wave functions. Comment on why the $\langle r \rangle$ keeps on decreasing with increasing Z.

3.25 Plot the radial functions $R_{n,l}(r)$, $R^2_{n,l}(r)$ and $4\pi^2 R^2_{n,l}(r)$ for (*i*) $2s$, (*ii*) $3s$, (*iii*) $3p$, (*iv*) $2p$ and (*v*) $3d$. See that these plots agree with those shown in Figs. 3.4, 3.5 and 3.6 respectively.

3.26 Plot the angular plot for (*i*) $2p_x$, (*ii*) $2p_y$, (*iii*) $3d_{z^2}$ as well as square of the angular wave functions. Compare the two plots and comment.

3.27 Plot the polar plot for: (*i*) $3d_{x^2-y^2}$, (*ii*) $3d_{xy}$, (*iii*) $3d_{yz}$ and (*iv*) $3d_{xz}$ orbitals of hydrogen like atoms in *xy*-plane. Plot also the square of the angular wave functions in *xy*-plane. Compare and comment.

3.28 Repeat the problem 3.27 in *yz*-plane and in *xz*-plane, *i.e.*, $\theta = 0^0$. Comment on the shapes of the orbitals obtained.

3.29 Using variation method, calculate the ground state energy of hydrogen atom using $\Phi(r) = e^{-Cr}$, where C is the variation constant. Show that the energy obtained by variation method is greater than the true value. (Hint: Since $l = 0$ for the ground state of hydrogen atom

$$\hat{H} = \frac{h^2}{8\pi^2\mu}\frac{d}{r^2\,dr}\left(r^2\frac{d}{dr} - \frac{e^2}{4\pi\varepsilon_0 r}\right).$$

3.30 Considering the approximate wave function $\Phi = x(1 - x)$ for a particle of mass m moving in a one-dimensional box of length a, calculate the ground state energy of the particle. Comment on your answer.

3.31 Knowing that oscillations of a particle around its mean position is symmetrical, take ground state variation function of this motions as $\Phi = \cos \alpha x$, where the limits of α are $-\frac{\pi}{2\alpha}$ and $\frac{\pi}{2\alpha}$. Calculate the value of E_0 and compare it with the actual value.

3.32 Using the linear combinations of functions, take the variation function as: $\Phi = C_1\, x(x - a) + C_2\, x^2 (x - a)^2$, where C_1 and C_2 are variation constants for a particle moving in a one dimensional box of length a, calculate the ground state energy and see how much it is different from the actual one.

3.33 If we consider the ground state trial wave functions for H-atom as, $\psi_1 = e^{-\frac{ar}{2}}$ and $\psi_1 = e^{-2ar}$, calculate the energies corresponding to each wave function and compare it with actual value.

3.34 Using the variation method similar to He-atom, calculate the energies of Li^+, Be^{2+}, B^{3+} and C^{4+}. Compare these values with those determined experimentally, *i.e.*, 24.47, 75.28, 153.10, 258.1 ±0.2 and 389.9 ± 0.4 eV respectively.

3.35 The atomic unit of energy is hartree and is denoted by H. Its value is E = 27.095eV. Calculate its value in Joules, kJ mol^{-1}, eV and wave numbers.

3.36 Show that the speed of light in au is equal to 137 times au.

3.37 The speed of electron in Bohr's first orbit is $\frac{e^2 2\pi}{4\pi\varepsilon_0 h}$ = 1 (natural unit). Show that it is equal to 2.88×10^6 ms^{-1}. This value of speed is a unit of speed in atomic units.

3.38 Show that the ground state energy of hydrogen atom is $\frac{4\pi^2 m e^4}{(4\pi\varepsilon_0)^2 h^2} = \frac{e^2}{4\pi\varepsilon_0 a_0}$ = 1 hartree. Show that it is equal to 4.35944×10^{-18} J.

3.39 Define orbital and spin orbital. Write the expression for the first two spin orbital of hydrogen atom.

3.40 Show that the wave functions given by equations (3.242) and (3.243) are symmetric and anti-symmetric under the interchange of the two electrons.

3.41 Show that the average value of energy obtained by using orbital wave functions and spin orbital wave functions are the same. Take helium atom as the case.

3.42 Write the normalized wave function of: (*i*) Li, (*ii*) Be, (*iii*) C and (*iv*) N atoms in the Slater's determinant form.

3.43 In Li atom, change either row or column and get a new determinant form of the wave function. Expand both the determinants and show that these are the criterion of symmetry.

3.44 Set up the ground state wave function of Li atom in Slater's determinant form. Now try to manipulate with either columns or rows in such a way that these become equal. Expand the new determinant to show that the wave function of Li atom becomes zero.

3.45 If the ground state wave function for He atom is written as: $\Psi = 1s\,\alpha(1)\,1s\,\beta\,(2) - 1s\,\alpha\,(2)\,1s\,\beta(1)$, show that $\int \Psi^*(1,2)\,\Psi(1,2)\,d\tau_1\,d\tau_2 = 2$ provided $1s$ wave functions are normalized.

3.46 For two electrons systems the total spin operator can be written as: $S_T = S_1 + S_2$. The four possible spin states can be written as:

$$\alpha\,(1)\,\alpha\,(2)$$

$$\beta\,(1)\,\beta\,(2)$$

$$1/\sqrt{2}\,[(\alpha\,(1)\,\beta\,(2) + \beta\,(1)\,\alpha\,(2)]$$

$$1/\sqrt{2}\,[(\alpha\,(1)\,\beta\,(2) - \beta\,(1)\,\alpha\,(2))]$$

Show that all these wave functions are eigen functions to S_T. For the first three states the eigen value S = 1 and for the fourth state it is zero.

3.47 The determinant form of wave function for Li atom can also be written as:

$$\Psi = 1/\sqrt{6}\,\{[1s(1)\,1s(2)\,2s(3) - 2s(1)\,1s(2)\,1s(3)]\,\alpha(1)\,\beta(2)\,\alpha(3) + [1s(1)\,2s(2)\,1s(3) - 1s(1)\,1s(2)\,2s(3)]\,\beta(1)\,\alpha(2)\,\alpha(3) + [2s(1)\,1s(2)\,1s(3) - 1s(1)\,2s(2)\,1s(3)]\,\alpha(1)\alpha(2)\,\beta(3)]\}.$$

Prove that Ψ is an eigen function of total $S_z = s_z\,(1) + s_z\,(2) + s_z\,(3)$ and calculate the eigen value.

3.48 In the Hartree-Fock method J_{ij} and K_{ij} are called Coulomb and Exchange integrals respectively. Explain why are these names given to these integrals?

3.49 Show that the term symbols for the closed shell configurations, *i.e.*, np^6 and nd^{10} are $^1S_0$.

3.50 Term symbol associated with nd^1 and nd^9 electronic configurations are same. Assign the ground state term symbol in each case.

3.51 Repeat the problem 3.52 for nd^2 and nd^8. Similar term symbols are obtained. Arrange them in order of increasing energy. *One can generalize that the term symbols obtained for the electronic configuration* d^n *or* p^n *and* d^{10-n} *or* p^{6-n} *are similar. The only difference can be in arranging the energies of states involving value of* J.

3.52 Show that the number of quantum states (sets) of magnetic quantum number (m_l) and spin quantum number (m_s) belonging to any term symbol is equal to (2L + 1) (2S + 1). Use this relation to find out the total number of quantum states for electronic configuration d^2. Does this value agrees with that obtained from equation (3.282).

3.53 Find the term symbols for the electronic configuration: (*i*) $1s^1\ 2s^1$, (*ii*) $2s^1 2p^1$, (*iii*) $3p^1\ 3d^1$, (*iv*) $3p^2\ 3d^1$ and (*v*) $2p^1\ 3p^1\ 3d^1$.

3.54 Calculate the term symbols for the electronic configuration: (*i*) d^3, (*ii*) d^4, and (*iii*) d^5. Arrange them in order of increasing energy.

3.55 The ground state term symbols for chromium and copper atoms are $^7S_3$ and $^2S_{1/2}$ respectively. Write the electronic configurations of each atom compatible with the term symbol.

3.56 The ground state term symbols for Nb, Ru, Rh and Pd are ^6D, ^5F, ^4F and ^1S. Write the electronic configurations of each atom compatible with the term symbol.

3.57 The ground state term symbols for Ce, Pr, Gd, W, and Pt are ^3H, ^4I, ^9D, ^5D and ^3D respectively. Write the electronic configurations of each atom compatible with the term symbol.

3.58 For the following term symbols, calculate the value of L, S and J. (*i*) $^9D_6$, (*ii*) $^4F_{\frac{3}{2}}$, (*iii*) $^4S_{\frac{3}{2}}$, (*iv*) $^7S_3$ and (*v*) $^6D_{9/2}$.

3.59 Write the electronic configurations of helium atom in its first ground and first excited states. Find out the term symbols for each case and indicate whether the spin states are the same or different. If not why?

3.60 An α particle is trapped in a nucleus whose radius is $r_0 = 1.4 \times 10^{-15}$ m. Show that the probabilities that an α particle will escape from the nucleus if its energy is (*a*) 2.0 MeV, or (*b*) 1.0 eV is (*a*) 0.124 or (*b*) 0.0423 respectively. The potential barrier at the surface of the nucleus is 4.0 MeV.

CHAPTER 4

Molecular Structure

From our earlier knowledge we know that elements except rare gases are always present as molecules. This means that molecules are more stable than the atoms of these major groups of elements. So the points come into our mind are:

(*i*) What causes the extra stability in the molecule?

(*ii*) How the electrons are distributed around the nuclei in the molecules and to be, more precise, whether all kind of electrons (core and valence) or only valence (outer most shell) electrons take part in holding the atoms together in molecules?

(*iii*) How are these atoms arranged in molecules in three dimensional space, *i.e.*, what is the shape of a molecule and what are the factors which govern the shapes of the molecules?

(*iv*) Knowing the distribution of electrons around the nuclei, *i.e.*, molecules, can we predict other properties of the molecules, for example, bond length, bond strength, bond moment etc.

These are the sorts of questions for which we shall be seeking answers in this chapter. We shall begin by giving a brief history and then discussing each point in detail.

Historically, soon after the discovery of electron, it was realized that the description of the bonds in the molecules is simply the description of electron distribution in them. J. J. Thomson (after the discovery of electron in 1897) was the first to develop an electron theory of valence. He was followed by G.N. Lewis, Irving Langmuir and, Kossel. The first qualitative picture of chemical bond was given by G.N. Lewis (1916). He discussed this kind of bonding as covalent in nature and postulated the idea of a shared electron bond or electron pair bond. In other words, it was thought that pairs of electrons, held together by two atoms, are responsible for the extra stability of the molecules, in comparison to atoms. Although it is true that this theory describes many aspects of chemical bond, but it was realized soon that it is a purely formal description and tells nothing what the sharing of electrons means or how it arises? Whether the electrons are in static equilibrium or in dynamic equilibrium? If it is former, it is in trouble because Earn Shaw's theorem of electrostatics clearly states that no system of charges can be in stable equilibrium when it is at rest. It is the dynamic equilibrium, which accounts for the stability of atoms or molecules. If it is so then the questions arise in our mind are: What kind of dynamic motion we are talking about and how this dynamic motion can explain the existence of binding energy, a saturation of combining power and stereochemistry or structure of a molecule.

Soon after the successful attempt of solving hydrogen atom problem with the help of Schrödinger wave equation, Heitler and London were the first to use the idea of quantum mechanics to explain the nature of chemical bonding between hydrogen molecules. This method is called valence bond approach or theory (v.b.t.). In thirties, another theory known as molecular orbital theory (m.o.t.), was developed by Mulliken. This theory gained more importance because it is simpler than valence bond. To determine the interactions of the method of molecular orbital and the method of valence bond, we correlate these two methods by applying them to the hydrogen molecule. For a heteroatomic diatomic molecule of type A-B, the wave function corresponding to stable state of AB molecule may be written, (to a first approximation) in the valence bond method as:

$$\Psi\,(A-B) = C_0\,\Psi_{cov} + \sum_i C_i\,\Psi^i_{ion}$$

where the first term denotes the covalent component of the wave function, *i.e.,* those particular solutions of the wave equation (with appropriate coefficients) which corresponds to finding the electrons near different nuclei, the second term denotes the ionic component, which corresponds to a simultaneous location of a pair of electron near a single nucleus. In the absence of second term, the valence bond method takes the status of ion-pair method.

$$\Psi_{cov} = [\varphi_A\,(1)\,\varphi_B\,(2) + \varphi_A\,(2)\,\varphi_B\,(1)]$$
$$\Psi^1_{ion} = \varphi_A\,(1)\,\varphi_A\,(2)$$
$$\Psi^2_{ion} = \varphi_B\,(1)\,\varphi_B\,(2)$$

φ's are atomic orbitals. Since the total wave function has to be anti-symmetric, a symmetric v.b. space-function has to be combined with an anti-symmetric spin function, $[\varphi_A\,(1)\,\varphi_B\,(2) + \varphi_A\,(2)\,\varphi_B\,(1)]$ goes with the spin function $[\alpha\,(1)\,\beta(2) - \beta\,(1)\,\alpha\,(2)$, singlet] where $\alpha(1)$ refers to the spin function of electron 1 in orbital A with $m_s = \pm\,\frac{1}{2}$. Similarly $[\varphi_A\,(1)\,\varphi_B\,(2) - \varphi_A\,(2)\,\varphi_B\,(1)]$ goes with the spin functions $\{\alpha\,(1)\,\alpha\,(2),\ \beta\,(1)\,\beta(2),\ \alpha\,(1)\,\beta(2) + \beta\,(1)\,\alpha\,(2)$; triplet$\}$. For hydrogen molecule in a stable state, we write

$$\psi(A-B) = \varphi_A\,(1)\,\varphi_B\,(2) + \varphi_A\,(2)\,\varphi_B\,(1) \qquad ...(4.1)$$

Expressing the molecular orbitals of both electrons in terms of atomic orbitals and assuming the motion of one electron as independent of that the other, *i.e.,* disregarding interaction of electrons, we have, in the method of molecular orbitals (to a first approximation),

$$\Psi_{m.o.} = C_1\,\varphi_A + C_2\,\varphi_B$$

where φ_A and φ_B are atomic orbitals combining to yield m.o. and C_1 and C_2 are the coefficients. Molecular orbital function for polyatomic molecules is $\psi_{m.o.} = \sum_i C_i\,\phi_i$, where φ_i's are the atomic orbitals and C_i are coefficients that minimize the energy of the m.o.

For hydrogen molecule

$$\psi_{m.o.} = [\varphi_A(1) + \varphi_B(1)]\,[\varphi_A(2) + \varphi_B(2)] \qquad ...(4.2)$$

$$= \varphi_A(1)\,\varphi_B(2) + \varphi_A(2)\,\varphi_B(1) + \varphi_A(1)\,\varphi_A(2) + \varphi_B(1)\,\varphi_B(2)$$

Note that for the case of hydrogen molecule we have $C_1 = \pm C_2$ and $\varphi_A = \varphi_B = \varphi_{1s}$, the m.o.; $\varphi_{1s}^A + \varphi_{1s}^B$ has a component of the total angular momentum along the molecular axis value equal to zero and hence is a σ orbital. The other combination $(\varphi_{1s}^A - \varphi_{1s}^B)$ is an anti-bonding σ orbital, usually designated as σ^*. Thus, we have in general, σ, π, δ,... , as well as, σ^* π^*, δ^* etc., molecular orbitals. It is customary to refer to σ m.o. formed from 1s atomic orbitals as σ_{1s} or $1s\sigma$ and similarly σ_{2p} or $2p\sigma$ and so on. Comparison of the expressions in (4.1) and (4.2) shows that the wave function calculated by the method of m.o., contains two terms other than those correspond to the location of one electron near nucleus A and the other near nucleus B. These terms, corresponding to finding both electrons near one or the other of the nuclei, *i.e.,* two possible ionic structures of molecule: $H^+ H^-$ and $H^- H^+$, which are not taken into account in the model of ion pair method, which is one of the forms of the valence bond. In the expression of the wave function calculated by the method of m.o.'s, the terms corresponding to ionic structures appear with the same coefficient (1) as the first term.

Strictly speaking, these formulae hold only for the case when atoms A and B are the same. In general case of different atoms A and B, each of the functions φ_A and φ_B is multiplied by the factors C_A or C_1 and C_B or C_2 respectively.

In this chapter, we will develop the basic concepts of both the theories, how these theories can be used to explain the following mentioned points, *i.e.,* electronic structure, energy levels, bond angles, bond lengths, dipole moment and spectra of simple molecules and draw backs of each theory in its simple form.

4.1 BORN OPPENHEIMER APPROXIMATION

The Schröndinger wave equation for a molecule can be written as:

$$\hat{H}\psi(r, R) = E\,\psi(r, R) \qquad ...(4.3)$$

where $\psi(r, R)$ is a complete wave function describing the behaviour of electrons and nuclei, r represents the coordinates of electrons and R those of nuclei, E is the total energy and $\hat{H}$ is the Hamiltonian operator and can be written as:

$$\hat{H} = -\frac{h^2}{8\pi^2}\sum_j \frac{1}{M_j}\nabla_j^2 - \frac{h^2}{8\pi^2 m_i}\sum_i \nabla_i^2 + \frac{1}{4\pi\varepsilon_0}\left[\sum_{i,i'}\frac{e^2}{r_{i,i'}} + \sum_{j,j'}\frac{Z Z' e^2}{R_{j,j'}} - \sum_{i,j}\frac{Z_j\, e^2}{r_{i,j}}\right] \qquad ...(4.4)$$

where the first term and second terms represent the kinetic energy terms of the nuclei and electrons respectively, third and fourth terms represent the repulsive potential between the electron-electron and nucleus-nucleus and fifth term the attraction between the electrons and nuclei, $r_{i,i'}$, $R_{j,j'}$ and $r_{i,j}$ represent the distance between electron i and i', nucleus j and j' and electrons and nucleus i and j respectively. The solution of equation (4.3) is not known even for simplest molecule, like, H_2^+. Born-Oppenheimer introduced approximation based on the fact that the mass of even the simplest atomic nucleus (hydrogen) is nearly two thousand times that of electron, but the forces acting on each other are equal. Thus it is reasonable to assume that: (*i*) the nuclei are stationary when solving the wave function for the electrons alone and (*ii*) the nuclei can be considered to move in the average electronic configuration, when solving a wave equation for nuclei alone. In the latter case, a characteristic energy value of the electronic wave equation, regarded as a function of internuclear distances occurs as a potential energy function. The above discussion leads to the fact that motion of electrons and nuclei can be treated independently. Following similar treatment as described in section (2.6.3), the total wave function $\psi(r, R)$ can be written as:

$$\psi_{n,v}(r, R) = \psi_n(r, R)\,\psi_{n,v}(R) \qquad ...(4.5)$$

where n and v are the quantum numbers associated with the motion of the electrons and the nuclei respectively. The Hamiltonian operator associated with the motion of electrons can be written as:

$$\hat{H} = -\frac{h^2}{8\pi^2 m_i}\sum_i \nabla_i^2 + \frac{1}{4\pi\varepsilon_0}\left[\sum_{i,i'}\frac{e^2}{r_{i,i'}} + \sum_{j,j'}\frac{ZZ'e^2}{R_{j,j'}} - \sum_i \frac{Z_j e^2}{r_{i,j}}\right] \qquad ...(4.6)$$

where $\psi_n(r, R)$ are the electronic wave functions corresponding to different set of values of the electronic quantum numbers n only, but it depends on the coordinates of electrons and nuclei. This is because the motion of the electrons is solved by taking a particular fixed framework of nuclei. The Schrödinger wave equation for the motion of electrons, after following the procedure of separation of coordinates, can be written as:

$$\sum_{i=1}^{s}\nabla_i^2\psi_n(r,R) + \frac{8\pi^2 m_i}{h^2}[E_n(R) - V(r, R)]\,\psi_n(r, R) = 0 \qquad ...(4.7)$$

where
$$V(r, R) = \sum_{i,i'}\frac{e^2}{4\pi\varepsilon_0 r_{i,i'}} + \sum_{j,j'}\frac{Z_j Z_{j'} e^2}{4\pi\varepsilon_0 R_{j,j'}} - \sum_{i,j}\frac{Z_j e^2}{4\pi\varepsilon_0 r_{i,j}} \qquad ...(4.8)$$

where s is the total number of electrons. $E_n(R)$ is the electronic energy for a given quantum state n and at a given distance from the nuclei. Using the approximation of fixed framework, the term $\sum_{j,j'}\frac{Z_j Z_{j'} e^2}{4\pi\varepsilon_0 R_{j,j'}}$ in the potential energy term will be constant. Thus equation (4.7) is an ordinary Schrödinger wave equation involving s electrons. Wave function $\psi_n(r, R)$ and its corresponding energy $E_n(R)$ for a given state n can be obtained from the solution of equation (4.7). In reality it gives many values of electronic energies and the corresponding electronic wave functions. The energy corresponding to the lowest value of n is known as the ground state and the other values are known as excited states. Next step is to solve the wave equation (4.7) for all the configurations of the nuclei by taking different values of $R_{j,j'}$ in the molecule. For example, in diatomic molecule the value of $R_{j,j'}$ can vary from zero (when the two atoms merge with each other) to infinity (when the atoms are separately held). It has been found that the value of electronic energy E_n (R) varies as a continuous function of nuclear coordinates. For example, in a diatomic molecule, there is only one nuclear coordinate (*i.e.,* internuclear distance, R) and the value of electronic energy varies continuously when R changes. The plot is shown in Fig. 4.1.

Similarly, the acceptable wave equation for the motion of nuclei becomes:

$$\sum_{j=1}^{t}\frac{1}{M_j}\nabla_j^2\psi_{n,v}(R) + \frac{8\pi^2}{h^2}[E_{n,v} - E_n(R)]\,\psi_{n,v}(R) = 0 \qquad ...(4.9)$$

where t is the number of nuclei. It was shown by Born and Oppenheimer that the wave functions $\psi_{n,\,v}$ (R) for the motion of nuclei are the acceptable solutions of wave equation (4.9) and are function of only nuclear coordinates R, in which the electronic energy function E_n (R) plays the role of potential energy, because the nuclei feel the average electronic configuration at each R.

The main aim of this chapter is to discuss and solve the wave equation (4.3), which is not possible. Even though the wave equation (4.3) has been reduced to two simple equations (4.7) and

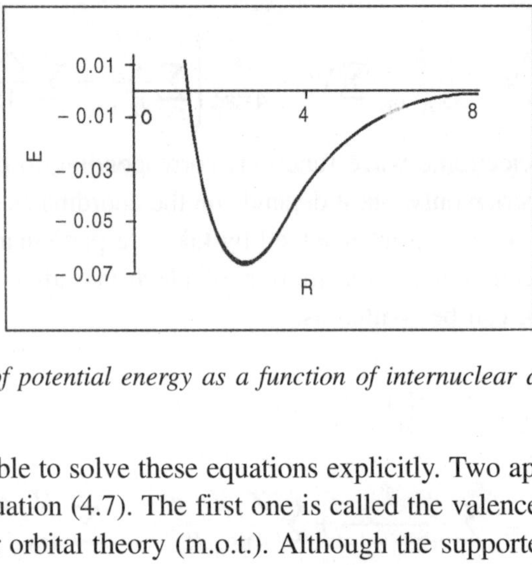

Figure 4.1: *Variation of potential energy as a function of internuclear distance in terms of Bohr's radius*

(4.9), but it is still not possible to solve these equations explicitly. Two approximate methods have been developed to solve equation (4.7). The first one is called the valence bond theory (v.b.t.) and the second one is molecular orbital theory (m.o.t.). Although the supporters of one or the other of these theories claim that the theory of their choice gives better insight of the chemical bonding, but it is pity to mention that none of these methods give complete picture of the structure of molecules. Though v.b.t. was developed earlier, m.o.t. gained more importance because of its simplicity. Thus first we will start with the m.o.t. and this will be followed by v.b.t.

4.2 MOLECULAR ORBITAL THEORY

The molecular orbital theory commonly known as m.o.t., was developed by Mulliken and was extended by Hund and Born. The m.o.t. attempts to treat a molecule from the same point of view as atom, the theory discussed in chapter 3 for atoms, i.e., m.o.t. assumes that basic features of self consistent field method for atoms can also be applied to molecules. So in the beginning we shall give general features, followed by a general treatment for diatomic molecules. The basic assumptions in the m.o.t. are:

(*i*) Each electron in a molecule, similar to atoms, can be described by a wave function ψ. Since it describes the motion of an electron in a molecule, it is called m.o. wave function. Unlike atomic orbitals, which are monocentric, molecular orbitals are polycentric. ψ^2 or $\psi^* \psi$ describes the charge density function and $\psi^2\, d\tau$ or $\psi^* \psi\, d\tau$ gives the relative probability of the electron in a small volume element $d\tau$. If ψ is normalized, (*i.e.*,$\int \psi^* \psi d\tau = 1$), then $\psi^2\, d\tau$ gives the exact probability of the electron in the volume element $d\tau$.

(*ii*) Each wave function is associated with a set of quantum numbers, which give the energy of the m.o. as well as describes its shape.

(*iii*) Each wave function is associated with a characteristic energy of the m.o. This energy is nearly equal to the ionization energy, *i.e.*, energy required to remove an electron to infinity. The total energy of the molecule can be calculated by adding the energy corresponding to each electron and subtracting the energy due to repulsion between electrons. Precaution should be observed that repulsive energy between electrons should not be counted twice.

(*iv*) Each electron is associated with a spin and it can either have a value of $+\frac{1}{2}$ (α) or $-\frac{1}{2}$ (β).

(*v*) Aufbau principle is used in filling the total number of electrons into different m.o.'s, keeping in mind the Pauli's Exclusion Principle and Hund's rules.

Before we use the above assumptions in understanding the structure of the molecules, let us try to get some ideas of molecular orbitals, their shapes and energies.

4.3 LINEAR COMBINATION OF ATOMIC ORBITALS (LCAO)

We know that the explicit solution the Schrödinger wave equation (4.7) is not possible. In order to get any meaningful picture about the molecules, variation theorem is used. This means we have to select trial wave functions. The concept involved in selecting the trial wave function is as follows: Let us consider a diatomic molecule. When an electron is very close to the nucleus A, it will experience the forces due to the nucleus A, attractive potential and the repulsive forces because of the electrons around the nucleus A. The interaction forces (both attractive and repulsive) of this electron with the nucleus B will be very small. The repulsive energy between the nuclei A and B is constant for a given configuration in a given space and can be taken into account later on. This means that the Hamiltonian operator and the Schrödinger wave equation resemble with those of an isolated atom. The solutions of this wave equation will give the molecular orbitals, which resemble the atomic orbitals around the nucleus A (*i.e.*, φ_A). In a similar fashion, the same electron when present near the nucleus B, the m.o.'s obtained from the wave equation will resemble the atomic orbitals around the nucleus B (*i.e.*, φ_B). Since the complete m.o. has the characteristics separately possessed by φ_A and φ_B, it is obvious that the complete m.o. will be obtained by the linear combination of a.o.'s of both the atoms. This method is known as LCAO-M.O. and m.o. so formed is generally expressed as:

$$\psi(i) = C_1\varphi_1 + C_2\varphi_2 \ldots \qquad + C_n\varphi_n = \sum_j C_j\varphi_j \qquad \ldots(4.10)$$

where φ_j are the atomic orbitals and C_j are the coefficients to be determined with the help of variation method (3.6) The a.o.'s used are called the basis functions. Larger the number of basis functions used the better is the trial function. If only the ground state occupied orbitals are used, the set is called the minimal basis set.

4.3.1 CONDITIONS ON *a. o.* TO FORM *m. o.*

In the last section we have mentioned that we should involve greater number of a.o.'s to form m.o.'s, but in practice it is not possible. Thus to have maximum participation of a.o.'s it is necessary that:

(*i*) *energies of the combining a.o.'s should be comparable.*

(*ii*) *overlap of charge cloud of the combining a.o.'s should be as much as possible.*

(*iii*) *a.o.'s should have the proper symmetry.*

The above conditions can be understood properly by considering the involvement of only one a.o. from each atom A and B in a diatomic molecule AB. So the m.o. can be expressed as:

$$\psi = C_1\varphi_A + C_2\varphi_B$$

Using variation theorem, where

$$E_A = \int \varphi_A \hat{H} \varphi_A\, d\tau,\ E_B = \int \varphi_B \hat{H} \varphi_B\, d\tau;$$

$$\beta = \int \varphi_A \hat{H} \varphi_B\, d\tau;\ S = \int \varphi_A \varphi_B\, d\tau;\ \int \varphi_A \varphi_A\, d\tau = \int \varphi_B \varphi_B\, d\tau = 1$$

the secular equations obtained will be:

$$(E_A - E)\, C_1 + (\beta - ES)\, C_2 = 0 \qquad \ldots(4.11)$$

$$(\beta - ES)\, C_1 + (E_B - E)\, C_2 = 0 \qquad \ldots(4.12)$$

(obtained using the expression (3.197) and minimizing the energy with respect to C_1 and C_2). As discussed earlier, E_A and E_B represent the energies of the electrons in the orbital φ_A and φ_B respectively. These are not the energies of the isolated atoms, but as said in the last section, they do correspond quite closely to the energies of the isolated atoms. S is the overlap integral of orbitals φ_A and φ_B. β is the resonance integral between orbitals φ_A and φ_B. E denotes the energy of the best linear combination of φ_A and φ_B. It is also clear from the above equations (4.11) and (4.12) that there will be two roots. It will be shown in section (4.4) that one of the values of E will be lower than those of E_A and E_B and the other value of E will be greater than those of E_A and E_B. The condition (*i*) can be understood from the following treatment. Rearranging equations (4.11) and (4.12) and eliminating C_1 and C_2, we get

$$\frac{E_A - E}{\beta - ES} = \frac{\beta - ES}{E_B - E} \qquad \ldots(4.13)$$

or

$$(E_A - E)\,(E_B - E) - (\beta - E\,S)^2 = 0 \qquad \ldots(4.14)$$

Assume $E_A \ll E_B$, *i.e.,* the energy of the atomic orbital φ_A is much smaller than that of φ_B. We have learnt in the last chapter that size of the wave function is determined by the corresponding energy of that orbital. This means that wave function φ_A will be smaller than φ_B and will lead to the smaller values of resonance integral and overlap integrals. The second term in equation (4.14) will be quite small. This will be possible only if either $(E_A - E)$ is small or $(E_B - E)$ is small. If we assume that $(E_A - E)$ is small, it means that $E \sim E_A$. Substituting $E \sim E_A$ in equation (4.14)

$$E = E_A - \frac{(\beta - E_A\ S)^2}{E_B - E_A} \qquad \text{...(4.15)}$$

Similarly substituting $E \sim E_B$ in equation (4.14) and rearranging:

$$E = E_B + \frac{(\beta - E_B\ S)^2}{E_B - E_A} \qquad \text{...(4.16)}$$

As we have said above $(\beta - ES)$ is small and $(E_B - E_A)$ is large. This means that the second term in equations (4.15) and (4.16) are quite small. The energies of the m.o. thus will be nearly equal to those of the atomic orbitals E_A and E_B. Substituting the values of E from equation (4.15) in equation (4.11)

$$\left(E_A - E_A + \frac{(\beta - E_A S)^2}{E_B - E_A}\right) C_1 + \left(\beta - E_A\ S + \frac{(\beta - E_A S)^2 S}{E_B - E_A}\right) C_2 = 0$$

$$\frac{(\beta - E_A S)^2}{E_B - E_A}\ C_1 + (\beta - E_A\ S)\left(1 + \frac{(\beta - E_A S)\ S}{E_B - E_A}\right) C_2 = 0 \qquad \text{...(4.17)}$$

Neglecting $\frac{(\beta - E_A S)\ S}{E_B - E_A}$ as it is much less than 1, equation (4.17) becomes:

$$\frac{C_2}{C_1} = -\frac{\beta - E_A\ S}{E_B - E_A} \ll 1.$$

Thus the m.o. corresponding to the energy given by equation (4.15) will be practically equal to the atomic orbital φ_A. Similarly the m.o. corresponding to the energy given by equation (4.16) will represent the atomic orbital φ_B. In other words there is practically no effective combination of atomic orbitals φ_A and φ_B.

Criterion (*ii*) demands that the overlap of atomic orbitals φ_A and φ_B should be large. If it is not so, S, as well as β will be small. It is clear from equations (4.15) and (4.16) that E will either be equal to E_A or E_B, meaning that φ_A and φ_B do not combine with each other. This criterion is also known as the criterion of maximum overlap.

To prove the criterion (*iii*) consider the molecular axis of diatomic molecule AB as the *z*-axis and a.o.'s used to form m.o. are *s* and p_x, as shown in Fig. 4.2. a.o.'s is totally symmetric, whereas one lobe of p_x is

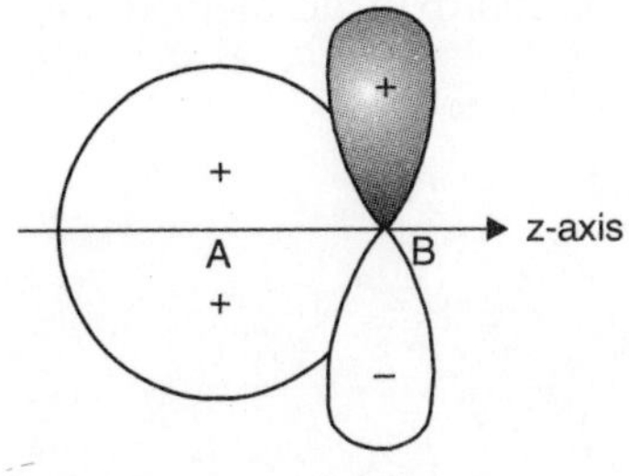

Figure 4.2: *Overlap of s and p_x a.o.'s*

having a positive sign and the other one is having a negative value. It is obvious from Fig. 4.2 that whatever constructive overlap of the two atomic functions takes place in the upper half, the lower half will have destructive overlap of exactly same magnitude. On the average the overlap integral S will have zero value. It is evident that p_x on one atom cannot combine with the s orbital on the other atom if z-axis is the molecular axis. Similarly it can also be concluded that p_z on one atom cannot combine with the p_y or p_x on the other atom because their overlap integrals S = 0 as each one is perpendicular to one and other. Table 4.1 gives the allowed combination of a.o.'s to m.o. using LCAO method.

Table 4.1: *Allowed and forbidden combination of a.o.'s to form m.o.'s considering z-axis as the molecular axis*

ψ_A	Allowed	ψ_B Forbidden
s	s, p_z, d_{z^2}	$p_y, p_x, d_{xy}, d_{yz}, d_{x^2-y^2}$
p_x	p_x, d_{z^2}	$s, p_y, p_z, d_{xy}, d_{yz}, d_{xz}, d_{x^2-y^2}$
p_y	p_y, d_{xy}	$s, p_x, p_z, d_{yz}, d_{xz} . d_{x^2-y^2}, d_{z^2}$
d_{z^2}	s, p_x, d_{z^2}	$p_y, p_z, d_{xy}, d_{yz}, d_{xz}, d_{x^2-y^2}$
d_{xy}	p_x, d_{xy}	$s, p_x, p_y, p_z, d_{xy}, d_{yz}, d_{xz}, d_{z^2}$
$d_{x^2-y^2}$	$d_{x^2-y^2}$	$s, p_x, p_y, p_z, d_{xz}, d_{yz}, d_{z^2}$

4.4 THE H_2^+ MOLECULE ION

The simplest molecular species for which the Schrödinger wave equation has been solved is H_2^+. This system consists of two nuclei, having positive charge of 1 unit each and only one electron. Thus, H_2^+ molecule ion has the same status in molecules as hydrogen ator [illegible] atoms. The major difference between hydrogen atom and H_2^+ is that the former is monocentric system and thus spherical polar coordinates have been used to understand its structure, whereas the latter is bicentric system and needs elliptical coordinates, with two nuclei at the two focuses of ellipse. The Schrödinger wave equation for H_2^+ ion has been solved exactly using the elliptical coordinates and exact nature of ψ is known for a given energy. However the results so obtained have little physical significance, because it does not have any physical picture to chemical bonding. So we will use only the approximate method, *i.e.,* LCAO-m.o.

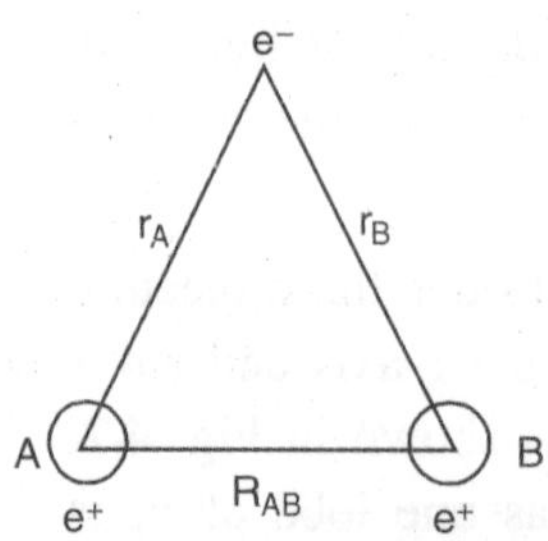

Figure 4.3: *Schematic representation of H_2^+ molecule ion*

The molecule H_2^+ has been shown in Fig. 4.3. Using Born-Oppenheimer approximation, the Hamiltonian can be written as:

$$\hat{H} = -\frac{h^2}{8\pi^2 m}\nabla^2 - \frac{e^2}{4\pi\varepsilon_0 r_A} - \frac{e^2}{4\pi\varepsilon_0 r_B} + \frac{e^2}{4\pi\varepsilon_0 R_{AB}} \quad ...(4.18)$$

and the trial LCAO-MO function for H_2^+ is as:

$$\psi = C_1\,\varphi_A + C_2\,\varphi_B \quad ...(4.19)$$

where C_1 and C_2 are the coefficients to be determined using variation method, φ_A and φ_B the ground state a.o.'s of hydrogen atom (*i.e.,* 1*s*), are given as:

$$\varphi_A = \frac{1}{\sqrt{\pi}}\, e^{-r_A/a_0}, \qquad \varphi_B = \frac{1}{\sqrt{\pi}}\, e^{-r_B/a_0} \quad ...(4.20)$$

The variation energy is given by the expression:

$$E = \frac{\int \psi^* \hat{H}\,\psi\, d\tau}{\int \psi^*\,\psi\, d\tau} \quad ...(2.125)$$

Using variation method (section 3.6) and minimizing the energy with respect to C_1 and C_2, *i.e.,* $\frac{\partial E}{\partial C_1} = \frac{\partial E}{\partial C_2} = 0$ (section 3.6) following secular equations are obtained:

$$(E_A - E)\, C_1 + (\beta - ES)\, C_2 = 0 \quad ...(4.21)$$

$$(\beta - ES)\, C_1 + (E_B - E)\, C_2 = 0 \quad ...(4.22)$$

where $E_A = \int \varphi_A\, \hat{H}\, \varphi_A\, d\tau,\ E_B = \int \varphi_B\, \hat{H}\, \varphi_B\, d\tau$

$$H_{AB} = \int \varphi_A\, \hat{H}\, \varphi_B\, d\tau;\ H_{BA} = \int \varphi_B\, \hat{H}\, \varphi_A\, d\tau \text{ and } S = \int \varphi_A\, \varphi_B\, d\tau$$

Since H^+_2 is comprised of similar atoms and is also symmetrical, we can conclude that $E_A = E_B$ and $H_{AB} = H_{BA} = \beta$. S is the overlap integral of the atomic orbitals of A and B (see better physical picture in v.b.t.). Further equations (4.21) and (4.22) can have non-trivial solution (*i.e.,* $C_1 = C_2 \neq 0$) only if the secular determinant:

$$\begin{vmatrix} E_A - E & \beta - ES \\ \beta - ES & E_B - E \end{vmatrix} = 0 \quad \text{or} \quad \begin{vmatrix} E_A - E & \beta - ES \\ \beta - ES & E_A - E \end{vmatrix} = 0 \qquad (\because E_A = E_B) \quad ...(4.23)$$

On expanding, we get

$$(E_A - E)^2 = (\beta - ES)^2 \quad ...(4.24)$$

or

$$E_A - E = \pm(\beta - ES) \quad ...(4.25)$$

i.e.,

$$E_1 = \frac{E_A + \beta}{1+S} \quad ...(4.26)$$

and $$E_2 = \frac{E_A - \beta}{1 - S} \quad ...(4.27)$$

***Evaluation of Coefficients*:** The coefficients C_1 and C_2 can be obtained by substituting the values of energy E in equation (4.21). Substitute E_1 for E in equation (4.21).

$$\left(E_A - \frac{E_A + \beta}{1+S}\right) C_1 + \left(\beta - \frac{E_A + \beta}{1+S} S\right) C_2 = 0$$

$$\left(\frac{E_A S - \beta}{1+S}\right) C_1 + \left(\frac{\beta - E_A S}{1+S}\right) C_2 = 0$$

or $$C_1 = C_2 = C_+ \text{ (say)} \quad ...(4.28)$$

Similarly using the value of E_2, the relation between the coefficients is:

$$C_1 = - C_2 = - C_- \quad ...(4.29)$$

So corresponding to energies E_1 and E_2, the wave functions are:

$$\psi_1 = C_+ (\varphi_A + \varphi_B) \quad ...(4.30)$$

$$\psi_2 = C_- (\varphi_A - \varphi_B) \quad ...(4.31)$$

Using the condition of normalization the values of C's can be determined, *i.e.*,

$$\int \psi_1^* \psi_1 \, d\tau = C_+^2 \int (\varphi_A + \varphi_B)^2 \, d\tau = 1 \quad ...(4.32)$$

$$C_+^2 \left[\int \varphi_A \varphi_A \, d\tau + \int \varphi_B \varphi_B \, d\tau + 2 \int \varphi_A \varphi_B \, d\tau\right] = 1$$

Since φ_A and φ_B are normalized wave functions, the first two terms will be equal to one.

$$C_+^2 (1 + 1 + 2S) = 1 \Rightarrow C_+ = \frac{1}{\sqrt{2 + 2S}} \quad ...(4.33)$$

Similarly $$C_- = \frac{1}{\sqrt{2 - 2S}} \quad ...(4.33a)$$

Thus, the normalized m.o.'s of H_2^+, along with energies are:

$$\psi_1 = \frac{1}{\sqrt{2 + 2S}} (\varphi_A + \varphi_B), \qquad E_1 = \frac{E_A + \beta}{1+S} \quad ...(4.34)$$

$$\psi_2 = \frac{1}{\sqrt{2 - 2S}} (\varphi_A - \varphi_B), \qquad E_2 = \frac{E_A - \beta}{1-S} \quad ...(4.34a)$$

It may come in some ones mind that in case of atoms, we have an infinite set of a.o.'s (*e.g.*, hydrogen like a.o.'s), whereas in case of molecule we are getting only two m.o.'s for H_2^+. It is important to realize that we obtained only two m.o's as we used a linear combination of two a.o.'s. This procedure was carried out for simplicity. It may be mentioned here that better results can be obtained by using more a.o.'s in the trial function and *thus the number of m.o.'s obtained will be equal to the number of total a.o.'s used.*

Evaluation of Integral, S, E_A and β

The energies of m.o.'s can be represented by a single equation:

$$E_{1,2} = E_A \pm \frac{\beta - E_A S}{1 \pm S} \qquad ...(4.35)$$

where plus and minus signs give the energies of m.o.'s ψ_1 (E_1) and ψ_2 (E_2) respectively. If we neglect the overlap integral S, the approximate values of E_1 and E_2 can be equal to:

$$E_{1,2} = E_A \pm \beta \qquad ...(4.36)$$

Equations (4.35) and (4.36) indicate that the energies of atomic orbitals of two hydrogen atoms which were equal before H_2^+ is formed, are split into two energy levels. The energy of one of the m.o. (ψ_1) is lower and that of other m.o. (ψ_2) is greater than the energy of the separated atomic orbitals. Since the sign of E_A and β are negative, the lowest energy is associated with plus sign and higher energy is associated with negative sign. The approximate relation (4.36) shows that the m.o.'s are equally placed around the a.o.'s, whereas equation (4.35) shows that the gap between the upper energy level and E_A is greater than the gap between the lowest energy level and E_A. This is because of 1 ± S term in the denominator. Figure. 4.4 represents the energies of the a.o.'s before forming m.o.'s towards the sides and those of m.o's in the middle.

The energies (E_1 and E_2) of m.o.'s (ψ_1 and ψ_2) can be calculated accurately if we can evaluate the integrals E_A, β an S. This can be done, by substituting the value of $\hat{H}$.

$$E_A = \int \varphi_A \left[-\frac{h^2}{8\pi^2 m}\nabla^2 - \frac{e^2}{4\pi\varepsilon_0 r_A} - \frac{e^2}{4\pi\varepsilon_0 r_B} + \frac{e^2}{4\pi\varepsilon_0 R_{AB}} \right] \varphi_A \, d\tau$$

$$= \int \varphi_A \left[-\frac{h^2}{8\pi^2 m}\nabla^2 - \frac{e^2}{4\pi\varepsilon_0 r_A} \right] \varphi_A \, d\tau - \int \varphi_A \frac{e^2}{4\pi\varepsilon_0 r_B} \varphi_A \, d\tau$$

$$+ \int \varphi_A \frac{e^2}{4\pi\varepsilon_0 R_{AB}} \varphi_A \, d\tau \qquad ...(4.37)$$

The first integral in equation (4.37) can be identified with that of energy of hydrogen atom. Since φ_A ($1s_A$) is normalized, the integral

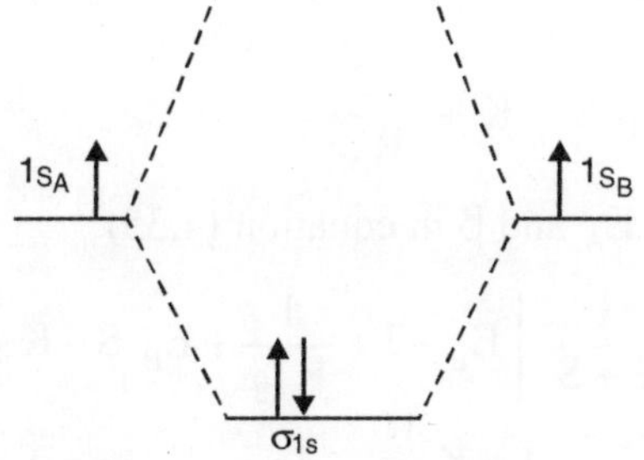

Figure 4.4: *Energy level diagram of H_2^+ molecule ion*

$$\int \varphi_A \left[-\frac{h^2}{8\pi^2 m}\nabla^2 - \frac{e^2}{4\pi\varepsilon_0 r_A}\right] \varphi_A \, d\varphi = E_H \qquad ...(4.38)$$

where E_H is the energy of a single H atom in its ground state and its value in a.u. is 0.5 (13.6 eV). The second integral describes the attraction of electron near nucleus A with the nucleus B at a distance r_B. If r_B is expressed in a.u., the integral gives the total attractive energy (Coulombic) between the electron and the nucleus B. Equating this letter to J,

$$J = \int \varphi_A \frac{e^2}{4\pi\varepsilon_0 r_B} \varphi_A \, d\tau \qquad ...(4.39)$$

The third integral

$$\int \varphi_A \frac{e^2}{4\pi\varepsilon_0 R_{AB}} \varphi_A \, d\tau = \frac{e^2}{4\pi\varepsilon_0 R_{AB}} \int \varphi_A \varphi_A \, d\tau = \frac{e^2}{4\pi\varepsilon_0 R_{AB}} \qquad ...(4.40)$$

represents the repulsive energy between nucleus A and nucleus B since φ_A is normalized. For a given configuration of nuclei A and B, the repulsive energy between two nuclei is constant. Thus

$$E_A = E_H - J + \frac{1}{R_{AB}} \qquad ...(4.41)$$

Similarly

$$\beta = \int \varphi A \left[-\frac{h^2}{8\pi^2 m}\Delta^2 - \frac{e^2}{4\pi\varepsilon_0 r_B}\right] \varphi_B \, d\tau - \int \varphi_A \frac{e^2}{4\pi\varepsilon_0 r_A} \varphi_B \, d\tau$$

$$+ \int \varphi_A \frac{e^2}{4\pi\varepsilon_0 R_{AB}} \varphi_b \, d\tau$$

$$= \int \varphi_A E_H \varphi_B \, d\tau - \frac{e^2}{4\pi\varepsilon_0} \int \varphi_A \frac{1}{r_A} \varphi_B \, d\tau + \frac{e^2}{4\pi\varepsilon_0 R_{AB}} \int \varphi_A \varphi_B \, d\tau$$

$$= E_H S - K + \frac{S}{R_{AB}} \qquad ...(4.42)$$

where $$K = \frac{e^2}{4\pi\varepsilon_0} \int \varphi_A \frac{1}{r_A} \varphi_B \, d\tau \qquad ...(4.43)$$

Equation (4.43) gives the energy, which arises due to exchange of electron between wave functions φ_A and φ_B. Thus

$$\beta = E_H S - K + \frac{S}{R_{AB}} \qquad ...(4.42)$$

Substituting the values of E_A and β in equation (4.35)

$$E_1 = \frac{1}{1+S}\left[E_H - J + \frac{1}{R_{AB}} + E_H S - K + \frac{S}{R_{AB}} \right]$$

$$= E_H - \frac{J+K}{1+S} + \frac{1}{R_{AB}} \qquad ...(4.44)$$

$$E_2 = \frac{1}{1-S}\left[E_H - J + \frac{1}{R_{AB}} - E_H S + K - \frac{S}{R_{AB}}\right]$$

$$= E_H - \frac{J-K}{1-S} + \frac{1}{R_{AB}} \qquad ...(4.45)$$

Solutions of the integrals J, K and S are beyond the scope of this book, but their values expressed as a function of R_{AB}, can be written as:

$$J = \frac{1}{R_{AB}}[1 - (1 + R_{AB})e^{-2R_{AB}}] \qquad ...(4.46)$$

$$K = (1 + R_{AB})e^{-R_{AB}} \qquad ...(4.47)$$

$$S = \left(1 + R_{AB} + \frac{R_{AB}^2}{3}\right) e^{-R_{AB}} \qquad ...(4.48)$$

Using equations (4.46), (4.47), and (4.48) as a function of R_{AB} the values of E_1 and E_2 can be calculated and are compiled in Table 4.2. For convenience the value of E_H has been taken as zero. The values of E_1 and E_2 so obtained are plotted in Figs. (4.5*a*) and (4.5*b*) respectively as a function of R_{AB}. The results of Table 4.2 and Figs. (4.5*a*) and (4.5*b*) bring forth certain very useful informations which are given below.

(*i*) Values of E_2 are always positive, *i.e.,* greater than E_H and keeps on decreasing as the value of R_{AB} increases. On the other hand, the values of E_1 are always greater than E_H at $R_{AB} \leq 1.5\ a_0$, becomes less than E_H at $\geq 1.5\ a_0$.

Table 4.2: *Values of E_1 and E_2 at different internuclear distances*

R_{AB}	E_1	E_2	R_{AB}	E_1	E_2
0			3.2	−0.054	0.112
0.2	4.01	6.83	3.4	−0.050	0.090
0.4	1.55	2.88	3.6	−0.045	0.078
0.6	0.78	2.24	3.8	−0.042	0.064
0.8	0.4	1.35	4.0	−0.037	0.053
1.0	0.211	1.04	4.2	−0.033	0.046
1.2	0.01	0.81	4.4	−0.029	0.038
1.4	0.002	0.63	4.6	−0.025	0.032
1.6	−0.01	0.51	4.8	−0.023	0.028
1.8	−0.04	0.41	5.0	−0.019	0.023
2.0	−0.054	0.336	5.5	−0.014	0.016
2.2	−0.062	0.278	6.0	−0.008	0.011
2.4	−0.064	0.230	6.5	−0.006	0.063
2.6	−0.0637	0.189	7.0	−0.004	0.004
2.8	−0.0628	0.160	7.5	−0.0026	0.0027
3.0	−0.06	0.133	8.0	−0.0017	0.002

(*ii*) There is no minimum in the value of E_2 and asymptotically approach zero at $R_{AB} = \infty$. In case of E_1, it deceases with increase of R_{AB}, reaches minimum and then starts increasing, reaching zero at $R_{AB} = \infty$.

(*iii*) At a given value of R_{AB}, the value of E_1 is always less than that of E_2, indicating that *m.o.* corresponding to E_1 is always stable than the m.o. corresponding to E_2.

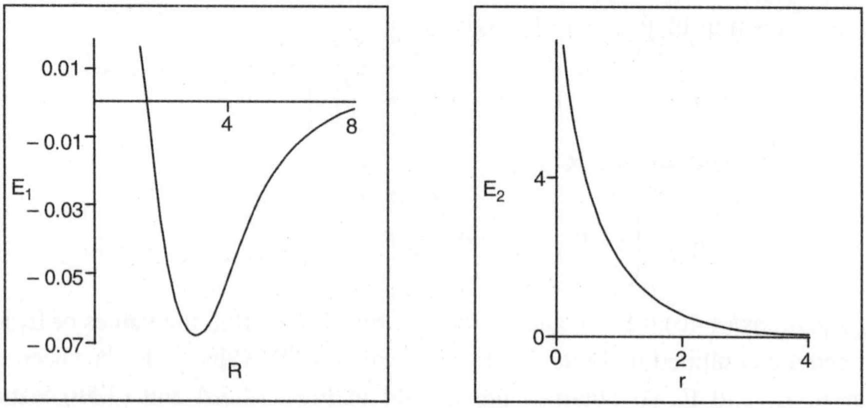

(a) (b)

Figure 4.5. *Plot of (a) E_1 versus R_{AB} and (b) E_2 versus R_{AB}. R_{AB} is in terms of Bohr's radius*

(*iv*) The value of E_1 and R_{AB} at the minimum value of energy are: 1.77 eV (170.8 kJ mol^{-1}) and 2.5 a_0 = 0.132 nm. It is clear from the above discussion that H_2^+ is a stable species with respect to H and H^+ system, the dissociation energy is 1.77 eV (170.8 kJ mol^{-1}) and the bond length of H_2^+ at the equilibrium position is 0.132 nm. The experimental evidence for the existence of H_2^+ can be seen in the discharge tube and mass spectrometer. The experimental value of E_1 (dissociation energy) and the equilibrium distance are found to be 2.8 eV (270.2 kJ mol^{-1}) and 0.106 nm respectively. Although the agreement between theoretical results and experimental observations is not good but one thing is clear that the above approach is leading in the right direction. As mentioned in section 3.6, the success of the variation method depends on the choice of the proper and correct wave function. In other words we need better wave function to describe the H_2^+ molecule.

The improvements in the approximate wave functions ($\varphi_A \pm \varphi_B$), were carried out by Dickenson, Coulson and many others. These are as follows:

(1) The first step is that 1*s* wave function used in the formation of H_2^+ must shrunk in comparison to 1*s* of hydrogen atom as the electron in H_2^+ is bound to two nuclei. In other words the value of Z in 1s orbital wave function can be taken as 1 when H and H^+ are infinite apart and this can be two when H and H^+ fuse together to form one nucleus, *i.e.*, He^+. Therefore at intermediate distances, similar to H_2^+, the value of Z will lie between 1 and 2 and thus the wave function for 1*s* orbital can be expressed in term of a variable parameter Z′, *i.e.*, $\sqrt{\dfrac{Z'^3}{\pi a_0^3}}\, e^{-\frac{Z'r}{a_0}}$. The value of Z′ can be calculated by minimizing the

energy with respect to Z′, *i.e.,* $\frac{\partial E}{\partial Z'} = 0$. The value of Z′ has been found to be 1.24 and the dissociation energy (D_e) = 2.25 eV (217.1 kJ mol^{-1}).

(2) The second improvement is about the shape of 1*s* wave function, *i.e.,* in presence of H^+, the 1*s* wave function cannot have spherical charge distribution as found in case of isolated hydrogen atom. It will be polarized along the direction of approach of H^+ (*i.e.,* along the molecular axis) to facilitate the attraction of the electron towards the second proton. In other words we may discard the LCAO form ($\varphi_A \pm \varphi_B$) altogether and may use the basis functions, which may have some *p* character, *i.e.,*

$$\psi = [1s_A + \lambda\, 2p_z\,(A)] + [1s_B + \lambda\, 2p_z\,(B)] \qquad ...(4.49)$$

where λ is the variable parameter and can be obtained by minimizing the energy with respect to λ, *i.e.,* $\frac{\partial E}{\partial \lambda} = 0$. Similarly the exponential part of 1*s* and $2p_z$ a.o. will have different Z′ parameters and can be determined similarly as: $\frac{\partial E}{\partial Z'} = \frac{\partial E}{\partial Z''} = \frac{\partial E}{\partial \lambda} = 0$.

The values λ, Z′ and Z″ are found to be 0.145, 1.247 and 2.868 respectively. The value of D_e found using these parameters is 2.65 eV (255.7 kJ mol^{-1}). Better value of D_e can be obtained further by using more refined wave functions, *i.e.,* by including more basis sets.

Charge Density and Bonding in H_2^+

As mentioned in section 2.5, the charge density of finding the electron in the molecule can be given by either ψ_1^2 or ψ_2^2 corresponding to the energy E_1 or E_2. Thus:

$$\psi_1^2 = \frac{1}{2(1 + S)}\left[1s_A^2 + 1s_B^2 + 2\,1s_A\,1s_B\right] \qquad ...(4.50)$$

$$\psi_2^2 = \frac{1}{2(1 - S)}\left[1s_A^2 + 1s_B^2 - 2\,1s_A\,1s_B\right] \qquad ...(4.51)$$

In both these equations the first two terms are identical, whereas the third term is also similar except that their signs are different. The first and second terms are similar to the charge cloud density of the isolated hydrogen atoms and are thus having spherical distribution, whereas the third

$$2 1s_A\, 1s_B \approx 2\, e^{-(r_A + r_B)} \qquad ...(4.52)$$

term represents a cloud that is eliptical in shape. The simplest way to get the physical picture of the charge cloud density is to study the variation of the charge density along the internuclear or molecular axis. When two species are infinite apart (*i.e.,* separate H and H^+) the average charge density is given by $\frac{1}{2}\,[1s_A^2 + 1s_B^2]$. As the two atoms approach towards each other, the third term also becomes finite. In case of E_1, ψ_1^2 will always be larger than the simple addition of the first two terms. For example, in the middle of the bond, $1s_A^2 = 1s_B^2$ and thus $\psi_1^2 = \frac{1}{2(1+S)}(4\ 1s_A^2)$, by equation 4.50, which will be always larger than either $1s_A^2$ or $1s_B^2$. This is shown in Fig. 4.6*a*. In other words, charge density is concentrated more in between the two nuclei and thus leads to the stability of the m.o. ψ_1. Due to this ψ_1 is known as bonding m. o. and as shown in Fig. 4.5, its energy is lower than that of $1s_A$.

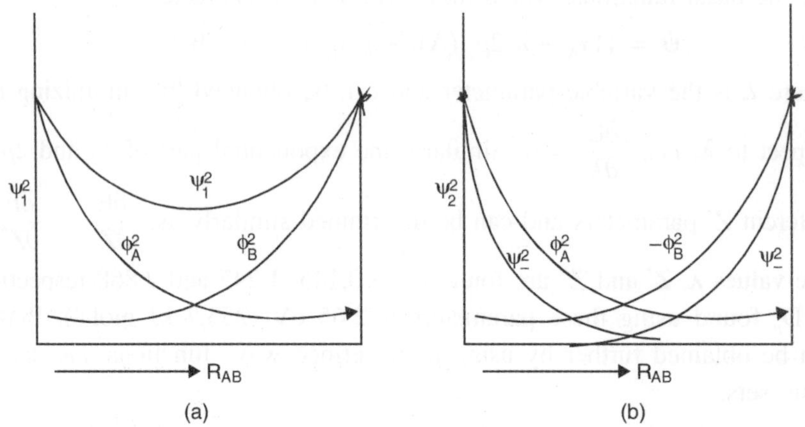

Figure 4.6: *(a) Plot of ψ_1^2 versus R_{AB} and (b) plot of ψ_2^2 versus R_{AB}*

A similar picture can also be arrived at, by analyzing equation (4.44). Taking $E_H = 0$, the first term depicts the attractive potential energy, due to the attraction of the electron to two nuclei, whereas the second term is the repulsion between the two nuclei. The minimum in energy E_1 is obtained at $R_{AB} = 0.132$ nm. If the calculations are carried out, using equation (4.44), it is found that at R_{AB} 0.132 nm, the repulsive energy $\frac{1}{R_{AB}}$ is nearly equal to the term $-\frac{J}{1+S}$, *i.e.,* thus, the decrease in the energy of E_1 (or m.o. ψ_1) when compared with that of pure a.o. is $-\frac{K}{1+S}$.

In other words, the binding force behind $H + H^+$ is the force of attraction, represented by the overlap of the electron cloud $1s_A\ 1s_B$, concentrated between the two nuclei. This suggests that the bonding is due to the overlap of two a.o.'s and exchange of electrons between the orbitals.

Figure 4.6*b* depicts the plot of Ψ_2^2 along the internuclear distance. The effect is opposite to what has been observed in case of Ψ_1^2, that is, the electron density decreases within the bond and reduces to zero in the middle. Thus it is called *anti-bonding* m.o., which has a higher energy relative to H or H^+. Physically it means that electron cloud is withdrawn from within the bond and thus increases repulsion between the two nuclei.

Similar to a.o.'s, physical picture of m.o.'s can also be represented in many ways. The one way we have already shown in Fig. 4.6*a* and 4.6*b*. The other way is taking the contour diagram (Fig. 4.7). This means that we join all those points of equal ψ^2 in three dimensional space and boundary of this three dimensional space is controlled by the value of the total electron density contained within this geometry. It varies between 90 to 95%.

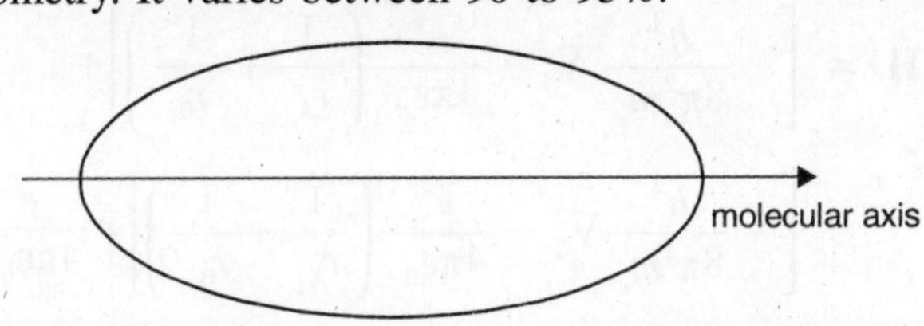

Figure 4.7: *Contour diagram of σ_{1s} molecular orbital*

One important difference between the bonding and anti-bonding m.o.'s is the presence of *nodal point* or *nodal plane* in the anti-bonding m.o. The bonding and anti-bonding m.o's thus formed are also known as σ and σ^* m.o.'s (details are given in section 4.6).

From the above discussion, it is clear that the concept of LCAO-m.o.'s, although not very accurate, but provides a simple and useful method to explain the bonding between H and H^+. The most important result that comes out is the concentration of charge cloud density between the chemical bond or the overlap of atomic orbitals involved in the bonding. Greater the overlap stronger is the chemical bond. m.o.'s thus can be formed by LCAO method and bonding nature of m.o.'s will be controlled by the magnitude of the attractive and repulsive potential enegy terms present in the Hamiltonian. Different m.o.'s obtained by LCAO will be discussed in section 4.6.

4.5 HYDROGEN MOLECULE (H_2)

The next simple molecule is hydrogen (H_2) and it contains two electrons. Figure 4.8 gives the coordinates of H_2. Using Born-Oppeneheimer approximation, the electronic Hamiltonian can be written as:

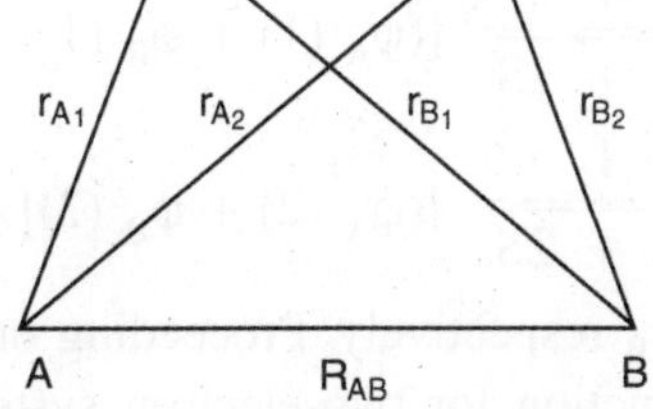

Figure 4.8: *Schematic diagram of hydrogen molecule*

$$\hat{H} = -\frac{h^2}{8\pi^2 m_e}[\nabla_1^2 + \nabla_2^2] - \frac{e^2}{4\pi\varepsilon_0}\left[\frac{1}{r_{A_1}} + \frac{1}{r_{A_2}} + \frac{1}{r_{B_1}} + \frac{1}{r_{B_2}}\right] + \frac{e^2}{4\pi\varepsilon_0 r_{12}} + \frac{e^2}{4\pi\varepsilon_0 R_{AB}} \qquad ...(4.53)$$

The first term in equation (4.53) gives the kinetic energy of both the electrons, second term is the attractive potential between the two electrons and the nuclei H_A and H_B, third and fourth terms are the repulsive potential between the electron-electron and nuclei H_A and H_B respectively. Equation (4.53) can be rearranged as:

$$\hat{H} = \left[-\frac{h^2}{8\pi^2 m_e} \nabla_1^2 - \frac{e^2}{4\pi\varepsilon_0}\left(\frac{1}{r_{A_1}} + \frac{1}{r_{B_1}}\right)\right] +$$

$$\left[-\frac{h^2}{8\pi^2 m_e} \nabla_2^2 - \frac{e^2}{4\pi\varepsilon_0}\left(\frac{1}{r_{A_2}} + \frac{1}{r_{B_2}}\right)\right] + \frac{e^2}{4\pi\varepsilon_0 r_{12}} + \frac{e^2}{4\pi\varepsilon_0 R_{AB}} \quad ...(4.54)$$

The first and second terms are similar to the electronic Hamiltonian for H_2^+ minus the fourth term and thus can be viewed as two separate H_2^+ systems. Equation (4.54) can be written as:

$$\hat{H} = \hat{H}_1(H_2^+) + \hat{H}_2(H_2^+) + \frac{e^2}{4\pi\varepsilon_0 r_{12}} - \frac{e^2}{4\pi\varepsilon_0 R_{AB}} \quad ...(4.55)$$

where $$\hat{H}(H_2^+) = -\frac{h^2}{8\pi^2 m_e} \nabla_1^2 - \frac{e^2}{4\pi\varepsilon_0}\left(\frac{1}{r_{A_1}} + \frac{1}{r_{B_1}}\right) + \frac{e^2}{4\pi\varepsilon_0 R_{AB}} \quad ...(4.56)$$

$$\hat{H}_2(H_2^+) = -\frac{h^2}{8\pi^2 m_e} \nabla_2^2 - \frac{e^2}{4\pi\varepsilon_0}\left(\frac{1}{r_{A_2}} + \frac{1}{r_{B_2}}\right) + \frac{e^2}{4\pi\varepsilon_0 R_{AB}} \quad ...(4.57)$$

***Wave functions*:** H_2 molecule can be viewed as the extension of H_2^+ by an addition of one further electron, on the same lines as done for helium from hydrogen atom. m.o. wave function for H_2, $\psi(1, 2)$, can be considered as the product of two one electron ground state wave functions of H_2^+ species, *i.e.*, m.o.'s $\sigma_g(1)$ and $\sigma_g(2)$. Now,

$$\psi(1, 2) = \sigma_g(1)\,\sigma_g(2) \quad ...(4.58)$$

where $\sigma_g(1)$ and $\sigma_g(2)$ are:

$$\sigma_g(1) = \frac{1}{\sqrt{2 + 2S}} [(\varphi_A(1) + \varphi_B(1)] \quad ...(4.59)$$

$$\sigma_g(2) = \frac{1}{\sqrt{2 + 2S}} [(\varphi_A(2) + \varphi_B(2)] \quad ...(4.60)$$

where φ_A and φ_B are $1s_A$ and $1s_B$ respectively. Proceeding on the same lines, as done for helium atom, the ground state wave function for two-electron system demands that the wave function has to be anti-symmetric (Pauli Exclusion Principle). It is obvious from the resultant equation after substitution of equations (4.59) and (4.60) that the space part of equation (4.58) is a symmetric function and can be made anti-symmetric by multiplying it with anti-symmetric spin combination of two-electron system, *i.e.*,

$$\psi(1, 2) = \sigma_g(1)\,\sigma_g(2)\,\frac{1}{\sqrt{2}} [\alpha(1)\,\beta(2) - \beta(1)\,\alpha(2)] \quad ...(4.61)$$

Equation (4.61) depicts that both the electrons in H_2 molecule are occupied in σ_g m.o., with the difference that electrons 1 and 2 are having anti-parallel spins α and β respectively. The electronic configuration can be written as $\sigma_g(1)\,\alpha(1)\,\sigma_g(2)\,\beta(2)$.

Electronic energy: The electronic energy of each m.o. can be calculated using average value postulate, *i.e.*,

$$E = \int \psi(1, 2)\, \hat{H}\, \psi(1, 2)\, d\tau \qquad ...(4.62)$$

as $\psi(1, 2)$ is normalized wave function. Substituting the value of $\psi(1, 2)$ in equation (4.62).

$$E = \frac{1}{2} \iint \sigma_g(1)\,\sigma_g(2)\,[\alpha(1)\,\beta(2) - \beta(1)\,\alpha(2)]\,\hat{H}$$
$$\sigma_g(1)\,\sigma_g(2)\,[\alpha(1)\,\beta(2) - \beta(1)\,\alpha(2)]\, d\tau_s\, d\tau_{spin} \qquad ...(4.63)$$

Since Hamiltonian operator will only be operating on the space part, equation (4.63) can be written as:

$$E = 1/2 \iint \sigma_g(1)\,\sigma_g(2)\,\hat{H}\,\sigma_g(1)\,\sigma_g(2)\, d\tau_s$$
$$\iint [\alpha(1)\,\beta(2) - \beta(1)\,\alpha(2)]^2\, d\tau_{spin} \qquad ...(4.64)$$

The integral containing spin part can be expanded as:

$$\iint \alpha^2(1)\,\beta^2(2)\, d\tau_{spin}(1)\, d\tau_{spin}(2) + \iint \beta^2(1)\,\alpha^2(2)\, d\tau_{spin}(1)\, d\tau_{spin}(2)$$
$$- 2\iint \alpha(1)\,\beta(1)\, d\tau_{spin}(1)\,\alpha(2)\,\beta(2)\, d\tau_{spin}(2)$$
$$= \int \alpha^2(1)\, d\tau_{spin}(1) \int \beta^2(2)\, d\tau_{spin}(2) + \int \beta^2(1)\, d\tau_{spin}(1)$$
$$\int \alpha^2(2)\, d\tau_{spin}(2) - 2 \times 0 \times 0$$
$$= 1 \times 1 + 1 \times 1 = 2$$

Thus, equation (4.64) can be written as:

$$E = \iint \sigma_g(1)\,\sigma_g(2)\,\hat{H}\,\sigma_g(1)\,\sigma_g(2)\, d\tau_s \qquad ...(4.65)$$

In other words it may be mentioned here that energies of m.o.'s can be calculated only by using the space part of m.o. Now substituting the value of $\hat{H}$ in equation (4.65), we get:

$$E = \iint \sigma_g(1)\,\sigma_g(2)\,[\hat{H}_1(H_2^+) + \hat{H}_2(H_2^+) + \frac{e^2}{4\pi\varepsilon_0 r_{12}} - \frac{e^2}{4\pi\varepsilon_0 R_{AB}}]$$
$$\sigma_g(1)\,\sigma_g(2)\, d\tau_1 d\tau_2$$
$$= \iint \sigma_g(1)\,\sigma_g(2)\,\hat{H}_1(H_2^+)\,\sigma_g(1)\,\sigma_g(2)\, d\tau_1 d\tau_2 + \iint \sigma_g(1)\,\sigma_g(2)\,\hat{H}_2$$
$$(H_2^+)\,\sigma_g(1)\,\sigma_g(2)\, d\tau_1\, d\tau_2 + \frac{e^2}{4\pi\varepsilon_0} \iint \sigma_g^2(1)\,\frac{1}{r_{12}}\,\sigma_g^2(2)\, d\tau_1 d\tau_2$$
$$- \frac{e^2}{4\pi\varepsilon_0 R_{AB}} \int \sigma_g(1)\,\sigma_g(1)\, d\tau_1 \int \sigma_g(2)\,\sigma_g(2)\, d\tau_2 \qquad ...(4.66)$$

$$= \int \sigma_g(1)\, \hat{H}_1\, \sigma_g(1)\, d\tau_1 \int \sigma_g(2)\, \sigma_g(2)\, d\tau_2 + \int \sigma_g(1)\, \sigma_g(1)\, d\tau_1 \int \sigma_g(2)\, \hat{H}_2\, \sigma_g(2)\, d\tau_2 + \frac{e^2}{4\pi\varepsilon_0} \iint \sigma_g^2(1)\, \frac{1}{r_{12}}\, \sigma_g^2(2)\, d\tau_1\, d\tau_2 - \frac{e^2}{4\pi\varepsilon_0 R_{AB}} \int \sigma_g(1)\, \sigma_g(1)\, d\tau_1 \int \sigma_g(2)\, \sigma_g(2)\, d\tau_2 \quad ...(4.67)$$

Since m.o. σ_g are normalized, the first and second integrals represent the electronic energy of electrons H_2^+ (1) and H_2^+ (2) respectively, *i.e.*, 2E(H_2^+), third integral represents a two centre electron integral, *i.e.*, it involves two centres, nuclei A and B and coordinates of both the electrons and thus cannot be separated, since $\frac{1}{r_{12}}$ is just a multiplier. Third integral can be written as equation (4.67). Since $\sigma_g^2(1)\, \sigma_g^2(2)$ represents the probability density of electron or charge density, the third integral represents the repulsive interaction between the electrons 1 and 2. Since the Hamiltonian $\hat{H}$ has been written for a given positions of nuclei A and B, R_{AB} is constant, integral will reduce to unity. Equation (4.67) reduces to:

$$E = 2E(H_2^+) + \frac{e^2}{4\pi\varepsilon_0} \iint \sigma_g^2(1)\, \frac{1}{r_{12}}\, \sigma_g^2(2)\, d\tau_1 d\tau_2 - \frac{e^2}{4\pi\varepsilon_0 R_{AB}} \quad ...(4.68)$$

Substituting the values of E (H_2^+) from equations (4.44) and (4.45) in equation (4.68), the energies of two m.o.'s can be written as:

$$E_1 = 2E_H - 2\,\frac{J + K}{1 + S} + \frac{1}{R_{AB}} + \frac{e^2}{4\pi\varepsilon_0} \iint \sigma_g^2(1)\, \frac{1}{r_{12}}\, \sigma_g^2(2)\, d\tau_1 d\tau_2 \quad ...(4.69)$$

$$E_2 = 2E_H - 2\,\frac{J - K}{1 - S} + \frac{1}{R_{AB}} + \frac{e^2}{4\pi\varepsilon_0} \iint \sigma_g^2(1)\, \frac{1}{r_{12}}\, \sigma_g^2(2)\, d\tau_1 d\tau_2 \quad ...(4.70)$$

and represent the bonding and anti-bonding m.o.'s respectively. Evaluation of fourth integral is quite involved and is beyond the scope of this book. Potential energy curve (*i.e.*, plot of energy versus internuclear distance) for H_2 is similar to that of H_2^+ except that R_{min} = 0.0732 nm, close to the true value of 0.0742 nm. The value of D_e at R_{min} comes out to be 3.49 eV (336.8 kJ mol^{-1}), which is not very good as compared to the true value 4.75 eV (458.4 kJ mol^{-1}).

Similar to H_2^+, simple m.o. theory does not represent the true wave function of H_2 molecule and thus needs refinement. The improvements on the same lines, as carried out for H_2^+ has been performed, *e.g.*, by allowing for more flexible basis functions, use of self consistent field method, Z = 1.193 in the orbital exponent instead of Z = 1 for 1*s* a.o. of hydrogen atom, inclusion of excited states of proper symmetry etc. These improvements have been carried out by a number of scientists. Best agreement between the calculated ($\frac{De}{hc}$ = 38,397 cm^{-1}) and experimental value of 38,293 cm^{-1} has been achieved by Kolos and Wolniewicz by including relativistic corrections and corrections to Born-Oppenheimer approximation. Their 100–term wave function is nearly indistinguishable from true wave function, but its great complexity allows no simple physical interpretation. (W. Kolos and C.C.J. Roothan, Rev. Mod. Phys., **32**, 219, 1960; W. Kolos and L. Wolniewicz, J. Chem. Phys., **41**, 3663, 1964; **49**, 404, 1968; R. J. Leroy and R. B. Bernstein, J. Chem. Phys., **49**, 4312, 1968).

4.6 SHAPES AND ENERGIES OF VARIOUS *m. o.'s*

4.6.1 SHAPES OF *m. o.'s*

Having discussed about the wave functions and energies of the corresponding m.o.'s of species, we will try to draw some physical picture of the bonding and anti-bonding m.o.'s, limiting ourselves to the period two elements only. In the period one, there are two elements only, involving 1s valence a.o.'s. The simplest diatomic species is H^+_2, containing two nuclei $1s_A$ and $1s_B$ and one electron, as shown in Fig. 4.9. Two 1*s* valence orbitals can overlap with each other and shaded area represents the region of overlap. The magnitude of the shaded area plays a major role in deciding the strength of the bond. The overlap between the two 1*s* a.o. can occur by two ways. One is the constructive overlap as shown in Fig. 4.9*a* and represents a zone of higher electron density lying in between the two nuclei. It can be interpreted as saying that the pair of electrons on an average is spending more time in between the two nuclei in comparison to the time spent in the isolated nuclei. This will increase the attractive potential energy between the electron and the nuclei in comparison to the energy of the isolated atoms. Greater the overlap greater will be the attractive potential and greater is the stabilization of the m.o. formed in this manner. This is also consistent with equations (4.35) and (4.44) respectively. Two points which emerge from Fig. 4.9*a* are:

(*i*) m.o. thus formed has a cylindrical symmetry around the molecular axis, *i.e.,* if we insert a needle passing through the molecular axis and rotate the m.o. around this needle, we get equivalent arrangements. The m.o. possessing this symmetry element is termed as sigma (σ) m.o. Since this m.o. is formed by the linear combination of 1*s* a.o., the m.o. so formed can be written as σ_{1s}.

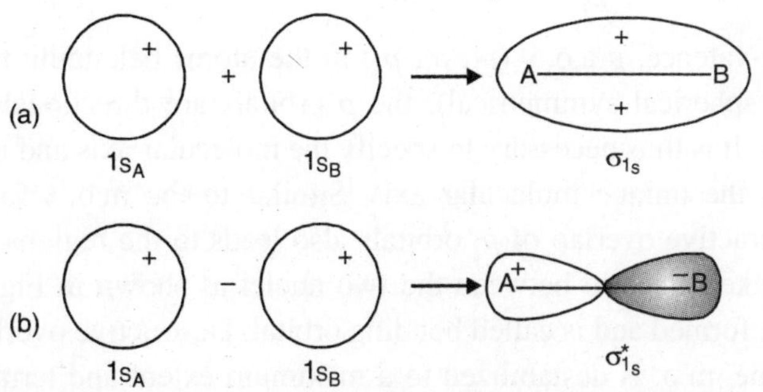

Figure 4.9: *Formation of (a)* σ_{1s} *bodning and (b)* σ^*_{1s} *anti-bonding from the LCAO of 1s orbitals*

(*ii*) This m.o. so formed is also having a centre of symmetry, *i.e.*, if we choose a point in the centre of a bond, connect this point to any point on the surface of m.o. (let this point be x at a distance l from the centre) and then proceed in opposite direction from the centre by a distance l. Since no change is observed in the wave function, it is symmetric with respect to centre of inversion symmetry element, denoted by g (gerade); σ_{1s} (g).

The overlap can also lead to negative (or destructive) combination, as shown by equation (4.36). In this combination, the charge density of 1*s* a.o.'s are subtracted from each other along the molecular axis and this creates a nodal region in between the nuclei. In other words, the electrons in this kind of m.o. spend less time on the average in this region than when present in isolated a.o.'s. This will lead to more repulsive rather attractive potential energy between the bonding region and is destabilized more in comparison to isolated a.o.'s and termed as anti-bonding m.o. Since this m.o. is also symmetrical around the bond axis, it is also termed as σ and is denoted as σ^*_{1s}. This m.o. does not possess a centre of inversion as a symmetry element and is called as un-gerade (u), *i.e.,* σ^*_{1s} (u). The equations (4.36) and (4.45) and Fig. 4.9*b* showing the energy levels correspond to these m.o.'s are consistent with this explanation.

The formation of m.o.'s from the LCAO of the second period elements can be proceeded on the same lines. The a.o.'s present are 1*s*, 2*s* and 2*p*. The size of 1*s* a.o. is quite small in these elements and is close to nucleus. Because of this the overlap of 1*s* orbitals of these atoms will be nearly negligible and does not add to the stabilization or destabilization of 1*s* a.o.'s. Due to this reason we generally involve the valence electrons only to form m.o.'s and leaves the core electrons localized on the respective atoms. Shape of 2*s* orbital (in reality of all the a.o.s') is similar to 1*s* a.o. and this will lead to the constructive and destructive overlap of 2*s* a.o.'s in the same manner as discussed for 1*s* a.o. The m.o.' so formed are also symmetrical around the bond axis and are termed as σ (bonding) and σ^* (anti-bonding). To distinguish these σ m.o.'s from those formed by the linear combination of 1*s* a.o., these are termed as $\sigma_{2s}(g)$ and $\sigma^*_{2s}(u)$ respectively. The normalized wave functions can be written as:

$$\psi\,(\sigma_{2s}) = \frac{1}{\sqrt{2}}\,(2s_A + 2s_B) \qquad ...(4.71)$$

$$\psi\,(\sigma^*_{2s}) = \frac{1}{\sqrt{2}}\,(2s_A - 2s_B) \qquad ...(4.72)$$

There are three-valence, p-a.o.'s (p_x, p_y, p_z) in the atoms belonging to second period. Unlike s orbital (which is spherical symmetrical), the p orbitals are directional and are located mostly along the three axes. It is thus necessary to specify the molecular axis and in general by convention, z-axis is chosen as the unique molecular axis. Similar to the m.o.'s formed by the overlap of s-orbitals, the constructive overlap of p_z orbitals also leads to the regions where the accumulation of charge density takes place in between the two nuclei as shown in Fig. 4.10. This leads to the stabilization of m.o. formed and is called bonding orbital. Destructive overlap leads to region which forms the nodal zone, m.o. is destabilized to a maximum extent and termed as anti-bonding m.o.

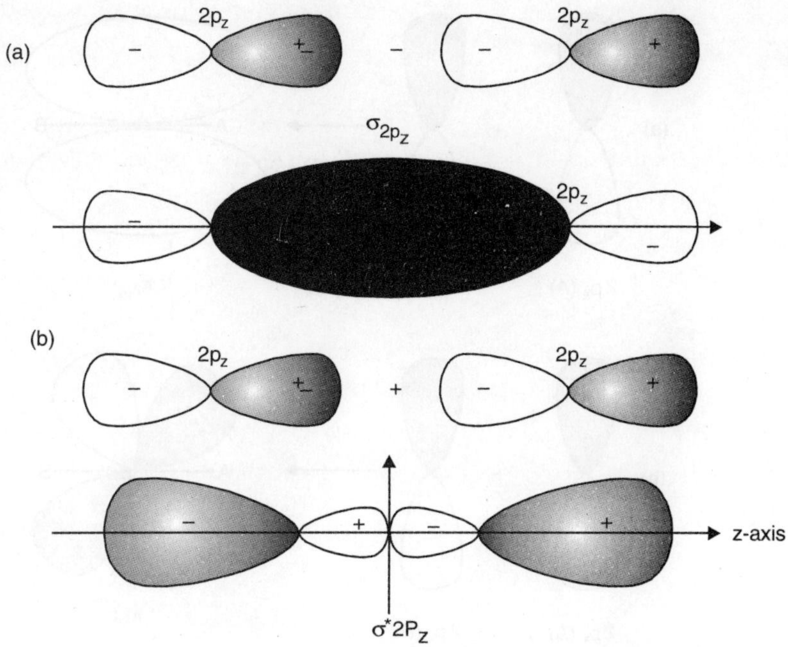

Figure 4.10: *Formation of σ bonding and anti-bonding molecular orbitals from $2p_z$ a.o.'s*

The bonding and anti-bonding m.o.'s also possess cylindrical symmetry around the molecular axis and are termed as σ_{2p_z} (*g*) and $\sigma^*_{2p_z}$ (*u*). The wave functions for these m.o.'s are:

$$\psi\ (\sigma(2p_z) = \frac{1}{\sqrt{2}}\ [2p_z\ (A) - 2p_z\ (B)] \qquad ...(4.73)$$

$$\psi\ (\sigma^*(2p_z) = \frac{1}{\sqrt{2}}\ [2p_z\ (A) + 2p_z\ (B)] \qquad ...(4.74)$$

The other two *p*-orbitals (p_x and p_y) are perpendicular to the molecular axis and thus can only have positive and negative overlap laterally. The constructive overlap leads to accumulation of charge density in between the nuclear region, with the difference from the earlier constructive overlap that it is concentrated more off the molecular axis rather than along the molecular axis and there is positive sign on one side and negative sign on the other side of the molecular axis. The plane passing through the molecular axis and perpendicular to *x*-axis (*yz*-plan) is a nodal plane. This is called bonding m.o.. On the other hand, the destructive overlap causes the loss of charge density even in the region off the molecular axis and creates nodal plane as shown in Fig. 4.11. This is called anti-bonding m.o.. The m.o.'s so formed, unlike σ ones, are not symmetrical around the molecular axis. By analogy with atomic orbitals, orbitals with one nodal plane are called *p*-orbitals, and so in case of molecules, orbitals with one nodal plane are called π (*pi*) orbital. The π bonding m.o. does not possess centre of symmetry and is termed as π_u, whereas the anti-bonding m.o. possesses the centre of inversion symmetry element and is termed as π^*_g. The former bonding m.o. is having less energy and the latter anti-bonding π^* m.o. is having greater energy in comparison to the a.o.'s of

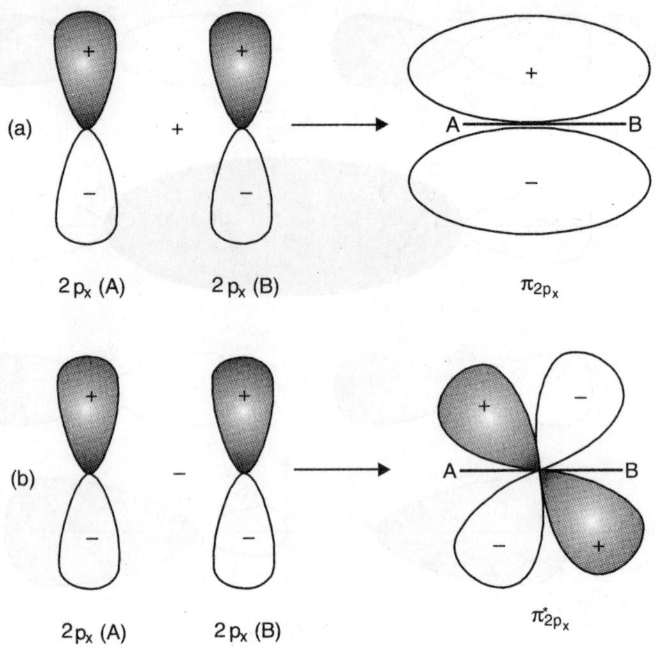

Figure 4.11: *Formation of (a) pi (π) bonding and (b) pi (π*) anti-bonding m.o.'s from the overlap of $2p_x$ a.o. of atoms A and B. Overlap of $2p_y$ a.o.'s will lead to the formation of pi (π) bonding and (b) p_y (π*) anti-bonding m.o.s perpendicular to these two m.o.'s*

the combining atoms. The shape of p_y orbital in comparison to z-axis is similar to p_x orbital. The overlap of p_y (A) and p_y (B) will also lead to the π-bonding and π^*-anti-bonding m.o.'s. These π orbitals are equivalent to the π orbitals formed from the combination of p_x a.o. and thus forms a degenerate set with degeneracy of two. These m.o.'s can be labeled as $\pi_{p_x} = \pi_{p_y}$ and $\pi^*_{p_y} = \pi^*_{p_x}$. The normalized wave functions for the π m.o.'s can be written as:

$$\psi\,(\pi_{2p_x}) = \frac{1}{\sqrt{2}}\,[2p_x\,(A) + 2p_x\,(B)] \qquad ...(4.75)$$

$$\psi\,(\pi_{2p_y}) = \frac{1}{\sqrt{2}}\,[2p_y\,(A) + 2p_y\,(B)] \qquad ...(4.76)$$

$$\psi\,(\pi^*_{2p_x}) = \frac{1}{\sqrt{2}}\,[2p_x\,(A) - 2p_x\,(B)] \qquad ...(4.77)$$

$$\psi\,(\pi^*_{2p_y}) = \frac{1}{\sqrt{2}}\,[2p_y\,(A) - 2p_y\,(B)] \qquad ...(4.78)$$

4.6.2 ENERGIES OF m. o.'s

After discussing the shapes and wave functions of m.o.'s formed from the a.o.'s we turn ourselves to the relative energies of these m.o.'s in comparison to respective a.o.'s. It has been established from the emission spectroscopy, *i.e.,* by observing transition from $2p \rightarrow 2s$, that 2*s* a.o. is more stable than 2*p* a.o. and this gap keeps on increasing with increase of atomic number along the second period. For example, the gap between 2*s* and 2*p* orbitals is 1.85 eV (178.5 kJ mol^{-1}) in lithium and 20 eV (1929.7 kJ mol^{-1}) in fluorine. Thus we place 2*s* orbital below 2*p* orbital in energy level diagram. Knowing that the overlap between a.o.'s to form σ m.o.'s are larger than that to form π m.o's, the possible energy level diagram is depicted in Fig. 4.12.

Lot of uncertainty existed about the position of σ_{2p_z} bonding orbital. It is only through experimental observations, *i.e.,* molecular spectroscopy, magnetic properties etc. that the position of σ_{2p_z} m.o. is ascertained in the energy level diagram. If the gap between 2*s* and 2*p a.o.*'s is

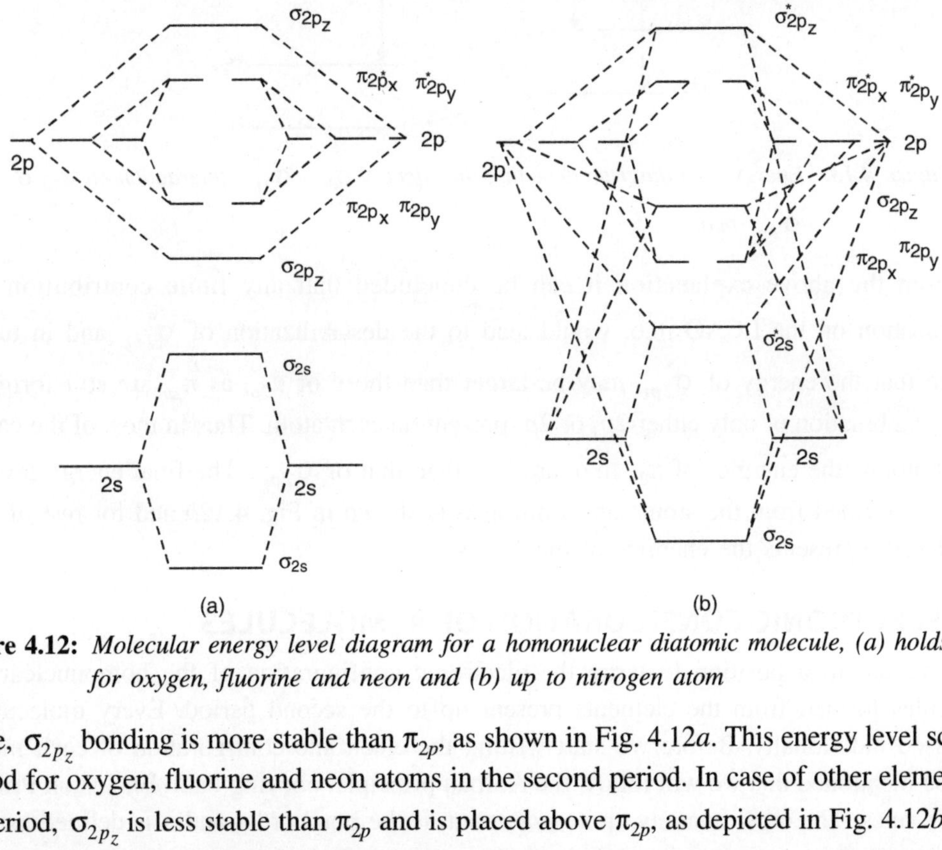

Figure 4.12: *Molecular energy level diagram for a homonuclear diatomic molecule, (a) holds good for oxygen, fluorine and neon and (b) up to nitrogen atom*

very large, σ_{2p_z} bonding is more stable than π_{2p}, as shown in Fig. 4.12*a*. This energy level scheme holds good for oxygen, fluorine and neon atoms in the second period. In case of other elements of second period, σ_{2p_z} is less stable than π_{2p} and is placed above π_{2p}, as depicted in Fig. 4.12*b*. This instability of σ_{2p_z} in comparison to π_{2p} has nothing to do with the poor overlap of $2p_z$ a.o. to form σ m.o., because if this would have been so then $\sigma^*_{2p_z}$ should have been less unstable than π^*_{2p}, which is not. This can be explained as follows. For 2*s* and 2*p* a.o.'s where the gap between them is small, it is better if we consider the formation of m.o.'s from the linear combination of two 2*s* and $2p_z$ together. The most stable m.o. from LCAO will be:

$$\psi = \frac{1}{\sqrt{2 + 2x^2}} [2s\ (A) + x\ 2p_z\ (A) + 2s\ (B) + x\ 2p_z\ (B)] \quad ...(4.789)$$

where the coefficient x is less than unity and represents the proportion of $2p_z$ a.o. included in the σ_{2s} m.o. In other words, *s-p* hybridization concept is used in forming the σ_{2s} and σ^*_{2s} m.o. and leads to their stability. But on the other hand, it will lead to the destabilization of σ_{2p_z} and σ^*_{2pz} m.o., suggesting that some 2*s* character will be present in the σ_{2p_z} and σ^*_{2pz} m.o. This effect is shown in Fig. 4.13.

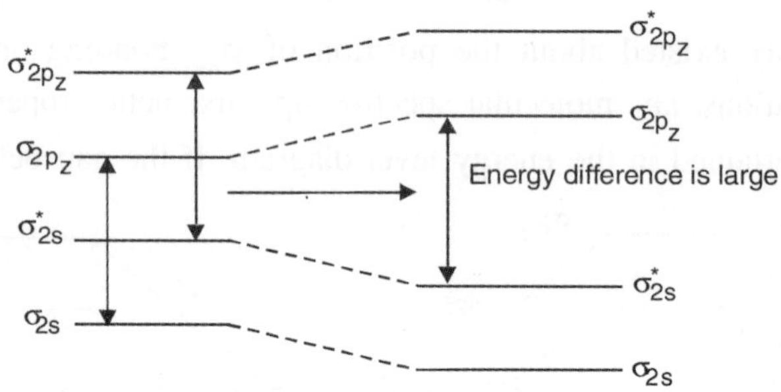

Figure 4.13: *Energy level diagram depicting the effect of $\sigma_{2s} - \sigma_{2p_z}$ interaction on σ_{2s}, σ^*_{2s}, σ_{2p_z} and $\sigma^*_{2p_z}$ m.o.*

From the above explanation it can be concluded that any finite contribution from *s-p* hybridization on the LCAO m.o. would lead to the destabilization of σ^*_{2pz} and in turn it may happen that the energy of σ^*_{2pz} may be larger than those of π_{2p}, as π_{2p} are still formed by the linear combination of only either $2p_x$ or $2p_y$ present on each atom. Thus in most of the cases, if not for all atoms, the energies of π_{2p} m.o. are less than that of $\sigma^*_{2p_z}$. The final energy level scheme of m.o.'s formed from the atoms up to nitrogen is shown in Fig. 4.12*b* and for rest of the atoms Fig. 4.12*a* represents the energies of the m.o.'s

4.7 ELECTRONIC CONFIGURATION OF A_2 MOLECULES

Now we are in a position to write the electronic configuration of the homonuclear diatomic molecules formed from the elements present up to the second period. Every molecule will be discussed individually. Before we start writing the electronic configuration of each molecule, it may be mentioned that we will follow the Aufbau principle, obeying Pauli Exclusion Principle and Hund's rules. The other quantity, quite important, is the bond order and it is defined as:

$$\text{Bond order} = \frac{1}{2} [\text{Number of electrons in the bonding m.o.} - \text{Number of electrons in the anti-bonding m.o.}] \quad ...(4.79)$$

Bond order can be related to the bond strength and bond length. Greater the bond order, greater will be the dissociation energy and smaller the bond length.

H_2^+ : It is the simplest molecule, which contains only one electron and can be placed in σ_{1s} m.o., *i.e.,* σ^1_{1s}. Bond order will be $\frac{1}{2}$. Its bond length is 0.106 nm and bond dissociation energy is 255.5 kJ mol^{-1}. H_2^+ species shows paramagnetic behaviour.

H_2: Each hydrogen atom subscribes one electron and two electrons can be placed in σ_{1s} m.o. as σ^2_{1s} following Pauli Exclusion Principle, *i.e.,* two electrons in σ_{1s} m.o. will have opposite spin. Thus the ground state of hydrogen molecule is formed by an electron pair bond and it may be emphasized here that electron pair bond which comes out of the necessity to place the electron in their most favourable arrangement is the backbone of the chemical bond theory.

The bond length of H_2 is 0.074 nm and dissociation energy is 432 kJ mol^{-1}. H_2 molecule with one bond has the smaller bond length and larger bond dissociation energy in comparison to H_2^+, which has only $\frac{1}{2}$ bond order. It may be generalized here that *when molecules comprised of nuclei of approximately the same atomic number are compared, the bond length decreases and bond dissociation energy increases with increase of bond order or number of bonds between A_2.*

He_2: The next element in the first period is helium and contains two 1*s* electrons in each atom. Two electrons will go to σ_{1s} with anti-parallel spin. The third electron cannot be placed in σ_{1s} m.o., because of Pauli Exclusion Principle. Next two electrons will be placed in σ^*_{1s} with anti-parallel spin, *i.e.,* σ^2_{1s} σ^{*2}_{1s}. This will weaken the He—He bond because the energy of σ^*_{1s} is higher than that of σ_{1s}, as well as, that of the individual He-atoms. This leads to a situation where He_2 has two electrons in the bonding and two in anti-bonding m.o.'s. This will cancel the stabilization gained by forming one bond (in reality the anti-bonding effect is more predominant than bonding one, equation 4.34) and thus, He_2 molecule is less stable in comparison to the individual helium atoms. This may also be stated as: "*This is because the electrons in bonding orbital tend to draw the nuclei together, whereas electrons in the anti-bonding orbital tend to draw the nuclei apart*". These two effects cancel each other and the result is that an anti-bonding electron cancels the effect of bonding electron. Thus in reality He_2 molecule is never observed experimentally.

Recently rare gas (noble gas) excimer lasers have been developed. Small stability of helium excimers are formed by the reaction:

$$He^* \ (S^1) + He \ (^0S^0) \rightarrow (He \ He)^*$$

where He^* (S^1) and He $(^0S^0)$ atoms are in their first excited and ground singlet states, with electronic configurations $1s^12s^1$ and $1s^2$ respectively. The electronic configuration for He-excimer will be σ^2_{1s} σ^{*1}_{1s} σ_{2s}. This configuration (common to all noble gas excimers) is stable than σ^2_{1s} σ^{*1}_{1s}, because it contains more bonding character than the anti-bonding character and will exist unless the excimer looses its energy.

Li_2: The electronic configuration of Li-atom is $1s^22s^1$ and Li_2 will be σ^2_{1s} σ^{*2}_{1s} σ^2_{2s}. The stability gained by placing two electrons in bonding σ_{1s} m.o. is cancelled by placing next two electrons in σ^{*2}_{1s}. The only pair of electrons, which explains the stability of Li_2, in comparison to the combining atoms, is due to σ^2_{2s} and leads to the formation of one bond between Li-Li. The above electronic configuration can also be written as $KK\sigma^2_{2s}$. The reason being, as stated earlier, the overlap between the core orbitals is generally very small because of smaller size. Theoretical calculations have shown (not given here) that there is hardly any difference between the electron densities in the two σ_{1s} and σ^*_{1s} m.o. of Li_2 and the electron densities of the two 1*s* a.o.'s of the individual Li-atoms. This justifies in considering only the valence electrons for bonding purposes and leaving the core electrons localized on the individual atom. The electronic configuration of Li_2 is consistent with the experimental results, *i.e.*, there is a single bond between Li-Li and molecule is diamagnetic in nature. The bond length and bond dissociation energy of Li_2 are 0.267 nm and 104.6 kJ mol^{-1} respectively. These parameters for Li_2 are much different from H_2 (0.074 nm, 432 kJ mol^{-1}), which also contains one bond. The larger bond length and smaller bond dissociation energy of Li_2 is due to the shielding of two σ_{2s} electrons by the electrons present in the $1s^2$ a.o. at the individual atom. This reduces the attractive potential between the electrons and the nuclei. The mutual repulsion of $1s^2$ electrons may also reduce the total attractive potential.

Be_2: The electronic configuration for Be atom is $K2s^2$ and that for Be_2 will be KK $\sigma^2_{2s}\ \sigma^{*2}_{2s}$. This electronic configuration can further be justified by saying that the nuclear charges of the atom beyond Li-atom in the periodic table are greater than the nuclear charge of Li-atom, and so the 1*s* electrons in these atoms will be held more tightly than the 1*s* electrons of the Li-atom. This configuration leads to (2 – 2 = 0) zero bond order and is consistent with the fact that Be_2 does not exist in the family of A_2 molecules.

B_2: The boron possesses $1s^22s^22p^1$ electronic configuration. In this case the problem starts because of the uncertainty whether σ_{2p_z} is lower in energy or $\pi_{p_x} = \pi_{p_y}$ are lower in energy. The experimental observations show that B_2 is a paramagnetic, with the indication of two unpaired electrons. The electronic configuration is: KK $\sigma^2_{2s}\ \sigma^{*2}_{2s}\ \pi^1_{2p_x} = \pi^1_{2p_y}$. This electronic configuration is also consistent with Hund's rules that the electrons occupy the degenerate orbitals by one each till all the degenerate orbitals are occupied by one electron. The bond order in Be_2 is one, a net π bond. The bond length is 0.159 nm and bond dissociation energy is 289 kJ mol^{-1}.

C_2: Carbon is $1s^22s^22p^2$ and electronic configuration for C_2 consistent with experimental results (*i.e.*, C_2 is diamagnetic) is: KK $\sigma^2_{2s}\ \sigma^{*2}_{2s}\ \pi^2_{2p_x}\pi^2_{2p_y}$. This gives two π bonds and all the electrons are paired, unlike B_2. Theoretical calculations have shown that the configurations KK $\sigma^2_{2s}\ \sigma^{*2}_{2s}\ \pi^2_{2p_x}\pi^2_{2p_y}$ and KK$\sigma^2_{2s}\ \sigma^{*2}_{2s}\ \pi^2_{2p_x}\pi^1_{2p_y}\sigma^1_{2p_z}$ have nearly the same energy, but the latter one can be rejected on the bases that this configuration leads to two unpaired electrons which will result in paramagnetic character in C_2, which actually is contrary to experimental results. In other words, it suggests that π_{2p_x} and π_{2p_y} m.o.'s are more stable than σ_{2p_z} m.o. The bond length in C_2 is 0.131 nm (consistent with the bond length observed in C=C bond in other molecules) and bond dissociation energy is 627.6 kJ mol^{-1}.

N_2: Electronic configuration of nitrogen atom is $1s^2 2s^2 2p^3$ and that of N_2 is KK σ^2_{2s} σ^{*2}_{2s} $\pi^2_{2p_x}\pi^2_{2p_y}\sigma^2_{2p_z}$. This leads to the presence of three bonds between, $N \equiv N$, maximum in A_2 family and no unpaired electrons. The bond length is 0.11 nm and bond dissociation energy is, 941.4 kJ mol^{-1}. All these facts agree with the experimental observations that N_2 is diamagnetic and the most inert diatomic molecule. Detailed analysis of electronic absorption spectrum has provided that the longest wavelength transition is due to excitation of electron from σ (highest occupied molecular orbital, HOMO) to π^* (lowest unoccupied molecular orbital, LUMO) orbital. Another proof comes from the fact that the electronic configuration of N_2^+ is KKσ^2_{2s} σ^{*2}_{2s} $\pi^2_{2p_x}\pi^2_{2p_y}\sigma^1_{2p_z}$.

O_2: Electronic configurations of oxygen atom and oxygen molecule are $1s^2 2s^2 2p^4$ and KK σ^2_{2s} σ^{*2}_{2s} $\sigma^2_{2p_z}\pi^2_{2p_x}\pi^2_{2p_y}$ $\pi^{*1}_{2p_x}$ $\pi^{*1}_{2p_y}$ respectively. Accordingly this suggests the formation of two bonds (one σ and one π) and the presence of two unpaired electrons. Paramagnetic character to the extent of two unpaired electrons is observed experimentally, further strengthening the use of m.o. theory. This is based on the fact that Lewis dot structure and simple valence bond theory (see later) could not explain the paramagnetic nature of O_2. Electronic absorption spectrum observed in case of O_2 also supports the presence of unpaired electrons in π^* anti-bonding m.o.

Bond length and bond dissociation energy in O_2 are 0.121 nm and 494 kJ mol^{-1}. As pointed out earlier that if a.o.'s of the combining atoms are nearly of the same atomic number, the addition of electron or removal of electron from anti-bonding m.o. will lead to a regular change in the bond length and bond dissociation energy of the species so formed. For example, if an electron is removed from π^* m.o., the bond order increases from 2 to 2.5 and there is a decrease in the bond length of O^+_2 (0.11227 nm). Similarly addition of electrons to π^* gives the species as O^-_2 and O_2^{2-} respectively. The bond order decreases from 2 to 1.5 in O^-_2 and 1.0 in $O^{2-}{}_2$, whereas the increase in bond length is observed as 0.126 nm and 0.149 nm respectively.

F_2: The electronic configurations of fluorine atom and F_2 molecule are represented as $1s^2 2s^2 2p^5$ and KK σ^2_{2s} σ^{*2}_{2s} $\sigma^2_{2p_z}\pi^2_{2p_x}\pi^2_{2p_y}$ $\pi^{*2}_{2p_x}$ $\pi^{*2}_{2p_y}$ respectively. There is no unpaired electron (showing diamagnetic character) and presence of one σ bond. The bond length is 0.142 nm and bond dissociation energy is 150.6 kJ mol^{-1}.

Ne_2: Ne atom is $1s^2 2s^2 2p^6$ is a closed shell structure and a hypothetical Ne_2 will be KK σ^2_{2s} σ^{*2}_{2s} $\sigma^2_{2p_z}\pi^2_{2p_x}\pi^2_{2p_y}$ $\pi^{*2}_{2p_x}$ $\pi^{*2}_{2p_y}$ $\sigma^{*2}_{2p_z}$. This configuration leads to no unpaired electrons and zero bond formation between Ne_2. This prediction conforms to the fact that so far Ne_2 has not been prepared.

4.8 OTHER A_2 TYPES OF MOLECULES

The above procedure can be generalized to A_2 molecules formed by the combination of two A-atoms present in other periods. In forming m.o.'s care should be taken to use only valence shell orbitals. Instead of discussing all the cases, we will take up the A_2 molecules belonging to alkali metals and halogen family.

Alkali metals: Diatomic molecules of alkali metals are Li_2, Na_2, K_2, Rb_2 and Cs_2. All of them possess σ_{ns}^2 electronic configuration, where *n* stands for the principal quantum number of the particular alkali atom. This indicates that each alkali metal atom is joined to other by a single σ bond possessing diamagnetic character as there is no unpaired electron. Bond lengths and bond dissociation energies of A_2 are given in Table 4.3. As expected, bond length increases and bond dissociation energy decreases with increase of principal quantum number. This is due to the shielding of σ_{ns}^2 m.o. from nuclei by the core electrons and repulsion among the core electrons.

Halogen family: All the four-halogen molecules (F_2, Cl_2, Br_2, I_2) possess the electronic configuration as $KK\sigma_{ns}^2\ \sigma^{*2}_{ns}\ \sigma_{np_z}^2\pi_{np_x}^2\pi_{np_y}^2\ \pi^{*2}_{np_x}\ \pi^{*2}_{np_y}$, where *n* is the principal quantum number of halogen atom. All the molecules are diamagnetic in nature with one σ bond. Bond lengths and bond energies of halogens are complied in Table 4.3. Regular increase in the bond length is consistent with the increase in principal quantum number *n*. On the other hand, no regular trend is observed in the bond dissociation energies, *i.e.,* it increases from F_2 to Cl_2 and then decreases regularly. The atomic electron absorption spectrum of fluorine has shown that the interelectronic repulsion in the 2*p* orbital of fluorine atom is more than that in 3*p* orbital of chlorine atom.

Table 4.3: *Bond lengths and bond dissociation energies of alkali metals and halogens.*

Molecule	Bond length (nm)	Bond dissociation energy (kJ mol^{-1})
Li_2	0.2672	104.6
Na_2	0.3078	72.4
K_2	0.3928	49.4
Rb_2	–	45.2
Cs_2	–	43.5
F_2	0.1418	150.6
Cl_2	0.1988	238.6
Br_2	0.2283	190.2
I_2	0.2667	148.7

4.9 MOLECULAR ORBITALS FORMED BY THE OVERLAP OF DIFFERENT ATOMIC ORBITALS

In the homonuclear diatomic molecules, m.o.'s are formed by the overlap of a.o.'s of the same quantum number and similar energies. The reason being that according to the results given in section 4.3.1, a strong overlap of a.o.'s (*i.e.,* formation of m.o.'s) is achieved if the a.o.'s have the similar energies and are of proper symmetry. So there is no need of taking any other combination. Actually in principle a.o. can combine with any other a.o. if the above two conditions are satisfied. In case of hetero-nuclear diatomic molecules or polyatomic molecules, we may need the overlap of a.o.'s having different quantum numbers and energies to form m.o., for example, *s* and *p*; *s* and *d*; *p* and *d* etc., We will consider the simplest case to understand the formation of m.o.'s from the overlap of *s* and *p* a.o.'s. Let us define the unique molecular axis as the *z*-axis as per convention. As shown in Fig. 4.14, *s* and p_z a.o.'s can combine to form m.o. as there is a positive overlap. The m.o. so obtained has accumulation of charge density in the middle of the bond and thus called bonding m.o. Similarly a negative overlap will lead to the depletion of charge density in the middle of bond, called anti-bonding m.o. In both cases the m.o.'s so formed are symmetrical around the molecular axis and thus called σ m.o. The wave function for the m.o. can be described by:

$$\psi\ (\sigma) = C_1 ns + C_2\ mp_z \quad ...(4.80)$$

where C_1 and C_2 are the coefficients and can be determined with the help of variation theorem, *n* and *m* are the principal quantum numbers of the combining a.o.'s, which may be same or different.

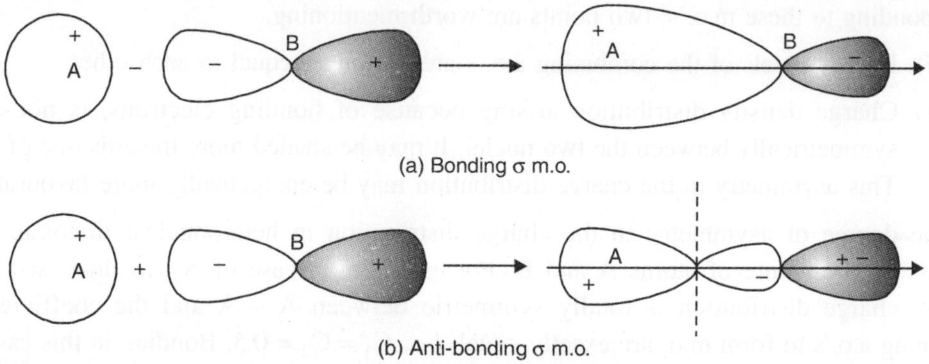

Figure 4.14: *Formation of bonding and anti-bonding m.o.'s by the overlap of s and* p_z *a.o.'s*

There are two *p* (p_x and p_y) orbitals which are perpendicular to molecular axis (*z*) and can interact to form π m.o. Figure 4.2 represents the overlap of *s* and $2p_x$ a.o.'s. It is obvious from Fig. 4.2 that in the upper half there is a constructive overlap between *s* and upper *lob* of p_x, which can increase the stability of m.o. On the other hand in lower half because of different sign of wave function of p_x orbital, there is a destructive overlap between *s* and p_x a.o.'s. The m.o. so formed will be destabilized. In other words, by whatever amount the m.o. so formed by the constructive overlap in the upper half is destabilized by equal amount due to destructive overlap in the lower half. The net effect is zero, *i.e., s* and p_x (similarly p_y) cannot combine to from m.o. One can also arrive at the similar conclusion by using the symmetry of combining a.o.'s. Table 4.1 provides the chart indicating which a.o.'s can combine to form m.o's.

As learnt earlier (4.4) the strength of the overlap of a.o.'s is given by the magnitude of overlap integral, *i.e.,*

$$S = \int \psi_A\ (r)\ \psi_B\ (r)\ d\tau$$

The magnitude of S depends on the size of the a.o.'s A and B. If the atomic orbital on A, ψ_A is small and that on B, ψ_B, is large and vice versa, the product of their amplitudes is small everywhere and the integral (summation) will also be small. On the other hand if the size of these combining a.o.'s are large in the region of combination space, there product is large in this region, as depicted in Fig. 4.14. The magnitude of the overlap integral is a good guide about the strength of the bond formed from the overlap of various a.o.'s. The value of S, lying between 0.2–0.3, is a good measure of the bond formed effectively.

4.10 HETERONUCLEAR DIATOMIC MOLECULES

Heteronuclear diatomic molecules are represented by AB, where A, and B are different atoms, having either different principal quantum number or different azimuthal quantum number. Conventionally A is less electronegative than B. Many important molecules like HCl, CO etc., belong to this class and lot of other physical properties; for example, dipole moment, electronegativity of the respective atoms etc., can be derived from the structure of these molecules. Before we start deriving the wave functions of the m.o.'s and the energy levels corresponding to these m.o.'s, two points are worth mentioning.

(*i*) Energy levels of the combining a.o.'s are no longer equal to each other.

(*ii*) Charge density distribution arising because of bonding electrons, is not distributed symmetrically between the two nuclei. It may be shaded more towards one of the nuclei. This asymmetry in the charge distribution may be energetically more favourable.

The degree of asymmetry in the charge distribution in heteronuclear diatomic molecules depends on the nature of atoms A and B. For example, in case of A_2, as discussed in the last section, charge distribution is totally symmetric between A – A and the coefficients of the combining a.o.'s to form m.o. are exactly equal, *i.e.*, $C_1 = C_2 = 0.5$. Bonding in this case is called pure covalent. The other extreme case can happen when $C_1 = 0$, $C_2 = 1$, *i.e.*, one or more than one electron has completely migrated from atom A (having lower ionization energy than B) to atom B to form ions $A^{n+}B^{n-}$. This is called pure ionic bond and the interactions between two ions are purely electrostatic in nature when using the first order approximation. In this section we are not going to discuss the two extreme kinds of molecules, but somewhat in between the extreme limits. This kind of bond is called polar bond, *i.e.*, pair of bonding electrons spend more time around one of the nuclei B in comparison to other nucleus A. This creates a partial (+δ) positive charge (less than one electron unit if it is a single bond) on nucleus A and partial (–δ) negative chrge but equal to that on A in magnitude on atom B. For example: $H^{\delta+}F^{\delta-}$, $N^{\delta+}O^{\delta}$ etc.

In quantum mechanics, the description of wave function and energy of m.o. formed from a.o.'s can also be described in the same way as has been carried out for homonuclear diatomic molecules, *i.e.*,

$$\psi = C_1\ \varphi_A + C_2\ \varphi_B \qquad \text{...(4.19)}$$

where ψ is a m.o. and φ_A and φ_B are the a.o.'s localized on A and B respectively before forming m.o.'s, C_1 and C_2 are coefficients and square of these give the contribution of a.o.'s in forming the m.o. Values of the energies of m.o.'s and their corresponding wave functions can be obtained with the help of variation theorem. In this section we will only discuss the physical picture of heteronuclear diatomic molecules rather than using variation theorem to do the detailed calculation of wave functions and energies.

4.10.1 LiH

The simplest heteronuclear diatomic molecule is LiH. The valence electrons on H and Li are $1s^1$ and $2s^1$. In the simplest m.o. picture, overlap of 1*s* and 2*s* orbitals leads to σ bonding and σ^* anti-bonding is shown in Fig. 4.15. The energies of a.o.'s 1*s* and 2*s* are shown on the left and right hand side respectively of the m.o.'s (Fig. 4.16). The wave function can be written as:

$$\psi = C_1\,\varphi_{1s}\,(\text{H}) + C_2\,\varphi_{2s}\,(\text{Li}) \qquad ...(4.81)$$

The values of the coefficients and energies of the m.o.'s so formed can be obtained using variation theorem. The molecular parameters; bond length, bond energies etc., so obtained do not agree with the experimental data. As mentioned earlier, due to unequal energies of a.o.'s of different atoms in heteronuclear diatomic molecules, it becomes difficult to choose the a.o.'s for proper combination. Knowing well that 2*s* a.o. is more stable than that of $2p_z$ of Li, the overlap of 2*s* of Li with 1*s* of H will be more effective. But we also know that the gap between 2*s* and $2p_z$ in Li atom is not very large (178.5 kJ mol^{-1}).

Participation of $2p_z$ a.o. in forming m.o. with 1*s* of H will also be possible. Thus the wave function of the bonding m.o. (σ) will be:

$$\psi = C_1\,\varphi_{2s}\,(\text{Li}) + C_2\,\varphi_{2p_z}\,(\text{Li}) + C_3\,\varphi_{1s}\,(\text{H}) \qquad ...(4.82)$$

Although not discussed in detail here, above equation (4.82), using variation theorem, can be written as:

$$\psi = 0.323\,\varphi_{2s}\,(\text{Li}) + 0.231\,\varphi_{2p_z}\,(\text{Li}) + 0.685\,\varphi_{1s}\,(\text{H}) \qquad ...(4.83)$$

indicating that $C_3 > C_1 > C_2$. The values of the coefficients are consistent with the energies of a.o.'s. The ionization enrgy of 1*s* of H is 1312.2 kJ mol^{-1} and that of Li is 521 kJ mol^{-1}, suggesting that bonding σ m.o. will be having more character of 1*s* (H) and bonding electrons will be spending more time around the hydrogen nucleus (74.85%). In other words, the pair of electrons will be nearly localized around the hydrogen atom. The contributions of 2*s* and $2p_z$ of Li to σ m.o. are 16.64% and 8.54% respectively. Similarly two anti-bonding m.o.'s so formed from the above combination will be having more character of 2*s* and $2p_z$ a.o.'s of Li. The wave functions representing these anti-bonding m.o.'s are:

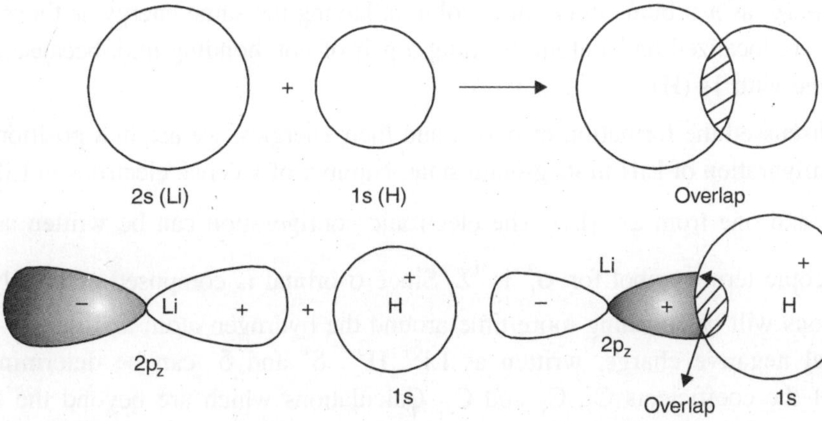

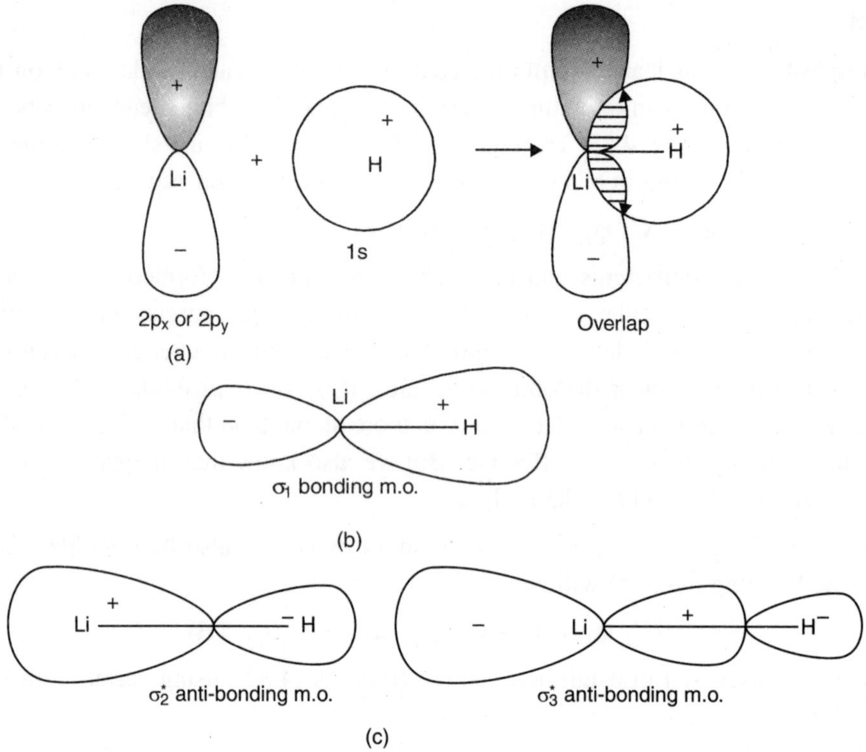

Figure 4.15: *(a) Overlap of 1s orbital of hydrogen a.o.with different a.o.'s of lithium atom. (b) Boundary surface of σ_1 bonding m.o. of LiH (equation 4.81). (c) Boundary surface of σ^*_2 and σ^*_3 anti-bonding m.o. of LiH (equations 4.82 and 4.83)*

$$\psi(\sigma^*_2) = C_4\, 2s\,(\text{Li}) - C_5\, 1s\,(\text{H}) \qquad C_4 > C_5 \qquad \text{...(4.84)}$$

$$\psi(\sigma^*_3) = C_6\, 2p_z\,(\text{Li}) - C_7\, 1s\,(\text{H}) \qquad C_6 > C_7 \qquad \text{...(4.85)}$$

As mentioned earlier $1s$ (H) cannot combine with $2p_x$ and $2p_y$ of Li and thus acts as non-bonding m.o. localized on Li atom.

The energy level diagram for LiH is depicted in Fig. 4.16, *i.e.*, $\sigma_1 < \sigma^*_2 < \sigma^*_3$. Figure 4.16 places $2p_x$ and $2p_y$ as π orbitals in the m.o. column, having the same energy as those of a.o.'s and these orbitals are localized on Li atom, forming a pair of non-bonding m.o. because these orbitals do not combine with $1s$ (H).

Having discussed the formation of m.o.'s and their energies, we are in a position to write the electronic configuration of LiH in its ground state. Number of valence electrons in LiH is two, one from $1s^1$ (H) and one from $2s^1$ (Li). The electronic configuration can be written as K (Li) σ_1^2. The spectroscopic term symbol for σ_1^2 is $^1\Sigma$. Since σ orbital is composed of mainly $1s$ (H), the pair of electrons will be spending more time around the hydrogen atom and thus H-atom will be having partial negative charge, written as $\text{Li}^{\delta+}\,\text{H}^{\delta-}$. δ^+ and δ^- can be determined from the knowledge of the coefficients C_1, C_2 and C_3. Calculations which are beyond the scope of this book have shown that the value of δ is 0.8, indicating that LiH is having 80% ionic character.

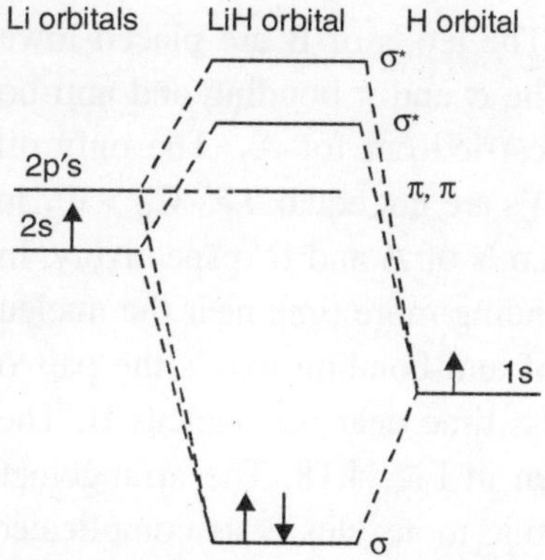

Figure 4.16: *Energy level diagram of LiH molecule*

In the end it may be pointed out that although the participation of $2p_z$ (Li) a.o. in the m.o. formation has improved the energies of m.o.'s and better picture has been obtained about the other physical properties like dipole moment etc., but it has lost the physical picture of chemical bond. This is more so in case of polyelectronic molecules, where it is more important to know the geometry of a molecule (*i.e.*, arrangements of atoms in a three dimensional space, the structure of molecule) than the energies of m.o.'s. Under these situations valence bond theory (and idea of hybridization) will be more helpful (to be discussed later).

4.10.2 AB MOLECULES IN GENERAL

We can now extend our discussion to AB molecules which involve the overlap of *s* and *p* orbitals of both the atoms A and B. Figure 4.17 represents the energy level diagram for m.o.'s of AB molecules. The a.o.'s of atom A are depicted on the left hand side, those of B on the right hand

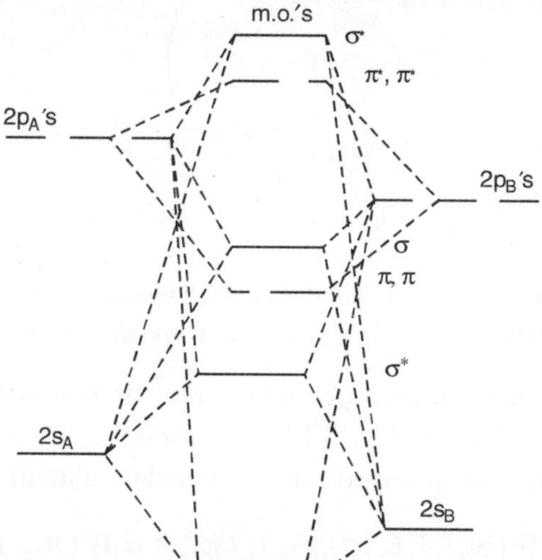

Figure 4.17: *Molecular orbital energy level diagram for heterocyclic diatomic molecules with appreciable σ – σ interaction. This generally happens when the heteroatoms differ by quantum numbers one or two. For other cases, Fig. 4.12 (a) holds*

side and m.o.'s in the middle. The a.o.'s of B are placed lower to the *a.o.*'s of A, as atom B is more electronegative than A. The σ and π bonding and anti-bonding m.o.'s of AB are formed in the same manner as has been carried out for A_2. The only difference is that the coefficients of a.o.'s taking part in forming m.o's are not equal, *i.e.*, $C_B > C_A$ in case of bonding m.o.'s, where C_A and C_B are the coefficients of a.o.'s of A and B respectively. In other words, the pair of electrons in the bonding m.o. will be spending more time near the nucleus B and less time near the nucleus A. On the other hand in case of anti-bonding m.o.'s the pair of electrons will be spending more time near the nucleus A and less time near the nucleus B. The bonding surface of the m.o.'s for general AB molecules are given in Fig. 4.18. The arrangement of m.o.'s in order of increasing energies in heteronuclear diatomic molecules is as complicated as for the homonuclear diatomic molecules, except that symbols *g* and *u* can not be used. Qualitatively it may be mentioned that if the energy difference between corresponding orbitals is less than the difference between the orbitals on a given atom or ion, (*in other words it may be said that as long as the atomic*

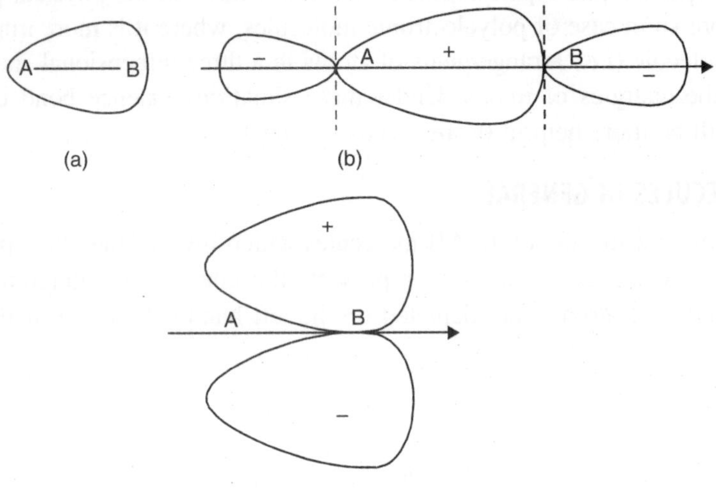

Figure 4.18: *Contour diagrams of the m.o.'s of heterodiatomic molecules. (a) Overlap of s-s orbitals, (b) overlap of s-p orbitals and (c) overlap of p-p orbitals. (a) and (b) depict σ m.o. and (c) depicts π m.o. All the three diagrams show that B is more electron negative than A. Thus the charge density lobes are shifted more towards B*

number of two atoms in the molecule are within one or two units) the orbitals and energies levels are similar to those given in Fig. 4.12. For example, NO, CO, NO^+, CO^- etc. Thus, the energy ordering most frequently observed for heteronuclear diatomic molecules is:

$$E(\sigma_{1s}) < E(\sigma^*_{1s}) < E(\sigma_{2s}) < E(\sigma^*_{2s}) < E(\pi_{2p_x}) = E(\pi_{2p_y}) < E(\sigma_{2p_z}) < E(\pi^*_{2p_x}) = E(\pi^*_{2p_y}) < E(\sigma^*_{2p_z}) < E(\sigma_{3s}) < E(\sigma^*_{3s}) < E(\pi_{3p_x}) = E(\pi_{3p_y}) < E(\sigma_{3p_z}) < E(\pi^*_{3p_x}) = E(\pi^*_{3p_y}) < E(\sigma^*_{3p_z}).$$

Since in case of AB molecules, the a.o.'s used in the formation of m.o.'s cannot be specified properly, the arbitrary numbering has also been assigned to m.o.'s. The lower most m.o. either σ or π starts with 1 and numbering keeps on increasing with increase of energy of the m.o.'s, *i.e.*,

$$\sigma_1 < \sigma_2 < \pi_1 = \pi_2 < \sigma_3 < \pi_3^* = \pi_4^* < \sigma_4$$

Examples 4.1*a* and 4.1*b* provide the use of diagram shown in Fig. 4.17 to write the electronic configuration and spectroscopic term symbols for couple of molecules.

Example 4.1: Write the electronic configuration and spectroscopic term symbols for (*a*) CO and (*b*) ClO molecules.

Solution.

(*a*) CO: Total number of valence electrons in CO is 10, (four from C and six from O-atoms). The electronic configuration is: $\sigma_1^2\ \sigma_2^{*2}\ \pi_1^2 = \pi_2^2\ \sigma_3^2$, *i.e.*, one σ bond and two π bonds. Since this is a completely closed shell structure, the term symbol is $^1\Sigma$. Any AB system with 10 electrons will have the electronic configuration and term symbol similar to CO. The bond length and bond dissociation energy of CO is 0.1128 nm and 1070 kJ mol^{-1} respectively. The bond dissociation energy of CO is larger than that of N_2 (941. 4 kJ mol^{-1}) even though both these molecules possess the triple bond.

(*b*) ClO: Total number of valence electrons are 13 (7 + 6). Electronic configuration is $\sigma_1^2\ \sigma_2^{*2}\ \pi_1^2 = \pi_2^2\ \sigma_3^2\ \pi_3^{*2} = \pi_4^{*1}$ and predicts bond order as $\frac{1}{2}$ [8 – 5] = $\frac{3}{2}$, less than the molecules containing ten electrons. Spectroscopic term symbol will be $^2\Pi$, because all the closed shell structures will have Λ = 0 and contribution to Λ is only from the partially filled π^* m.o.

4.10.3 DIFFERENT TYPES OF m.o.'s

From the above discussion it may be generalized here that the number of electrons in heteronuclear diatomic molecules can be assigned to four kinds of orbitals. These can be: (*i*) inner shell or core electrons, (*ii*) valence shell non-bonding electrons, (*iii*) valence shell bonding electrons and (*iv*) valence shell anti-bonding electrons. It will be easier to understand the above classification by taking two examples, HF and O_2.

HF: In HF, hydrogen can have only one electron in 1*s* orbital in its stable configuration, whereas in fluorine K electrons can be classified as of type (*i*). The same is the case of 2*s* electrons (This assignment can be different if hybridized atomic orbitals are considered for fluorine (section 4.19)). In this case, these will not be pure 2*s* electrons but a mixture of 2*s* and $2p_z$ (if *z* is the molecular axis). But still these will be localized around fluorine atom. If *z*-axis is the molecular axis of HF, p_z a.o. of fluorine and 1*s* of hydrogen atom will form two m.o.'s, having σ character. Bonding m.o. can be classified as (*iii*) and anti-bonding as (*iv*). The remaining two a.o.'s of fluorine ($2p_x$ and $2p_y$) can not form m.o. with hydrogen atom, because the a.o. of hydrogen atom which can form m.o. with fluorine atom by LCAO are $2p_z$ orbital and energies of these a.o.'s are very high. In other words, there will, effectively, be no combination among themselves. Although $2p_x$ and $2p_y$ a.o.'s are the valence electrons of fluorine, but these a.o.'s will remain localized around their original nucleus and may suffer a very small polarization. Two electrons with opposite spin in these orbitals are called lone pair electrons.

O_2: In O_2, both the oxygen atoms can have six valence electrons in its stable configuration and both of them possess K electrons. These can be classified as of type (*i*). The same is the case of 2*s* electrons If *z*-axis is the molecular axis of O_2, p_z a.o. of each oxygen atom will form two m.o.'s, having σ character. Bonding m.o. can be classified as (*iii*) and anti-bonding as (*iv*). The remaining one unpaired electron of oxygen $2p_x$ (or $2p_y$) can form π m.o.'s, one bonding and one anti-bonding, because these a.o.'s of oxygen atoms are perpendicular to the molecular axis and thus can have lateral overlap. These molecular orbitals will be placed between the two σ m.o.'s. The rest of the two electrons $2p_y$ (or $2p_x$) a.o.'s are the valence electrons of each of the oxygen atoms, but these a.o.'s will remain localized around their original nucleus and may suffer a very small polarization. Two electrons with opposite spin in these orbitals of each atom are called lone pair electrons.

4.11 VALENCE BOND THEORY (v.b.t.)

Although the large majority of molecular electronic structure calculations are carried out using some form of m.o.t., most of our qualitative thinking about chemical bonding employs v.b. model. This model considers the combining atoms as a whole and a chemical bond is viewed as a pairing of electrons between the atomic orbitals localized on the bonded atoms. The pairing of electrons can be achieved by the unpaired electrons on each atom combining to form an electron pair with opposite spin functions (a covalent bond) or by the donation of an electron pair on one atom to an empty atomic orbital on a second atom (a coordinate covalent bond). Under both circumstances the pair of electrons can be viewed as present between the bonded atoms. Thus it differs from the m.o. theory where only the nuclei (nuclei + core electrons) are first arranged in space and the valence electrons are then filled up in the poly-centric m.o.'s (*i.e.*, electron pair is spread over the entire molecule in m.o.). It is quite clear that the v.b. description represents more closely the conventional picture of a chemical bond. Heitler and London were the first to explain the chemical bonding between the simplest molecules (H_2).

Similar to m.o.t., the simplest molecule (H_2) will be considered in v.b.t. also. The model will start with simple mathematical picture and the results so obtained will be improved upon by taking new mathematical functions. But it must be kept in mind that all the mathematical functions used will be based on chemical intuition, experience and experimental results. In the following sections, results already discussed, will be used without proving them again.

4.11.1 HYDROGEN MOLECULE (H_2)

H_2 molecule consists of two electrons and each hydrogen atom contributes only one electron, (say H_A and H_B). Similar to helium atom, we will consider only the space part of the a.o.'s first and then spin part will be added. To start with, consider two hydrogen atoms separated by a large distance. Under these conditions, each electron will be present in the ground state and can be represented by:

$$\varphi_A(1) = 1s_A(1) \text{ and } \varphi_B(2) = 1s_B(2) \qquad ...(4.86)$$

and each electron will be having the same ground state energy equal to E_H. If the internuclear distance R is so large between two atoms that practically there is no interaction between them, then the total wave function ψ and the total energy E can be written as:

$$\psi = \varphi_A(1)\,\varphi_B(2) \qquad \text{...(4.87)}$$

and

$$E = 2\,E_H \qquad \text{...(4.88)}$$

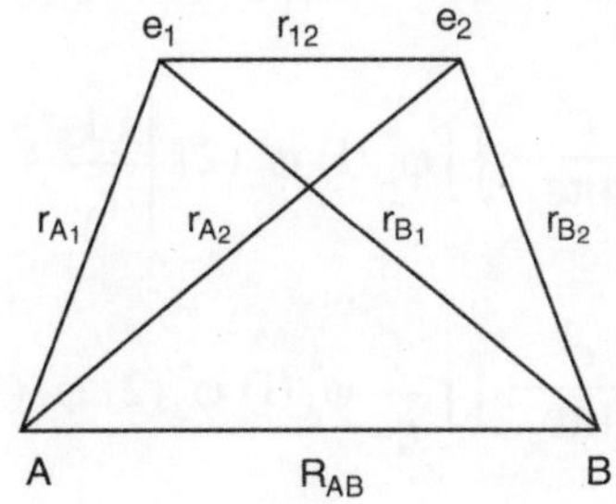

Figure 4.19: *Schematic diagram of hydrogen molecule*

If you see the wave function (4.87), we have proceeded on the same lines as we did for He-atom, *i.e.,* considering only the space part and adding spin part later on. Further when the two atoms are brought together, interactions will take place and molecule may be represented as shown in Fig. 4.19. Complete Hamiltonian operator can be written as:

$$\hat{H} = -\frac{h^2}{8\,\pi^2 m_e}(\nabla_1^2 + \nabla_2^2) - \frac{e^2}{4\pi\varepsilon_0}\left(\frac{1}{r_{A_1}} + \frac{1}{r_{A_2}} + \frac{1}{r_{B_1}} + \frac{1}{r_{B_2}} - \frac{1}{r_{12}} - \frac{1}{R_{AB}}\right) \qquad \text{...(4.89)}$$

and the average energy E can be calculated using equation (2.126) and variation method.

$$E = \frac{\int \psi^*\,\hat{H}\,\psi\,d\tau}{\int \psi^*\,\psi\,d\tau} \qquad \text{...(2.126)}$$

The simplest approximate wave function for variation method can be considered as (4.87), hoping that it may give useful information. On the similar lines as done earlier (4.6), value of E have been calculated by taking fixed position of the nuclei and then determining the E (R) for different values of R. Some of the integrals are difficult to solve, but can be carried out. Here we will only give the integrals and their results. Simplification can be made by using earlier results that φ_A (1) is an eigen function to the operator

$$\hat{H} = -\frac{h^2}{8\pi^2 m_e}\nabla_1^2 - \frac{e^2}{4\pi\varepsilon_0}\frac{1}{r_{a_1}} \qquad \text{...(4.90)}$$

with corresponding eigen value E_H (1). Assuming that φ_A and φ_B are normalized, equation (2.126) can be written as:

$$E = \iint \varphi_A^*(1)\,\varphi_B^*(2)\left[-\frac{h^2}{8\pi^2 m_e}(\nabla_1^2 + \nabla_2^2) - \frac{e^2}{4\pi\varepsilon_0}\left(\frac{1}{r_{A_1}} + \frac{1}{r_{A_2}} + \frac{1}{r_{B_1}} + \frac{1}{r_{B_2}} - \frac{1}{r_{12}} - \frac{1}{R_{AB}}\right)\right]$$
$$\varphi_A(1)\,\varphi_B(2)\,d\tau_1\,d\tau_2$$

$$= \int \varphi_A^*(1) \left[-\frac{h^2}{8\pi^2 m_e} \nabla_1^2 - \frac{e^2}{4\pi\varepsilon_0} \frac{1}{r_{A_1}} \right] \varphi_A(1)\, d\tau_1 \int \varphi_B^*(2)\, \varphi_B(2)\, d\tau_2$$

$$+ \int \varphi_B^*(2) \left[-\frac{h^2}{8\pi^2 m_e} \nabla_2^2 - \frac{e^2}{4\pi\varepsilon_0} \frac{1}{r_{B_2}} \right] \varphi_B(2)\, d\tau_2 \int \varphi_A^*(1)\, \varphi_A(1)\, d\tau_1$$

$$- \frac{e^2}{4\pi\varepsilon_0} \iint \varphi_A^*(1)\, \varphi_B^*(2) \left[\frac{1}{r_{A_2}} + \frac{1}{r_{B_1}} \right] \varphi_A(1)\, \varphi_B(2)\, d\tau_1\, d\tau_2$$

$$+ \frac{e^2}{4\pi\varepsilon_0} \iint \frac{1}{r_{12}} \varphi_A^*(1)\, \varphi_B^*(2)\, \varphi_A(1)\, \varphi_B(2)\, d\tau_1\, d\tau_2$$

$$+ \frac{e^2}{4\pi\varepsilon_0 R_{AB}} \int \varphi_A^*(1)\, \varphi_B^*(2)\, \varphi_A(1)\, \varphi_B(2)\, d\tau_1 d\tau_2$$

$$= 2E_H - 2\frac{e^2}{4\pi\varepsilon_0} \int \varphi_A^*(1) \frac{1}{r_{B_1}} \varphi_A(1)\, d\tau_1 + \frac{e^2}{4\pi\varepsilon_0}$$

$$\iint \frac{1}{r_{12}} \varphi_A^*(1)\, \varphi_B^*(2)\, \varphi_A(1)\, \varphi_B(2)\, d\tau_1\, d\tau_2 + \frac{e^2}{4\pi\varepsilon_0 R_{AB}}$$

$$= 2E_H + J \qquad \text{...(4.91)}$$

where

$$J = -2\frac{e^2}{4\pi\varepsilon_0} \int \varphi_A^*(1) \frac{1}{r_{B_1}} \varphi_A(1)\, d\tau_1 + \frac{e^2}{4\pi\varepsilon_0}$$

$$\iint \frac{1}{r_{12}} \varphi_A^*(1)\, \varphi_B^*(2)\, \varphi_A(1)\, \varphi_B(2)\, d\tau_1\, d\tau_2 + \frac{e^2}{4\pi\varepsilon_0 R_{AB}} \qquad \text{...(4.92)}$$

The terms in equation (4.91) can be explained as follows: the first term gives the energy of individual hydrogen atom when separated apart, the second term gives the attractive potential (or energy) between the charge cloud of electron (1) and the nucleus B, factor 2 appears because of the similar attractive potential between the charge cloud of electron 2 and nucleus A; third term represents the repulsive potential (or energy) between the charge clouds of two electrons (*i.e.*, $e\,\varphi_A^2(1)$ and $e\,\varphi_B^2(2)$) and fourth term gives the repulsive energy expression between the nuclei. From the above expression, it is clear that a chemical bond will be formed between two hydrogen atoms if the attractive energy terms will be greater than both the repulsive energy terms. Figure 4.20 shows the variation of E as a function of internuclear distance R. The curve I undergoes a minimum at R = 0.09 nm (or 1.7 a_0) with E = 0.25 eV (24.2 kJ mol^{-1}), indicating that a chemical bond can be formed between the two hydrogen atoms. The agreement between the observed values of R (0.0741 nm) is not that bad with the calculated one, but the stabilization energy is very far off from the experimental one (4.747 eV, 458 kJ mol^{-1}). The only satisfactory point is that the above treatment is on the right track but needs further refinement.

The further modification was proposed by Heitler and London. In the last paragraph we assumed that electron 1 is present around the nucleus A and electron 2 is present around the nucleus B. We also know that all the electrons are identical and thus it is not possible to distinguish that the electron around the nucleus A is 1 and around the nucleus B it is 2. It could also be possible that electron 2 is around the nucleus A and electron 1 is around the nucleus B, *i.e.,* the wave functions in both cases can be written as:

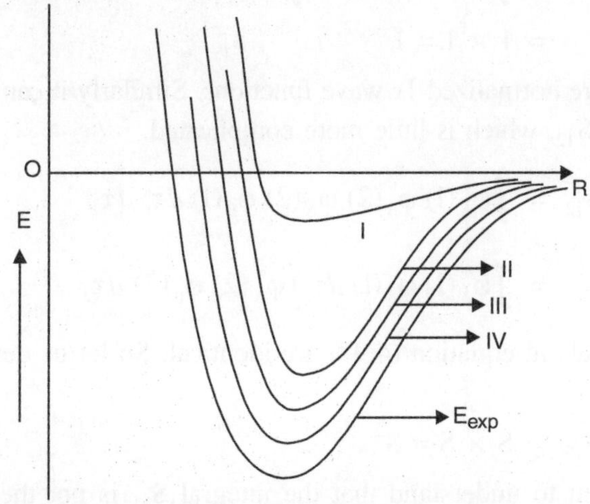

Figure 4.20: *Plot of energy (E) versus internuclear distance for different models. (I) using equation (4.87); (II) using equation (4.93); (III) using equation (4.111); (IV) using equation (4.114); E_{expt}, experimental results*

$$\psi_I = \varphi_A(1)\,\varphi_B(2),\quad \psi_{II} = \varphi_A(2)\,\varphi_B(1) \qquad ...(4.87a)$$

and the actual wave function may be the linear combination of ψ_I and ψ_{II}.

$$\psi = C_1\,\psi_I + C_2\,\psi_{II} \qquad ...(4.93)$$

where C_1 and C_2 are coefficients to be determined from the secular equations, as in section 3.6. Square of these coefficients give the weighted average of these wave functions in the total wave function. Since the wave functions ψ_I and ψ_{II} are identical, the secular equations will give $H_{11} = H_{22}$, $H_{12} = H_{21}$, $S_{11} = S_{22}$, as well as, $C_1^2 = C_2^2$. Thus

$$C_1 = \pm C_2 = C \qquad ...(4.94)$$

Substituting the values of coefficients in the secular equations, we get the two wave functions and their corresponding energies, *i.e.,*

$$\psi_+ = C\,[\varphi_A(1)\,\varphi_B(2) + \varphi_A(2)\,\varphi_B(1)],\quad E_+ = \frac{H_{11}+H_{12}}{S_{11}+S_{12}} \qquad ...(4.95)$$

$$\psi_- = C\,[\varphi_A(1)\,\varphi_B(2) - \varphi_A(2)\,\varphi_B(1)],\quad E_- = \frac{H_{11}-H_{12}}{S_{11}-S_{12}} \qquad ...(4.96)$$

Since both H_{11} and H_{12} have negative signs, E_+ represents the stabilized or lowest energy state with corresponding wave function ψ_+. In order to get the better quantitative picture, we must now evaluate the integrals, H_{11}, H_{12}, S_{11} and S_{12}.

S_{11} and S_{12}: We have been using the overlap integral word very commonly earlier but its physical meaning can be understood only here.

$$S_{11} = \int \varphi_A^*(1)\, \varphi_B^*(2)\, \varphi_A(1)\, \varphi_B(2)\, d\tau_1\, d\tau_2$$

$$= \int \varphi_A^*(1)\, \varphi_A(1)\, d\tau_1 \int \varphi_B^*(2)\, \varphi_B(2)\, d\tau_2$$

$$= 1 \times 1 = 1$$

Since φ_A and φ_B are normalized 1*s* wave functions. Similarly it can also be shown that $S_{22} = 1$. Now we can evaluate S_{12}, which is little more complicated.

$$S_{12} = \int \varphi_A^*(1)\, \varphi_B^*(2)\, \varphi_A(2)\, \varphi_B(1)\, d\tau_1\, d\tau_2$$

$$= \int \varphi_A^*(1)\, \varphi_B(1)\, d\tau_1 \int \varphi_A^*(2)\, \varphi_B(2)\, d\tau_2 \qquad ...(4.97)$$

Both of the integrals in equation (4.97) are identical. So let us denote each of these by S, so that

$$S_{12} = S \times S = S^2 \qquad ...(4.97a)$$

It is very important to understand that the integral S_{11} is not the same as integral S_{12}. It is because in the above integral both the 1*s* orbitals are present on the same atom A, whereas in the (4.97), one of 1*s* orbital is centred on nucleus A and the other 1*s* orbital is present on nucleus B. We have depicted these orbitals in Fig. 4.6 and can be realized that this integral will be non-zero only when both 1*s* (A) and 1*s* (B) are non-zero simultaneously. This is only possible in the region of the Fig. 4.6 where two orbitals overlap with each other. The magnitude of the integral S, as defined above, depends on the extent of the overlap of these two orbitals it is called as *overlap integral*. It is also obvious that magnitude of the overlap integral gives the extent of overlap of two orbitals centred on two different nuclei. As mentioned in m.o.t., it is not easy to solve this integral explicitly, but one thing is clear that it depends on the internuclear distance and its magnitude as a function of internuclear distance is given by equation (4.50).

H_{11} and H_{12}

$$H_{11} = \int \varphi_A^*(1)\, \varphi_B^*(2)\, \hat{H}\, \varphi_A(1)\, \varphi_B(2)\, d\tau_1\, d\tau_2 \qquad ...(4.98)$$

where $\hat{H}$ is defined by the equation (4.89). H_{11} is exactly similar to equation (4.91), *i.e.*,

$$H_{11} = 2\, E_H + J \qquad ...(4.91)$$

As explained earlier, each term in equation (4.92) can be interpreted as a classical Coulomb interaction and so J is called Coulomb integral. The value of J depends on the internuclear separation and its evaluation is a tedious exercise. The value of J can be evaluated and can be expressed in terms of atomic units as:

$$J = e^{-2R}\left[\frac{1}{R}+\frac{5}{8}-\frac{3}{4}R-\frac{R^2}{6}\right] \quad ...(4.99)$$

Similarly H_{12} can also be evaluated as

$$H_{12} = \iint \varphi_A^*(1)\,\varphi_B^*(2)\,H\,\varphi_A(2)\,\varphi_B(1)\,d\tau_1\,d\tau_2 \quad ...(4.100)$$

$$= \iint \varphi_A^*(1)\,\varphi_B^*(2)\left[-\frac{h^2}{8\pi^2 m_e}\left(\nabla_1^2+\nabla_2^2\right)-\frac{e^2}{4\pi\varepsilon_0}\right.$$

$$\left.\left(\frac{1}{r_{A_1}}+\frac{1}{r_{A_2}}+\frac{1}{r_{B_1}}+\frac{1}{r_{B_2}}-\frac{1}{r_{12}}-\frac{1}{R_{AB}}\right)\right]\varphi_A(2)\,\varphi_B(1)\,d\tau_1\,d\tau_2$$

Using the following relations,

$$\left[-\frac{h^2}{8\pi^2 m_e}\nabla_1^2-\frac{e^2}{4\pi\varepsilon_0}\frac{1}{r_{B_1}}\right]\varphi_B(1) = E_H\,\varphi_B(1) \quad ...(4.101)$$

and $$\left[-\frac{h^2}{8\pi^2 m_e}\nabla_2^2-\frac{e^2}{4\pi\varepsilon_0}\frac{1}{r_{A_2}}\right]\varphi_A(2) = E_H\,\varphi_A(2) \quad ...(4.102)$$

H_{12} can be written as:

$$= E_H \int \varphi_A^*(1)\,\varphi_B(1)\,d\tau_1 \int \varphi_A^*(2)\,\varphi_B(2)\,d\tau_2$$

$$+ E_H \int \varphi_A^*(2)\,\varphi_B(2)\,d\tau_2 \int \varphi_A^*(1)\,\varphi_B(1)\,d\tau_1 + \frac{e^2}{4\pi\varepsilon_0}$$

$$+ \iint \varphi_A^*(1)\,\varphi_B^*(2)\left[-\frac{1}{r_{A_1}}-\frac{1}{r_{B_2}}\right]\varphi_B(1)\,\varphi_A(2)\,d\tau_1 d\tau_2$$

$$+ \frac{e^2}{4\pi\varepsilon_0}\iint \frac{1}{r_{12}}\varphi_A^*(1)\,\varphi_B^*(2)\,\varphi_A(2)\,\varphi_B(1)\,d\tau_1\,d\tau_2$$

$$+ \frac{e^2}{4\pi\varepsilon_0\,R_{AB}}\int \varphi_A^*(1)\,\varphi_B^*(2)\,\varphi_A(2)\,\varphi_B(1)\,d\tau_1 d\tau_2$$

Thus, $$H_{12} = 2E_H\,S^2 + K \quad ...(4.103)$$

where

$$K = \frac{e^2}{4\pi\varepsilon_0}\iint \varphi_A^*(1)\,\varphi_B^*(2)\left[-\frac{1}{r_{A_1}}-\frac{1}{r_{B_2}}\right]\varphi_B(1)\,\varphi_A(2)\,d\tau_1\,d\tau_2$$

$$+ \frac{e^2}{4\pi\varepsilon_0} \iint \frac{1}{r_{12}} \varphi_A^*(1)\, \varphi_B^*(2)\, \varphi_A(2)\, \varphi_B(1)\, d\tau_1\, d\tau_2$$

$$+ \frac{e^2}{4\pi\varepsilon_0\, R_{AB}} \int \varphi_A^*(1)\, \varphi_B^*(2)\, \varphi_A(2)\, \varphi_B(1)\, d\tau_1 d\tau_2 \qquad ...(4.104)$$

Difference between J and K is that we have a factor φ_A (1) φ_B (1) in K and φ_A (1) φ_A (1) in J. Following the similar treatment as carried out for H_2^+, the secular determinant can be written as:

$$\begin{vmatrix} 2E_H + J - E & 2E_H S^2 + K - ES^2 \\ 2E_H S^2 + K - ES^2 & 2E_H + J - E \end{vmatrix} = 0$$

or $\quad (2E_H + J - E)^2 = (2E_H S^2 + K - ES^2)^2$

By taking square root of both the sides, we get

$$2E_H + J - E = \pm (2E_H S^2 + K - ES^2)$$

or
$$E_+ = 2E_H + \frac{J + K}{1 + S^2} \qquad ...(4.105)$$

$$E_- = 2E_H + \frac{J - K}{1 - S^2} \qquad ...(4.106)$$

Considering $2E_H$ as the energy of the two isolated hydrogen atoms, the energy of H_2 levels in comparison to isolated hydrogen atoms are:

$$\frac{J \pm K}{1 \pm S^2} \qquad ...(4.107)$$

Similar to H_2^+ ion, stability of H_2 bond is due to the term K, which arises only because of the use of wave function (4.93). This wave function clearly shows the presence of indistinguishability of two electrons. In other words K is the outcome of applying quantum mechanics to the molecular structure. Since it arises when the electrons are exchanged between the two nuclei, K is called *exchange integral.* Since exchange integral K is strictly an outcome of quantum mechanical treatment, we can say that the stability or even the existence of chemical bond is a quantum mechanical effect. The value of K can be calculated in terms of internuclear distance R, but it is beyond the scope of this book. However, variation of K as a function of R is plotted in Fig. 4.20.

Having calculated the energies of the states, we can normalize the corresponding wave functions using Born-relation, *i.e.*,

$$\int \psi^* \psi\, d\tau = 1. \text{ Taking } \psi = \psi_+$$

$$C^2 \iint (\psi_I + \psi_{II})^2 d\tau_1 d\tau_2 = 1$$

or
$$C^2 \iint [\varphi_A(1)\, \varphi_B(2) + \varphi_A(2)\, \varphi_B(1)]^2 d\tau_1 d\tau_2 = 1$$

$$C^2 \iint [\varphi_A^2(1)\,\varphi_B^2(2)\,d\tau_1\,d\tau_2 + \iint \varphi_A^2(2)\,\varphi_B^2(1)\,d\tau_1 d\tau_2$$

$$+ 2\iint [\varphi_A(1)\,\varphi_B(2)\,\varphi_A(2)\,\varphi_B(1)\,d\tau_1 d\tau_2] = 1$$

$$C^2\,[\,1 + 1 + 2S.S] = \text{ or } C = \frac{1}{\sqrt{2+2S^2}}$$

So
$$\psi_+ = \frac{1}{\sqrt{2+2S^2}}\,[\varphi_A(1)\,\varphi_B(2) + \varphi_A(2)\,\varphi_B(1)] \qquad ...(4.108)$$

Similarly
$$\psi_- = \frac{1}{\sqrt{2-2S^2}}\,[\varphi_A(1)\,\varphi_B(2) - \varphi_A(2)\,\varphi_B(1)] \qquad ...(4.109)$$

Note of caution: Before proceeding further, one point is worth considering. In the above discussion it has been pointed out that stability or formation of chemical bond is due to exchange integral H_{12} and energy released when the electrons exchange positions over the two nuclei (4.93). Sometimes even exchange frequency of ~ $7 \times 10^{14}\ s^{-1}$ (obtained from $\frac{H_{12}}{h}$) is assigned. It may be mentioned here that these words should be treated cautiously. Since we cannot even point out which electron is which, then how we can say that a particular electron has moved from one nucleus to another nucleus. Further the value of energy obtained for bonding H_2 by v.b.t. is similar to that obtained by m.o.t., in which no such terms appear in this manner. As it is known that quantum mechanics is totally silent about the motion of particle, there is no such thing like exchange phenomenon. As mentioned by Coulson, there is a wave function ψ, which closely resembles with (4.93) and the decrease in energy is observed by allowing either of the electrons near either of nucleus. The lowering of energy is because (*i*) of the attraction of both the electrons by both the nuclei rather than from one and (*ii*) indistinguishable nature of electrons compels us to chose the wave function of type (4.93) and allows greater freedom to the electrons. Such a large freedom is followed generally by the lowering of energy of the system.

E_+ as a function of R is also plotted in Fig. 4.20. Value of E_+ is 3.14 eV (303 kJ mol^{-1}) and is much better than 0.25 eV (24.1 kJ mol^{-1}). R_{min} = 1.6 a.u. = 0.087 nm. Although better agreement with the experimental value of minimum energy and equilibrium bond length is observed, still some refinement in the treatment is required.

In the conventional Heitler-London treatment, as given above, the molecular wave function has been obtained by taking ground state wave functions of hydrogen atom, *i.e.,* 1*s* with Z =1. We have observed that pair of electrons is attracted by both the nuclei. In other words, the value of Z will differ from 1 (when two hydrogen atoms are far apart) to 1.625 (when two hydrogen atoms will fuse to form helium atom). Note that we have not taken the value of Z = 2 in the latter case as one electron shields the nucleus from other. Molecular wave function (ψ) can be constructed from atomic functions by taking nuclear charge as the variable parameter (Z′), *i.e.,*

$$\varphi_A = N e^{-\frac{Z' r_A}{a_0}} \qquad \text{...(4.110)}$$

Minimization of energy with respect to Z′ shows that Z′ = 1.166 and the binding energy increases to 3.76 eV (362.8 kJ mol^{-1}). R_{min} is found to be 1.39 a.u. (0.076 nm). Number of improvements in the wave functions has been carried out to get better agreement with the experimental results and the comparison has been compiled in Table 4.4. Here we will list two more refinements in the wave function, which are based on chemical evidence. First one is the polarization of the atomic charge by the presence of another atom. Due to this, the charge cloud of atom will no longer be spherical around the nucleus. Similar to that mentioned in m.o.t. the wave function can be written as:

$$\varphi_A = a_1 e^{-\frac{a_3 r_A}{a_0}} + a_2 Z e^{-\frac{a_3 r_A}{a_0}} \qquad \text{...(4.111)}$$

where z-axis is directed from A to B. Wave function φ_B can also be represented in the similar manner. Variables a_1, a_2 and a_3 are varied in the similar manner, as discussed earlier, to get the

Table 4.4: *Energy values for ground state of H_2 using various approximate wave functions*

Type of Wave Function	Maximum Binding Energy	Equilibrium Separation (nm)
Heitler-London	3.14 eV	0.0869
Heitler-London with screening (Wang)	3.78	0.0743
Wang + Polarization (Roson)	4.04	0.074
Wang +ionic (Weinbaum)	4.02	0.074
Wang + ionic + Polarization (Weinbaum)	4.12	0.0749
Molecular Orbital $\{\psi_A(1)+\psi_B(1)\}\{\psi_A(2)+\psi_B(2)\}$	2.70	0.085
Molecular orbital, plus screening	3.49	0.0732
Correlated m.o. $\{\psi_A(1)+\psi_B(1)\}\{\psi_A(2)+\psi_B(2)\}(1+pr_{12})$	4.11	0.71
Best possible molecular orbitals (self consistent field)	3.62	0.074
Coolidge-James, without r_{12} terms	4.29	0.074
Coolidge-James, with r_{12} terms (13 term function)	4.72	0.074
Coolidge-James, with r_{12} terms (50 term function)	4.747	0.0741
Oberved values	4.747	0.0741

(Copied from Valence by C. A. Coulson, Oxford University Press, Oxford, 1963.

minimum energy. The value obtained for binding energy is 4.02 eV (387.9 kJ mol^{-1}) instead of 3.76 eV (362. 8 kJ mol^{-1}) obtained at third stage. Chemical evidence for polarization can be looked up from the Stark effect. Earlier theories of chemical bonding did involve the mutual polarization of both atoms. Loose molecular complexes like $(O_2)_2$ and $(NO)_2$ are few examples.

So far we have assumed that charge clouds of both the combining atoms are concentrated between the nuclei and are polarized. But chances are that the charge clouds may be localized on either of the atom howsoever small it may be. This will lead to the following arrangements (leading to the presence of ionic character).

$$H_A^- \, H_B^+ \qquad H_A^+ \, H_B^- \qquad ...(4.112)$$

having equal probability of each. The wave function, which gives rise to the ionic character, can be written as:

$$\psi_{ion} = \varphi_A (1) \, \varphi_A (2) + \varphi_B (1) \, \varphi_B (2) \qquad ...(4.113)$$

The complete wave function will comprise the contributions from the pure covalent (discussed earlier) and pure ionic (4.113) to give (4.114).

$$\psi = a_3 \, \psi_{cov} + a_4 \, \psi_{ion} \qquad ...(4.114)$$

Following similar procedure as explained earlier and minimizing the energy for different parameters, the binding energy has been found to be 4.1 eV (395.7 kJ mol^{-1}). Value of a_4 is found to be 0.26 and for a_3 = 1.166, suggesting that major contribution to chemical bonding is from covalent character rather than from ionic one.

The above discussion can be summed up by mentioning that the total wave function of H_2 molecule can be written as the combination of the different proportions of pure covalent (ψ_{cov}) and pure ionic (ψ_{ion}, $H^- H^+$, $H^+ H^-$) wave functions. So far importance is not given to the individual structures (covalent or ionic), it seems all right to write structures depicting different ways of pairing the electrons. Chemical evidence for ionic structures will be better understood when dealing with heteronuclear diatomic molecules, which exhibit polar character. Thus, the best way of describing the final wave function is that ψ must have characteristics of (*i*) covalent, (*ii*) polar and (*iii*) ionic structures at the same time. The final wave function is different for different molecules.

4.11.2 HOMONUCLEAR DIATOMIC MOLECULES

We have learnt from the last section that v.b.t. puts the electrons of a molecule in a.o.'s and constructs the molecular wave function by allowing for exchange of valence electrons pairs between the a.o.'s of bonding atoms. This method is referred to as pairing of electrons or electron pairs. In other words the pairing between those electrons of the combining atoms will be effective which are unpaired. This suggests the presence of multiple bonding between the combining atoms if these atoms possess more than one unpaired electrons. The only difficulty, as we have seen in case of H_2 molecule, is the indistinguishability of the electrons and will lead to number of structures depicting electron-pair formation. Knowing fully well that some of these structures will have poor overlap integral (obtained by using $\int \varphi_A \, \varphi_B \, d\tau$), but their presence cannot be neglected. This will increase the problem of handling these calculations and thus quantitative information is available only for simple molecules. The main reasons can be summed up as:

(*i*) All the refinements listed for H_2 molecule have to be employed to get the satisfactory wave functions. The number will increase with increase of number of electrons.

(*ii*) The presence of core electrons as well as the valence electrons which are paired have to be considered as they exert significant forces.

(*iii*) As mentioned above, every kind of electron pairing is to be considered, although some of them may be improbable. For simplicity some of the highly improbable structures can be neglected, but will lead to bad results.

Few examples can be considered to show the complexity of the problem.

(*a*) The v.b. function for He_2 has each electron paired with another electron in an orbital on the same atom and hence predicts no bonding.

(*b*) Now consider Li_2. The ground state configuration of Li atom is $1s^2 2s^1$. The Lewis structure of Li_2 is Li – Li, with 2*s* Li electrons paired and 1*s* electrons remaining in the inner shell of each atom. The part of v.b. wave function involving the 1*s* electrons will be similar to that used for He_2 and the other part of v.b. wave function involving bonding 2*s* electrons will be similar to the Heitler-London H_2 wave function. Since each electron cannot be distinguished, we must allow every electron to be in every orbital and thus ground state v.b. wave function for Li_2 will be 6 × 6 determinant (three from each atom).

(*c*) The next example can be N_2 molecule. In the ground state, the electronic configuration for N-atom is $1s^2 2s^2 2p^3$. Hund's rule gives the ground state term symbol as $^4S_{3/2}$, indicating that one electron is present in all the three *p*-orbitals. Thus electron pair can be formed between $2p_x$ electrons, $2p_y$ electrons and $2p_z$ electrons to form a triple bond with Lewis structure as $N \equiv N$. The electron pair along the molecular axis (say *z*) is the σ bond and along other two axes will lead to 2π bonds. 2*s* electrons can be left because they are already paired internally in the separate atoms. Leaving aside the 2*s* electrons, in v.b.t., opposite spins are given to electrons bonded together. We have three such pairs of electrons and two ways to give opposite spins to the electrons of each bonding pair of a.o.'s. Thus, there are $2^3 = 8$ possible ways of writing structures and this leads to complexity.

(*d*) As regards to bonding, v.b.t. predicts correctly double bond in O_2 molecule. Electron pairing will lead to total spin S = 0 and suggests a singlet ground state, which is not correct. It is triplet as predicted by simple m.o.t. Thus needs a better picture based on v.b.t.

4.11.3 HETERONUCLEAR DIATOMIC MOLECULES

Heteronuclear diatomic molecules can also be treated by v.b.t. as done for homonuclear diatomic molecules. The only difference can be the weightage of ionic structures as compared to covalent ones. For example, consider the heteronuclear diatomic molecule AB with B more electronegative than A. The possible structures can be covalent (A – B) and ionic structures as A^+B^- and A^-B^+. The possible wave functions can be:

$$\underbrace{\varphi_A(1)\,\varphi_B(2) + \varphi_A(2)\,\varphi_B(1)}_{\psi_{cov}} \qquad \underbrace{\varphi_A(1)\,\varphi_A(2)}_{\psi_{ion}} \qquad \underbrace{\varphi_B(1)\,\varphi_B(2)}_{\psi_{ion}} \qquad ...(4.115)$$

and the v.b. wave function can be written as:

$$\psi = C_1 \psi_{cov} + C_3 \varphi_A(1)\varphi_A(2) + C_2 \varphi_B(1)\varphi_B(2) \qquad ...(4.116)$$

Rest of the treatment is the same, *i.e.,* using the variation theorem to minimize the energy with respect to C_1, C_2 and C_3. As usual the values of C_1, C_2 and C_3 give the relative contributions of each structure. The expression (4.116) becomes simple if electronegativity of B is much greater than that of A. Under this situation the weightage of structure $\varphi_A(1)\varphi_A(2)$ will be negligible and can be omitted. The wave function (4.116) can be written as:

$$\psi = C_1 \psi_{cov} + C_2 \varphi_B(1)\varphi_B(2) \qquad ...(4.117)$$

Results show that $C_2 \gg C_1$ for these kinds of molecules, suggesting the presence of asymmetry in the charge distribution and thus leading to polar character in the heteronuclear diatomic molecules. Knowing the values of C_1 and C_2, one can calculate the percentage ionic character in AB by the expression

$$\frac{C_2^2 \times 100}{C_1^2 + C_2^2} \qquad ...(4.118)$$

We shall now consider couple of examples. The simplest example can be HF. Using the valence electron approximation we will be considering the bonding electron for the wave function. A single bond is expected to be formed by the pairing of the hydrogen 1*s* electron and unpaired fluorine $2p_z$ electron leading to σ bond. Pure covalent or Heitler-London wave function can be written as:

$$\psi_{cov} = \frac{1}{\sqrt{2+2S^2}}\left[\varphi_H(1)\varphi_F(2) + \varphi_H(2)\varphi_F(1)\right] \qquad ...(4.119)$$

High electronegativity of fluorine will suggest the contribution from the ionic structures as well (just for information some ionic character can also be visualized from the covalent wave function because the charge density distribution will show asymmetric distribution). The ionic part of the wave function is $\varphi_F(1)\varphi_F(2)$ and the wave function for the bonding electrons can be written as:

$$\psi = C_1 \varphi_{cov} + C_2 \varphi_F(1)\varphi_F(2) \qquad ...(4.120)$$

Following the usual procedure the values of C_1 and C_2 can be calculated. We have ionic-covalent resonance involving the structures H – F and H^+F^-. The true molecular structure is intermediate between the covalent and ionic structures. The structure H^-F^+ having the term $C_3\ \varphi_H(1)\ \varphi_H(2)$ will contribute very little. The reason for this can easily be understood. The ionization energies of H- and F-atoms are 13.6 and 17.4 eV respectively. One requires 13.6 – 3.6 = 10 eV to take an electron from H-atom to give to F-atom (*i.e.,* H^+F^-), whereas to form H^-F^+, one needs 17.4 – 0.72 = 16.68 eV. Since the Coulombic attractive energy between H^+F^- and H^-F^+ would be nearly similar at the same value of R, it is clear that H^+F^- will be 6.68 eV more stable than H^-F^+. In other words even if we neglect H^-F^+ structure, error will not be very large. These calculations have shown that the most ionic (CsF) compound contains nearly 9% of covalent character. But it should be kept in mind that for less ionic system, both ionic structures could be important.

The above picture is better understood in case of highly ionic compounds. For example, in case of NaCl, which is highly ionic, even $C_2 \gg C_1$ and thus no need of taking Na^-Cl^+. One point worth mentioning here is that NaCl when dissociates in vapour phase, separates out as Na and Cl atoms rather than Na^+ and Cl^- ions. This is because the latter process is endothermic by 1.5 eV, because the first ionization energy of Na is 5.1 eV and electron affinity of Cl-atom is 3.6 eV. This means that in vapour phase, when R increases from R_e, the ratio of $\frac{C_2}{C_1}$ will decrease, becoming zero at $R = \infty$. For intermediate values of R, the Coulombic attraction energy between the ions is less than 1.5 eV and the molecule is largely covalent. Similarly even the most ionic compound (CsF) dissociates into Cs and F atoms in gas phase rather than Cs^+ and F^- ions. The latter process is also 3.9 – 3.45 = 0.045 eV endothermic than the former process and should be included if very detailed calculations are required. However it is clear that when ionic compounds are dissolved in water they separate into ions. This is due to the fact that solvation energies of these ions are quite large and thus the hydrated separate ions are more stable than even the diatomic NaCl molecule. Lastly it may be mentioned that there are cases of excited states of diatomic molecules whose potential energy curves correspond to dissociation of ions.

4.12 COMPARISON OF V.B. AND M.O. THEORIES

After giving the preliminary details of both the theories, we are now in a position to compare the results of both the theories. It may be pointed out that both the theories are based on certain approximations and even applying these approximations, one needs sophisticated computers to get quantitative results. Let us first start with features, which are common to both the theories.

Similarities

(*a*) Both the theories consider the bonding electrons to be associated with both the nuclei, *i.e.,* pairing of electrons.

(*b*) Both the theories predict an increase in charge density between the nuclei when a bond is formed.

(*c*) Both the theories distinguish between bonds of σ and π type.

It may be mentioned here that though both the approaches are different but the end results are nearly similar. We will consider each point separately.

(*a*) Considering the m.o.t. first, it is clear from equation (4.19), *i.e.,* $C_1\,\varphi_A + C_2\,\varphi_B$, that an electron belongs to both the nuclei A and B. Further in m.o.t. energy levels are determined first and then filling of electrons is carried out. We have observed that each electron tries to occupy the m.o. of minimum energy and would have kept on doing so. It is the Pauli Exclusion Principle which has put a restriction that not more than two electrons can go to any m.o. (*i.e.,* same space wave function) with opposite spins. In other words, it is clear that electrons are present in pairs with anti-parallel spins. Exceptions are there when degenerate m.o.'s are present, for example, O_2 etc. In these cases each m.o. is occupied by a single electron till all are occupied and this arises because to avoid greater Coulombic repulsion between electrons in the same space orbital. In other words the pairing arises completely because of Pauli Exclusion Principle and the desire to have as low energy as possible.

In v.b.t. wave function, φ_A (1) φ_B (2) + φ_A(2) φ_B (1), implies that electron 1 may be present around either nucleus A or nucleus B. Pairing of electrons is little less obvious in case of v.b.t., *i.e.*, pairing of orbitals is clear but pairing of spins needs some explanation. As discussed in case of H_2 molecule, Pauli Exclusion Principle states that the total wave function including spin part of H_2 molecule should be anti-symmetric. This can be obtained by taking either, space part symmetric and spin part anti-symmetric or vice versa. We have seen earlier that symmetric space wave function of form φ_A (1) φ_B (2) + φ_A(2) φ_B (1) leads to increase of charge between the nuclei and a tendency of forming a bond. Whereas the anti-symmetric space wave function of form φ_A (1) φ_B (2) – φ_A(2) φ_B (1) leads to decrease of charge between the nuclei and tendency not to form a bond. This suggests that bonding between the atoms requires that spin part of the wave function to be anti-symmetric and this is only possible if the pair of electrons are anti-parallel. Thus, it may be summed up that spin pairing does not arise directly but forced on us by a combination of symmetry requirements of Pauli Exclusion Principle and need for a certain choice of possible space functions in order to give bonding. In other words, it is the space part, which determines the spin alignments in v.b.t. and not other way.

(*b*) Figures (4.6*a*) and (4.6*b*) have clearly depicted that the charge density in case of bonding m.o. is concentrated in the middle of the two nuclei. This can easily be shown, by using the electron density or probability density function. In m.o.t. the wave function is give by:

$$\psi_+ = \frac{1}{\sqrt{2 + 2S}} [\varphi_A + \varphi_B] \qquad ...(4.35)$$

where S is the overlap integral ($\int \varphi_A \varphi_B \, d\tau$) and the probability density function for each electron is:

$$\psi_+^2 = \frac{1}{2(1 + S)} [\varphi_A + \varphi_B]^2 \qquad ...(4.50)$$

and total charge density for two electrons is obtained by multiplying it by a factor of 2,

$$i.e., \quad \frac{1}{1 + S} [\varphi_A + \varphi_B]^2 = \frac{\varphi_A^2 + \varphi_B^2 + 2\varphi_A\varphi_B}{1+S} = \varphi_A^2 + \varphi_B^2 + \frac{2\varphi_A\varphi_B - S(\varphi_A^2 + \varphi_B^2)}{1 + S} \qquad ...(4.121)$$

It is obvious from equation (4.121) that electron density in the molecule is larger than the sum of two isolated atomic charge densities in the region where S is large. By definition S is large where neither φ_A nor φ_B is small and this will occur only in the middle of the nuclei. In other regions either φ_A or φ_B is small. In other words, formation of bond between two atoms leads to the flow of the charge in the middle of the two nuclei. Similarly one can also show that in case of anti-bonding m.o. the charge density is smaller than the sum of the individual atomic charge density. Third term in equation (4.121), which suggests the flow of charge in the middle of bond, is called *bond charge.*

The v.b.t. also gives similar results except that approach is different. For example, if we are interested only in the charge density distribution of electron 1, it can be obtained by calculating it for all possible positions of electron 2 and taking the sum of all the calculated values. The final expression is:

$$\rho(i) = \frac{2\,\varphi_A\,\varphi_B + \varphi_A^2 + \varphi_B^2}{2\,(1+S^2)}. \qquad \text{...(4.122)}$$

Expressions (4.121) and (4.122) are similar to each other except that (4.121) involves S and (4.122) involves S^2. Since $S < 1$, expression (4.122) predicts greater accumulation of charge than the expression (4.121).

From the above results it may be concluded that a single bond is formed, by pairing of two electrons between the nuclei. This will result in the increase in the charge density in the middle of the two nuclei and decrease in the regions out side the two nuclei. To your surprise, the increase of charge density is not very large and its absolute magnitude is $\frac{1}{16}$ of electron. This magnitude is enough to decrease the potential energy to such a large extent that stable bond is formed. The decrease in potential energy is due to the attraction of electrons from both the nuclei rather than from one.

Differences

Having discussed similarities between the two approaches, let us turn ourselves to some properties when discussed by either method, which can give better insight into the physical basis of chemical bond. For example:

(*a*) ***Covalent and Ionic character*:** The ground state wave function for H_2 molecule, according to m.o.t., leaving the normalization constant and spin part, can be written as:

$$[\varphi_A(1) + \varphi_B(1)]\,[\varphi_A(2) + \varphi_B(2)] \qquad \text{...(4.61)}$$

which on expansion becomes:

$$\varphi_A(1)\,\varphi_B(2) + \varphi_A(2)\,\varphi_B(1) + \varphi_A(1)\,\varphi_A(2) + \varphi_B(1)\,\varphi_B(2) \qquad \text{...(4.123)}$$

If one analyzes these terms, the first two terms tell that each electron in an a.o. is localized on different nucleus and we have termed them as covalent terms, *i.e.*, equal sharing of the electrons between the atoms. On the other hand, the last two terms tell that both the electrons in a.o. are localized only on one nucleus and we have termed them as ionic terms having the structures as given earlier I ($H^+\,H^-$) and II (($H^-\,H^+$). This means that according to m.o.t., the weightage of each term is equal, *i.e.*, 50% covalent and 50% ionic. This may be looked from different angle. If we decompose H_2 moleucle, it may dissociate equally as two neutral hydrogen atoms or one of them as proton (H^+) and other as hydride ion (H^-) with equal porbabilities. But we know that H_2 is a non-polar molecule and dissociates into two neutral atoms. In other words, m.o.t gives wrong picture about the decomposition of H_2 and over emphasizes the contribution of the ionic structures to the total wave function. *This anamoly can be removed by including the configuration interaction (i.e., the contribution from the anti-bonding* m.o.) *in the above wave function.* The modified m.o. wave function is:

$$[\varphi_A (1) + \varphi_B (1)] [\varphi_A (2) + \varphi_B (2)] + \lambda [\varphi_A (1) - \varphi_B (1)] [\varphi_A (2) - \varphi_B (2)] \quad ...(4.124)$$

where λ is the coefficient and gives the weightage of this term in the total wave function. On expanding and rearranging the terms, the above equation can be written as:

$$(1 - \lambda) [\varphi_A (1) \varphi_B (2) + \varphi_A (2) \varphi_B (1)] + (1 + \lambda)$$

$$[\varphi_A (1) \varphi_A (2) + \varphi_B (1) \varphi_B (2)] \quad ...(4.125)$$

or

$$C_{cov} [\varphi_A (1) \varphi_B (2) + \varphi_A (2) \varphi_B (1)] + C_{ion} [\varphi_A (1) \varphi_A (2) + \varphi_B (1) \varphi_B (2)] \quad ...(4.126)$$

Since coefficients of covalent and ionic wave functions are different, thus it suggests different weightage of each term. This equation is exactly similar to that given in v.b.t treatment, *i.e.,* (4.117). *So it may be concluded here that valence bond theory plus the ionic terms are formally identical to molecular orbital theory including configuration interaction terms.* If more number of configuration interactions terms is included in m.o. wave function, better picture of molecular structure will be obtained from m.o.t.

Equation (4.123) can be looked from other point also. This m.o. wave function underestimates the electron correlation, *i.e.* it tells us that the structures having both the electrons on the same atom are having equal weightage as the structures having electrons on different atoms. On the other hand, the v.b.t. (4.117) overestimates electron correlation as it totally neglects the structures where both the electrons are localized on the same atom. *Thus, in* v.b.t., *the electron correlation is reduced by adding the ionic terms and in* m.o.t., *electron correlation is added by including the configuration interaction.* Thus it may be mentioned here that simple v.b.t is more well behaved at large internuclear distance than simple m.o.t., because the latter gives wrong picture about the dissociation products. Leave aside simple H_2 molecule, even highly ionic compounds only give neutral atoms on dissociation, *e.g.,* NaCl, CsF etc.,

***(b) Resonance*:** Resonance is extremely useful in getting qualitative picture of the actual molecular structure in terms of Lewis dot structures, which seems to be a reasonable description in chemical common sense. The valence bond method introduces the idea of resonance very satisfactorily. The idea of resonance is very useful in describing organic molecules and also helps qualitatively in predicting chemical reactions. For example, electron-donating groups attached to benzene, is known to direct electrophilic groups mainly into ortho and para positions in substitution reactions. This can be easily explained if the molecule is considered as a resonance hybrid of these canonical forms.

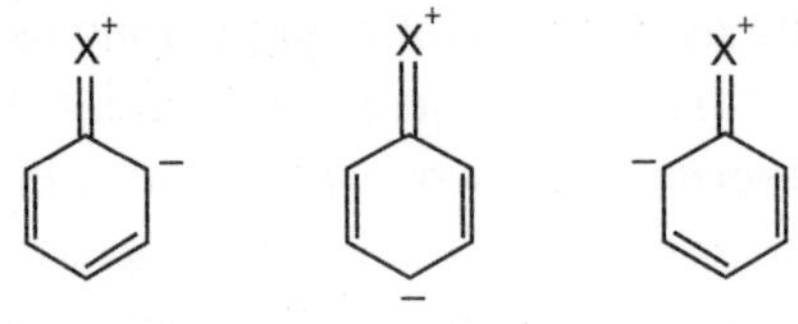

(*ii*) m.o.t. explains very nicely the paramagnetic characters of O_2 and S_2 molecules and many inorganic molecules, whereas one has to modify the v.b.t.

(*iii*) The presence of anti-bonding molecular orbitals finds better physical significance in m.o.t. than in v.b.t. Anti-bonding in v.b.t. is limited to the repulsion between the pairs of electrons in different a.o.'s, for example, in F_2, there is repulsion between the lone pairs of electrons in the $2p_x$ and $2p_y$ orbitals.

(*iv*) The excited states of molecules are not readily described by the v.b.t. and again m.o.t. may be used to explain the behaviour of individual electrons when dealing with electronic excitation. This point can be illustrated by referring once again to the hydrogen molecule. Out of many possible transitions, two transitions can be represented as:

(*i*) $\sigma^2{}_{1s} \rightarrow \sigma_{1s}\, \sigma^*{}_{1s}$

(*ii*) $\sigma^2{}_{1s} \rightarrow \sigma_{1s}\, \pi_{2p}$

In the first case an electron is excited from σ_{1s} to $\sigma^*{}_{1s}$ and in the second case to π_{2p} m.o. This is because m.o.t. treats the electrons as independent of one another. It considers only the possibility of one electron being excited, unlike in case of v.b.t. where we are obliged to consider wave functions intimately involving both electrons. On the other hand v.b.t. does not distinguish between the electrons (any one can be promoted). In the first case, based on v.b.t., it can be represented by transforming the normal ground state function to purely ionic type wave function $[\varphi_A(1)\, \varphi_B(2) - \varphi_A(2)\, \varphi_B(1)]$. Whereas the second case needs a wave function involving no less than four terms of type $\varphi_{s_A}(1)\, \varphi_{\pi B}(2)$. This is because we do not know whether π orbital is associated with A or B and correspondingly the *s* orbital with B or A. The wave function becomes even more elaborate when more complex molecules are considered.

(*v*) Lastly let us consider another class of stable molecules, which form either one electron or three electrons chemical bond, also known as odd electron molecules. The simplest molecule to be considered is H_2^+, as described earlier and the electronic configuration is σ_{1s}. Similarly, Li_2^+, can be represented as KK σ_{2s}. The only difference is that in one electron chemical bond, m.o. is half filled and bond dissociation energies of these species are nearly half that of H_2 and Li_2 respectively. Another point to be observed in going from H_2 to H_2^+ or Li_2 to Li_2^+ is that repulsion energy between the two electrons is lost and is balanced by increase in bond length.

Most common and important molecule containing three electrons bond is NO. The electronic configuration, as mentioned earlier, can be written as:

$KK(z\sigma)^2(y\sigma)^2(x\sigma)^2(w\pi)^4(v\pi)^1$ or may be $KK(2s_O)^2(2s_N)^2(\sigma_{2p_z})^2(\pi_{2p_x})^2(\pi_{2p_y})^2(\pi^*_{2p_x})$

Because of this single electron in the anti-bonding m.o., the NO is paramagnetic and bonding is weaker than that in N_2 molecule. NO is unexpectedly stable for a molecule with unpaired electron and this stability must be attributed to the electron being distributed over the complete m.o. Even though NO is essentially a free radical, the unpaired electron is distributed over the complete molecule. Thus, the sensitivity and tendency for pairing of this electron is diminished. A dimmer $(NO)_2$ is formed but only at low temperature and its structure is:

I N----O II $N_I\ O_I = N_{II}\ O_{II}$ = 0.11nm (triple bond)

||| |||

I O----N II $N_I\ O_{II} = N_{II}\ O_I$ = 0.238 nm (much less than a single bond)

Description of one electron bond by v.b.t. can be explained as follows. Consider again H_2^+ molecule. According to v.b.t., one electron can be localized either on H_A or it can be on H_B. This can be represented by two structures, I and II. Resonance between the

I H^+ $^{\bullet}H$

II $H^{\bullet}$ ^{+}H

the structures leads to the wave function like I ± II. These combinations can further be treated on the similar lines as is carried out for m.o.t.

Simplest three electrons system can be written as:

III He: $^{\bullet}He^+$

IV $^{+}He^{\bullet}$:He

Thus, wave functions can be written as III ± IV, because of resonating structures. These structures only tell us that electrons are paired when the atoms are isolated. The final wave functions can only tell that one of the electrons (which one is not sure) can be paired either with the atom A or it may be paired with the atom B. This kind of bond is depicted as A-----B. In case if A and B are two different atoms, the resonating structures I and II or III and IV will not have the same energies, suggesting that these are not strong conditions from resonance, Thus, one electron and three electron bonds are not very stable according to v.b.t., except between the atoms having similar electronegativity. From the above arguments, it is clear that both the theories will not give reliable results unless these theories are suitably modified. Each theory has certain limitations and when improved upon, they converge and finally become equivalent.

4.13 MOLECULAR TERM SYMBOLS

Similar to atomic term symbols, the molecular term symbols for the diatomic molecules are designated as:

$$^{2S+1}\Lambda_{\Omega} \quad ...(4.127)$$

where

(*i*) S is the total spin quantum number given as: $S = \Sigma s_i$ ...(4.128) (*ii*) Λ is the sum of the components of the orbital angular momentum of the individual electron along the molecular axis (λ) and is given by: $\Lambda = \Sigma\lambda_i$ ...(4.129)

This is different from the total orbital angular momentum term for atoms, because in case of diatomic molecules, only one quantum number is uniquely defined by solving the space part of the Schrödinger equation, *i.e.*, the component of the orbital angular momentum along the unique molecular axis. It is much easier to calculate the resultant of λ_i's because λ_i can have only two possible directions. One is towards the positive side of the axis and the other is towards the negative side as shown in Fig. 4.21. In other words, for $\lambda_i \neq 0$, the degeneracy of λ_i is only two, irrespective of the value of λ_i.

For $\lambda = 1$

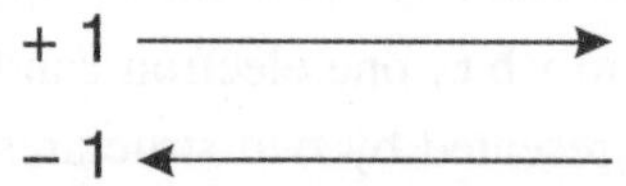

Figure 4.21: *Directions of the components of the orbital angular momentum along the molecular axis*

(*iii*) $\Omega = \Lambda + S, \Lambda - S - 1, \ldots \quad .|\Lambda - S|$ ...(4.130)

where Ω is the total angular momentum equivalent to J. Before we take up any particular case and apply the above rules, let us try to designate the molecular orbital corresponding to the value of Λ.

For one electron		More than one electron (Total)	
$\lambda = 0$	σ	$\Lambda = 0$	Σ
$\lambda = 1$	π	$\Lambda = 1$	Π
$\lambda = 2$	δ	$\Lambda = 2$	Δ

This is similar to the designation of atomic orbitals (*s*, *p*, *d*, *f*,... for l = 0, 1, 2, 3.....) and total angular momentum (S, P, D, F, ... for L = 0, 1, 2, 3... .).

The other thing commonly observed for homonuclear diatomic molecular orbitals is the presence of sub-script *g* (gerade) and *u* (un-gerade). As discussed earlier (4.6), σ bonding and π^* anti-bonding m.o.'s are having the parity *g*, whereas σ^* anti-bonding and π bonding m.o. are having the parity *u*. In many electron systems, note the parity of each electron present in the respective m.o. and total parity of the system, can be obtained by the product of the parity of each electron. Care should be taken that $g \times g = g$, $g \times u = u \times g = u$, $u \times u = g$ and so on. The similar use of parity as sub-script cannot be used for heteronuclear diatomic molecules because of the presence of two different atoms and thus absence of centre of inversion.

Example 4.2: Consider a very simple molecule H_2 with electron configuration σ_{1s}^2. Calculate the term symbol.

Solution: Here $\lambda_1 = 0$, $\lambda_2 = 0$, $\Lambda = 0$

Since both the electrons are present in the same orbital, the electrons must have anti-parallel spin, according to Pauli Exclusion Principle, *i.e.*,

$$s_1 = \frac{1}{2},\ s_2 = -\frac{1}{2},\ S = \frac{1}{2} - \frac{1}{2} = 0 \text{ and } 2S + 1 = 2 \times 0 + 1 = 1$$

$$\Omega = \Lambda + S = 0 + 0 = 0.$$

So the spectroscopic term symbol is $^1\Sigma_0$. Further the parity of σ m.o. is g and total parity will be $g \times g = g$. Final spectroscopic term symbol is $^1\Sigma_0\ (g)$.

From the results of H_2 molecule, it may be concluded that the spectroscopic term symbol for any molecule having close shell structure (or completely filled orbital) is always $^1\Sigma_0(g)$. This is because for every value of λ, there will be a value of λ which is equal in magnitude but opposite in direction. Thus for completely filled orbital, the total value of $\Lambda = 0$. Similarly is the case for the total value of spin, because in each orbital pair of electrons present have anti-parallel spin. This leads to $S = \Sigma i\left(\frac{1}{2}\right) - \Sigma i\left(-\frac{1}{2}\right) = 0$. In other words the above discussion saves us from a lot of labour and we have to do the above exercise for only partially filled orbitals. This is taken in the next example.

Example 4.3: Consider O_2 molecule with electronic configuration $KK\sigma_{2s}^2\ \sigma_{2s}^{*2}\ \sigma_{2p_z}^2\ \pi_{2p_x}^2\ \pi_{2p_y}^2\ \pi_{2p_x}^{*1}\ \pi_{2p_y}^{1}$. Calculate the term symbols.

Solution: O_2 molecule has two unpaired electrons. Here $\lambda_1 = +1$ or -1, $\lambda_2 = +1$ or -1.

$\Lambda = +2, -2, 0$. These values of Λ are obtained as shown below pictorially.

Two values of Λ obtained (2 or -2) are consistent with the fact that $\Lambda \neq 0$, are doubly degenerate. $\Lambda = 0$ also corresponds to two different ways of arranging λ_i. In other words $\Lambda = 0$ is also doubly degenerate, which is not possible. Though it is not easy but plane through internuclear axis, *i.e.*, plane of symmetry can help to remove degeneracy for $\Lambda = 0$ is also doubly degenerate,

λ_1	$\rightarrow$	+1	$\leftarrow$	-1		λ_1	$\rightarrow$	+1	$\leftarrow$	-1
λ_2	$\rightarrow$	+1	$\leftarrow$	-1		λ_2	$\leftarrow$	-1	$\rightarrow$	+1
Λ	$\rightarrow$	+2		-2		Λ		0		0

Figure 4.22: *Different ways of depicting the components of the orbital angular momentum in O_2 molecule*

which is not possible. The electronic eigen functions unchange or change sign when reflected at any plane passing through both nuclei. The former is called Σ^+ state and the latter is called Σ^- state.

For $\Lambda = +2$ or -2, both the electrons are present in the same m.o. Thus, $s_1 = +\frac{1}{2}$, and $s_2 = -\frac{1}{2}$, leading to $S = \frac{1}{2} - \frac{1}{2} = 0$, $2S + 1 = 2 \times 0 + 1 = 1$

$$\Omega = 2 + 0,\ 2 - 0 = 2$$

The term symbols are $^1\Delta_2$.

For $\Lambda = 0$, $s_1 = +\frac{1}{2}$, $s_2 = +\frac{1}{2}$ and $S = 1$, $2S + 1 = 3$

$$\Omega = 0 + 1 = 1$$

$$\Omega = 0 + 0 = 0$$

$$^1\Sigma_g^+ \text{ ———————}$$

$$^1\Delta_g \text{ ———————}$$

$$^3\Sigma_g^- \text{ ———————}$$

Figure 4.23: *Energy level diagram of different term symbols of O_2*

Term symbols are $^3\Sigma^-$ and $^1\Sigma^+$. Since the total wave function ($\psi_T = \psi_{space}\ \psi_{spin}$) should be anti-symmetric, anti-parallel spins of electrons has to be combined with $^1\Sigma^+$ and parallel spins with $^3\Sigma^-$. The energy level diagram or spectroscopic states can be based on Hund's rule and are depicted in Fig. 4.23.

Example 4.4: Write the electronic configuration of NO and calculate the term symbols.

Solution: Electronic configuration of N = $K2s^22p^3$ and that of O = $K2s^22p^4$

Total number of valence electrons = 5 + 6 = 11

Electronic configuration of NO is; KK $\sigma_1^2\ \sigma_2^{*2}\ \sigma_3^2\ \pi_1^2 = \pi_2^2\ \pi_3^{*1}$

As mentioned earlier only those m.o.'s will be considered which are partially occupied. So in this case π_3^{*1} is the only orbital to be considered.

$$\lambda = 1 = \Lambda,\ s = \frac{1}{2} = S,\ 2S + 1 = 2 \times \frac{1}{2} + 1 = 2$$

$$\Omega = 1 + \frac{1}{2},\ 1 - \frac{1}{2} = \frac{3}{2},\ \frac{1}{2}$$

Term symbols can be $^2\Pi_{\frac{3}{2}}$, $^2\Pi_{\frac{1}{2}}$.

4.14 DIPOLE MOMENT OF AB MOLECULES

Charge density calculations in AB molecules have shown that it is asymmetrically distributed in AB molecules and shifted towards the atom, which is more electronegative. This will create partial positive charge and partial negative charge on the respective atoms. When two charges q^+ and q^- are separated by a distance r, the magnitude of the electrical dipole moment (μ) is given by:

$$\mu = qr \qquad ...(4.131)$$

In diatomic molecules we can consider the partial charges as the point charges. Thus the value of r will be the equilibrium bond length of AB molecule. If q is measured in electrostatic units (esu, $e^- = 4.8 \times 10^{-10}$ esu, charge on one electron) and bond length in cm (typical bond length is of the order of 10^{-8} cm, Å) the μ will be a multiple of $10^{-10} \times 10^{-8} = 10^{-18}$ esu cm and is denoted by a unit debye (D). If $q = e = 4.8 \times 10^{-10}$ esu, $r = 1$ Å $= 10^{-8}$ cm, then

$$\mu = 4.8 \times 10^{-10} \text{ esu} \times 1 \times 10^{-8} \text{ cm} = 4.8 \times 10^{-18} \text{ esu cm} = 4.8 \text{ D}.$$

In S.I. units, $e = 1.6022 \times 10^{-19}$ C, r in meters (1 Å $= 1 \times 10^{-10}$ m),

$$\mu = 1.6022 \times 10^{-19} \text{ C} \times 1 \times 10^{-10} \text{ m} = 1.6022 \times 10^{-29} \text{ C m}.$$

$$1 \text{ D} = \frac{1.6022 \times 10^{-29}}{4.8} = 3.336 \times 10^{-30} \text{ C m}.$$

Dipole moment is a vector quantity. Thus, it has a magnitude and direction. There are two conventions. According to Physicists, the dipole is directed from negative end to the positive end, whereas the Chemists consider the direction in opposite manner, *i.e.,* dipole is directed from positive end to negative end, as shown in Fig. 4.24.

$+\delta \longrightarrow -\delta$

Figure 4.24: *Pictorial view of the direction of dipole moment*

Two kinds of information can be drawn from the magnitude of dipole moments.

(*i*) ***Percentage ionic character***: Dipole moments of AB molecule can be measured experimentally. Knowing the equilibrium bond length (determined from other methods) one can calculate the partial charge actually present on the respective atoms. Theoretical dipole moment (μ_{the}), assuming transfer of one complete unit of electronic charge, can also be calculated knowing the value of r. This value provides the maximum value of μ. Since the experimental value of μ provides a partial value of electrical charge, the percentage ionic character (IC) in AB bond can be obtained as:

$$\% \text{ IC} = \frac{\mu_{exp}}{\mu_{the}} \times 100 = \frac{q \times r}{e \times r} \times 100 = \frac{q}{e} \times 100 \quad ...(4.132)$$

In other words, dipole moments are used to get information about the molecular structure (section 4.6) and to test the calculated wave function.

Example 4.5: Calculate the percentage ionic character in LiH if the equilibrium bond length is 0.16 nm and $\mu_{exp} = 5.9$ D.

Solution: Using equation (4.131):

$$\mu_{the} = (1.6 \times 10^{-19} \text{ C}) (0.16 \times 10^{-9} \text{ m}) = 0.256 \times 10^{-28} \text{ cm}.$$

$$\% \text{ IC} = \frac{\mu_{exp}}{\mu_{the}} \times 100 = \frac{(5.9\text{D})\ (3.336 \times 10^{-30} \text{ C m D}^{-1})}{0.256 \times 10^{-28} \text{ C m}} \times 100 = 76.88$$

This agrees quite nicely with the value of 80% IC, mentioned in section 4.10.

(*ii*) ***Solubility of ionic compounds in polar solvents***: Dipole moments or polarity of the solvent molecules can be used to find the suitability of a solvent to dissolve an ionic compound. Qualitatively two kinds of parameters are involved in affecting the solubility of ionic solids. (*i*) Lattice energy. Energy has to be supplied and (*ii*) solvation energy,

energy released during solvation of ions. Negative value of the resultant of two processes decides the solubility of ionic compounds. Parameter (*ii*) can be initiated by two ways. (*a*) One end of the polar solvent molecule will be oriented towards the oppositely charged ion and thus surrounds the solute molecules in a particular manner. Due to this potential energy of the ions will be lowered. (*b*) Whenever an ionic solid is dissolved in a polar solvent, it reduces the electrostatic interactions between the charged ions and can be explained as follows. If in a solution, two ions are separated by a distance *r*, the electrostatic interaction is proportional $\frac{1}{4\pi\varepsilon_0 r}$ in vacuum, where ε_0 is the permittivity. This electrostatic interaction is reduced to $\frac{1}{4\pi\varepsilon_0 r D}$, in a medium of relative permittivity D (or dielectric constant). The value of dielectric constant depends on the magnitude of dipole moment or polarity of solvent molecules. Depending on the value of D, the magnitude of electrostatic interaction can be reduced significantly. For example, for water the value of D is ~ 80. In other words in water as solvent, the electrostatic interaction will be reduced nearly by two orders of magnitude. From the above discussion it is evident that polar solvent molecules really disrupt the order or arrangements of ions in ionic solids and helps in getting solubilized in polar solvents.

4.15 ELECTRONEGATIVITY

In section 4.10, we have learnt that charge density of the bonding electrons in heteronulcear diatomic molecules AB does not lie in the middle of the bond, but it is shifted towards one of the atoms, *i.e.,* electron-pair is spending more time around one of the atoms as compared to other atom. This property of unequal sharing of electrons by the atoms was designated as *electronegativity* (χ) by Pauling in 1931. According to Pauling, *the power of an atom in a molecule to attract electrons towards it self,* is called electronegativity. It is evident from the definition that electronegativity is not a property of an isolated atom, although it has been defined and related to atomic property. In reality it is a property of an atom in a molecule and depends on the environments around that particular atom, although it is invariably assumed that it remains constant. The degree of asymmetry of the electronic charge in the structure of the molecules and the direction of shift of its electron cloud are determined by the relationship of the electronegativities of the atoms that make the molecule.

Many methods were developed to calculate and relate the electronegativity of different atoms in a molecule. Here we will discuss only two methods, Pauling's method and Mulliken's method.

4.15.1 PAULING'S METHOD

Pauling's method of calculating electronegativity is based on the thermochemical data. It is found that bond dissociation energy of the mclecule AB is greater than the bond dissociation energies (BE) of either individual A_2 or B_2 molecules and even the average of the bond dissociation energies of A_2 and B_2, *i.e.,* $BE_{AB} > BE_{A_2} > BE_{B_2}$ or $\frac{1}{2}(BE_{A_2} + BE_{B_2})$. The extra stability in the bond strength of AB, according to Pauling is due to the presence of ionic structure in the molecule AB and this leads to the ionic resonance energy. Thus the wave function of AB molecule is defined as:

$$\psi_{AB} = C_1\,\psi_{cov} + C_2\,\psi_{A^+B^-} + C_3\,\psi_{A^-B^+} \quad ...(4.133)$$

If the molecule AB is purely covalent, then $C_2 = C_3 = 0$ and charge density of the bond pair will be present more in the middle. If B is more electronegative than A, the energy due to the structure A^+B^- will contribute quite large to the total energy and will be approaching to the energy stabilization by the covalent structure. This will increase the ionic resonance energy. On the other hand contributions from the structure A^-B^+ to the total energy will be nearly negligible because the formation energy of structure A^-B^+ is very large. Thus, it may be concluded that for a covalent but polar molecule, $C_1 > C_2 \gg C_3$. In other words, greater the contribution from ionic structure A^+B^-, greater will be the contribution to the resonance ionic energy. Pauling suggested that this ionic resonance energy is related to the difference in the electronegativity of the atoms A and B. Thus to calculate the actual values of χ, Pauling took hydrogen atom as a reference atom and assigned the value of 2.05 to it. The reason to choose this value was to have positive values of χ, as well as, near whole number. The procedure to calculate the electronegativity is as follows.

Let BE of A_2 and B_2 are BE_{A_2} and BE_{B_2} respectively. The covalent bond dissociation energy of AB can be calculated by taking the average of BE_{A_2} and BE_{B_2}:

$$BE_{cov} = \frac{1}{2}\left[BE_{A_2} + BE_{B_2}\right] \quad ...(4.134)$$

$$\chi_A - \chi_B = \left[\frac{BE_{exp}(AB) - \frac{1}{2}\left\{BE_{A_2} + BE_{B_2}\right\}}{23}\right]^{\frac{1}{2}}$$

or

$$BE_{exp}(AB) = 23\,(\chi_A - \chi_B)^2 + \frac{1}{2}\left[BE_{A_2} + BE_{B_2}\right] \quad ...(4.134a)$$

provided BE's are in kcal mol^{-1}.

Although BE_{cov}, calculated by this procedure, was always less than the actual BE (expected), but in some cases it is found that BE_{cov} calculated by this method was larger than BE_{exp}. So it was suggested by Pauling to take the geometric mean rather than arithmetic mean of BE_{A_2} and BE_{B_2}, *i.e.*,

$$BE_{cov} = \left[BE_{A_2}\;BE_{B_2}\right]^{\frac{1}{2}} \quad ...(4.135)$$

The resonance ionic energy contribution to the BE (Δ) is given as:

$$\Delta = BE_{exp} - BE_{cov} \quad ...(4.136)$$

If χ_A and χ_B are the electronegativities of atoms A and B respectively,

$$\chi_A - \chi_B = \Delta^{\frac{1}{2}} \text{ provided } \Delta \text{ is expressed in eV.} \quad ...(4.137)$$

Thus, using hydrogen atom as a reference, we can calculate the electronegativity of other atoms. These values are compiled in Table 4.5. Few things have emerged from the data of Table 4.5.

(*i*) Atoms with smaller size are more electronegative as these atoms have higher tendency

Table 4.5: *Electronegativities of some elements*

Element	χ	Element	χ	Element	χ	Element	χ	Element	χ
H	2.1								
Li	1.0	Na	0.9	K	0.8	Rb	0.8	Cs	0.7
Be	1.5	Mg	1.2	Ca	1.0	Sr	1.0	Ba	0.9
B	2.0	Al	1.5						
C	2.5	Si	1.8	Ge	1.7	Sn	1.7		
N	3.0	P	2.1	As	2.0	Sb	1.8		
O	3.5	S	2.5	Se	2.4	Te	2.1		
F	4.0	Cl	3.0	Br	2.8	I	2.4		

to attract the electron pair to themselves, *i.e.,* in a given group the electronegativity decreases if we go down a group.

(*ii*) Atoms with nearly filled valence shell are more electronegative than the less filled valence shells. For example, electronegativity increases if we go from left to right in a given period.

(*iii*) It has been observed that if the atoms of nearly same electronegativity form molecules these molecules are less polar, whereas the polarity of molecules increases with increase in difference of the electronegativities of the combining atoms. Polarity in a diatomic molecule can be expressed as a percentage ionic character in that bond. Percentage ionic character can be determined by comparing the dipole moment determined experimentally with that calculated theoretically, as described in section 4.14.

Pauling electronegativity and percentage ionic character of single bonds are correlated as:

$$\% \text{ IC} = 100 \left[1 - e^{-\{0.25 (\chi_A - \chi_B)^2\}}\right]^2 \qquad ...(4.138)$$

This formula is of an approximate nature and is especially not valid for large $\chi_A - \chi_B = \Delta\chi$ values. Hannay-Smyth has suggested the following equation (4.139) to relate the percentage ionic character (IC) to the difference in electronegativities of the combining atoms.

$$\% \text{ IC} = 16 |\chi_A - \chi_B| + 3.5 |\chi_A - \chi_B|^2 \qquad ...(4.139)$$

Pauling electronegativity and bond distance are correlated by the Schomaker-Stevenson equation as:

$$r\,(A - B) = r_A + r_B + C\,|\chi_A - \chi_B| \qquad ...(4.140)$$

r_A and r_B are covalent radii of atoms A and B in Å, r (A – B) is the bond distance between A and B in Å, C = Schomaker-Stevenson coefficient in Å units.

Nature of A and B in bond A – B	C in Å
Either A or B or both A and B belong to first row in the periodic table	0.08
A = Si, P or S and $\chi_A > \chi_B$	0.06
A = Ge, As or Se and $\chi_A > \chi_B$	0.04
A = Sn, Sb or Te and $\chi_A > \chi_B$	0.02
A = C and B is an atom in the fifth, sixth or seventh group (not in the first row)	

4.15.2 MULLIKEN METHOD

Mulliken proposed alternative method to calculate the electronegativity of a particular atom using ionization energy (IP) and electron affinity (EA) of the atom. It is based on the fact that in a molecule A, the formation of structure A^+B^-, which involves the transfer of an electron from A to B, is accompanied by the energy change as [IP (A) – EA (B)], where IP and EA are expressed in eV. Similarly the formation of structure A^-B^+, which involves the transfer of an electron from B to A is accompanied by the energy change as [IP (B) – EA (A)]. If in a given molecule AB, A^+B^- is the predominant structure, then

$$IP(A) - EA(B) < IP(B) - EA(A)$$

or $$IP(A) + EA(A) < IP(B) + EA(B)$$

Mulliken, thus suggested that the electronegativity could be regarded as the average of IP and EA,

$$\chi = \frac{1}{2}[IP + EA] \qquad ...(4.141)$$

In other words, the value of χ can be calculated by substituting the measured values of IP and EA of a respective atom. It is observed that the value of χ obtained by using Mulliken method is nearly 2.8 times larger than that obtained by using Pauling method. Equation (4.141) can be modified as:

$$\chi = \frac{IP + EA}{5.6} \qquad ...(4.142)$$

This method, although has a better theoretical justification, found to be less popular. Because it is difficult to measure the electron affinity for an atom and data for EA's are not available. Further, this method has better explanation for the fact that different oxidation states of atom have different electronegativities.

The following is a table of the electronegativities for several elements in kJ/g-atom (after Mulliken). These values are obtained by means of the IP, which corresponds to the removal of the sole unpaired electron.

Element	F	Cl	Br	I	H	Li
$(IP + EA) \times 10^{-3}$	2.452	1.891	1.749	1.598	1.372	0.565

The Pauling and Mulliken scales are related approximately:

$$\chi(\text{Mulliken}) = 2.78\chi\ (\text{Pauling}).$$

It is seen that electronegativities of the halogens and hydrogen fall off in the order F > Cl > Br > I > H. From this it follows that electron clouds in molecules of hydrogen halides or halogens hydrides (HX) must be displaced towards the halogen atoms and the more considerably the lighter halogen. Consequently, the dipole moments of HX molecules should be related as HF > HCl > HBr > HI. Indeed, experiment gives the following values of dipole moments of these compounds: 1.74 (HF), 1.03 (HCl), 0.79 (HBr), and 0.38 (HI) D. In contrast to halogens, the electronegativity of alkaline elements is less than that of hydrogen. From this we conclude that the electron cloud in the hydride molecules of alkali elements must be shifted towards the hydrogen atom. This is evidenced or evident from *x*-ray studies of crystal lattices of hydrides of alkaline elements in which the hydrogen atom is negative. Related to this is also the large dipole moment of hydride molecules; thus, the computed moment of LiH comes out to be 6.04 D.

The difference in electronegativities is particularly great in the case of halogens (X) and alkaline elements (Me). For this reason, in MeX molecules, the electron cloud of pair of valence electrons is so much shifted towards the X atom that it particularly localizes on this atom, so that grounds appear for speaking of the ionic structure of the molecule Me^+Cl^-. An experimental confirmation of this conclusion is close to agreement of measured values of the bond energy and other molecular constants of alkaline –halogen salts with these values computed on the assumption that MeX molecules are built up from ions, the electrostatic interaction of which is responsible for the chemical bond in the molecule.

From the foregoing discussion, we can conclude that ionic or electrovalent, bonding is a limiting case of chemical bonding, when the electron cloud of a pair of electrons that are responsible for chemical bond is shifted to such an extent towards the electronegative atom that these electrons will actually belong to the latter. But in the overwhelming majority of cases, the cloud of valence electrons is more or less shifted towards one of the atoms from the symmetrical position that it would occupy if the electrical properties of atoms comprising, the molecules are the same. Bearing in mind both the limits of cases, (the symmetric electron cloud and the ionic structure), one can speak of the real structure of a molecule as a certain intermediate structure between these limiting structures.

Due to the difference in physical and chemical properties of molecules with covalent and electrovalent bonds, and also in so far as the limiting cases of purely ionic or purely covalent bonds are comparatively rare, it is meaningful to speak of degree of electrovalence or degree of polarity of chemical bonds. The natural measure of this latter quantity is the electrical dipole moment.

Better methods were also suggested to account for the electronegativity of an atom, but these methods are beyond the scope of this book.

Example 4.6: The bond dissociation energies of H_2, F_2 and HF are 435.29 kJ mol^{-1}, 154.82 kJ mol^{-1} and 564.84 kJ mol^{-1} respectively. If the electronegativity of hydrogen is 2.1, calculate the electronegativity of fluorine atom.

Solution: Using equation (4.135), BE_{cov} can be calculated as:

$$BE_{cov} = \left[BE_{A_2}\ BE_{B_2}\right]^{\frac{1}{2}} = [435.21 \times 154.82]^{\frac{1}{2}} = 259.55 \text{ J mol}^{-1}$$

Resonance ionic energy (Δ) can be calculated with the help of (4.136)

$$\Delta = 564.84 - 259.55 = 305.29 \text{ kJ mol}^{-1} = \frac{305.29 \text{ kJ mol}^{-1}}{96.5 \text{ kJ eV}^{-1} \text{ mol}^{-1}} = 3.16 \text{ eV}$$

$$\chi_A - \chi_B = [3.16]^{\frac{1}{2}} = 1.77$$

$$\chi_A = 2.1 + 1.77 = 3.87.$$

4.16 SATURATION OF CHEMICAL FORCES

The saturation of chemical forces in chemistry is a well-known established fact and is manifested in a constancy of composition of chemical compounds and in the stoichiometry of reactions. This fact first received substantiation in quantum mechanics. From the result of a quantum mechanical analysis (London) of a system of three hydrogen atoms it is concluded that saturation is due to the mutual exchange repulsion of electrons with parallel spins. To illustrate the correctness of this conclusion let us examine the following examples. From the fact that molecules like HeH are not observed, it follows that the H-atom and He atom, in the ground states, mutually repel. Indeed, in the ground state of the He ($^1S_0$) atom its two electrons have anti-parallel spins. When the atoms, He and H approach, hydrogen electron can exchange with each of the He electrons. However as a result of exchange with an electron, there should arise a state of helium atom that is forbidden by the Pauli Exclusion Principle because in this state both electrons have the same four quantum numbers. Consequently only the exchange of electrons with parallel spins is possible; but this leads to exchange repulsion of He and H atoms. Similar reasoning suggests the mutual repulsion of H and H_2 molecule, which can form to the conclusion from quantum mechanical calculations on a system with three hydrogen atoms. Further as the electronic structures of the inert gas atoms are built up of pairs of electrons with anti-parallel spins, these atoms (like the He atom) repel the hydrogen atom (like any other atom). This accounts for the experimentally observed inertness of Ne, Ar, Kr and other inert gases. Of the same nature is the mutual repulsion of ions (at small distances) that have the same electronic structures as those of inert gases. For example, the Na^+ and Cl^- ions, by virtue of opposite charges, these ions attract at large distances and repel at short distances when the exchange forces of repulsion (with a small radius of action) become predominant. Of the same very nature is the mutual repulsion of saturated molecules that finds expression in the extreme difficulty with which they interact chemically.

4.17 TRIATOMIC MOLECULES

In this section we shall discuss heteronuclear molecules of type AB_2. AB_2 molecules can be linear (*e.g.*, BeH_2, CO_2) and non-linear (*e.g.*, H_2O, NO_2). Although BeH_2 and CO_2 are linear (and similarly H_2O and NO_2 are non-linear), these belong to different classes in the sense that in BeH_2 (and so in H_2O), the terminal atoms (H) are using only 1*s*-a.o.'s to form m.o.'s with the central atoms, whereas in CO_2 and (NO_2) the terminal atoms will be using *p*-a.o.'s to form chemical bonds with the central atoms.

4.17.1 TRIATOMIC LINEAR MOLECULES

BeH_2:

Molecular orbital theory: Both the hydrogen atoms (H_A and H_B) possess 1*s* valence electrons, whereas the valence orbitals on Be are 2*s* and degenerate $2p_x$, $2p_y$ and $2p_z$, possessing two electrons placed in 2*s* a.o. in its ground state configuration. Fig. (4.25*a*) represents the coordinate axis of BeH_2, *i.e.,* H_BBeH_A are placed along the *z*-axis. As discussed in case of AB molecules, m.o.'s will be formed from the linear combination of the a.o.'s of Be and two 1*s* orbitals of H-atoms having proper symmetry. 1*s* (H_A) and 1*s* (H_B) can have the following linear combinations.

$$1s^+ = 1s\,(H_A) + 1s\,(H_B) \qquad ...(4.143)$$

$$1s^- = 1s\,(H_A) - 1s\,(H_B) \qquad ...(4.144)$$

Since both the *s* orbtials are symmetrical and equivalent, the coefficients of both the orbitals will be same. $1s^+$ m.o. is symmetrical and can combine with 2*s* a.o. of Be, which is also symmetrical. The constructive overlap will give the bonding m.o. and destructive overlap will give the anti-bonding m.o. (Fig. 4.25*b*). The wave functions for the m.o.'s are:

$$\psi\,(\sigma_{2s}) = C_1\,2s\,(Be) + C_2\,[1s\,(H_A) + 1s\,(H_B)] \qquad ...(4.145)$$

$$\psi\,(\sigma^*_{2s}) = C_3\,2s\,(Be) - C_4\,[1s\,(H_A) + 1s\,(H_B)] \qquad ...(4.146)$$

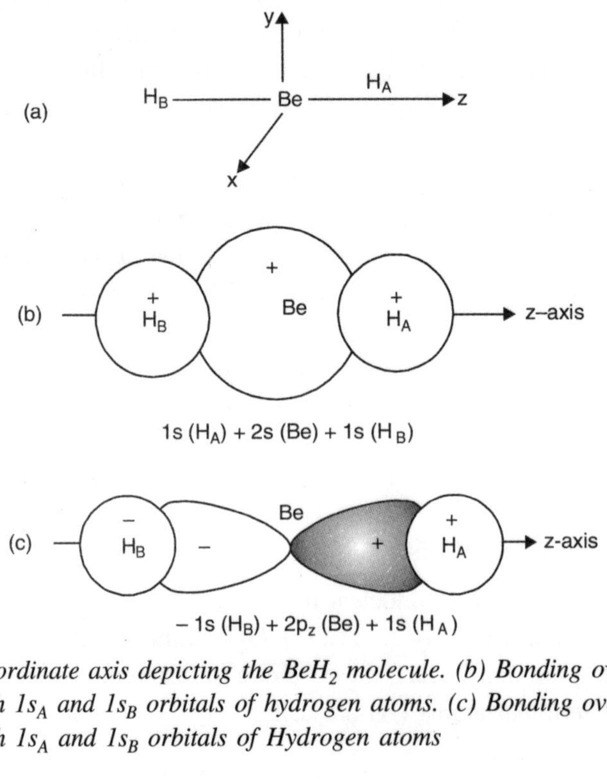

Figure 4.25: *(a) Coordinate axis depicting the BeH_2 molecule. (b) Bonding overlap of 2s orbital of Be with $1s_A$ and $1s_B$ orbitals of hydrogen atoms. (c) Bonding overlap of $2p_z$ orbital of Be with $1s_A$ and $1s_B$ orbitals of Hydrogen atoms*

The second a.o. on Be, which can form σ bond with hydrogen atom is $2p_z$. $2p_z$ orbital has a positive lobe on the increasing side of z-axis. The shape of p_z is such that it can combine with H^- as shown in Fig. 4.25*c*. The wave functions for the bonding and anti-bonding m.o.'s are:

$$\psi\,(\sigma_{2p_z}) = C_5\, 2p_z\,(\text{Be}) + C_6\,[1s\,(\text{H}_A) - 1s\,(\text{H}_B)] \qquad ...(4.147)$$

$$\psi\,(\sigma^*_{2p_z}) = C_7\, 2p_z\,(\text{Be}) - C_8\,[1s\,(\text{H}_A) - 1s\,(\text{H}_B)] \qquad ...(4.148)$$

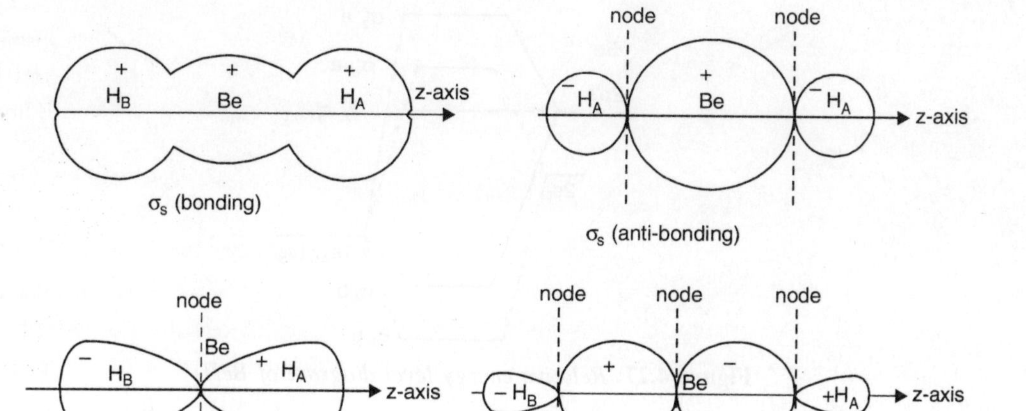

Figure 4.26: *Contour diagrams of the boundary surface of the BeH$_2$ molecular orbitals*

$2p_x$ and $2p_y$ orbitals on Be are placed perpendicular to the bond axis, thus these orbitals cannot combine with the $1s$ orbital of hydrogen atoms, because these are the π-orbitals. These are labeled as non-bonding m.o.'s, localized on Be atom. The coefficients C_1 to C_8 can be calculated using variation theorem and are beyond the scope of this book. The m.o.'s given by the equations (4.145) and (4.147) are bonding m.o.'s and m.o.'s given by equations (4.146) and (4.148) are anti-bonding m.o.'s respectively, because the signs of nearby lobes of the a.o.'s are similar in Fig. (4.25 *b*, *c*) and although not shown these signs will be different in anti-bonding m.o.'s. The contour diagrams or the boundary surfaces of the various m.o.'s formed are depicted in Fig. 4.26.

Figure (4.27) depicts the energy level diagram. Towards the left hand side are the energy levels of the atom Be, towards right hand side are those of 1*s* of the H-atoms and on the middle are those of m.o.'s of BeH_2. H^+ will be of slightly lower in energy in comparison to H^-, because former is symmetrical and latter is asymmetrical. Further hydrogen is more electronegative than

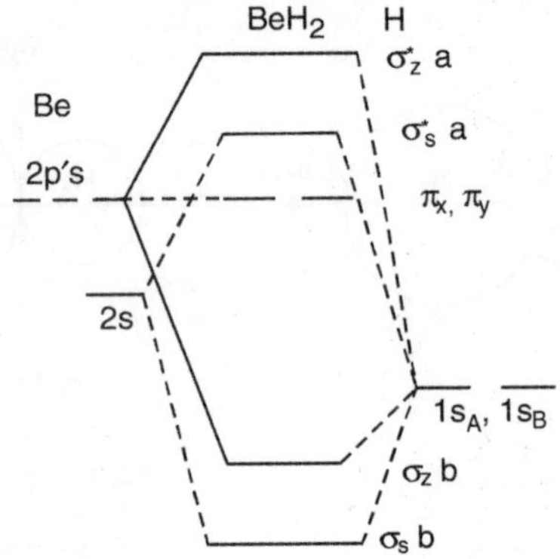

Figure 4.27: *Relative energy level diagram of* BeH_2

Be, 2*s* and 2*p* a.o.'s of Be will be placed higher than the 1*s* of H-atoms. ψ (σ_{2s}) and ψ ($2p_z$) are bonding m.o. formed by the overlap of H^+ and H^- with 2*s* and $2p_z$ a.o. of Be atom respectively. ψ (σ_{2s}) will be of lower energy than ψ (σ_{2p_z}). Similarly ψ ($\sigma^*_{2p_z}$) and will be of higher energy than ψ (σ^*_{2s}). Since $2p_x$ and $2p_y$ are non-bonding m.o., their energies will not change and are localized on Be.

From the energy level diagram (Fig. 4.27) and from the boundary surfaces of m.o. (Fig. 4.26) it is obvious that percentage of 1*s* (H) orbital present in ψ (σ_{2s}), as well as, in ψ ($2p_z$) is more than that of 2*s* (Be) and $2p_z$ (Be). Similarly in ψ (σ^*_{2s}) and ψ (σ^*_{2pz}), the contribution of 2*s* (Be) and $2p_z$ (Be) are much larger than 1*s* (H). In other words, $2C_2 > C_1$; $2\,C_6 > C_5$; $C_3 > 2C_4$; and $C_7 > 2C_8$. The ground state electronic configuration of BeH_2 can be obtained by placing the valence electrons, 2 from Be and 1 each from H-atoms, (*i.e.*, $\sigma^2_{2s}\ \sigma^2_{2p_z}$).

Valence bond theory (localized m.o.'s): In the last section, we have learnt that the m.o's are formed by the use of four a.o.'s and delocalized over all the three atoms. This gives little diffuse picture of chemical bonding. A better picture of chemical bonding in BeH_2 can be obtained if we use the localized bonding. This can be done by first forming two equivalent *sp* hybrid orbitals, from the linear combination of 2*s* and $2p_z$ a.o.'s, obtained by promoting one electron from 2*s* to 2*p* orbitals of Be. The wave functions for the hybrid orbitals (will be discussed in more details later, 4.19) are:

$$h_1\ (\text{Be}) = C_1\ 2s\ (\text{Be}) + C_2\ 2p_z\ (\text{Be}) \qquad ...(4.149)$$

$$h_2\ (\text{Be}) = C_3\ 2s\ (\text{Be}) - C_4\ 2p_z\ (\text{Be}) \qquad ...(4.150)$$

The boundary surfaces of h_1 and h_2 are shown in Fig. (4.28*a* and *b*). The lobes of h_1 and h_2 are much larger than those of either 2*s* or $2p_z$ and are directed along the molecular axis (*z*). Now 1*s* (H) orbital of each hydrogen atom approaching from each direction can overlap to form directed m.o. The wave functions for the bonding orbitals can be written as (Fig. 4.28*c*, *d*).

$$\psi_{1,l} = C_1\, h_1 + C_2\, 1s\,(H_A) \qquad ...(4.151)$$

$$\psi_{2,l} = C_3\, h_2 + C_4\, 1s\,(H_B) \qquad ...(4.152)$$

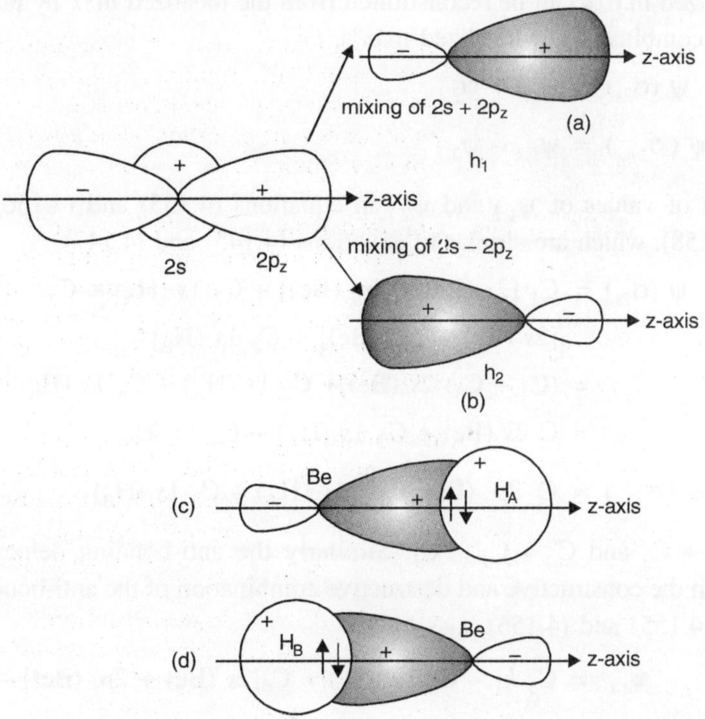

Figure 4.28: *Formation of two hybrid orbitals by the combination of s and $2p_z$ atomic orbitals. (a) is the h_1 hybrid orbital and (b) is the h_2 hybrid orbital. (c) depicts the formation of m.o. by the combination of h_1 and $1s_A$ and (d) depicts the formation of m.o. by the combination of h_2 and $1s_B$. Double arrows depict the presence of electron pair for most of the time*

and thus explains the directional property of σ orbital. Substituting the value of h_1 and h_2, the wave functions ψ_1 and ψ_2 can be written as:

$$\psi_{1,l} = C_1\,[2s\,(Be) + 2p_z\,(Be)] + C_2\, 1s\,(H_A) \qquad ...(4.153)$$

$$\psi_{2,l} = C_3\,[2s\,(Be) - 2p_z\,(Be)] + C_4\, 1s\,(H_B) \qquad ...(4.154)$$

Comparison of delocalized and localized m.o.'s: The m.o.'s described by equations (4.145) and (4.147) give a very diffuse picture of a chemical bond involving all the three atoms, *i.e.*, all the four electrons are delocalized over the complete molecule, whereas the localized m.o.'s describe the classical picture of a chemical bond involving sharing of two electrons in a bond and possessing directional property. We know from experimental data that both the Be-H bonds have similar characteristics. These are also called as *equivalent bonds or equivalent orbitals* (EO).

Delocalized m.o.'s give a better description of a spectral transition, because the molecular energy levels are very well defined (Fig. 4.27) and lowest energy transition is defined as $\sigma^b_z \rightarrow \sigma^*_s$. In case of localized m.o.'s formation, it is difficult to assign spectral transition, *i.e.*, excitation of electron, because both the bonds (Be–H) are equivalent. It may be pointed out here that the delocalized m.o.'s describe spectroscopic properties nicely, whereas the localized m.o's give a better picture of chemical bond and thus geometry of the molecule.

The delocalized m.o.'s can be reconstituted from the localized m.o. by taking the constructive and destructive combination of localized m.o.'s, *i.e.*,

$$\psi\,(\sigma_{2s}) = \psi_{1,l} + \psi_{2,l} \qquad \text{...(4.155)}$$

$$\psi\,(\sigma_{2p_z}) = \psi_{1,l} - \psi_{2,l} \qquad \text{...(4.156)}$$

Substitution of values of $\psi_{1,l}$ and $\psi_{2,l}$ in equations (4.155) and (4.156), leads to equations (4.157) and (4.158), which are similar to equations (4.145) and (4.147).

$$\begin{aligned}\psi\,(\sigma_{2s}) &= C_1\,[2s\text{ (Be)} + 2p_z\text{ (Be)}] + C_2\,1s\,(H_A) + C_3\\ &\quad [2s\text{ (Be)} - 2p_z\text{ (Be)}] + C_4\,1s\,(H_B)\\ &= (C_1 + C_3)\,2s\text{ (Be)} + C_2\,1s\,(H_A) + C_4\,1s\,(H_B)\\ &= C\,2s\text{ (Be)} + C_2\,1s\,(H_A) + C_4\,1s\,(H_B)\end{aligned} \qquad \text{...(4.157)}$$

$$\psi\,(\sigma_{2p_z}) = C'\,2p_z\text{ (Be)} + C_2\,1s\,(H_A) - C_4\,1s\,(H_B) \qquad \text{...(4.158)}$$

where $C = C_1 + C_3$ and $C' = C_1 - C_3$. Similarly the anti-bonding delocalized m.o.'s can be constructed from the constructive and destructive combination of the anti-bonding localized m.o.'s, *i.e.*, equations (4.155) and (4.156), *i.e.*,

$$\begin{aligned}\psi^*_{1,l} &= C_5\,h_1 - C_6\,1s\,(H_A) = C_5[2s\text{ (Be)} + 2p_z\text{ (Be)}] - C_6\,1s\,(H_A)\\ \psi^*_{2,l} &= C_7\,h_2 - C_8\,1s\,(H_B)\\ &= C_7\,[2s\text{ (Be)} - 2p_z\text{ (Be)}] + C_8\,1s\,(H_B)\end{aligned} \qquad \text{...(4.159)}$$

$$\psi\,(\sigma^*_{2s}) = \psi^*_{1,l} + \psi^*_{2,l} = C''\,2s\text{ (Be)} - C_6\,1s\,(H_A) - C_8\,1s\,(H_B)$$

$$\psi^*\,(\sigma_{2p_z}) = \psi^*_{1,l} - \psi^*_{2,l} = C'''\,2p_z\text{ (Be)} - C_6\,1s\,(H_A) + C_8\,1s\,(H_B) \qquad \text{...(5.160)}$$

where $C'' = C_5 + C_7$ and $C''' = C_5 - C_7$

In equations (4.157) and 4.159), the contributions from $(C_1 - C_3)\,2p_z$ (Be) and $(C_5 - C_7)\,2p_z$ (Be) are very small and have been neglected. Similarly the terms $(C_1 - C_3)\,2s$ (Be) and $(C_5 - C_7)$ $2s$ (Be) in equations (4.158) and (4.160) have been neglected.

Carbon dioxide (CO_2)

***Delocalized m. o.'s for CO_2*:** CO_2 is also a triatomic linear molecule and can be treated similarly as BeH_2. The only difference between BeH_2 and CO_2 is that in the latter molecule, both $2s$ and $2p$ a.o.'s are placed on both carbon and oxygen atoms. Because of these a.o.'s, both atoms contain the a.o.'s which are capable of forming π m.o.'s. As per the normal convention followed in this book, bond axis in CO_2 is depicted along the *z*-axis, as shown in Fig. 4.29. Since the gap between the 2*s* and 2*p* orbitals of oxygen atom is very large, the probability of participation of 2*s* orbital of oxygen will be very small when the mixing of a.o.'s is considered on oxygen atom and thus can be neglected, *i.e.*, we will only consider the linear combination of 2*p* a.o.'s on the oxygen atoms. As mentioned earlier, (4.6), $2p_z$ will be used for forming σ bonds and $2p_x$ and $2p_y$ will be used for π-bonds. The proper linear combinations for the σ bonding m.o.'s are obtained by writing the combinations of $2p_z$ (O_A) and $2p_z$ (O_B), that matches the algebraic signs on the lobes of 2*s* and $2p_z$ a.o.'s of the central carbon atom respectively as shown in Fig. 4.29 *b* and *c*. These are:

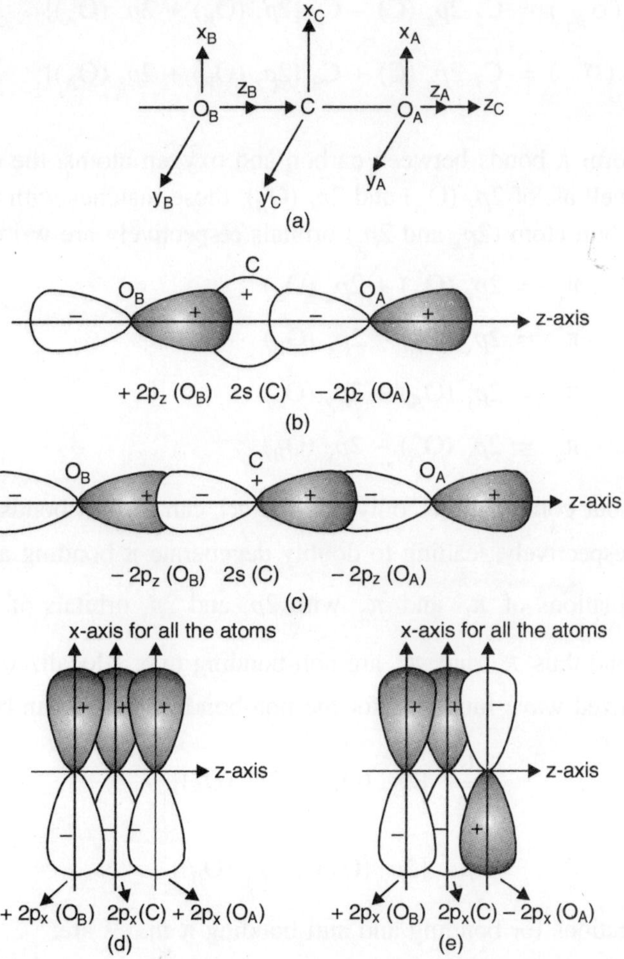

Figure 4.29: *(a) Coordinate axis for CO_2 showing the directions how each atom is depicted in space. (b) overlap of 2s (C) with $2p_z(O_A)$ and $2p_z(O_B)$ to form σ m.o. (c) overlap of $2p_z$(C) with $2p_z(O_A)$ and $2p_z$ (O_B) to form σ m.o. (d) overlap of $2p_x$ (C) with $2p_x(O_A)$ and $2p_x(O_B)$ to lead π m.o. (e) overlap of $2p_x$ (C) with $-2p_x(O_A)$ and $2p_x(O_B)$ will not lead to π m.o because the $2p_x$ (O_B) leads to constructive overlap and $2p_x(O_A)$ leads to destructive overlap. This net result is zero overlap*

$$2p_z^+ (O) = 2p_z (O_B) + 2p_z (O_A) \qquad ...(4.161)$$

$$2p_z^-\,(\text{O}) = 2p_z\,(\text{O}_\text{B}) - 2p_z\,(\text{O}_\text{A}) \quad ...(4.162)$$

The m.o.'s for σ bonds can be formed by the LCAO of $2s$ and $2p_z$ of carbon atom with the $2p_z^+$ (O) and $2p_z^-$ (O) having proper symmetry respectively. The wave functions for bonding and anti-bonding σ orbitals are:

$$\psi\,(\sigma_s) = C_1\,2s\,(\text{C}) + C_2\,[2p_z\,(\text{O}_\text{B}) - 2p_z\,(\text{O}_\text{A})] \quad ...(4.163)$$

$$\psi\,(\sigma_s^*) = C_3\,2s\,(\text{C}) - C_4\,[2p_z\,(\text{O}_\text{B}) - 2p_z\,(\text{O}_\text{A})] \quad ...(4.164)$$

$$\psi\,(\sigma_{p_z}) = C_5\,2p_z\,(\text{C}) - C_6\,[2p_z\,(\text{O}_\text{B}) + 2p_z\,(\text{O}_\text{A})] \quad ...(4.165)$$

$$\psi\,(\sigma_{p_z}^*) = C_7\,2p_z\,(\text{C}) + C_8\,[2p_z\,(\text{O}_\text{B}) + 2p_z\,(\text{O}_\text{A})] \quad ...(4.166)$$

Similarly to form π bonds between carbon and oxygen atoms, the combinations of $2p_x\,(\text{O}_\text{A})$ and $2p_x\,(\text{O}_\text{B})$, as well as, of $2p_y\,(\text{O}_\text{A})$ and $2p_y\,(\text{O}_\text{B})$, those matches with the algebraic signs on the lobs of central carbon atom ($2p_x$ and $2p_y$) orbitals respectively are written as:

$$\pi_1^+ = 2p_x\,(\text{O}_\text{A}) + 2p_x\,(\text{O}_\text{B}) \quad ...(4.167)$$

$$\pi_1^- = 2p_x\,(\text{O}_\text{A}) - 2p_x\,(\text{O}_\text{B}) \quad ...(4.168)$$

$$\pi_2^+ = 2p_y\,(\text{O}_\text{A}) + 2p_y\,(\text{O}_\text{B}) \quad ...(4.169)$$

$$\pi_2^- = 2p_y\,(\text{O}_\text{A}) - 2p_y\,(\text{O}_\text{B}) \quad ...(4.170)$$

Out of these four combinations, only π_1^+ and π_2^+ can form π bonds with $2p_x$ and $2p_y$ orbitals on carbon atom respectively, leading to doubly degenerate π bonding and π-anti-bonding m.o.'s. Overlap of combinations of π_1^- and π_2^- with $2p_x$ and $2p_y$ orbitals of carbon atom respectively gives zero value and thus π_1^- and π_2^- are non-bonding m.o.'s localized on the respective oxygen atom. The normalized wave functions for the non-bonding m.o.'s can be written as:

$$\frac{1}{\sqrt{2}}\,[2p_x\,(\text{O}_\text{A}) - 2p_x\,(\text{O}_\text{B})] \quad ...(4.171)$$

and

$$\frac{1}{\sqrt{2}}\,[2p_y\,(\text{O}_\text{A}) - 2p_y\,(\text{O}_\text{B})] \quad ...(4.172)$$

The wave functions for bonding and anti-bonding π m.o.'s are:

$$\begin{aligned}\psi\,(\pi_x) &= C_9\,2p_x\,(\text{C}) + C_{10}\,\pi_1^+ \\ &= C_9\,2p_x\,(\text{C}) + C_{10}\,[2p_x\,(\text{O}_\text{A}) + 2p_x\,(\text{O}_\text{B})] \quad ...(4.173)\end{aligned}$$

$$\begin{aligned}\psi\,(\pi_y) &= C_9\,2p_y\,(\text{C}) + C_{10}\,\pi_2^+ \\ &= C_9\,2p_y + C_{10}\,[2p_y\,(\text{O}_\text{A}) + 2p_y\,(\text{O}_\text{B})] \quad ...(4.174)\end{aligned}$$

$$\begin{aligned}\psi\,(\pi_x^*) &= C_{11}\,2p_x\,(\text{C}) - C_{12}\,\pi_1^+ \\ &= C_{11}\,2p_x\,(\text{C}) - C_{12}\,[2p_x\,(\text{O}_\text{A}) + 2p_x\,(\text{O}_\text{B})] \quad ...(4.175)\end{aligned}$$

$$\psi\,(\pi_y^*) = C_{11}\ 2p_y\,(C) - C_{12}\ \pi_2^+$$
$$= C_9\ 2p_y - C_{12}\ [2p_y\,(O_A) + 2p_y\,(O_B)] \qquad ...(4.176)$$

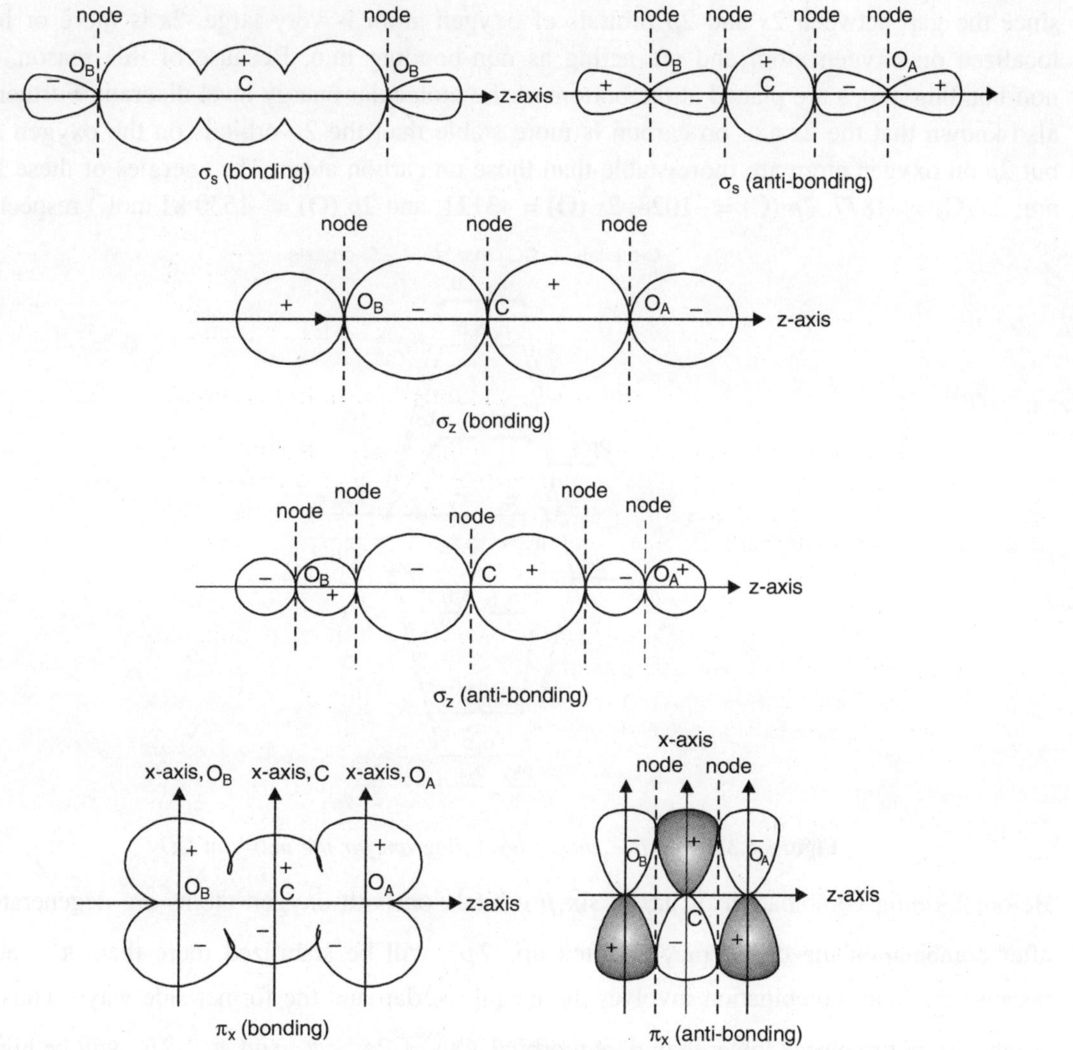

Figure 4.30: *Contour diagrams of the m.o's of CO_2*

As mentioned earlier, the coefficients of bonding and anti-bonding σ and π orbitals can be calculated using approximate methods, but these are outside the scope of this book. Qualitative picture is shown in Fig. 4.29 *d* and *e* respectively. The bonding surfaces of the bonding and anti-bonding m.o.'s has been depicted in Fig. 4.30. The nature of a m.o., as described earlier, can be determined from the signs of the nearby lobes of the combining atoms. Thus we can see easily from the contour diagrams (Fig. 4.30) that $\psi\,(\sigma_s)$, $\psi\,(\sigma_{p_z})$, $\psi\,(\pi_x)$ and $\psi\,(\pi_y)$ are bonding m.o.'s and $\psi(\sigma_s^*)$, $\psi(\sigma^*_{p_z})$, $\psi(\pi^*_x)$ and $\psi(\pi^*_y)$ are anti-bonding m.o.'s.

The energy level diagram for the a.o.'s and m.o.'s of CO_2 with their respective atoms are shown in Fig. 4.31. Similar to the other energy level diagram of m.o.'s, left hand side depicts the energy levels of carbon-atom, right hand side depicts the energy levels of oxygen atoms and the middle ones are those of m.o.'s of CO_2. Since oxygen atom is more electronegative than carbon atom, 2*s* energy level of oxygen atom is placed much lower than that of carbon atom. Further, since the gap between 2*s* and 2*p* orbitals of oxygen atom is very large, 2*s* is more or less is localized on oxygen atom and are acting as non-bonding m.o. Because of this reason, these non-bonding m.o.'s are placed at the bottom of the molecular energy level diagram. Further, it is also known that the 2*s* a.o. on carbon is more stable than the 2*p* orbitals on the oxygen atom, but 2*p* on oxygen atom are more stable than those on carbon atom. The energies of these levels are: 2*s* (C) = –1877, 2*p* (C) = –1028, 2*s* (O) = –3121, and 2*p* (O) = –1530 kJ mol^{-1} respectively.

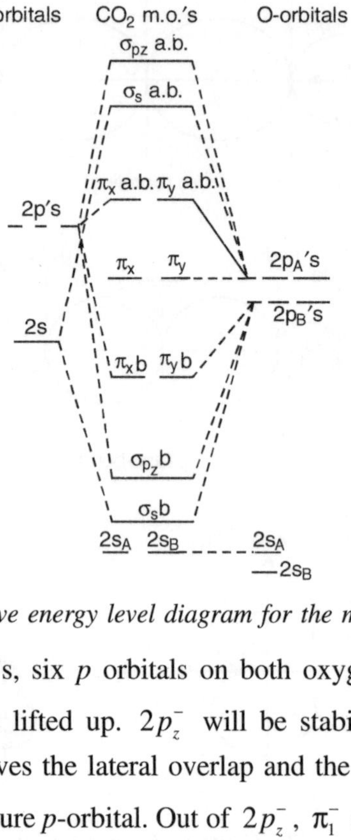

Figure 4.31: *Relative energy level diagram for the m.o.'s of* CO_2

Before forming combination of a.o.'s, six *p* orbitals on both oxygen atoms are degenerate, but after combination the degeneracy is lifted up. $2p_z^-$ will be stabilized more than π_1^+ and π_2^+, because the latter combination involves the lateral overlap and the former side ways. Thus these combinations are placed lower than pure *p*-orbital. Out of $2p_z^-$, π_1^- and π_2^-, $2p_z^-$ will be higher in energy than the other two, but is destabilized in comparison to pure *p*-orbitals and thus is placed above pure 2*p*-orbitals.(This internal combination of energy level on oxygen a.o.'s are not shown in the energy level diagram). Out of the lower set of the combinations, orbitals on oxygen atoms, $2p_z^-$ forms σ bonding m.o. with 2*s* orbital of carbon and the other two (π_1^+ and π_2^+) form the π m.o.'s with the overlap of $2p_x$ and $2p_y$ a.o.'s on the carbon atom. Similarly out of the three other combinations, ($z_{p_z}^+$, π_1^- and $\bar{\pi}_2$), $2p_z^+$ forms σ m.o. with $2p_z$ a.o. on carbon atom and the other two acts as non-bonding m.o.'s. Since in general the overlap integral with 2*s* orbital will be larger than that with $2p_z$ orbital, the energy of σ_s m.o. will be less than that of σ_{p_z} and vice versa of those of anti-bonding ones. Overlap integrals in π orbitals are not very large and thus are stabilized less than those of σ ones. These are placed lower than the a.o.'s of the respective atom, but higher than σ ones. The values of the energies for non-bonding are same as those one on oxygen atom.

The electronic configuration of CO_2 can be written by placing the valence electrons in the m.o.'s of increasing energies following Pauli's Exclusion Principle and Hund's rule. There are sixteen valence electrons (four from carbon atom and six each from oxygen atoms). The electronic configuration for the ground state of CO_2 is:

$$(1s_A)^2(1s_B)^2(\sigma_s)^2(\sigma_{p_z})^2(\pi_x)^2(\pi_y)^2(\pi_1)^2(\pi_2)^2 \qquad ...(4.177)$$

and the molecular term symbol is $^1\Sigma$. There are thus four electrons in two σ bonds (ψ_{σ_s} and $\psi_{\sigma_{p_z}}$) and four electrons in two π bonds (ψ_{π_x} and ψ_{π_y}), in agreement with Lewis dot structure. $(2s_A)^2$, $2s_B$, $\psi_{\bar{\pi_1}}$ and $\psi_{\bar{\pi_2}}$ are non-bonding orbitals. The qualitative estimates of the coefficients in all the wave functions for the bonding and anti-bonding orbitals can be evaluated similarly as discussed earlier, *i.e.*,

$$C_2 > C_1, C_3 > C_4, C_6 > C_5, C_7 > C_8, C_{10} > C_9 \text{ and } C_{11} > C_{12}.$$

The charge densities in ψ_{σ_1}, ψ_{σ_2}, ψ_{π_1} and ψ_{π_2} are distributed more towards the oxygen atoms, because of their higher electronegativity, whereas in $\psi_{\sigma_1^*}$, $\psi_{\sigma_2^*}$, $\psi_{\pi_1^*}$ and $\psi_{\pi_2^*}$, the electrons are spending more time near the carbon atom.

Localized m. o. for CO_2: Proceeding exactly on the same lines as carried out for BeH_2, we can form sp_z hybrid orbitals on carbon atom, projecting their lobes as shown in Fig. 4.28*a* and *b*. The wave functions for the hybrid orbitals are:

$$h_1\ (C) = C_1\ 2s\ (C) + C_2\ 2p_z\ (C) \qquad ...(4.178)$$

$$h_2\ (C) = C_3\ 2s\ (C) - C_4\ 2p_z\ (C) \qquad ...(4.179)$$

The localized σ m.o.'s, can be formed by combining the $2p_z$ orbital on the respective oxygen atom as:

$$\psi\ (\sigma_{1,l}) = C_1\ h_1 - C_2\ 2p_z\ (O_A)$$

$$= C_1\ [2s\ (C) + 2p_z\ (C)] - C_2\ 2p_z\ (O_A) \qquad ...(4.180)$$

$$\psi\ (\sigma_{2,l}) = C_1\ h_2 + C_2\ 2p_z\ (O_B)$$

$$= C_1\ [2s\ (C) - 2p_z\ (C)] + C_2\ 2p_z\ (O_B) \qquad ...(4.181)$$

p_x on carbon atom can combine with p_x on O_A (or O_B) and p_y on carbon can combine with p_y on O_B (or O_A) to form π orbitals as shown in Fig. 4.32. This will leave 2*s* and $2p_y$ on O_A (or 2*s* and $2p_x$) and 2*s* and $2p_x$ on O_B (or 2*s* and $2p_y$) as non-bonding orbitals on the oxygen atoms. The wave functions for the π orbitals are:

$$\psi\ (\pi_1) = C_3\ 2p_x\ (C) + C_4\ 2p_x(O_A) \qquad ...(4.182)$$

$$\psi\ (\pi_2) = C_3\ 2p_y\ (C) + C_4\ 2p_y\ (O_B) \qquad ...(4.183)$$

The wave functions for the localized anti-bonding m.o.'s can be written by the negative combination of the respective atomic wave functions, *i.e.*,

$$\psi(\sigma^*_{1,l}) = C_1\, h_1 + C_2\, 2p_z\,(O_A)$$
$$= C_1\,[2s\,(C) + 2p_z\,(C)] + C_2\, 2p_z\,(O_A) \quad ...(4.184)$$
$$\psi(\psi^*_{2,l}) = C_1\, h_2 - C_2\, 2p_z\,(O_B)$$
$$= C_1\,[2s\,(C) - 2p_z\,(C)] - C_2\, 2p_z\,(O_B) \quad ...(4.185)$$
$$\psi(\pi^*_1) = C_3\, 2p_x\,(C) - C_4\, 2p_x(O_A) \quad ...(4.186)$$
$$\psi(\pi^*_2) = C_3\, 2p_y\,(C) - C_4\, 2p_y\,(O_B) \quad ...(4.187)$$

***Comparison of delocalized and localized m.o.'s*:** Proceeding on the same lines as done for BeH_2, spectroscopic transitions are well defined by the delocalized m.o.'s and structural properties

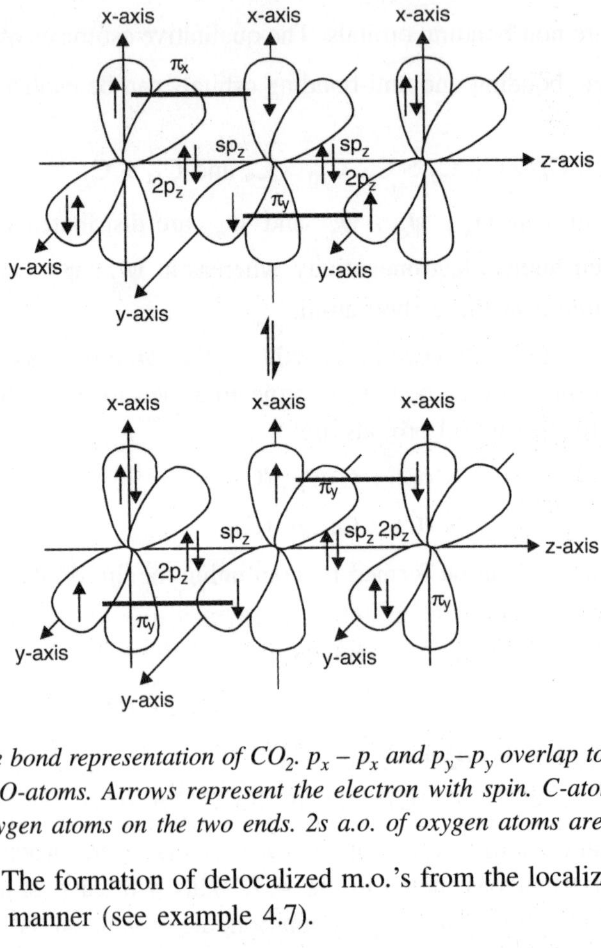

Figure 4.32: *Valence bond representation of CO_2. $p_x - p_x$ and p_y–p_y overlap to form π bonds between C and O-atoms. Arrows represent the electron with spin. C-atom is in the middle and two oxygen atoms on the two ends. 2s a.o. of oxygen atoms are not shown*

by the localized m.o.'s. The formation of delocalized m.o.'s from the localized m.o.'s can also be carried out in the same manner (see example 4.7).

Now let us try to compare some of the properties of these bonds. The carbon oxygen bond length in CO_2 is found to be 0.115 nm. On the other hand, the normal carbon oxygen double bond (such as in ketones) is 0.122 nm and that of triple bond is 0.110 nm. In other words the bond in CO_2 possesses some intermediate character. CO_2 molecule possesses two kinds of bond energies. One is dissociation energy of C — O bond, obtained by breaking CO_2 into CO and oxygen atom, *i.e.*,

$$O - C - O \rightarrow CO + O$$

The dissociation energy refers to the breaking of one of CO bonds in CO_2 and its value is 531.4 kJ mol^{-1}. The other way of breaking CO_2 molecule is to have one carbon and two oxygen atoms, *i.e.*,

$$O - C - O \rightarrow C + 2\,O$$

and the dissociation energy is found to be 1602.5 kJ mol^{-1}. This can also be obtained from the process

$$O - C - O \rightarrow CO + O \rightarrow C + O + O$$

and total dissociation energy of CO_2 can be calculated by adding the dissociation energy of each step, *i.e.*,

$$E = 531.4 + 1071.\,1 = 1602.5 \text{ kJ mol}^{-1}$$

The term regarding dissociation energy can be defined in multi-atomic molecules containing similar bonds, as the bond energy, obtained by dividing the total dissociation energy by the number of bonds. This may also be called as the average bond energy. In case of CO_2, the average bond energy of C — O bond is:

$$BE\,(C - O) = \frac{1}{2}\,(1602.5) = 801.25 \text{ kJ mol}^{-1}.$$

Let us try to compare these values with those obtained by considering the CO_2 to be made up of two C ═ O bonds. The heat of formation of C ═ O bond is known to be 723.8 kJ mol^{-1}. Thus the total energy of CO_2 molecule will be twice the C ═ O value, *i.e.*, 2 × 723.8 = 1447.6 kJ mol^{-1}, whereas measured heat of formation of CO_2 is 1602.5 kJ mol^{-1}. The difference between these two (1602.5 –1447.6 = 154.9 kJ mol^{-1}) represents the extent to which the actual structure is more stable than that represented by the simple formula I. This suggests that the actual structure of CO_2 cannot be represented by the simple formula I. The difference in the above properties (*i.e.*, shortening of bond length, difference in energy) can be explained if we describe the actual structure as the combination of the following structures also called as resonance hybrid formed from the three canonical structures.

I	$O = C = O$	ψ_I
II	$O^+ \equiv C - O^-$	ψ_{II}
III	$O^- - C \equiv O^+$	ψ_{III}

so that $\psi = C_I\,\psi_I + C_{II}\,\psi_{II} + C_{III}\,\psi_{III}$, describes the actual state of CO_2, which is more stable than any of the contributing forms. This may be seen as that each structure generates a wave front which is obtained when all the interactions leading to the formation of individual bond in each structure. The actual structure then arises from the interaction of these stationary waves, leading to a frequency splitting, a new stationary wave and consequently of lower energy state for the molecule to result.

One thing, which we have learnt from the above discussion, as we have mentioned in diatomic molecules, is that simple structure does not describe the true picture of polyatomic molecules. Different canonical forms have to be considered for each molecule and actual structure is described by the hybrid of all the canonical forms. This means that the calculations will be more tedious as the number of canonical structures increases in the polyatomic molecules.

Example 4.7: Form the delocalized m.o.'s from the localized m.o.'s of CO_2.

Solution: The localized wave functions of σ bonding and anti-bonding m.o.'s are:

$$\psi(\sigma_{1,l}) = C_1 h_1 - C_2\, 2p_z(O_A)$$

$$= C_1 [2s(C) + 2p_z(C)] - C_2\, 2p_z(O_A) \quad ...(4.180)$$

$$\psi(\sigma_{2,l}) = C_1 h_2 + C_2\, 2p_z(O_B)$$

$$= C_1 [2s(C) - 2p_z(C)] + C_2\, 2p_z(O_B) \quad ...(4.181)$$

$$\psi(\sigma^*_{1,l}) = C_1 h_1 + C_2\, 2p_z(O_A)$$

$$= C_1 [2s(C) + 2p_z(C)] + C_2\, 2p_z(O_A) \quad ...(4.184)$$

$$\psi(\sigma^*_{2,l}) = C_1 h_2 - C_2\, 2p_z(O_B)$$

$$= C_1 [2s(C) - 2p_z(C)] - C_2\, 2p_z(O_B) \quad ...(4.185)$$

Taking the linear combination of the bonding wave functions (4.180) and (4.181), we get

$$\psi(\sigma_{1,l}) + \psi(\sigma_{2,l}) = [C_1 [2s(C) + 2p_z(C)] - C_2\, 2p_z(O_A)] + [C_1 [2s(C) - 2p_z(C)] + C_2\, 2p_z(O_B)]$$

$$= 2\, C_1\, 2s(C) + C_2 [2p_z(B) - 2p_z(A)]$$

$$= C_1\, 2s(C) + C_2 [2p_z(B) - 2p_z(A)] \quad ...(4.180')$$

$$\psi(\sigma_{1,l}) - \psi(\sigma_{2,l}) = [C_1 [2s(C) + 2p_z(C)] - C_2\, 2p_z(O_A)] - [C_1 [2s(C) - 2p_z(C)] + C_2\, 2p_z(O_B)]$$

$$= 2\, C_1\, 2p_z(C) - C_2 [2p_z(B) + 2p_z(A)]$$

$$= C_1\, 2p_z(C) - C_2 [2p_z(B) + 2p_z(A)] \quad ...(4.181')$$

Similarly taking the linear combination of the anti-bonding wave functions (4.184) and (4.185), we get

$$\psi(\sigma^*_{1,l}) + \psi(\sigma^*_{2,l}) = [C_1 [2s(C) + 2p_z(C)] + C_2\, 2p_z(O_A)] + [C_1 [2s(C) - 2p_z(C)] - C_2\, 2p_z(O_B)]$$

$$= 2C_1\, 2s(C) - C_2 [2p_z(B) - 2p_z(A)]$$

$$= C_1\, 2s(C) - C_2 [2p_z(B) - 2p_z(A)] \quad ...(4.184')$$

$$\psi(\sigma^*_{1,l}) - \psi(\sigma^*_{2,l}) = [C_1 [2s(C) + 2p_z(C)] + C_2\, 2p_z(O_A)] - [C_1 [2s(C) - 2p_z(C)] - C_2\, 2p_z(O_B)]$$

$$= 2C_1\ 2p_z\ (C) + C_2\ [2p_z\ (B) + 2p_z\ (A)]$$

$$= C_1\ 2p_z\ (C) + C_2\ [2p_z\ (B) + 2p_z\ (A)] \qquad ...(4.185')$$

Thus, it is clear that the equations (4.184′) and (4.185′) are similar to (4.181) and (4.185) respectively.

4.17.2 TRIATOMIC BENT MOLECULES

Water

***Delocalized* m.o.'s:** The simplest and most familiar triatomic bent molecule is water (H_2O). The only difference between BeH_2 and H_2O is that the central oxygen atom possesses three *p* orbitals besides the 2*s* one and thus can be treated in the similar manner. The valence shell electronic configuration of oxygen atom is $2s^2\,2p_x^2\,2p_y^1\,2p_z^1$. As mentioned in case of CO_2, the 2*s* electrons or orbital will not participate in LCAO of oxygen a.o.'s as the gap between 2*s* and 2*p* orbitals of oxygen is very large and thus, will act as non-bonding m.o. Assume that the oxygen atom is present at the origin of the three dimensional *xyz*-plane and hydrogen atoms are placed in *y-z* plane (Fig. 4.33). To start with, assume that water is a linear molecule and is placed along the *z*-axis, as the molecular axis. The actual water molecule is generated if we bent the H — O — H bond symmetrically with respect to *y*-axis till the angle HOH is 105°. This will mean that *y*-axis divides the HOH angle equally and this can happen with all the similar triatomic molecules. Now proceeding on the same lines, as carried out in case of BeH_2, 1*s* (H_A) and 1*s* (H_B) can have the following linear combinations.

$$1s^+ = 1s\ (H_A) + 1s\ (H_B) \qquad ...(4.188)$$

$$1s^- = 1s\ (H_A) - 1s\ (H_B) \qquad ...(4.189)$$

p_x, p_y and p_z orbitals on oxygen are shown in Fig. (4.34 *a, b*) respectively. Considering the signs presents on the lobes of p_y and p_z orbitals of oxygen atom, it is obvious that $1s^-$ will overlap constructively with $2p_z$ atoms of oxygen (Fig. 4.34*a*) and $1s^+$ will overlap with $2p_y$ of oxygen (Fig. 4.34*b*) to form bonding orbitals. The wave functions for the bonding and anti-bonding m.o.'s can be written as:

$$\psi\left(\alpha_{2p_y}\right) = C_1\ 2p_y\ (O) + C_2\ [1s\ (H_A) + 1s\ (H_B)] \qquad ...(4.190)$$

$$\psi\left(\alpha_{2p_y}\right) = C_3\ 2p_z\ (O) + C_4\ [1s\ (H_A) - 1s\ (H_B)] \qquad ...(4.191)$$

$$\psi\left(\alpha_{2p_y}\right) = C_5\ 2p_y\ (O) - C_6\ [1s\ (H_A) + 1s\ (H_B)] \qquad ...(4.192)$$

$$\psi\left(\alpha_{2p_y}\right) = C_7\ 2p_z\ (O) - C_8\ [1s\ (H_A) - 1s\ (H_B)] \qquad ...(4.193)$$

Since $2p_x$ orbital on oxygen is perpendicular to *yz*-plane and thus, there will not be any overlap between the hydrogen atoms. However $2p_x$ orbital has a tendency to form π orbital, but 1*s* orbital of hydrogen atom cannot do so. So $2p_x$ orbital is acting as a non-bonding m.o.

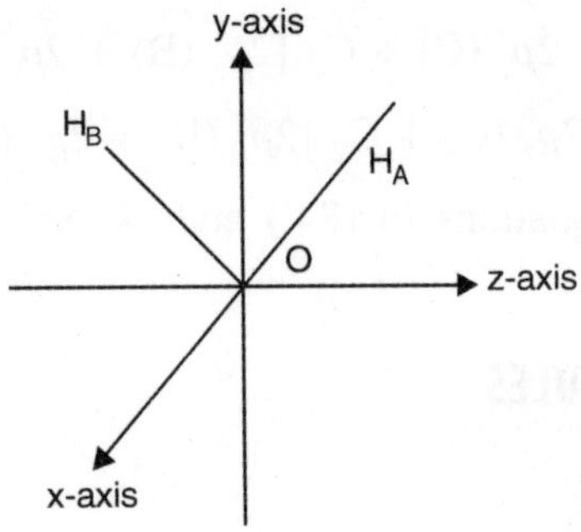

Figure 4.33: *Coordinate system showing water molecule*

Although it is mentioned above that 2*s* orbital of oxygen is too low in energy to mix with 1*s* orbitals of hydrogen atoms, the detailed calculations have shown some overlap of these orbitals. Figure 4.35 indicates that $1s^+$ combination of hydrogen atoms will have constructive overlap with the 2*s* orbital of oxygen atom, leading to the formation of σ_s. This means that σ_{p_y} and σ_s can mix

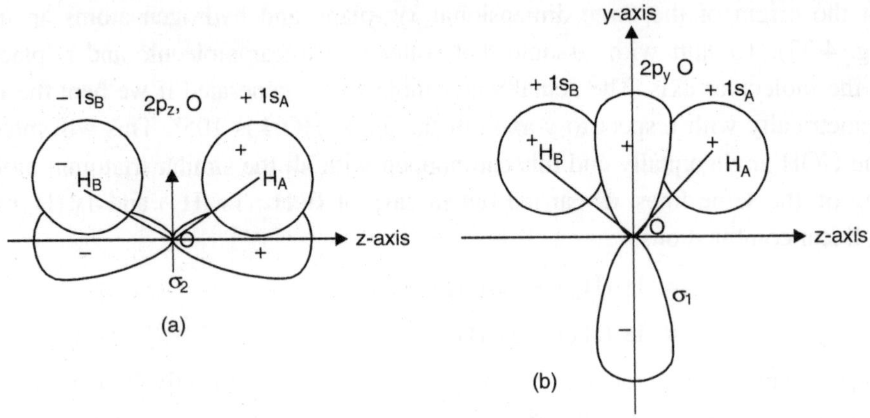

Figure 4.34: *Formation of water m.o.'s by the overlap of (a) $2p_z$ (O) with $1s_A$ and $1s_B$ of hydrogen atoms, (b) $2p_y$ (O) with $1s_A$ and $1s_B$ of hydrogen atoms*

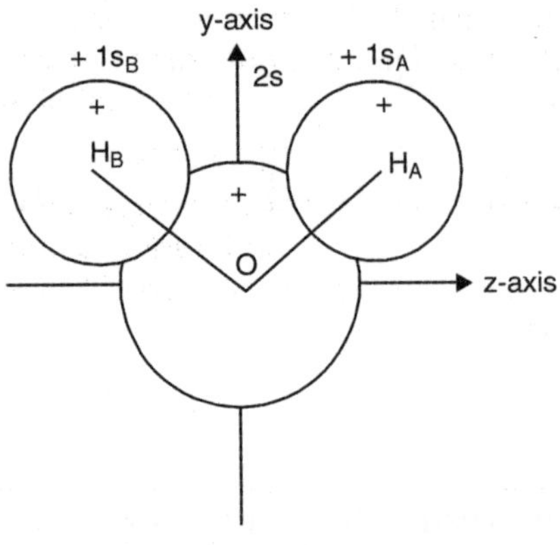

Figure 4.35: *Overlap of the 2s atomic orbital of oxygen atom with 1s atomic orbitals of the hydrogen atoms*

with each other. This will lead to the formation of three m.o.'s, one bonding, one nearly non-bonding and one anti-bonding. The energy level diagram shows that the order is σ_s, σ_{p_y} and $\sigma^*_{p_y}$. The molecular energy level diagram is depicted in Fig. 4.36. 1*s* orbitals of hydrogen atoms are placed above the 2*s* and 2*p* a.o.'s of oxygen. Due to interactions between σ_s and σ_{p_y}, the energy of σ_{p_y} is higher than that of σ_{p_z} and vice versa of the anti-bonding m.o.'s. The ground state configuration of H_2O can be written as: $\sigma_s^2\ \sigma_{p_z}^2\ \sigma_{p_y}^2\ \sigma_{p_z}^2$ because there are six electrons from oxygen and one each form hydrogen atoms. All the m.o.'s are paired, thus the total S = 0, *i.e.*, water molecule is diamagnetic. There are four electrons in bonding orbitals (σ_s and σ_{p_z}) and thus, will form two σ bonds. There are four non-bonding obitals ($\sigma^2_{p_y}$ and p_x^2).

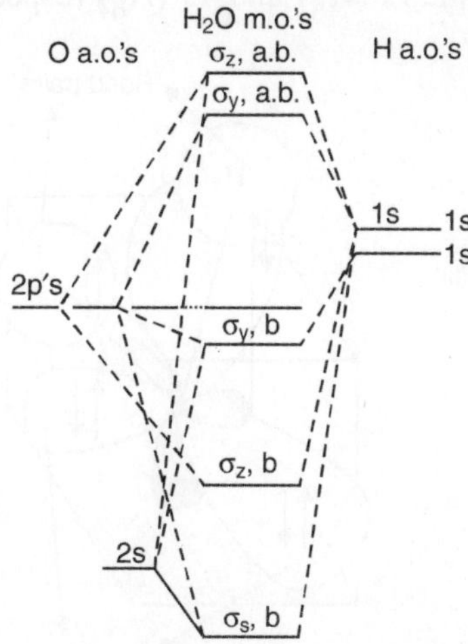

Figure 4.36: *Molecular energy level diagram of water. The scale is arbitrary*

***Localized molecular orbitals*:** If we consider bonding between $2p_y$ and $2p_z$ orbitals of oxygen and 1*s* of hydrogen atoms, the angle between H — O — H will be 90° as the two orbitals $2p_y$ and $2p_z$ are perpendicular to each other. This can be viewed as by placing the hydrogen atoms along the *y*- and *z*-axis. Further similar to the molecules discussed earlier, here also the above treatment assumes that complete delocalization of electrons over the whole molecule does take place and thus if a molecule of water loses an electron, the electron will come from the complete molecule. In other words each m.o. will lose some charge density. Whereas we know that the bond angle (H — O — H) is 105° and both the O — H bonds are equivalent. Some of the characteristics of O — H bond are so unique that even if one of the H-atoms of water molecule is replaced by say alkyl group, these remain nearly same, *e.g.*, bond length, bond energy, infrared stretching frequency. Partly the interaction of σ_s with p_y m.o. may lead to the extension of bond angle H — O — H by 15°, but the other properties may be explained as follows.

Similar to BeH_2 or CO_2, four equivalent or hybrid (sp^3) orbitals can be constructed by mixing or from the linear combination of 2*s* and three-2*p* a.o.'s. These equivalent orbitals, unlike m.o.'s do not belong to any symmetry type but can be transformed to one another by symmetry operations of the group to which the molecules belong and are orthogonal to each other. Although different procedures are there to construct these equivalent or hybrid orbitals, one of the methods is given in Fig. 4.37. Thus, the four equivalent orbitals formed are:

$$h_1 = 1/2\,[s + p_x + p_y + p_z] \qquad ...(4.194)$$

$$h_2 = 1/2\,[s - p_x + p_y - p_z] \qquad ...(4.195)$$

$$h_3 = 1/2\,[s - p_x - p_y + p_z] \qquad ...(4.196)$$

$$h_4 = 1/2\,[s + p_x - p_y - p_z] \qquad ...(4.197)$$

and are having an angle of 109° between each bond. Two of these orbitals (h_3 and h_4) will have two electrons each with unparallel spins and the other two (h_1 and h_2) will have one electron each. Former are labeled as non-bonding ones. Two σ bonds, with each H-atoms can be formed with the overlap of h_1 and h_2 with 1*s* (H_A) and 1*s* (H_B) respectively.

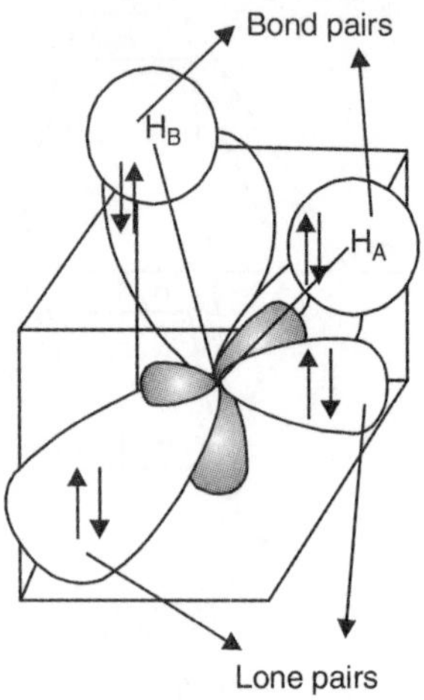

Figure 4.37: *Valence bond description of H_2O molecule using sp^3 hybrid orbitals on oxygen-atom*

The wave functions for the bonding orbitals can be written as:

$$\psi_{1,l} = C_1\, h_1 + C_2\, 1s\,(H_A) \qquad ...(4.198)$$

$$\psi_{2,l} = C_3\, h_2 + C_4\, 1s\,(H_B) \qquad ...(4.199)$$

These wave functions for the equivalent orbitals are fairly well localized in bonds or which represent lone pairs of electrons. Experimentally the angle observed between two O — H bonds (105°) is less than that observed in tetrahedron (109°) and can be explained as follows. Water molecule contains two non-bonding (lone pairs) and two bonding pairs. Mellish has pointed out that the lone pair of electrons in water will occupy more volume than will the electrons in the O — H bonds. In other words, lone pair-lone pair repulsion is larger than the bond pair-bond pair, thus, squeezing the two O — H bonds to an angle of 105°.

Sulphur and selenium also form dihydrides as XH_2 and bond angles in SH_2 and SeH_2 are 92° and 90° respectively, much closer to or equal to the angle between the two *p* orbitals. We can also consider these X — H bonds to be formed from tetrahedral orbitals. The sulphur atom is less electronegative than oxygen (H = 2.1, O = 3.5, S = 2.5, Se = 2.4, on Pauling scale), so that the electrons in S — H bonds will be further pulled away from the sulphur atom than those in the O — H bonds from the lone pairs and thus, smaller lone pair-lone pair interactions in SH_2 as compared to that in HO_2. The redistribution of electrons required for minimum configuration energy then accounts for the closing of S — H bonds.

The other property, which can also be accounted for from the above discussion is polarity of bonds. Since oxygen atom is more electronegative than the hydrogen atom, the electrons in the bonding orbitals will spend more time near the oxygen atom than around the hydrogen atom. This will lead to partial positive charge on the hydrogen atom and making the O — H bond polar in the ground state. This can be associated with the dipole moment of the individual bond and resultant dipole moment of H_2O. *The former is also called bond moment*. The dipole moment of water is 1.84 D. This dipole moment is the resultant of two bond moments and two lone pair moments as shown in Fig. 4.38. The dipole moments of some of the triatomic molecules are given in Table 4.6.

Table 4.6: *Dipole moments of some bent triatomic molecules*

Molecule	H_2O	H_2S	SO_2	NO_2	O_3
Dipole moment (D)	1.844	0.92	1.633	0.39	0.52

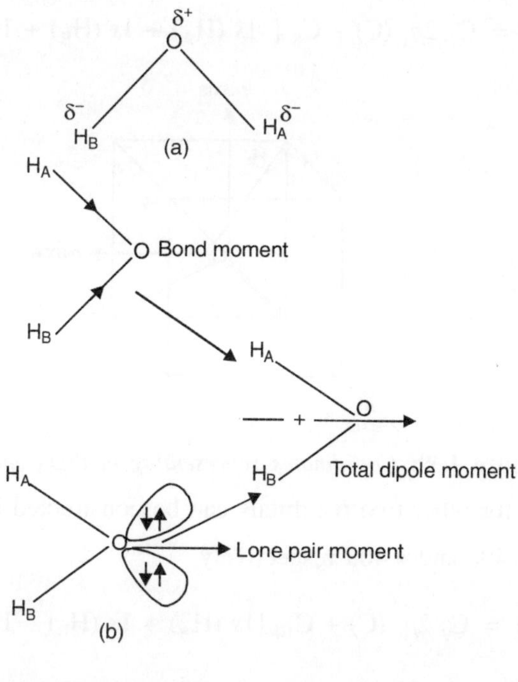

Figure 4.38: *(a) Depicts the separation of charges in the water molecule in the ground state. (b) Depicts the bond moments and lone pair dipole moment and thus leading to resultant total dipole moment*

4.18 CARBON COMPOUNDS

4.18.1 SATURATED HYDROCARBONS

Because of its good catenation property, carbon forms large number of compounds, (may be equal to the compounds formed by all the elements put together), either with hydrogen, with itself or with other elements. The simplest molecule is methane (CH_4). It has a tetrahedral structure, as depicted in Fig. 4.39. The carbon atom is placed at the centre of a cube and the four hydrogen atoms at the corners of a cube. Since the carbon atom has four σ bonds in CH_4 and this can be only explained if all the four orbitals of carbon ($2s$, $2p_x$, $2p_y$, $2p_z$) are used to form bonds as shown in Fig. (4.40*a*, *b*, *c*, *d*) with all the four hydrogen atoms. $2s$ orbital of carbon is spherical symmetrical with positive sign (or lobe) and it can form bonding and anti-bonding σ orbitals by taking positive combination of all the $1s$ hydrogen atom orbitals.

$$\psi(\sigma_s) = C_1\ 2s\ (C) + C_2\ [1s\ (H_A) + 1s\ (H_B) + 1s\ (H_C) + 1s\ (H_D)] \quad ...(4.200)$$

$$\psi(\sigma_s^*) = C_3\ 2s\ (C) - C_4\ [1s\ (H_A) + 1s\ (H_B) + 1s\ (H_C) + 1s\ (H_D)] \quad ...(4.201)$$

$2p_x$ orbital has one positive lobe (coming out) and one negative lobe (going in) in the cube. From Fig. 4.40*b*, it is clear that H_C and H_B will combine with the positive lobe of $2p_x$ and H_A and H_D with the negative lobe. The wave functions for the bonding and anti-bonding m.o.'s are:

$$\psi(\sigma_{p_x}) = C_5\ 2p_x\ (C) + C_6\ [-1s\ (H_A) + 1s\ (H_B) + 1s\ (H_C) - 1s\ (H_D)] \quad ...(4.202)$$

$$\psi(\sigma_{p_x}^*) = C_7\ 2p_x\ (C) - C_8\ [-1s\ (H_A) + 1s\ (H_B) + 1s\ (H_C) - 1s\ (H_D)] \quad ...(4.203)$$

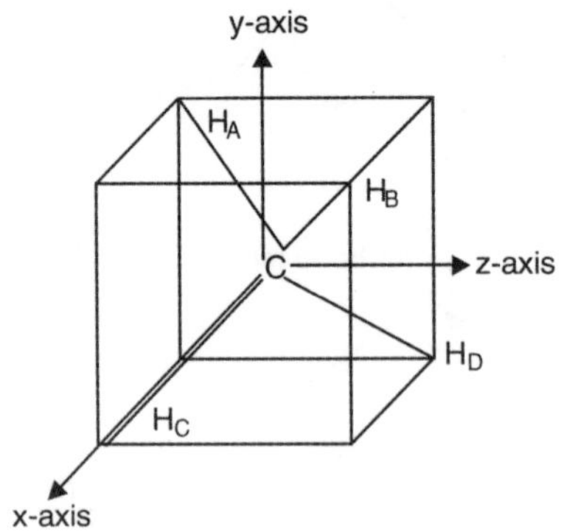

Figure 4.39: *Coordinates representing methane molecule*

The wave functions for other two σ orbitals can be constructed in the same manner as done for σ_{p_x}, shown in Fig. 4.40*c* and 4.40*d* respectively.

$$\psi(\sigma_{p_y}) = C_9\ 2p_y\ (C) + C_{10}\ [1s\ (H_A) + 1s\ (H_B) - 1s\ (H_C) - 1s\ (H_D)] \quad ...(4.204)$$

$$\psi(\sigma_{p_y}^*) = C_{11}\ 2p_y\ (C) - C_{12}\ [1s\ (H_A) + 1s\ (H_B) - 1s\ (H_C) - 1s\ (H_D)] \quad ...(4.205)$$

$$\psi(\sigma_{p_z}) = C_{13}\ 2p_z\ (C) + C_{14}\ [-1s\ (H_A) + 1s\ (H_B) - 1s\ (H_C) + 1s\ (H_D)] \quad ...(4.206)$$

$$\psi(\sigma^*_{p_z}) = C_{15}\ 2p_z\ (C) - C_{16}\ [-1s\ (H_A) + 1s\ (H_B) - 1s\ (H_C) + 1s\ (H_D)] \quad ...(4.207)$$

The molecular energy level diagram for the ground state of methane is shown in Fig. 4.41. As observed from the Figs. (4.40*b, c, d*) that overlap of all the four hydrogen atoms are similar with p_x, p_y and p_z orbitals of carbon, these σ m.o.'s will have the same energies and are degenerate. Now there are eight valence electrons, (four from carbon atom and one each from four hydrogen atoms), the ground state electronic configuration of methane is:

$$\sigma_s^2\sigma_{p_x}^2\ \sigma_{p_y}^2\ \sigma_{p_z}^2\ ,\ S = 0 \quad ...(4.208)$$

The total spin value is zero and thus it is diamagnetic. The average C — H bond energy is 413.5 kJ mol^{-1} and C — H bond length in CH_4 is 0.1093 nm.

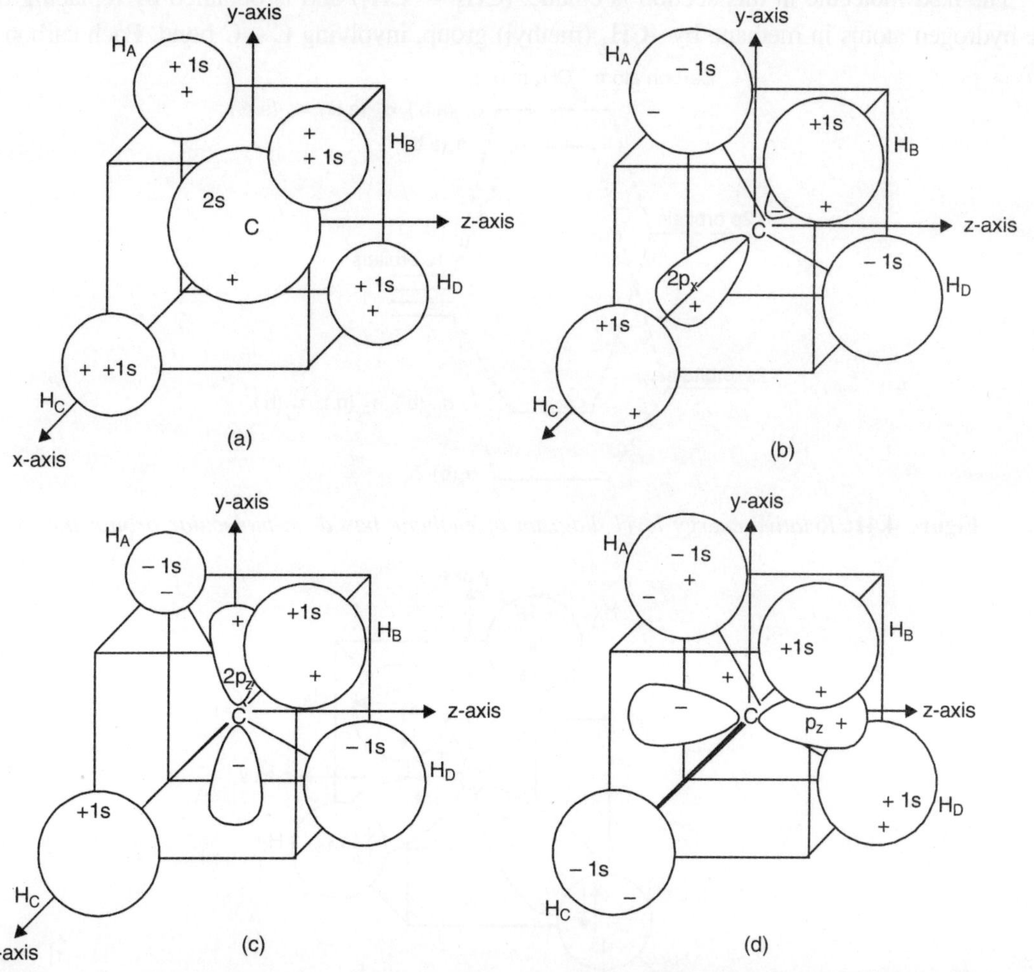

Figure 4.40: *(a) Formation of σ_s by the overlap of 2s orbital of C-atom and 1s orbitals of hydrogen atoms, (b) formation of σ_{px} by the overlap of $2p_x$ orbital of C-atom and 1s orbitals of hydrogen atoms, (c) formation of σ_{py} by the overlap of $2p_y$ orbital of C-atom and 1s orbitals of hydrogen atoms and (d) formation of σ_{pz} by the overlap of $2p_z$ orbital of C-atom and 1s orbitals of hydrogen atoms*

In the v.b.t., four equivalent or hybrid orbitals on carbon can be constructed by taking linear combination of 2*s*, $2p_x$, $2p_y$, $2p_z$ a.o.'s (obtained by exciting one electron from 2*s* a.o. to one of the vacant 2*p* a.o.) and is shown in Fig. 4.42. The wave functions for the four sp^3 hybrid orbitals can be constructed in the same manner as done for oxygen atom, except with the difference that in carbon atom each hybrid orbital is singly occupied, whereas in oxygen, two of the hybrid orbitals are doubly occupied and two with one electron each. The construction of hybrid orbitals and the evaluation of the coefficients are discussed in section 4.19. These hybrid orbitals are arranged towards the corners of tetrahedron and each hybrid orbital will have 1/4th share of *s* orbital and 3/4th from *p* orbitals. Four hydrogen atoms can be placed at the corners of the tetrahedron and localized σ orbitals can be easily formed by the overlap of 1*s* orbital of each hydrogen atom with each hybrid orbital.

The next molecule in this section is ethane. (CH_3 — CH_3) and is obtained by replacing one of the hydrogen atoms in methane by $-CH_3$ (methyl) group, involving C – C bond. Each carbon atom

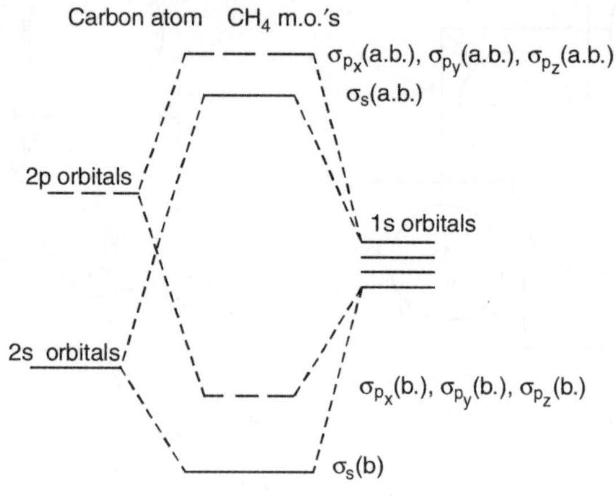

Figure 4.41: *Relative energy level diagram of methane based on molecular orbital theory*

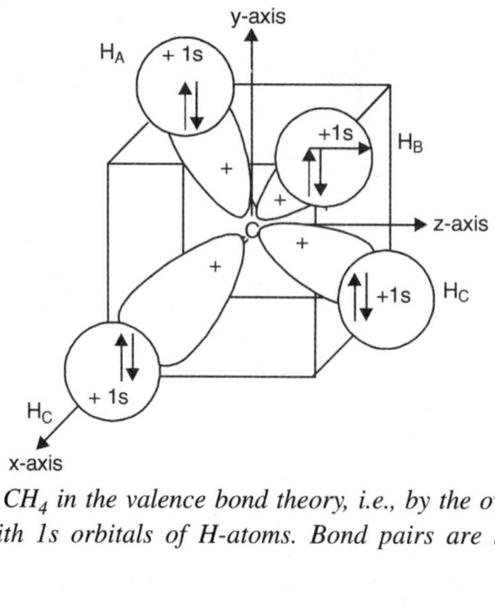

Figure 4.42: *Formation of CH_4 in the valence bond theory, i.e., by the overlap of sp^3 hybrid orbitals of C-atom with 1s orbitals of H-atoms. Bond pairs are shown by the two opposite arrows*

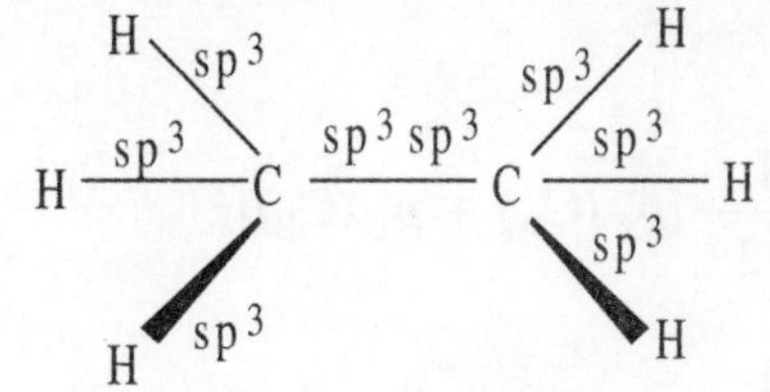

Figure 4.43: *Formation of ethane in the valence bond theory, i.e., sp^3 – sp^3 bond between C and C, sp^3 – 1s bond between each hydrogen atom*

is sp^3 hybridized, utilizes three of the four sp^3 hybrid orbitals on each carbon atom in forming σ bonds with three hydrogen atoms and fourth sp^3 hybrid orbital in forming C — C bond as shown in Fig. 4.43. Similarly many other carbon compounds, commonly known as *saturated hydrocarbons,* are formed, by replacing C — H hydrogen by — CH_3 group. Because of directional property and large value of overlap integral, the gap between the bonding σ and anti-bonding σ* orbitals is very large. The transitions observed between these orbitals lie in the vacuum U.V. region and thus these bonds or electrons are not much of practical use.

4.18.2 UNSATURATED HYDROCARBONS

It has been observed that in many organic compounds, carbon atom does not use its all the four hybrid orbitals in forming σ bonds but utilizes either three or two σ bonds. In the former case, carbon utilizes sp^2 hybridization and in the latter case it is *sp* hybridization. This leaves one or two pure *p*-orbitals on each carbon atom, perpendicular to the σ framework for forming π-orbitals. In the coming sections, we will discuss the characteristics of π bonding in various kinds of these molecules, also commonly known as *unsaturated hydrocarbons.*

In describing the structure and chemical bonding of this class of molecules, it is a common practice to treat separately the σ bonding and π bonding. σ bonding is described by the interaction of sp^2 or *sp* hybrid orbitals of the carbon atom with the hybrid orbitals or pure a.o.'s of the combining atoms. The reason for separating the σ and π bonding m. o. treatment is that π orbitals are perpendicular to the σ frame work and thus, the overlap of $p\pi$ (p_x a.o.) will be zero with either sp^2 or *sp* hybrid orbtials or with the *s*, p_z and p_y a.o.'s as bond axis is *z*-axis. π bonding will involve the overlap of $p\pi$ orbitals to form π bonding and π^* anti-bonding m.o.'s. On the energy level diagram, π bonding m.o. will be placed much above the σ bonding and π^* anti-bonding much below the σ^* anti-bonding m.o., because of the smaller value of the overlap integral in the π bonding. Thus, the gap between the π-π^* m.o. is not large. Due to this the π-π^* spectral transition lies in the region of spectrum, (*i.e.,* u.v. visible) which is chemically very useful.

***Ethylene*:** Ethylene (C_2H_4) molecule is the simplest molecule belonging to this class of molecules and is depicted in Fig. 4.44. C_2H_4 is a planar molecule, each carbon atom is sp^2 hybridized and thus forms two σ bonds with the overlap of 1*s* orbitals of two hydrogen atoms at each carbon atom and one σ bond with the overlap of one of the sp^2 hybrid orbitals of each carbon atom. These are commonly depicted as C — H and C — C bonding respectively. If we assume that C_2H_4 molecule is lying in *yz*-plane, then each carbon atom will be left with one p_x *a.*

o. Proceeding exactly on the same lines as done for H_2^+, the π and π* m.o. formed by the overlap of p_x orbitals are described by:

$$\psi_\pi = \frac{1}{\sqrt{2}} [p_x (C_A) + p_x (C_B)] \quad ...(4.209)$$

$$\psi_{\pi^*} = \frac{1}{\sqrt{2}} [p_x (C_A) - p_x (C_B)] \quad ...(4.210)$$

where $\frac{1}{\sqrt{2}}$ is the normalization constant. The bonding surfaces of π and π* are shown in Fig. 4.44. The energies of these m.o.'s can be obtained as described in section 4.20, *i.e.*,

$$E_H = \int \psi_\pi \hat{H} \psi_\pi^* d\tau$$

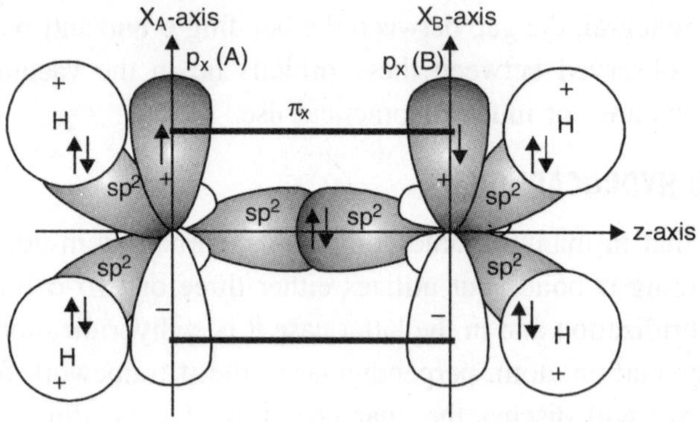

Figure 4.44: *Formation of ethylene molecule. Overlap of sp^2–sp^2 of carbon atoms, sp^2–1s overlap of carbon atom with each hydrogen atom. p_x– p_x overlap of each carbon atom to form π_x bond*

where $\hat{H}$ is the Hamiltonian operator for π-electrons. Substituting the values of ψ_π and ψ_{π^*}, we can obtain E_π and E_{π^*} respectively, *i.e.*,

$$E_\pi = \frac{1}{2} [\int\{(p_x (C_A) + p_x (C_B)\} \hat{H} \{(p_x (C_A) + p_x (C_B)\} d\tau$$

$$= \frac{1}{2} [\int p_x (C_A) \hat{H} p_x (C_A) d\tau + \int p_x (C_A)] \hat{H} p_x (C_B) d\tau$$

$$+ \int p_x (C_B)) \hat{H} p_x (C_A) d\tau + \int p_x (C_B) \hat{H} p_x (C_B) d\tau]$$

$$= \int p_x (C_A) \hat{H} p_x (C_A) d\tau + \int p_x (C_A) \hat{H} p_x (C_B) d\tau = \alpha + \beta \quad ...(4.211)$$

As p_x a.o. on both the carbon atoms are similar, integrals 1st and 4th, 2nd and 3rd are similar.

Here
$$\alpha = \int p_x (C_A) \hat{H} p_x (C_A) d\tau = \int p_x (C_B) \hat{H} p_x (C_B) d\tau \quad ...(4.212)$$

$$\beta = \int p_x (C_A) \hat{H} p_x (C_B) d\tau = \int p_x (C_B) \hat{H} p_x (C_A) d\tau \quad ...(4.213)$$

α and β are also known as Coulombic and exchange (or resonance) integrals respectively, as

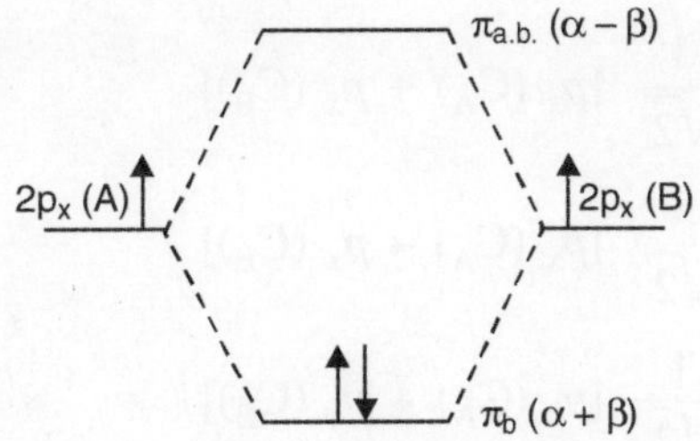

Figure 4.45: *Relative energy level diagram of ethylene molecule, showing only π orbitals*

discussed in section 4.20. Similarly

$$E_{\pi^*} = \alpha - \beta$$

Since both α and β are negative, the π-bonding will be stabilized and π^*-anti-bonding will be destabilized in comparison to p_x a.o. as shown in Fig. 4.45. This diagram is exactly similar to that observed for H_2 molecule.

***Electronic configuration*:** The total number of electrons in C_2H_4 molecule is 12, *i.e.,* four each from carbon atom and 1 each from four hydrogen atoms. Out of these 12 electrons, 10 electrons are used for forming σ orbitals, as shown in Fig. 4.45. The remaining two can be placed in π bonding (Fig. 4.45). The ground state electronic configuration for π system is π^2, *i.e.,* there is one π bond between C — C aroms. Ethylene molecule possesses two kinds of bonds. σ bond corresponds to C — H single bond and a C ═ C bond, which describes a double bond (one σ and one π). The bond length and bond strength of C — H is 0.108 nm and 413 kJ mol^{-1} respectively nearly similar to that present in C — H bond of CH_4. The bond length and bond strength of C ═ C double bond are 0.135 nm and 619 kJ mol^{-1} respectively. The former is smaller than C — C single bond (0.154 nm) and the latter is nearly double the C — C bond strength (346 kJ mol^{-1}).

***Values of α and β in C_2H_4*:** Values of α and β in C_2H_4 molecule can be determined from the ionization energy and the absorption band maximum of the longest wavelength transition, determined experimentally. It is evident from Fig. 4.45 that ionization energy of C_2H_4 corresponds to α + β and the absorption band maximum of the longest wavelength transition to ΔE (–2β) corresponds to the excitation of an electron from bonding π m.o. to anti-bonding π^* m.o. ($\pi \to \pi^*$). The energy gap between $\pi \to \pi^*$ is:

$$\Delta E = \alpha - \beta - (\alpha + \beta) = -2\beta = 728 \text{ kJ mol}^{-1}.$$

So
$$\beta = -\,364 \text{ kJ mol}^{-1} \quad \text{...(4.214)}$$

$$IE = 10.24 \text{ eV} = 1{,}003.6 \text{ kJ mol}^{-1}$$

Thus,
$$\alpha = -\,(\text{I.P.} + \beta) = -\,1{,}003.6 + 364 = -\,639.6 \text{ kJ mol}^{-1} \quad \text{...(4.215)}$$

***Ethyne molecule*:** The structure of ethyne molecule (C_2H_2) is given in Fig. 4.46, taking *z*-axis as the molecular axis. The σ bonds in C_2H_2 involve the *sp* hybridization on each carbon atom. This leaves two *p*-a.o.'s (p_x and p_y) on each carbon atom for π bonding, perpendicular to molecular axis or σ framework. The wave functions for the bonding and anti-bonding π orbitals are:

$$\psi_{\pi_1} = \frac{1}{\sqrt{2}} [p_x (C_A) + p_x (C_B)] \quad ...(4.216)$$

$$\psi^*_{\pi_1} = \frac{1}{\sqrt{2}} [p_x (C_A) - p_x (C_B)] \quad ...(4.217)$$

$$\psi_{\pi_2} = \frac{1}{\sqrt{2}} [p_y (C_A) + p_y (C_B)] \quad ...(4.218)$$

$$\psi^*_{\pi_2} = \frac{1}{\sqrt{2}} [p_y (C_A) - p_y (C_B)] \quad ...(4.219)$$

Since p_x and p_y a.o.'s on each carbon atom are identical, thus the π bonding and π^* anti-bonding m.o.'s are of equal energies, *i.e.,* these are doubly degenerate. The energy of π and π^* m.o.'s are:

$$E_{\pi_1} = E_{\pi_2} = \alpha + \beta \quad ...(4.220)$$

$$E^*_{\pi_1} = E^*_{\pi_2} = \alpha - \beta \quad ...(4.221)$$

The boundary surface of the π and π^* m.o.'s are depicted in Fig. 4.46 and the energy level diagram of π orbitals is shown in Fig. 4.47. There are 10 electrons in C_2H_2 (four each from carbon atoms and 1 each from hydrogen atoms). Six electrons are placed in σ orbitals, *i.e.,* 3 σ bonds are present in ethyne. Two σ bonds are obtained by the overlap of 1*s* a.o. of hydrogen atoms with one of the *sp* hybridized orbitals of carbon atom and third σ bond is formed by the overlap of one of the *sp* hybridized orbital of each carbon atom, leading to two C—H bonds in the former and one C — C bond in the latter case. The remaining four electrons are placed in two π

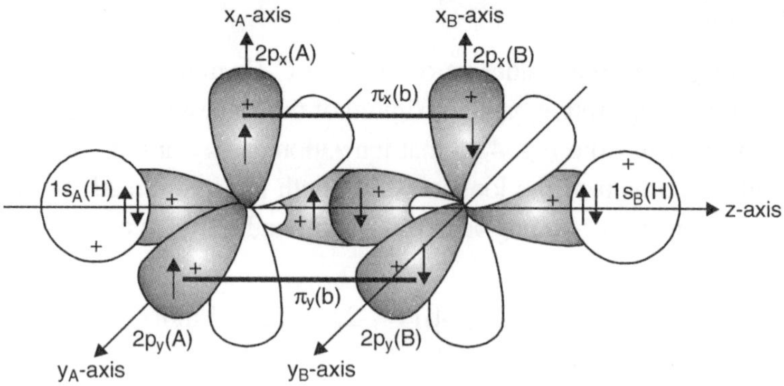

Figure 4.46: *Formation of ethyne molecule, showing the overlap of sp – sp overlap of C-atoms, sp – 1s overlap of C-atom and each hydrogen atom, $p_x - p_x$ overlap of each C-atoms to form π_x and $p_y - p_y$ overlap of each C-atoms to form π_y m.o.*

bonding orbitals. The ground state electronic configuration for π orbitals in ethylene is $\pi_1^2 \, \pi_2^2$. This indicates that there are three bonds between two carbon atoms, one σ bond and two π bonds. The bond length of C ≡ C bond

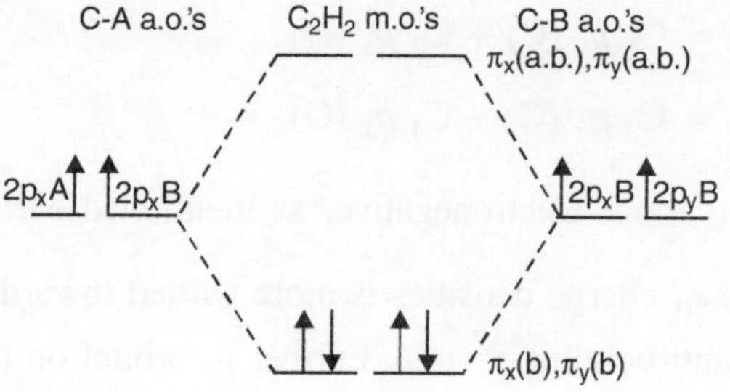

Figure 4.47: *Relative energy level diagram for ethyne molecule*

is 0.123 nm, smaller than those of C — C and C = C bonds. Whereas the bond strength or energy of C ≡ C bond is 835 kJ mol^{-1}, larger than those of C — C and C = C bonds.

4.18.3 CARBONYL GROUP

The other very important group in organic chemistry is the carbonyl group. This group is present in aldehydes, ketones, carboxylic acids and their derivatives. The simplest molecule containing carbonyl group is formaldehyde, as shown in Fig. 4.48, depicted in *yz*-plane. This molecule is planar and thus, sp^2 hybridized orbitals are utilized by carbon atom. The σ bonds on carbon atom are depicted by the sp^2 hybridization on carbon. This leaves one p_x orbital on carbon atom to form π orbital with the p_x orbital on oxygen atom. Since oxygen atom is more electronegative than carbon atom, 2*s* orbital on oxygen is largely stabilized, both in comparison to sp^2 hybrid orbitals on carbon atom and *p* orbitals on oxygen atom. This pair of electrons or 2*s* electrons more or less are localized on the oxygen atom and is not involved either in forming sp^2 hybrid orbitals on oxygen atom or forming m.o. with σ bond with carbon atom. So $2p_z$ orbital on oxygen atom is forming σ bond with one of the three sp^2 hybrid orbitals. The other two sp^2 hybrid orbitals on carbon atom are used to form σ bonds with 1*s* orbitals of two hydrogen atoms. π and π^* orbitals between oxygen and carbon atoms are formed by the overlap of p_x (C) and p_x (O) orbitals. The wave functions for these orbitals are given as:

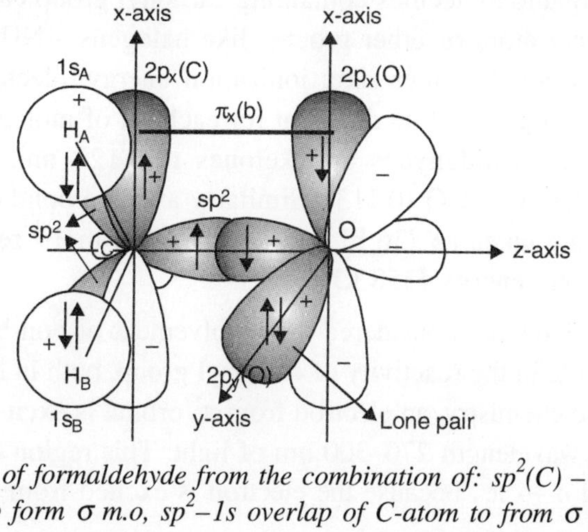

Figure 4.48: *Formation of formaldehyde from the combination of: sp^2(C) – $2p_z$(O) overlap of C and O-atoms to form σ m.o, sp^2–1s overlap of C-atom to from σ m.o.'s and each H-atom, $2p_x$ (C) – $2p_x$ (O) to from π m.o., $2p_y$ is the lone pair. 2s a.o. of oxygen is not involved*

$$\psi_{\pi_x} = C_1\, p_x\,(C) + C_2\, p_x\,(O) \quad ...(4.222)$$

$$\psi^*_{\pi_x} = C_3\, p_x\,(C) - C_4\, p_x\,(O) \quad ...(4.223)$$

Since oxygen atom is more electronegative, as mentioned earlier, $C_2^2 > C_1^2$ in π orbital and $C_4^2 < C_3^2$ in π* orbital, *i.e.,* charge densities is more shifted towards oxygen bonding π m.o. and towards carbon atom in anti-bonding π^* m.o. Further p_y orbital on oxygen is not involved in any kind of bonding with carbon atom, it is present as non-bonding electrons on oxygen atom. The energy level diagram for the π bonding is shown in Fig. 4.49. There are twelve electrons in H_2CO, *i.e.,* six from oxygen atom, four from carbon atom and one each from two hydrogen atoms. Six electrons are used in forming three σ bonds, two with hydrogen atoms and one with oxygen atom. Two electrons are placed in π bond and remaining four electrons are placed as non-bonding electrons on oxygen atom, one pair on $2s^2$ (O) and the other pair on $2p_y^2$ (O). The ground state electronic configuration, depicting only π configuration is π^2, thus, having double bond character between carbon and oxygen atoms, one σ and one π bond.

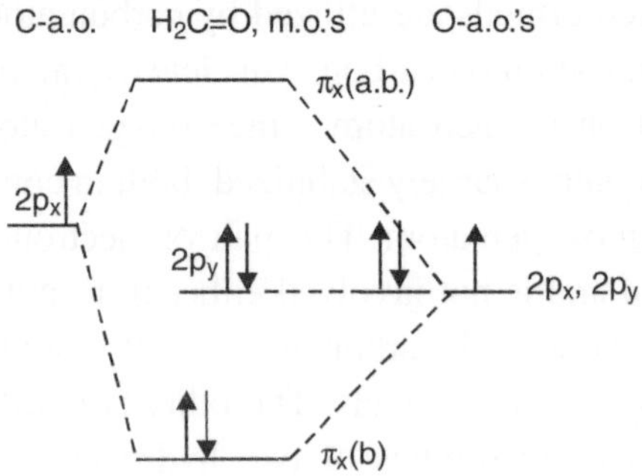

Figure 4.49: *Relative π energy levels only of formaldehyde*

Different kind of organic molecules containing carbonyl group can be obtained by replacing hydrogen atom with other atom or other groups, like halogens, $-NH_2$, $-OH$, $-OCH_3$ etc. Thus, the properties of the carbonyl groups like, ionization energy, electronic and infrared spectra, bond length and bond strength, will be different for each set of molecules. But the average bond length of C = O bond in aldehydes and ketones is 0.122 nm, smaller than C—O bond (0.143 nm) and larger than C ≡ O (0.113). Similarly average bond energy of C = O bond in aldehydes and ketones are taken as 736 kJ mol^{-1} and 749 kJ mol^{-1} respectively, which is nearly twice that of C — O bond energy (358 kJ mol^{-1}).

Although so far we have not considered the involvement of non-bonding electrons, but these electrons play a major role in the reactivity of carbonyl group, both in the ground state and excited state. In the excited state chemistry, an electron from π_y orbital is excited to π^* and thus absorption occurs in the region of wavelength 270–300 nm of light. This region of light is easily accessible. This transition is called $n \rightarrow \pi^*$, because the electron is excited from the non-bonding (n) orbital to π^* anti-bonding m.o.

4.19 HYBRIDIZATION

In the last section, we have tried to use mixing of a.o.'s on Li atom to form m.o.'s of LiH. This picture enabled to explain the energy levels and other physical parameters of molecular structure of LiH. Although we did mention in the formation of LiH that its bond strength can be better understood using the hybrid a.o.'s on Li, we postponed our discussion on hybridization. Similarly in many other cases, theoreticians working on m.o. theory feel that it is not necessary and there is no justification to introduce the concept of hybridization on mathematical or physical reasons. On the other hand, scientists using v.b.t. feel that hybridization is the backbone of v.b.t. Introduction of the concept of hybridization simplifies the description of chemical bond, explains the arrangement of atoms in three dimensional network to give better picture about the geometry of the molecules, as well as, the bond strength of the chemical bond. In this section, we will discuss first why there is a need to introduce this idea, followed by the physical picture of hybridization, its mathematical formulation and applications.

There are many molecules where simple theories fail to explain the bonding. For simplicity, let us consider the well-known example of methane (CH_4). Carbon in methane is tetravalent. The electronic configuration of carbon $1s^2 2s^2 2p_x^1 2p_y^1$, is compatible with the ground state spectroscopic term symbol ^3P. This configuration shows that carbon is divalent because there are only two unpaired electrons. The molecules having divalent nature of carbon are known, for example, carbene ($:CH_2$), ketene ($CH_2 = C = O$) etc., but these are not very stable and have been predicted only as intermediates in the chemical reactions. Thus, to explain the tetravalent character of carbon, it is necessary to promote an electron from $2s$ orbital to $2p_z$ orbital. This process is endothermic and requires 406 kJ mol^{-1} of energy. The electronic configuration of this state is $1s^2 2s^1 2p_x^1 2p_y^1 2p_z^1$ and spectroscopic term symbol is ^5S, compatible with four unpaired electrons to form four bonds with four hydrogen atoms. The compatible process (*i.e.,* promotion of an electron and formation of four bonds) is an exothermic process, with the release of energy equivalent to –74.8 kJ mol^{-1}. In other words it explains that overall process is compatible with the heat of reaction observed in the formation of methane from carbon and hydrogen. Let us see the second aspect, *i.e.,* the geometry of methane molecule. Having formed the carbon atom with the electronic configuration, $1s^2 2s^1 2p_x^1 2p_y^1 2p_z^1$, three hydrogen atoms can form bonds with three p orbitals. To have the maximum overlap and stability, $1s$ orbital of hydrogen atoms must approach each of the p orbitals along their orientation. Since three p orbitals are equivalent, the three bonds formed with p orbitals will also be equivalent. Fourth bond will be formed by the overlap of $2s$ of C-atom with $1s$ of hydrogen atom. s orbital, being spherically symmetrical, the overlap of $1s$ and $2s$ will adjust itself to have minimum repulsion from other bonding orbitals. This means that in methane three bonds should be of similar in nature and the fourth one different from the other set of three bonds. This kind of methane is never observed and methane we know has four equivalent, C — H bonds and angle between each bond is 109.5°.

The above discussion clearly explains that it is not only the promotion of an electron to the vacant orbital to create the desired number of unpaired electrons but some thing more than that is needed to explain the structure of methane, in particular and other similar molecules in general, *e.g.*, divalent, trivalent, pentavlent etc. of Be, B, P, atoms respectively and bond angles H—O—H in H_2O, H—N—H in NH_3 to be 105° and 107° respectively.

The concept of hybridization involves the mixing or linear combination of pure a.o.'s such as *s*, *p*, *d*, etc. of the same atom in every respect. Before we proceed further, it may be made clear here that though the hybridization also involves the linear combination of atomic orbitals, the a.o.'s used belong to the same atom, unlike in molecules where m.o.'s are also formed from LCAO but in latter case a.o.'s belong to different atoms combining to form molecules.

For simplicity, consider 2*s* and one of the 2*p* orbitals, as shown in Fig. 4.50. If we take the linear combination of *s* and *p* orbitals it is clear from the Fig. (4.50 *a*) that on right hands side of origin, there will be a positive combination of two wave functions and on the left hand side a negative

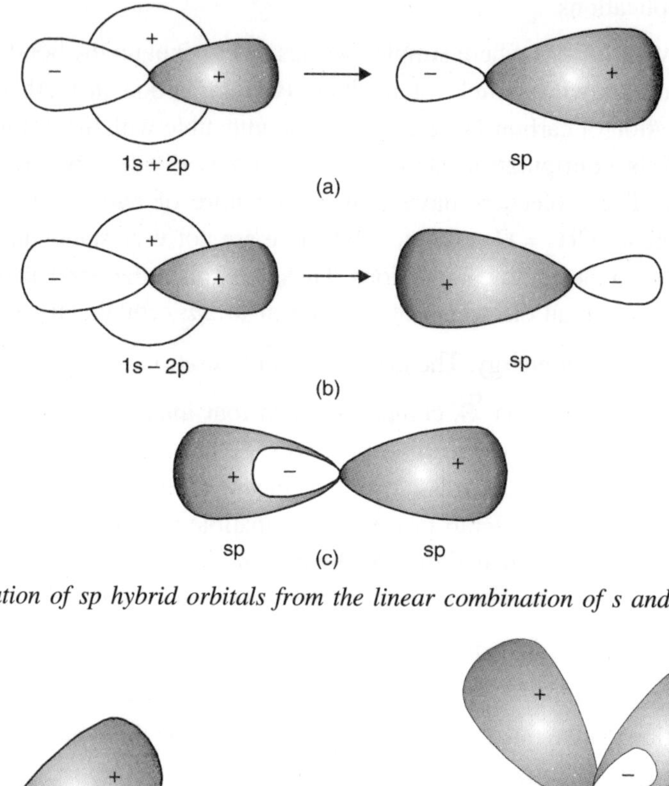

Figure 4.50: *Formation of sp hybrid orbitals from the linear combination of s and one of the p-a.o*

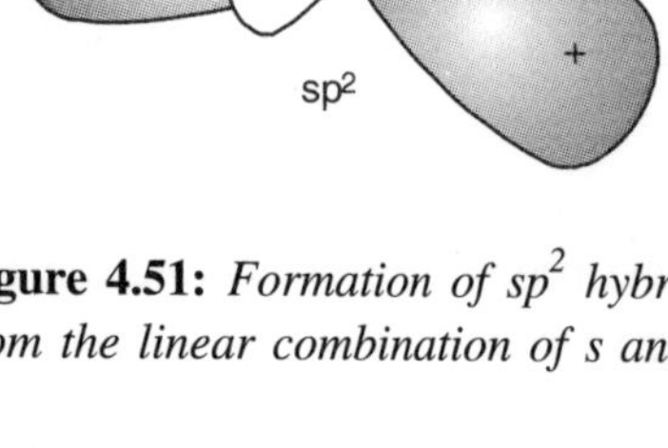

Figure 4.51: *Formation of sp^2 hybrid orbitals from the linear combination of s and 2p a.o.'s*

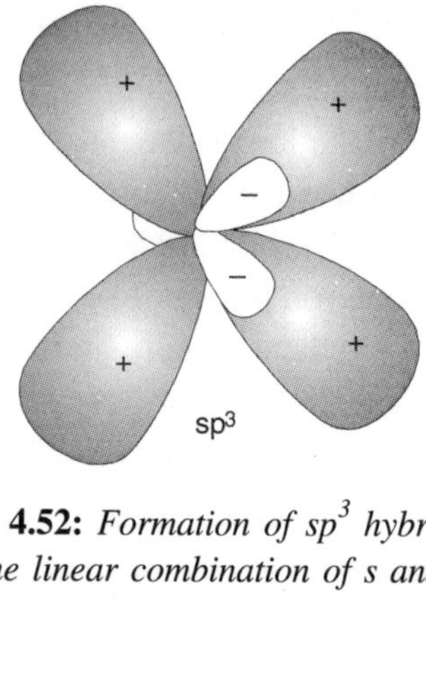

Figure 4.52: *Formation of sp^3 hybrid orbitals from the linear combination of s and 2p a.o.'s*

combination. The former leads to an increase in size of the lobe, larger than *s* and *p* separately. Latter leads to decrease in the size of the lobe, smaller than *s* and *p* separately. Similarly the combination of *s* and negative *p* leads to exactly the similar description as described above with a difference that bigger lobe is on the left hand side and smaller lobe on the right hand side (Fig. 4.50*b*). The new orbitals formed from the linear combination of *s* and *p* a.o.'s are known as hybrid atomic orbitals or simply hybrid orbitals. To place these hybrid orbitals in space so that there is minimum repulsion between the negative charges, there is only one way as shown in Fig. (4.50*c*), *i.e.,* the bigger lobes are pointing opposite to each other with an angle of 180° between them. This combination is also known as *sp* hybridization.

The other combinations involving *s* and *p* orbitals are: by taking *s* and two *p* and three *p* orbitals. The former is known as sp^2 and the latter is known as sp^3. Proceeding exactly on the same lines, sp^2 combination will lead to the formation of three equivalent hybrid orbitals, each having nearly the similar shapes as described for *sp* combination. These three hybrid orbitals can be arranged pointing towards the corners of equilateral triangle and angle between each orbital is 120°. This arrangement will lead to the most stable or minimum energy arrangement. On the same lines, sp^3 combination will lead to the four equivalent hybrid orbitals, pointing towards the corners of the tetrahedron and the angles between each orbital is 109.5°. These are depicted in Fig. 4.51 and 4.52 respectively.

Before proceeding for mathematical description of hybrid orbital, it may be pointed out that hybridization does not limit only to the combination of *s* and *p* orbitals. It can involve the combination of *s, p* and *d* orbitals also. The main condition for forming hybrid orbitals is that the radial wave function for the combining orbitals should be the combination of *s, p* and *d* orbitals also. The main condition for forming hybrid orbitals is that the radial wave function for the combining orbitals should be similar. If this is not so, the charge density distribution in the various orbitals will be different and overlap will be poor. The above criterion can also be described, as the energies of the combining orbtials should be similar. But in practice hybrid orbitals can also be formed if the energies of the orbitals are adjacent to each other, *e.g.,* the participation of *d* orbitals with *s* and *p* orbitals. The hybrid orbitals formed by involving *d* orbitals are dsp^2, dsp^3 and d^2sp^3 and lead to square planar, trigonal bipyramidal and octahedral geometries. Table 4.7 represents the geometries and bond angles between hybrid orbitals. Further, the shape of the molecule depends on the number of bond pairs and number of lone pairs present in the molecule. This can also be looked up in Table 4.7.

Table 4.7

Structure SN_x	Geometrical shape	Hybridization	Molecular shape possibilities
1	Spherical	–	–
2	Linear	*sp*	LP = 0, BP = 2, Linear LP = 1, BP = 1, Linear
3	Triangular	sp^2	LP = 0, BP = 3, Triangular LP = 1, BP = 2, Bent, Angle = 120° LP = 2, BP = 1, Linear

4	Tetrahedron	sp^3	LP = 0, BP = 4, Tetrahedron LP = 1, BP = 3, Trigonal pyramidal LP = 2, BP = 2, Bent (Angle 109°) LP = 3, BP = 1, Linear
5	Square planar	dsp^2 or sp^2d	LP = 0, BP = 4, Square planar
6	Trigonal bipyramidal	dsp^3 or sp^3d	LP = 0, BP = 5, Trigonal bipyramidal LP = 1, BP = 4, Distortedsquare planar LP = 2, BP = 3, T-shaped LP = 3, BP = 2, Linear LP = 4, BP = 1, Linear
7	Octahedral	d^2sp^3 or sp^3d^2	LP = 0, BP = 6, Octahedral LP = 1, BP = 5, Square pyramidal LP =2, BP = 4, Square planar

4.19.1 WAVE FUNCTIONS

The hybrid orbitals are obtained by LCAO belonging to the same atom. The wave function for the hybrid orbitals in terms of pure a.o.'s can be written as:

$$\psi_{h,i} = \Sigma C_{ij}\ \varphi_j \qquad ...(4.224)$$

where φ_j is the wave function of pure a.o. taking part in the formation of hybrid atomic orbital, ψ_{hi} and C_{ij} give the proportion of pure a.o. contributing to the hybrid orbital. The summation is carried out over the number of a.o.'s taking part in the combination. Thus one thing is certain that the number of hybrid orbitals formed is equal to the number of pure a.o.'s. The values of coefficients can be determined by many methods. Here we will give only one method, given by Hsu and Orchin (C. Y. Hsu and M. Orchin, J. Chem. Edu., 80 (1973) 114), based on geometrical calculation. Following rules are involved in calculating the coefficients C_{ij}.

(*i*) Each hybrid wave function is normalized, *i.e.*,

$$\sum_{j=1}^{n} C_{ij}^2 = 1 \qquad ...(4.225)$$

(*ii*) Each hybrid wave function in the set of hybrid orbitals must be orthogonal to other hybrid orbitals.

$$\sum_{j=1}^{n} C_{ij}C_{kj} = 0 \qquad ...(4.226)$$

(*iii*) Square of the coefficients of any component atomic function summed over all the hybrid orbitals in which it participates, must be equal to unity, *i.e.*,

$$\sum_{i=1}^{n} C_{ij}^2 = 1 \qquad ...(4.227)$$

Keeping the above rules in mind, the following procedure can be used to generate the equivalent hybrid orbitals.

(*a*) Specify a Cartesian coordinate system and place one of the *i*th hybrid orbitals along one of the coordinate axes and as many of the other hybrid orbitals as possible in the plane containing this axis and other coordinate axes.

(*b*) Since *s* orbital has a spherical symmetry, each equivalent hybrid orbital of a set contains $1/n$ of the total of the unit *s* orbital distributed in the *n* hybrid orbitals. Therefore coefficient in each hybrid orbital for *s* orbital is $\frac{1}{\sqrt{n}}$.

These rules and procedure can be described by taking particular examples as discussed in the examples 4.8 and 4.9 give below.

Example 4.8: Calculate the wave functions for the sp^3 hybrid orbitals.

Solution: Four hybrid orbitals can be formed by the linear combination of sp^3 a.o.'s. The wave functions can be written as:

$$h_1 = C_{11}\, s + C_{12}\, p_x + C_{13}\, p_y + C_{14}\, p_z \qquad ...(4.228)$$

$$h_2 = C_{21}\, s + C_{22}\, p_x + C_{23}\, p_y + C_{24}\, p_z \qquad ...(4.229)$$

$$h_3 = C_{31}\, s + C_{32}\, p_x + C_{33}\, p_y + C_{34}\, p_z \qquad ...(4.230)$$

$$h_4 = C_{41}\, s + C_{42}\, p_x + C_{43}\, p_y + C_{44}\, p_z \qquad ...(4.231)$$

where h_1, h_2, h_3 and h_4 are the wave functions of the four hybrid orbitals. According to rule (*iii*) *b*,

$$C_{11}^2 = C_{21}^2 = C_{31}^2 = C_{41}^2 = \frac{1}{4}$$

or

$$C_{11} = C_{21} = C_{31} = C_{41} = \frac{1}{2}$$

Using rule (*iii*) *a*, arrange hybrid orbital h_1 along the *z*-axis, h_2 hybrid orbital in such a manner that it lies on *xz*-plane, whereas the *xz*-plane divides the hybrid orbitals h_3 and h_4 equally as shown in Fig. 4.53. It is clear from the Fig. 4.53, that the contribution to h_1 hybrid orbital by p_x and p_y orbitals is zero. Thus

$$C_{12} = C_{13} = 0$$

Now using rule (*i*) on h_1 hybrid orbital and substituting the values of C_{11}, C_{12} and C_{13} in (4.228), we get

$$C_{11}^2 + C_{12}^2 + C_{13}^2 + C_{14}^2 = 1$$

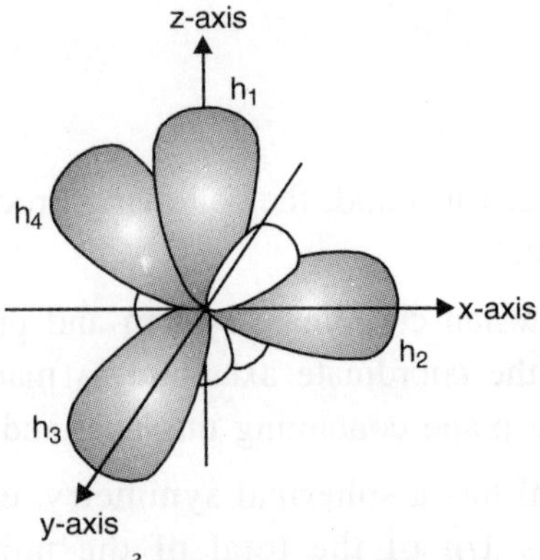

Figure 4.53: *Representation of sp^3 hybrid orbitals in Cartesian coordinate system*

Which finally gives $C_{14}^2 = 1 - \frac{1}{4} = \frac{3}{4}$, *i.e.,* $C_{14} = \frac{\sqrt{3}}{2}$

Now consider the hybrid orbital h_2. The hybrid orbitals h_2, h_3 and h_4 are placed symmetrically along the z-axis. This shows that the remaining portion of p_z orbital will be equally contributing to these three hybrid orbitals. Thus,

$$C_{24}^2 = C_{34}^2 = C_{44}^2 = C^2 \text{ (say)}$$

Using rule number (*iii*) on p_z and substituting the values of respective coefficient in (4.226)

$$C_{14}^2 + C_{24}^2 + C_{34}^2 + C_{44}^2 = 1$$

i.e., $\frac{3}{4} + 3C^2 = 1$ gives $C = \frac{1}{\sqrt{12}}$.

Thus $C_{24} = C_{34} = C_{44} = \frac{1}{\sqrt{12}}$

Further all these hybrid orbitals are pointing towards the direction where the wave function of p_z is negative. Thus

$$C_{24} = C_{34} = C_{44} = -\frac{1}{\sqrt{12}}$$

As h_2 hybrid orbital lies along the x-axis or in the xz-plane, the contribution by p_y orbital to this orbital will be zero, *i.e.,* $C_{23} = 0$

Using the rule (*i*) on h_2 and substituting the values of respective coefficients in (4.228):

$$C_{21}^2 + C_{22}^2 + C_{23}^2 + C_{24}^2 = 1$$

i.e., $\frac{1}{4} + C_{22}^2 + 0 + \frac{1}{12} = 1$ will give $C_{22} = \sqrt{\frac{2}{3}}$.

Since it lies along the positive side of the p_x orbital, $C_{22} = \sqrt{\frac{2}{3}}$.

It is clear from Fig. 4.53 that h_3 and h_4 orbitals are placed equally around the *x*-axis, but towards the negative side of p_x orbital, *i.e.*, $C_{32} = C_{42} = -C$ and using rule (*iii*) again to p_x orbital,

$$C_{12}^2 + C_{22}^2 + C_{32}^2 + C_{42}^2 = 1$$

$0 + \frac{2}{3} + C^2 + C^2 = 1$ will give $C = \frac{1}{\sqrt{6}}$. Thus $C_{32} = C_{42} = -\frac{1}{\sqrt{6}}$.

Lastly using any rule, either (*i*) or (*iii*) and locations of h_3 and h_4, it is clear that p_y orbital is equally placed around h_3 and h_4. Further h_3 is towards the positive direction and h_4 towards negative direction. This will give $C_{33} = \frac{1}{\sqrt{2}}$ and $C_{43} = -\frac{1}{\sqrt{2}}$. The complete set of wave functions of all the four hybrid orbitals can be written as:

$$h_1 = \frac{1}{2} s + \frac{\sqrt{3}}{2} p_z$$

$$h_2 = \frac{1}{2} s + \sqrt{\frac{2}{3}} p_x - \frac{1}{\sqrt{12}} p_z$$

$$h_3 = \frac{1}{2} s - \sqrt{\frac{1}{6}} p_x + \sqrt{\frac{1}{2}} p_y - \frac{1}{\sqrt{12}} p_z$$

$$h_4 = \frac{1}{2} s - \sqrt{\frac{1}{6}} p_x - \sqrt{\frac{1}{2}} p_y - \frac{1}{\sqrt{12}} p_z .$$

It may be pointed out here that the values of coefficients of the pure a.o.'s, obtained by this method do not represent the unique values of the coefficients, *i.e.*, the values of the coefficients obtained depends on the orientations of the hybrid orbitals along the *x*-, *y*- and *z*-axis. But one thing is certain that the wave functions of the hybrid orbitals obtained by using any geometrical representation give the equivalent set and it is easy to obtain the relation between the coefficients of any two sets of sp^3 hybrid orbitals using matrix transformation.

For example, four hybrid orbitals from sp^3 combination can also be pointed towards the four corners of the cubes as shown in Fig. 4.52. This arrangement of the hybrid orbitals clearly shows that all the four hybrid orbitals are placed symmetrically around the coordinate axis. In other words all the four sp^3 orbitals will be contributing equally to the hybrid orbitals. Only thing different can be the sign of the coefficients. From the Fig. 4.52, it is obvious:

$$C_{11} = C_{21} = C_{31} = C_{41} = \frac{1}{2}$$

$$-C_{12} = C_{22} = C_{32} = -C_{42} = \frac{1}{2}$$

$$-C_{13} = C_{23} = -C_{33} = C_{43} = \frac{1}{2}$$

$$C_{14} = C_{24} = -C_{34} = -C_{44} = \frac{1}{2}$$

And the wave functions for the hybrid orbitals are:

$$h_1 = \frac{1}{2}s - \frac{1}{2}p_x - \frac{1}{2}p_y + \frac{1}{2}p_z$$

$$h_2 = \frac{1}{2}s + \frac{1}{2}p_x + \frac{1}{2}p_y + \frac{1}{2}p_z$$

$$h_3 = \frac{1}{2}s + \frac{1}{2}p_x - \frac{1}{2}p_y - \frac{1}{2}p_z$$

$$h_4 = \frac{1}{2}s - \frac{1}{2}p_x + \frac{1}{2}p_y - \frac{1}{2}p_z$$

Example 4.9: Calculate the coefficients of dsp^3 hybrid orbitals forming trigonal bipyramidal structure.

Solution: Based on symmetry, out of five d orbitals d_{z^2} will be the most effective in forming hybrid orbitals. The five hybrid orbitals can be written as:

$$h_1 = C_{11}\,s + C_{12}\,p_x + C_{13}\,p_y + C_{14}\,p_z + C_{15}\,d_{z^2}$$

$$h_2 = C_{21}\,s + C_{22}\,p_x + C_{23}\,p_y + C_{24}\,p_z + C_{25}\,d_{z^2}$$

$$h_3 = C_{31}\,s + C_{32}\,p_x + C_{33}\,p_y + C_{34}\,p_z + C_{35}\,d_{z^2}$$

$$h_4 = C_{41}\,s + C_{42}\,p_x + C_{43}\,p_y + C_{44}\,p_z + C_{45}\,d_{z^2}$$

$$h_5 = C_{51}\,s + C_{52}\,p_x + C_{53}\,p_y + C_{54}\,p_z + C_{55}\,d_{z^2}$$

Proceeding on the same lines as described in the last example (4.8), the coefficients of s orbital in the five hybrid orbitals are equal and are given as:

$$C_{11} = C_{21} = C_{31} = C_{41} = C_{51} = \frac{1}{\sqrt{5}}$$

Arrangement of five hybrid orbitals can be divided into two sets, h_1 and h_2 are placed along the z-axis representing as axial hybrid orbitals. The other set represents three hybrid orbitals in the xy-plane, h_3 pointing towards x-axis, h_4 and h_5 placed symmetrically around the negative side of x-axis. These hybrid orbitals are shown in Fig. 4.54. It is clear from Fig. 4.54 that p_z orbital will be contributing equally to h_1 and h_2 but with different sign. Using rule (*iii*) and $C_{34} = C_{44} = C_{54} = 0$ and $C_{14} = -C_{24} = C$

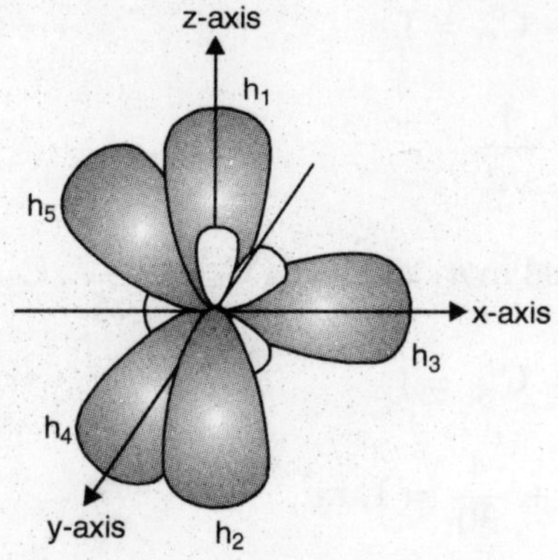

Figure 4.54: *Representation of dsp³ hybrid orbitals in the Cartesian coordinate system*

$$C_{14}^2 + C_{24}^2 = 1, \textit{ i.e.}, C_{14} = -C_{24} = \frac{1}{\sqrt{2}}$$

Applying rule (*i*) to h_1, *i.e.*,

$$C_{11}^2 + C_{12}^2 + C_{13}^2 + C_{14}^2 + C_{15}^2 = 1$$

and $$C_{11} = \frac{1}{\sqrt{5}}, C_{14} = \frac{1}{\sqrt{2}}, C_{12} = C_{13} = 0, \textit{ i.e.},$$

$$\frac{1}{5} + \frac{1}{2} + C_{15}^2 = 1, \text{ we get } C_{15} = \sqrt{\frac{3}{10}},$$

because h_1 is oriented towards the positive side of d_{z^2} of *z*-axis. The shape of d_{z^2} orbit in *xy*-plane is such that it will be contributing equally to the hybrid orbitals h_3, h_4 and h_5. Further the sign of wave function for d_{z^2} in *xy*-plane is negative therefore all the coefficients will be negative. Thus, using the rule (*iii*) to d_{z^2} and $C_{35} = C_{45} = C_{55} = -C$, equation

$$C_{15}^2 + C_{25}^2 + C_{35}^2 + C_{45}^2 + C_{55}^2 = 1 \text{ is reduced to:}$$

$$\frac{3}{10} + \frac{3}{10} + 3C^2 = 1 \Rightarrow C = \frac{2}{\sqrt{30}}$$

$$\textit{i.e.}, C_{35} = C_{45} = C_{55} = -\frac{2}{\sqrt{30}}.$$

Having evaluated the coefficients of *s*, p_z and d_{z^2} in all the hybrid orbitals, let us turn our attention towards h_3, h_4 and h_5. Description of Fig. 4.54 demands that p_y will not contribute to h_3 at all, but contributes equally to h_4 and h_5, positive contribution to h_4 and negative contribution to h_5. Using rule (*iii*) to p_y orbital and $C_{13} = C_{23} = C_{33} = 0$ and $C_{43} = -C_{53}$, equation

$$C_{13}^2 + C_{23}^2 + C_{33}^2 + C_{43}^2 + C_{53}^2 = 1$$

reduces to give $C_{43} = -C_{53} = \frac{1}{\sqrt{2}}$

Applying rule (*i*) to p_x orbital in h_3 and using $C_{31} = \frac{1}{\sqrt{5}}$, $C_{33} = C_{34} = 0$, $C_{35} = \frac{2}{\sqrt{30}}$, equation

$$C_{31}^2 + C_{32}^2 + C_{33}^2 + C_{34}^2 + C_{35}^2 = 1$$

reduces to: $\frac{1}{5} + C_{32}^2 + 0 + 0 + \frac{4}{30} = 1$, *i.e.*,

$$C_{32} = \sqrt{\frac{2}{3}}$$

Lastly p_x orbital will be contributing equally to h_4 and h_5 with negative value to coefficient as both of these orbitals are pointing towards the negative side of *x*-axis, *i.e.*, $C_{42} = C_{52} = -C$. Using rule (*i*) on p_x and $C_{12} = C_{22} = 0$, $C_{32} = \sqrt{\frac{2}{3}}$, equation

$$C_{31}^2 + C_{32}^2 + C_{33}^2 + C_{34}^2 + C_{35}^2 = 1$$

reduces to: $0 + 0 + \frac{2}{3} + 2C^2 = 1$

and gives $C = \frac{1}{\sqrt{6}}$, *i.e.*, $C_{42} = C_{52} = -\frac{1}{\sqrt{6}}$

Now the complete wave functions for all the five hybrid orbitals can be written as:

$$h_1 = \frac{1}{\sqrt{5}} s + \frac{1}{\sqrt{2}} p_z + \sqrt{\frac{3}{10}} d_{z^2}$$

$$h_2 = \frac{1}{\sqrt{5}} s - \frac{1}{\sqrt{2}} p_z + \sqrt{\frac{3}{10}} d_{z^2}$$

$$h_3 = \frac{1}{\sqrt{5}} s + \sqrt{\frac{2}{3}} p_x - \frac{2}{\sqrt{30}} d_{z^2}$$

$$h_4 = \frac{1}{\sqrt{5}} s - \frac{1}{\sqrt{6}} p_x + \frac{1}{\sqrt{2}} p_y - \frac{2}{\sqrt{30}} C_{45}\, d_{z^2}$$

$$h_5 = \frac{1}{\sqrt{5}} s - \frac{1}{\sqrt{6}} p_x - \frac{1}{\sqrt{2}} p_y - \frac{2}{\sqrt{30}} d_{z^2}$$

Using above procedure as described in examples 4.7 and 4.8, we can find the wave functions of hybrid orbitals from the linear combination of sp^2, d^2sp^3 and dsp^2 a.o.'s commonly used in chemistry.

4.19.2 ANGULAR DEPENDENCE OF HYBRID ORBITALS

In the last section, we have learnt that LCAO of the orbitals of the same atom leads to hybrid orbitals and they possess strong charge density orientations along particular directions, *i.e., these hybrid orbitals have strong directional properties*. Qualitatively we have learnt that *sp* hybrid orbitals will be having 180° angle between them, whereas sp^2 and sp^3 will have 120° and 109.5° respectively. Mathematically these angles can also be derived. We have learnt that normalized wave functions for *sp* hybrid a.o. are:

$$\psi_1 = \frac{1}{\sqrt{2}}(s + p) \qquad ...(4.232)$$

$$\psi_2 = \frac{1}{\sqrt{2}}(s - p) \qquad ...(4.233)$$

Angles between the hybrid orbitals and their shapes can be determined if we consider the radius of the spherical contour surface, representing the magnitude of *s* orbital as 1. Since the *p* orbitals possess angular dependence, its magnitude relative to *s* orbital can be written as:

$$p_x = \sqrt{3}\sin\theta\cos\varphi \qquad ...(4.234)$$

$$p_y = \sqrt{3}\sin\theta\sin\varphi \qquad ...(4.235)$$

$$p_z = \sqrt{3}\cos\theta \qquad ...(4.236)$$

Using normalized angular functions of $\Theta(\theta)$ and $\Theta(\varphi)$ for 2*s* and 2*p* orbitals and selecting $2p_z$ as the orbital taking part in *sp* hybridization, using equation (4.236), ψ_1 and ψ_2, can be written as

$$\psi_1 = \frac{1}{\sqrt{2}}\left[\frac{1}{\sqrt{4\pi}} + \sqrt{\frac{3}{4\pi}}\cos\theta\right]$$

$$\psi_2 = \frac{1}{\sqrt{2}}\left[\frac{1}{\sqrt{4\pi}} - \sqrt{\frac{3}{4\pi}}\cos\theta\right]$$

If we take out $\frac{1}{\sqrt{4\pi}}$ out of these terms, the functions, which we get, determine the directions of the two hybrid orbitals.

$$f_1 = \frac{1}{\sqrt{8\pi}}\left[1 + \sqrt{3}\cos\theta\right]$$

$$f_2 = \frac{1}{\sqrt{8\pi}}\left[1 - \sqrt{3}\cos\theta\right]$$

Since these functions do not depend on the angle φ, it means that these functions will concentrate alone *z*-axis. The maximum value of f_1 can be for θ = 0°, *i.e.*, 1.932 and of f_2 for θ = 180°, without taking $1/\sqrt{4\pi}$ into account again 1.932. These values are greater than pure *s* orbital, which is 1 and pure *p* orbitals, which is 1.732. Further pure p_z orbital is directed along both sides of *z*-axis equally, whereas *sp* hybrid orbitals are directed mainly along one side of *z*-axis, *i.e.*, either along the + *z*-axis or – *z*-axis and the angle between them is 180°. The polar plots of f_1 and f_2 for different values of θ are drawn in Figs. 4.55*a* and 4.55*b*.

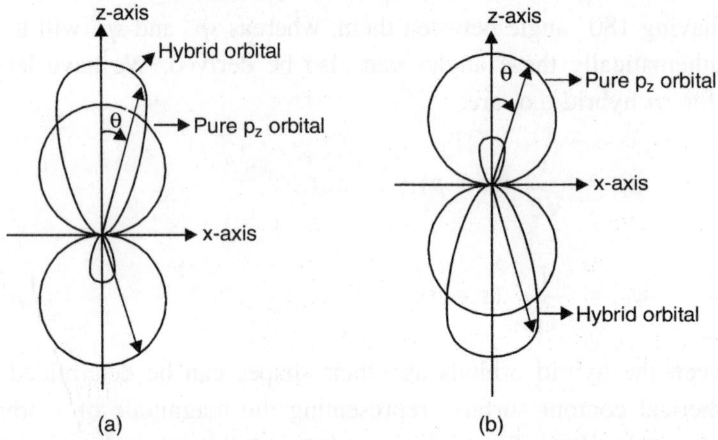

Figure 4.55: *(a) Plot of function f_1 as a function of angle θ. (b) Plot of function f_2 as function of θ. Pure p_z orbital is also shown. The scale for both the functions is not same*

Linear structure of ethyne has been explained by involving the *sp* hybridization on each carbon atom and π-m.o. formed by the linear combination of p_x–p_x and p_y–p_y atomic orbitals.

***sp^2-Hybrid Orbitals*:** Using the procedure given in section 4.19.1 , the wave functions for sp^2 hybrid orbitals, using *s*, p_x and p_y a.o.'s can be written as:

$$\psi_1 = \frac{1}{\sqrt{3}}\, s + \sqrt{\frac{2}{3}}\, p_x \qquad ...(4.237)$$

$$\psi_2 = \frac{1}{\sqrt{3}}\, s - \sqrt{\frac{1}{6}}\, p_x + \frac{1}{\sqrt{2}}\, p_y \qquad ...(4.238)$$

$$\psi_3 = \frac{1}{\sqrt{3}}\, s - \sqrt{\frac{1}{6}}\, p_x - \frac{1}{\sqrt{2}}\, p_y \qquad ...(4.239)$$

Combining equations (4.234-236) and (4.237-239), the above wave functions can be written as:

$$\psi_1 = \frac{1}{\sqrt{3}} + \sqrt{\frac{2}{3}}\, \sqrt{3}\, \sin\theta\, \cos\varphi \qquad ...(4.240)$$

$$\psi_2 = \frac{1}{\sqrt{3}} - \sqrt{\frac{1}{6}}\ \sqrt{3}\sin\theta\cos\varphi + \frac{1}{\sqrt{2}}\sqrt{3}\ \sin\theta\sin\varphi \qquad ...(4.241)$$

$$\psi_3 = \frac{1}{\sqrt{3}} - \sqrt{\frac{1}{6}}\ \sqrt{3}\sin\theta\cos\varphi - \frac{1}{\sqrt{2}}\sqrt{3}\ \sin\theta\sin\varphi \qquad ...(4.242)$$

One thing is evident from equations (4.240-442) that all the three hybrid orbtials do not contain p_z-a.o., *i.e.*, sp^2 hybrid orbitals are planar and are lying in xy-plane. In this situation $\theta = 90°$ and sin 90 = 1. Equations (4.240–242) reduce to

$$\psi_1 = \frac{1}{\sqrt{3}} + \sqrt{2}\ \cos\varphi$$

$$\psi_2 = \frac{1}{\sqrt{3}} - \sqrt{\frac{1}{2}}\ \cos\varphi + \sqrt{\frac{3}{2}}\ \sin\varphi$$

$$\psi_3 = \frac{1}{\sqrt{3}} - \sqrt{\frac{1}{2}}\ \cos\varphi - \sqrt{\frac{3}{2}}\ \sin\varphi$$

Arranging ψ_1 along x-axis, means that its maximum value will be when $\varphi = 0°$. On the other hand, the maximum value of ψ_2 and ψ_3 can be obtained by setting $\frac{d\psi_2}{d\varphi} = \frac{d\psi_3}{d\varphi} = 0$, *i.e.*,

$$\frac{d\psi_2}{d\varphi} = \sqrt{\frac{1}{2}}\ \sin\varphi + \sqrt{\frac{3}{2}}\ \cos\varphi = 0$$

$$\tan\varphi = -\sqrt{3}. \text{ Thus } \varphi = 120°$$

Similarly, $$\frac{d\psi_3}{d\varphi} = \sqrt{\frac{1}{2}}\ \sin\varphi - \sqrt{\frac{3}{2}}\ \cos\varphi = 0$$

$$\tan\varphi = \sqrt{3} \text{ will give } \varphi = 240°$$

Thus, each hybrid orbital is aligned to other at an angle of 120°. The maximum value of f_1, f_2 and f_3 will be:

$$f_1 = \frac{1}{\sqrt{3}} + \sqrt{2} = 0.578 + 1.414 = 1.992$$

$$f_2 = \frac{1}{\sqrt{3}} - \sqrt{\frac{1}{2}}\ \cos 120 + \sqrt{\frac{3}{2}}\ \sin 120 = \psi_2$$

$$= \frac{1}{\sqrt{3}} - \sqrt{\frac{1}{2}}\ (-1/2) + \sqrt{\frac{3}{2}}\sqrt{\frac{3}{2}}$$

$$= 1.992.$$

Similarly the value of f_3 will also be 1.992. In other words all the three sp^2 hybrid orbitals are equivalent and form angles of 120° with each other in *xy*-plane. The main examples belonging to this class of molecules are olefins or aromatic compounds, which are planar and where the bond angles are 120°. Further there remains a pure *p* orbital on each carbon atom, which forms a π-m.o. The other example is BF_3.

Similarly using the wave functions for sp^3 hybrid orbitals, as arrived at in the example 4.8, one can find out that the angle between the two hybrid orbitals is 109.5°. All the saturated hydrocarbons conform to sp^3 hybrid orbitals on each carbon.

4.19.3 STRENGTH OF HYBRID BONDS

Two main things, which we have learnt from the last section, are: (*i*) Strong directional property and, (*ii*) the relatively larger magnitude of the hybrid orbital in comparison to the constituent pure a.o.'s. Thus according to the principle of maximum overlap, bonding involving hybrid orbitals will be stronger than those arising from pure a.o.'s. This can be better understood by considering the following three examples: (*i*) alkali metals, (*ii*) halogens and (*iii*) C_2 as present in C_2H_6.

Alkali metals possess the valence electrons as ns^1 and thus will involve the overlap of *ns* orbitals in diatomic molecules, like say Li_2. Because of spherically symmetrical distribution of charge density of *s* orbital (and particularly it is quite diffuse in alkali metals), the overlap of similar *s* orbtitals between the adjacent atoms is quite small and leads to weaker bond. In case of halogens, which involve the bonding between *p* orbitals on the adjacent halogen atoms and possess better directional character as compared to pure *s* orbital, overlap between *p-p* a.o.'s is larger and thus a stronger bond. Lastly, on the basis of relative magnitude of hybrid orbitals, the bonding between *s-p* hybrid orbitals will be larger as the orbitals will have considerable overlap. This is quite clear from the bond energies of Li_2, F_2 and C_2 in (ethane) as 108.8, 159 and 334.7 kJ mol^{-1} respectively.

Before we stop discussion on the hybrid orbitals and their characteristics, few points demand attention, even though qualitatively at this stage. Although so far we have emphasized, and it is reasonable to expect that the strength of the bond between two identical atoms depends on the overlap of the two a.o.'s, which are paired together, and is one of the most significant terms determining the bond strength. But there are other factors also.

(*i*) ***Weightage of Each a.o.'s in the hybrid orbitals***: We have seen in the last section that weightage of each a.o. in each hybrid orbital depends on their participation. For example, taking *sp*, sp^2 and sp^3 hybrid orbitals, values of coefficients of a.o.'s in each hybrid orbital varies between 0.5 to 0.25 respectively and thereby decreasing its participation as we go from *sp* to sp^2 to sp^3. Calculation of overlap integral (S) in each case between two adjacent carbon atoms is maximum for *sp* and decreases when we consider sp^2 and sp^3. In other words *sp* bond is the strongest and shortest as compared to sp^2 and sp^3. Table 4.8 lists the characteristics in support of the above discussion.

Table 4.8: *Characteristics of C — H bonds involving differing hybridization*

Hybridization	Molecule	C — H bond length(nm)	Stretching force constant (10^3 N m^{-1})	Bond energy (kJ mol^{-1})
sp	Ethyne	0.0106	6.397	~ 506
sp^2	Ethene	0.1069	6.126	~ 444
sp^3	Methane	0.1090	5.387	~ 431
(*p*)	CH radical	0.012	4.490	335

(*ii*) Coulomb repulsion between the attached atoms, which carry net positive or negative charges.

(*iii*) Coulomb and exchange repulsion between the two bonds.

(*iv*) Involvement of lone pair of electrons present on the particular atom.

(*v*) Configuration interactions between different configurations are discussed in section 3.14. This will correspond to different pairing schemes (or structures) in the language of valence bond theory.

4.20 HÜCKEL MOLECULAR ORBITAL THEORY (HMO)

In m.o.t., we have learnt that these m.o.'s can be obtained by taking LCAO of the atoms of the molecules. The energies of the various m.o.'s are obtained by using variation method, *i.e.,* by solving the secular determinant and are given in terms of diagonal elements (H_{ii}) and off diagonal (H_{ij}) elements, as well as, overlap integrals (S_{ij}), defined in chapter 3 as:

$$\alpha = H_{ii} = \int \varphi_i \ \hat{H} \ \varphi_i \ d\tau_i, \ \beta = H_{ij} = \int \varphi_i \ \hat{H} \ \varphi_j \ d\tau_i, \ S_{ij} = \int \varphi_i \ \varphi_j \ d\tau_i$$

where φ_i and φ_j are a.o.'s, $\hat{H}$ is the Hamiltonian operator. The values of the coefficients of a.o.'s, used in forming m.o.'s, are calculated by substituting the values of each energy in the secular simultaneous equations one by one and solving them. The solution of the secular determinant and secular equations become more and more complicated as the number of atoms increases in a molecule. These calculations may become imposible for complicated molecules. Similar difficulties are also faced when v.b.t. is applied to complex molecules. To overcome these difficulties, simple theory was developed for unsaturated planar organic molecules, which is described here briefly.

In 1930, Hückel developed the most celebrated semi-empirical π-electron theory, known as Hückel molecular orbital (HMO) method. In this method, m.o.'s of planar unsaturated organic molecules are divided into σ and π m.o.'s. The σ electrons are described in terms of localized hybrid bonding orbitals. Thus σ m.o.'s possess maximum probability charge density in molecular plane. On the other hand, π-electrons are described by molecular wave functions those extend over each of the atom that contributes a π-electron. In other words, π-electrons are delocalized. These m.o.'s change sign on reflection in the molecular plane, *i.e., molecular plane acts as a nodal plane for these* m.o.'s *and probability charge density lies above and below the* σ *molecular framework.*

In other words, the π-electrons can be treated separately from σ-electrons and approximate theory can be developed as given below.

Assumptions

1. The main assumption in this theory is that the π-electrons move in a fixed electrostatic effective potential due to the σ-framework of the molecule (*i.e.,* σ-electrons and the nuclei). The approximate Hamiltonian for π-electrons in its simple form is given as:

$$\hat{H}_\pi = \sum_{j=1}^{n_\pi} \hat{H}^{eff}(j) \qquad ...(4.243)$$

where $\hat{H}^{eff}(j)$ is the effective Hamiltonian for the jth π-electron. (*The commonly used symbol 'i' is avoided here because it may create confusion with other meaning of i = √−1. a quantity which will appear in some other expression of HMO*). It includes the kinetic energy operator for π-electron, the average effect of σ-electrons on the π-electron and the average effect of π-electron repulsion. It is not defined explicitly in HMO. Since total HMO Hmiltonian is the sum of one electron Hamiltonian, separation of variables can be carried out as described in (2.6.3). The Schröndinger wave equation for the jth electron can be written as:

$$\hat{H}^{eff}(j)\,\varphi_j = E_J\,\varphi_j \qquad ...(4.244)$$

Since $\hat{H}^{eff}(j)$ is not defined explicitly, as mentioned above, thus, there is no need to solve this equation (4.244). On the other hand one can use the variational method.

The above assumption leads to the following results:

(*a*) The total wave function of the molecule, including σ and π-electrons, can be written as:

$$\psi = \psi_\sigma\,\psi_\pi \qquad ...(4.245a)$$

(*b*) The total energy of the molecule will be

$$E = E_\sigma + E_\pi \qquad ...(4.245b)$$

(*c*) Since the π-electron Hamiltonian is defined as the sum of the one π-electron effective Hamiltonian, the total wave function describing even π-electron framework can be written as:

$$\psi_\pi = \varphi_1\,(1)\,\varphi_1\,(2)\,\varphi_2\,(3)\,\varphi_2\,(4)\ldots\ldots\,\varphi_{n/2}\,(n-1)\,\varphi_{n/2}\,(n) \qquad ...(4.246)$$

where n is the even number of π-electrons. The total energy

$$E_\pi = n_1\,E_1 + n_2\,E_2 + \ldots\ldots\ldots\ldots\ldots\ldots\ldots\ldots + n_n\,E_n \qquad ...(4.247)$$

where E_j is the energy of the one electron in φ_j orbital, as obtained from equation (4.244). The coefficients n_j can be either 0, 1 or 2, depending on whether the particular orbital is unoccupied, singly occupied or doubly occupied.

2. π m.o.'s (φ_j) is obtained by taking the linear combination of only those a.o.'s of carbon atom which are having π symmetry and are perpendicular to the molecular plane. These are $2p\pi$ orbitals of carbon atom. Thus φ_j can be written as:

$$\varphi_j = \sum_{k=1}^{n} C_{jk}\, p_k = C_{j1}\, p_1 + C_{j2}\, p_2 + \ldots\ldots\ldots + C_{jn}\, p_n \qquad \ldots(4.248)$$

φ_j is the jth HMO of the molecule, p_k is the $2p_x$ a.o. of the kth carbon atom (yz is considered as molecular plane), n is the number of carbon atoms and C_{jn} is the coefficient of the nth a.o. in the jth HMO. This method is mainly applied to the systems containing carbon atoms and not hydrogen atoms, but later on it was modified to include other atoms also, known as extended HMO method.

3. The functions φ_j, defined by equation (4.248) is the linear variation function. Thus the energy and the corresponding wave function of π-electron can be calculated using variation method (3.6). The secular equations obtained are:

$$(H_{11} - ES_{11})\, C_1 + (H_{12} - ES_{12})\, C_2 + \ldots\ldots\ldots + (H_{1n} - ES_{1n})\, C_n = 0$$
$$(H_{21} - ES_{21})\, C_1 + (H_{22} - ES_{22})\, C_2 + \ldots\ldots\ldots + (H_{2n} - ES_{2n})\, C_n = 0$$
$$(H_{31} - ES_{31})\, C_1 + (H_{32} - ES_{32})\, C_2 + \ldots\ldots\ldots + (H_{3n} - ES_{3n})\, C_n = 0 \qquad \ldots(4.249)$$
$$\ldots\ldots\ldots\ldots\ldots\ldots\ldots\ldots\ldots\ldots\ldots\ldots\ldots\ldots\ldots$$
$$\ldots\ldots\ldots\ldots\ldots\ldots\ldots\ldots\ldots\ldots\ldots\ldots\ldots\ldots\ldots$$
$$(H_{n1} - ES_{n1})\, C_1 + (H_{n2} - ES_{n2})\, C_2 + \ldots\ldots\ldots + (H_{nn} - ES_{nn})\, C_n = 0$$

Similarly the nth order secular determinant will be

$$\begin{vmatrix} H_{11} - ES_{11} & H_{12} - ES_{12} + \ldots\ldots\ldots & H_{1n} - ES_{1n} \\ H_{21} - ES_{21} & H_{22} - ES_{22} + \ldots\ldots\ldots & H_{2n} - ES_{2n} \\ \ldots\ldots & \ldots\ldots\ldots\ldots & \ldots\ldots \\ \ldots\ldots & \ldots\ldots\ldots\ldots & \ldots\ldots \\ H_{n1} - ES_{n1} & H_{n2} - ES_{n2} \ldots\ldots\ldots & H_{nn} - ES_{nn} \end{vmatrix} = 0 \qquad \ldots(4.250)$$

where $H_{JJ} = \alpha = \int p_j\, \hat{H}_{\text{eff}}(j)\, p_j\, d\tau_j$, $H_{Jk} = \beta = \int p_j\, \hat{H}_{\text{eff}}(j)\, p_k\, d\tau_j$, $S_{jk} = \int p_j\, p_k\, d\tau$

4. The other major assumptions in HMO are: (*a*) that H_{jj} has the same value for every carbon atom in the molecule. There is no problem in this assumption for benzene molecule, where all the six carbon atoms are equivalent, but the problem can arise in butadiene molecule where the terminal carbon atoms and middle carbon atoms have different behaviour. It is not only that H_{jj} is the same for each carbon atom in butadiene but the integral, commonly known as Coulomb integral (α), is assumed to be same for all the carbon atoms, *i.e.*,

$$H_{11} = H_{22} = H_{33} = \ldots\ldots\ldots\ldots H_{nn} = \alpha$$

(*b*) H_{jk}, represents the energy of interaction between two p_j (π) a.o.'s and is assumed to be same for any two carbon atoms bonded to each other (*i.e., adjacent or neighbouring carbon atoms*) and is zero if the two carbon atoms are not bonded to each other (*i.e., not adjacent or neighbouring carbon atoms*), *i.e.,*

$$H_{jk} = \beta \quad \text{if } C_j \text{ and } C_k \text{ are adjacent atoms.}$$
$$H_{jk} = 0 \quad \text{if } C_j \text{ and } C_k \text{ are not adjacent atoms.}$$

This is commonly denoted as β and often called resonance integral (although this term is more appropriate to v.b.t. than that for m.o.t.). Since the carbon atoms other than neighbouring ones, are well separated in space, this assumption is reasonable.

(*c*) The overlap integral of type S_{jj} is assumed to be equal to unity, since the a.o. are normalized, whereas the overlap integral S_{jk} is assumed to be equal to zero, *i.e.,*

$$S_{11} = S_{22} = S_{33} = \ldots\ldots\ldots\ldots = S_{jj} = 1$$

and

$$S_{12} = S_{23} = \ldots\ldots\ldots\ldots = S_{jk} = 0$$

The assumption that overlap integral for carbon atoms bonded to each other is equal to zero, is a very poor or drastic assumption. In reality if one uses Slator's *p* orbitals, the value of overlap integral for adjacent carbon atoms lies between 0.2 to 0.3, depending on bond distance, which is far from zero. Finite value of S_{jk} has been incorporated into the simple HMO, but will not be discussed in this book. However the values of energy obtained for each HMO by the simple HMO method is not very different from those obtained by more accurate methods.

These assumptions certainly simplify the secular determinant. Equation (4.249), *i.e.,* secular determinant for linear (4.251) and cyclic (4.252) conjugated planar hydrocarbons can be written as:

$$\begin{vmatrix} \alpha - E & \beta & 0 & 0 \ldots\ldots\ldots 0 \\ \beta & \alpha - E & \beta & 0 \ldots\ldots\ldots 0 \\ 0 & \beta & \alpha - E & \beta \ldots\ldots\ldots 0 \\ \ldots\ldots\ldots\ldots & & & \\ \ldots\ldots\ldots\ldots & & & \\ \ldots\ldots\ldots\ldots & & & \beta \quad \alpha - E \end{vmatrix} = 0 \qquad \ldots(4.251)$$

and

$$\begin{vmatrix} \alpha - E & \beta & 0 & 0 \ldots\ldots\ldots\ldots\beta \\ \beta & \alpha - E & \beta & 0 \ldots\ldots\ldots\ldots 0 \\ 0 & \beta & \alpha - E & \beta \ldots\ldots\ldots\ldots 0 \\ \ldots\ldots\ldots\ldots\ldots\ldots\ldots\ldots \\ \ldots\ldots\ldots\ldots\ldots\ldots\ldots\ldots \\ \beta \ldots\ldots\ldots\ldots\ldots\ldots\ldots\ldots & & & \beta \quad \alpha - E \end{vmatrix} = 0 \qquad \ldots(4.252)$$

These secular determinants can also be written as:

$$\begin{vmatrix} x & 1 & 0 & 0 \ldots\ldots\ldots\ldots 0 \\ 1 & x & 1 & 0 \ldots\ldots\ldots\ldots 0 \\ 0 & 1 & x & 1 \ldots\ldots\ldots\ldots 0 \\ \ldots\ldots\ldots\ldots\ldots\ldots\ldots\ldots \\ \ldots\ldots\ldots\ldots\ldots\ldots\ldots\ldots \\ 0\ldots & 0\ldots\ldots & 0\ldots\ldots & 0 \ldots\ldots\ldots 1 \quad x \end{vmatrix} = 0 \qquad \ldots(4.253)$$

and

$$\begin{vmatrix} x & 1 & 0 & 0 \ldots\ldots\ldots\ldots 1 \\ 1 & x & 1 & 0 \ldots\ldots\ldots\ldots 0 \\ 0 & 1 & x & 1 \ldots\ldots\ldots\ldots 0 \\ \ldots\ldots\ldots\ldots\ldots\ldots\ldots\ldots \\ \ldots\ldots\ldots\ldots\ldots\ldots\ldots\ldots \\ 1\ldots & 0\ldots & 0\ldots & 0 \ldots\ldots\ldots 1 \quad x \end{vmatrix} = 0 \qquad \ldots(4.254)$$

$$\text{where } x = \frac{\alpha - E}{\beta} \qquad \ldots(4.255)$$

Solution of equations (4.251) and (4.253) or (4.252) and (4.254) gives '*n*' values of energies of *n*-HMO's. Relative values of coefficients of HMO's (*i.e.*, C_{jk}), can be calculated by substituting the values of energies, one by one, in *n* secular simultaneous equations and solving them. The actual values can be obtained by using the normalization condition. Thus for each value of energy E, we will get a set of coefficients of *p* orbital involved in HMO. For a conjugated sytem, consisting of *n* carbon atoms, we get *n* HMO's. Pauli Exclusion Principle and Hund's rules can be used to fill $n\pi$ electrons in *n* HMO's, giving electron configuration for the ground state. The above principles and assumptions involved in HMO method can be better understood by studying the following examples.

Applications

1. *Ethylene molecule*: Ethylene is the simplest molecule to be treated by HMO. The σ framework has been shown in Fig. 4.56. Each carbon atom contributes $2p_x$ orbital to form delocalized π m.o. This is similar to the hydrogen molecule problem and can be treated on the same lines. The π-m.o. can be written as:

$$C_1 \text{———} C_2$$

Figure 4.56: *σ framework of ethylene molecule*

$$\varphi_j = C_{j1}\, p_1 + C_{j2}\, p_2 \qquad ...(4.256)$$

where p_1 and p_2 are the $2p_x$ orbital of carbon belonging to atoms 1 and 2 respectively. The secular determinant can be written as:

$$\begin{vmatrix} \alpha - E & \beta \\ \beta & \alpha - E \end{vmatrix} = 0 \quad \text{or} \quad \begin{vmatrix} x & 1 \\ 1 & x \end{vmatrix} = 0$$

Expanding the determinant,

$$x^2 - 1 = 0 \quad \text{or } x = \pm 1$$

Thus, the two HMO energies corresponding to $x = +1$ and $x = -1$ are:

$$E_1 = \alpha + \beta \qquad ...(4.257)$$

$$E_2 = \alpha - \beta \qquad ...(4.258)$$

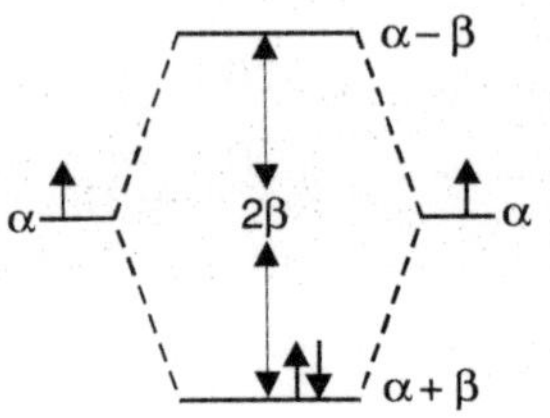

Figure 4.57: *Relative energy level of ethylene molecule. Single arrows show px π-electron at each carbon atom and double arrow shows π-electron in m.o. 2β is the gap between the π-bonding and π-anti bonding m.o.*

Since both α and β are negative, E_1 will give rise to bonding and E_2 will give rise to anti-bonding HMO's. To know the exact values of α and β, we must know the effective Hamiltonian operator. Luckily we do not need to know α and β as some empirical values are assigned to them. Further since α is nearly the energy of an electron in an isolated $2p_x$ orbital of carbon atom, a zero value can be assigned. Thus, the quantity β defines the stability of bonding and unstability of anti-bonding HMO. The energy level diagram of HMO is shown below in Fig. 4.57.

***π-Bond energy*:** Ethylene molecule has two π-electrons and the ground state electronic configuration will be π_1^2. Both the electrons are placed in the π_1 HMO. Thus the total energy will be:

$$E = 2(\alpha + \beta) \qquad ...(4.258)$$

In the absence of a chemical bond between the two carbon atoms, electrons will be placed on the two isolated carbon atoms and the total energy will be 2α. Thus the π bond energy of ethylene molecule is $2(\alpha + \beta) - 2\alpha = 2\beta$.

Coefficients and HMO wave functions: The secular equations for *j*th HMO of ethylene molecule can be written as:

$$(H_{11} - ES_{11})\, C_{j1} + (H_{12} - ES_{12})\, C_{j2} = 0$$
$$(H_{21} - ES_{12})\, C_{j1} + (H_{22} - ES_{22})\, C_{j2} = 0 \qquad ...(4.259)$$

Under the approximations of HMO, these equations reduce to:

$$(\alpha - E)\, C_{j1} + \beta\, C_{j2} = 0$$
$$\beta\, C_{j1} + (\alpha - E)\, C_{j2} = 0 \qquad ...(4.260)$$

The values of coefficients for each value of E can be calculated by substituting the particular value of E. Thus, for $E_1 = \alpha + \beta$, the above equations become:

$$(\alpha - \alpha - \beta)\, C_{11} + \beta\, C_{12} = 0$$

or $$C_{11} = C_{12}$$

Thus $$\varphi_1 = C_{11}\,(2p_1 + 2p_2) \qquad ...(4.261)$$

Value of C_{11} can be obtained by normalizing φ_1, *i.e.*,

$$1 = \int \varphi_1\, \varphi_1\, d\tau = C_{11}^2 \int [2p_1^2 + 2p_2^2 + 2\; 2p_1\; 2p_2]\, d\tau$$
$$= 2C_{11}^2, \text{ since } \int 2p_1\, 2p_2\, d\tau = S_{12} = 0 \text{ and } \int 2p_1,\, 2p_1 d\tau = \int 2p_2 2p_2 = 1$$

Thus $$C_{11} = \frac{1}{\sqrt{2}} \qquad ...(4.262)$$

and $$\varphi_1 = \frac{1}{\sqrt{2}}\,[2p_1 + 2p_2] \qquad ...(4.263)$$

Similarly by using $E_2 = \alpha - \beta$ in equations (4.260), one can easily find out

$$\varphi_2 = \frac{1}{\sqrt{2}}\,[2p_1 - 2p_2] \qquad ...(4.264)$$

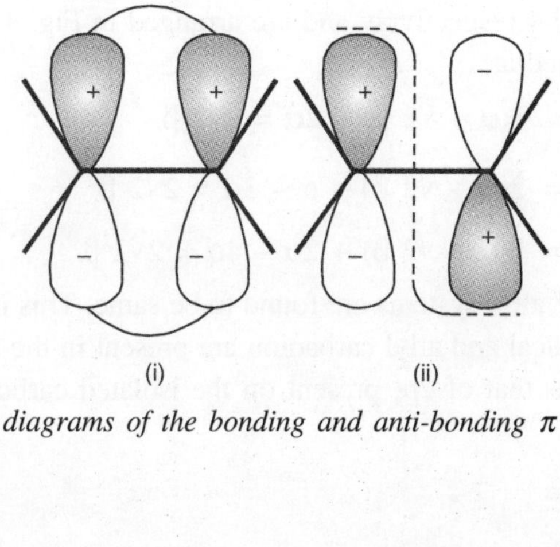

Figure 4.58: *Contour diagrams of the bonding and anti-bonding π molecular orbitals*

In case of φ_1, the magnitude of φ_1 increases, reaches a maximum value at carbon atom C_1, decreases slightly in between carbon atoms C_1 and C_2, reaches maximum again at C_2 and then decreases continuously. In case φ_2, the behaviour is similar to that of φ_1 up to C_1, then decreases and changes sign in the middle of C_1 and C_2, reaches continuously to reach zero. Similar to m.o. treatment of H_2, φ_2 has zero electron density in the middle, called a node and thus known as anti-bonding m.o. The three dimensional contour diagrams of φ_1 and φ_2 are shown in Fig. 4.58. These are similar to those depicted in Fig. 4.11. Both these HMO's are having *yz*-plane as a nodal plane. φ_1 is asymmetric with respect to centre of inversion, called ungerade (*u*) and φ_2 symmetric with respect to centre of inversion, called gerade (*g*).

Allyl system: Allyl system consists of three carbon atoms and σ framework can be written as:

$$C_1 \text{------} C_2 \text{------} C_3$$

The wave function can be written as:

$$\varphi_j = C_{j1}\, 2p_1 + C_{j2}\, 2p_2 + C_{j3}\, p_3 \qquad \text{...(4.265)}$$

and secular determinant as:

$$\begin{vmatrix} x & 1 & 0 \\ 1 & x & 1 \\ 0 & 1 & x \end{vmatrix} = 0$$

which on expansion can be written as:

$$x^3 - 2x = 0 \quad \text{or } x(x^2 - 2) = 0$$

Three roots are: $x_1 = 0$, $x_2 = -\sqrt{2}$, $x_3 = +\sqrt{2}$. The corresponding HMO energy levels are:

$$E_1 = \alpha + \sqrt{2}\beta \qquad \text{...(4.266)}$$

$$E_2 = \alpha \qquad \text{...(4.267)}$$

$$E_3 = \alpha - \sqrt{2}\beta \qquad \text{...(4.268)}$$

Since α and β, both have negative signs, the relation between E_1, E_2 and E_3 is $E_1 < E_2 < E_3$. The energy level diagram is depicted in Fig. 4.59. Three different allyl systems can be considered, *i.e.*, allyl carbonium ion, allyl radical and allyl carbanion. The maximum numbers of π-electrons in these systems are 2, 3 and 4 respectively and are arranged in Fig. 4.59. π bond energies for each allyl system can be obtained as:

$$E_\pi\, (C_3H_5^+) = 2(\alpha + \sqrt{2}\,\beta) - 2\alpha = 2\sqrt{2}\,\beta$$

$$E_\pi\, (C_3H_5^\bullet) = 2(\alpha + \sqrt{2}\,\beta) + \alpha - 3\alpha = 2\sqrt{2}\,\beta$$

$$E_\pi\, (C_3H_5^-) = 2(\alpha + \sqrt{2}\,\beta) + 2\alpha - 4\alpha = 2\sqrt{2}\,\beta$$

π-bond energies for all the allyl systems are found to be same. This is because the third and fourth π-electrons in the allyl radical and allyl carbanion are present in the non-bonding orbital, which is having the same energy as that of $2p_x$ present on the isolated carbon atom.

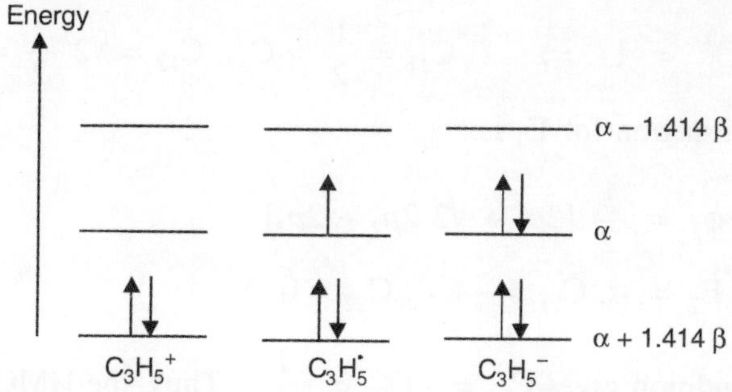

Figure 4.59: *Relative energy level diagram for π-electrons in allyl carbonium ion, allyl system and allyl carbanion*

Delocalized or resonance energy: *Delocalized or resonance energy is defined as the difference between the π-bond energy (obtained by HMO) and the localized π-bond energy (obtained by classical* v.b. *structures).* Now all the three-allyl systems in the valence bond or classical picture can be written as:

$C = C - C^+$	$C = C - C^{\bullet}$	$C = C - C^-$
Carbonium ion	Radical	Carbanion

which contain one π bond and one single bond. As we have learnt earlier that m.o. is spread over the whole molecule (*i.e.,* comprising all the atoms of the molecule), it may be assumed that in the allyl systems, π-electrons are spread over all the three carbon atoms or it may be mentioned that π-electrons are delocalized over the three corbon atoms. The delocalized structures can be written as:

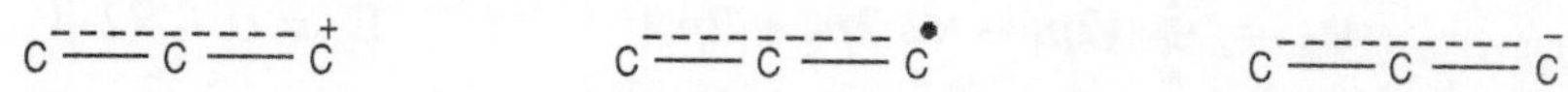

The π-bond energies in the v.b. or localized allyl structures are the same, *i.e.,* 2β (one π bond). Thus the delocalized π-energy is $2\sqrt{2}\beta - 2\beta = 0.828\beta$, which is a negative quantity. This means that delocalized allyl radical structures are more stable than the localized or v.b. structures by 0.828 β.

HMO coefficients: The simplified secular equations for allyl systems as are:

$$(\alpha - E)\, C_{j1} + \beta\, C_{j2} = 0$$

$$\beta\, C_{j1} + (\alpha - E)\, C_{j2} + \beta\, C_{j3} = 0$$

$$\beta\, C_{j2} + (\alpha - E)\, C_{j3} = 0$$

Substituting the value of E_1, in the above equations, we get

$$(\alpha - \alpha - \sqrt{2}\,\beta)\, C_{11} + \beta\, C_{12} = 0 \quad \Rightarrow \quad C_{12} = \sqrt{2}\, C_{11}$$

$$\beta\, C_{11} + (\alpha - \alpha - \sqrt{2}\,\beta)\, C_{12} + \beta\, C_{13} = 0 \quad \Rightarrow \quad C_{11} = C_{13}$$

$$\beta\, C_{12} + (\alpha - \alpha - \sqrt{2}\,\beta)\, C_{13} = 0 \quad \Rightarrow \quad C_{12} = \sqrt{2}\, C_{13}$$

Using normalization condition on φ_1, we get

$$C_{11}^2 + C_{12}^2 + C_{13}^2 = 1$$

or $$C_{11}^2 + 2C_{11}^2 + C_{11}^2 = 1 \Rightarrow C_{11} = \frac{1}{2} = C_{13}, C_{12} = \sqrt{2}\,\frac{1}{2} = \frac{1}{\sqrt{2}}$$

The HMO wave function for E_1 is:

$$\varphi_1 = \frac{1}{2}[2p_1 + \sqrt{2}\,2p_2 + 2p_3]$$

Similarly for $E_2 = \alpha, C_{21} = -C_{23}, C_{22} = 0.$

Normalization condition gives $C_{21} = -C_{23} = \frac{1}{\sqrt{2}}$. Thus, the HMO wave function for

$$\varphi_2 = \frac{1}{\sqrt{2}}(2p_1 - 2p_3)$$

For $E_3 = \alpha - \sqrt{2}\,\beta, C_{31} = C_{33}, C_{32} = -\frac{\sqrt{2}}{2}$ and

$$\varphi_3 = \frac{1}{2}[2p_1 - \sqrt{2}\,2p_2 + 2p_3]$$

Thus, the energy levels and corresponding wave functions are:

$$\varphi_1 = \frac{1}{2}[2p_1 + \sqrt{2}\,2p_2 + 2p_3] \qquad E_1 = \alpha + \sqrt{2}\,\beta$$

$$\varphi_2 = \frac{1}{\sqrt{2}}(2p_1 - 2p_3) \qquad E_2 = \alpha$$

$$\varphi_3 = \frac{1}{2}[2p_1 - \sqrt{2}\,2p_2 + 2p_3] \qquad E_3 = \alpha - \sqrt{2}\,\beta$$

Graphical pictures of these HMO's can be represented in many ways and we have depicted the contour diagram of π-system only. Figure 4.60 shows the contour diagram of each π HMO. It is clear from Fig. 4.60 that the wave functions φ_1 and φ_3 are symmetric with respect to point C_2 whereas φ_2 is anti-symmetric. As expected the number of nodes in the function increases with

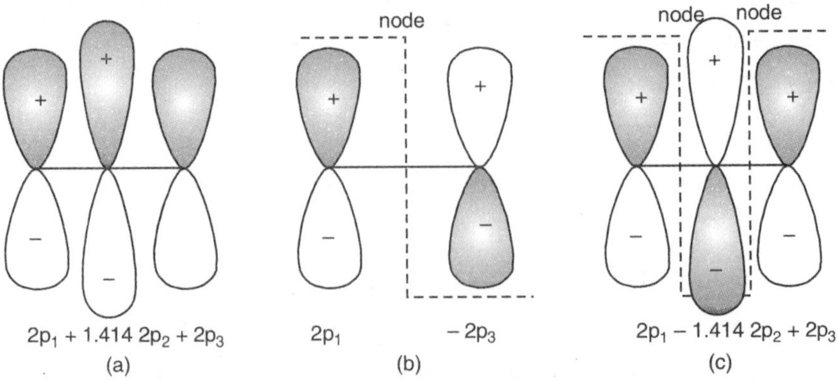

Figure 4.60: *Graphical representation of the wave functions of the allyl system. Size of the orbitals represents the contribution of each p_x-orbital. Dotted lines give the nodal points*

increase in energy, *i.e.,* zero in φ_1, one in φ_2 and two in φ_3. Figure 4.60*a* shows that the participation of p_x orbital on C_2 is 1.414 times larger than those of p_x orbital on C_1 and C_3 in φ_1 and φ_3, except that in φ_3, p_x (2) is having negative amplitude, whereas in φ_2, only contribution from p_x orbitals are found from C_1 and C_3 with opposite sign.

Butadiene: Butadiene molecule is more interesting as compared to the molecules discussed so far. The reason being that; (*i*) delocalization of π-electrons leads to formation of cis- and trans-isomers, suggesting that bond $C_2 - C_3$ has partial double bond character and (*ii*) knowledge of wave functions of different HMO's in the ground state and excited singlet state plays a major role in the Diels-Alder reaction. We will ignore the cis-trans isomers and instead take butadiene molecule as simply a linear sequence of four carbon atoms in *yz*-plane. Each carbon atom contributes a $2p_x$ orbtital to the π-electron orbital as below. In butadiene, HMO will be formed, by taking LCAO of four atomic orbitals. It will lead to 4 × 4 secular determinant, four different energy levels and four different HMO. The wave function for *j*th HMO can be given as:

$$\varphi_j = C_{j1}\,2p_1 + C_{j2}\,2p_2 + C_{j3}\,2p_3 + C_{j4}\,2p_4$$

Using the approximation of HMO method, the four simultaneous secular equations are:

$$(\alpha - E)\,C_{j1} + \beta\,C_{j2} = 0$$
$$\beta\,C_{j1} + (\alpha - E)\,C_{j2} + \beta\,C_{j3} = 0$$
$$\beta\,C_{j2} + (\alpha - E)\,C_{j3} + \beta\,C_{j4} = 0$$
$$\beta\,C_{j3} + (\alpha - E)\,C_{j4} = 0$$

and secular determinant

$$\begin{vmatrix} x & 1 & 0 & 0 \\ 1 & x & 1 & 0 \\ 0 & 1 & x & 1 \\ 0 & 0 & 1 & x \end{vmatrix} = 0$$

where $x = \dfrac{\alpha - E}{\beta}$. On expanding, this determinant becomes:

$$x^4 - 3x^2 + 1 = 0$$

On solving this equation for x^2, we get

$x^2 = \dfrac{3 \pm \sqrt{5}}{2}$. Therefore four roots of the above equation are:

$$x = \pm 1.618, \pm 0.618$$

Using the value of *x*, the energies of four HMO's are:

$$E_1 = \alpha + 1.618\,\beta$$
$$E_2 = \alpha + 0.618\,\beta$$
$$E_3 = \alpha - 0.618\,\beta$$
$$E_4 = \alpha - 1.618\,\beta$$

and are depicted in Fig. 4.61. As we know that α and β have negative sign, E_1 and E_2 are π bonding HMO's and E_3 and E_4 are π anti-bonding HMO's. Thus four π-electrons are accommodated in E_1

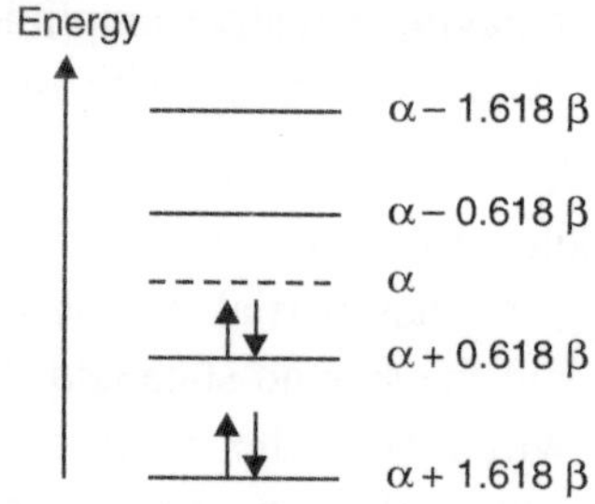

Figure 4.61: *Relative energy level diagram for π-electrons in butadiene molecule*

and E_2, with ground state electronic configuration as $\pi_1^2\ \pi_2^2$. Thus the total energy is:

$$E_\pi = 2(\alpha + 1.618\ \beta) + 2(\alpha + 0.618\ \beta) = 4\ \alpha + 4.472$$

and π-bond energy $= 4\alpha + 4.472\beta - 4\ \alpha = 4.472\ \beta$

The delocalization energy will be $= 4\alpha + 4.472\beta - 4(\alpha + \beta) = 0.472\beta$

Similar to allyl systems, in butadiene also, it is the π bond energy calculated with the help of HMO is smaller in magnitude than that of the butadiene structure based on localized bonding (*i.e.*, there are two π bonds between C_1 and C_2; C_3 and C_4 and a single bond between C_2 and C_3). In other words, the latter structure appears as if two ethylene molecules are joined to form butadiene. The stability that butadiene derives relative to two isolated doble bonds is due to the delocalization of π-electrons over the entire length of the molecule, rather than these are localized over two π bonds.

***Coefficients*:** Substituting the values of HMO energies one by one in the secular equations, we can calculate the coefficients of the HMO's. Starting with $E_1 = \alpha + 1.618\ \beta$, the secular equations give:

$$(\alpha - \alpha - 1.618\ \beta)\ C_{11} + \beta\ C_{12} = 0 \quad \Rightarrow \quad C_{12} = 1.618\ C_{11}$$

$$\beta\ C_{11} + (\alpha - \alpha - 1.618\ \beta)\ C_{12} + \beta\ C_{13} = 0$$

$$C_{13} = 1.618\ C_{12} - C_{11} = (1.618)\ (1.618)\ C_{11} - C_{11} = 1.618\ C_{11}$$

In other words $C_{12} = C_{13}$

Further,

$$\beta\ C_{13} + (\alpha - \alpha - 1.618\ \beta)\ C_{14} = 0 \quad \Rightarrow \quad C_{13} = 1.618\ C_{14}$$

$$C_{13} = C_{12} = 1.618\ C_{11} = 1.618\ C_{14}.$$

Hence $C_{11} = C_{14}$

Using the condition of normalization,

$$C_{11}^2 + C_{12}^2 + C_{13}^2 + C_{14}^2 = 1$$

gives $C_{11}^2 + (1.618)^2 C_{11}^2 + (1.618)^2 C_{11}^2 + C_{11}^2 = 1$

or $\qquad C_{11} = 0.372 = C_{14}$ and $C_{12} = C_{23} = 1.618 \times 0.372 = 0.602$

Thus, $\qquad \varphi_1 = 0.372\ 2p_1 + 0.602\ 2p_2 + 0.602\ 2p_3 + 0.372\ 2p_4$

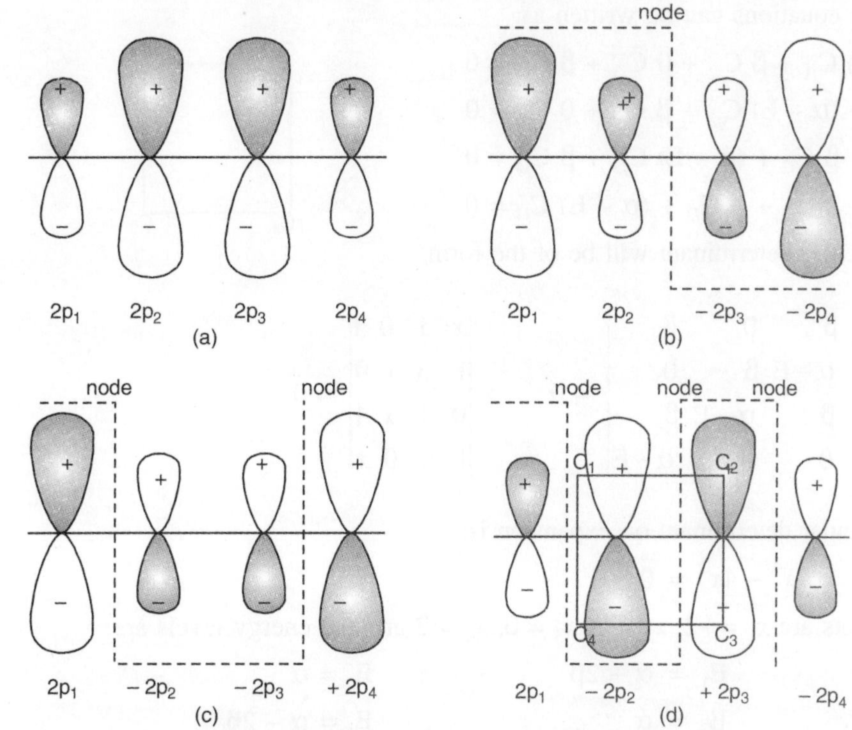

Figure 4.62: *Graphical representation of wave functions: (a) φ_1 with no nodes, (b) φ_2 with one node in between points 2 & 3, (c) φ_3 with two nodes in between points 1 & 2 and 3 & 4 and (d) φ_4 with three nodes in between points 1 & 2, 2 & 3 and 3 & 4*

Similarly using the other values of energies, E_2, E_3 and E_4, φ_2, φ_3 and φ_4 respectively can be evaluated. Four HMO's and the corresponding energies of butadiene are found to be:

$$\varphi_1 = 0.372\ 2p_1 + 0.602\ 2p_2 + 0.602\ 2p_3 + 0.372\ 2p_4 \qquad E_1 = \alpha + 1.618\ \beta$$

$$\varphi_2 = 0.602\ 2p_1 + 0.372\ 2p_2 - 0.372\ 2p_3 - 0.602\ 2p_4 \qquad E_2 = \alpha + 0.618\ \beta$$

$$\varphi_3 = 0.602\ 2p_1 - 0.372\ 2p_2 - 0.372\ 2p_3 + 0.602\ 2p_4 \qquad E_3 = \alpha - 0.618\ \beta$$

$$\varphi_4 = 0.372\ 2p_1 - 0.602\ 2p_2 + 0.602\ 2p_3 - 0.372\ 2p_4 \qquad E_4 = \alpha - 1.618\ \beta$$

Simlar to allyl system, the HMO's of butadiene can be presented in different manners. Figure 4.62 gives the contour diagram of the π-m.o.'s. One thing is obvious that the number of nodes increase with increase of π-energy. These are 0, 1, 2 and 3 for φ_1, φ_2, φ_3 and φ_4 respectively.

***Cyclobutadiene*:** We are discussing this molecule from two points. (*i*) Comparing the results of cyclobutadiene with linear butadiene will provide us the information regarding the relative stability of each system. (*ii*) Our aim is to carry out HMO calculations for the very important molecule, benzene, which is the starting point of aromatic systems. Comparing the results of cyclobutadiene and benzene will provide the information regarding the size of the ring and stability of the system.

Carbon skeleton of cyclobutadiene is given below and HMO's can be written as:

$$\varphi_j = C_{j1}\, 2p_1 + C_{j2}\, 2p_2 + C_{j3}\, 2p_3 + C_{j4}\, 2p_4$$

and secular equations can be written as:

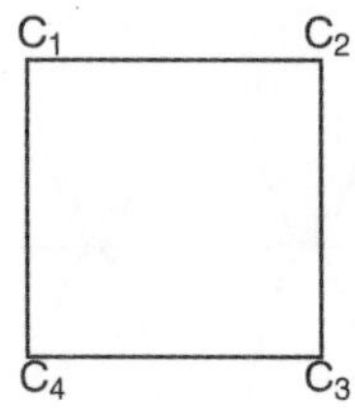

$$(\alpha - E)\, C_{j1} + \beta\, C_{j2} + 0\, C_{j3} + \beta\, C_{j4} = 0$$
$$\beta\, C_{j1} + (\alpha - E)\, C_{j2} + \beta\, C_{j3} + 0\, C_{j4} = 0$$
$$0\, C_{j1} + \beta\, C_{j2} + (\alpha - E)\, C_{j3} + \beta\, C_{j4} = 0$$
$$\beta\, C_{j1} + 0\, C_{j2} + \beta\, C_{j3} + (\alpha - E)\, C_{j4} = 0$$

The secular determinant will be of the form:

$$\begin{vmatrix} \alpha - E & \beta & 0 & \beta \\ \beta & \alpha - E & \beta & 0 \\ 0 & \beta & \alpha - E & \beta \\ \beta & 0 & \beta & \alpha - E \end{vmatrix} = 0 \Rightarrow \begin{vmatrix} x & 1 & 0 & 1 \\ 1 & x & 1 & 0 \\ 0 & 1 & x & 1 \\ 1 & 0 & 0 & x \end{vmatrix} = 0$$

The secular determinant on expansion is:

$$x^4 - 4x^2 = 0$$

The roots are $x_1 = -2$, $x_2 = 0$, $x_3 = 0$, $x_4 = 2$ and the energy levels are:

$$E_1 = \alpha + 2\beta \qquad E_2 = \alpha$$
$$E_3 = \alpha \qquad E_4 = \alpha - 2\beta$$

These energy levels are depicted in Fig. 4.63. On the similar grounds as discussed earlier,

E_1 is bonding, E_4 is an anti-bonding, E_2 and E_3 are non-bonding doubly degenerate. Four π-electrons in the ground state of cyclobutadiene are arranged as $\pi_1^2\, \pi_2\, \pi_3$. Total energy (E_π) is obtained as:

$$E_\pi = 2(\alpha + 2\beta) + \alpha + \alpha = 4\alpha + 4\beta$$
$$\pi\text{-bonding energy} = 4\alpha + 4\beta - 4\alpha = 4\beta$$
$$\text{Delocalization energy} = 4\alpha + 4\beta - 4\alpha - 4\beta = 0$$

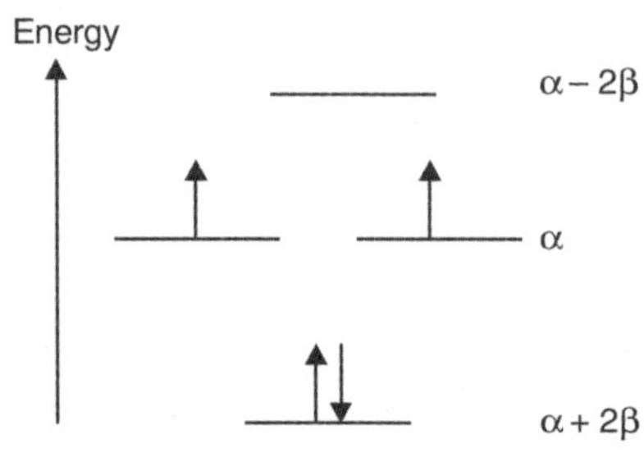

Figure 4.63: *Relative energy level diagram of the π m.o.'s of cyclobutadiene*

Calculation of localized energy is obtained by using the localized structure of cyclobutadiene as shown in Fig. 4.63. Zero value of delocalization energy suggests that there is no extra stability in considering that all the four π-electrons are delocalized or distributed over the complete molecule. This could be due to the small size of cyclobutadiene molecule. The strain present in the molecule is not overcome by the extra energy due to delocalization. Thus the molecule is unstable and difficult to synthesize.

Coefficients of HMO's: As mentioned earlier, coefficients of HMO's can be calculated by substituting the values of HMO's energies one by one in the secular equations. For example, starting with E_1, the secular equations become:

$$-2\beta\, C_{11} + \beta C_{12} + 0\, C_{13} + \beta C_{14} = 0 \qquad ...(4.269)$$

$$\beta C_{11} - 2\beta\, C_{12} + \beta C_{13} + 0 C_{14} = 0 \qquad ...(4.270)$$

$$0 C_{11} + \beta C_{12} - 2\beta\, C_{13} + \beta C_{14} = 0 \qquad ...(4.271)$$

$$\beta C_{11} + 0 C_{12} + \beta C_{13} - 2\beta\, C_{14} = 0 \qquad ...(4.272)$$

Since each equation contains three unknowns, one has to use the method of elimination. For example, multiplying equation (4.270) by 2 and add to equation (4.269), we get

$$-3\beta\, C_{12} + 2\beta\, C_{13} + \beta C_{14} = 0 \qquad ...(4.273)$$

Now add equation (4.271) to equation (4.273), one gets

$$-2\beta\, C_{12} + 2\beta\, C_{14} = 0 \quad \Rightarrow \quad C_{12} = C_{14}$$

Combining $C_{12} = C_{14}$, with equation (4.273), we get

$$-3\beta\, C_{12} + 2\beta\, C_{13} + \beta\, C_{12} = 0 \quad \Rightarrow \quad C_{12} = C_{13}$$

Using $C_{12} = C_{13} = C_{14}$ in equation (4.269), we obtain

$$-2\beta\, C_{11} + 2\beta\, C_{12} = 0 \qquad \Rightarrow \qquad C_{11} = C_{12}$$

i.e., $C_{11} = C_{12} = C_{13} = C_{14} = C_{11}$. Using normalization relation,

$$C_{11}^2 + C_{12}^2 + C_{13}^2 + C_{14}^2 = 1,$$

we get $4C_{11}^2 = 1 \Rightarrow C_{11} = C_{12} = C_{13} = C_{14} = \frac{1}{2}$ Thus, the HMO corresponding to $E_1 = \alpha + 2\beta$, is:

$$\varphi_1 = \frac{1}{2}\,[2p_1 + 2p_2 + 2p_3 + 2p_4]$$

Similarly, HMO's φ_2, φ_3 and φ_4 can be calculated using E_2, E_3 and E_4 respectively and these are:

$$\varphi_1 = \frac{1}{2}\,[2p_1 + 2p_2 + 2p_3 + 2p_4] \qquad E_1 = \alpha + 2\beta$$

$$\varphi_2 = \frac{1}{\sqrt{2}}\,[2p_1 - 2p_3] \qquad E_2 = \alpha$$

$$\varphi_3 = \frac{1}{\sqrt{2}} [2p_2 - 2p_4] \qquad E_3 = \alpha$$

$$\varphi_4 = \frac{1}{2} [2p_1 - 2p_2 + 2p_3 - 2p_4] \qquad E_4 = \alpha - 2\beta$$

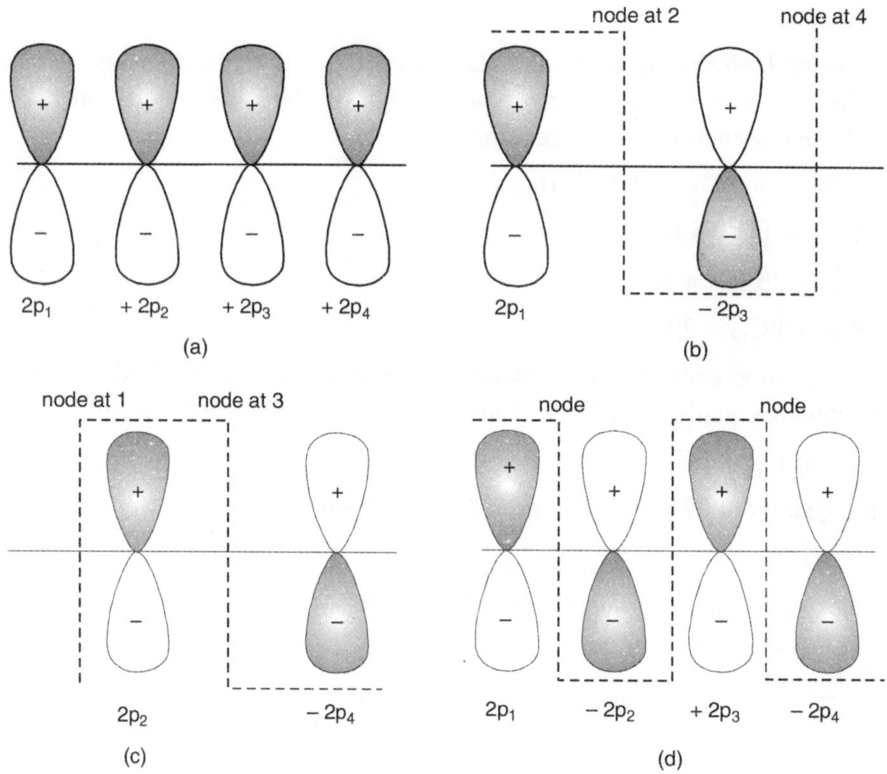

Figure 4.64: *Graphical representation of wave functions: (a) φ_1 with no nodes, (b) φ_2 with one node at points 2 & 4, (c) φ_3 with one node at points 1 & 3 & (d) φ_4 with two nodes at points in between 1 & 2 & 3 & 4*

Figure 4.64 gives the pictorial view of the HMO's of cyclobutadiene. As usual the number of nodes increases with increase in energy, no node in φ_1, one node in φ_2 and φ_3 at points 2 and 4; at points 1 and 3 respectively and two nodes in φ_4 present between the two adjacent carbon atoms.

***Benzene*:** The carbon skeleton of benzene molecule, which comprises of six-carbon atoms connencted to form cyclic structure, is given below. The HMO wave function will be a LCAO of six p_x orbitals of carbon atom as:

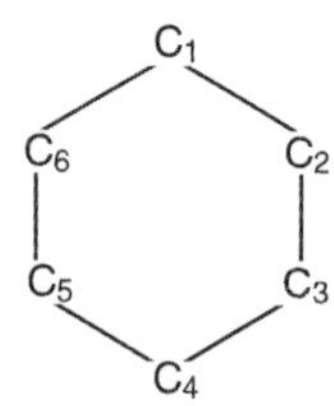

$$\varphi_j = C_{j1}\, 2p_1 + C_{j2}\, 2p_2 + C_{j3}\, 2p_3 + C_{j4}\, 2p_4 + C_{j5}\, 2p_5 + C_{j6}\, 2p_6$$

The secular equations and secular determinant will be:

$$(\alpha - E)\, C_{j1} + \beta\, C_{j2} + \beta\, C_{j6} = 0$$

$$\beta C_{j1} + (\alpha - E) C_{j2} + \beta C_{j3} = 0$$
$$\beta C_{j2} + (\alpha - E) C_{j3} + \beta C_{j4} = 0$$
$$\beta C_{j3} + (\alpha - E) C_{j4} + \beta C_{j5} = 0$$
$$\beta C_{j4} + (\alpha - E) C_{j5} + \beta C_{j6} = 0$$
$$\beta C_{j1} + \beta C_{j5} + (\alpha - E) C_{j6} = 0$$

and

$$\begin{vmatrix} \alpha - E & \beta & 0 & 0 & 0 & \beta \\ \beta & \alpha - E & \beta & 0 & 0 & 0 \\ 0 & \beta & \alpha - E & \beta & 0 & 0 \\ 0 & 0 & \beta & \alpha - E & \beta & 0 \\ \beta & 0 & 0 & 0 & \beta & \alpha - E \end{vmatrix} = 0$$

$$\begin{vmatrix} x & 1 & 0 & 0 & 0 & 1 \\ 1 & x & 1 & 0 & 0 & 0 \\ 0 & 1 & x & 1 & 0 & 0 \\ 0 & 0 & 1 & x & 1 & 0 \\ 1 & 0 & 0 & 0 & 1 & x \end{vmatrix} = 0$$

which on expansion gives an equation of degree 6.

$$x^6 - 6x^4 + 9x^2 - 4 = 0$$

This equation on factorization reduces to:

$$(x + 2)(x - 2)(x^2 - 1)^2 = 0$$

The values of six roots of this equation and the corresponding energy levels are given in the tabular form as under:

Root	No.	Energy
$x = -2$	1	$E_1 = \alpha + 2\beta$
$x = -1$	2	$E_2 = \alpha + \beta$
$x = -1$	3	$E_3 = \alpha + \beta$
$x = +1$	4	$E_4 = \alpha - \beta$
$x = +1$	5	$E_5 = \alpha - \beta$
$x = +2$	6	$E_6 = \alpha - 2\beta$

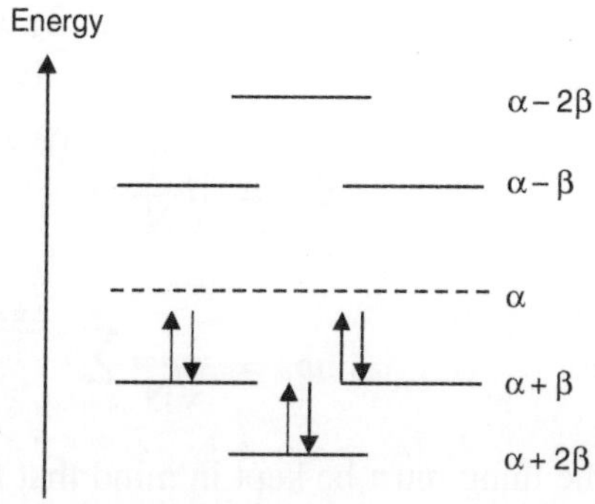

Figure 4.65: *Relative energies of the π m.o.'s of benzene molecule*

It is clear from the table that the first and sixth HMO are non-degenerate, whereas second, third, fourth and fifth are doubly degenerate. HMO's 1, 2 and 3 are π-bonding orbitals and 4, 5 and 6 are anti-bonding π-orbitals. The energy level diagram is depicted in Fig. 4.65. Six π-electrons are placed in three π-bonding HMO's and the ground state electronic configuration is $\pi_1^2\ \pi_2^2\ \pi_3^2$. The total π energy for the six electrons is:

$$E_\pi = 2(\alpha + 2\beta) + 4(\alpha + \beta) = 6\alpha + 8\beta$$

If six π-electrons were localized into three localized double bonds, then the π-electronic energy would be equal to the energy of three ethylene molecules. The π-electronic energy of three ethylene molecules is:

$$E_\pi = 3(2\alpha + 2\beta) = 6\alpha + 6\beta$$

Therefore the delocalization or resonance energy in benzene is: $6\alpha + 8\beta - 6\alpha - 6\beta = 2\beta$

Thus, HMO method predicts that benzene is stabilized by 2β.

***Coefficients of HMO's*:** We have observed in the previous applications that solution of secular determinant, as well as, that of secular equations to get the HMO coefficients is becoming tedious and time consuming with increase of complexity of the molecule. We can get the HMO coefficients for benzene molecule also by elimination method, as performed on cyclobutadiene, but it will be cumbersome. There are many other methods, but we will use the following equations (without going into the details of their derivations) to calculate the energies and their corresponding HMO's of linear and cyclic conjugated hydrocarbons.

For linear conjugated polyenes:

$$E_j = \alpha + 2\beta \cos \frac{j\pi}{N + 1} \qquad \text{...(4.274)}$$

where $j = 1, 2, 3, \ldots\ldots\ldots, N$ and N is the number of carbon atoms contributing π-electrons.

$$C_{jk} = \sqrt{\frac{2}{N + 1}} \sin \frac{jk\pi}{N + 1}, \; k = 1, 2, 3, \ldots\ldots\ldots, N \qquad \text{...(4.275)}$$

For cyclic conjugated polyenes:

$$E_j = \alpha + 2\beta \frac{2j\pi}{N}, \text{ where } j = 0, 1, 2, 3, \ldots\ldots\ldots, N - 1 \qquad \text{...(4.276)}$$

$$C_{jk} = \sqrt{\frac{1}{N}}\, e^{\frac{2\pi i j (k - 1)}{N}}, \; j = 0, 1, 2, 3, \ldots\ldots\ldots, N - 1 \qquad \text{...(4.277)}$$

$$\varphi_j = \frac{1}{\sqrt{N}} \sum_{k=1}^{N} e^{\frac{2\pi i j (k-1)}{N}}\, 2p_k \qquad \text{...(4.278)}$$

One thing must be kept in mind that the index k in these equations does not correspond to the actual order of the m.o.'s. The m.o. with lowest energy has $k = 0$, next one in energy are $k = 1$, and $k = N - 1$, next m.o.'s are with $k = 2$, and $k = N - 2$ and so on.

For linear conjugated polyenes, these equations will be tested on in examples 4.10, whereas the coefficients and m.o.'s of benzene molecule will be shown below.

Use equation (4.277) and take different values of j we can get

For first level $j = 0$,

$$C_{11} = \frac{1}{\sqrt{6}} e^{\frac{2\pi i.0\,(1-1)}{6}} = \frac{1}{\sqrt{6}} e^{0} = \frac{1}{\sqrt{6}}\,;\; C_{12} = \frac{1}{\sqrt{6}} e^{\frac{2\pi i.0\,(2-1)}{6}} = \frac{1}{\sqrt{6}} e^{0} = \frac{1}{\sqrt{6}}$$

$$C_{13} = \frac{1}{\sqrt{6}} e^{\frac{2\pi i.0\,(3-1)}{6}} = \frac{1}{\sqrt{6}} e^{0} = \frac{1}{\sqrt{6}}\,;\; C_{14} = \frac{1}{\sqrt{6}} e^{\frac{2\pi i.0\,(4-1)}{6}} = \frac{1}{\sqrt{6}} e^{0} = \frac{1}{\sqrt{6}}$$

$$C_{15} = \frac{1}{\sqrt{6}} e^{\frac{2\pi i.0\,(5-1)}{6}} = \frac{1}{\sqrt{6}} e^{0} = \frac{1}{\sqrt{6}}\,;\; C_{16} = \frac{1}{\sqrt{6}} e^{\frac{2\pi i.0\,(6-1)}{6}} = \frac{1}{\sqrt{6}} e^{0} = \frac{1}{\sqrt{6}}$$

For second level $j = 1$

$$C_{21} = \frac{1}{\sqrt{6}} e^{\frac{2\pi i.1\,(1-1)}{6}} = \frac{1}{\sqrt{6}} e^{0} = \frac{1}{\sqrt{6}}\,;\; C_{22} = \frac{1}{\sqrt{6}} e^{\frac{2\pi i.1\,(2-1)}{6}} = \frac{1}{\sqrt{6}} e^{\frac{\pi i}{3}}$$

$$C_{23} = \frac{1}{\sqrt{6}} e^{\frac{2\pi i.1\,(3-1)}{6}} = \frac{1}{\sqrt{6}} e^{\frac{2\pi i}{3}}\,;\; C_{24} = \frac{1}{\sqrt{6}} e^{\frac{2\pi i.1\,(4-1)}{6}} = \frac{1}{\sqrt{6}} e^{\frac{6\pi i}{6}} = -\frac{1}{\sqrt{6}}$$

$$C_{25} = \frac{1}{\sqrt{6}} e^{\frac{2\pi i.1\,(5-1)}{6}} = \frac{1}{\sqrt{6}} e^{\frac{4\pi i}{3}}\,;\; C_{26} = \frac{1}{\sqrt{6}} e^{\frac{2\pi i.1\,(6-1)}{6}} = \frac{1}{\sqrt{6}} e^{\frac{5\pi i}{3}}$$

For third level $j = 5$, but the coefficients can be written as:

$$C_{31} = \frac{1}{\sqrt{6}} e^{\frac{2\pi i.5\,(1-1)}{6}} = \frac{1}{\sqrt{6}} \mathrm{e}^{0} = \frac{1}{\sqrt{6}}$$

$$C_{32} = \frac{1}{\sqrt{6}} e^{\frac{2\pi i.5\,(2-1)}{6}} = \frac{1}{\sqrt{6}} e^{\frac{5\pi i}{3}} = \frac{1}{\sqrt{6}} e^{2\pi i} e^{-\frac{\pi i}{3}} = \frac{1}{\sqrt{6}} e^{-\frac{\pi i}{3}}$$

$$C_{33} = \frac{1}{\sqrt{6}} e^{\frac{2\pi i.5\,(3-1)}{6}} = \frac{1}{\sqrt{6}} e^{\frac{10\pi i}{3}} = \frac{1}{\sqrt{6}} e^{4\pi i} e^{-\frac{2\pi i}{3}} = \frac{1}{\sqrt{6}} e^{-\frac{2\pi i}{3}}$$

$$C_{34} = \frac{1}{\sqrt{6}} e^{\frac{2\pi i.5\,(4-1)}{6}} = \frac{1}{\sqrt{6}} e^{5\pi i} = -\frac{1}{\sqrt{6}}$$

$$C_{35} = \frac{1}{\sqrt{6}} e^{\frac{2\pi i.5\,(5-1)}{6}} = \frac{1}{\sqrt{6}} e^{\frac{20\pi i}{3}} = \frac{1}{\sqrt{6}} e^{8\pi i} e^{-\frac{4\pi i}{3}} = \frac{1}{\sqrt{6}} e^{-\frac{4\pi i}{3}}$$

$$C_{36} = \frac{1}{\sqrt{6}} e^{\frac{2\pi i.5\,(6-1)}{6}} = \frac{1}{\sqrt{6}} e^{\frac{25\pi i}{3}} = \frac{1}{\sqrt{6}} e^{10\pi i} e^{-\frac{5\pi i}{3}} = \frac{1}{\sqrt{6}} e^{-\frac{5\pi i}{3}}$$

For for fourth level $j = 2$ but the coefficients can be written as:

$$C_{41} = \frac{1}{\sqrt{6}} e^{\frac{2\pi i.2\,(1-1)}{6}} = \frac{1}{\sqrt{6}} e^{0} = \frac{1}{\sqrt{6}};\ C_{42} = \frac{1}{\sqrt{6}} e^{\frac{2\pi i.2\,(2-1)}{6}} = \frac{1}{\sqrt{6}} e^{\frac{2\pi i}{3}}$$

$$C_{43} = \frac{1}{\sqrt{6}} e^{\frac{2\pi i.2\,(3-1)}{6}} = \frac{1}{\sqrt{6}} e^{\frac{4\pi i}{3}};\ C_{44} = \frac{1}{\sqrt{6}} e^{\frac{2\pi i.2\,(4-1)}{6}} = \frac{1}{\sqrt{6}} e^{2\pi i} = \frac{1}{\sqrt{6}}$$

$$C_{45} = \frac{1}{\sqrt{6}} e^{\frac{2\pi i.2\,(5-1)}{6}} = \frac{1}{\sqrt{6}} e^{\frac{8\pi i}{3}} = \frac{1}{\sqrt{6}} e^{\frac{2\pi i}{3}};\ C_{46} = \frac{1}{\sqrt{6}} e^{\frac{2\pi i.2\,(6-1)}{6}} = \frac{1}{\sqrt{6}} e^{\frac{4\pi i}{3}}$$

For fifth level $j = 4$ and coefficients can be written as:

$$C_{51} = \frac{1}{\sqrt{6}} e^{\frac{2\pi i.4\,(1-1)}{6}} = \frac{1}{\sqrt{6}};\ C_{52} = \frac{1}{\sqrt{6}} e^{\frac{2\pi i.4\,(2-1)}{6}} = \frac{1}{\sqrt{6}} e^{2\pi i} e^{-\frac{2\pi i}{3}} = \frac{1}{\sqrt{6}} e^{-\frac{2\pi i}{3}}$$

$$C_{53} = \frac{1}{\sqrt{6}} e^{\frac{2\pi i.4\,(3-1)}{6}} = \frac{1}{\sqrt{6}} e^{4\pi i} e^{-\frac{4\pi i}{3}} = \frac{1}{\sqrt{6}} e^{-\frac{4\pi i}{3}}$$

$$C_{54} = \frac{1}{\sqrt{6}} e^{\frac{2\pi i.4\,(4-1)}{6}} = \frac{1}{\sqrt{6}} e^{4\pi i} = \frac{1}{\sqrt{6}}$$

$$C_{55} = \frac{1}{\sqrt{6}} e^{\frac{2\pi i.4\,(5-1)}{6}} = \frac{1}{\sqrt{6}} e^{6\pi i} e^{-\frac{2\pi i}{3}} = \frac{1}{\sqrt{6}} e^{-\frac{2\pi i}{3}}$$

$$C_{56} = \frac{1}{\sqrt{6}} e^{\frac{2\pi i.4\,(6-1)}{6}} = \frac{1}{\sqrt{6}} e^{8\pi i} e^{-\frac{4\pi i}{3}} = \frac{1}{\sqrt{6}} e^{-\frac{4\pi i}{3}}$$

For sixth level, $j = 3$

$$C_{61} = \frac{1}{\sqrt{6}} e^{\frac{2\pi i.3\,(1-1)}{6}} = \frac{1}{\sqrt{6}} e^{0} = \frac{1}{\sqrt{6}};\ C_{62} = \frac{1}{\sqrt{6}} e^{\frac{2\pi i.3\,(2-1)}{6}} = \frac{1}{\sqrt{6}} e^{i\pi} = -\frac{1}{\sqrt{6}}$$

$$C_{63} = \frac{1}{\sqrt{6}} e^{\frac{2\pi i.3\,(3-1)}{6}} = \frac{1}{\sqrt{6}} e^{2\pi i} = \frac{1}{\sqrt{6}};\ C_{64} = \frac{1}{\sqrt{6}} e^{\frac{2\pi i.3\,(4-1)}{6}} = \frac{1}{\sqrt{6}} e^{3\pi i} = -\frac{1}{\sqrt{6}}$$

$$C_{65} = \frac{1}{\sqrt{6}} e^{\frac{2\pi i.3\,(5-1)}{6}} = \frac{1}{\sqrt{6}} e^{4\pi i} = \frac{1}{\sqrt{6}};\ C_{66} = \frac{1}{\sqrt{6}} e^{\frac{2\pi i.3\,(6-1)}{6}} = \frac{1}{\sqrt{6}} e^{5\pi i} = -\frac{1}{\sqrt{6}}$$

Thus the Hückel m.o.'s can be written as:

$$\varphi_1 = \frac{1}{\sqrt{6}} \left[2p_1 + 2p_2 + 2p_3 + 2p_4 + 2p_5 + 2p_6\right]$$

$$\varphi_2 = \frac{1}{\sqrt{6}} \left[2p_1 + e^{\frac{\pi i}{3}}\, 2p_2 + e^{\frac{2\pi i}{3}}\, 2p_3 - 2p_4 + e^{\frac{4\pi i}{3}}\, 2p_5 + e^{\frac{5\pi i}{3}}\, 2p_6\right]$$

$$\varphi_3 = \frac{1}{\sqrt{6}} \left[2p_1 + e^{-\frac{\pi i}{3}}\, 2p_2 + e^{-\frac{2\pi i}{3}}\, 2p_3 - 2p_4 + e^{-\frac{4\pi i}{3}}\, 2p_5 + e^{-\frac{5\pi i}{3}}\, 2p_6\right] \quad ...(4.279)$$

$$\varphi_4 = \frac{1}{\sqrt{6}} \left[2p_1 + e^{\frac{2\pi i}{3}}\, 2p_2 + e^{\frac{4\pi i}{3}}\, 2p_3 + 2p_4 + e^{\frac{2\pi i}{3}}\, 2p_5 + e^{\frac{4\pi i}{3}}\, 2p_6\right]$$

$$\varphi_5 = \frac{1}{\sqrt{6}} \left[2p_1 + e^{-\frac{2\pi i}{3}}\, 2p_2 + e^{-\frac{4\pi i}{3}}\, 2p_3 + 2p_4 + e^{-\frac{2\pi i}{3}}\, 2p_5 + e^{-\frac{4\pi i}{3}}\, 2p_6\right]$$

$$\varphi_6 = \frac{1}{\sqrt{6}} \left[2p_1 - 2p_2 + 2p_3 - 2p_4 + 2p_5 - 2p_6\right]$$

The degenerate Hückel m.o.'s depicted in equation (4.279) are of complex in nature. Since the two m.o.'s of each degenerate levels are complex conjugate of each other, the real forms of these m.o.'s can be obtained by taking the plus and minus combinations of the complex ones. The real form of these m.o.'s are:

$$\varphi_{2,\text{real}} = \frac{1}{\sqrt{2}} [\varphi_2 + \varphi_3], \; \varphi_{3,\text{real}} = -\, i\, \frac{1}{\sqrt{2}} [\varphi_2 - \varphi_3]$$

$$\varphi_{2,\text{real}} = \frac{1}{\sqrt{12}} [2\; 2p_1 + 2p_2 - 2p_3 - 2\; 2p_4 - 2p_5 + 2p_6]$$

$$\varphi_{3,\text{real}} = \frac{1}{2} [2p_2 + 2p_3 - 2p_5 - 2p_6]$$

Similarly taking the linear combination of the 4th and 5th orbitals, we can form the real form of these m.o.'s and are given below.

$$\varphi_{4,\text{real}} = \frac{1}{\sqrt{12}} [2\; 2p_1 - 2p_2 - 2p_3 + 2\; 2p_4 - 2p_5 - 2p_6]$$

$$\varphi_{5,\text{real}} = \frac{1}{2} [2p_2 - 2p_3 + 2p_5 - 2p_6] \; \varphi\varphi$$

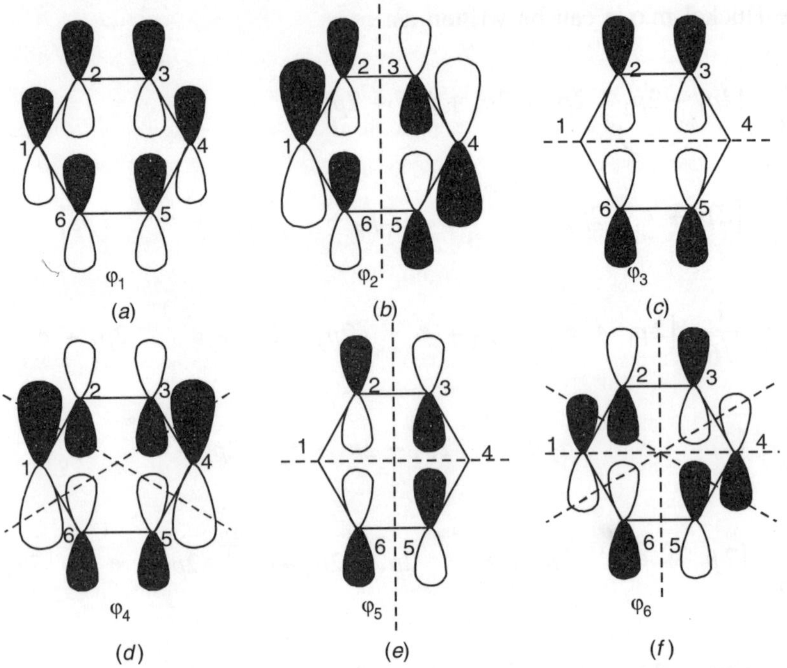

Figure 4.66: *Graphical representation of wave functions Hückel molecular orbitals of benzene molecule: (a) φ_1 with no nodes, (b) φ_2 with one nodal plane passing through bonds 2-3 and 5-6, (c) φ_3 with one nodal plane passing through bonds points 1 and 4, (d) φ_4 with two nodal planes passing through bonds 1-2 and 4-5 and bonds 3-4 and 5-6, (e) φ_5 with two nodal planes passing through bonds 1 and 4 and bonds 2-3 and 5-6 and (f) φ_6 with three nodal planes passing through bonds 1-2 and 4-5, 2-3 and 5-6 and 3-4 and 6-1. Dark lobes depict the positive sign and light lobes negative signs*

It may be pointed out here that the equation (4.277), used here to calculate the coefficients, is the solution of the secular determinant based on symmetry considerations without using the Hückel approximation. Thus the m.o.'s depicted in equation (4.279), leaving aside the normalization constants, are the minimal-basis-set SCF π-electrons m.o.'s for benzene. On the other hand the Hückel energies obtained do not depict the true SCF orbital energies because Hückel approximation does not take into account the electron-electron repulsions and the total π-electron energy is obtained by taking the sum of orbital energies. As mentioned in the chapter 3, the SCF method does take into account the electron repulsions in an average manner and the total SCF energy is not the sum of the orbital energies. Figure 4.66 shows the real form of the π-orbitals of benzene.

π-Electron Charges

According to the first principle of quantum mechanics, the probability of locating the electron in a small volume element $d\tau$ can be obtained by the relation $\psi^* \psi \, d\tau$, where ψ^* and ψ have their usual meanings. In case of HMO, the π-electron wave function is given as:

$$\varphi_j = C_{j1}\, 2p_1 + C_{j2}\, 2p_2 + C_{j3}\, 2p_3 + \ldots\ldots\ldots = \sum_{k=1} C_{jk}\, 2p_k \qquad \ldots(4.250)$$

Thus, total probability of locating the electron will be:

$$\int \varphi_j^* \varphi_j \, d\tau = \int \sum C_{jk}^* \, 2p_k \sum C_{jk} \, 2p_k \, d\tau$$

$$= \int \sum C_{jk}^2 \, 2p_k^2 \, d\tau + \sum_k \sum_{k \neq l} C_{jk} C_{kl} \int 2p_k \, 2p_l d\tau \qquad ...(4.280)$$

In case of HMO, we have assumed that:

$\int 2p_k \, 2p_l \, d\tau = 0$ if $k \neq l$ and $\int 2p_k \, 2p_k \, d\tau = 1$, since $2p_k$ orbitals are normalized. Thus the expression (4.280) reduces to:

$$\int \varphi_j^* \varphi_j \, d\tau = \sum_{k=1} C_{jk}^2 = 1$$

$$= C_{j1}^2 + C_{j2}^2 + C_{j3}^2 + \cdots\cdots\cdots = 1 \qquad ...(4.281)$$

as φ_j is the normalized π-electron HMO wave function. Since the relation (4.281) is obeyed by the HMO's coefficients, it is quite natural to say that the coefficients C_{jk}^2 give the probability of an electron near the kth carbon atom in the jth HMO. Since the number of electrons (n_j) present in jth π HMO (φ_j) can be either 0, 1 or 2, the total π-electron density q_k in the region of carbon atom k is defined as:

$$q_k = \sum n_j \, C_{jk}^2 \qquad ...(4.282)$$

It may be made clear that the name charge density for q_k is a misnomer, because the dimensions for q_k are neither statcol cm^{-3} nor electrons cm^{-3}. q_k is basically a pure number and gives the approximate number of π-electrons close to carbon atom k. In a π system neutral atom is associated with an electron density of one, and net charge density is defined as

$$\xi_k = 1 - q_k \qquad ...(4.283)$$

π-Electron bond order (p_{kl}): It is also called mobile bond order. Consider carbon atoms j and k, which are bonded to each other in a planar conjugated molecule. Then the quantity $C_{jk} \, C_{kl}$ gives the π-electron probability density in a bond region between atoms k and l. p_{kl} is defined as:

$$p_{kl} = \sum_j n_j \frac{1}{2}(C_{jk}^* C_{jl} + C_{jl}^* C_{jk}) \qquad ...(4.284)$$

where the sum is over the HMO's. Further it may be mentioned here that each π HMO (φ_j) contributes a real quantity to the bond order p_{kl}. If the σ-electrons of a single bond are added to p_{kl}, then the total bond order (ρ_{kl}) is given as:

$$\rho_{kl} = 1 + p_{kl} \qquad ...(4.285)$$

Bond length (R_{kl}): We have learnt earlier that there is some correlation between the bond length and bond order (*i.e.,* greater the bond order, smaller the bond length). A similar correlation is also expected naturally between bond order and bond length in case of HMO method. The simplest relation is to assume as a linear relation:

$$R_{kl} = a + b \, p_{kl} \qquad ...(4.286)$$

where a and b are constants. In case of linear conjugated molecules, the values of a, and b are obtained by using the bond order and bond length of ethylene molecule, whereas in case of cyclic conjugated systems, the similar data for benzene are used. Thus, the empirical relations for linear conjugated systems and cyclic systems are equations (4.287) and (4.288) respectively.

$$R_{kl} = 1.707 - 0.186\, P_{kl} \qquad ...(4.287)$$

$$R_{kl} = 1.521 - 0.186\, P_{kl} \qquad ...(4.288)$$

Example 4.10: (*a*) Using equation (4.275), calculate the coefficients of all the four π HMO's of butadiene and compare them with those determined otherwise. (*b*). Calculate the bond order and bond lengths for the bond 1-2, 2-3 and 3-4 for butadiene.

Solution:

(*a*) Using equation (4.275), *i.e.*, $C_{jk} = \sqrt{\dfrac{2}{N+1}}\sin\dfrac{j\,k\,\pi}{N+1}$

For HMO 1, $j = 1$

$$C_{11} = \sqrt{\frac{2}{5}}\sin\frac{1\times 1\pi}{5} = 0.632 \times 0.588 = 0.372$$

$$C_{12} = \sqrt{\frac{2}{5}}\sin\frac{1\times 2\pi}{5} = 0.632 \times 0.951 = 0.602$$

$$C_{13} = \sqrt{\frac{2}{5}}\sin\frac{1\times 3\pi}{5} = 0.632 \times 0.951 = 0.602$$

$$C_{14} = \sqrt{\frac{2}{5}}\sin\frac{1\times 4\pi}{5} = 0.632 \times 0.588 = 0.372$$

For HMO 2, $j = 2$

$$C_{21} = \sqrt{\frac{2}{5}}\sin\frac{2\times 1\pi}{5} = 0.602$$

$$C_{22} = \sqrt{\frac{2}{5}}\sin\frac{2\times 2\pi}{5} = 0.372$$

$$C_{23} = \sqrt{\frac{2}{5}}\sin\frac{2\times 3\pi}{5} = -\,0.372$$

$$C_{24} = \sqrt{\frac{2}{5}}\sin\frac{2\times 4\pi}{5} = -\,0.602$$

For HMO 3, $j = 3$

$$C_{31} = \sqrt{\frac{2}{5}}\sin\frac{3\times 1\pi}{5} = 0.602$$

$$C_{32} = \sqrt{\frac{2}{5}} \sin\frac{3 \times 2\pi}{5} = -0.372;$$

$$C_{33} = \sqrt{\frac{2}{5}} \sin\frac{3 \times 3\pi}{5} = -0.372;$$

$$C_{34} = \sqrt{\frac{2}{5}} \sin\frac{3 \times 4\pi}{5} = 0.602$$

For HMO 4, $j = 4$

$$C_{41} = \sqrt{\frac{2}{5}} \sin\frac{4 \times 1\pi}{5} = 0.372$$

$$C_{42} = \sqrt{\frac{2}{5}} \sin\frac{4 \times 2\pi}{5} = -0.602$$

$$C_{43} = \sqrt{\frac{2}{5}} \sin\frac{4 \times 3\pi}{5} = 0.602$$

$$C_{44} = \sqrt{\frac{2}{5}} \sin\frac{4 \times 4\pi}{5} = -0.372$$

These coefficients are similar to those obtained in earlier section.

(*b*) π-bond order can be calculated using equation (4.284)

$$p_{kl} = \sum_j n_j \frac{1}{2}(C^*_{jk}C_{jl} + C^*_{jl}C_{jk})$$

$$\begin{aligned} p_{12} &= 2/2\,(0.372 \times 0.602 + 0.602 \times 0.372) + 2/2\,(0.602 \times 0.372 \\ &\quad + 0.372 \times 0.602) + 0/2\,[(0.602) \times (-0.372) + (-0.372)(0.602)] \\ &\quad + 0/2[(0.372) \times (-0.602) + (-0.602 \times 0.372)] \\ &= (0.224 + 0.224) + (0.224 + 0.224) = 0.896 \\ p_{23} &= 2/2[0.602 \times 0.602 + 0.602 \times 0.602] + 2/2\,[(0.372) \times (-0.372) \\ &\quad + (-0.372)(0.372)] + 0/2\,[.....] + 0/2\,[.......] \\ &= (0.362 + 0.362) - (0.138 + 0.138) = 0.448 \\ p_{34} &= 2/2\,[0.602 \times 0.372 + 0.372 \times 0.602] + 2/2\,[(-0.372) \times (-0.602) \\ &\quad + (-0.602) \times (-0.372)] + 0/2\,[.....] + 0/2\,[......] \\ &= (0.224 + 0.224) + (0.224 + 0.224) = 0.896 \end{aligned}$$

The above calculations have shown that the π bond order between the terminal carbon atoms ($C_1 - C_2$) and ($C_3 - C_4$) is the same and is larger than the middle bond ($C_2 - C_3$). The total bond order between the bonds, 1-2, 2-3 and 3-4 are:

$$\rho_{12} = \rho_{34} = 1.0 + 0.896 = 1.896$$
$$\rho_{23} = 1.0 + 0.448 = 1.448$$

Similarly, the bond lengths between different bonds can be obtained using equations (4.287 and 4.288).

$$R_{12} = R_{34} = 1.707 - 0.186 \times 1.896 = 1.354\ \text{Å} = 0.1354\ \text{nm}$$
$$R_{23} = 1.707 - 0.186 \times 1.448 = 1.438\ \text{Å} = 0.1438\ \text{nm}$$

The solution of example 4.10 has given us very useful and beautiful results:

(*a*) Total bond orders between 1-2 and 3-4 are equal to each other, whereas between 2-3 is shorter than both 1-2 and 3-4. This suggests that the middle bond is tightly bound than the terminal ones.

(*b*) Total bond orders between each bond is either less than two or greater than one. This means that none of the bond is either a full double bond or full single bond. All of them

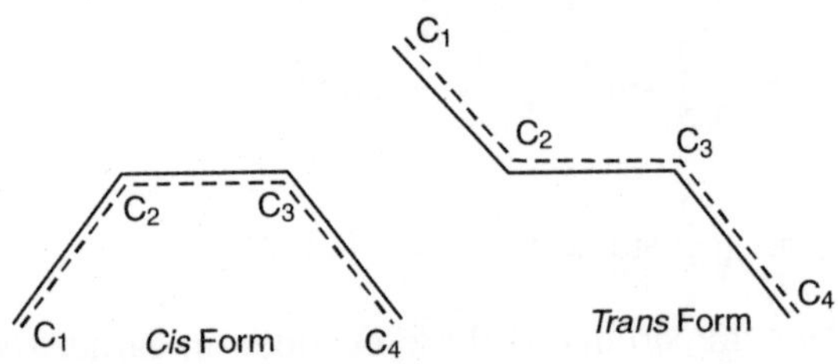

have partial double bond character. This means that cis-trans isomerism can be observed in butadiene, as shown above. Experimentally both the forms have been synthesized in the laboratory.

(*c*) Correlation between the bond order and bond length suggests that the bonds 1-2 and 3-4 will be shorter than 2-3. The presence of *cis*- and *trans*-butadiene has been observed experimentally and thus agrees with above calculations.

Example 4.11: Calculate the bond orders and bond lengths of all the six bonds in benzene molecule.

Solution: Bond order can be calculated using equation 4.284 and real coefficients of each m.o. obtained earlier.

$$p_{12} = \frac{2}{2}\,[C_{11}\,C_{12} + C_{12}\,C_{11}] + \frac{2}{2}\,[C_{21}\,C_{22} + C_{22}\,C_{21}]$$
$$+ \frac{2}{2}\,[C_{31}\,C_{32} + C_{32}\,C_{31}]$$

In the summation the last three terms have been neglected as there will not be any electron in orbitals 4, 5 and 6. Substituting the values of the coefficients, we get

$$p_{12} = \left[\frac{1}{\sqrt{6}}\frac{1}{\sqrt{6}} + \frac{1}{\sqrt{6}}\frac{1}{\sqrt{6}}\right] + \left[\frac{2}{\sqrt{12}}\frac{1}{\sqrt{12}} + \frac{1}{\sqrt{12}}\frac{2}{\sqrt{12}}\right] + \left[0\,\frac{1}{\sqrt{6}} + \frac{1}{\sqrt{6}}\,0\right]$$

$$= \frac{1}{6} + \frac{1}{6} + \frac{2}{12} + \frac{2}{12} = \frac{8}{12} = 0.66$$

$$p_{23} = \frac{2}{2} [C_{12}\, C_{13} + C_{13}\, C_{12}] + \frac{2}{2} [C_{22}\, C_{23} + C_{23}\, C_{22}] + \frac{2}{2} [C_{32}\, C_{33} + C_{33}\, C_{32}]$$

$$= \left[\frac{1}{\sqrt{6}}\frac{1}{\sqrt{6}} + \frac{1}{\sqrt{6}}\frac{1}{\sqrt{6}}\right] + \left[\frac{1}{\sqrt{12}}\left(\frac{1}{\sqrt{12}}\right) + \left(\frac{1}{\sqrt{12}}\right)\frac{1}{\sqrt{12}}\right] + \left[\frac{1}{2}\frac{1}{2} + \frac{1}{2}\frac{1}{2}\right]$$

$$= \frac{1}{6} + \frac{1}{6} - \frac{1}{12} - \frac{1}{12} + \frac{1}{4} + \frac{1}{4} = \frac{8}{12} = 0.66$$

$$p_{34} = \frac{2}{2} [C_{13}\, C_{14} + C_{14}\, C_{13}] + \frac{2}{2} [C_{23}\, C_{24} + C_{24}\, C_{23}] + \frac{2}{2} [C_{33}\, C_{34} + C_{34}\, C_{33}]$$

$$= \left[\frac{1}{\sqrt{6}}\frac{1}{\sqrt{6}} + \frac{1}{\sqrt{6}}\frac{1}{\sqrt{6}}\right] + \left[\left(-\frac{1}{\sqrt{12}}\right)\left(-\frac{1}{\sqrt{12}}\right) + \left(-\frac{1}{\sqrt{12}}\right)\left(-\frac{1}{\sqrt{12}}\right)\right]$$

$$+ \left[0\,\frac{1}{2} + \frac{1}{2}\,0\right]$$

$$= \frac{1}{6} + \frac{1}{6} + \frac{2}{12} + \frac{2}{12} = \frac{8}{12} = 0.66$$

Similarly we also get the value of p_{45}, p_{56} and p_{61} and all are equal to 0.666. In other words all the values of p_{kl} are equal. Using equation (4.285), we can calculate the total bond order:

$$\rho_{kl} = 1 + p_{kl} = 1.0 + 0.66 = 1.666$$

It shows that bond order of each bond in benzene is the same.

The bond length of each bond will also be equal and can be calculated by using equation (4.288), *i.e.*,

$$R_{kl} = 1.521 - 0.186\, p_{kl} = 1.521 - 0.186 \times 1.66$$

$$= 1.521 - 0.309 = 1.21 \text{ Å.} = 0.121 \text{ nm.}$$

The results obtained in examples 4.10 and 4.11 are in excellent agreement with the experimental observations involving the reactivity of these bonds in butadiene and so on. Hückel m.o.t. has been applied to a large number of organic molecules with some success. Further refinements to simple Huckel m.o.t have also been made and extended in a number of ways so that it is now a useful tool and guide in organic chemistry.

4.21 BONDING IN TRANSITION METALS

Interest in the synthetic and understanding of the transition elements chemistry started since the pioneer work of Werner in 1893. Even after one century, the interest in understanding the chemistry of these elements is as alive as in 1893 and poses a very challenging problems for chemists, because the compounds of these elements have and are finding lot of applications as catalysts in the laboratory as well as in the industry.

As the heading conveys, this class of molecules represents the interaction of transition metals or transition metal ions with anions or neutral molecules, possessing lone pairs of electrons. In other words the participation of *d* orbitals is involved in the binding of these molecules. In this section brief description of each theory will be given starting from Werner.

Werner's understanding regarding, the structural aspects of transition metal chemistry is totally based on synthetic aspects of these complexes because at that time even the electron was not discovered by Thompson. Werner assigned two kinds of bonding in the coordination complexes.

(*i*) ***Primary valences*:** These are non-directional. In the light of modern theories, these can be explained as follows. The complexes can be both positive and negative ions. The primary valence is the number of charges on the complexes. Thus, in the compounds, these charges are counter-balanced by the ions of oppositely charged ions. For example, in $K_4[Fe(CN)_6]$, the primary valence on $[Fe(CN)_6]^{4-}$ is four negative, in $[Co(NH_3)_6]Cl_2$ the primary valence on $[Co(NH_3)_6]^{2+}$ is two positive. These valences are similar to those found in ionic compounds.

(*ii*) ***Secondary valences*:** Secondary valence is the number of ligands surrounding the central metal atom or metal ion. This is also called as the *coordination number* of metal. Since the ligands around the metal atom can be arranged only in certain manner in space, these valences are having directional properties. The most common coordination numbers are 4 and 6. The former forms tetrahedral or square planar complexes and the latter forms octahedral complexes, depending on the nature of the ligand. As mentioned earlier the ligands can either be anions (halide ions) or neutral molecules possessing the lone pair of electrons ($\ddot{N}H_3$, $H_2\ddot{O}$ ....). For example, in $K_4[Fe(CN)_6]$, the secondary valence of Fe(II) is six, arranged in octahedral manner.

As mentioned above, the secondary valences have directional properties and these ligands can be arranged in space around the transition element. Werner made these assignments by preparing the number of isomers and then comparing them with the theoretically possible isomers those can exist in the particular arrangements of ligands in space. For example, consider the complex $[Co(NH_3)_4Cl_2]Cl$. There are six equivalent secondary valences and these can be arranged in space as shown in Fig. 4.67. The number of possible isomers can be prepared based on each structure and actually prepared in laboratory are compiled in Table 4.9 . The results given clearly indicate that the best way of arranging six ligands around the transition elements is octahedral. Although the recent studies, using *x*-ray crystallography have confirmed that the octahedral symmetry is correct one in the above case, but Werner at that time had no concrete proof. This may be because it might have been possible that the synthetic techniques were not so developed properly to prepare all the possible isomers. Based on this procedure Werner was able to assign different geometries to coordinate complexes, but the final proof was lacking. In the coming sections, more refined theories will be discussed to have a better picture regarding the geometries and bonding in the transition metal complexes.

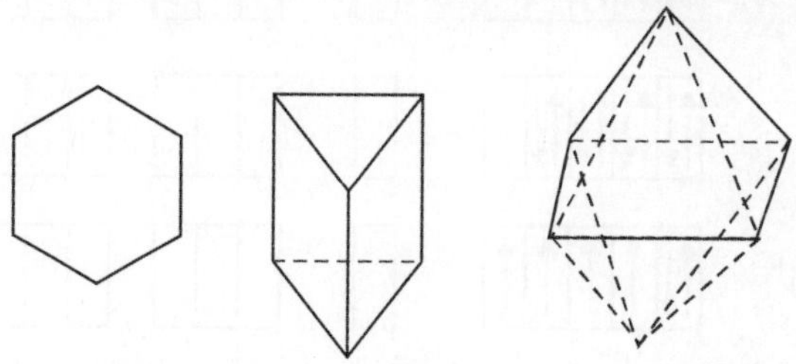

Figure 4.67: *Octahedral geometry for MX_6 complex*

Table 4.9: *Number isomers predicted and actually prepared*

Complex	Predicted planar hexagom	Trigonal prism	Octahedral	Actually prepared
MX_6	1	1	1	1
MX_5Y	1	1	1	1
MX_4Y_2	3	3	2	2
MX_3Y_3	3	3	2	2

Valence Bond Theory (v.b.t.)

Linus Pauling was the first to apply the v.b.t. to explain the structure and nature of bonding involved in the coordination complexes. It is also commonly known as the v.b.t. of coordination compounds. In this theory, it is assumed that coordinate covalent or a dative bond is formed between the transition metal or its ion and the ligand. The ligand acts as a Lewis base (*i.e.,* it supplies the pair of electrons) and metal or metal ion acts as Lewis acid (*i.e.,* it accepts the pair of electrons). Further the basic concept in this theory is very closely related to the idea of hybridization and geometry of the main group elements. For example, as mentioned in section 4.19, the hybridization at the carbon atoms in methane and ethane can be predicted only if we know the structure of these molecules. Similarly using very simple concept of v.b.t. it is also necessary that we should know the structure of the complex. If we want to know the precise nature of the atomic orbitals involved in the hybridization (*i.e.,* $(n-1)d^a\ ns\ np^b$ or $ns\ np^b\ nd^a$), we should know also the value of magnetic moment of the complex. To understand the above concepts, consider the two complexes of Fe(II), $K_4[Fe(CN)_6]$ and $K_4[FeF_6]$. In both cases, the Fe is in two plus state and is surrouned by six ligands. The results of the following Scheme, show that electrons present in the *d* orbitals are 6. The arrangement of electrons in the metal ion can be represented as below:

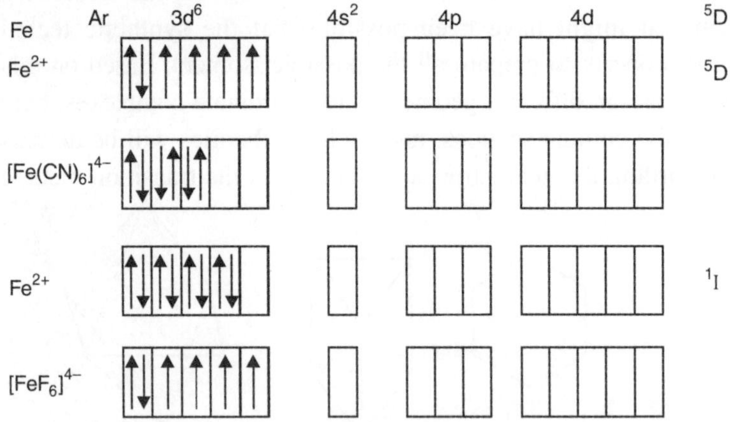

From the molecular formula of the individual anion, it is obvious that each ligand is present at the corner of the octahedron. Six ligands connected to the central Fe^{2+} ion to have an octahedron arrangement, one needs $3d^2 4s4p^3$ or $4s4p^3 4d^2$ hybridization at the central ion. Selection between the two arrangements can be made only if we know the magnetic moment (*i.e.,* qualitatively whether the complex is paramagnetic or diamagnetic). The hybridization involving 3*d* orbitals will lead to the diamagnetic species, whereas involvement of 4*d* orbitals in hybridization will give rise to paramagnetic species. Since $[Fe(CN)_6]^{4-}$ is a diamagnetic and $[FeF_6]^{4-}$ is a paramagnetic, the former will involve 3*d* orbitals and the latter 4*d* orbitals. This is also known as using inner *d* orbitals or using outer *d* orbitals. From the above discussion one can make the following conclusion. *Octahedral complexes having central atom or metal ion with* d^6 *electronic configuration will involve inner d orbitals if these are diamagnetic or outer d orbitals if these are paramagnetic*. Examples are: $[Co(NH_3)_6]^{3+}$ and $[CoF_6]^{3-}$; Co(III) is a d^6 system. The former is a diamagnetic and the latter is paramagnetic, with magnetic moment equivalent to four unpaired electrons.

More complications arise with d^8 systems when the number of ligands attached to central metal atom ion is four. The metal ions can be Ni^{2+}, Pd^{2+} and Pt^{2+}. In general the complexes of Pd^{2+} and Pt^{2+} are four coordinated and are diamagnetic in nature. Similar magnetic characteristic is also found for Ni^{2+}. Based on this arrangement of electrons in various orbitals of neutral metal, metal cation hybridized state can be written as follows: In few cases (only limited to Ni^{2+}) paramagnetic 4-centred complexes of Ni^{2+} are known.

To explain the paramagnetic behaviour of these complexes, it is essential that the inner *d* orbitals should be occupied by d^8 electrons of Ni^{2+}. Thus, the four ligands must be connected to Ni^{2+} forming tetrahedral arrangement, involving sp^3 hybridization.

From the above discussion it is possible to predict the involvement of inner *d* or outer *d* orbitals in case of octahedral complexes or geometry of the complexes in case of four cornered complexes, if the magnetic moment (susceptibility) of the complex is known. Thus, four-cornered complex will be tetrahedral if the complex is paramagnetic and square planar if the complex is diamagnetic. As mentioned earlier this is the weakest point in the simplified picture of v.b.t when

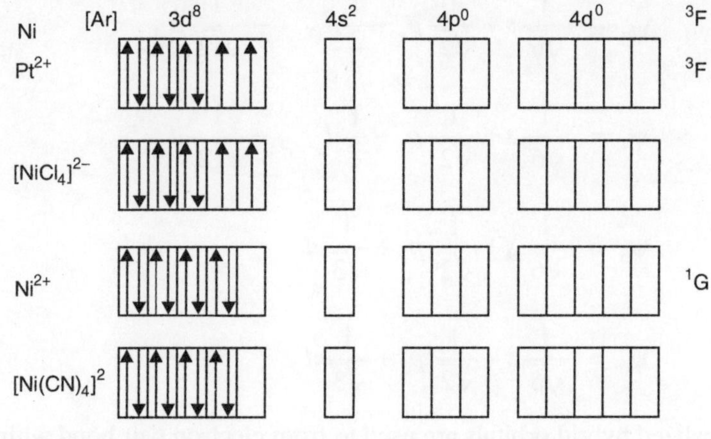

it is used to explain the bonding in the coordinate complexes. In order to interpret the experimental data observed in coordinate complexes, better theories were developed.

Using simplified version of v.b.t (*i.e.,* localized bonding), the pictorial representation of the bonding in octahedral complexes, $[Fe(CN)_6]^{4+}$ is shown in Fig. 4.68. As mentioned earlier the six hybrid orbitals in the octahedral geometry can be constructed by taking the linear combination of, d_{z^2}, $d_{x^2-y^2}$ (neglecting for time being whether these are inner *d* orbitals or outer *d* orbitals), *s*, p_x, p_y and p_z atomic orbitals of the central metal ion. The wave functions are so obtained that the larger lobes of these hybrid atomic orbitals project towards the direction of ligand. Using the ideas developed in section 4.19 the wave functions of the hybrid orbitals are:

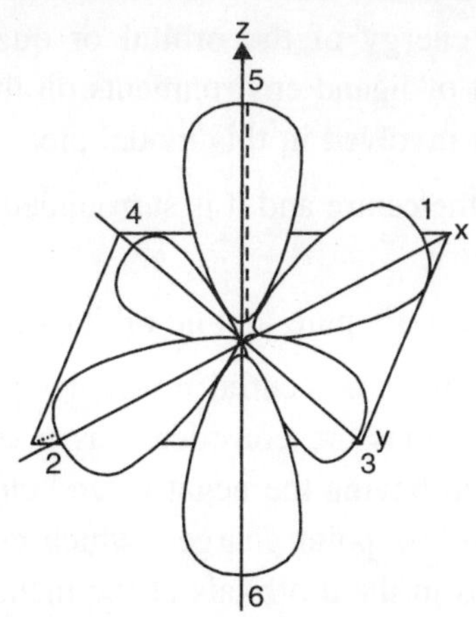

Figure 4.68: *Octahedral arrangements of six CN⁻ ligands around the Fe^{2+} ion*

$$\psi_1 = \frac{1}{\sqrt{6}}s + \frac{1}{\sqrt{2}}p_x + \frac{1}{2}d_{x^2-y^2} - \frac{1}{\sqrt{12}}d_{z^2}$$

$$\psi_2 = \frac{1}{\sqrt{6}}s + \frac{1}{\sqrt{2}}p_x - \frac{1}{2}d_{x^2-y^2} - \frac{1}{\sqrt{12}}d_{z^2}$$

$$\psi_3 = \frac{1}{\sqrt{6}}s + \frac{1}{\sqrt{2}}p_y + \frac{1}{2}d_{x^2-y^2} - \frac{1}{\sqrt{12}}d_{z^2} \quad \text{...(4.288)}$$

$$\psi_4 = \frac{1}{\sqrt{6}}s + \frac{1}{\sqrt{2}}p_y - \frac{1}{2}d_{x^2-y^2} - \frac{1}{\sqrt{12}}d_{z^2}$$

$$\psi_5 = \frac{1}{\sqrt{6}}s + \frac{1}{\sqrt{2}}p_z + \frac{1}{\sqrt{3}}d_{z^2}$$

$$\psi_6 = \frac{1}{\sqrt{6}}s - \frac{1}{\sqrt{2}}p_z + \frac{1}{\sqrt{3}}d_{z^2}$$

These six localized hybrid orbitals are used to from electron pair bond with the six CN^- ions as shown in Fig. 4.68. The rest of the d^6 electrons present in Fe^{4+} are placed in *d* orbitals which have not been used to construct hybrid bonding orbitals.

Crystal Field Theory (c.f.t)

The crystal field theory, commonly known as c.f.t. was developed by Bethe in 1929 and later on by Van Vleck. Even though this theory and v.b.t. proposed by Pauling were developed at the same time, the c.f.t. till 1950 was only used by physicists. It is only in fifties that the chemists started using it when they failed to explain the experimental results on the basis of v.b.t. because of its limitations.

As pointed out earlier that the fundamental aims of the chemists are only two fold: to calculate the (*i*) wave function and (*ii*) energy of the orbital or quantum state. The c.f.t. model was developed to calculate the effect of ligand environments on the energies of five *d* orbitals which are degenerate. The assumptions involved in this model are:

(*i*) Metal ion is present in the centre and it is surrounded by the ligands arranged in space to have the proper geometry.

(*ii*) *d* orbitals of the metal ion are pure atomic orbitals and are localized on the metal ion.

(*iii*) Ligands can either be anions or neutral molecules possessing at least one non-bonding pair of electrons. The anions are considered as point charges and the Lewis bases as point dipole. Each ligand having the negative end close to the cation is further assumed as structure-less, orbital-less point charges, which creates electrostatic field. The effect of this field on electrons in the *d* orbitals of the metal will be studied.

Consider an independent free metal ion containing *d* orbitals. It has been shown in Chapter 3.1.4 that for $n \geq 3$ and $l = 2$, there can be only five independent wave functions, called orbitals and their pictorial representation is given in Fig. 3.9. These five orbitals $d_{z^2}, d_{x^2-y^2}, d_{xy}, d_{yz}, d_{xz}$ are degenerate. Now assume that the ligands are arranged around the metal ion in a spherical symmetry and negative end of the charge is lying on the surface of the sphere. Because of spherical symmetry, each electron in each orbital will experience the similar repulsive interaction and thus, each orbital gets destabilized equally in comparison to free ion as shown in Fig. 4.69. Now suppose that instead of spherical symmetry, the ligands are placed at the corners of the octahedron and metal ion is placed in the centre. The total spherical charge is redistributed into six ligands.

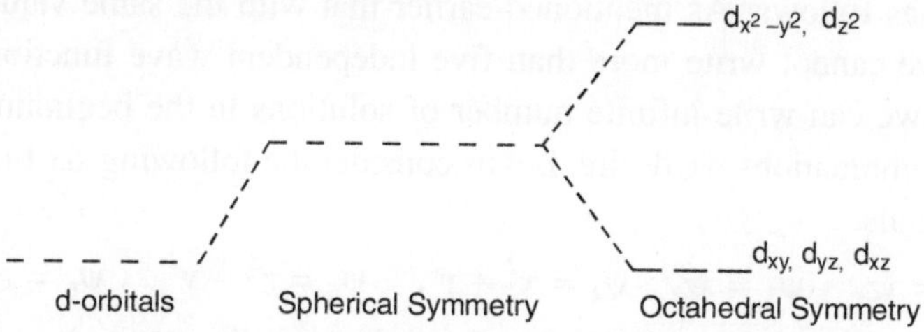

Figure 4.69: *Splitting of d-orbitals in the presence of spherical and octahedral symmetries*

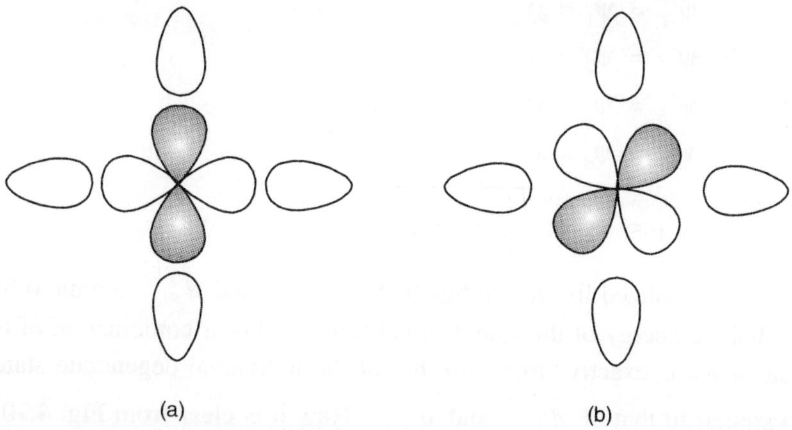

Figure 4.70: *a; Arrangements of ligands around the (a) $d_{x^2-y^2}$ (b) d_{xy} orbitals. The section d-orbitals by xy-plane is shown*

Although the shape of all the d orbitals except d_{z^2} are exactly similar, but their lobes are distributed along different directions in space and thus, will experience different electrostatic repulsion. Consider $d_{x^2-y^2}$ and d_{xy} orbitals in the octahedral field of ligands, as shown in Fig. 4.70. It is apparent from Fig. 4.70*a* that electrostatic repulsion felt by an electron in $d_{x^2-y^2}$ orbital from the negative end of the ligand placed along x- and y-axis will be more than that felt by the electron in d_{xy} orbital (Fig. 4.70*b*). This is because the charges of ligands are concentrated more towards the lobes of $d_{x^2-y^2}$ orbital that the lobes of d_{xy} orbital. In other words, d_{xy} orbital will be stabilized (lower energy) and $d_{x^2-y^2}$ orbital destabilized (higher energy) in comparison to the two orbitals placed in the spherical symmetrical field. The interaction of electronic charge with the charges of the ligands placed along the z-axis will be the same for $d_{x^2-y^2}$ and d_{xy} orbitals.

Similarly, we can also consider the electronic effect of the charges of the ligands present in other d orbitals. It is obvious from Fig. 4.70 that d_{xz} and d_{yz} orbitals are identical to d_{xy} except that the lobes for each of these three orbitals are lying in different planes. Thus, it can be concluded that d_{yz} and d_{xz} orbitals will be stabilized to the same extent as d_{xy} in comparison to spherical symmetrical field and thus, have the same energy. On the other hand, similar conclusion that d_{z^2} orbital will have the same energy as $d_{z^2-y^2}$ in the octahedral field is not so obvious, but it is true and can be shown as follows: As mentioned earlier that with the same value of quantum number $n \geq 3$ and $l = 2$, we cannot write more than five independent wave functions representing five d orbitals. However we can write infinite number of solutions in the beginning and then select any five independent combinations we desire. Let us consider the following six functions for the angular part of the nd orbitals.

$$\psi_1 = xy, \quad \psi_2 = yz, \quad \psi_3 = xz, \quad \psi_4 = x^2 - y^2, \quad \psi_5 = z^2 - y^2, \quad \psi_6 = z^2 - x^2. \qquad ...(4.289)$$

Out of these six functions the following five are generally selected.

$$\begin{aligned} \psi'_1 &= \psi_1 = xy \\ \psi'_2 &= \psi_2 = yz \\ \psi'_3 &= \psi_3 = xz \\ \psi'_4 &= \psi_4 = x^2 - y^2 \\ \psi'_5 &= \psi_5 = \sqrt{\frac{1}{2}}(\psi_5 + \psi_6) = 2z^2 - x^2 - y^2 \end{aligned} \qquad ...(4.290)$$

i.e., d_{z^2} orbital is normalized linear combination of $d_{z^2-y^2}$ and $d_{z^2-x^2}$ atomic orbitals. It has been shown earlier that the energy of the state formed from the linear combination of two or more than two degenerate states is exactly similar to that of the individual degenerate state, *i.e.*, energy of d_{z^2} orbital is similar to that of $d_{z^2-y^2}$ and $d_{z^2-x^2}$. Now it is clear from Fig. 4.70 that the orbitals $d_{z^2-y^2}$ and $d_{z^2-x^2}$ (and d_{z^2}) will be destabilized to the same extent as $d_{x^2-y^2}$ in octahedron crystal field. Fig. 4.69 represents the enrgy level diagram of five d orbitals in octahedron field. Five fold degeneracy of d orbitals, is lifted up and broken into two fold. d_{xy}, d_{yz} and d_{xz} form the one set and d_{z^2} and $d_{x^2-y^2}$ form the other set. The former has the lower energy than the latter. Although it will not be discussed here but based on the group theory representations d_{xy}, d_{yz} and d_{xz} are designated as t_{2g} and d_{z^2} and $d_{x^2-y^2}$ are called e_g. The energy difference between the e_g and t_{2g} levels is commonly represented by 10 Dq or Δ where Dq are the symbols commonly used in literature.

Another point worth considering is the magnitude of the stabilization and destabilization energy of t_{2g} and e_g states respectively. Let A be the lowering of the energy of t_{2g} and B be the raising of the energy of e_g state with respect to spherical symmetry. Then:

$$A + B = 10\ Dq\ (\text{or } \Delta) \qquad ...(4.291)$$

Now consider a d electron and is surrounded by $6q$ units of negative charge. If this charge is uniformly spherically distributed around the central metal ion and the charge is lying only at the periphery of sphere, the energy of d electron will be raised (let us say by E_s with respect to the ground state of the free ion) because of the electronic repulsion between the electronic charge and charges of the ligands. Since d orbitals can maximum have 10 electrons, the repulsive energy will be 10 E_s. Now if we assume that $6q$ units of charge is redistributed in the octahedron form, with q units of charge present at each corner, the five fold degeneracy of d-orbitals will be removed, three of d orbitals (d_{xy}, d_{yz} and d_{xz}) are stabilized and two of the d orbitals (d_{z^2} and $d_{x^2-y^2}$) are destabilized with respect to the spherical symmetrical field. Let the stabilization and destabilization energy of each electron is A and B respectively. Since t_{2g} set and e_g set can occupy 6 and 4 electrons respectively, the total energy of 10 d electrons in spherical symmetry will be equal to the total energy of the electrons in the octahedral symmetry, *i.e.*,

$$6(E_s - A) + 4(E_s + B) = 10E_s \quad ...(4.292)$$

Also $$A + B = 10Dq \text{ (or } \Delta) \quad ...(4.291)$$

Solving for A and B,

$$A = 4Dq \text{ or } \frac{2}{5}\Delta \quad ...(4.293)$$

$$B = 6Dq \text{ or } \frac{3}{5}\Delta \quad ...(4.294)$$

Tetrahedral Field

In this case, four ligands (negative ions or dipoles) are arranged around the central metal ion in such a way that it forms tetrahedron, the splitting of d orbitals in a tetrahedral electrostatic field can be calculated by similar treatment as done for octahedral field. The above treatment can be simplified if we consider a cubic symmetry. The four ligands are placed at the four corners of the cube as shown in Fig. 4.71 and metal ion in the middle. From Fig. 4.71 and the shapes of the d orbitals, it is obvious that d_{z^2} and $d_{x^2-y^2}$ are not oriented directly towards the directions of the charges of the ligands, whereas the other set of d orbitals (d_{xy}, d_{yz} and d_{xz}) also do not point towards the direction of the ligands as in the case of octahedral complexes, but still close to the ligands than d_{z^2} and $d_{x^2-y^2}$.

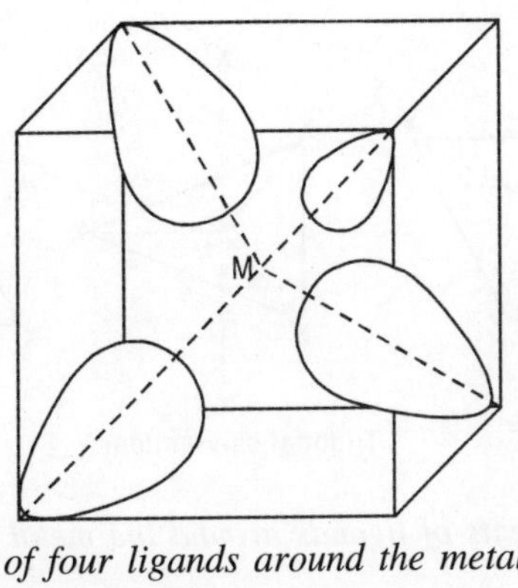

Figure 4.71: *Arrangement of four ligands around the metal ion in tetrahedral manner*

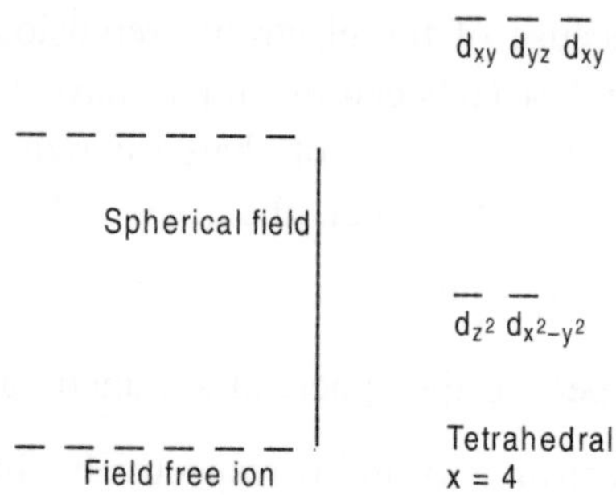

Figure 4.72: *Splitting of d orbitals in tetrahedral crystal field*

Thus, the electron in d_{z^2} and $d_{x^2-y^2}$ will have a lower potential energy than one in d_{xy}, d_{yz} and d_{xz} orbitals is shown in Fig. 4.72. Since tetrahedron point group does not have the centre of inversion, the set of orbitals d_{xy}, d_{yz} and d_{xz} and d_{z^2} and $d_{x^2-y^2}$ are termed as simply t_2 and e respectively. If Δ_t is the energy separation between t_2 and e, then following the similar procedure as carried out for octahedron symmetry, it can be shown that the energies of t_2 and e orbitals relative to their energies in a spherical shell of the same total energy will be:

$$t_2 = A = \frac{2}{5}\Delta_t,\ e = \frac{3}{5}\ \Delta_t \qquad ...(4.295)$$

Considering pure electrostatic interaction in c.f.t., it can be shown that the gap between the two set of d orbitals for the same metal ion and for the same set of ligands in octahedral and tetrahedral symmetries is:

$$\Delta_t = \frac{4}{9}\Delta_0 \qquad ...(4.296)$$

It may be pointed out here that the ratio between $\frac{\Delta_t}{\Delta_0}$ is observed to a good approximation by the experimental results even though it is known that c.f.t. model grossly fails to calculate the energies of d orbital splitting in the coordinate complexes.

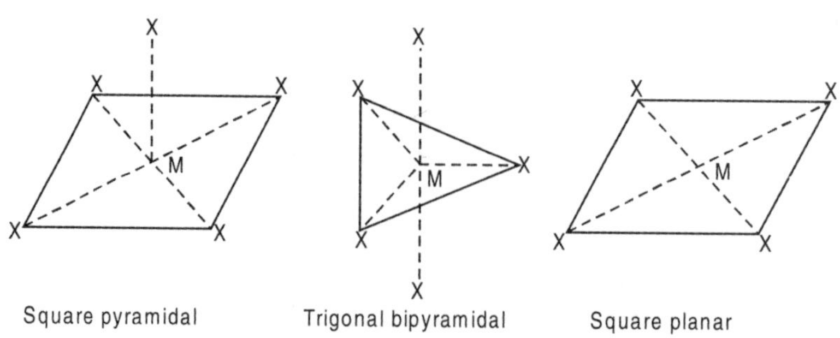

Figure 4.73: *Arrangements of ligands around the metal ions in different manner*

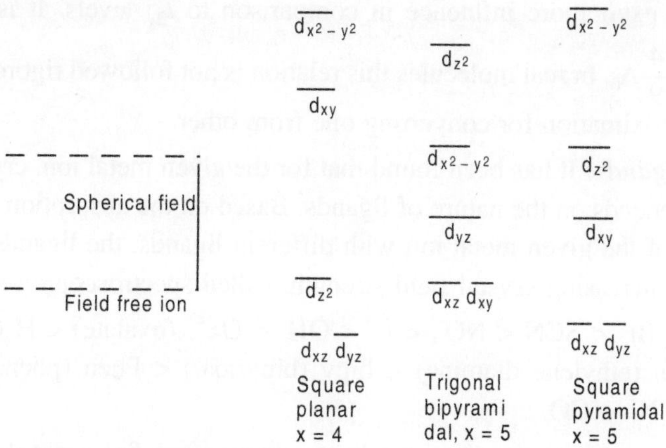

Figure 4.74: *Splittings of five d orbitals in spherical, square planar, trigonal bipyramidal and square pyramidal arrangements of the ligand around the metal ion*

The other kind of arrangements of ligands around the central metal atom or cation can be square planar and trigonal bipyramidal. Following exactly the similar treatment as has been done for octahedral and tetrahedral complexes, the splitting between the *d* orbitals can be calculated in case of square planar and trigonal bipyramidal complexes. These have been depicted in Fig. 4.73 and 4.74.

Factors Affecting the Magnitude of 10 Dq

There are many factors which affect the splitting of *d* orbitals in c.f.t. Here we will discuss only the four major ones.

(i) ***Charge on the central atom***: Keeping the same ligands, the charge on the central metal ion plays a major role on the c.f.t. splitting or the magnitude of 10 D*q*. This is because if the oxidation state of metal atom is changed from +2 to +3, it will bring the ligands close to central metal ion. Since c.f.t. involves the interaction between the electronic charge and the charges on the ligands, the repulsive potential energy of e_g levels in octahedral complexes will increase. This increase is predicted to be nearly 50% if the charge on the central metal ion is increased from +2 to +3. For example, 10 Dq for $[Fe(H_2O)_6]^{2+}$ and $[Fe(H_2O)_6]^{3+}$ are 1000 D*q* and 1400 D*q* cm^{-1} respectively, whereas for $[V(H_2O)_6]^{2+}$ and $[V(H_2O)_6]^{3+}$ are 1180 and 1800 D*q* cm^{-1} respectively.

(ii) ***Number and geometry of ligands***: The number of ligands around the metal ion reflects the total negative charge or total dipole moment, *i.e.*, the increase in the number of ligands will increase the value of 10 D*q*. Thus, changing the number of ligands from four in tetrahedral to six in octahedral complexes will increase the 10 D*q* by 33%, whereas the increase in the 10 D*q* in octahedral complexes is little larger than twice of tetrahedral. The rest of the difference is because of the placement of ligands in tetrahedral and octahedral complexes. In tetrahedral complexes, the ligands are not aimed directly at any of the *d* orbitals but exert some what greater influence on t_2 orbitals than on *e* orbitals, whereas in case of octahedral complexes, e_g orbitals are pointing directly towards the

ligands and exert more influence in comparison to t_{2g} levels. It is shown that in c.f.t. model $\Delta_t = \frac{4}{9}\Delta_0$. In real molecules this relation is not followed rigorously, but this is good enough approximation for converting one from other.

(*iii*) ***Nature of ligands***: It has been found that for the given metal ion, crystal field splitting of *d* orbitals depends on the nature of ligands. Based on the absorption spectra of coordinate complexes of the given metal ion with different ligands, the ligands have been arranged in order of increasing crystal field strength, called spectroscopic series.

I^- <, S^{2-} < Br^- < SCN^- < NO_3^- < F^- < OH^- < Ox^{2-} (oxalate) < H_2O < NCS^- < CH_3CN < NH_3 < en (ethylene diamine) < bipy (bipyridal) < Phen (phenanthroline) < NO_2^- < Phosph < CN^- < CO.

This spectroscopic series has not been constructed for a single metal ion forming complexes of the above ligands, but it is possible to synthesize the coordinate complexes of the different metal ions and to construct the series from the overlapping sequences, each depicting a portion of the series.

On inspection of the series the shortcomings of c.f.t. are quite obvious, *i.e.,* the pure electronic repulsion between electronic charge and the charged dipole of ligand. If one considers the ligands like S^{2-}, it should not be listed at the lower end in comparison to other monovalent ligands, ion OH^- causes less splitting in comparison to H_2O as ligands. Similarly NH_3 comes towards higher side of the series in comparison to water even though the dipole moment of NH_3 (4.0×10^{-30} C m) is less than that of water (6.17×10^{-30} Cm).

(*iv*) ***Nature of metal ion***: The size of the metal atom (in term of charge density) plays a big role in calculating the crystal field splitting energies. It is observed that the atomic radius of the transition elements does not change significantly while moving across a given transition series. Thus, for given combination of ligands, crystal field splittings are not very much different in a given transition element series. Whereas for a given transition metal, the size of the element decreases in going from 3*d* to 4*d*. Because of this the value of 10 D*q* will increase in a given group of elements with increase of Principal quantum number of the transition element. For example, ingoing from Cr to Mo or from Co to Rh, the value of 10 D*q* increases by 50%. Similarly 10 D*q* increases by nearly 25% in going from Rh to Ir. Because of this reason, depending on the nature of ligand, low spin and high spin (see next section) complexes are observed for the first transition series, but for the second and third transition series only low spin complexes are observed.

Electronic Configuration in Transition Metal Complexes

Having established the energies of the given *d* orbitals we are in a position to place the electrons in the *d* orbitals. Considering the octahedral crystal field, there is no problem in placing the first three electrons in *d* orbitals (*i.e.,* up to d^3), in t_{2g} as it is lower in energy and it is triply degenerate. Following Hund's rules, each orbital will have one electron. In d^4 system, there are two alternatives, *i.e.,* fourth electron can either go to e_g state or t_{2g} state. This is decided by two factors: (*i*) 10 D*q* value and (*ii*) pairing energy (energy required to pair two electrons in the same *d* orbital).

Table 4.10: *Electronic configuration of octahedral and tetrahedral complexes*

Elecronic configuration of metal ion	Electronic structure of the complex	Example
$3d^1$	$[\pi(d)]^1$	$[Ti(H_2O)_6]^{3+}$
$3d^2$	$[\pi(d)]^2$	$[V(H_2O)_6]^{3+}$
$3d^3$	$[\pi(d)]^3$	$[Cr(H_2O)_6]^{3+}$
$3d^4$ (low spin)	$[\pi(d)]^4$	$[Mn(CN)_6]^{3-}$
(high spin)	$[\pi(d)]^3\ [\sigma^*\ (d)]^1$	$[Cr(H_2O)_6]^{2+}$
$3d^5$ (low spin)	$[\pi(d)]^5$	$[Fe(CN)_6]^{3-}$
(high spin)	$[\pi(d)]^3\ [\sigma^*\ (d)]^1$	$[Mn(H_2O)_6]^{2+}$
$3d^6$ (low spin)	$[\pi(d)]^6$	$[Co(NH_3)_6]^{3+}$
(high spin)	$[\pi(d)]^4\ [\sigma^*\ (d)]^2$	$[CoF_6]^{3-}$
$3d^7$ (low spin)	$[\pi(d)]^6$	$[Co(NO_2)_6]^{4-}$
(high spin)	$[\pi(d)]^6\ [\sigma^*\ (d)]^1$	$[Co(H_2O)_6]^{2+}$
$3d^8$	$[\pi(d)]^6\ [\sigma^*\ (d)]^2$	$[Ni(NH_3)_6]^{2+}$
$3d^9$	$[\pi(d)]^6\ [\sigma^*\ (d)]^3$	$[Cu(H_2O)_6]^{2+}$

Case I*: *Low spin complexes: In this case 10 D*q* value is much higher than the pairing energy of electrons. So the 4th electron in d^4 system will prefer to pair up with other electron rather than going to less stable e_g state. Due to pairing of electrons, the total resultant spin will decrease if the number of electrons increases beyond three. Because of low spin values, these are called *low spin complexes*. This situation is commonly observed for ligands on the right hand side of the spectrochemical series or for the elements of 4th and 5th *d* series.

Case II*: *High spin complexes: In this case the 10 D*q* value is smaller than the pairing energy. Thus, the 4th electron in d^4 system will prefer to occupy e_g state rather pairing up with t_{2g} state electrons. This type of configuration will lead to high resultant spin value and are called high spin complexes. This is mostly observed for the ligands lying on the left hand side of the spectrochemical series and the first transition elements in general. In case of 4-cornered tetrahedral complexes, the 10 D*q* value is much less than that of octahedral complexes. Due to this reason most of the tetrahedral complexes are high spin complexes, though some exceptions might be observed. Table 4.10 lists the electronic configuration for octahedral complexes.

Ligand Field Theory (l.f.t.)

c.f.t. represents the interactions between the electronic charge of the *d* orbitals and charges on the ligands, considering the ligands either point charges or point dipoles and *d* electrons as the pure orbitals localized on the metal ion. This is an over simplification and is not true representation. It is now widely accepted and proved by the experimental results that even in the complexes which are highly ionic; one would expect the interaction to be pure electrostatic, for example, $[FeF_6]^{4-}$ and $[CoF_6]^{3-}$ but fair degree of overlap (even though small) between the *d* orbitals and ligand orbitals is observed. The use of physical methods, like EPR and NMR, has proved that spin densities of *d* electrons of the metal atom have a finite value at the nuclei of the ligands and fair degree of delocalization of the *d* electrons are found in the ligand orbitals. Theoretical calculations have concluded that *d* orbitals in the coordinate complexes represents only 80 ± 10% *d* character and the rest 10 to 30% character is due to the ligand electrons. Thus, the degree of the *d* orbitals, based on c.f.t. model in coordinate complexes, will be higher as compared to the true value. It is found from the experimental observations that the inter-electronic repulsion energies between the d^n electrons in coordinate complexes are nearly 70% of the value obtained using c.f.t. for free ion.

Two things have emerged from the above discussion. Firstly, it is not possible to calculate a priori the crystal field splittings of the *d* orbitals by using pure *d* orbital wave functions and the point charge or point dipole of the ligands. It has been found that in general we can still use the same values of *d* orbital energies in general form but we have to express: (*i*) the effective charge or dipole of the ligand, (*ii*) distance between the ligand and (*iii*) radial part of the *d* orbital wave function in terms of some fictitious parameters.

Molecular Orbital Theory (m.o.t.)

In the previous section we have learnt that c.f.t. has failed to account for the position of certain ligands in the spectroscopic series. Few examples have been given in that section, but most glaring one is that of CO molecule. This molecule is neutral and virtually has no dipole moment (0.374×10^{-30} C m) and still causes the maximum crystal field splittings. l.f.t. does take care of the above by incorporating some overlap of the ligand and *d* orbitals (*i.e.,* covalent character) by defining the effective charges on the metal *d* orbitals in terms of some fictitious parameters. Still l.f.t. is not enough to take care of the failures of c.f.t. Thus, one needs a better theory than l.f.t., which takes care of the salient features of c.f.t. and overlap of the charges of the ligands with electronic charges of the metal orbital. m.o.t. has been developed to remove this descripency. We will still be discussing only the salient features of this theory by taking six cornered octahedral complexes, four cornered tetrahedral and square planar complexes.

Octahedral complexes: The simplest example forming only σ bonds can be the hexa-aqua complex of Ti^{3+}, *i.e.,* $[Ti(H_2O)_6]^{3+}$. The complex contains one electron in the *d* orbital of Ti and twelve electrons of six water molecules as ligands. Each water molecule donates a lone pair of electrons present in sp^3 hybridized orbitals on oxygen and forms a σ bond with the titanium ion. Our job is now to construct the m.o.'s formed by the LCAO of the ligands and metal ion. We have learnt this in section 4.3 that (*i*) m.o. can be formed from the LCAO of atoms (or species) if the atomic orbitals of the combining atoms are of comparable energies and are of proper symmetry and (*ii*) energies of the m.o.'s are determined by the overlap of the combining orbitals.

The orientation of the ligand around Ti^{3+} is depicted in Fig. 4.75. Although the valence orbitals available on Ti^{3+} are nine, *i.e.,* five 3*d*, one 4*s* and three 4*p* orbitals but we need only six to form σ bond with the ligands. Thus, out of five *d* orbitals, orbitals which can form σ bonds with ligands can be d_{z^2} and $d_{x^2-y^2}$. Qualitatively m.o.'s can be easily formed if the proper symmetries of the combining atomic orbitals are considered. For example, let us start with, d_{z^2}. Now the shape and symmetry of d_{z^2} is such that the proper combination of the ligand orbitals is:

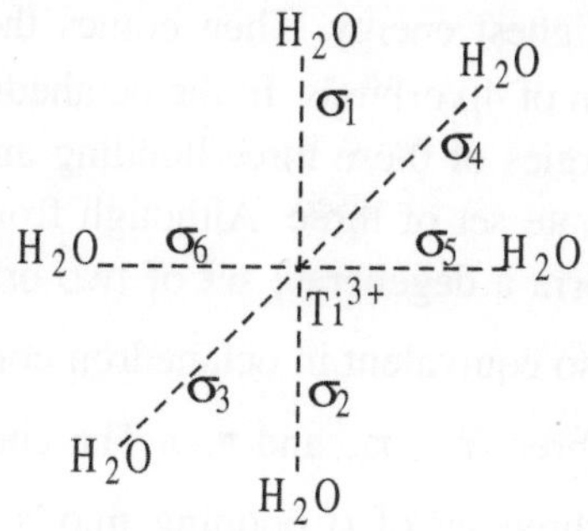

Figure 4.75: *Depiction of octahedron arrangement of six water ligands around Ti^{3+}*

$2\sigma_1 + 2\sigma_2 - \sigma_3 - \sigma_4 - \sigma_5 - \sigma_6$

(neglecting the normalization factor) and is shown in Fig. 4.76*a*. The wave function for the m.o. involving d_{z^2} is thus can be written as:

$$\psi_1(\sigma_z^2) = C_1\, 3d_{z^2} + C_2\,(2\sigma_1 + 2\sigma_2 - \sigma_3 - \sigma_4 - \sigma_5 - \sigma_6) \quad ...(4.297)$$

Similarly the wave functions for the other m.o.'s can be written by matching the metal orbital lobes and signs with the σ orbitals of the ligands those having the proper sign and symmetry. This has been shown in Fig. 4.76*b-f*. The wave functions of the m.o.'s thus, formed are:

$$\psi_2(\sigma_{x^2-y^2}) = C_3(3d_{x^2-y^2}) + C_4(\sigma_3 + \sigma_4 - \sigma_5 - \sigma_6) \quad ...(4.298)$$

$$\psi_3(\sigma_s) = C_5\, 4s + C_6(\sigma_1 + \sigma_2 + \sigma_3 + \sigma_4 + \sigma_5 + \sigma_6) \quad ...(4.299)$$

$$\psi_4(\sigma_{p_x}) = C_7\;4p_x + C_8(\sigma_3 - \sigma_4) \quad ...(4.300)$$

$$\psi_5(\sigma_{p_y}) = C_9\;4p_y + C_{10}(\sigma_5 - \sigma_6) \quad ...(4.301)$$

$$\psi_6(\sigma_{p_z}) = C_{11}\;4p_z + C_{12}(\sigma_1 - \sigma_2) \quad ...(4.302)$$

Pictorially the overlap of the metal orbitals and the ligand orbitals are shown in Fig. 4.76*b-f*. It is quite evident now that the symmetry of the other set of *d* orbitals (d_{xy}, d_{yz} and d_{xz}) is such that it cannot form σ orbitals with the ligands arranged in the octahedral complexes. These are thus, called non-bonding orbitals. Later on we will learn that these orbitals can form π m.o.'s with the ligands of proper symmetry and thus, can be termed as π_{xy}, π_{yz} and π_{xz}. Since water is not a good π bonding ligand, it will not form π orbitals.

From the above discussion it is now possible to set up qualitatively the energy level diagram for the above complex. Before we start discussing the energies of the m.o.'s we must know the relative energies of the starting orbitals; 3*d*, 4*s* and 4*p* and σ (H_2O). In this case, σ (H_2O) orbitals are more stable than any of the combining orbitals on Ti^{3+}. But in general it is known that the σ orbitals of the ligands are always of less energy than the metal valence orbitals. Similarly, it is also true that the order of increasing energy for the metal valence orbitals in transition metal complexes is $nd < (n+1)s < (n+1)p$. The shape of 4*s* orbital is such that it will have maximum overlap with the ligand orbitals. In other words, the σ_s bonding m.o. (ψ_3) will have the minimum energy and $\sigma_s^*(\psi_3^*)$ anti-bonding m.o. of highest energy. Then comes the bonding and anti-bonding m.o.'s formed by the linear combination of 4*p* orbitals. In the octahedral complex, the wave functions ψ_4, ψ_5 and ψ_6 are such that the energies of these three bonding and three anti-bonding m.o.'s will be the same and thus form degenerate set of three. Although from the wave functions ψ_1 and ψ_2, it is not obvious that these will form a degenerate set of two orbitals but it can also be shown that $3d_{z^2}$ and $3d_{x^2-y^2}$ orbitals are also equivalent in octahedron complex. Similarly, $3d_{xy}$, $3d_{yz}$ and $3d_{xz}$ also froms a degenerate set of three (π_{xy}, π_{yz} and π_{xz}). The energy level diagram is shown in Fig. 4.77. It is clear that there are three set of σ bodning m.o.'s: σ_s; σ_x, σ_y, σ_z and σ_{z^2}, $\sigma_{x^2-y^2}$ and three set of σ anti-bonding orbitals, which are having energy more than σ_d orbtials and also less than σ_d^*. The ground state electronic configuration of $[Ti(H_2O)_6]^{3+}$ can be written using Aufbau principle and Hund's rule. It is:

$$(\sigma_s)^2(\sigma_x, \sigma_y, \sigma_z)^6(\sigma_{z^2}, \sigma_{x^2-y^2})^4(\pi_d)^1$$

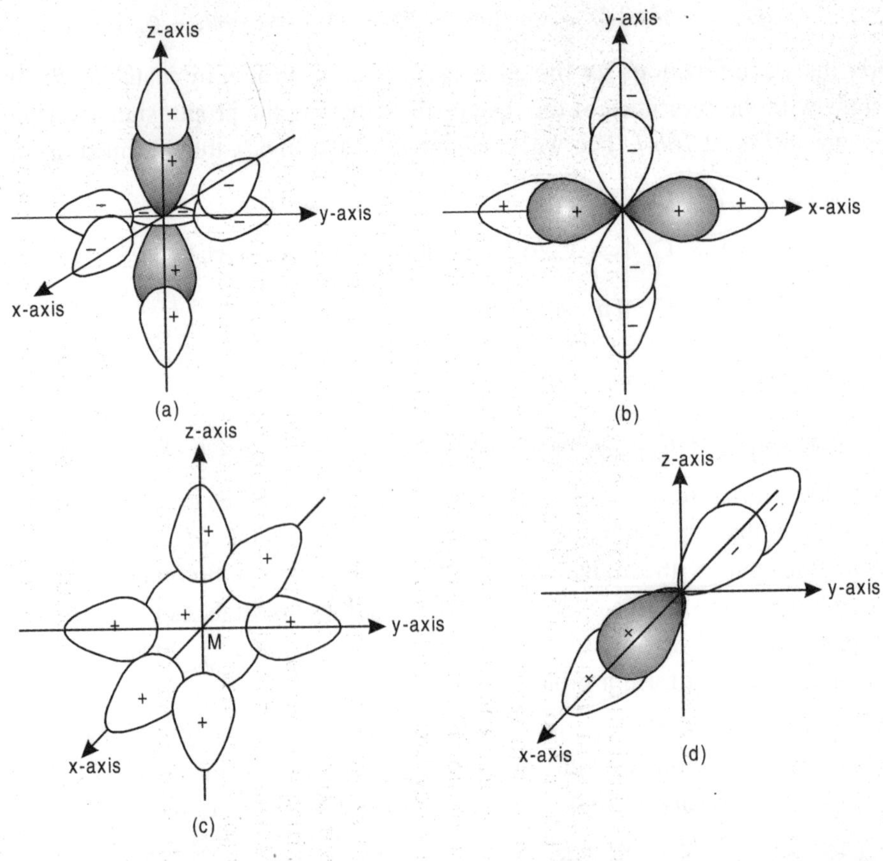

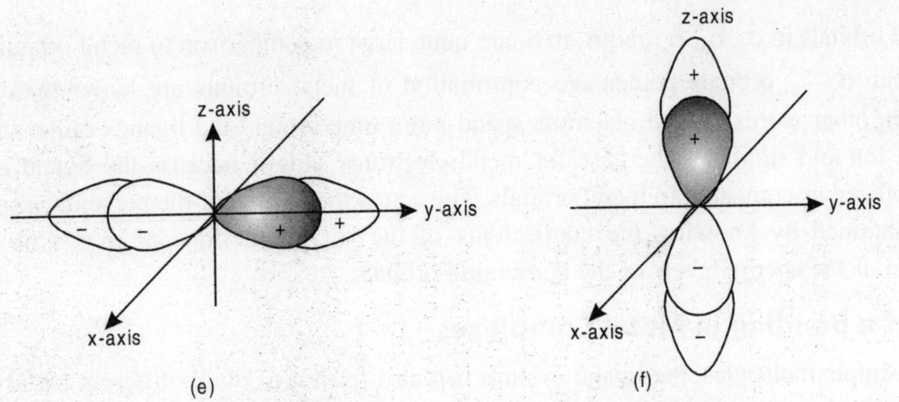

Figure 4.76: *(a) Overlap of* $3d_{z^2}$ *and (b)* $3d_{x^2-y^2}$*, (c) 4s, (d) 4px, (e) 4py and (f) 4pz orbitals of metal atom or ion with* σ *orbitals of the ligands*

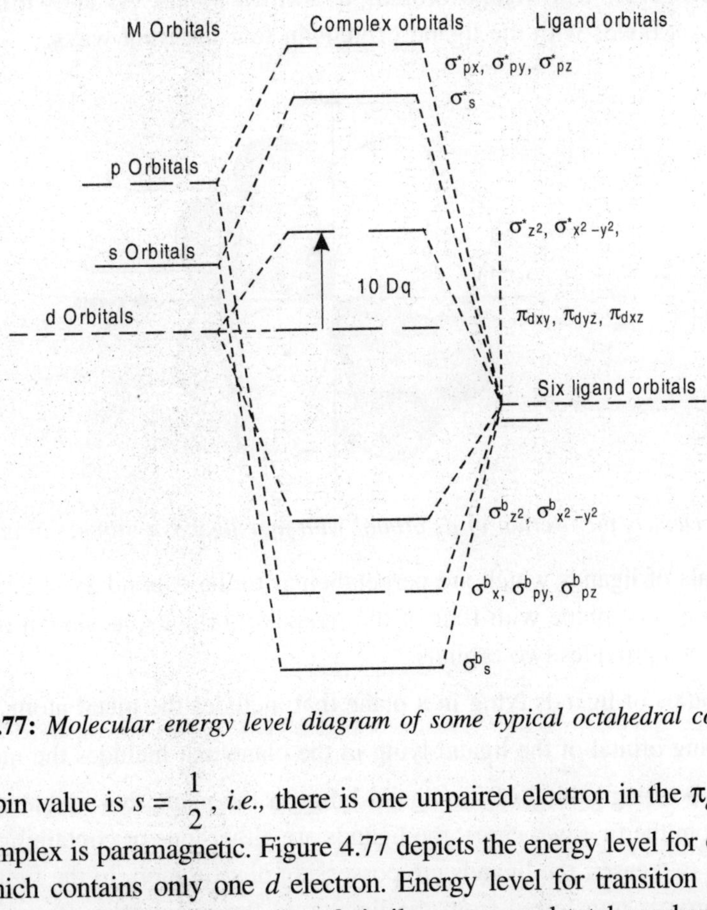

Figure 4.77: *Molecular energy level diagram of some typical octahedral complex*

The resultant spin value is $s = \frac{1}{2}$, *i.e.*, there is one unpaired electron in the π_d, consistent with the fact that this complex is paramagnetic. Figure 4.77 depicts the energy level for σ bonded ligands in $[Ti(H_2O)_6]^{3+}$, which contains only one *d* electron. Energy level for transition metal complexes containing more than one *d* electrons are also of similar nature and total number of electrons can also be arranged according to Aufbau principle and Hund's rule. This will be discussed in the next section. Energy level diagram as shown in Fig. 4.77 also reveals that the weighted percentage of the ligand orbitals in σ_s, σ_x, σ_y and σ_z m.o. are quite large in comparison to metal orbitals, whereas in σ_{z^2} and $\sigma_{x^2-y^2}$ orbitals percentage contribution of metal orbitals are larger than the ligand orbitals. In other words, ligand electrons spend more time around the ligands rather than around the metal ion and similar is the case for metal electrons. This is because the ligand orbitals are more stabilized as compared to metal orbitals. The actual magnitude of the percentage contribution can be obtained by knowing the coefficients of the wave functions of m.o.'s or the actual magnitude of the energy levels of the combining orbitals.

Types of π Bonding in Metal Complexes

Like simple molecules, the ligand orbitals can also form π.o.'s with different metal *d* orbitals, depending on the geometry of the coordinate complex. For simplicity we will again consider the octahedral complex, *i.e.*, $[ML_6]^{n+}$. In the last section we have found that in octahedral complex, the lobes of d_{xy}, d_{yz} and d_{xz} orbitals lie in between the axes and thus cannot form σ bonds with the ligands, but can interact with ligand orbitals to form π bonds. As shown in Fig. 4.78, these *d* orbitals can form π bonds with the ligand orbitals in four different ways.

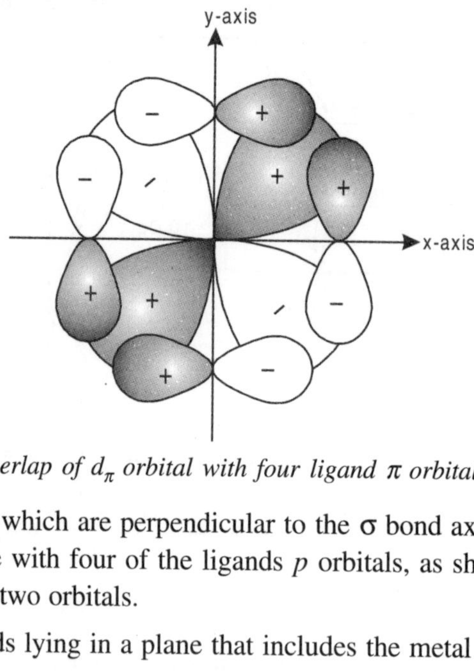

Figure 4.78: *Depicts the overlap of d_π orbital with four ligand π orbitals in octahedral complex*

(*i*) *p* orbitals of ligands which are perpendicular to the σ bond axis. Each of d_{xy}, d_{yz} and d_{xz} orbitals can combine with four of the ligands *p* orbitals, as shown in Fig. 4.78. Each of the ligands provides two orbitals.

(*ii*) π *d* orbitals of ligands lying in a plane that includes the metal atom.

(*iii*) π bonding orbital of the ligand lying in the plane that includes the metal atom.

(*iv*) π^* anti-bonding orbital of the ligand lying in the plane that includes the metal atom. In general in the first two cases, the ligands are either ions or containing saturated bonds. In the last two cases, the ligands are possessing unsaturation. In the first three cases, mostly the charge is transferred from the ligands to the metal atom (termed as L → M). This will increase the charge density on the metal atom and thus the energies of the degenerate set of d_{xy}, d_{yz} and d_{xz} are raised in comparison to the situation when these atomic orbitals do not form π bonds. This decreases the gap between π_d and $\sigma^*_{z^2}$ and

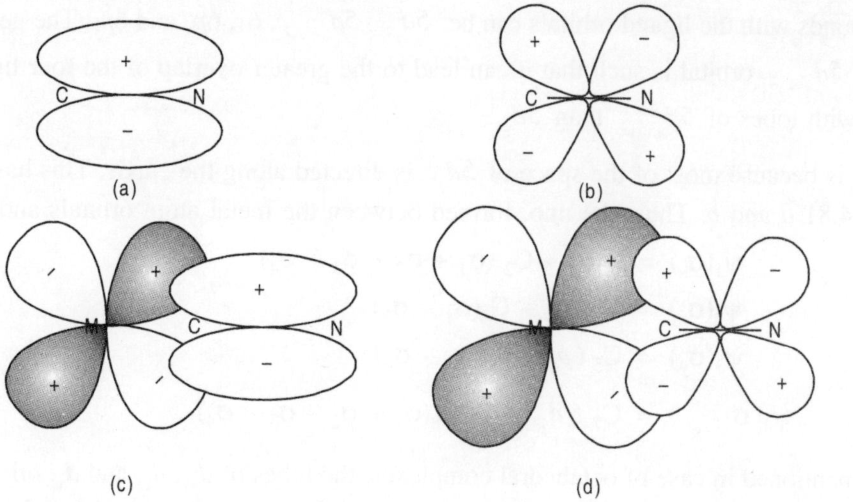

Figure 4.79: *(a) Shows the bonding π bonding and (b) shows the π* anti-bonding orbitals of CN⁻. (c) Shows the overlap one of the d_{xy}, d_{yz} and d_{xz} π orbitals with π bonding orbital of CN⁻. This is an example of L → M type charge transfer. (d) Shows the overlap of the d_{xy}, d_{yz} and d_{xz} π orbitals with π* anti-bonding orbital of CN⁻. This is an example of M → L type charge transfer*

$\sigma^*_{x^2-y^2}$ orbital, *i.e.*, the value of 10 Dq or Δ_0. In the last case charge is transferred from metal to ligand (M → L), thereby decreasing the charge density on the metal atom. Thus, M → L process stabilizes the π_d orbitals or increases the value of 10 Dq or Δ_0. The examples of the ligands in the first two cases can be halides, S^{2-} etc., whereas in the last two cases, it can be CN^-, CO etc., The formation of the m.o.'s by the overlap of orbitals in the first case is shown in Fig. 4.78, whereas in the last two cases it is shown in Fig. 4.79.

Square Planar Complexes

Square planar complexes are mostly formed by d^8 system and particularly by the 5d elements, *i.e.*, Pt^{2+}. Simple square planar complex thus to be considered is, $PtCl_4^{2-}$. The coordinate system used to depict this complex is shown in Fig. 4.80. The possible orbitals of the metal which can from

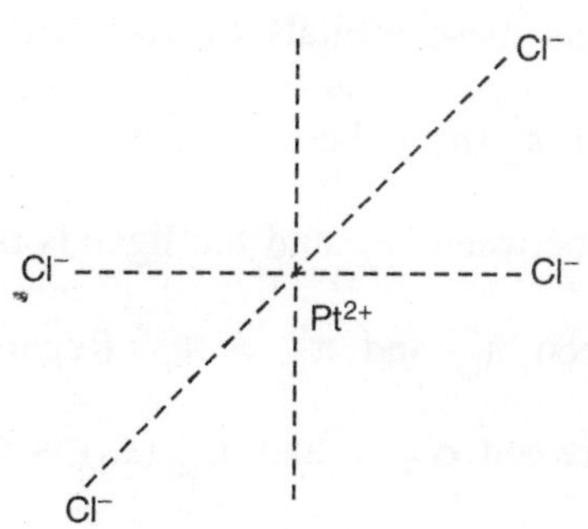

Figure 4.80: *Arrangement of four Cl⁻ ions around the platinum ion in the square planar complex*

form σ bonds with the ligand orbitals can be, $5d_{z^2}$, $5d_{x^2-y^2}$, 6*s*, $6p_x$ and $6p_y$. The geometry and the shape of $5d_{x^2-y^2}$ orbital is such that it can lead to the greater overlap of the four ligand σ valence orbitals with lobes of $5d_{x^2-y^2}$ than $5d_{z^2}$.

This is because most of the space of $5d_{z^2}$ is directed along the *z*-axis. This has been depicted in Figs. 4.81 *a* and *b*. Thus, the m.o. formed between the metal atom orbitals and ligands are:

$$\psi_1(\sigma_s) = C_1 6s + C_2(\sigma_1 + \sigma_2 + \sigma_3 + \sigma_4) \quad ...(4.304)$$

$$\psi_2(\sigma_x) = C_3 6p_x + C_4(\sigma_1 - \sigma_2) \quad ...(4.305)$$

$$\psi_3(\sigma_y) = C_5\, 6p_y + C_6(\sigma_3 - \sigma_4) \quad ...(4.306)$$

$$\psi_4(\sigma_{x^2-y^2}) = C_7\, 5d_{x^2-y^2} + C_8(\sigma_1 + \sigma_2 - \sigma_3 - \sigma_4) \quad ...(4.307)$$

As mentioned in case of octahedral complexes, the lobes of d_{xy}, d_{yz} and d_{xz} orbitals are located in between the axes rather than along the axes and thus, cannot form σ bonds with the ligand orbitals. But these are good for forming π bonds with the p_x and p_y orbitals of Cl^- ions. From the coordinate axis, it is eveident that d_{xy} orbital can interact with π valence orbitals of all the four ligands, whereas d_{yz} and d_{xz} can interact with only two ligands and thus, form the equivalent set. Figures 4.81*c* and 4.81*d* gives the pictorial depiction of the overlap of 5*d* metal orbitals with the valence orbitals of the four ligands. Figure 4.82 depicts the approximate energy level diagram of m.o.'s obtained by combining atomic orbitals and m.o.'s of the ligands from their linear combination. Though it is possible to calculate the energies of each m.o., we have tried to depict the region of the energies. Since the maximum overlap will be between the σ orbital, σ bonding m.o.'s are the lowest and σ^* will be highest in the energy level scheme. The electronegativity of Cl^- ion is much larger than the metal atom, so the ligand orbitals are more stabilized than the metal atom orbitals. Due to this, percentage contribution by the Cl^- orbitals in the bonding orbitals will be larger in comparison to metal orbitals. Thus, it can be said that σ bonding m.o.'s are more or less localized on Cl^- ligands and σ* anti-bonding on the metal atom. The orbitals next in the sequence of increasing energy are the π bonding m.o., also localized on Cl^- ligands. This is followed by the metal *d* orbitals. Since the σ overlap between $d_{x^2-y^2}$ is very large with the ligands, $\sigma^*(d_{x^2-y^2})$ anti-bodning m.o. will be highest in energy in the metal *d* orbitals. The next in the decreasing energy will be $\pi^*_{xy}(d_{xy})$ m.o. as these orbitals overlap with π orbitals of the four ligands in comparison to the $\pi^*_{yz}(d_{yz})$ and $\pi^*_{xz}(d_{xz})$, because these orbitals overlap with only π orbitals of the two ligands. Since overlap between d_{z^2} and the ligands is weak, it is believed that $\sigma^*_{z^2}(d_{z^2})$ anti-bonding m.o. lies in between π^*_{xy} and $\pi^*_{yz} = \pi^*_{xz}$. Regardless of the placement of d_{z^2}, one thing is certain that the gap between $\sigma_{x^2-y^2}$ and π^*_{xy} (Δ_1) is very large in comparison to the gap between π^*_{xy} and σ_{z^2} and $\pi^*_{yz} = \pi^*_{xz}$ (Δ_2 and Δ_3) but $\Delta_1 + \Delta_2 + \Delta_3$ is greater than Δ_0.

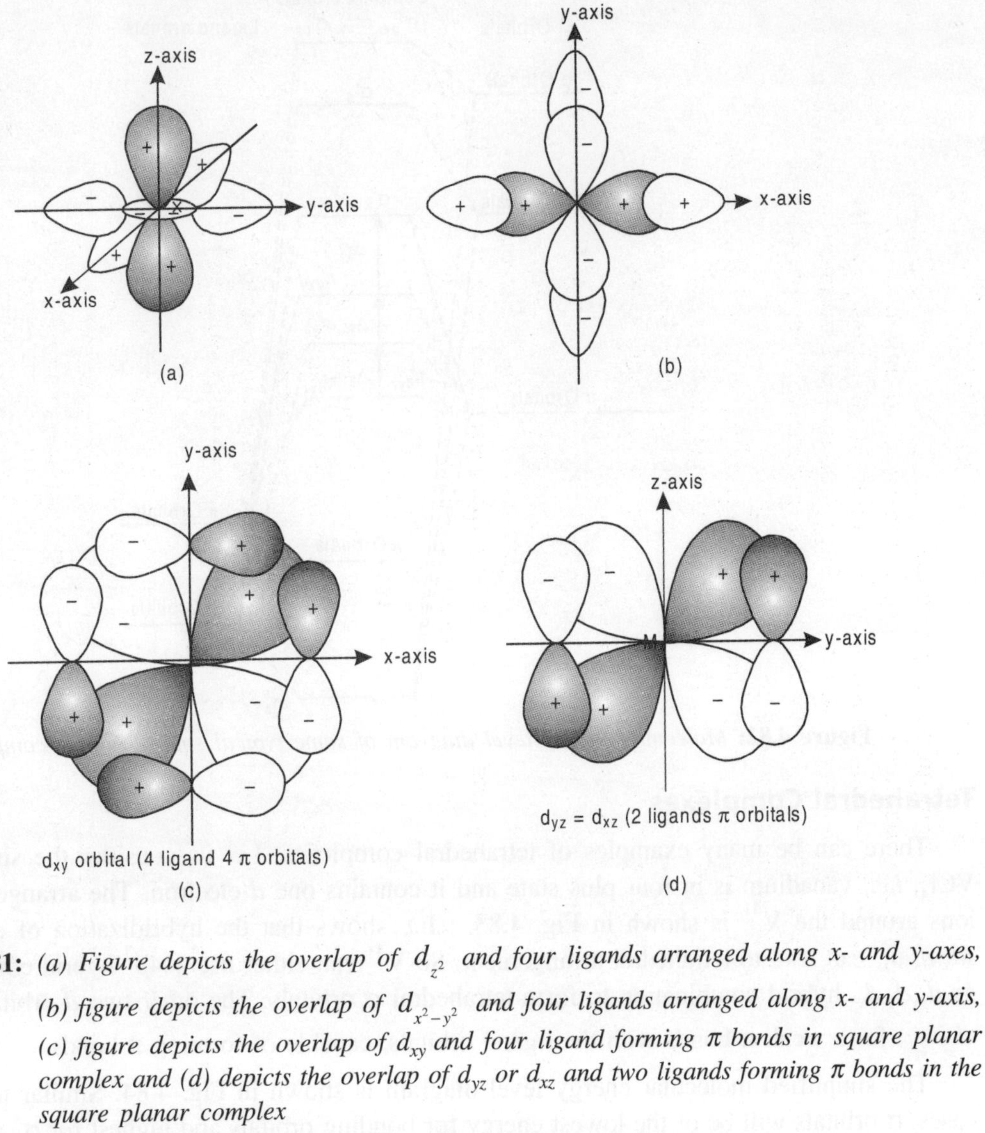

Figure 4.81: *(a) Figure depicts the overlap of d_{z^2} and four ligands arranged along x- and y-axes, (b) figure depicts the overlap of $d_{x^2-y^2}$ and four ligands arranged along x- and y-axis, (c) figure depicts the overlap of d_{xy} and four ligand forming π bonds in square planar complex and (d) depicts the overlap of d_{yz} or d_{xz} and two ligands forming π bonds in the square planar complex*

The ground state electronic configuration for $PtCl^{4-}$, which contains eight *d* electrons, eight σ and 16 π electrons is expressed as:

$$(\sigma)^8(\pi)^{16}(\pi^*_{xz}\ \pi^*_{yz})^4(\sigma^*_{z^2})^2(\pi^*_{xy})^2 \qquad ...(4.308)$$

Each set of m.o.'s contains even number of electrons and are completely paired, the total spin S = 0, is compatible with the experimental results. The other examples of transition metals forming mostly square planar complexes are those of Ni^{2+}, Pt^{2+}, Pd^{2+} and Au^{3+}. All these metal ions of d^8 system are compatible with energy level diagram shown in Fig. 4.82 and show diamagnetic behaviour.

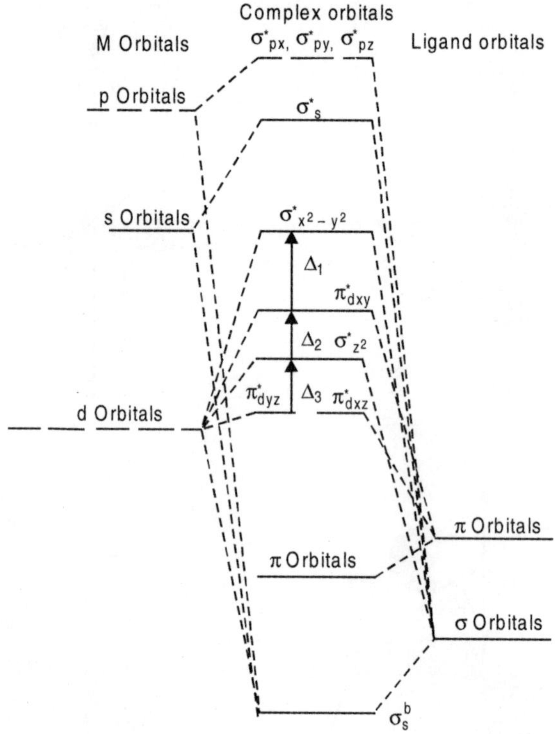

Figure 4.82: *Molecular energy level diagram of some typical square planar complex*

Tetrahedral Complexes

There can be many examples of tetrahedral complexes. Let us consider the simple one as VCl_4, *i.e.*, vanadium is in four plus state and it contains one *d* electron. The arrangement of Cl^- ions around the V^{4+} is shown in Fig. 4.83. v.b.t. shows that the hybridization of either sp^3 or $sd_{xy}d_{yz}d_{xz}$ can lead to tetrahedral arrangement. So V^{4+} can either use $4s4p^3$ hybrid combination or $4s\ d_{xy}d_{yz}d_{xz}$ hybrid combination to form tetrahedral σ orbitals. The other two *d* orbitals (d_{z^2} and $d_{x^2-y^2}$) form weak π bonds with the ligand orbitals, capable of forming π bonds.

The simplified molecular energy level diagram is shown in Fig. 4.84. Similar to the earlier cases, σ orbitals will be of the lowest energy for bonding orbitals and highest for σ^* anti-bonding m.o.'s. Next in the increasing energy are the π bonding orbitals. As mentioned in the lastparagraph,

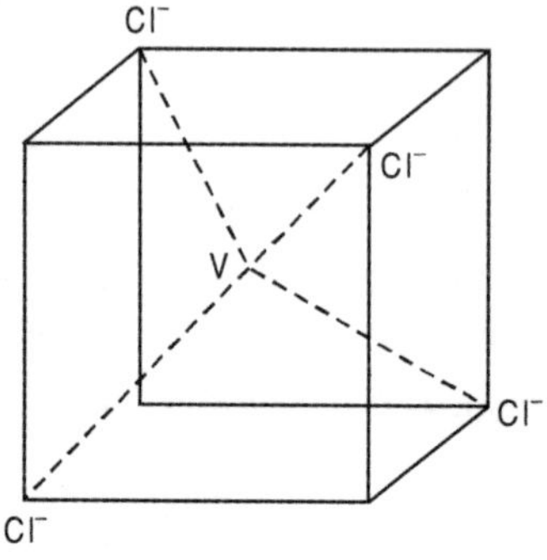

Figure 4.83: *Arrangement of the 4 Cl^- ions around the V^{4+} metal ion*

degeneracy of five d orbitals is lifted because d_{xy}, d_{yz} and d_{xz} can form σ orbitals with ligand orbitals. These orbitals will be part of σ^*, whereas $d_{x^2-y^2}$ and d_{z^2} will be part of π* anti-bonding m.o.'s. Since these orbitals in each set are equivalent, they form the degenerate set of two, former being of higher energy in comparison to the latter. This is similar to t_2 and e levels obtainbed in c.f.t. and difference between t_2 and e is known as 10 Dq or Δ_t.

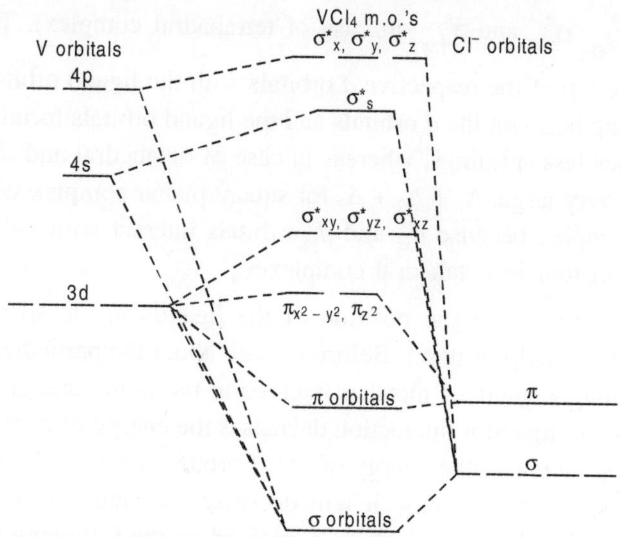

Figure 4.84: *Relative energy levels of different m.o.'s of* VCl_4

The total number of electrons in VCl_4 is: one in 3d; 8 σ electrons and 16 π electrons. The ground state electronic configuration will be:

$$(\sigma_b)^8(\pi_b)^{16}(\pi_d^*)^1$$

and S = $\frac{1}{2}$. This is consistent with experimental results that VCl_4 is paramagnetic in nature. Similarly, molecular energy level diagram can be drawn for other tetrahedral complexes and electron can be arranged according to Aufbau principle and Hund's rules. As mentioned in case of c.f.t., Δ_t is smaller than the pairing energy of electrons, the electrons will be filled first one each in orbital and then pairing will be done.

Splitting of *d* Orbitals

One thing which has emerged from m.o.t. of transition metal complexes and common to c.f.t. is the splitting of d orbitals. m.o.t. provides a better picture about this splitting, called 10 Dq or Δ and removed doubts raised earlier regarding the dependence on the value of Δ casued by various parameters.

We have seen earlier that other things being kept same, the value of Δ decreases in the following order:

Square planar	> Octahedral	> tetrahedral
$1.3\Delta_0$	Δ_0	$0.45\ \Delta_0$

In m.o.t., Δ_0 is defined as the difference between the non-bonding matel *d*-orbitals (i.e. t_{2g}; $\pi_{d_{xy}}, \pi_{d_{yz}}$ and $\pi_{d_{xz}}$ in case of octahedral complex as and e; $\pi_{x^2-y^2}, \pi_{dz^2}$ in case of tetrahedral complexes) and metal anti-bonding *d*-orbitals (*i.e.*, eg; $\sigma^*_{d_{x^2-y^2}}, \sigma^*_{d_{z^2}}$ in case of octahedral complexes and *t*; $\sigma^*_{d_{xy}}, \sigma^*_{dyz}$ and $\sigma^*_{d_{zx}}$ in case of tetrahadral complex). This in turn depends on how strong is the overlap of the respective *d* orbitals with the ligand orbital. In case of tetrahedral complexes, the overlap between the *d* orbitals and the ligand orbitals forming σ orbitals is not very strong and thus causes less splittings, whereas in case of octahedral and square planar complexes, the above overlap is very large. $\Delta_1 + \Delta_2 + \Delta_3$ for square planar complex will be always larger than that of octahedral complex because d_{xz} and d_{yz} orbitals interact with only two ligands in square planar as compared to four in octahedral complexes.

The discrepancies regarding the position of the ligands in the spectroscopic series can be nicely explained with the help of m.o.t. Before we talk about the particular case, let us make some generalization. A strong Ligand → metal σ interaction raises the energies of the σ* (*d*) orbitals. Similarly strong metal to ligand π interaction decreases the energy of π(*d*) orbital. A strong Ligand → metal π interaction increases the energy of –π(*d*) orbital. Thus, in the first two cases, the value of Δ will increase and in the last case it will decrease the value of 10 D*q*. In the light of this discussion, the spectrochemical series can be explained on the following lines. A good π-accpetor ligand (which causes strong metal-ligand π bonding) causes large splitting, whereas the good π donor ligand (those are capable of strong ligand to metal π bonding) casues small splitting. The ligands with intermediate Δ values have little or no π bonding capabilities. Thus, the spectrochemical series can be written as:

$$I^- < Br^- < S^{2-} < SCN^- < NO_3^- < F^- < OH^- < Ox^{2-} < H_2O < NCS^- < CH_3CN < NH_3$$

π-donors   weak π-donors   non-π-donors

$< en <$ *o*-phen $< NO_2^- < CN^- < CO$

π-acceptors

Increase of Δ value with increase in the charge on the metal atom can be explained nicely with m.o.t. In case of complexes, containing ligands which are not good π-acceptors, the increase in the charge on metal ion will increase the value of 10 D*q*. This is because the increase in the charge on metal ion will increase the σ overlap between the ligands and metal orbitals, thereby increasing the gap between t_{2g} (π_d) and e_g (σ^*_d). For example, the values of 10 D*q* in case of $[V(H_2O)_6]^{3+}$ and $[V(H_2O)_6]^{2+}$ are 17,850 and 11,800 cm^{-1} respectively. On the other hand, Δ values do not change substantially with increase in the charge on central metal atom if the ligands present are good π acceptors. For example, the value of Δ observed for $[Fe(CN)_6]^{4-}$ and $[Fe(CN)_6]^{3-}$ are nearly the same (34,000 cm^{-1}). This could be because of a decrease in the metal-ligand π bonding when the positive charge is increased on the metal ion and thus, the π(*d*) orbital is stabilized just as much as the $\sigma^*(d)$ orbital.

Lastly the increase in the Δ observed with increase in the values of principal quantum number may be because 4*d* and 5*d* valence orbitals are better in forming σ bonds with the ligands than 3*d* orbitals.

PROBLEMS

1. Considering *x*-axis as the molecular axis of diatomic molecule, suggest which combination will lead to bonding m.o's. Depict the a.o.'s and arrive at the results. (*i*) *s* and p_z, (*ii*) p_x and p_x, (*iii*) p_x and d_{xy}, (*iv*) p_x and p_y, (*v*) *s* and d_{xy}.
2. Considering *z*-axis as the molecular axis of diatomic molecule, indicate by using diagrams, which of the following combinations will lead to bonding m.o's. (*i*) *s* and p_z, (*ii*) p_x and p_x, (*iii*) p_z and d_{xy}, (*iv*) p_z and d_{xz}, (*v*) *s* and d_{xz}.
3. Considering *y*-axis as the molecular axis of diatomic molecule, find out by using diagrams which of the following combinations will lead to bonding m.o's. Depict the a.o.'s and arrive at the results. (*i*) *s* and p_z, (*ii*) *s* and p_y, (*iii*) p_x and p_x, (*iv*) p_y and $d_{x^2-y^2}$, (*v*) p_y and d_{yz}.
4. In valence bond theory for hydrogen molecule, show that $\hat{H}_{11} = \hat{H}_{22}$.
5. Write the total Hamiltonian operator for hydrogen molecule.
6. If we use the wave function like (4.87) for hydrogen molecule, the exchange integral, K, defined as:

$$K = \int 1s_A(1)1s_B(2)\left[-\frac{1}{r_{1A}} - \frac{1}{r_{2B}} + \frac{1}{r_{12}} + \frac{1}{R}\right]1s_A(2)1s_B(1)dr_1dr_2$$

does not arise. The integral *k* only arises when we use the wave function (4.95).

7. Prove that the wave function corresponding to $E_2 = 2E_H + \frac{J-K}{1-S^2}$ is

$$\Psi_- = \frac{1}{\sqrt{2-2S^2}}\,[\varphi_{1s_A}(1)\,\varphi_{1s_B}(2) + \varphi_{1s_B}(1)\,\varphi_{1s_A}(2)]$$

8. Show that linear combination of the two degenerate spin states, *i.e.*, α(1) β(2) and α(2) β(1) leads to the symmetric and anti-symmetric normalized wave functions as:

$$\frac{1}{\sqrt{2}}\,[\alpha(1)\,\beta(2) + \alpha(2)\,\beta(1)\}\text{ and }\frac{1}{\sqrt{2}}\,[\alpha(1)\,\beta(2) - \alpha(2)\,\beta(1)]$$

9. Show that the total wave function for Ψ_+ has a singlet character ($S_T = 0$) and for Ψ_- has a triplet character ($S_T = 1$).
10. Suppose that in hydrogen molecule ion both H_A and H_B use four a.o.'s each in forming LCAO m.o.'s. Indicate how many total number of m.o.'s will be formed. How many will be bonding ones and how many will be anti-bonding ones?
11. In v.b.t. Ψ_1 and Ψ_2 are represented in the matrix form as

$$\psi_1 = \begin{vmatrix} 1s_A(1)\,\alpha(1) & 1s_B(1)\,\beta(1) \\ 1s_A(2)\,\alpha(2) & 1s_B(2)\,\beta(2) \end{vmatrix} \text{ and } \psi_2 = \begin{vmatrix} 1s_A(1)\,\beta(1) & 1s_B(1)\,\alpha(1) \\ 1s_A(2)\,\beta(2) & 1s_B(2)\,\alpha(2) \end{vmatrix}$$

Then show that

$$\psi_+ = \frac{1}{\sqrt{2(1+S^2)}}\left[\frac{1}{\sqrt{2}}\psi_1 - \frac{1}{\sqrt{2}}\psi_2\right]$$

$$\psi_- = \frac{1}{\sqrt{2(1-S^2)}}\left[\frac{1}{\sqrt{2}}\psi_1 + \frac{1}{\sqrt{2}}\psi_2\right]$$

12. Show that both Ψ_+ and Ψ_- are eigen functions of S_z (total) and find the eigen values.

13. Write Hamiltonian operator (in a.u.) for C_2H_2 molecule.

14. Write the wave function of hydrogen molecule in the Slater determinant form and show that on expanding it resembles:

$[\varphi_{1s_A}(1)\, \varphi_{1s_B}(2)\, \alpha(1)\, \beta(2) + \varphi_{1s_B}(1)\, \varphi_{1s_A}(2)\, \beta(1)\, \alpha(2)]$

15. Write the total wave functions of hydrogen molecule in m.o.*t.* as:

$$\psi_{m.o} = \frac{1}{(2+2\,S)}[1s_A(1)+1s_B(1)]\,[1s_A(2)+1s_B(2)]$$

If we substitute this function in $E = \int \psi\, \hat{H}\, \psi\, d\tau$, the expression which we get for the energy is similar to that obtained by using v.b.t.

16. Show that the contribution from ionic and covalent character to the total wave function in m.o.t. are the same, whereas in v.b.t. we have to add the ionic terms and their contributions are less than those from covalent ones.

17. If *z*-axis is the molecular axis of hydrogen molecule, why the polarized a.o.'s will give better results of energy. The polarized a.o. used is $\varphi = 1s + C\, 2p_z$. Why cannot we use p_x or p_y a.o.'s.

18. Diatomic helium does not exist, but He_2^+ has been observed. Explain briefly.

19. Write the electronic configuration of N_2, O_2 and F_2. Find the order and arrange the bond strength and bond length in these molecules.

20. Is B_2 paramagnetic or diamagnetic? Predict using the electronic configuration.

21. Write the electronic configuration of O_2^+, O_2 O_2^- and O_2^-. Calculate the bond order in each case and arrange their bond dissociation energies and bond lengths.

22. Both C_2 and O_2 have bond order two. The bond dissociation energies are 599 kJ mol^{-1} and 494 kJ mol^{-1} respectively, whereas the bond lengths of C_2 and O_2 are 124 and 121 pm respectively. Explain the two observations.

23. The excimer of rare gases are formed by the combination of ns^2 and ns^1 $(n + 1)s^1$. Write the electronic configuration of the excimer and explain their stability relative to diatomic rare gas molecules.

24. (*a*) Construct energy level diagram for m.o. of NO from the a.o.'s of nitrogen and oxygen atoms. Write the electronic configuration of NO^+, NO and NO^-, (*b*) Arrange the species according to increasing bond dissociation energies and bond lengths, (*c*) Also construct the m.o.'s qualitatively showing the charge densities, (*d*) Write the wave functions for each m.o. and indicate which a.o. is participating more than the other.

25. Repeat the problem 24 for ClO.

26. The electronic configuration for CO [KK $(2\sigma_g)^2$ $(2\sigma_u)^2$ $(1\pi_u)^4$ $(3\sigma_g)^2$] and N_2 [$\sigma_{2s}^2\sigma_{2s}^{*2}\pi_u^4\sigma_{2p_z}^{*2}$] are similar. Bond orders are also similar, but the bond dissociation energies of CO and N_2 are 1071 and 942 kJ mol^{-1} respectively. Explain briefly.

27. Draw the energy level for SO. Write the electronic configuration and calculate the bond order. Compare the electronic configuration with S_2 and suggest qualitatively which molecule will have higher dissociation energy and why. Draw the shapes of m.o for each species.

28. Calculate the molecular term symbols for (*i*) B_2, (*ii*) O_2^-, (*iii*) O_2, (*iv*) N_2^+, (*v*) N_2 and arrange them according to increasing energy.

29. Calculate the molecular term symbols for (*i*) CO, (*ii*) SO, (*iii*) NO, (*iv*) NO^-, (*v*) ClO, (vi) OF^-. Arrange them according to increasing energy levels.

30. Write the electronic configuration of the first excited state of C_2 and calculate the molecular term symbols.

31. The longest wavelength transition in N_2 is observed when an electron is excited from σ_g^2 to π_g^1 m.o. with electronic configuration σ_g^1 π_g^1. Calculate the molecular term symbols.

32. Repeat the problem 31 for (*i*) NO, (*ii*) OF and (*iii*) ClO.

CHAPTER 5

Spectroscopy

All the spectroscopic techniques involve the interaction of radiation with matter. During this interaction: (*i*) the energy in the form of radiation can be absorbed, (*ii*) energy can be emitted in the form of radiation, and (*iii*) the photon can be scattered from its original path either with less, same or more energy. The first one is called the absorption spectroscopy, the second one is called emission spectroscopy and the third one is called Raman scattering or spectroscopy. In the case of Raman scattering, the photon either absorbs the energy from the molecules, called Anti-Stokes line, gives energy to the molecules, called Stokes line or energy is not exchanged at all and is called as Rayleigh scattering.

Basic purpose of using and studying spectroscopy is to understand the structure of matter, as well as, the dynamic behaviour of the system at atomic or molecular level. Before each branch of spectroscopy is discussed it is nice to understand some general features of spectroscopy. A spectroscopic experiment generally involves the measurement of two parameters; the energy of the electromagnetic radiation absorbed or emitted by the matter and intensity of the absorbed or emitted radiation. This chapter describes first some factors related to the intensity of absorption spectrum. These are nearly, the same for emission spectrum. This will be followed by the survey about the various types of spectroscopy.

5.1 INTENSITY

Intensity of absorption or emission in molecular or atomic spectrum is as important as the frequency difference between the two energy states. The practical measure of intensity of absorption spectrum at any wavelength is the value of molecular extinction coefficient, ε, also called the *molar absorbance,* on molar decadic coefficient. In this section, we will discuss first Beer-Lambert's law and it will be followed by the theoretical treatment of the subject.

5.1.1 BEER-LAMBERT'S LAW

Lambert was the first to derive this law. He only considered the changes in the intensity of radiation passing through the absorbing material of the system. Further modification, was carried out by Beer. He proposed that the changes in the intensity of radiation also depend on the concentration of the absorbing material. This is known as Beer-Lambert's law.

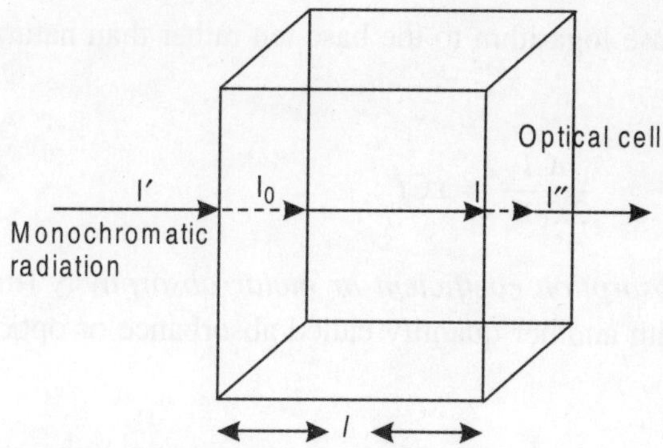

Figure 5.1: *Derivation of Beer-Lambert law*

Consider the situation as shown in Fig. 5.1. I′ is the intensity of monochromatic radiation (radiation of one wavelength) just before entering the cell, I_0 is the intensity of radiation just after passing through the cell material of certain thickness, I is the intensity of radiation passing through the absorbing material and I″ is the intensity of radiation just after coming through another thickness of the absorbing material of the cell. Change in the intensity of above arrangement is defined in two ways.

(*i*) The ratio of I″/I′ is called transmittance (T)

$$T = I''/I' \qquad \text{...(5.1}a\text{)}$$

(*ii*) Ratio of I/I_0 is defined as the internal transmittance, *i.e.*,

$$T' = I/I_0 \qquad \text{...(5.1}b\text{)}$$

The physical quantity T′ is more important to chemists because this quantity tells about the amount of radiation absorbed by the material in which we are interested. In general T and T′ are not very different from each other because the amount of radiation absorbed or scattered by the cell material is negligibly small. Even this difference in absorption of radiation by the cell material can be eliminated by carrying out the experiment with two matched cells, one containing the solution of the absorbing material in a given solvent and the other without the absorbing material. Thus, the Beer-Lambert's law states that the change in the intensity of radiation depends on: (*a*) concentration of the absorbing material (C, mol dm^{-3}), (*b*) intensity of monochromatic radiation and (*c*) thickness of the absorbing layer, *i.e.*,

$$-\,dI = aIC\,dl \qquad \text{...(5.2)}$$

where a is the absorbing constant and dl is the small thickness. The above expression can be integrated, after separation of variables and with the condition that when $l = 0$, $I = I_0$.

$$-\frac{dI}{I} = aC\,dl$$

$$-\ln\frac{I}{I_0} = aCl \qquad \text{...(5.3)}$$

or

$$I = I_0\, e^{-aCl}$$

Since it is easier to use logarithm to the base ten rather than natural logarithm, the expression (5.3) can be written as:

$$-\log\frac{I}{I_0} = \frac{aCl}{2.303} = \varepsilon Cl \qquad ...(5.4)$$

where ε is the *molar absorption coefficient or molar absorptivity* (molar extinction coefficient). A chemist is familiar with another quantity called absorbance or optical density (A). It is defined as:

$$A = \log\frac{I_0}{I} = \varepsilon Cl \qquad ...(5.5)$$

The optical density has no units. These days the spectrophotometers are calibrated in the absorbance scale rather than transmittance scale. The unit of ε is $dm^3\ mol^{-1}\ cm^{-1}$. In SI system, concentration is measured in mol m^{-3} and l in *m*, so the units for ε (expressed as κ) will be m^2 mol^{-1} and equation (5.5) can be written as:

$$\ln\frac{I_0}{I} = \kappa\, C'l \qquad ...(5.5a)$$

In molecular dimensions equation (5.3) can also be written as:

$$I = I_0\, e^{-\sigma N x} \qquad ...(5.5b)$$

where N is the number of molecules per m^3, x is the thickness in *m*, so the unit for σ is m^2, also known as *absorption cross section.* It may be mentioned here that σ, although has the units of area it should not be interpreted as the area of the absorbing molecule. Equations (5.5*b*), (5.5) and (5.5*a*) have been derived for a particular monochromatic radiation. Thus the expression tells that molar absorption coefficient ε not only depends on the wavelength of excitation but also on the nature of the absorbing material. The value of ε quoted generally in literature is that of the absorption band maximum, denoted by ε_{max}. It will be explained later that ε_{max} also varies with the nature of solvent and temperature, because the absorption band maximum also depends on these two parameters. The most intense electronic spectrum or fully allowed transition of a compound has the value of ε of the order of $10^5\ dm^3\ mol^{-1}\ cm^{-1}$. Values of ε_{max} below the order of $10^3\ dm^3\ mol^{-1}\ cm^{-1}$ correspond to the quasi-forbidden or probable transition and ε_{max} below 100 $dm^3\ mol^{-3}\ cm^{-1}$ refers to the forbidden transition.

The other quantity, which is also commonly used in the absorption spectrum, is the *integrated molar absorption coefficient, i.e.,*

$$A = \int_{\bar{v}_2}^{\bar{v}_1} \varepsilon\, d\bar{z} \qquad ...(5.6)$$

where v_1 and v_2 (frequencies) are the limits of the absorption spectrum. As seen from this relation, many absorption spectra are spread over a wide range of frequencies, and so monitoring the molar absorption coefficient for a single frequency may not depict a true indication of intensity of absorption of a molecular transition. The unit for integrated absorption coefficient is $L mol^{-1} cm^{-2}$. If the shape of absorption band is of Gaussian type, the integrated absorption coefficient can be approximated to the relation (5.6*a*).

$$\int_{\bar{\nu}_1}^{\bar{\nu}_2} \varepsilon \, d\bar{\nu} = 1.06 \, \varepsilon \, \Delta\bar{\nu}_{1/2} \qquad ...(5.6a)$$

where $\Delta\bar{\nu}_{1/2}$ is the band width at half the maximum height. In general ν_{max} depends on temperature and thus the value of ε_{max} will also be a function of temperature. But the value of integrated absorption coefficient does not depend on temperature. However bandwidth ($\Delta\bar{\nu}_{1/2}$) does depend on temperature.

When more than one kind of absorbing species is present in the solution, the absorbance is given as:

$$A = l \, \Sigma \, \varepsilon_i \, c_i \qquad ...(5.7)$$

where ε_i is the molar absorption coefficient and c_i is the concentration of the *i*th species. The additivity law is of great importance to chemist, because it allows the spectroscopists to carry out the analysis of solution containing several species.

Deviation from Beer-Lambert's law may be real or apparent. Since in the derivation of expression (5.5) radiation used is assumed to be a perfectly monochromatic, the law can be true only for monochromatic radiation. Exit beam from the spectrophotometer, whose energy is measured or compared with some standard in the calculation of molar absorption coefficient always consists of a range of wavelengths, determined by the dispersive power of the grating or prism and the slit width used. Thus for this law to be valid over this range of wavelengths, molar absorption coefficients of the species should not vary too much. Although this law is valid over large range of concentration of the species, still a non-linear relation is observed between absorbance and concentration. This happens when a molecule undergoes association or dissociation, *e.g.*, aqueous solutions of dyes, which are prone to dimerize or polymerize, rarely follow Beer-Lambert's law over an extended concentration range.

Example 5.1: (*a*) Light of wavelength 400 nm is passed through a cell of 1 mm path length containing 10^{-3} mol dm^{-3} of compound A. If the absorbance of this solution is 0.25, calculate the molar absorption coefficient and the transmittance. (*b*) Derive the expression relating absorption cross-section (σ) and molar absorption coefficient (ε).

Solution: (*a*) Using equation (5.5), A = 0.25, C = 10^{-3} mol dm^{-3}, l = 0.1 cm

$$\varepsilon = \frac{0.25}{(10^{-3} \text{ mol dm}^{-3})\,(0.1 \text{ cm})} = \frac{0.25}{10^{-4} \text{ mol dm}^{-3} \text{ cm}}$$

$$= 2.5 \times 10^{3} \text{ dm}^{3} \text{ mol}^{-1} \text{ cm}^{-1}$$

Now $$A = \log \frac{I_0}{I} = 0.25 \tag{A}$$

$$\frac{I_0}{I} = 1.778 \text{ and } \frac{I}{I_0}\ 0.56$$

Transmittance $$T = \frac{I}{I_0} = 0.56 \times 100 = 56\%$$

Note: The molar absorption coefficient is some times expressed in $mol^{-1}\ cm^2$ unit. This choice of unit suggests that ε can be interpreted as a molar cross-section for photon absorption. Thus the present example gives the value of ε as $2.5 \times 10^6\ mol^{-1}\ cm^2$ as $1\ dm^3 = 1000\ cm^3$.

(*b*) Equation (5.5*b*) can be written as:

$$\log \frac{I_0}{I} = \frac{\sigma N x}{2.303} \tag{B}$$

Comparing equations (B) and (5.5) we get:

$$\frac{\sigma N x}{2.303} = \varepsilon C l \tag{C}$$

Now in the above equation, before we compare any thing, we must express the quantities in proper units.

$$C\ (mol\ dm^{-3}) = C\ (mol\ dm^{-3})\ (10^3\ dm^3\ m^{-3})$$

$$x(m) = 1\ (cm)\ (10^{-2}\ m\ cm^{-1})$$

$$N = \text{number of molecules per } m^3 \text{ in the concentration C}$$

$$= C\ (mol\ dm^{-3})\ (10^3\ dm^3\ m^{-3})\ (N_{av}).$$

Now substituting the above quantities in equation (C).

$$\frac{\sigma\ (C\ mol\ dm^{-3})\ (10^3\ dm^3\ m^{-3})\ (N_{av})\ (1 \times 10^{-2}\ m)}{2.303} = \varepsilon C l$$

$$\sigma = \frac{(2.303)\ (\varepsilon)\ (10^2\ cm\ m^{-1})}{(N_{av})\ (10^3\ dm^3\ m^{-3})}$$

Substituting the values of the respective quantities, we get

$$\sigma = \frac{(2.303)\ (2.5 \times 10^3\ dm^3\ mol^{-1}\ cm^{-1})\ (10^2\ m^{-1} cm)}{(6.022 \times 10^{23}\ mol^{-1})\ (10^3\ dm^3\ m^{-3})} = 9.56 \times 10^{-22}\ m^2$$

Example 5.2: A solution contains a mixture of two substances A and B, whose absorption spectra overlap. The absorbances of the given solution at wavelength 300 and 350 nm are 0.24

and 0.66 respectively. Calculate the concentration of species A and B if ε_{300} (A) = 2 × 10^3 dm^3 mol^{-1} cm^{-1}, ε_{350} (A) = 10^2 dm^3 mol^{-1} cm^{-1}, ε_{300} (B) = 10^5 dm^3 mol^{-1} cm^{-1}, ε_{350} (B) = 5 × 10^5 dm^3 mol^{-1} cm^{-1}. The path length is 1 cm.

Solution: Using equation (5.7)

$$A_{300} = l\,\varepsilon_{300}\,(A)\,C_A + l\,\varepsilon_{300}\,(B)\,C_B \qquad ...(5.8)$$

$$A_{350} = l\,\varepsilon_{350}\,(A)\,C_A + l\,\varepsilon_{350}\,(B)\,C_B \qquad ...(5.9)$$

where C_A and C_B are the concentrations of species A and B. Substituting the respective values and taking l = 1 cm, we get

$$0.24 = (2 \times 10^3\ dm^3\ mol^{-1}\ cm^{-1})\,(1\ cm)\,(C_A) + (10^5\ dm^3\ mol^{-1}\ cm^{-1})\,(1\ cm)\,(C_B)$$

$$= 2 \times 10^3\,C_A + 10^5\,C_B \qquad ...(5.10)$$

and

$$0.64 = 10^3\,C_A + 5 \times 10^5\,C_B \qquad ...(5.11)$$

Solving equations (5.10) and (5.11), one gets: C_A = 5.45 × 10^{-5} mol dm^{-3} and C_B = 1.31 × 10^{-6} mol dm^{-3}.

Note: When there are more than one absorbing species (n) in the solution and their absorption spectra overlap, the concentration of each species can be calculated by generating as many equations as the number of species. This is done by measuring the absorbances at n wavelengths and independently calculating the molar absorption coefficients of each species at these wavelengths. For example, consider a solution containing two species A and B. Measure the molar absorption coefficients of A and B at two wavelengths λ_1 and λ_2. Thus the absorbance of this solution containing mixture of A and B for l = 1 cm is given as:

$$A\,(\lambda_1) = C_A\,\varepsilon_A\,(\lambda_1) + C_B\,\varepsilon_B\,(\lambda_1) \qquad ...(5.8a)$$

$$A\,(\lambda_2) = C_A\,\varepsilon_A\,(\lambda_2) + C_B\,\varepsilon_B\,(\lambda_2) \qquad ...(5.8b)$$

Solving these equations, one gets

$$C_A = \frac{A\,(\lambda_1)\,\varepsilon_B\,(\lambda_2) - A\,(\lambda_2)\,\varepsilon_B\,(\lambda_1)}{\varepsilon_A\,(\lambda_1)\,\varepsilon_B(\lambda_2) - \varepsilon_B\,(\lambda_1)\,\varepsilon_A\,(\lambda_2)} \qquad ...(5.12)$$

$$C_B = \frac{A\,(\lambda_1)\,\varepsilon_A(\lambda_2) - A\,(\lambda_2)\,\varepsilon_A(\lambda_1)}{\varepsilon_B\,(\lambda_1)\,\varepsilon_A\,(\lambda_2) - \varepsilon_A\,(\lambda_1)\,\varepsilon_B\,(\lambda_2)} \qquad ...(5.13)$$

5.2 THEORIES OF ABSORPTION OR EMISSION SPECTRUM

In this section we will briefly discuss the outlines of the results of classical and quantum mechanical theories of intensities of absorption of radiation. In both theories, electrons are displaced from their mean position by the interaction of electrical field of the radiation, when the frequency of the radiation matches with the characteristic frequency of the electronic transition of the atoms or molecules. Due to this, the induced electrical dipole moment is created in the molecule. Thus intensity of absorption depends on the strength of this induced electrical dipole moment. Only difference between two theories is the procedure of calculating the induced dipole moment and the interpretation of the characteristic frequencies of electrons in the atoms or molecules.

5.2.1 CLASSICAL THEORY

In classical theory, the electron bound to an atom or a molecule is considered to oscillate in a simple harmonic motion with respect to the positive centre, called nucleus, with a characteristic frequency of the molecule. This characteristic frequency can be calculated from the mass of electron and the binding force between the nucleus and electron. Now if the electron is displaced from its mean position by a distance x_0, the electric dipole moment of magnitude, ex_0, is created. In this theory, the acceleration represented in this oscillating motion caused the electron to loose some energy in the form of emission and this is proportional to $(ex_0)^2$, where x_0 is the amplitude of the oscillatory motion. It is well understood in classical theory that this steady loss of energy by the electron in the form of emission of light, during the harmonic oscillation, cannot predict the stability of the arrangement of electron around the positively charged nucleus. This theory is inconsistent with the behaviour of electrons in atoms and molecules. However, this theory helps in defining certain terms quantitatively. The quantity, which is quite commonly used, is called as *oscillator strength* (f). It is defined as the effective number of dispersion electrons, which are oscillating with a characteristic frequency ν_i. For example, spectral lines emitted by an atom can be associated with the corresponding electronic oscillator strength with a particular value of f. It turns out that intense transitions have oscillator strengths of *about unity.*

The oscillating strength can be expressed in many ways and their relation to each other will be given in example 5.4. The oscillator strength can be related to the experimentally known quantity ε, determined at each frequency by the expression (5.14) in cgs units:

$$f = \frac{2303\ \mathrm{m}_e\, c^2}{\pi\, \mathrm{N}_{av}\, e^2} \int \varepsilon(\bar{\nu})\, d\bar{\nu} = 4.39 \times 10^{-9} \int \varepsilon(\bar{\nu})\, d\bar{\nu} \qquad ...(5.14)$$

where m_e is the mass of an electron in gram, e is the charge of the electron in electrostatic unit, c the velocity of light, N_{av} is the Avogadro number, $\bar{\nu}$ is the frequency expressed in wave numbers (cm^{-1}), factor 2303 involves the conversion from natural logarithms as well as the number of cubic centimeters in a liter required by the introduction of the molar absorption coefficient. The advantage of introducing the oscillator strength is that it can be interpreted theoretically and can be calculated with the help of quantum mechanics. The integral $\int \varepsilon(\bar{\nu})\, d\bar{\nu}$ gives the area of the transition band under consideration. The approximate value of the area under the band can be determined from the relation:

$$\int \varepsilon(\nu)\, d\nu = 1.06 \times \varepsilon \times \text{width of the band at half the maximum height} \qquad ...(5.6b)$$

The oscillator strength in SI units can be expressed as:

$$f = \left[\frac{(4\pi\varepsilon_0)\, \mathrm{m}_e\, c}{\pi\, \mathrm{N}_{av}\, e^2}\right] \int k d\nu \qquad ...(5.14a)$$

where $4\pi\varepsilon_0$ is the permittivity factor. After substituting the values of the constants, the above relation can be written as:

$$f = 6.27 \times 10^{-19} \int kd\,\bar{\nu} \qquad ...(5.14b)$$

A point to be noted here is that although the expression for f contains so many constants but the units of each quantity is such that the product of all gives rise to a dimensionless quantity and for three dimensional oscillator its value is 1.

Example 5.3: In the longest wavelength band of 5-aminoindazole in water as a solvent, molar absorption coefficient at the band maximum is 5.4×10^3 dm^3 mol^{-1} cm^{-1} and the bandwidth at half the maximum height is 4047 cm^{-1}. Calculate the oscillator strength of this band.

Solution: Using equation (5.6*a*)

$$\int \varepsilon(\bar{\nu})\, d\bar{\nu} = 1.06\ (5.4 \times 10^3 \text{ dm}^3 \text{ mol}^{-1} \text{ cm}^{-1})\ (4047 \text{ cm}^{-1})$$

$$= 2.317 \times 10^7 \text{ dm}^3 \text{ mol}^{-1} \text{ cm}^{-2}$$

Using equation (5.14)

$$f = 4.39 \times 10^{-9} \int \varepsilon\ (\nu)\ d\nu = 4.39 \times 10^{-9} \times 2.317 \times 10^7 = 0.1$$

5.2.2 QUANTUM THEORY

Before quantum mechanics was developed, Einstein explained the intensities of the transitions involved between two quantum states. His theory is based on the following facts:

(*i*) Energies of the states are quantized, as suggested by Bohr.

(*ii*) Molecules in the two states are present according to the Maxwell-Boltzmann's distribution.

(*iii*) Equilibrium distribution of energy among the frequencies of radiation is given by Planck's equation for black body radiation.

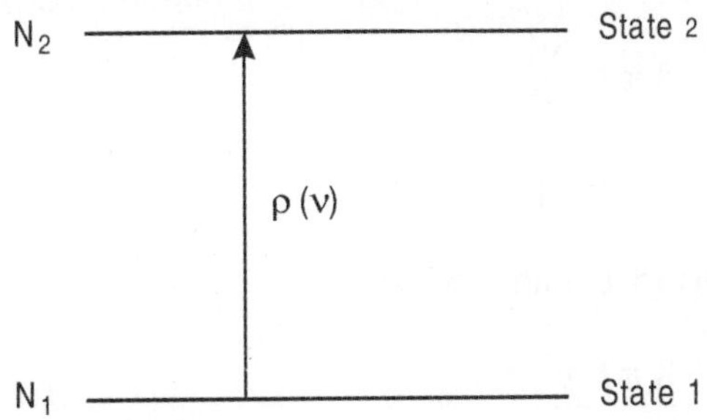

Figure 5.2: *Absorption of radiation between two quantum states depicting Einstein's theory of absorption of radiation*

Consider two states of atom: state 1 is of lower energy and state 2 of higher energy, as depicted in Fig. 5.2. Difference between the energies of these states is related by frequency ν. N_1 is the number of atoms in state 1 and N_2 in state 2. ρ (ν) erg cm^{-3} per unit interval of frequency unit (s^{-1}) is the density of the radiation at the frequency ν, which is capable of inducing transition between the states 1 and 2. Now if the atom is placed in the isotropic radiation of above characteristics, number of atoms absorbing radiation per second is given by the relation $B_{12} N_1 \rho (\nu)$, where B_{12} is the constant, characteristic of atom and is called Einstein's coefficient of induced absorption $1 \rightarrow 2$. By successful interpretation, Einstein suggested that the atoms from state 2 can revert back to the state 1 by two paths: (*i*) by spontaneous emission and (*ii*) by emission induced by the radiation of density $\rho (\nu)$. Thus the number of atoms coming to ground state is given by the expression: $N_2 [A_{21} + B_{21} \rho (\nu)]$, where A_{21} and B_{21} are the Einstein's coefficient of spontaneous and induced emission from state 2 to state 1 respectively. State of equilibrium can be determined by equating rate of emission with rate of absorption.

$$N_1 B_{12} \rho (\nu) = N_2 B_{21} \rho (\nu) + N_2 A_{21} \quad \text{...(5.15)}$$

$$\frac{N_2}{N_1} = \frac{B_{12} \rho(\nu)}{B_{21} \rho(\nu) + A_{21}} \quad \text{...(5.16)}$$

According to Maxwell-Boltzmann distribution,

$$\frac{N_2}{N_1} = \frac{g_2 e^{-\Delta\varepsilon / k_b T}}{g_1} = \frac{g_2 e^{-h\nu / k_b T}}{g_1} \quad \text{...(5.17)}$$

g_1 and g_2 the are degeneracy of the states 1 and 2 respectively; k_b is the Boltzmann's constant. Equating (5.16) and (5.17) and rearranging the resulting equation, it reduces to:

$$\rho (\nu) = \frac{\dfrac{A_{21}}{B_{21}}}{\dfrac{g_1}{g_2} \dfrac{B_{12}}{B_{21}} e^{h\nu / k_b t} - 1} \quad \text{...(5.18)}$$

According to Planck's radiation law, which considers a large number of identical systems in equilibrium with radiation at temperature T, density of radiant energy is given by

$$\rho (\nu) = \frac{8 \pi h \dfrac{\nu^3}{c^3}}{e^{\frac{h\upsilon}{k_b T}} - 1} \quad \text{...(5.19)}$$

Comparing the equations (5.18) and (5.19)

$$\frac{A_{21}}{B_{21}} = \frac{8 \pi h \nu^3}{c^3} \quad \text{...(5.20)}$$

and $$B_{12}\, g_1 = B_{21}\, g_2 \quad \text{...(5.21)}$$

The beauty of this formulation by Einstein is that after 40 years, the idea of stimulated emission helped in the invention of radiations amplifying devices, the MASER and LASER. In other words the intensity of transition is related to the calculation of Einstein's coefficients and can be nicely discussed and understood with the help of quantum mechanics.

5.2.3 QUANTUM MECHANICAL FORMULATION

After the development of quantum mechanics, intensities of transitions were explained with the help of time dependent Schrödinger wave equation. This is because in the act of absorption or emission, the system no longer stays in the stationary states, as defined by Bohr's theory of atomic structure but is forced to go from one state to another state if the proper amount of energy is supplied. As explained in earlier sections, the electron keeps moving in a non-radiating stationary states and its motion is described by a wave like distribution of probability density function. In the presence of radiation, the electron is forced to jump from one stationary state to another, under the influence of radiation-induced perturbation. This will be possible if the frequency of radiation matches with the frequency of transition, as defined by Bohr, *i.e.*,

$$\nu = \frac{E_2 - E_1}{h}$$

where E_1 and E_2 are the energies two different stationary states respectively. Thus one can visualize that quantum mechanical description also involves the terms like characteristic frequencies and induced dipole moment, as described in the classical mechanics. But the only difference is that characteristic frequencies and induced dipole moment is due to effect of displacement of electron charge accompanying the transition.

The first point is very trivial, *i.e.*, the electron will only be displaced from the state 1 to state 2 if the characteristic frequency matches with the transition frequency. Let us try to understand the second quantity, *i.e.*, induced dipole moment. Consider the two states with energies E_1 and E_2, described by the wave functions ψ_1 and ψ_2 respectively. For simplicity let us assume that oscillations of the electric field of the radiation are parallel to the x-axis. Under the influence of radiation field, the probability of radiation-induced transition is found to depend on the square of the induced dipole moment, ($e\,x_{12}$), called transition moment, where x_{12} is the average displacement of electron along the x-axis. The average displacement is considered because this displacement is a periodic function of time, due to the wave nature of radiation. In the above case, x_{12} is measured from the centre of gravity of positive charge, which is considered as a fixed point. The average displacement of electron, *i.e.*, charge over two states is described by:

$$\bar{x}_{12} = \langle x_{12} \rangle = \int \psi_2^* \, x \, \psi_1 \, dx \qquad \text{...(5.22)}$$

where ψ_2^* is a complex conjugate of ψ_2 and thus the transition moment is defined by the equation (5.23).

$$e \langle x_{12} \rangle = \int \psi_2^* \, \varepsilon \hat{x} \, \psi_1 \, dx = e \int \psi_2^* \, x \, \psi_1 \, dx \qquad \text{...(5.23)}$$

Now one can calculate the similar terms for all possible stationary states and interactions possible between these stationary states. These quantities can be arranged as the components of elements of matrix. In this matrix some of the terms will be like $x_{11} = \int \psi_1^* \, x \, \psi_1 \, dx$, representing the average value of x coordinate of the electrons in the same stationary state and the other terms will be like $x_{12} = \int \psi_2^* \, x \, \psi_1 \, dx$, representing the time dependent average displacement of the electron undergoing transitions. Thus x_{12} is therefore nothing but the effective length of induced dipole. Since the square of dipole moment integral defines the probability of transition, the square of dipole length is called *dipole strength* as D_{12}.

The above relations were developed with the help of quantum mechanics to relate them with the Einstein's coefficients A_{21} and B_{21}. These equations are:

$$A_{21} = \frac{64\,\pi^4\,\nu^3\,M_{12}^2}{3\,h\,c^3} \qquad \text{...(5.24)}$$

and $$B_{12} = B_{21} = \frac{8\,\pi^3\,M_{12}^2}{3\,h^2} \qquad \text{...(5.25)}$$

where M_{12} and M_{21} are the quantum mechanical description of absorbance and emission processes, called transition moment. As defined earlier:

$$M_{12} = M_{21} = e\,r_{12} = e \int \psi_2^* \;\hat{r}\; \psi_1\, dr \qquad \text{...(5.26)}$$

Since dipole moment is a vector quantity, so is the transition moment and can be resolved into three different components, *i.e.*, M_x, M_y and M_z. Three components are related to M_{12} or $M_{21} = M$ by the relation (5.27).

$$M_{21}^2 = M_{12}^2 = M_x^2 + M_y^2 + M_z^2 \qquad \text{...(5.27)}$$

where

$$M_x = \int \psi_2^* \;\hat{x}\; \psi_1\, dx \qquad \text{...(5.27}a\text{)}$$

$$M_y = \int \psi_2^* \;\hat{y}\; \psi_1\, dy \qquad \text{...(5.27}b\text{)}$$

$$M_z = \int \psi_2^* \;\hat{z}\; \psi_1\, dz \qquad \text{...(5.27}c\text{)}$$

If the wave functions are known, M_{12} or M_{21} can be calculated by evaluating the integral. The connection between the transition moment and the observed intensity is:

$$f = \frac{8\,\pi^2\,m_e\,c\bar{\nu}}{3\,h\,e^2}\,|M_{12}|^2 \qquad \text{...(5.28)}$$

where m_e and e are the mass and charge of the electron respectively and ν is the frequency (s^{-1}) of absorption band and M is measured in electrostatic units (debyes). If the value of f is obtained from equation (5.14), one can calculate the value of M_{12}^2 by the relation:

$$M_{12} = \left(\frac{3\,h\,e^2\,f}{8\,\pi^2 m_e\,c\bar{\nu}}\right)^{\frac{1}{2}} = 0.0958\left(\frac{\int \varepsilon d\bar{\nu}}{\bar{\nu}}\right)^{\frac{1}{2}} \qquad \text{...(5.28}a\text{)}$$

Knowing wave functions, the intensity of the line can be calculated and thus it is an important link between theory and experiment.

There is also an interesting relationship between the intensity of absorption band and the mean lifetime of the molecule in its excited state before it returns to the lower state with emission of radiation. This lifetime is called *radiative lifetime*. This is derived from Einstein spontaneous emission coefficient (A_{21}), which means as how many times per second the molecule in the excited state returns to the lower state with emission of radiation. The mean radiative lifetime thus in the excited state is defined as:

$$\tau_{FM} = \frac{1}{A_{21}}$$

The value of A_{21} (without derivation), can be related to the experimental quantity oscillator strength by the relation (in cgs units):

$$A_{21} = \frac{64\,\pi^4\bar{\nu}^3 G_1 M_{21}^2}{3\,h} = \frac{8\,\pi^2 e^2 \bar{\nu}^2 G_1 f_{12}}{mc\,G_2} \qquad \text{...(5.24}a\text{)}$$

Thus: $$\tau_{FM} = \frac{m_e c}{8\pi^2 e^2 \bar{\nu}^2 f_{12}}\left(\frac{G_2}{G_1}\right) = \frac{1.5 G_2}{\bar{\nu}^2 f_{12}\cdot G_1} = \frac{Nav}{8\pi\, 2303\bar{\nu}^2 c \int \varepsilon\, d\bar{\nu}}\left(\frac{G_2}{G_1}\right)$$

$$= \frac{3.47\times10^8}{\nu^2 \int \varepsilon\, d\nu}\left(\frac{G_2}{G_1}\right) \qquad \text{...(5.29)}$$

It is obvious from equation (5.29) that relatively shorter mean lifetime of the excited state is associated with an intense absorption band and longer mean lifetime with weak absorption band. If the molecule is deactivated to the lower state by only emission process, *i.e.*, fluorescence quantum yield is unity the value of the τ_{FM} from equations (5.24*a*) or (5.29) is the actual and maximum possible mean lifetime of the excited state. This is also called *intrinsic lifetime* of the excited state. If the molecule is being deactivated by other processes, also *e.g.*, non-radiative processes, collisional deactivation, energy transfer etc., the mean lifetime of the molecule in its excited state will be shorter than the intrinsic lifetime. In general the intrinsic lifetime of the molecule is of the order of 10^{-8} to 10^{-9} s for the stronger absorption band. In the end, it may be mentioned here that this is the most simple and approximate method to calculate the mean lifetime of the molecule in the excited state. More sophisticated methods are available but still the experimental procedure gives the actual value of mean lifetime.

Example 5.4: (*a*) The oscillator strength of the electronic transition in 5-aminoindazole at 316.7 nm is 0.1. Calculate the transition moment as well as dipole length of this transition. (*b*) Calculate the mean lifetime of this molecule in its first excited state using the data of example (5.3). ($\Delta\bar{\nu}$ = 4047 cm^{-1}, $\int \varepsilon\, d\bar{\nu}$ = 2.317 × 10^7).

Solution: (*a*) We can make use of the equation (5.28). Here

$$\bar{\nu} = \frac{1}{\lambda} = \frac{1}{(316.7\times10^{-9}\,\text{m})\,(10^2\ \text{cm m}^{-1})} = 31575.6\ \text{cm}^{-1}$$

$$M = 0.0958\left(\frac{2.317\times10^7}{31575.6}\right)^{1/2} = 2.595\ \text{D}$$

$$D_{12} = \text{dipole length} = \frac{M}{e} = \frac{2.59\times10^{-18}\ \text{cgs cm}}{4.8\times10^{-10}\ \text{cgs}} = 0.54 \times 10^{-8}\ \text{cm.}$$

(*b*) Using equation (5.29):

$$\tau_{FM} = \frac{3.47\times10^{8}}{(31575.6)^2\ (2.317\times10^{7})} = 1.50 \times 10^{-8}\ \text{s} = 15\ \text{ns}$$

5.2.4 SELECTION RULES

In the last section, we have learnt how to calculate the intensity of each transition by evaluating the transition moment integral. The equations (5.24) and (5.25) give the accurate values. But the evaluation of these integrals is a quite cumbersome job. If we are only interested in knowing whether these transition moment integrals are zero or non-zero, this can be easily carried out. There are certain rules by which we can easily find out qualitatively whether the equations (5.27, *a, b, c*) are finite or zero. These rules are called *'selection rules'. If the transition moment integral is zero, the transition is called forbidden and if the dipole moment integral is non-zero or finite it is called 'allowed transition'*. In the following section, we will discuss certain rules. These rules help us in finding out whether a particular transition will be allowed or forbidden, *i.e.,* whether the transition moment integral will be zero or non-zero.

Spin Selection Rules: The dipole moment operator works only on the space part of the wave function. Thus equation (5.23) can be written as:

$$M = \int \Psi_2^{space}(er)\Psi_1^{space}(er)dr^{space} \int \Psi_2^{spin}\Psi_1^{spin}\, d\tau^{spin} \qquad ...(5.30)$$

Since the second integral of equation (5.30) only involves the dot product of the spin wave functions, it is quite evident that this integral will be zero until the wave functions for the spin belong to the same state, *i.e.,*

$$\Delta S = 0 \qquad ...(5.31)$$

This means that for allowed transitions, the change in the spin quantum number should be zero.

Equation (5.30) has been written with the assumption that the interaction between the wave function for space part and spin part is negligibly small, *i.e.,* orbital motion and spin motion of an electron are independent of each other. Thus, the total wave function can be written as the product of two wave functions depicting the space part and spin part. This is only true for atoms and molecules of light masses. For heavy particles, the interaction between the space part and spin part is finite. Thus the total wave function cannot be written as the product of wave functions depending only on one variable. Therefore for heavy particles (atoms or molecules) the selection rule ($\Delta S = 0$) is not valid and the transitions are observed even for $\Delta S \neq 0$. For example, the strongest transition in mercury atom corresponding to 253.7 nm is due to 6 ($^3P_1$) $\rightarrow$ 6 ($^1S_0$).

Example 5.5: Show that the transition between the singlet and triplet state is forbidden.

Solution: The wave functions for the spin part of the singlet and triplet states are:

$$\Psi_s = [\alpha(1)\ \beta(2) - \alpha(2)\ \beta(1)] \qquad ...(5.32)$$

and

$$\Psi_T = \alpha(1)\ \alpha(2) \qquad ...(5.32a)$$

$$= \beta(1)\ \beta(2) \qquad ...(5.32b)$$

or

$$= [\alpha(1)\ \beta(2) \pm \alpha(2)\ \beta(1)] \qquad ...(5.32c)$$

The normalization constants are omitted for sake of simplicity. Now considering one of the triplet state wave functions, equation (5.23) becomes:

$$\int \Psi_1^{spin}\ \Psi_2^{spin}\ d\tau^{spin} = \int [\alpha(1)\ \beta(2) - \alpha(2)\ \beta(1)]\ [\alpha(1)\ \beta(2) + \alpha(2)\ \beta(1)]\ d\tau^{spin}$$

$$= \int \alpha(1)\ \alpha(1)\ d\tau_1^{spin} \int \beta(2)\ \beta(2)\ d\tau_2^{spin} + \int \alpha(1)\ \beta(1)\ d\tau_1^{spin} \int \beta(2)\ \alpha(2)\ d\tau_2^{spin} - \int \alpha(2)\ \beta(2)\ d\tau_2^{spin} \int \alpha(1)\ \beta(1)\ d\tau_1^{spin} - \int \alpha(2)\ \alpha(2)\ d\tau_2^{spin} \int \beta(1)\ \beta(1)\ d\tau_1^{spin}$$

Since the wave functions are normalized, the first and fourth integral will be unity, whereas the second and third integrals will be zero. Therefore

$$\int \Psi_1^{spin}\ \Psi_2^{spin}\ d\tau^{spin} = 1 + 0 + 0 - 1 = 0$$

Thus, this transition is forbidden. It can also be shown similarly that the integrals

$$\int [\alpha\ (1)\ \beta\ (2) - \alpha\ (2)\ \beta\ (1)]\ \alpha\ (1)\ \alpha\ (2)\ d\tau^{spin}$$

and

$$\int [\alpha\ (1)\ \beta\ (2) - \alpha\ (2)\ \beta\ (1)]\ \beta\ (1)\ \beta\ (2)\ d\tau^{spin}$$

are also zero, indicating that the singlet to triplet transition is not allowed.

Symmetry Selection Rules

Transitions are forbidden for which the transition moment integral vanishes, because of the symmetry properties of the integrand. To be non-zero, dipole moment integral ($\int \psi_2\ M\ \psi_1\ d\tau$) must be totally symmetric. Since ψ_1 in general is symmetric, the transition will be allowed only if the product M ψ_2 is symmetric. Because the three components of the dipole moment operator transform differently under the covering operations of the various point groups, a transition may be allowed with polarization along one or two axis, but forbidden with polarization along the other axis. In molecules, which have centre of symmetry, the general selection rule is that only $u \leftrightarrow g$ and $g \leftrightarrow u$ transitions are allowed. It is also known as *Laporte's rule*. Transitions between states having similar parity, *i.e.*, $g \leftrightarrow g$ and $u \leftrightarrow u$ are forbidden and are sometimes called as *parity forbidden*.

There are also many selection rules but these are very specific and will depend on the branch of spectroscopy and will be discussed at the proper places rule of this chapter.

$u \leftrightarrow g$, $g \leftrightarrow u$ are allowed and $u \leftrightarrow u$, $g \leftrightarrow g$ are forbidden ...(5.33)

If we consider that $\pi\ \pi^*$ is the fully allowed transition of oscillator strength (F_A) of the order of unity, then any other transition will have its oscillator strength F as to order of magnitude given by:

$$F = f_s\ f_0\ f_m\ f_p\ F_A \qquad ...(5.34)$$

where f factors differ from unity as follows for the different kinds of forbiddenness which have been described so far:

f_s = spin forbidden transition involving change of multiplicity, 10^{-5} for second row elements.

f_o = overlap forbidden involving change of position of electronic charge, 10^{-2} for $n \leftrightarrow \pi^*$ transition of second row heteroatom.

f_m = momentum forbidden or orbitally forbidden involving large change of linear or angular momentum, 10^{-1} to 10^{-3} for condensed ring system.

f_p = parity forbidden, initial and final states both are either even or odd, 10^{-1} for condensed systems.

Example 5.6: In octahedral complexes of transition elements, the five degenerate orbitals split into a set of two, t_{2g} (three orbitals) and e_g (two orbitals). Will the transition from d to d orbital be allowed or forbidden?

Solution: According to Lapport's rule, the transition is allowed between the orbitals of parity g and u. On this basis the d-d transition is forbidden.

Example 5.7: The ground state and the first three excited states of oxygen molecule have the following symbols: $^3\Sigma_g$, $^1\Sigma_g$, $^1\Delta_g$, and $^3\Delta_u$. Based on the above selection rules, indicate the allowed and forbidden transitions starting from the ground state.

Solution: Above selection rules indicate that the allowed transitions starting from the ground state are:

(*i*) $^3\Sigma_g$ to $^1\Sigma_g$ transition is forbidden, based on spin and parity selection rules.

(*ii*) $^3\Sigma_g$ to $^1\Delta_g$ transition is forbidden, based on parity, spin and component of orbital angular momentum along the molecular axis ($\Delta\lambda = 0, \pm 1$). Here $\lambda = 0$ and 2 for Σ and Δ states respectively.

(*iii*) $^3\Sigma_g$ to $^3\Sigma_u$, is an allowed transition, based on all counts. $\Delta S = 0$, $u \leftrightarrow g$, and $\Delta\lambda = 0$.

5.3 DIFFERENT KINDS OF SPECTROSCOPY

This section will briefly describe the various types of spectroscopic techniques utilized to know the structure of molecules and have been described in order of increasing size of the quantum of radiation used. Classification of electromagnetic radiation into different Regions and branches of spectroscopy associated with them are compiled in appendix A-6.

***Nuclear Magnetic Resonance (NMR) Spectroscopy*:** NMR spectroscopy detects the transitions between the nuclear states of the system in presence of applied magnetic field and utilizes the radio frequency of the electromagnetic radiation. This part of radiation is of very small energy. The commonly used NMR spectrophotometers employ 100 MHz corresponding to 6.626×10^{-26} J (although very sensitive NMR spectrophotometers employing 800 MHz or of higher frequencies are available) and the energy of photon is much less than the thermal energy, k_bT (4.02×10^{-21} J at 300 K). Thus all the nuclear states are nearly equally populated. Detection of the signals requires very sensitive apparatus and large amount of sample. The latter difficulty has been over come by using Fourier transform NMR spectrophotometers (section 5.8).

***Electron Spin Resonance (ESR) Spectroscopy*:** Similar to NMR, ESR detects the transitions between the various spin states of the unpaired electrons of the systems when placed in the external magnetic field and utilizes the microwave frequency of the radiation. Commonly used frequencies are 9.5×10^3 MHz (X-band) or 1.2×10^4 MHz (K-band), but any frequency in the range of 2×10^3 to 3.6×10^3 MHz can be used. The energy of this part of the radiation is greater than radio wave but still considerably less than k_b T, *e.g.*, 10^4 MHz photon corresponds to 6.62×10^{-24} J and k_b T at 300 K is 4.02×10^{-21} J. Thus higher spin states will be still nearly equally populated, but as compared to NMR, the lower level will be populated more than the upper ones. Thus, the *saturation effects* (*when the populations of two states become equal*) are not so pronounced as in case of NMR.

***Rotational Spectroscopy*:** In rotational spectroscopy, the transitions are observed between the different rotational states of the molecules. The spacing between the rotational states corresponds to the microwave region of electromagnetic radiation, unless the molecules are of low molecular weights (H_2, D_2) or hydrogen halides. Thus similar remarks can be made as were mentioned in case of ESR.

***Vibrational Spectroscopy*:** Quantum states are evolved due to the vibrational motion of the nuclei around their mean positions and transitions occur between these states. This part of the spectrum consists of infrared radiation and thus commonly known as infrared spectroscopy. Commonly used infrared spectrophotometers scan the range from 4000 to 400 cm^{-1}, but lower than this wavenumbers can also be scanned. The energy of the photon in this part of the radiation is much greater than the thermal energy k_b T and thus the higher vibrationals levels are scarcely populated unless the far infrared part of the spectrum is involved. In this case, transitions generally take place from lower most energy state. The source of radiation is based on the principle of blackbody radiations and is produced by hot glowing body. The optics part consists of large single crystals of sodium chloride, lithium chloride or potassium bromide etc.

***Raman Spectroscopy*:** Raman spectroscopy also gives information about the rotational and vibrational energy levels, not by absorbing but by scattering of the monochromatic beam of radiation. The monochromatic radiation generally used is of visible region (to avoid the direct absorption of radiation because of electronic transitions) and it is allowed to strike the sample. The scattered light is measured at right angle to the incident radiation and consist mostly radiation of the same frequency, but small amount of lower and higher frequency radiations are also detected. The energy difference between these weak lines (or closely spaced to the incident radiation) and the energy difference between the strong and weak lines (or placed at greater distance from the incident light) gives the information about the rotational and vibrational spectra of the molecules. A new branch different from normal Raman Effect known as Resonance Raman Spectroscopy (RRS) came into being with the advent of lasers. Due to high intensity of the scattered radiation observed in RRS, very low concentration of analytical sample can be observed in contrast to normal Raman Effect. This phenomenon has made smooth roads in the study of biological systems.

***Electronic Spectroscopy*:** This branch of spectroscopy involves the transitions between the electronic states of the systems and the radiation involved varies from visible to far ultra violet region.

5.4 ROTATIONAL SPECTRUM

In this section we shall like to deduce molecular structure from the observation of radiation, which when absorbed increases the rotational energy of molecule. In the beginning the equations based on classical mechanics will be derived, to explain the rotation of diatomic molecules. Then quantum restrictions will be imposed to get the expression for rotational energy. This will be followed by the interaction of the radiation with the molecules to change its rotational energy and the comparison of the rotational spectrum with analytical expression to get information regarding molecular structure.

5.4.1 DIATOMIC MOLECULES

***Angular Momentum and Kinetic Energy*:** Consider a diatomic molecule AB, with mass of atom A as m_A and that of atom B as m_B. These two atoms are separated by a distance r as shown in Fig 5.3. To the first approximation we can assume that the internuclear distance r remains constant and molecule is rotating around the axis passing through centre of gravity of the molecule AB. This problem is also known as *rigid rotor* problem. The kinetic energy of the molecule can be written as:

$$\text{K.E.} = \frac{1}{2}m_A v_A^2 + \frac{1}{2}m_B v_B^2 \quad ...(5.35)$$

where v_A and v_B are the velocities of the atoms A and B respectively. It is nice to introduce certain quantities, which can describe the rotatory motion of the molecule AB, for example:

(*i*) Frequency of rotation (*i.e.,* the number of rotations that a particle makes per second) and can be denoted by $\nu_{rot.}$

(*ii*) Angular velocity: number of angular units or radians swept out per second and denoted as ω. Since the angle swept in one rotation is 2π, angle swept in ν_{rot} will be $2\pi\,\nu_{rot}$ and this will be the angular velocity, *i.e.,*

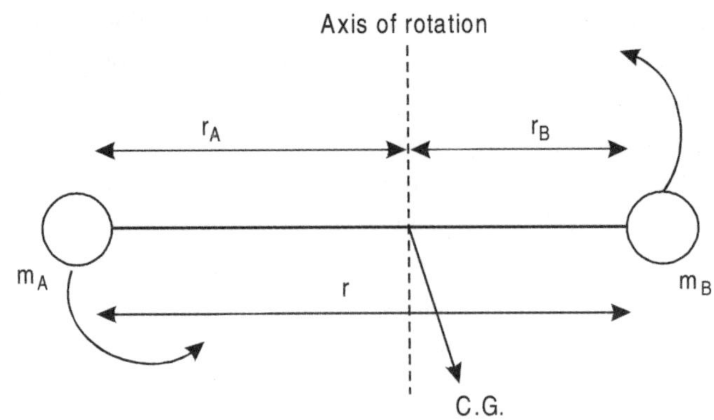

Figure 5.3: *Schematic diagram depicting the rotational motion of diatomic molecule*

$$\omega = 2\pi\, \nu_{rot} \qquad ...(5.36)$$

If r_A and r_B are the distances of atoms A and B from the centre of gravity, the relation between the linear velocity and the angular velocity will be given by the expression:

$$v_A = r_A\,\omega;\ v_B = r_B\,\omega \qquad ...(5.37)$$

Substituting these values in the equation (5.35),

$$\text{K.E.} = \frac{1}{2} m_A r_A^2 \omega^2 + \frac{1}{2} m_B r_B^2 \omega^2 = \frac{1}{2}(m_A r_A^2 + m_B r_B^2)\omega^2 = \frac{1}{2}\,\text{I}\omega^2 \qquad ...(5.38)$$

where
$$\text{I} = m_A r_A^2 + m_B r_B^2 \qquad ...(5.39)$$

is called the moment of inertia and for many particle system this is defined as:

$$\text{I} = \Sigma\, m_i\, r_i^2 \qquad ...(5.39a)$$

where m_i is the mass of the ith particle and r_i is its distance from the centre of gravity or from the axis of rotation of the system.

The moment of inertia in terms of bond length r can be calculated as follows. Since the centre of gravity is located by r_A and r_B, the relation (5.40),

$$m_A\, r_A = m_B\, r_B \qquad ...(5.40)$$

must hold. Similarly

$$r = r_A + r_B \qquad ...(5.41)$$

Using equations (5.40) and (5.41), we get

$$r_A = \frac{m_B}{m_A + m_B}\, r, \qquad r_B = \frac{m_A}{m_A + m_B}\, r \qquad ...(5.42)$$

Substituting the values of r_A and r_B in the equation (5.39)

$$\text{I} = \frac{m_B^{\,2} m_A\, r^2}{(m_A + m_B)^2} + \frac{m_A^{\,2}\, m_B\, r^2}{(m_A + m_B)^2} = \frac{m_A\, m_B\, r^2}{(m_A + m_B)^2}\,(m_A + m_B)$$

$$= \frac{m_A\, m_B\, r^2}{m_A + m_B} = \mu\, r^2 \qquad ...(5.43)$$

where μ is the reduced mass. As mentioned earlier, reduced mass is a hypothetical molecule, which allows the two body problem to be represented by a single particle of mass μ and thus makes the expression simple. Substituting the value of I in equation (5.38),

$$\text{K.E.} = \frac{1}{2}\,\text{I}\omega^2 = \frac{1}{2}\,\mu\, r^2\, \omega^2 \qquad ...(5.44)$$

gives the expression for the rotational energy of the molecule. In the expression (5.44) the potential energy term has not been included. This is because we have assumed that the distance between two atoms does not change during rotation and thus, the potential energy will be a constant quantity and can be taken as zero. Even if we add a constant term to the expression (5.44), the total energy of each rotational level will be increased by a constant amount, but the successive difference between the two energy levels will not change. So it will not make any difference to the rotational lines or transitions.

Here one point, which should be mentioned, is the introduction of angular momentum, as said in section (2.4.4). Angular momentum is related to angular velocity and the moment of inertia in a similar way as the linear momentum is related to the product of mass and linear velocity, *i.e.*,

$$\mathbf{L} = \text{Angular momentum} = \text{I}\omega \qquad ...(5.45)$$

$$\text{K.E.} = \frac{1}{2}\frac{(\text{I}\omega)^2}{\text{I}} \qquad ...(5.46)$$

5.4.2 QUANTUM RESTRICTIONS

Though the allowed energies of different rotational states can be obtained by using Schrödinger equation (2.247), it involves complicated mathematics. The simplest procedure is to calculate the expression for the energy with the help of classical mechanics and impose the quantum restrictions. Further it is easy to impose quantum restrictions on the angular momentum rather than on rotational energy. The angular momentum for the rotational motion is given by the expression:

$$\mathbf{L} = \text{I}\omega = [\text{J}(\text{J}+1)]^{\frac{1}{2}}\frac{h}{2\pi} \qquad ...(5.47)$$

where J = 0, 1, 2, ..., is called rotational quantum number. It can be shown that each level is (2J + 1) times degenerate except for J = 0 (equation 2.272). Substituting the value of **L** in equation (5.46), we get:

$$\text{E}_\text{J} = \frac{h^2}{8\,\pi^2\,\text{I}}\,(\text{J} + 1) \qquad ...(5.48)$$

where E_J is the rotational energy of the *j*th rotational level. Equation (5.48) gives the most important expression for the rotational spectroscopic studies. It shows how the rotational energies depend on the molecular property I and the rotational quantum number J. Figure 5.4 represents the energies of various rotational energy levels. One thing is quite evident that the rotational levels are not equally spaced.

In spectroscopy, the energies of the various quantum states are generally expressed in cm^{-1}, because, similar to frequency, energy is also directly proportional to wave number ($\bar{\nu}$, inverse of wave length). This can be achieved by dividing equation (5.48) by $h\,c$ (h is a Planck's constant and c is the velocity of light), *i.e.*,

$$\text{F (J)} = \frac{\text{E}_\text{J}}{h\,c} = \frac{h \times 10^{-2}}{8\pi^2\,\text{I}\,c}\,\text{J (J + 1)} \qquad ...(5.49)$$

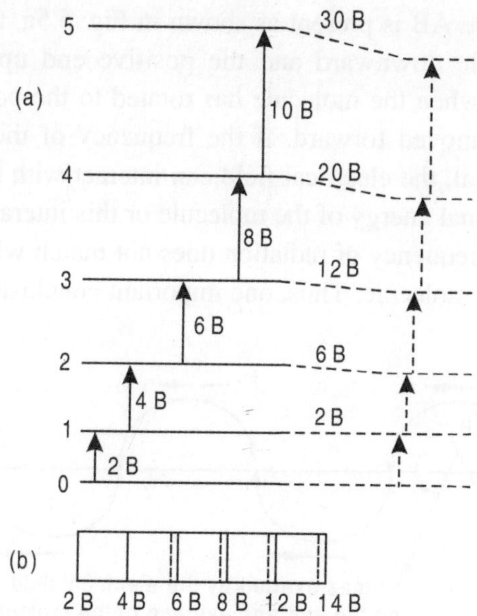

Figure 5.4: *Energy level diagram of rigid rotator and transitions. (a) The energy level diagram. (b) The resulting spectrum. The dotted lines show the effect of centrifugal distortion. The gap between the successive lines decreases as J increases. The gap shown may be larger as compared to actually observed*

where F (J) is known as *rotational term* and the factor 10^{-2} has been introduced as a constant because of the introduction of the SI unit for the respective constants in equation (5.49). For a given molecule, I is constant and a new term, known as *rotational constant,* B can be defined as:

$$B = \frac{h \times 10^{-2}}{8 \pi^2 I c} = \frac{h \times 10^{-2}}{8 \pi^2 \mu r^2 c} \quad \text{...(5.50)}$$

B is a constant for a given diatomic molecule. The equation (5.49) can be written as:

$$F(J) = B J (J + 1) \quad \text{...(5.51)}$$

5.4.3 INTERACTION OF RADIATION WITH MOLECULE

It has been seen that all the diatomic molecules do not absorb the radiation to increase their respective rotational energies. To understand why all the molecules do not interact one has to understand the fundamental nature of radiation and the molecular system. We have learnt in the previous chapter that molecules consist of positively charged nuclei and negatively charged electrons. Similarly the radiation is also electrical and magnetic in nature. Thus the electrostatic interactions between the molecules and the radiation can play a major role in the rotation of all the molecules one way or the other. For this, consider the wave nature of the radiation as shown in Fig. 5.5 *a*. Now consider the molecule which is having a permanent dipole moment, *i.e.,* one end is more negatively charged and the other end is more positively charged, *e.g.,* HCl, CO, ClO etc. whereas homonuclear diatomic molecules, like H_2, O_2 etc. have identical atoms at the two ends and thus do no possess permanent dipole moment.

If the position of molecule AB is present as shown in Fig. 5.5a, the electrical field pushes the negative end of the molecule downward and the positive end upward. It will tend to make the molecule to rotate. Now when the molecule has rotated to the position as shown in Fig 5.5*b*, the radiation wave has also moved forward. If the frequency of the radiation wave and that of molecular rotation are identical, the electrical field can interact with the molecular dipole and can keep on increasing the rotational energy of the molecule or this interaction will make the molecule to rotate faster and faster. If frequency of radiation does not match with that of molecular rotation, it will not be absorbed by the molecule. Thus, one important conclusion, which can be drawn from

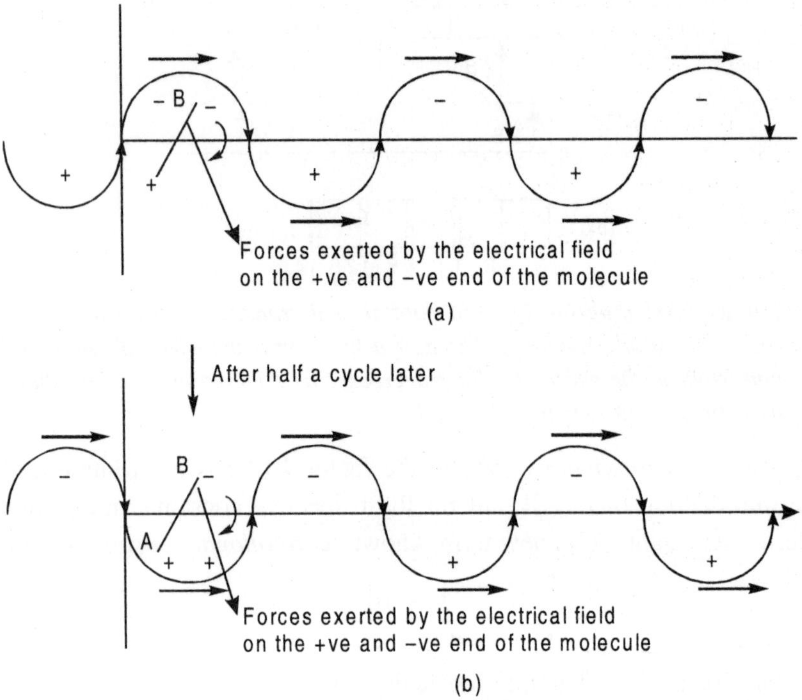

Figure 5.5: *(a) Interaction between the electrical field of the radiation with a molecule AB possessing a finite dipole moment. For the frequency of the radiation the same as the rotational frequency of the molecule, the effect of the interaction forces can be to make the molecule rotate faster. Y-axis represents the electrical field of the radiation. (b) After the radiation wave has moved half wave front forward*

the above discussion, is that *only those molecules will interact with the radiation and will give rise to rotational spectrum, which possesses permanent dipole moment. This is also known as gross selection rule for the rotational spectrum.* This means that homonuclear diatomic molecules, H_2, N_2, O_2 etc., will not give rise to rotational spectra and will be seen later that the internuclear bond lengths of these molecules cannot be determined with the help of rotational spectroscopy.

One other restriction is also imposed on the rotational spectroscopy. This can be derived from the transition moment integral relation (5.26). *To have allowed transitions, the transition moment integral should be non-zero.* This is only possible if:

$$\Delta J = \pm 1 \quad ...(5.52)$$

(see appendix 5.1) for those molecules where $\lambda = 0$. $\Delta J = +1$ corresponds to the absorption of energy and is known as absorption spectrum. $\Delta J = -1$ corresponds to the release of energy and is known as emission spectrum. These rules are also called as the *specific selection rules.*

5.4.4 SPECTRUM

The frequency of radiation can be calculated using Bohr's frequency rule, *i.e.,*

$$\bar{\nu} = F(J') - F(J'') \quad ...(5.53)$$

where single prime (′) refers to the final state and double prime (″) to the initial state. Substituting the values of respective term value in (5.53), we get

$$\bar{\nu} = BJ'(J'+1) - BJ''(J''+1) \quad ...(5.54)$$

In absorption spectrum, $J' = J'' + 1$.

$$\bar{\nu} = B[(J''+1)(J''+2) - J''(J''+1)]$$

$$= 2B(J''+1) = 2B(J+1), \text{ where } J = 0, 1, 2\ldots\ldots\ldots \quad ...(5.55)$$

For general expression J″ can be substituted for J. The spectral transitions are depicted in Fig. 5.4 and below the figure 5.4 are shown the various transitions observed with above selection rule. It is clearly evident that the successive difference between each rotational line is 2B. One thing must be clearly understood here that the successive rotational energy levels are not equally spaced but the rotational lines are equally spaced. This can also be observed from the analytical relation (5.55). The value of B can be determined from the difference between rotational lines and knowledge of B leads to the determination of bond length of the diatomic molecule. The relation is (5.56):

$$r = \sqrt{\frac{h \times 10^{-2}}{8\pi^2 \mu B c}} \quad ...(5.56)$$

The frequency of transition for pure rotational spectrum lies generally in the microwave region. That is why it is also commonly known as *microwave spectroscopy*. There are few lighter molecules, *e.g.,* H_2, D_2 etc. (note that the rotational constants of these kind of molecules are determined from either Raman spectroscopy or electronic spectrum, see later) and hydrogen halides. The frequency of rotational spectrum of these molecules extends to far infrared region. In case of hydrogen halides it is not the small masses of the halogen atoms, but because of the small reduced mass which is close to hydrogen atom.

Example 5.8: The bond length of HF molecule is 0.093 nm. Calculate (*a*) moment of inertia, (*b*) rotational constant (*c*) rotational energy of the 4th and 5th rotational levels and (*d*) the frequency of the transition between J = 4 to J = 5 in s^{-1} and in cm^{-1}.

Solution: (*a*) Expression for the reduced mass of HF is:

$$\mu_{HF} = \frac{m_H\, m_F}{m_H + m_F} = \frac{\frac{M_H}{N_{av}} \frac{M_F}{N_{av}}}{\frac{M_H + M_F}{N_{av}}}$$

$$= \frac{M_H \times M_F}{M_H + M_F} \frac{1}{N_{av}}$$

where M stands for molar mass, m is the mass of one atom and N_{av} the Avogadro's number.

$$\mu_{HF} = \frac{1 \times 19}{(1 + 19) \times 6.02 \times 10^{23}} = 1.578 \times 10^{-24} \text{ g} = 1.578 \times 10^{-27} \text{ kg}$$

Moment of inertia can be calculated as:

$$I = \mu\, r^2 = (1.578 \times 10^{-27} \text{ kg})\,(0.093 \times 10^{-9} \text{ m})^2$$

$$= 1.365 \times 10^{-47} \text{ kg m}^2$$

(*b*) $B = \frac{h \times 10^{-2}}{8\,\pi^2 \mu\, r^2\, c} = \frac{(6.62 \times 10^{-34} \text{ Js})(10^{-2} \text{ m cm}^{-1})}{(8)\,(\pi)^2\,(1.365 \times 10^{-47} \text{ kg m}^2)\,(3 \times 10^8 \text{ ms}^{-1})} = 20.46 \text{ cm}^{-1}$

(*c*) $F(4) = 20.46 \text{ cm}^{-1} \times 4 \times 5 = 409.2 \text{ cm}^{-1}$

$F(5) = 20.46 \text{ cm}^{-1} \times 5 \times 6 = 613.8 \text{ cm}^{-1}$

(*d*) $\bar{\nu}(4 \rightarrow 5) = 2\,B\,(J + 1) = 2 \times 20.46 \text{ cm}^{-1} \times (4 + 1) = 204.6 \text{ cm}^{-1}$.

This can also be calculated by taking the difference between F (5) and F (4), *i.e.*

$$613.8 - 409.2 = 204.6 \text{ cm}^{-1}$$

$$\nu = c\, \bar{\nu} = (204.6 \text{ cm}^{-1})\,(3 \times 10^8 \text{ ms}^{-1})\,(10^2 \text{ cm m}^{-1})$$

$$= 6.138 \times 1012 \text{ s}^{-1}$$

Example 5.9. The successive difference between the rotational lines of NO molecule is 3.41 cm^{-1}. Calculate the bond length of NO.

Solution:

$$\mu_{NO} = \frac{m_O\, m_N}{m_N + m_O} = \frac{14 \times 16}{(14 + 16)\,(6.02 \times 10^{23} \text{ mol}^{-1})}$$

$$= 1.24 \times 10^{-23} \text{ g} = 1.24 \times 10^{-26} \text{ kg}$$

$$2\,B = 3.41 \text{ cm}^{-1};\; B = 1.705 \text{ cm}^{-1}$$

Substituting the values of respective terms in equation (5.56)

$$r = \sqrt{\frac{(6.62 \times 10^{-34} \text{ Js})\,(10^{-2} \text{ m cm}^{-1})}{(8\,\pi^2)\,(1.24 \times 10^{-26} \text{ kg})\,(1.705 \text{ cm}^{-1})\,(3 \times 10^8 \text{ ms}^{-1})}}$$

$$= 0.1149 \text{ nm}$$

The above example has illustrated a method that can be used to calculate the bond length of diatomic molecules. The rotational constants and the bond lengths of few molecules are compiled in Table 5.1.

Table 5.1: *Rotational constants and bond lengths of few selected molecules*

Molecule	B_e (cm^{-1})	r_e(nm)	Molecule	B_e (cm^{-1})	r_e(nm)
HBr	8.473	0.1413	$^{79}Br^{19}F$	0.357	0.1756
HCl	10.591	0.1275	$^{2}H^{35}Cl$	5.445	0.1274
HF	20.939	0.0917	$^{2}H^{19}F$	11.007	0.0917
HI	6.551	0.1604	$^{127}I^{35}Cl$	0.114	0.2301
CO	1.931	0.1128	$^{16}O^{1}H$	18.872	0.09706
NO	1.705	0.1151	$^{16}O^{2}H$	10.01	0.0977

5.4.5 THERMAL DISTRIBUTION OF ROTATIONAL QUANTUM STATES (INTENSITIES OF THE ROTATIONAL LINES)

Besides the values of transition moment integral, the intensities of the rotational lines depend on the number of molecules present in the initial states from where the particular transition originates. The spectrum in general is recorded when the system is in thermal equilibrium. Thus one has to know the distribution of the molecules to various quantum states. This can be determined with the help of Maxwell-Boltzmann distribution, *i.e.,*

$$\frac{N_J}{N_0} = \frac{g_J\, e^{-\frac{E_J}{k_b T}}}{g_0\, e^{-\frac{E_0}{k_b T}}} = (2\,J + 1)\; e^{-\frac{B\,h\,c\,100\,(J^2+J)}{k_b T}} \qquad ...(5.57)$$

where N_J, N_0 are the number of molecules, g_J and g_0 is the degeneracy in the Jth and 0th states respectively. g_J is given by the relation 2J + 1, k_b is the Boltzmann constant and T is the temperature in degree Kelvin. (*It may be noted that* $\frac{N_J}{N_0}$ *is not the fraction of molecules in the* J^{th} *energy level, but it is the number of molecules in the* J^{th} *level relative to the number of molecules in the ground state*). Since $\frac{N_J}{N_0}$ is a linear function of J as well as of the exponential function, the ratio $\frac{N_J}{N_0}$ will increase first, reach a maximum value and will decrease with increase of rotational quantum number. The maximum value of J, at any given temperature, can be calculated as:

$$\frac{d\frac{N_J}{N_0}}{dJ} = 2e^{-\frac{B\,h\,C\,100\,(J^2+J)}{k_b\,T}} - (2J+1)^2\frac{100\,B\,h\,c}{k_b T}\,e^{-\frac{B\,h\,C\,100\,(J^2+J)}{k_b\,T}} = 0$$

and this will be equal to zero for $J = J_{max}$, *i.e.*,

$$\left[2-(2J_{max}+1)^2\frac{100\,B\,h\,c}{k_b T}\right]e^{-\frac{B\,h\,C\,100\,(J^2+J)}{k_b\,T}} = 0$$

or $$J_{max} = \frac{1}{2}\left[\frac{2\,k_b\,T}{100\,B\,h\,c}\right]^{\frac{1}{2}} - 1/2 \qquad ...(5.58)$$

This relation indicates that J_{max} will increase with increase of temperature and decrease of B. This kind of distribution for HF molecule has been shown in Fig. 5.6 at 300 and 1000 K. The population ratio suggests also why number of rotational lines is observed in absorption spectroscopy. This is due to the fact that at room temperature the thermal energy $k_b\,T = 4.1 \times 10^{-22}$ J at 300 K. Thus many rotational levels will be populated at room temperature to give the rotational transitions. The other information one can get from this plot is that as the thermal energy increases higher levels will be more populated and maximum shifts to higher J values, *e.g.*, J_{max} at 300 K and 1000 K are 2 and 4 respectively.

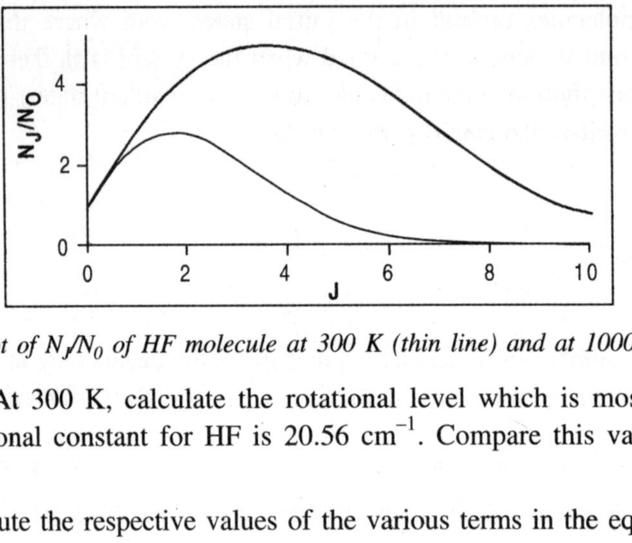

Figure 5.6: *Plot of N_J/N_0 of HF molecule at 300 K (thin line) and at 1000 K (thick line)*

Example 5.10: At 300 K, calculate the rotational level which is most populated for HF molecule. The rotational constant for HF is 20.56 cm^{-1}. Compare this value that obtained in figure 5.6.

Solution: Substitute the respective values of the various terms in the equation (5.58).

$$J_{max} = \frac{1}{2}\left[\frac{2\,(1.38\times10^{-23}\,\mathrm{J\,K^{-1}})\,(300\,\mathrm{K})}{(20.56\,\mathrm{cm^{-1}})\,(6.62\times10^{-34}\,\mathrm{J\,s})\,(3\times10^{8}\,\mathrm{m\,s^{-1}})\,(100\,\mathrm{cm\,m^{-1}})}\right]^{\frac{1}{2}} - 1/2$$

$$= 2.25 - 0.5 = 1.75 \approx 2$$

The rotational level with J = 2 will be most populated and same results are obtained from Fig. 5.6.

Example 5.11: Calculate the most populated rotational level at 1000 K. B(HF) = 20.56 cm^{-1}. Compare this value that obtained in Fig. 5.6.

Solution: Substitute the respective values of the various terms in the equation (5.58).

$$J_{max} = \frac{1}{2}\left[\frac{2(1.38\times10^{-23}\,JK^{-1})(1000K)}{(20.56cm^{-1})(6.62\times10^{-34}\,Js)(3\times10^{8}\,ms^{-1})(10^{2}\,cm\,m^{-1})}\right]^{1/2} - \frac{1}{2}$$

$$= 4.11 - 0.5 = 3.65 \approx 4$$

The value of J_{max} agrees with the result that the value of J incleases with increase of temperature.

5.4.6 EXTENSION TO POLYATOMIC MOLECULES

Similar to the study of the structure of polyatomic molecules, the extension of the spectroscopic techniques to the polyatomic molecules also give rise to complicated spectra. Microwave spectroscopy is no exception. But still large useful information can be obtained from the complicated spectrum. In this section only some of the difficulties, those occur in the study of polyatomic molecules other than diatomic molecules, will be mentioned and some of the results obtained for these molecules will be indicated.

Let us first consider the linear triatomic molecules. The simplest molecules those come into mind are CO_2, CS_2. But these are symmetric linear molecules, *i.e.*, O=C=O, S=C=S and thus will have zero permanent dipole moment. Due to this microwave spectroscopy will not be observed for this kind of molecules. Next simplest molecule to be considered will be O=C=S, carbonyl sulphide. Since this molecule is similar to diatomic molecule (thin rod type) the values of rotational energy levels will be given by the similar expression, *i.e.*,

$$E_J = \frac{h^2}{8\pi^2 I} J(J+1) \qquad ...(5.48)$$

$$J = 0, 1, 2, 3, 4 \ldots\ldots\ldots\ldots\ldots\ldots\ldots\ldots$$

Due to the similarity with the diatomic molecules, the rotational energy levels or spectral lines observed for linear triatomic molecules are also equally spaced and can be seen from the data compiled in Table 5.2. Comparison of the spacing between the rotational lines with the theoretical relations, one can calculate the rotational constant, B, in the similar manner as obtained for diatomic molecules. This will lead to the calculation of moment of inertia, *i.e.*,

$$I\,(^{16}O^{12}C^{32}S) = 138.0 \times 10^{-47}\ kg\ m^2 \qquad ...(5.59)$$

One difference, which is very apparent from the above data, is that mass of each atom is mentioned clearly. The reason is clear from discussion. On seeing the structure of the molecule, it is quite evident that the molecule possesses two bond lengths; one $r_{C=O}$ and the other $r_{C=S}$. Thus it is impossible to calculate two unknown parameters from one equation. This problem is generally overcome by the use of isotopic substitution. Since the chemical bond involves the overlap of the

Table 5.2: *Rotational spectral data of* $^{16}O^{12}C^{32}S$ *molecule*

Transition	Frequency of radiation observed × 10^{-9} Hz	Calculated value of [$h^2/8\pi^2 I \times 10^{-9}$]Hz	Calculated value of I × 10^{40} g cm^2
J = 1 → J = 2	24.3259	6.08149	137.9450
J = 2 → J = 3	36.4882	6.08147	137.9454
J = 3 → J = 4	48.6516	6.08145	137.9459
J = 4 → J = 5	60.8141	6.08141	137.9468

electronic charge clouds of the two atoms and it is independent of the masses of the nuclei, it can be assumed that the bond lengths remain unchanged with change of isotopic substitution, *i.e.*, $r\,(^{12}C^{32}S) = r\,(^{12}C^{34}S)$. Form the rotational spectrum of $^{16}O^{12}C^{34}S$, one can calculate

$$I\,(^{16}O^{12}C^{34}S) = 141.4 \times 10^{-47}\ \text{kg m}^2 \qquad ...(5.60)$$

Now to calculate $r_{C=O}$ and $r_{C=S}$, one has to set up the expressions for the moment of inertia of the two isotopic molecules. This needs the location of the centre of gravity of the molecule and then the moment of inertia is expressed in terms of r_{CO} and r_{CS}. This procedure leads to an expression for the moment of inertia in terms of the above parameters for each isotopic molecule. Comparison of the calculated and experimental values will lead to the determination of r_{CO} and r_{CS}, *i.e.*,

$$r_{CO} = 1.165 \times 10^{-10}\ \text{m} = 0.1165\ \text{nm}$$

$$r_{CS} = 1.558 \times 10^{-10}\ \text{m} = 0.1558\ \text{nm}$$

The study of linear molecules has revealed that for each isotopic molecule, we can calculate one moment of inertia. But there are many polyatomic molecules for which more than one moment of inertia can be calculated. For simplicity or in general all these non-linear polyatomic molecules can be divided into following three broad categories, based on the values of the principle moment of inertia.

(*i*) **Spherical top:** In these molecules, all the three principal moment of inertia are equal.

(*ii*) **Symmetric top:** Molecule having two of the three principal moment of inertia equal.

(*iii*) **Asymmetric top:** In these molecules all the three principal moment of inertia are different.

The molecules belonging to the above three categories are shown in Fig. 5.7 (*a*), (*b*) and (*c*) respectively. Unlike diatomic molecules, the rotational spectra of the last two categories are quite complicated and it is not easy to get the structural information for these molecules. In spite of this complexity one can analyze these data, as in the case of diatomic molecules to get two or three moment of inertia for the categories (*ii*) and (*iii*). Again one can use various isotopes to get some information regarding the additional values of moment of inertia. From these data the values of bond lengths and bond angles can be calculated. Some of the molecules belonging to these classes are shown in Fig. 5.8. One can visualize that microwave spectroscopy can be used to get a lot of information regarding the structure of the molecules.

5.4.7 LIMITATIONS OF SIMPLE MODEL

The manner in which the rotational spectrum is discussed so far gives an idea that it is very simple and information about the structure of the molecules can be obtained easily. But it may be stated here that it is not so simple even for diatomic molecules. Here two points will be discussed without going into the detailed discussion.

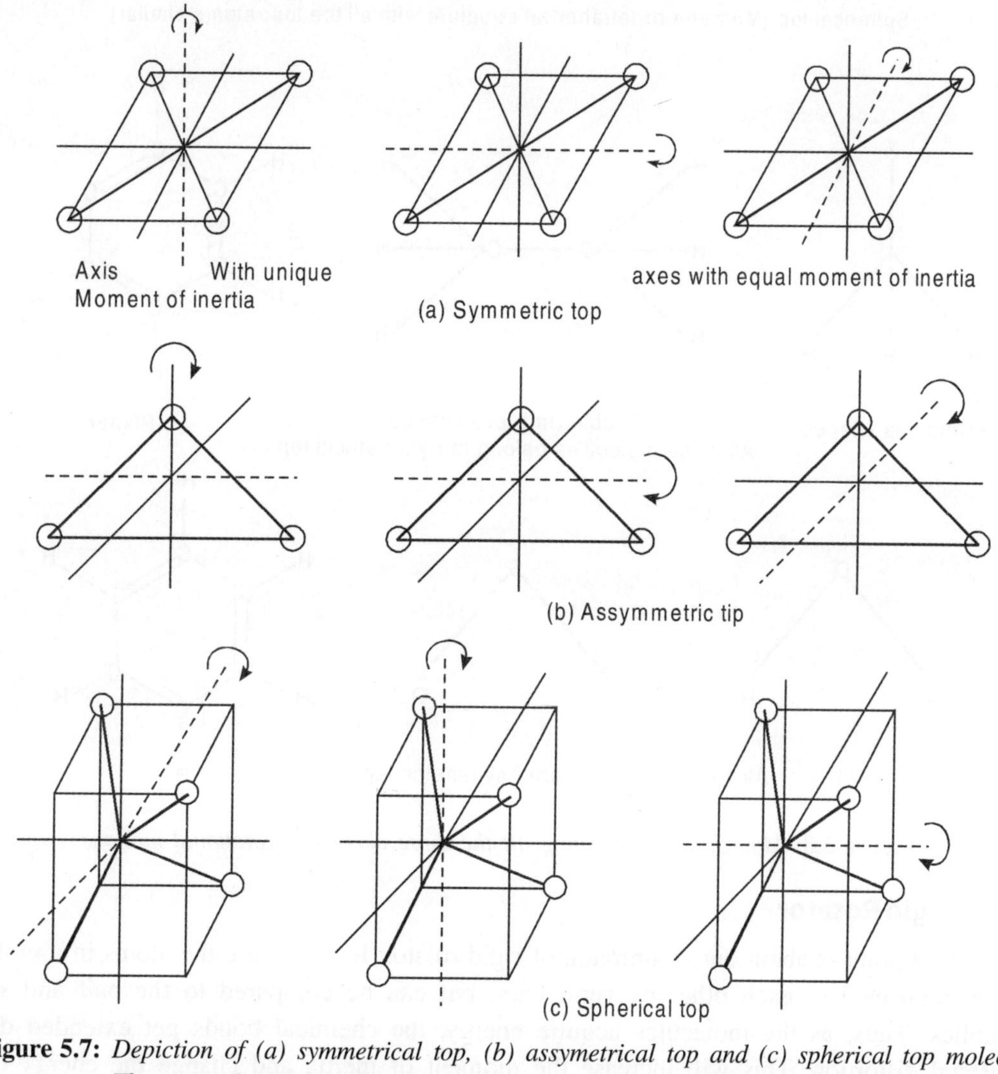

Figure 5.7: *Depiction of (a) symmetrical top, (b) assymetrical top and (c) spherical top molecules. The axes of moment of inertia are shown by thin dotted lines. Continuous lines for the normal axes and dotted line the axis around which the rotation is taking place. Thick lines show the distances that enter into the calculation of* $\Sigma\, m_i\, r_i^2$ *about these axes. For spherical top, all the three moment of inertia are equal, in symmetric top two are equal and in assyemtric top all the moment of inertia are different*

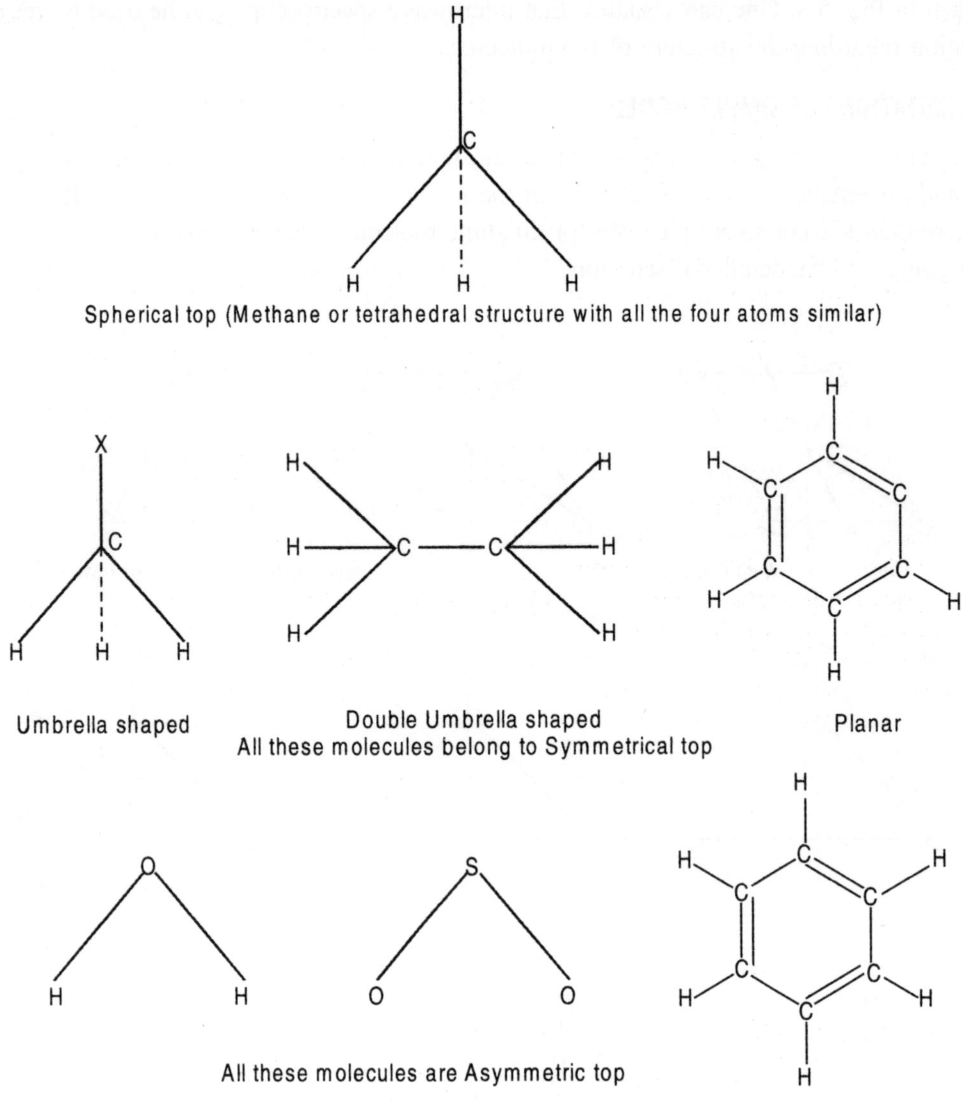

Figure 5.8: *Molecules representing all the three classes of rotational motions*

1. Non-rigid Rotator

The first point is about our assumption of rigid rotator. In real sense the atoms in a molecule are not connected to each other by rigid bars, but can be compared to the ball and spring assemblies. Thus, as the molecules acquire energy, the chemical bonds get extended due to centrifugal distortion. This will increase the moment of inertia and change the energy of the rotational level. Due to this the successive rotational lines are no longer equally spaced, *i.e.,* 2B, but the gap between the successive lines will keep on decreasing as J increases. In very precise work, one takes into account the effect of centrifugal distortion and thus one always lists the value of inertia for J = 0, where the centrifugal distortion is zero.

Before we discuss the effect of centrifugal distortion on the rotational spectrum of diatomic molecules, let us try to understand couple of things, though these will be more useful for the coming sections. Firstly, if we consider that the bonds are elastic, the energy gained can be used for vibrationl excitation. By this we mean that bond is either stretched or contracted periodically with certain fundamental frequency. It is dependent on the masses of atoms and the elasticity, *i.e.*, force constant of the bond. As shown by the equation (2.204) the force constant, is connected to the fundamental frequency by the relation:

$$k = 4\pi^2 \omega^2 c^2 \mu \qquad ...(5.61)$$

where k is the force constant, ω is the fundamental frequency expressed in cm^{-1} and other quantities have the usual meanings. The above relation clearly shows that weaker the bonds, smaller will be the force constant. This can be further extended to rotational motion, *i.e.*, smaller the force constant, larger the centrifugal distortion and smaller the value of B as J increases.

It will be seen later (section 5.5) that the vibrational frequency ($\sim 10^{14}\ s^{-1}$) is much larger than that of rotational frequency ($\sim 10^{11}\ s^{-1}$). This means that during one rotation, molecule undergoes many vibrations. In other words one can consider the average behaviour of the bond length. Although the average value of the bond length will not be different from the equilibrium value of bond length (r_e), but square of the average bond length in case of even harmonic oscillators are not exactly same as r_e. Since the rotational constant B, depends inversely on the square of the bond length ($B \propto 1/r^2$), the value of B and thus r will be different for different vibrational states of the molecule. In other words, one has to label the value of rotational constants, *i.e.*, B_e, B_0 and B_v represent rotational constants for equilibrium bond length, bond length for zero vibrational state and vibrational state v.

Though the distinction has been made between different rotational constants, in the last section, only one rotational constant will be considered. This is because the molecules always possess zero point energy even in the lowest vibrational state ($v = 0$). Thus, the bond length of the diatomic molecule, calculated from the rotational spectrum will be r_0 and not r_e. The energies of various rotational levels of the non-rigid rotator (though not derived here) can be calculated from the expression (5.62).

$$E_J = \frac{h^2}{8\pi^2 I} J(J+1) - \frac{h^4}{32\pi^4 I^2 r^2 k} J^2 (J+1)^2 \qquad ...(5.62)$$

or

$$F(J) = \frac{h}{8\pi^2 I c} J(J+1) - \frac{h^3}{32\pi^4 I^2 r^2 k c} J^2 (J+1)^2$$

$$F(J) = BJ(J+1) - DJ^2 (J+1)^2 \qquad ...(5.63)$$

where B has been defined earlier and the centrifugal distortion constant (D, cm^{-1}) is defined as:

$$D = \frac{h^3 \times 10^{-2}}{32\pi^4 I^2 r^2 k c} \qquad ...(5.64)$$

D is a positive quantity. The above expression for the rotational energy levels has been obtained for the molecule behaving as simple harmonic oscillator. For anharmonic oscillator (see later) the above expression can be further modified but will not be discussed here. The value of D is much smaller than the value of B (D < < B) and thus the centrifugal distortion will only affect the energy of rotational level for higher value of J. This can be seen from the following equation relating B, D and ω.

$$D = \frac{h^3}{32\,\pi^4\, I^2\, r^2\, k\, c} = \frac{16\, B^3\, \pi^2\, \mu\, c^3}{k} \qquad ...(5.65)$$

$$= \frac{4\, B^3}{\omega^2} \qquad ...(5.66)$$

The value of B is generally, of the order of 10 cm^{-1} and that of ω is 10^3 cm^{-1} (see section 5.5). Thus the value of D is of the order of ~ 10^{-3} cm^{-1}. The frequency of rotational lines can be calculated by using the selection rules, ΔJ = ± 1. For absorption spectrum, Δ J = + 1.

$$\bar{\nu} = F(J') - F(J'')$$

$$= B\,[\,J'(J'+1) - J''(J''+1)] - D\,[\,J'^2(J'+1)^2 - J''^2(J''+1)^2]$$

$$= 2B\,(J''+1) - 4\,D\,(J''+1)^3 \qquad ...(5.67)$$

or general expression will be

$$\bar{\nu} = 2B\,(J+1) - 4D\,(J+1)^3 \qquad ...(5.67a)$$

where J = 0, 1, 2, ...

The rotational spectrum thus obtained is similar to that obtained for rigid rotator but with only one difference, *i.e.*, in case of non-rigid rotator the rotational lines are no longer equally spaced but the gap between the successive lines keeps on decreasing with increase of J. Figure 5.4 shows the energy levels for the rigid and non-rigid rotators as well as the rotational lines. The difference between two cases has been exaggerated for clarity purposes. The analysis of the rotational lines can be carried out in a similar manner as has been done for rigid rotator (see example 5.12). Besides calculating the bond length, one can also calculate the approximate value of vibrational frequency knowing B and D from equation (5.66).

Example 5.12: The wave numbers for fourth and fifth rotational lines of HF molecule are at 164.0 and 204.62 cm^{-1} respectively. Calculate B, D, ω and *r*.

Solution: The fourth and fifth rotational lines will be observed for J = 3 and J = 4 respectively. Substituting the respective values in the equation (5.67*a*), we get

$$164 = 2B\,(3+1) - 4D\,(3+1)^3 = 8B - 256D \qquad ...(i)$$

$$204.62 = 2B\,(4+1) - 4D\,(4+1)^3 = 10B - 500D \qquad ...(ii)$$

Multiply the first equation by 10 and the second by 8 and subtracting them we get

$$1640 - 1636.96 = (-2560 + 4000)\,D$$

$$D = \frac{3.04}{1440} = 2.11 \times 10^{-3}\ cm^{-1}$$

Substituting D in the first equation we get

$$B = \frac{1}{8}\ (164 + 256 \times 2.11 \times 10^{-3}) = 20.568\ cm^{-1}$$

Equation (5.66) gives

$$\omega_e = \left[\left(\frac{4\ (20.568)^3}{2.11 \times 10^{-3}}\right)^{\frac{1}{2}}\right] = 4061.4\ cm^{-1}$$

Using equation (5.56)

$$r = \left[\frac{h \times 10^{-2}}{8\ \pi^2\ \mu\ B\ c}\right]^{\frac{1}{2}} = \left[\frac{(6.62\times10^{-34}\,Js)\ (10^{-2}\,m\ cm^{-1})}{(8\pi^2)\ (1.58\times10^{-27}\,kg)\ (20.568\ cm^{-1})\ (3\times10^8\,ms^{-1})}\right]^{½}$$

$$= 0.926 \times 10^{-10}\ m = 0.0926\ nm.$$

Example 5.13: Consider the molecule DF. Assuming that the bond distance does not change with change of isotope, calculate: *(a)* reduced mass, *(b)* moment of inertia, *(c)* value of B and *(d)* $\bar{\nu}$ for the transition 4 → 5. Compare this value with that obtained in example 5.8. The bond length is 0.093 nm.

Solution:

(a)
$$\mu = \frac{m_D\ m_F}{m_D + m_F} = \frac{2 \times 19}{(2 + 19) \times 6.02 \times 10^{23}} = 3.0 \times 10^{-24}\ g$$

$$= 3.0 \times 10^{-27}\ kg$$

(b)
$$I = \mu\ r^2 = (3.0 \times 10^{-27}\ kg)\ (0.093 \times 10^{-9}\ m)^2 = 2.6 \times 10^{-47}\ kg\ m^2$$

(c)
$$B = \frac{h}{8\ \pi^2\ I\ c} = \frac{(6.62 \times 10^{-34}\ Js)\ (10^{-2}\ m\ cm^{-1})}{(8\ \pi^2)\ (2.6 \times 10^{-47} kg\ m^2)\ (3 \times 10^8 m\ s^{-1})} = 10.74\ cm^{-1}$$

(d)
$$\bar{\nu}_{4\to5} = 2B\ (J + 1) = (2)\ (10.74)\ (4 + 1) = 107.4\ cm^{-1}$$

Note: If one compares the results of examples (5.8) and (5.13), it is clear that the differences are because of isotope masses, *i.e.*,

$$\frac{B}{B'} = \frac{h \times 10^{-2}}{8\ \pi^2\ \mu\ r^2\ c} \times \frac{8\ \pi^2\ \mu'\ r^2\ c}{h \times 10^{-2}} = \frac{\mu'}{\mu} \qquad ...(5.68)$$

where B′ and μ′ are the rotational constant and reduced mass of the isotopic molecule. Utilizing B and reduced masses μ and μ′ one can calculate B′ directly, *i.e.*, using the data of the example 5.8

$$B' = \frac{B\,\mu}{\mu'} = \frac{(20.46\ \text{cm}^{-1})\,(1.578 \times 10^{-27}\ \text{kg})}{3.0 \times 10^{-27}\ \text{kg}} = 10.76\ \text{cm}^{-1}$$

The results of example (5.13) are similar to that of the above value. One arrives at the *conclusion that rotational constant is inversely proportional to the reduced mass.*

Effect of Nuclear Spin

Second point is about the effect of rotation of nuclei on rotational spectra. Although effect is discussed in details in advanced treatment of rotational spectroscopy, here it will be discussed briefly to give a physical picture. So far we have taken nuclei as purely a point mass particles and this helps in calculating moment of inertia. But nuclei also rotate about the center of gravity, besides it forms a part of the molecule, which is rotating about the principal axis. One can consider the molecule as a rather complicated system. Actually this is an interaction of rotational angular momentum vector with the nuclear spin angular momentum. Due to this the rotational energy levels are getting perturbed, though by small amount, as these interactions are not very large. In other words one observes that for molecules which are containing the nuclei with finite spin the absorption lines are split into closely spaced lines as shown in the Fig. 5.9. Without going into details, it may be asserted here that one can obtain information about the rotating nuclei of the molecule by studying the finer detail of rotational spectroscopy.

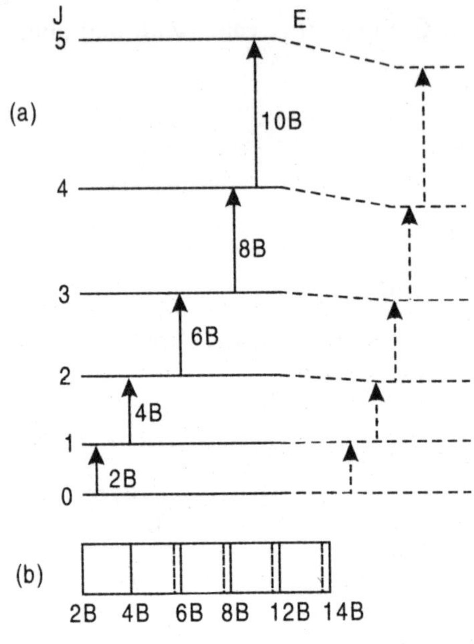

Figure 5.9: *Effect of nuclear spin on the rotational energy levels. Dotted lines show the effect of nuclear spin. The separation shown between the rotational lines are too much. In reality the separation is small*

5.5 VIBRATIONAL SPECTROSCOPY

The radiation of next higher energy than microwave is of infrared. As said earlier, this part of radiation activates the vibrational motion of the molecules. In other words, the interaction of the infrared radiation with the molecule gives information about the flexibility of the molecules, *i.e.*, how the bond lengths and bond angles can be distorted from their equilibrium values.

There is one problem, which should be mentioned before this branch is discussed. As mentioned in section 5.3, the rotational quantum is much smaller as compared to that of vibrational quantum. Due to this rotational motion also gets activated with the absorption of infrared radiation and leads to more complicated spectrum. This situation can be avoided by studying the vibarational spectrum in solution phase, where the rotational motion is hindered due to the presence of other molecules in the near vicinity. In other words, first of all only the pure vibrational spectrum will be discussed and this will be followed by the rotational-vibrational spectrum. The latter part can be studied when the molecules, especially diatomic ones, are investigated in the gas phase.

5.5.1 VIBRATIONAL SPECTRUM

In the same manner as discussed for the rotational spectrum, we should first apply the classical mechanics, *i.e.,* Newtonian mechanics, to study the oscillating motion of the molecules. In the beginning the oscillatory motion of a ball of mass *m*, attached to a spring will be discussed. This example will be proceeded by the two balls attached to a spring. The latter example is quite close the description of the diatomic molecule.

Consider a ball of mass *m*, attached to a spring and displaced from its mean position as shown in Fig. 5.10. The ball will follow oscillatory motion, *i.e.,* a restoring force in the direction opposite to the displacement will force the ball to return back to its equilibrium position. To the first approximation, this motion follows the Hook's law, *i.e.,*

$$f \propto x = (-k)\ x \qquad ...(5.69)$$

where f is the force, x is the displacement from its equilibrium position and k is known as force constant. The negative sign written in equation (5.69) explicitly shows that although the force acts in the direction opposite to that in which x increases, the proportionality constant k will have a positive

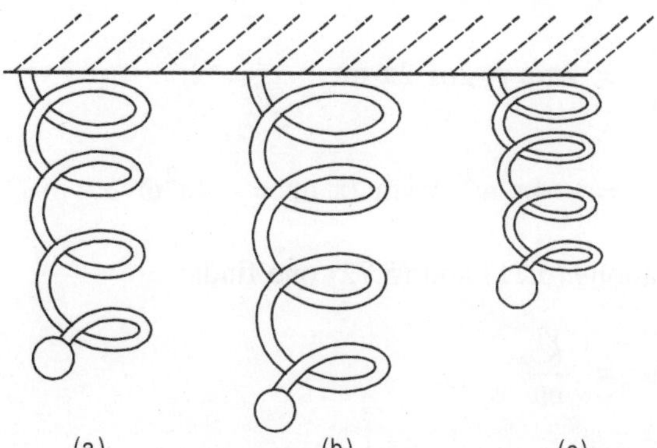

Figure 5.10: *Vibrational motion of a single ball attached to spring on one side and the other end of the string is attached to a wall, representing: (a) equilibrium position, (b) stretched position and (c) compressed position*

value. *The force constant k is equal to the force when unit displacement is made or it measures the stiffness of the spring.* Its unit in CGS system is dynes cm^{-1} and in SI units it is Newton m^{-1} ($N\ m^{-1}$). A strong inflexible spring will have higher value of k and easily extendable or compressible spring will have small value of k.

Let us now study the vibrational motion of the spring. This vibrational motion of the spring expects us to set up an equation, which describes the displacement of the ball from its equilibrium position cycelically. This kind of motion is described by either sine function or cosine function of time. But here we will only describe the sine function, *i.e.,*

$$x = A \sin (\text{const.})\, t \qquad ...(5.70)$$

where A is amplitude of vibration. The constant is a coefficient of time and can be identified as 2 $\pi\omega$, where ω is the frequency of oscillation, *i.e.,* the number of oscillations or cycles made by the ball in one second (ω has been used to represent the fundamental vibrational frequency instead of commonly used symbol ν or ν' to avoid confusion between the transition frequency and fundamental frequency). This means that the sine function will bring the particle at the same position when $2\pi\omega\, t$ is advanced by an amount 2π. This can only happen if t is increased by an amount $1/\omega$. In other words, $1/\omega$ gives the time in which the ball completes one full cycle or vibration. ω gives the number of cycles or oscillations completed by the ball in one second.

Having discussed the meaning of frequency ω, let us try to relate this frequency with the characteristics of the spring (k and mass of the ball m). According to Newton's second law of motion

$$f = -kx \qquad ...(5.69)$$

The restoring force can be expressed as a product of mass of the ball and the acceleration of the vibrating motion, given by the second derivative of displacement with respect to time.

$$m\frac{d^2x}{dt^2} = -kx \qquad ...(5.71)$$

If the equation (5.70) describes the motion of the ball with respect to time (x = A sin $2\pi\omega t$) it should satisfy the above equation (5.71).

$$\frac{dx}{dt} = 2\pi\omega\, A \cos 2\pi\, \omega t$$

$$\frac{d^2x}{dt^2} = -4\pi^2 w^2 A \sin 2\pi\, \omega t = -4\pi^2\omega^2 x \qquad ...(5.72)$$

On comparing equations (5.71) and (5.72) one finds:

$$4\pi^2\omega^2 = \frac{k}{m}$$

or

$$\omega = \frac{1}{2\pi}\left(\frac{k}{m}\right)^{\frac{1}{2}} \qquad ...(5.73)$$

Simple mathematics described above has shown that $x = A \sin 2\pi\omega t$ and indicates the vibratory motion of the ball as well as it defines the oscillatory frequency in terms of force constant and the mass of the ball. The relation (5.71) satisfies the criterion of the oscillatory motion. The force required to set the ball attached to spring, into vibratory motion will be large if the spring is very stiff or the ball attached to the spring is heavy. Lastly one important contribution coming out of this classical treatment of oscillatory motion is that for a given spring and mass of the ball there exits a characteristic or classical frequency (or natural frequency) of vibration and that can be calculated from equation (5.73).

Let us move one step closer to the diatomic molecule. Consider a spring attached with two balls of masses m_1 and m_2 respectively and lying on a frictionless table as shown in Fig. 5.11*a*. Stretch the spring and allow it to vibrate as shown in Fig. 5.11 *b* and *c*. The displacement from the equilibrium position is $x_2 - x_1$ and Newton's equations of motion for each particle can be written as:

$$k\,(x_2 - x_1) = m_1 \frac{d^2 x_1}{dt^2} \qquad \text{...(5.74)}$$

$$-\,k\,(x_2 - x_1) = m_2 \frac{d^2 x_2}{dt^2} \qquad \text{...(5.75)}$$

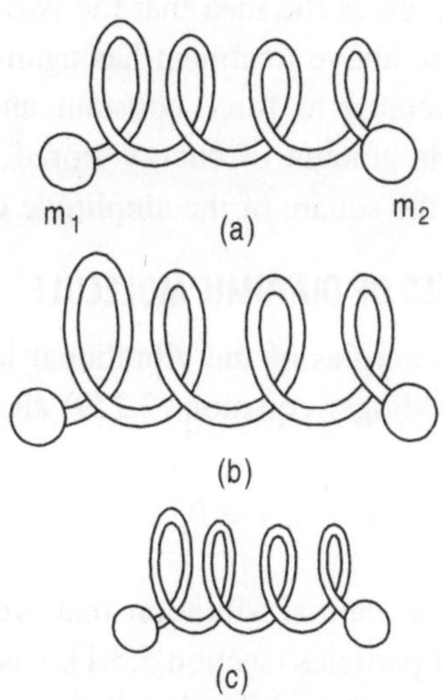

Figure 5.11: *Vibrational motion of two balls of mass m_1 and m_2 attached to spring representing: (a) Equilibrium position (b) stretched position and (c) compressed position*

where each term has the usual meaning as described earlier. The simple solution to the above equations can be written as:

$$x_1 = A_1 \sin 2\pi\,\omega t \qquad \text{...(5.76)}$$

$$x_2 = A_2 \sin 2\pi\,\omega t \qquad \text{...(5.77)}$$

where the amplitude terms for particles 1 and 2 can be different but the frequency ω must have the same value for both particles. Substituting the values of x_1 and x_2 in equations (5.74) and (5.75),

$$k\,A_2 \sin 2\pi\,\omega t - k\,A_1 \sin 2\pi\,\omega t = -\,4\pi^2\,\omega^2\,m_1\,A_1 \sin 2\pi\,\omega t \qquad ...(5.78)$$

$$-\,k\,A_2 \sin 2\pi\,\omega t - k\,A_1 \sin 2\pi\,\omega t = -\,4\pi^2\,\omega^2\,m_2\,A_2 \sin 2\pi\,\omega t \qquad ...(5.79)$$

The term sin 2π ωt can be cancelled and the above equations can be written after rearrangement as:

$$A_2\,k = (k - 4\pi^2\,\omega^2\,m_1)\,A_1 \qquad ...(5.80)$$

$$(k - 4\pi^2\,\omega^2\,m_2)\,A_2 = k\,A_1 \qquad ...(5.81)$$

A_1 and A_2 are eliminated, by dividing equation (5.80) by (5.81).

$$\frac{k}{k - 4\pi^2\,\omega^2\,m_2} = \frac{k - 4\pi^2\,\omega^2\,m_1}{k} \qquad ...(5.82)$$

After multiplication and rearranging, we get

$$\omega = \frac{1}{2\pi}\left(\frac{k}{\mu}\right)^{\frac{1}{2}} \qquad ...(5.83)$$

where μ is the reduced mass and gives the idea that the two particle motion can be expressed as one body problem of mass μ. The above treatment has again shown that the spring has a natural frequency of vibration and is related to force constant and the reduced mass of the system. Although not shown here but the amount of energy stored in the vibrations of such a ball and spring system would depend on the square of the amplitude of vibrational motion.

5.5.2 VIBRATIONAL ENERGY LEVELS OF DIATOMIC MOLECULE

In quantum mechanics, the energies of the vibrational levels of a diatomic molecule can be calculated with the help of Schrödinger equation (2.229) and is expressed as:

$$E_v = h\omega_e\,(v + \frac{1}{2}),\ v = 0, 1, 2............. \qquad ...(5.84)$$

where ω_e is the frequency of the natural vibration that would occur if the diatomic molecule behaved as do the ordinary sized particles (section 5.5.1), v is the vibrational quantum number and describes the quantized behaviour of the vibrational energy. The classical frequency ω_e can be calculated by the expression:

$$\omega_e = \frac{1}{2\pi}\left(\frac{k}{\mu}\right)^{\frac{1}{2}} \qquad ...(5.85)$$

as described earlier k and μ have the usual meanings. The wave functions describing the behaviour of vibrational motion, corresponding to each state, as well as the importance of zero point energy (energy of the molecules when $v = 0$ state) has been discussed in section 2.7. The vibrational energy in spectroscopy is commonly expressed in terms of cm^{-1} and is obtained by dividing equation (5.84) by *hc*, *i.e.*,

$$G(v) = \frac{E_v}{hc} = \overline{\omega}_e \left(v + \frac{1}{2}\right) \qquad ...(5.86)$$

where G (*v*) is called the vibrational term value and $\overline{\omega}_e$ is the classical frequency expressed in cm^{-1}. The values of vibrational energy levels are shown in Fig. 5.12 for diatomic molecule. On comparing with the rotational energy levels, two points are clearly evident.

1. The vibrational levels are equally spaced.
2. All the vibrational levels can be described by a single vibrational frequency ω_e (s^{-1}) or $\overline{\omega}$ (cm^{-1}).

5.5.3 MECHANISM OF ABSORPTION OF INFRARED RADIATION BY MOLECULES

The absorption of infrared radiation by the molecules can be explained on the same lines as has been done in case of rotational spectrum. This can be divided in two parts. One forms the gross or general selection rule and the other specific selection rule.

(*i*) **Gross Selection Rule:** This selection rule can be derived if we assume that absorption of infrared radiation by the vibrating molecules is based on the electrical interactions, *i.e.*, electrical field vector of radiation can interact with the diatomic molecules if we assume that the molecules consist of positive and negative charges at the two ends. Further if the

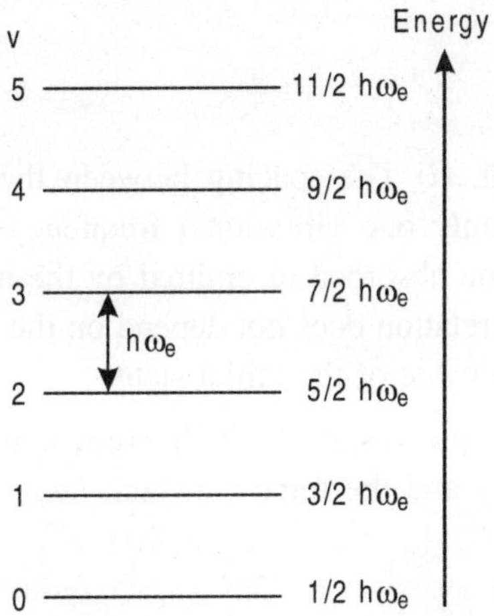

Figure 5.12: *Vibrational energy levels of a diatomic molecule*

charges vary with change of distance between two ends, the dipole moment of the molecule will be greater in the extended position than that in the compressed position. During vibration, the two charges will oscillate around their mean position with certain frequency. Consider that the oscillating charges are in phase with the electric field. During this act, the electrical field pushes the molecule to vibrate more if the radiation field has the same frequency as that of the molecular vibrations. This means that *the oscillating dipole moment should have a finite value or should have non-zero value.* The heteronuclcear diatomic molecules like CO, HCl will have infrared spectrum. On the other hand the homonuclear molecules, like N_2, H_2, O_2 etc., will not have spectrum in the infrared region, because these molecules do not possess permanent dipole moment and thus no oscillating dipole moment.

(*ii*) **Specific Selection Rule:** Similar to the rotational spectrum, the specific selection rule can be obtained from the transition moment integral:

$$\int \psi\,(v')\;\hat{M}\;\psi\,(v'')\;d\tau \qquad \text{...(5.26)}$$

where $\psi\,(v')$ and $\psi\,(v'')$ are the vibrational wave functions of the final and initial states respectively, M is the dipole moment operator. It has been shown in appendix (5.2) that for vibrations which follow harmonic motion, the above integral will be non-zero if $v' - v'' = \pm 1$, *i.e.*, $\Delta v = \pm 1$. As usual $\Delta v = +1$ is for absorption and $\Delta v = -1$ for emission spectrum.

5.5.4 SPECTRUM

According to the Bohr's frequency rule, the vibrational frequency is given by:

$$\bar{\nu} = G\,(v') - G\,(v'') = \bar{\omega}_e \left(v' + \frac{1}{2}\right) - \bar{\omega}_e \left(v'' + \frac{1}{2}\right)$$

$$= \bar{\omega}_e\;(v' - v'') \qquad \text{...(5.86}a\text{)}$$

For $\qquad \Delta v = +1,$

$$\bar{\nu} = \bar{\omega}_e \qquad \text{...(5.87)}$$

Two things are quite evident. (*i*) The spacing between the vibrational levels is the same. (*ii*) Unlike rotational spectrum only one vibrational frequency is observed. The above results indicate that frequency of radiation absorbed or emitted by the molecule is equal to the classical frequency of the molecules. This relation does not depend on the initial state of the molecules and holds good no matter what is the value of the initial state.

On comparing the spectroscopic results with theoretical results one can find the relation between the vibrational frequency and the force constant, *i.e.*,

$$\bar{\omega}_e = \frac{10^{-2}}{2\pi c}\left(\frac{k}{\mu}\right)^{\frac{1}{2}} \qquad \text{...(5.88)}$$

where factor of 10^{-2} is to convert CGS units to SI units. Thus knowing ω_e we can calculate the value of force constant.

Example 5.14: The vibrational frequency of NO is 1876 cm^{-1}. Calculate the value of force constant.

Solution: $$\mu = \frac{m_O\, m_N}{(m_N + m_O)\,(6.02\times10^{23})} = \frac{14\times16}{(14+16)\,(6.02\times10^{23})} = 1.23\times10^{-23}\,\text{g}$$

$$= 1.23\times10^{-26}\ \text{kg}$$

From equation (5.88)

$$k = \frac{\bar{\omega}_e^2\ 4\pi^2c^2\mu}{10^{-4}} = \frac{(1.24\times10^{-26}\ \text{kg})\,(4\pi^2)\,(3\times10^8\ \text{m s}^{-1})^2\,(1876\ \text{cm}^{-1})^2}{10^{-4}\ \text{m}^2\ \text{cm}^{-2}}$$

$$= 1.55\times10^3\ \text{N m}^{-1}$$

Example 5.15: The force constant for HI is 314 N m^{-1}. Calculate the frequency of radiation to excite the molecule from $v = 0$ to $v = 1$.

Solution: $$\mu_{HI} = \frac{127\times1}{(127+1)\,(6.02\times10^{23})} = 1.65\times10^{-24}\ \text{g} = 1.27\times10^{-27}\ \text{kg}.$$

Using equation (5.88):

$$\bar{\omega}_e = \frac{10^{-2}\ \text{m cm}^{-1}}{(2\pi)\,(3\times10^8\ \text{ms}^{-1})}\left[\frac{314\ \text{N m}^{-1}}{1.6\times10^{-27}\ \text{kg}}\right]^{\frac{1}{2}} = 2313\ \text{cm}^{-1}$$

The values of fundamental frequencies and the force constants of some of the molecules are compiled in Table 5.3. As said earlier, one can get the idea of stiffness of the chemical bond from the values of force constants, for example, one can say that among the hydrogen halides the HF bond is stiffer than HI and similarly triple bond is stiffer than double bond and double bond is stiffer than single bond. Though there is no direct relationship between the force constant and the dissociation energy of the chemical bond but it can be said that the greater the value of force constant greater will be bond dissociation energy.

5.5.5 INTENSITY OR THERMAL POPULATION DISTRIBUTION

The treatment is similar as carried out for rotational lines, *i.e.,* besides the value of transition moment integral (5.2), the intensities of the vibration transitions depend on the population (*i.e.,* number of molecules present in each level) of each vibrational level. This can be obtained from the Maxwell-Boltzmann distribution.

$$\frac{N_v}{N_0} = \frac{g_v\, e^{-\frac{E_v}{k_b T}}}{g_0\, e^{-\frac{E_0}{k_b T}}} = e^{\left(-\frac{h\bar{\omega}_e c v}{k_b\ T}\right)} \qquad ...(5.89)$$

where each term has the usual meaning, E_v and E_0 are the vibrational energies, g_v and g_0 are the degeneracy of the vth and zeroth vibrational levels respectively. Further the degeneracy of each vibrational level is unity. (*Similar to rotational motion, it may also be noted that* $\frac{N_v}{N_0}$ *is not* *the fraction of molecules in the vth energy level, but it is the number of molecules in the vth level relative to the number of molecules in the ground state*). Table 5.4 gives the ratio of $\frac{N_v}{N_0}$ for some of the molecules at 300 K and 1000 K. It can be seen that even at 1000 K the number of molecules present in higher levels are very small. Thus, for all practical purposes the infrared spectrum is observed only from the ground state vibrational level. This unlike rotational spectrum, is due to the fact that the successive difference between vibrational states is much larger than the thermal energy k_b T. Figure 5.13 depicts the plot of $\frac{N_v}{N_0}$ for $^{129}I^{35}Cl$ molecule.

Example 5.16: The oscillation frequency of $^{129}I^{35}Cl$ is 394.2 cm^{-1}. Assuming that the molecule follows simple harmonic motion, plot $\frac{N_v}{N_0}$ at 300 K and 1000 K.

Solution: Since $^{129}I^{35}Cl$ follows simple harmonic motion

$$E_v = h\,\omega_e\left(v + \frac{1}{2}\right)$$

and Maxwell-Boltzmann relation gives:

$$\frac{N_v}{N_0} = \frac{e^{-\frac{E_v}{k_b T}}}{e^{-\frac{E_0}{k_b T}}} = e^{-\frac{h\bar{\omega}_e c\, v}{k_b T}}$$

At 300 K

$$\frac{h\bar{\omega}_e c}{k_b\ T} = \frac{(6.62 \times 10^{-34}\ \text{Js})\ (354.2\ \text{cm}^{-1})\ (3 \times 10^8\ \text{ms}^{-1})}{(1.38 \times 10^{-23}\text{J K}^{-1})\ (300\ \text{K})\ (10^{-2}\text{m cm}^{-1})} = 1.843$$

Table 5.3: *Vibrational frequencies and force constants of few selected molecules*

Molecule	$\bar{\omega}_e$ (cm^{-1})	k (Nm^{-1})	Molecule	$\bar{\omega}_e$ (cm^{-1})	k (Nm^{-1})
$^{1}H^{19}F$	4138.5	960.4	$^{16}O^{1}H$	3735.2	775
$^{1}H^{35}Cl$	2989.7	513	$^{16}O^{2}H$	2720.9	772
$^{2}H^{35}Cl$	2144	513.4	$^{79}Br^{19}F$	671.0	407
$^{2}H^{19}F$	2998.3	960.2	$^{127}I^{35}Cl$	384.2	239
$^{1}H^{79}Br$	2649.7	311	$^{14}N^{16}O$	1904	1598
$^{1}H^{127}I$	2309.5	317	$^{12}C^{16}O$	2170	1906

Note: From the data of Table 5.3 it is evident that force constant of a bond containing elements of different isotopes is independent of reduced mass.

At 1000 K,

$$\frac{h\bar{\omega}_e c}{k_b T} = \frac{(6.62 \times 10^{-34} \text{ Js})(354.2 \text{ cm}^{-1})(3 \times 10^8 \text{ ms}^{-1})}{(1.38 \times 10^{-23} \text{J K}^{-1})(1000 \text{ K})(10^{-2} \text{m cm}^{-1})} = 0.553$$

Table 5.4: *Values of* $\frac{N_1}{N_0}$ *for some selected molecules at 300 K and 1000 K*

Molecule	$\frac{N_1}{N_0}$ **(300 K)**	$\frac{N_1}{N_0}$ **(1000 K)**	**Molecule**	$\frac{N_1}{N_0}$ **(300 K)**	$\frac{N_1}{N_0}$ **(1000 K)**
$^1H^{19}F$	2.39×10^{-9}	2.59×10^{-3}	$^{16}O^1H$	1.65×10^{-8}	4.63×10^3
$^1H^{35}Cl$	5.91×10^{-7}	1.35×10^{-2}	$^{16}O^2H$	2.14×10^{-6}	1.99×10^{-2}
$^2H^{35}Cl$	1.31×10^{-4}	6.84×10^{-2}	$^{79}Br^{19}F$	4.00×10^{-2}	0.38
$^2H^{19}F$	5.67×10^{-7}	1.34×10^{-2}	$^{127}I^{35}Cl$	0.16	0.57
$^1H^{79}Br$	3.02×10^{-6}	2.2×10^{-2}	$^{14}N^{16}O$	1.08×10^{-4}	6.46×10^{-2}
$^1H^{127}I$	1.54×10^{-5}	3.6×10^{-2}	$^{12}C^{16}O$	3.01×10^{-5}	4.4×10^{-2}

Values of $\frac{N_v}{N_0}$ at 300 K and 1000 K as a function of v are compiled below

v	0	1	2	3	4	5	6	7	8
$\frac{N_v}{N_0}$ 300 K	1	0.16	0.025	0.004	0.00063	0.0001	—	—	—
$\frac{N_v}{N_0}$ 1000 K	1	0.58	0.33	0.14	0.095	0.063	0.04	0.02	0.01

The above data are plotted in Fig. 5.13.

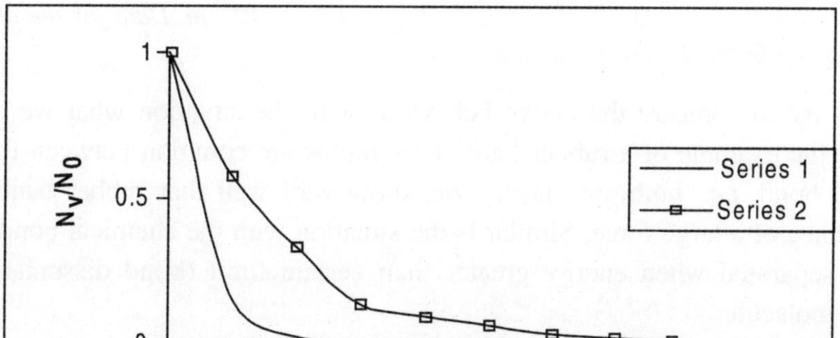

Figure 5.13: *Plot of N_v/N_0 for $^{129}I\ ^{35}Cl$ at 300 K and 1000 K versus v. Top line is at 1000 K and bottom line at 300 K*

5.5.6 POTENTIAL ENERGY AND ASYMMETRY IN CHEMICAL BOND (AN-HARMONIC MOTION)

In the last section we assumed that particles (atoms) when extended or compressed try to attain their equilibrium positions, under the influence of restoring forces. These periodic motion or oscillatory motion follows the so-called simple harmonic motion. In this case the restoring force is proportional to the distance moved by the particle from its mean position and acts in the direction opposite to the extension or compression, also called *Hook's law*, Under this assumption the molecules will keep on vibrating even if a large force is applied or even if the atoms are extended or compressed by a large amount. The potential energy for such kind of motion can be calculated from the equation (5.90).

$$\text{P.E.} = -\int_0^x (-F)\, dx = \int_0^x k\, x\, dx = \frac{1}{2}\, k\, x^2 \qquad \text{...(5.90)}$$

where each term has usual meaning. Potential energy as a function of x is plotted in Fig. 5.14 and energies of various vibrational levels calculated with the help of equation (5.84) are also plotted in Fig. 5.14. It is quite clear from Fig. 5.14 that potential energy increases very sharply when the chemical bond is extended or compressed and all the vibrational levels are equally spaced.

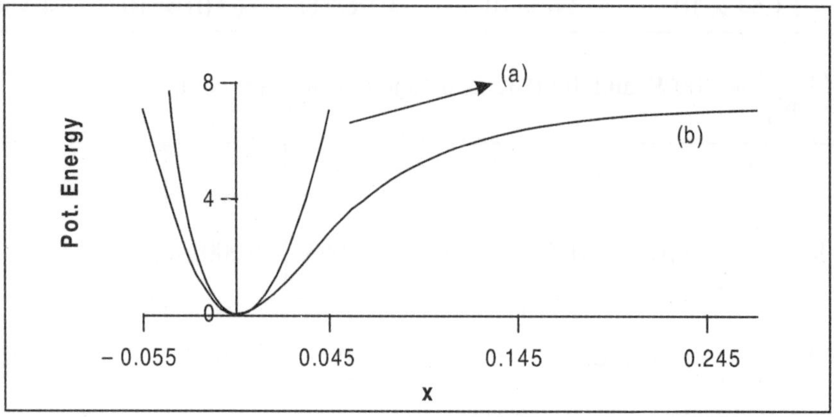

Figure 5.14: *(a) Potential energy curve for HCl molecule following Hook's law. (b) Potential energy curve following Morse potential. x is in units of nm and Pot. Energy in units of* 10^{-19} *J,* $k = 482\ N\ m^{-1}$. $D_e = 7.31 \times 10^{-19}$ *J,* $a = 1.82 \times 10^{10}$ *m. Data for the above curves are given in the Tables of example 5.18*

Now let us try to compare the above behaviour with the situation what we observe in our daily life. Take the example of a rubber band. Few things are common between the rubber band and a chemical bond, *i.e.,* both are elastic. We know very well that rubber band breaks down under the influence of a large force. Similar is the situation with the chemical bond, *i.e.,* atoms in molecules get separated when energy greater than certain limit (bond dissociation energy) is supplied to the molecule.

Let us try to study the beahviour of diatomic molecules in a little more detailed fashion. In Fig. 5.14, the actual potential energy curve of a diatomic molecule can be obtained by method as discussed in chapter 4.4 and shown in Fig. 4.5*a*. But the other curve given in Fig. 5.14 is drawn with the help of Morse potential (5.99), just for simplicity. It is quite obvious that the two curves drawn in Fig. 5.14 do not match with each other. We know that if a chemical bond is extended by a factor of two or three to its normal length, the bond between the two atoms is pretty well destroyed and further extension will be very easy. This means that for further extension the restoring force is no longer linearly proportional to the increase in the bond length, but decreases very quickly and becomes zero at infinite distance. In terms of potential energy this implies that it does not increase parabolically but it increases very slowly as the bond length increases and eventually decreases to zero. On the other hand when atoms are compressed the increase in the potential energy is much sharper than that observed by using equation 5.90. This is due to the repulsion between the positively charged nuclei as well as between the negatively charged electrons. Thus, it can be seen from Fig. 5.14 that agreement between the actual and theoretical potential energy curves is only near the equilibrium bond length. In other words the oscillations in the chemical bond of a diatomic molecule follow simple harmonic motion as long as the chemical bond is not extended or compressed by large amount. Figure 5.15 contains the plot of the vibrational energy levels obtained experimentally (from the infrared spectroscopy). It can be seen clearly that the vibrational energy levels are no longer equally spaced and the successive difference between them keeps on decreasing with increase of v, the vibational quantum number. This departure of oscillatory motion from simple harmonic motion is called *anharmonic motion or anharmonicity.* This behaviour of the chemical bond is called asymmetry in the molecule. One can relate this departure from the normal behaviour of the molecule with the following equation if we express the potential energy as a function of extension or compression of a chemical bond from its equilibrium position, *i.e.,*

$$V = \text{P.E.} = f(r - r_e) \qquad ...(5.91)$$

where $r - r_e = x$ is the change in the equilibrium bond length. Expanding this function as a Taylor series around the mean position one gets

$$V = V(r_e) + \left(\frac{\partial V}{\partial r}\right)_{r=r_e}(r-r_e) + \frac{1}{2!}\left(\frac{\partial^2 V}{\partial r^2}\right)_{r=r_e}(r-r_e)^2$$

$$+ \frac{1}{3!}\left(\frac{\partial^3 V}{\partial r^3}\right)_{r=r_e}(r-r_e)^3 + \frac{1}{4!}\left(\frac{\partial^4 V}{\partial r^4}\right)_{r=r_e}(r-r_e)^4 + \ldots\ldots\ldots \qquad ...(5.92)$$

If we measure the energy of the vibrational levels from the minimum energy of the system and take this value equal to zero [$V(r_e) = 0$], the first derivative of V with respect to r at $r = r_e$ will be zero. Thus equation (5.92) reduces to:

$$V = \frac{1}{2!}\left(\frac{\partial^2 V}{\partial r^2}\right)_{r=r_e}(r-r_e)^2 + \frac{1}{3!}\left(\frac{\partial^3 V}{\partial r^3}\right)_{r=r_e}(r-r_e)^3 + \frac{1}{4!}\left(\frac{\partial^4 V}{\partial r^4}\right)_{r=r_e}(r-r_e)^4 \qquad ...(5.93)$$

Further neglecting the higher terms, equation (5.93) reduces to:

$$V = \frac{1}{2!}\left(\frac{\partial^2 V}{\partial r^2}\right)_{r=r_e}(r-r_e)^2 \qquad ...(5.94)$$

This equation resembles the equation (5.90), derived earlier. On comparing equations (5.94) and (5.90),

$$k = \left(\frac{\partial^2 V}{\partial r^2}\right)_{r=r_e} \quad ...(5.95)$$

Thus the higher derivatives of the potential energy function are called anharmonicity constants, as these terms describe the anharmonic behaviour of the vibrational motion of the diatomic molecules. Following relations exists between the anharmonic constants and derivatives of Taylor series expansion

$$\bar{\omega}_e x_e = \frac{1}{3!}\left(\frac{\partial^3 V}{\partial r^3}\right)_{r=r_e} (r-r_e)^3 \quad ...(5.96)$$

$$\bar{\omega}_e y_e = \frac{1}{4!}\left(\frac{\partial^4 V}{\partial r^4}\right)_{r=r_e} (r-r_e)^4 \quad ...(5.97)$$

$$\bar{\omega}_e z_e = \frac{1}{5!}\left(\frac{\partial^5 V}{\partial r^5}\right)_{r=r_e} (r-r_e)^5 \quad ...(5.98)$$

and so on.

It has been shown in chapter 4 that calculation of the potential energy curve for any diatomic molecule is quite complicated and it is hard to express the desired potential energy curve by a suitable analytical expression. Many, empirical expressions have been given by many scientists but the simplest and more commonly used is given by P.M. Morse. This is called Morse function.

$$V = D_e\left[1-e^{-a(r-r_e)}\right]^2 \quad ...(5.99)$$

where a is a constant and it depends on the nature of the molecule, D_e is the equilibrium bond dissociation energy. This function gives nice picture of variation of potential energy towards the right hand side of the Fig. 5.14, but it does not explain the left hand side of the potential energy curve. The values of anharmonicity constants can be calculated in terms of the variables defined by Morse function.

Example 5.17: (*a*) Show that the potential energy given by the Morse function is minimum at $r = r_e$. (*b*) Calculate the constant a for hydrogen molecule. $D_0 = 4.476$ eV and $\bar{\omega} = 4395.2\ cm^{-1}$.

Solution:

(*a*) Expanded form of equation (5.99) is written as (5.100):

$$V = D_e\left[1-2\,e^{-a(r-r_e)}+e^{-2a(r-r_e)}\right] \quad ...(5.100)$$

$$\frac{dV}{dr} = D_e\left[2a\,e^{-a(r-r_e)}-2a\,e^{-2a(r-r_e)}\right] \quad ...(5.101)$$

From the above equation at $r = r_e$, $dV/dr = 0$

Further $$\frac{d^2V}{dr^2} = D_e\left[-2a^2\,e^{-a(r-r_e)}+4a^2\,e^{-2a(r-r_e)}\right] \quad ...(5.102)$$

Now at $r = r_e$, $\dfrac{d^2V}{dr^2} = 2\,a^2\,D_e$ ...(5.103)

$a^2\,D_e$ is a positive quantity because D_e, the dissociation energy is positive and the square of the constant a will also be positive. Thus the requirement for the minimum is satisfied.

(*b*) Equation (5.95) gives that

$$\left(\frac{\partial^2 V}{\partial v^2}\right)_{r=r_e} = k = 2\,a^2\,D_e$$

From equation (5.73), $k = 4\pi^2\,\omega_e^2\,\mu$

Combining the equations (5.73) and (5.95):

$$2\,a^2\,D_e = 4\,\pi^2\,\omega_e^2\,\mu$$

$$\text{or} \qquad a = \left[\frac{2\pi^2\mu}{D_e}\right]^{\frac{1}{2}} \omega_e \qquad \text{...(5.103}a\text{)}$$

$$\text{For} \qquad H_2,\ \mu = \frac{1 \times 1}{(1+1) \times 6.02 \times 10^{23}} = 8.31 \times 10^{-25}\ \text{g} = 8.31 \times 10^{28}\ \text{kg}$$

$$a = \left[\frac{(2\pi^2)(8.31\times10^{-28}\ \text{kg})}{(4.476\,\text{eV})(1.6\times10^{-19}\ \text{J eV}^{-1})}\right]^{1/2} (4395.2\ \text{cm}^{-1})(3.0 \times 10^{10}\ \text{cm s}^{-1})$$

$$= 1.006 \times 10^{10}\ \text{m}^{-1}$$

Example 5.18: Construct the potential energy curve for HCl based on (*i*) Hook's law and (*b*) Morse potential. $k = 482$ N m^{-1}, $D_e = 7.31 \times 10^{-19}$ J, $a = 1.82 \times 10^{10}$ m^{-1}.

Solution:

(*a*) Using equation $V(x) = \dfrac{1}{2}\,k\,x^2$(Hook's law) and taking different values of x (displacement from the mean position) one can calculate V and these values are listed in the following table.

Sr. No	**± *x* (nm)**	**V(*x*) (10^{-19} J)**	**Sr. No**	***x* (nm)**	**V(*x*) (10^{-19} J)**	**Sr. No**	**± *x* (nm)**	**V(*x*) (10^{-19} J)**
1	0	0	5	0.02	0.964	9	0.04	3.856
2	0.005	0.06	6	0.025	1.506	10	0.045	4.88
3	0.01	0.241	7	0.03	2.169	11	0.05	6.025
4	0.015	0.542	8	0.035	2.952	12	0.055	7.229

(*b*) Similarly using equation (5.99), $V(x) = D_e [1 - e^{-ax}]^2$, the values of V calculated for

Sr. No.	x (nm)	$V(x)$ (10^{-19} J)	Sr. No.	x (nm)	$V(x)$ (10^{-19} J)	Sr. No.	x (nm)	$V(x)$ (10^{-19} J)
1	−0.04	8.84	11	0.01	0.202	21	0.06	3.227
2	−0.035	5.80	12	0.015	0.417	22	0.08	4.30
3	−0.03	3.857	13	0.02	0.680	23	0.1	5.130
4	−0.025	2.427	14	0.025	0.9768	24	0.12	5.757
5	−0.02	1.409	15	0.03	1.294	25	0.14	6.211
6	−0.015	0.72	16	0.035	1.6225	26	0.16	6.536
7	−0.01	0.291	17	0.04	1.955	27	0.18	6.768
8	−0.005	0.0664	18	0.045	2.285	28	0.2	6.931
9	0	0.0	19	0.05	2.610	29	0.25	7.156
10	0.005	0.0554	20	0.055	2.924	30	0.3	7.248

different values of V are also listed in the following table.

In both cases V versus x are plotted in Fig. 5.14. Explanation is given in the text.

5.5.7 ENERGY LEVELS

The energy levels of different vibrational states of the anharmonic oscillator can be calculated with the help of Schrödinger wave equation, using Morse potential function for potential energy term. The expression so obtained is:

$$E_v = h\,\omega_e\,(v + 1/2) - h\,\omega_e x_e\,(v + 1/2)^2 + h\,\omega_e y_e\,(v + 1/2)^3 + h\,\omega_e z_e\,(v + 1/2)^4 \ldots \quad \ldots(5.104)$$

and the corresponding term value is:

$$G(v) = \bar{\omega}_e\,(v + 1/2) - \bar{\omega}_e\,x_e\,(v + 1/2)^2 + \bar{\omega}_e\,y_e\,(v + 1/2)^3 + \bar{\omega}_e\,z_e\,(v + 1/2)^4 \quad \ldots(5.105)$$

$$v = 0, 1, 2, 3 \ldots\ldots\ldots\ldots$$

where v is the vibrational quantum number, $\omega_e x_e$, $\omega_e y_e$ ... are called anharmonic constants, with $\omega_e x_e << \omega_e$ and $\omega_e y_e << \omega_e x_e$. Further the sign of $\omega_e x_e$ is positive and that of $\omega_e y_e$ can be positive or negative. It is a normal convention to write $\omega_e x_e$ and $\omega_e y_e$ together rather than writing only x_e and y_e. But some books do write the anharmonic constants only x_e and y_e. It is quite clear from the expression (5.104) that the successive gap between the vibrational energy levels will keep on decreasing as v increases and is shown in Fig. 5.15.

The zero point energy in case of anharmonic oscillator can be obtained by substituting $v = 0$ in equation (5.105)

$$G(0) = 1/2\ \bar{\omega}_e - 1/4\ \bar{\omega}_e\ x_e + 1/8\ \bar{\omega}_e\ y_e + \ldots\ldots\ldots\ldots \quad ...(5.106)$$

and if the energy levels are measured from zero vibrational level, the expression for energy levels can be written as:

$$G_0(v) = \bar{\omega}_e\ v - \bar{\omega}_e\ x_0\ v^2 + \bar{\omega}_e\ y_0\ v^3 + \quad ...(5.107)$$

where

$$\bar{\omega}_0 = \bar{\omega}_e + \bar{\omega}_e\ x_e + 3/4\ \bar{\omega}_e\ y_e + \ldots \quad ...(5.108)$$

$$\bar{\omega}_0 x_0 = \bar{\omega}_e\ x_e - 3/2\ \bar{\omega}_e\ y_e + \ldots \quad ...(5.109)$$

$$\bar{\omega}_0 y_0 = \bar{\omega}_e\ y_e + \ldots \quad ...(5.110)$$

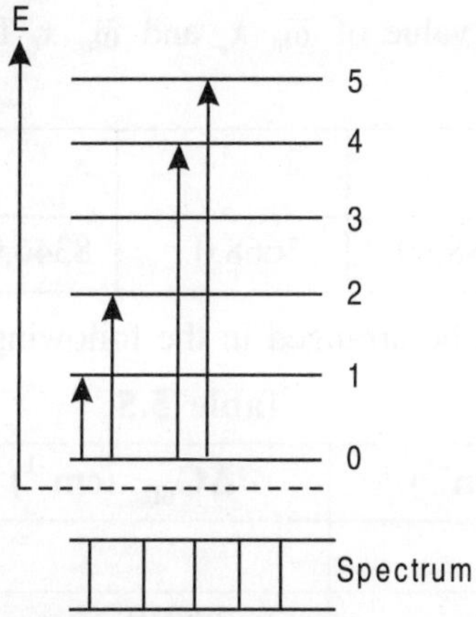

Figure 5.15: *Energy levels and vibrational transitions of anharmonic oscillator. The absorption spectrum is given schematically below. Dotted line shows zero of the energy scale*

5.5.8 SPECTRUM

The vibrational transitions are still observed from the lowest most energy level, *i.e.*, $v = 0$. Besides $\Delta v = \pm 1$, $\Delta v = \pm 2$, $\Delta v = \pm 3$, $\Delta v = \pm 4$ etc. (see appendix 5.2), transitions are also observed for anharmonic oscillator. $\Delta v = \pm 1$ *is called fundamental transition or band;* $\Delta v = \pm 2$ *first overtone;* $\Delta v = \pm 3$ *the second overtone;* and so on. The intensity of the fundamental band is the maximum and intensities of the overtones keep on decreasing with increase of Δv gap. The vibrational spectrum observed for HCl is given in Table 5.5 and the successive gaps between the vibrational levels are also listed in Table 5.5. It is quite clear that these successive gaps are not equal to each other and the frequencies of the overtones are not the integral multiple of the fundamental frequency. The values of fundamental frequency as well as those of anharmonic constants can be calculated as follows (neglecting the $\omega_0\ y_0$ anharmonic constant):

$$\Delta G_{v+\frac{1}{2}} = G_0(v') - G(v'')$$

$$= [\bar{\omega}_0\ v' - \bar{\omega}_0\ x_0\ v'^2] - [\bar{\omega}_0\ v'' - \bar{\omega}_0\ x_0\ v''^2] \quad ...(5.111)$$

For successive gap, $\Delta v = +1$, so

$$\Delta G_{v+\frac{1}{2}} = \bar{\omega}_0 (v'' + 1 - v'') - \bar{\omega}_0 x_0 [(v'' + 1)^2 - v''^2]$$

$$= \bar{\omega}_0 - \bar{\omega}_0 x_0 (1 + 2v'') \quad ...(5.112)$$

General expression can be written as:

$$\Delta G_{v+\frac{1}{2}} = \bar{\omega}_0 - \bar{\omega}_0 x_0 - 2\bar{\omega}_0 x_0 v \quad ...(5.113)$$

The second difference will lead to

$$\Delta^2 G_{v+1} = 2\bar{\omega}_0 x_0 \quad ...(5.114)$$

Example 5.19: Calculate the value of $\bar{\omega}_0 x_e$ and $\bar{\omega}_0 x_0$ from the following the vibrational spectrum of HCl.

v	0	1	2	3	4	5
$\bar{\omega}$ (cm^{-1})	0	2885.9	5668.0	8346.9	10923.1	13396.5

Solution: The above data can be arranged in the following Table 5.5.

Table 5.5

v	$\bar{\omega}_{obs}$ (cm^{-1})	ΔG_{obs} (cm^{-1})	$\Delta^2 G$ (cm^{-1})
0	0		
		2885.9	
1	2885.9		–103.7
		2782.1	
2	5668.0		–103.2
		2678.9	
3	8346.9		–102.8
		2576.1	
4	10923.1		–102.6
		2473.4	
5	13396.5		

The average value of second difference is – 103.075 cm^{-1}

According to equation (5.114)

$$\Delta^2 G_{v+1} = -2\bar{\omega}_0 x_0 = -2\bar{\omega}_e x_e = -103.075 \text{ cm}^{-1}$$

$$\bar{\omega}_0 x_0 = \bar{\omega}_e x_e = 51.538 \text{ cm}^{-1}$$

The value of $\bar{\omega}_0$ can be calculated either from the plot of $\Delta G_{v+½}$ versus v (*i.e.*, the intercept) or analytically by taking $\bar{\omega}_{obs}$ from different bands and using equation (5.107), *e.g.*, using the data of $0 \rightarrow 1$ band

$$2885.9 = \bar{\omega}_0 - 51.538\,(1)^2$$

$$\bar{\omega}_0 = 2885.9 + 51.538 = 2937.438 \text{ cm}^{-1}$$

and $\bar{\omega}_e$ from equation (5.108)

$$2937.438 = \bar{\omega}_e - 51.538$$

$$\bar{\omega}_e = 2937.438 + 51.538 = 2988.976 \text{ cm}^{-1}$$

5.5.9 ROTATIONAL-VIBRATIONAL SPECTRUM

In the previous section, we have considered rotational and vibrational motions separately. But it is well known that rotational and vibrational motions in the molecule take place simultaneously. Further we have also learnt that the rotational quantum is much smaller than the vibrational quantum. Due to this many rotational levels are present within each vibrational level. Thus, infrared spectrum does not only contain the vibrational line but a number of rotational lines will also be observed during the excitation of vibrational motion. This set of closely lying lines is called a *band*. This model will be developed, which can take care of vibrational and rotational motions simultaneously. This is called the vibrating rotator.

Energy Levels

According to Born-Oppenheimer approximation if the interaction between the rotational and vibrational motions is neglected (one can do so because the value of B is of the order of 10 cm^{-1} and that of vibrational frequency is of the order of 10^3 cm^{-1}), the total energy of the vibrating rotator can be written as the sum of the energy of anharmonic oscillator and the non-rigid rotator. However better results can be obtained about the energy levels if one takes into account the interaction of the two motions. As we have seen in the last section (5.5.8) that the internuclear distance between the two atoms changes when the molecule is excited vibrationally, this will change the rotational constant. Since hundreds of vibrations are completed during one rotation, we can take the average distance between the two atoms as the internuclear distance, as stated in section (5.4.1). Further the average bond length will be different in each vibrational state and thus the values of rotational constant will be different in each vibrational state. In reality it keeps on decreasing as the value of v increases and can be given by the expression:

$$B_v = \frac{h \times 10^{-2}}{8\,\pi^2\, c\, \mu \left\langle r^2 \right\rangle} \qquad ...(5.115)$$

where $\left\langle r^2 \right\rangle$ is the average value of r^2 and B_v is the rotational constant in the vth vibrational state. Theoretically, one can derive a relation between the various values of rotational constants and this is given by equation (5.116).

$$B_v = B_e - \alpha\left(v + \frac{1}{2}\right) \qquad ...(5.116)$$

where B_e is the rotational constant corresponding to the equilibrium bond length and α is the first order constant due to the rotational-vibrational interactions. The value of α is much smaller than B_e and α is always a positive quantity. As defined above, $B_v < B_e$. Now we can see that the rotational constant calculated in section 5.4 is B_0 (rotational constant in the zeroth vibrational state) and not B_e. This is because the molecule is always present in the zeroth vibrational state.

Neglecting the correction due to non-rigid rotator or centrifugal distortion, the expression for total energy is:

$$T = G(v) + F(J) \qquad ...(5.117)$$

$$= \bar{\omega}_e \left(v + \frac{1}{2}\right) - \bar{\omega}_e x_e \left(v + \frac{1}{2}\right)^2 + B_v J (J + 1) \qquad ...(5.118)$$

The energy level diagram corresponding to the expression (5.118) is shown in Fig. 5.16. As described earlier there are number of rotational levels in between successive vibrational levels.

SELECTION RULES

Similar to the treatment given for non-interacting particles (sec. 2.6.4), the total wave function for the vibrating-rotator can be written as a product of rotational and vibrational wave functions. Thus similar selection rules are also valid in this case also as were applicable to rotational and vibrational spectra separately, *i.e.*, $\Delta v = \pm 1$ and $\Delta J = \pm 1$. $\Delta v = 0$ is also valid but involves the

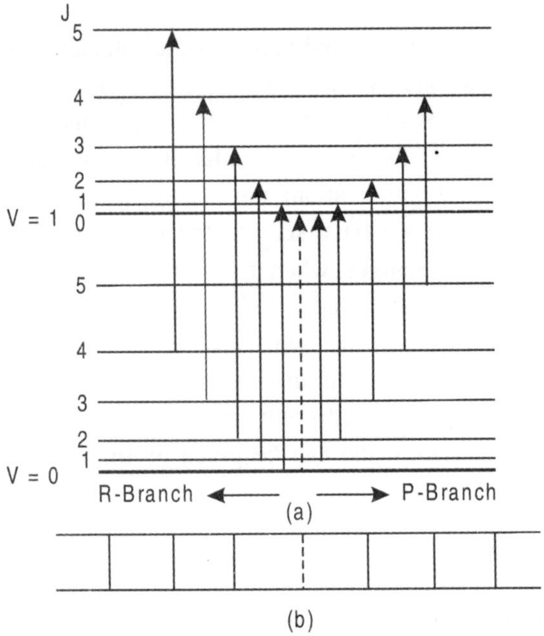

Figure 5.16: *(a) Rotational-Vibrational energy level diagram for v =0 and v = 1. Arrows show the particular transitions following the selection rules. Dotted line depicts the zero line (forbidden transition). Right hand side shows the P-Branch and left hand side shows the R-Branch. (b) Shows the rotational-vibrational spectrum*

promotion of molecules from one rotational level to another rotational level within similar vibrational state. This is just similar to pure rotational spectrum. Similar to earlier discussion $\Delta v = +1$ corresponds to absorption and $\Delta v = -1$ to emission spectrum. But $\Delta J \pm 1$ is valid for both absorption and emission spectroscopy.

Rotational-Vibrational Spectrum

With the above selection rules in mind, one can write the wave number of transition resulting in the excitation of vibrational level v'' to v', *i.e.*,

$$\begin{aligned}\bar{\nu} &= T_{v'} - T_{v''} = G(v') - G(v'') + F(J') - F(J'') \\ &= \bar{\omega}_0 + B_{v'} J'(J'+1) - B_{v''} J''(J''+1) \qquad \text{...(5.119)}\end{aligned}$$

where $\bar{\omega}_0 = G(v') - G(v'')$ is the frequency of pure vibrational transition, *i.e.*, $J' = J'' = 0$. Since both values of $\Delta J = \pm 1$ are valid, we can write the two separate expressions with $\Delta J = +1$, *i.e.*, $J' = J'' + 1$ and with $\Delta J = -1$, $J' = J'' - 1$.

Now with $J' = J'' + 1$

$$\begin{aligned}\bar{\nu}_R &= \bar{\omega}_0 + B_{v'}(J''+1)(J''+2) - B_{v''} J''(J''+1) \\ &= \bar{\omega}_0 + 2B_{v'} + (3B_{v'} - B_{v''}) J'' + (B_{v'} - B_{v''}) J''^2\end{aligned}$$

or we can write the general expression

$$\bar{\nu}_R = \bar{\omega}_0 + 2B_{v'} + (3B_{v'} - B_{v''}) J + (B_{v'} - B_{v''}) J^2 \quad J = 1, 2, 3, \ldots \qquad \text{...(5.120)}$$

with $\qquad J' = J'' - 1$

$$\begin{aligned}\bar{\nu}_P &= \bar{\omega}_0 + B_{v'}(J''-1)(J'') - B_{v''} J''(J''+1) \\ &= \bar{\omega}_0 - (B_{v'} + B_{v''}) J'' + (B_{v'} - B_{v''}) J''^2\end{aligned}$$

or general expression will be:

$$\bar{\nu}_P = \bar{\omega}_0 - (B_{v'} + B_{v''}) J + (B_{v'} - B_{v''}) J^2 \quad J = 1, 2, 3, \ldots\ldots \qquad \text{...(5.121)}$$

Since J can have a whole series of values a set of lines can be observed on both sides of pure vibrational frequency, $\bar{\omega}_0$. In $\Delta J = +1$, the rotational lines are on the higher frequency side of $\bar{\omega}_0$ and it is called the *R branch*. With $\Delta J = -1$ the rotational lines are on the decreasing side of $\bar{\omega}_0$ and it is called the P *branch*. Further in each vibrational state, the rotational levels involved for rotational lines of R branch start with $J = 0$ and P branch has to have the minimum value of $J = 1$. Thus the rotational-vibrational spectrum observed is shown in Fig. 5.16. The salient features emerging from the above infrared spectrum are:

(*i*) It is clear from equations (5.120) and (5.121) that the first line of R branch will be at $\bar{\omega}_0 + 2B_{v'}$ and the first line of P branch will be at $\bar{\omega}_0 - 2B_{v'}$. In other words, $\bar{\omega}_0$ line will not be observed in the rotational-vibrational spectrum and is a forbidden transition. This is called a *null line* or *zero line* and is also known as *band origin*. Thus the minimum gap observed between the first R line and P line is $4B_{v'}$.

(*ii*) Since $B_{v'} < B_{v''}$, the third term in the equation (5.120) will be positive, whereas the fourth term is negative. In other words the successive lines in the above spectrum are not equally spaced and the gap between them will keep on decreasing with increase of J. A situation will arise when for certain value of J the fourth term will outweigh the third term and thus the reversal of the rotational lines will take place. This means that after this value of J the frequency of rotational lines will start decreasing rather than increasing, a trend that is normally observed for R branch. Due to this there will be a lot of crowding of lines near this point. The point of reversal is called the *band head*. The value of J for the band head formation can be calculated from the relation when

$(B_{v'} - B_{v''})\, J^2 > -\, (3B_{v'} - B_{v''})\, J$, *i.e.*,

$$J > -\frac{3B_{v'} - B_{v''}}{B_{v'} - B_{v''}} \qquad \text{...(5.122)}$$

Since the difference between $B_{v'}$ and $B_{v''}$ is not very large for the vibrational levels involved in the rotational-vibrational spectrum, the value of J in general for band head formation is quite large. Due to this reason, band head formation is often not observed in the infrared spectrum. This is because the relative population of molecules up to such large rotational levels in a given vibrational level is quite small. On the other hand for P branch both the second and third terms are negative and also third term is proportional to J^2, the change in this term will be more as J increases. Thus, the gap between the successive lines of P branch increases with increase of J.

(*iii*) If one considers that there is no interaction between the rotational and vibrational motions, $B_{v'} = B_{v''} = B$. The equations (5.120) and (5.121) for R and P branches respectively become:

$$\bar{v}_R = \bar{\omega}_0 + 2B + 2BJ = \bar{\omega}_0 + 2B\,(J + 1) \qquad \text{...(5.123)}$$

$$\bar{v}_P = \bar{\omega}_0 - 2BJ \qquad \text{...(5.124)}$$

i.e., in these cases the rotational lines in both the branches are equally spaced and the gap between successive lines is 2B, similar to the pure rotational spectrum without involving the centrifugal distortion.

(*iv*) Value of α, defined in equation (5.116) is not very large. Thus, for small values of v, $B_{v'}$ and $B_{v''}$, are not much different from each other. The rotational-vibrational spectrum can be described nicely with a set of equations (5.123) and (5.124). But for overtones, when $\Delta v = 2, 3......$ the difference between $B_{v'}$ and $B_{v''}$ is large and one needs equations (5.120) and (5.121) to describe the true picture of infrared spectrum.

(*v*) Equations for P and R branches can be described by a single equation, *i.e.,*

$$\bar{\nu}_{P,R} = \bar{\omega}_0 + (B_{v'} + B_{v''})\, m + (B_{v'} - B_{v''})\, m^2 \qquad ...(5.125)$$

where $m = \pm 1, \pm 2, \pm 3, \pm 4....$ Positive m refers to R branch and negative m refers to P branch. Further for R branch, $m = J + 1$ $(J = m - 1)$ and for P branch $m = -J$. Also note that m cannot have value equal to zero, because it will lead to the values of J′ or J″ to be −1, which is not possible.

Example 5.20. The value of B_e and α_e for CO are 1.9313 and 0.0.01747 cm^{-1}. (*a*) Calculate the value of J for which the reversal of R branch will occur. (*b*) Calculate the relative population of this level with respect to J = 0 at 300 K.

Solution: Using equation (5.116)

$$B_0 = 1.9313 - (0.01747)\,(1/2) = 1.9226 \text{ cm}^{-1}.$$

$$B_1 = 1.9313 - (0.01747)\,(3/2) = 1.9051 \text{ cm}^{-1}$$

Now using the relation (5.122)

$$J \geq -\frac{3B_1 - B_0}{B_1 - B_0} = -\frac{(3)\,(1.9051) - (1.9226)}{1.9051 - 1.9226} = 217$$

$$(b)\quad \frac{N_{217}}{N_0} = (2J + 1)\exp\,(-\,BhcJ\,(J + 1/k_b\,T))$$

$$\frac{Bhc\,(J^2 + J)}{k_b T} = \frac{(1.9226\text{ cm}^{-1})\,(6.62\times10^{-34}\text{ J s})\,(3\times10^{10}\text{ cm s}^{-1})\,(217)\,(218)}{(1.38\times10^{-23}\text{ J K}^{-1})\,(300\text{ K})} = 436$$

$$\frac{N_{217}}{N_0} = (2 \times 217 + 1)\, e^{-436} = 0$$

i.e., band head will not be observed in case of rotational-vibrational spectrum of CO.

Calculation of the Rotational Constants

The following procedures can be used to calculate the rotational constants, corresponding to different vibrational levels.

1. **$B_{v'}$:** Consider the rotational lines of P and R branches, which originate from the same rotational quantum number of lower vibrational state. Take the difference of these two lines. Since both these lines are starting from the same J value, the difference of these lines will give the gap between the rotational levels of the upper vibrational state. Substitute the value of J in the equations (5.120) and (5.121) and take the difference.

$$\bar{\nu}_R(J) - \bar{\nu}_P(J) = \bar{\omega}_0 + 2B_{v'} + (3B_{v'} - B_{v''})\,J + (B_{v'} - B_{v''})\,J^2$$

$$-\,\bar{\omega}_0 + (B_{v'} + B_{v''})\,J + (B_{v'} - B_{v''})\,J^2$$

$$= 2B_{v'} + 4B_{v'}\,J = 2B_{v'}\,(2J + 1) \qquad ...(5.126)$$

For the proper value of J one can calculate the value of $B_{v'}$.

2. **$B_{v''}$:** Consider the rotational lines of P and R branches, which end on the same rotational level of the upper vibrational state. These transitions will correspond to J – 1 for R branch and J + 1 for P branch. Substitute the values of rotational quantum numbers in the equations (5.120) and (5.121) respectively and take the difference.

$$\begin{aligned}\bar{\nu}_R(J-1) - \bar{\nu}_P(J+1) &= \bar{\omega}_0 + 2B_{v'} + (3B_{v'} - B_{v''})(J-1) + (B_{v'} - B_{v''})(J-1)^2 \\ &\quad - \bar{\omega}_0 + (B_{v'} + B_{v''})(J+1) - (B_{v'} - B_{v''})(J+1)^2 \\ &= 2B_{v''} + 4B_{v''}J = 2B_{v''}(2J+1) \qquad ...(5.127)\end{aligned}$$

From the above relation, *i.e.*, equation (5.127), one can calculate, the rotational constant, $B_{v''}$ of the vibrational level v''.

The only disadvantage in this method is that it gives one and only one value of each rotational constant. Since there are some inherent errors in the measurement of the positions of the spectral lines, the best values of these rotational constants can be obtained by taking many such combinations and then taking the average of these values.

3. **Analytical Method:** Using this method one can calculate the rotational constants of both the vibrational states involved in the given transition. Since this method involves plotting, one can get the accurate value from the best fit of the graphical plot. The procedure is as follows: Arrange the rotational lines from maximum negative value of *m* to maximum positive value of *m* as shown in Table 5.6. The line for *m* = 0 is missing. Take the successive difference and it is listed in the third column of the Table 5.6. The second difference is also recorded in the fourth column. It can be seen that the data in the fourth column are pretty well constant with in experimental error. Using equation (5.129), one can calculate $B_{v'}$. Similarly using other data one can calculate $B_{v''}$ also. The analytical equations involved are:

$$\begin{aligned}\Delta\bar{\nu} &= \bar{\omega}_0 + (B_{v'} + B_{v''})(m+1) + (B_{v'} - B_{v''})(m+1)^2 - \hat{\omega}_0 \\ &\quad - (B_{v'} + B_{v''})m - (B_{v'} - B_{v''})m^2 \\ &= B_{v'} + B_{v''} + (B_{v'} - B_{v''})(2m+1) = B_{v'} + B_{v''} + B_{v'} - B_{v''} \\ &\quad + 2(B_{v'} - B_{v''})m \\ &= 2B_{v'} + 2(B_{v'} - B_{v''})m \qquad ...(5.128)\end{aligned}$$

$$\begin{aligned}\Delta^2\bar{\nu} &= 2B_{v'} + 2(B_{v'} - B_{v''})(m+1) - 2B_{v'} - 2(B_{v'} - B_{v''})m \\ &= 2(B_{v'} - B_{v''}) \qquad ...(5.129)\end{aligned}$$

Figure 5.17 gives the plot between *m* and Δν. It gives a straight line. Intercept gives the value of $B_{v'}$ and slope that of $B_{v'} - B_{v''}$. From these observations, one can calculate $B_{v'}$ and $B_{v''}$.

Calculation of α and B_e

Substituting the values of $B_{v''}$ and $B_{v'}$ in equation (5.116) and one gets the following set of equations:

$$B_{v'} = B_e - \alpha\,(v' + 1/2) \quad \text{...(5.130)}$$

$$B_{v''} = B_e - \alpha\,(v'' + 1/2) \quad \text{...(5.131)}$$

Substituting the values of $B_{v'}$ and $B_{v''}$ in these equations, we can get the values of B_e and α.

Calculation of r_e, $r_{v'}$ and $r_{v''}$: Knowing the values of B_e, $B_{v'}$ and $B_{v''}$, we can easily calculate r_e, $r_{v'}$ and $r_{v''}$ from equation (5.56).

Calculation of Null Line or Band Origin: Knowing the values of $B_{v'}$ and $B_{v''}$ the value of $\bar{\omega}_0$ can easily be calculated from the value of any of the P or R line of infrared spectrum.

Example 5.21 gives the complete procedure to calculate the various constants from the infrared spectrum of ClO.

Example 5.21: The following data are observed in infrared spectrum of ClO. Calculate, $\bar{\omega}_0$, B_e, B_0 B_1 and α_e.

Solution:

J	R (J) (cm^{-1})	P (J) (cm^{-1})	J	R (J) (cm^{-1})	P (J) (cm^{-1})
0	2906.25		4	2980.90	2798.78
1	2925.78	2865.09	5	2997.78	2775.79
2	2944.89	2843.56	6	3014.29	2752.03
3	2963.34	2821.49	7	3029.96	2727.75

The above data can be arranged in the following manner according to equation (5.125) in Table 5.6.

$\Delta\bar{v}_m$ has been plotted in Fig. 5.17. The slope gives: 2 ($B_{v'} - B_{v''}$) and intercept at $m = 0$ gives $2B_{v'}$

$$i.e., \quad 2\,(B_1 - B_0) = \frac{23.85 - 18.35}{-6 - 3} = -\,0.61$$

$$\text{or} \quad B_1 - B_0 = -\;0.305 \text{ cm}^{-1}$$

Intercept, $2B_1 = 20.2$ cm^{-1}, $B_1 = 10.1$ cm^{-1} and $B_0 = 10.1 + 0.305 = 10.405$ cm^{-1}

Table 5.6

m	$\bar{\nu}$ (cm^{-1})	$\Delta\bar{\nu}_m$ (cm^{-1-})	$\Delta^2\bar{\nu}_m$ (cm^{-1})
8	3029.96		
		15.67	
7	3014.29		– 0.84
		16.51	
6	2997.78		– 0.37
		16.88	
5	2980.90		– 0.78
		17.66	
4	2863.24		– 0.77
		18.43	
3	2944.89		– 0.63
		19.03	
2	2925.78		–0.50
		19.53	
1	2906.25		
0			
–1	2865.09		
		21.53	
–2	2843.56		–0.54
		22.07	
–3	2821.49		–0.28
		22.71	
–4	2798.78		–0.77
		22.99	
–5	2775.79		–0.41
		23.76	
–6	2752.03		–0.61
		24.24	
–7	2727.75		

Using equations (5.130) and (5.131)

$$10.405 = B_e - \alpha_e/2$$

$$10.10 = B_e - 3/2\ \alpha_e$$

gives $B_e = 10.557\ \text{cm}^{-1}$ and $\alpha_e = 0.305\ \text{cm}^{-1}$

Lastly, using equation (5.120) for any value of R line and substituting the value of B_0 and B_1, we get

$$R(2) = 2944.89 = \bar{\omega}_0 + 2 \times 10.1 + (3 \times 10.1 - 10.405)(2) + (10.1 - 10.405)(2)^2$$

$$= \bar{\omega}_0 + 20.2 + 39.7 - 1.22$$

or $\bar{\omega}_0 = 2944.89 - 58.68 = 2886.21\ \text{cm}^{-1}$.

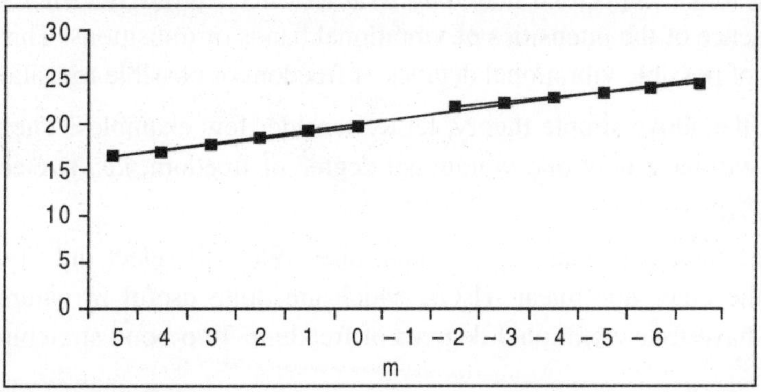

Figure 5.17: *Plot of $\Delta\bar{v}_m$ versus m. The value of $\Delta\bar{v}_m$ at m = 0 gives the intercept, i.e., $2B_1$ and slope gives $2(B_1 - B_2)$*

5.5.10 VIBRATION OF POLYATOMIC MOLECULES

In the last couple of sections we have learnt about the rotational and vibrational motions of the diatomic molecules. In this section we shall like to know about the vibrational motions of the polyatomic molecules and how this information can be used to determine the structure and geometry of the molecules. Similar to rotational spectra, before we proceed for this discussion it may be mentioned here that even though the vibrational spectra of the polyatomic molecules are quite complicated, only slight and logical extensions to the simple theory are adequate to give us the understanding of the spectra. It is necessary to know the followings before we start using the vibrational spectra to elucidate the structure of the molecules.

Vibrational Degrees of Freedom and Symmetry

Consider a molecule containing N atoms. As we know that we need three coordinates (*x, y, z*) to specify any point in space, the number of coordinates required to specify N atoms in space will be 3 N. These are also called degrees of freedom, because value of each coordinate may be specified independently of the others. The advantage of specifying these coordinates is that the distances between the respective atoms (bond lengths and angles) are fixed.

The polyatomic molecules can have three different kinds of motions. First one is the motion of the molecules as a whole and can be described by considering the centre of mass coordinates of the molecule and thus it can have only *three* independent values. These are called *translational degrees of freedom*. The second one is rotational degrees of freedom. As we have seen in the last section that the total numbers of axes around which the non-linear molecule can rotate are three, the *rotational degrees of freedom* for non-linear molecules should be thus *three*. On the other hand for linear molecules there is no rotation around the rotational axis coinciding the bond axis. Thus for linear molecules the rotational degrees of freedom are *two*. The third type of motion the molecules can have is the vibrational motion and the number of vibrational degrees of freedom will be 3N – 5 for linear molecules and 3N – 6 for non-linear molecules. Further the number of bonds between N-atom molecules are N – 1, N – 1 of vibrations are bond stretching motions and the rest 2N – 5, for non-linear and 2N – 4 for linear molecules are the bending motions. One thing may be specified here that 3N – 5 or 3N – 6 rule does not say anything about the presence or absence of the intensities of vibrational bands or transitions. This rule only tells the maximum number of possible vibrational degrees of freedom or possible transitions to be observed.

To understand the above simple theory, let us consider few examples. The simplest molecule is diatomic and it will have only one vibrational degree of freedom, *i.e.*, stretching motion of the bond.

The next set of molecules can be tri-atomic ones. We will select only two molecules (one linear, CO_2; and the other non-linear, H_2O), which are quite useful in nature. CO_2 is a linear molecule and will have four vibrational degrees of freedom. Two bond stretching and two will be

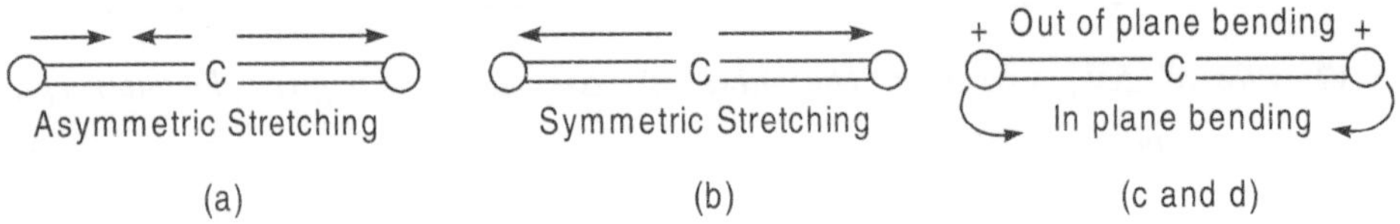

Figure 5.18: *Schematic representation of the different vibrational modes of* CO_2

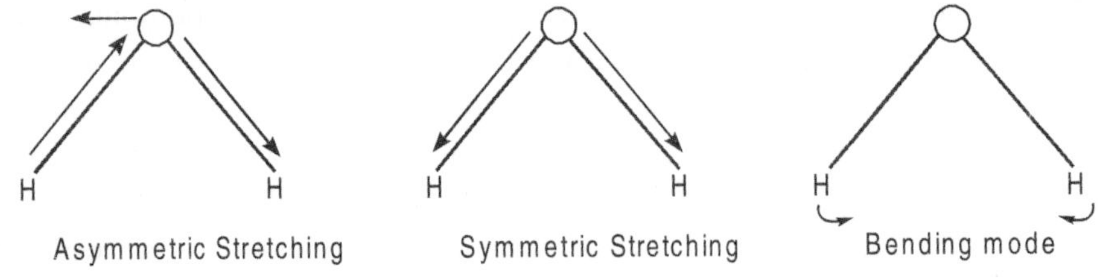

Figure 5.19: *Schematic representation of different vibrational modes of water*

bending modes. Figure 5.18 describes them pictorially as *a*, *b*, *c* and *d*. Vibrational motions as described in Fig. 5.18*a* and *b* involve the stretching of bonds, whereas Fig. 5.18*c* and *d* involves the bending of the bonds. Looking at the molecule, it is clear that the molecule can have two kinds of bending motions. One is the bending of the oxygen atoms in the plane of the molecule or plane of paper and the other in which oxygen atoms move in and out simultaneously perpendicular to the plane of the paper. Since these two vibrational motions are similar to each other in all aspects except direction, so these are termed as degenerate. These two motions should be considered separately. The distinction between the two different kind of stretching motions can be made by applying simple rules of symmetry. CO_2 can have two different kind of rotational axes, *i.e.,* one coincides with the bond axis and is termed as C_∞ (referred to as ∞ fold since rotation of the molecule around the bond axis through any angle gives the structure identical in all respect). The second axis is perpendicular to the bond axis, passing through centre of gravity of CO_2 (*i.e.,* through carbon atom) and is called C_2 axis. If we rotate the molecule by 180° around C_2 axis it will be seen from Fig. 5.18*a* that the motion depicted will be anti-phase to that prior to rotation and is called anti-symmetric stretching vibration. On the other hand, Fig. 5.18 *b* depicts the vibrational motion, which is identical before and after rotation through 180°. It is called symmetrical stretching. Out of four vibrational modes there is a change in the dipole moment of the modes depicted by 5.18 *a, c* and *d* and thus are infrared active, whereas there is no change in the dipole moment in the vibrational mode given in 5.18 *b* and it is infrared inactive. The vibrational frequency corresponding to this motion can be found by other technique (Raman spectroscopy or electronic spectroscopy to be discussed in next sections or mathematical tool of normal coordinate analysis).

Second tri-atomic molecule, H_2O, is a non-linear molecule and will have only three vibrational modes. Two are bond stretching and one is bond bending. These are shown in Fig. 5.19 *a, b* and *c.* Symmetry rules tell that water will have only one C_2 fold axis of rotation. Rotation of water molecule around this axis by 180° tells that the vibrational motion depicted in Fig. 5.19 *a*, will be anti-symmetric stretching and depicted in Fig 5.19 *b* will be symmetric stretching. In water all the motions will be infrared active because these are accompanied by a change in the dipole moment.

Finally the molecular vibrations depicted in Fig. 5.19 *a, b* and *c* can be labeled as $\bar{\omega}_1$, $\bar{\omega}_2$ and $\bar{\omega}_3$ respectively. By convention these are labeled according to the decreasing vibrational frequencies with their symmetry type. Thus, the symmetric stretching vibration of CO_2 is labeled as $\bar{\omega}_2$ for highest fully symmetric stretching frequency (1330 cm^{-1}) and $\bar{\omega}_3$ for the next highest (667 cm^{-1}). The anti-symmetric stretching vibration at, 2309.3 cm^{-1} is labeled as $\bar{\omega}_1$.

Overtones and Combination Bands

Based on the above discussion we can infer that we will be only observing either three (nonlinear molecule, H_2O) or two out of four (linear molecule, CO_2) vibrational bands in the infrared spectra of triatomic molecules. But in actual practice, the absorption spectra in the infrared region are more complicated and contain many other bands. It can be explained as follows. The above discussion is based on the fact that the molecules follow the simple harmonic motion. In reality it is no so. We have learnt that the molecules act as anharmonic oscillators and besides observing the fundamental transitions ($\Delta v = \pm 1$), the transition arising out of $\Delta v = \pm 2, \pm 3, \ldots$ (overtones) are also observed. The intensity of the overtones decreases with increase of Δv. Besides the observance of overtones bands, two other kinds of transitions are also observed. These are called combination bands *i.e.,* sum and difference bands. The former is arising out from the addition of two or more fundamental frequencies or overtones. For examples, $\bar{\omega}_1 + \bar{\omega}_2$, $\bar{\omega}_1 + 2\bar{\omega}_2$, $\bar{\omega}_1 + \bar{\omega}_2 + \bar{\omega}_3$ ..... The latter one can be $\bar{\omega}_1 - \bar{\omega}_2$, $2\bar{\omega}_1 - \bar{\omega}_2$, $\bar{\omega}_1 + \bar{\omega}_2 - \bar{\omega}_3$.

Similar to overtones the intensities of these combination and difference bands are also quite weak.

It may be pointed out here that in certain cases, the combination bands observed between the fundamental and the overtones etc., are more intense than the corresponding overtones or combination band (sum or difference). The increase in intensity of these bands is considerably enhanced by a resonance phenomenon. It may happen that two vibrational modes in a particular molecule may have energy levels very close to each other and thus are described as accidental degenerate. It may be pointed out here that accidental degeneracy does not point to two exactly degenerate modes of CO_2 ($\bar{\omega}_3$), but rather to the possibility of two quite different modes having similar energies. In general the fundamental vibrational modes are quite different from each other and thus will not have accidental degeneracy. But the accidental degeneracy can occur between the fundamental and some overtone or combination band. For example, $\bar{\omega}_2$ of CO_2 is approximately at 1330 cm^{-1}, is very close to the first overtone of $\bar{\omega}_3$ (*i.e.,* 1334 cm^{-1}). Further for information both these bands are not observed in infrared region, as explained earlier, but can be observed in Raman spectroscopy. It will be discussed later that law of quantum mechanics are equally applicable to both infrared and Raman spectroscopy. It can be shown by quantum mechanics that two such bands interfere with each other in such a way that the energy of the higher mode is raised further (1385 cm^{-1}) and that of lower one is lowered (1285 cm^{-1}), whereas the mean of two is at 1330 cm^{-1}. This kind of combination between the fundamental and overtone or combination modes is called **Fermi Resonance.** Normally one would expect the intensity of 1385 cm^{-1} band to be more than that of 1285 cm^{-1}, because the former arises from the fundamental band ($\bar{\omega}_2$) and the latter from the first overtone ($2\bar{\omega}_3$). Actually the intensities of both these bands are found to be nearly equal, *i.e.,* the overtone band has gained intensity at the expanse of fundamental band. The above situation can be compared to the two oscillating pendulums, tied to the same bar. The exchange of energy or phenomenon of resonance will be the maximum when these pendulums have similar vibrational frequencies. In the spectra of complex molecules, it is quite possible that the accidental degeneracy can occur between the fundamental and overtone and thus one would expect the Fermi resonance. It may be pointed out here that not all such degeneracies' lead to resonance. It is also necessary to consider the molecular symmetry and the type of degenerate vibrations. Further discussion is beyond the scope of this book and interested students can look up the advanced books.

Example 5.22: In polyatomic molecules, the vibrational transitions are generally represented in short hand notation as:

(*i*) (0, 0, 0) $\rightarrow$ (0, 2, 0)

(*ii*) (0, 0, 2) $\rightarrow$ (0, 1, 1). Explain their meanings.

Solution:

(*i*) Transition (0, 0, 0) → (0, 2, 0) means that the vibration $\bar{\omega}_2$ is excited from $v_2 = 0$ to $v_2 = 2$ and this is first overtone. In all, there are three distinct modes of vibrations.

(*ii*) The transition (0, 0, 2) → (0, 1, 1) means that $\bar{\omega}_2$ is excited from $v_2 = 0$ to $v_2 = 1$ and $\bar{\omega}_3$ is de-excited from $v_3 = 2$ to $v_3 = 1$, *i.e.*, it is combination band with frequency $\bar{\omega}_2 + \bar{\omega}_3$. Overall there are three modes of vibrations.

Example 5.23: The fundamental vibrations in water have the frequencies. $\bar{\omega}_1$ (asymmetric stretching) = 3756 cm^{-1}, $\bar{\omega}_2$ (symmetric stretching) = 3652 cm^{-1} and $\bar{\omega}_3$ (bending) = 1595 cm^{-1}. Besides these the following transitions at 3151.4 cm^{-1}, 5332 cm^{-1} and 6874 cm^{-1} are also observed. These are either weak or of medium intensities. Assign these transitions.

Solution: Since the last three transitions are either weak or medium, these must be either overtones or combination bands.

(*i*) 3151.4 cm^{-1} band seems to be the overtone of bending mode because 2 × 1595 cm^{-1} = 3190 cm^{-1} is quite close to 3151.4 cm^{-1}. Difference of 38.6 cm^{-1} could be due to anharmonicity in the molecular band.

(*ii*) 5332 cm^{-1} band system could be due to the combination band due to $\bar{\omega}_3 + \bar{\omega}_1$ = (1595 + 3756) cm^{-1} = 5351 cm^{-1}. The first overtones of $\bar{\omega}_3$ and $\bar{\omega}_3$ is having frequencies which are greater than 5332 cm^{-1}.

(*iii*) 6874 cm^{-1} could also be a combination band: 2 $\bar{\omega}_3 + \bar{\omega}_1$ = (2 × 1595 + 3755.8) cm^{-1} = 6945.8 cm^{-1}.

5.5.11 APPLICATIONS OF INFRARED SPECTROSCOPY

We have learnt in the last section that the number of vibrational modes in a molecule is either 3N – 5 or 3N – 6. Thus the infrared spectrum of the complex molecule will be quite complicated. But to the first approximation, the molecular vibrations can be divided broadly into two categories. One involves the molecular vibrations, to be associated with the vibrations of individual bond or fundamental groups (*localized vibrations*). The other involves the vibrations of more or less of all the atoms of a molecule. The latter is called *skeletal vibrations and involves* all the atoms to the same extent. The former involves the atoms of the characteristic group vibrations, which only comprises a small portion of the molecules and the other part of the molecule being more or less stationary.

The skeletal vibrations generally fall in 1400 to 700 cm^{-1} region of infrared and contain large number of bands. Though it is not possible to assign each band to a particular vibrational mode it still provides enough information to find the structure of molecules. Further the molecules in general possess more bending modes than the stretching modes and these vibrational modes lie in this region, (*i.e.*, 1400 – 700 cm^{-1}). The molecular vibrations, which have been analyzed in this region, are the most intense fundamental and few overtones or combination bands. For this reason, this region is also called the *Finger Print region* and is extremely useful in establishing conclusively the identity of two samples. It may be pointed out here and it is quite possible that the two molecules may have the similar vibrational spectra above the 1400 cm^{-1} region, whereas in the finger print region the spectral features are quite different.

Table 5.7 gives the vibrational frequencies data regarding the group frequencies. It is quite evident that the functional group containing light atoms (for example, —CH_3, —CH_2—, —NH_2, —OH, >C═O etc.) possess vibrational frequencies greater than 1400 cm^{-1}, whereas the functional groups containing heavier atoms, > C═S, —C—Cl, —C—Br etc., have molecular vibrations in the region below 700 cm^{-1}. The vibrational spectra arising out of these vibrations are highly characteristics of a group and thus can be used to establish the presence of a particular group in a molecule. For example, —CH_3 group gives rise to symmetric —C—H stretching transition generally falling between 2850 and 2890 cm^{-1}, symmetric stretching frequency in the

Table 5.7: *Characteristic stretching frequencies of certain groups of atoms*

Group	Band (cm^{-1})	group	Band (cm^{-1})
O—H	3650 – 3590 (*v*)	—CN	2260 – 2200 (*m*)
N—H	3500 – 3300 (*m*)	—C≡C—	2270 – 2150 (*v*, *m*, *w*)
≡C—H	3400 – 3300 (*m*)	>C═O	1850 – 1700 (*s*)
═C—H	3095 – 3075 (*m*)	>C═C<	1650 – 1550 (*m*)
C—H aromatic	3100 – 3050 (*v*)	>C═S	1200 (*s*)
C—H aliphatic C—H	3000 – 2800 (*m–s*) 2960 – 2850 (*s*)	—>C—F	1400 – 1000 (*s*) 780 – 680 (*w*)
>CH_2 asym	1470 – 1430 (*m*) deformation	—>C—Cl	800 – 600 (*s*)
—CH_3 Sym def.	1390 – 1370 (*m*)	—>C—Br	750 – 500 (*s*)
—S—H	2600 – 2550 (*w*)	—>C—I	~ 500 (*s*)

v = variable, *s* = strong, *m* = medium, *w* = weak.

region 2940–2980 cm^{-1}, asymmetric deformation (*i.e.*,opening and closing of the 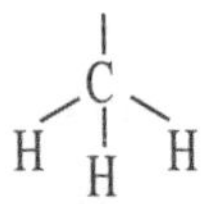 umbrella) at about 1375 cm^{-1} and asymmetric deformation approximately at 1470 cm^{-1}. Similarly the other common group is >C=O (carbonyl group) and its vibrational frequency lies between 1600–1950 cm^{-1}, depending on the other subsituents of the group.

Before we close this section let us try to analyze the frequencies of certain group present in different molecule:

(*i*) We have learnt in earlier sections that vibrational frequency of any oscillator is directly proportional to the square root of the force constant and inversely to the reduced mass. For example, the vibrational frequencies of the stretching mode of —C—H, —C—F, —C—Cl, —C—Br etc., decrease in this order, because of the increase in the atomic mass of the undergoing oscillations. Similarly it is also for the groups >C=O, >C=S. The increase in the bond order increases the force constant and thus increases the vibrational frequency of the following in the order mentioned, *i.e.*, —C —C — < C = C < C ≡ C etc.

(*ii*) Presence of interacting group also affects the vibrational frequency as well as the intensity of a band. For example, the >C=C< and >C=O groups, individually have the vibrational frequencies at 1650 and 1715 cm^{-1} respectively. However, if these groups are present as —C=C—C=O, the vibrational frequencies are affected and are observed at 1600 and 1675 cm^{-1}. The intensity of —C=C— absorption band increases to become comparable with that of the intensity of strong >C=O band. The decrease in the vibrational frequency is because of delocalization of the π cloud and thereby decreasing the bond order and hence the force constant of the individual —C=C— and >C=O groups.

(*iii*) Vibrational frequencies of the groups are affected due to the presence of interactions between the similar or different molecules. The most vulnerable groups are —OH and —NH_2 groups. The change in the frequencies is due to hydrogen bonding and it could be either intra- or intermolecular hydrogen bonding. The shift in the frequency depends on the nature of vibrational mode. For example, in case of —OH group the intermolecular hydrogen bonding decreases the stretching mode but increases the bending mode of —OH bond.

(*iv*) Change in the physical state of the molecule may also cause a shift in the vibrational frequency. In general it has been found that vibrational frequency decreases in going from gas phase to solid phase ($\nu_{gas} > \nu_{liq} > \nu_{solid}$) and especially if the molecule is polar. For example, $\bar{\omega}_{liq}$ for HCl is 100 cm^{-1} less than $\bar{\omega}_{gas}$ and $\bar{\omega}_{solid}$ is 20 cm^{-1} less than $\bar{\omega}_{liq}$, whereas the similar changes for CO_2 are quite small. ($\bar{\omega}_{gas} - \bar{\omega}_{solid} \approx 60\ cm^{-1}$).

5.6 RAMAN SCATTERING

Spectroscopic techniques, which are discussed in the previous sections, are only helpful in determining the molecular parameters of heteroatomic molecules, *i.e.,* the molecules, which are having permanent dipole moment or nonzero oscillating dipole moment. There are other classes of molecules, known as homonuclear diatomic or polyatomic molecules whose permanent dipole moment, as well as, oscillating dipole moment for certain oscillations is zero. In such cases the techniques discussed so far cannot be used because of certain selection rules discussed already.

For this class of molecules Raman scattering is very helpful. The basic difference between the infrared or microwave spectroscopy and Raman scattering is that in former cases radiation is absorbed by the molecules, whereas in the latter case the molecular parameters are determined from the changes observed in the frequency of monochromatic incident radiation by the molecules.

In Raman scattering the monochromatic radiation, mostly in the visible region is passed through the cell containing the desired sample. Small part of the incident radiation (nearly 10^{-4} of I_0) is scattered in all directions. Intensity of scattered light is determined by placing the detectors at right angle to the incident beam. The following changes can be observed:

(*i*) Frequency of the scattered radiation is equal to that of the incident radiation beam. This is known as Rayleigh scattering.

(*ii*) Frequency of the scattered radiation is smaller than that of the incident radiation beam. This is known as Stokes lines.

(*iii*) Frequency of the scattered radiation is greater than that of the incident radiation beam. This is known as anti-Stokes lines.

The above results can be summed as follows: Whenever monochromatic radiation is passed through a sample (may be a gas, liquid or solid), the light is scattered. Frequency of the scattered radiation may be equal to, less than or greater than that of the incident radiation and these can be termed as Rayleigh scattering, Stokes line or anti-Stokes line respectively. The effect is called Raman scattering. Change in the frequency of the incident radiation is due to the inelastic collisions between the molecules and photons. Since the rotational and vibrational energy levels are quantized, the energy transfer to or from the photon will be in the form of quantum. Thus changes observed in the frequency of the scattered radiation are a characteristic function of the molecules of the sample.

Before describing the Raman scattering and its applications to determine molecular parameters let us discuss the basic theory. As usual we will discuss first the classical picture and it will be followed by quantum theory.

5.6.1 CLASSICAL THEORY OF RAMAN SCATTERING

If an atom or a molecule is placed in a static electrical field of strength say F, the positive charge (nuclei) will be displaced towards the negative end and the negative charge (the electron) towards the positive end of the electrical field. This interaction will produce an induced dipole moment in the system. The relation between μ and F is given by

$$\mu \propto F$$

$$\mu = \alpha F \qquad ...(5.132)$$

where α is called polarisability, having the units of volume. Unless the molecule possesses a spherical symmetry, the magnitude of the induced dipole moment produced depends on the direction of the electrical field with the molecule. If the molecule does not possess a spherical symmetry, each component of the induced dipole moment can be related to each component of electrical field as:

$$\mu_x = \alpha_{xx} F_x + \alpha_{xy} F_y + \alpha_{xz} F_z \quad ...(5.133)$$

$$\mu_y = \alpha_{yx} F_x + \alpha_{yy} F_y + \alpha_{yz} F_z \quad ...(5.134)$$

$$\mu_z = \alpha_{zx} F_x + \alpha_{zy} F_y + \alpha_{zz} F_z \quad ...(5.135)$$

We can see from above equations that there are nine components of the polarisability and these depend on the electronic structure of the molecule. Out of nine, only six components are independent and these behave like their subscripts under the influence of symmetry operations. For example, α_{xz} will behave like xz. Now if we select the suitable or principal axis in such a way that the direction of the electrical field vector F coincides with that of induced dipole moment μ, the cross components of polarisability (*i.e.,* α_{xy}, α_{xz} etc.) will vanish. The above equations reduce to:

$$\mu_x = \alpha_{xx} F_x \quad ...(5.136)$$

$$\mu_y = \alpha_{yy} F_y \quad ...(5.137)$$

$$\mu_z = \alpha_{zz} F_z \quad ...(5.138)$$

In general for most of the molecules α_{xx}, α_{yy} and α_{zz} are different but in case of diatomic molecule, if the molecular axis is z-axis, then α_{xx} and α_{yy} will be equal. This is because it is difficult to distinguish, which one is x-axis and which one will be y-axis. This is similar to the rotational motion of the diatomic molecule, where moment of inertia of the molecule around the axis perpendicular to molecular axis is equal. In any way, one thing is clear that induced dipole moment is still proportional to the electrical field, similar to equation (5.132) and this is all that is important in most cases. In future we will only consider this effect, without giving emphasis on the direction of axis.

If the radiation of frequency ν falls on the molecules, it will have an alternating electrical field, given by the expression:

$$F = F_0 \sin 2\pi\nu t \quad ...(5.139)$$

where t is the time. This alternating field induces alternating dipole moment, given by the expression

$$\mu = \alpha F_0 \sin 2\pi\nu t \quad ...(5.140)$$

This alternating induced dipole moment in turn will emit radiation of frequency equal to the frequency of incident beam. This is called Rayleigh scattering (one can explain the Tyndall effect or the phenomenon of refraction in this way). If UV or visible monochromatic radiation is used, the induced dipole moment is only due to the rapid motion of the electrons, because the nuclei are too heavy to follow such rapid oscillations.

Since polarisability possesses the unit of volume, any change in the bond length will change the polarisability, even though slightly. Polarisability also depends on the direction of the electrical field with the molecular axis. Thus polarisability will change during the rotational motion of the molecules. In other words the magnitude of the dipole moment will change during the rotational and vibrational motion of the molecules. To a good degree of approximation, the polarisability during the vibrational motion can be expressed as:

$$\alpha = \alpha_{0v} + \alpha_{1v} \sin 2\pi\nu_{\text{vib}} t \quad ...(5.141)$$

where α_{0v} is the polarisability at the equilibrium position, α_{1v} is the amplitude of the change in polarisability during vibrational motion. $\alpha_{0v} >> \alpha_{1v}$ and ν is the frequency of vibration. In case of rotational motion,

$$\alpha = \alpha_{0r} + \alpha_{1r} \sin 2\pi\, 2\nu_{rot}t \qquad ...(5.142)$$

α_{0r} is the average polarisability and α_{1r} is the amplitude of change in polarisability during rotational motion frequency. In case of rotational motion, $\alpha_{0r} \gg \alpha_{1r}$ and ν_{rot} is the frequency of the rotational motion, the frequency with which the polarisability changes is twice that of rotational frequency and can be explained in number of ways. The physical picture of this fact is that the polarization of the molecules returns to its initial values twice on every revolution. Therefore half a full rotation corresponds to a complete oscillation of polarisability as shown in Fig. 5.20. The selection rule does not conflict with the earlier argument that a photon has a unit spin and overall conservation of angular momentum. Two photons are required in Raman process. One coming in and other are going out.

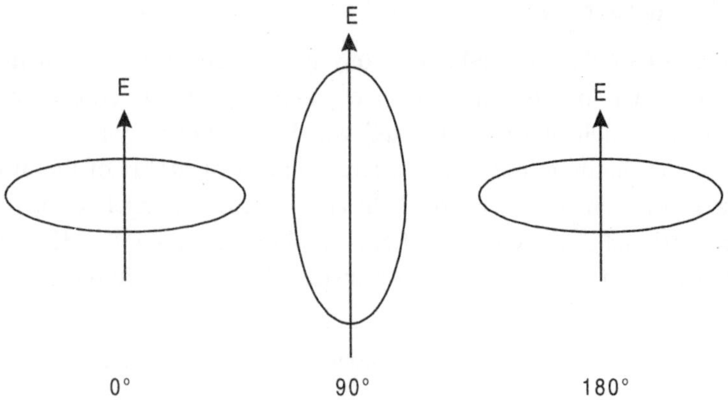

Figure 5.20: *The polarization of the molecule returns to its initial value after a rotation of only 180°*

Now combining equations (5.140), (5.141) and (5.142), one can write the expression for induced dipole moment for vibrational motion.

$$\mu_v = \alpha_{0v} F_0 \sin 2\pi\nu t + \alpha_{1v} F_0 \sin 2\pi\nu t \sin 2\pi\nu_v t \qquad ...(5.143)$$

Similarly for rotational motion:

$$\mu_r = \alpha_{0r} F_0 \sin 2\pi\nu t + \alpha_{1r} \sin 2\pi\nu t \sin 2\pi 2\nu_r t \qquad ...(5.144)$$

Using trigonometric relations:

$$\cos(\alpha + \beta) = \cos\alpha \cos\beta - \sin\alpha \sin\beta \qquad ...(5.145a)$$

$$\cos(\alpha - \beta) = \cos\alpha \cos\beta + \sin\alpha \sin\beta \qquad ...(5.145b)$$

The expressions (5.143) and (5.144) become:

$$\mu_v = \alpha_{0v} F_0 \sin 2\pi\nu t + \frac{1}{2} \alpha_{1v} F_0 [\cos 2\pi(\nu - \nu_v)t - \cos 2\pi(\nu + \nu_v)t] \qquad ...(5.146)$$

$$\mu_r = \alpha_{0v} F_0 \sin 2\pi v t + \frac{1}{2} \alpha_{1r} F_0[\cos 2\pi (v - 2v_r)t - \cos 2\pi (v + 2v_r)t] \quad ...(5.147)$$

It is clear from the equations (5.146) and (5.147) that the induced dipole moment changes in three different ways: (*i*) with the frequency of the incident radiation, (*ii*) with frequency $v - v_v$ for vibrational and $v - 2v_r$ for rotational motion and (*iii*) with frequency equal to $v + v_v$ for vibrational and $v + 2v_r$ for rotational motions. In other words, as said earlier, after the interaction of the molecule with the incident radiation, the emitted radiation will have the same frequency as that of incident radiation, as well as, two more frequencies; $v - v_v$ and $v + v_v$ or $v - 2\,v_r$ and $v + 2\,v_r$, *i.e.,* besides getting the undisplaced line, one gets two other lines, one of the higher frequency and one of lower frequency than the incident radiation. Since the molecular vibrations have fixed vibrational frequencies, one gets the displaced lines at fixed distance from the original line, where as in case of rotational motion (especially in condensed phase) all the rotational frequencies are possible, thus one get a continuous spectrum on both sides of the undisplaced line. The intensities of the displaced lines (*i.e.,* square of the amplitude) will be equal but much weaker than the undisplaced lines, because $\alpha_{1i} << \alpha_{0i}$. In case of vibrational motion, the intensities of the vibrational lines should also depend on the amplitude of vibrations because polarisability also depends on them.

From above discussion one can explain the qualitative features the Raman scattering but it lacks the following features of the observed spectrum: (*i*) Rotational Raman scattering is also discrete set of lines rather than continuous spectrum, (*ii*) The intensity of Raman scattering on higher frequency (anti-Stokes line) side of incident light is much weaker than that observed for the line on the lower frequency side (Stokes line).

5.6.2 QUANTUM THEORY OF RAMAN SCATTERING

The quantum theory of Raman scattering can be explained as follows. When a photon of frequency v strikes the molecule, the following results can be observed. (*i*) The collision between the photon and the molecule can be elastic, *i.e.,* energy is neither gained nor lost by the photon. In this case frequency of the radiation remains same and is called Rayleigh scattering. (*ii*) Energy can be lost by the photon to the molecule. Since the energy levels of molecules are quantized, only discrete amount of energy will be transferred to the molecule. If E'' and E' are the energies of the lower and upper states of the molecules respectively, $\Delta E = E' - E''$ will be gained by the molecule and lost by the photon. The energy or frequency of the photon after collision will be $hv - \Delta E$ or $v - \Delta E/h$ respectively. Thus, one observes a line displaced toward a longer wavelength or shorter frequency from the unscattered radiation. It is called Stokes line. (*iii*) The energy can be gained by the photon from the molecule possessing energy. Thus the energy or frequency of the photon will be $h\,v + \Delta E$ or $v + \dfrac{\Delta E}{h}$ respectively. This line will be displaced towards the shorter wavelength or higher frequency side of Rayleigh line. It is also known as anti-Stokes line. The above description clearly indicated that the Raman scattering or spectrum depends on the value of ΔE. Since there are number of states present in the molecule, one can observe a number of Raman lines.

Quantum theory also explains the intensity of Stokes and anti-Stokes lines very nicely. This is because the number of molecules present in the higher states is very small and thus the photons scattered after collisions with molecules having higher energies will also be small, as compared to the photons loosing energies to the molecules having lower or zero energies. In other words, the intensity of Stokes lines will be more than that of ant-Stokes line. We have also learnt in the earlier sections that the rotational quantum is much smaller than the vibrational quantum. Thus, rotational Raman lines (both Stokes and anti-Stokes) will be lying very close to the incident radiation (Rayleigh line), whereas the vibrational Raman Stokes and anti-Stokes lines will displaced at a longer distance from the Rayleigh line. Quantitatively both rotational Raman and vibrational Raman lines are shown in Fig. 5.21. This explains completely all the features of Raman scattering where the classical theory has failed to do.

5.6.3 VIBRATIONAL RAMAN SCATTERING

The gross or general selection rule for vibrational Raman scattering is that the oscillating polarisability should be finite. The specific selections rules for the vibrational Raman spectrum are the same as for vibrational spectrum, *i.e.*, $\Delta v = \pm 1$. In other words the molecule loses or gains energy by unit of vibrational quantum. These are based on the fact that molecular vibrations are following simple harmonic oscillations. Thus, one Stokes and one anti-Stokes line is observed on both sides of the Rayleigh (unscattered) line by an amount of classical frequency, *i.e.*,

$$\nu' = \nu_{inc} - \nu_{clas} \text{ or } \bar{\nu}' = \bar{\nu}_{inc} - \bar{\omega}_{clas} = \bar{\nu} - \bar{\omega}_e \quad ...(5.148)$$

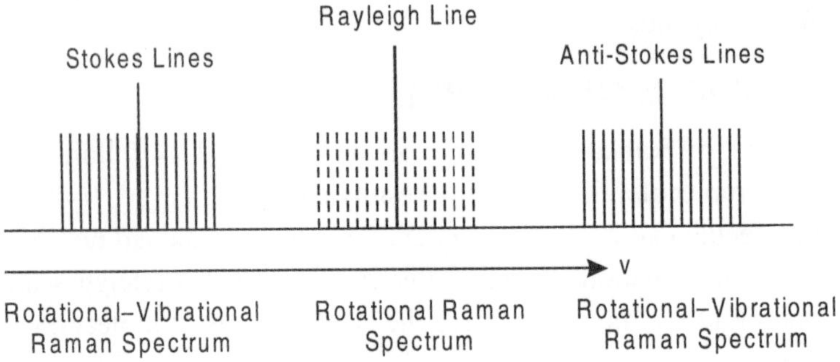

Figure 5.21: *Schematic diagram showing the presence of pure Rotational Raman line, Rotational-Vibrational lines observed in Stokes and Anti-Stokes lines. Rayleigh line is also depicted*

On comparing the scattering line with the infrared spectrum it is evident that both these spectra give the same information about the vibrational data of the molecules. But one thing must be kept in mind that the mechanism of scattering is very different from that of infrared spectrum, *i.e.*, in the latter case it is the absorption of photon of proper energy or frequency, whereas in the former case energy transfer is occurring between the photon of one-particular energy and molecules having many different quantum states. The vibrational constants can be obtained even for those molecules, which do not possess the permanent dipole moment.

As we know that the gap between the successive vibrational levels is quite large as compared to k_b T, especially for diatomic molecules, so the number of molecules present in $v \geq 1$ states will be very small. Actually in many cases anti-Stokes line observed is too weak to be observed. But in many polyatomic molecules the number of vibrational modes is quite large and some of these modes are of small frequencies. In these cases anti-Stokes lines are also relatively stronger than those in diatomic molecules.

Example 5.24: If $^1H^{35}Cl$ is irradiated with 435.8 nm mercury line, calculate the Raman line in nm if the fundamental vibrational frequency of $^1H^{35}Cl$ is $8.667 \times 0113\ s^{-1}$. Calculate the intensity ratio of Stokes to anti-Stokes line at room temperature.

Solution: The frequency of mercury line is:

$$\overline{\nu}_{inc} = \frac{c}{\lambda} = \frac{3 \times 10^8\ ms^{-1}}{435.8 \times 10^{-9}\ m} = 68.839 \times 10^{13}\ s^{-1}$$

Stokes Raman line is observed at $\nu_R = (68.839 - 8.667) \times 10^{13}\ s^{-1} = 60.172 \times 10^{13} s^{-1}$

Stokes Raman line in meter can be obtained as

$$\lambda_R = \frac{3.0 \times 10^8\ ms^{-1}}{60.172 \times 10^{13}\ s^{-1}} = 498.57\ nm.$$

$$\frac{I_{asm}}{I_{sm}} = \frac{g_1\, e^{-\varepsilon_1/k_b T}}{g_0\, e^{-\varepsilon_0/k_b T}} = \frac{g_1}{g_0} e^{-(\varepsilon_1 - \varepsilon_0)/k_b T}$$

Each vibrational level is non-degenerate, $g_1 = g_0 = 1$ and $\varepsilon_1 - \varepsilon_0 = h\nu$

$$\frac{h\nu}{k_b T} = \frac{(6.62 \times 10^{-34}\ Js)\ (8.667 \times 10^{13}\ s^{-1})}{(1.38 \times 10^{-23} J\ K^{-1})\ (298\ K)} = 13.95$$

$$\frac{I_{asm}}{I_{sm}} = e^{-13.95} = 8.74 \times 10^{-7}.$$

5.6.4 ROTATIONAL RAMAN SCATTERING

The gross selection rule for rotational Raman spectrum also involves polarisability, *i.e.*, anisotropic polarisability should be finite. Unlike rotational spectrum the rotational Raman scattering has the following specific selection rules:

$$\Delta J = 0, \pm 2 \qquad ...(5.149)$$

$\Delta J = 0$ corresponds to no change in the rotational quantum number and thus corresponds to the Rayleigh scattering. The frequency of the displaced lines on both sides of incident radiation can be explained with $J' - J'' = \Delta J = \pm 2$, where J' and J'' are the rotational quantum numbers of the upper and lower states respectively. $\Delta J = 2$ corresponds to gain of energy by the molecules from the photon. Thus, the energy of the scattered photon will be less than that of incident photon and one would observe line displaced towards longer wave length, called Stokes line. The frequency of the Raman Stokes line can be calculated as follows:

$$\bar{\nu}_{\text{Raman}} = \bar{\nu}_{\text{inc}} - \frac{\Delta E}{hc} \quad \text{...(5.150)}$$

where $$\Delta E = E_{J'} - E_{J''} \quad \text{...(5.150}a\text{)}$$

$E_{J'}$ and $E_{J''}$ are the energies of the rotational levels J' and J" respectively.

$$\Delta E = Bhc\,J'(J'+1) - B\,h\,c\,J''(J''+1)$$

Using $$J' = J'' + 2,$$

$$\Delta E = B\,[(J''+2)(J''+3) - J''(J''+1)]\,h\,c$$

$$= 4B\left(J'' + \frac{3}{2}\right) \text{ where } J'' = 0, 1, 2.....$$

$$\bar{\nu}_{\text{Raman}} = \bar{\nu}_{\text{inc}} - 4B\left(J'' + \frac{3}{2}\right) \quad \text{...(5.151)}$$

$\Delta J = -2$, *i.e.*, corresponds to gain of energy by the photon from the molecule and rotational Raman line will be at higher frequency. This is also known as anti-Stokes line. The rotational anti-Stokes line can be calculated as:

$$\bar{\nu}_{\text{Raman}} = \bar{\nu}_{\text{inc}} + \frac{\Delta E}{hc} \quad \text{...(5.152)}$$

$$\Delta E = Bhc\,J''(J''+1) - B\,h\,c\,J'(J'+1),\ J = 0, 1, 2, 3,..... \quad \text{...(5.153)}$$

Using $$J' = J'' - 2$$

$$\Delta E = Bhc\,[J''(J''+1) - (J''-2)(J''-1)] = 4\,Bhc\,(J'' - 1/2)$$

$$\bar{\nu}_{\text{Raman}} = \bar{\nu}_{\text{inc}} + 4B\left(J'' - \frac{1}{2}\right),\ J = 2, 3, 4,.......... \quad \text{...(5.154)}$$

As we know from our earlier studies the spacing between the rotational levels is quite small and thus, both the Stokes and anti-Stokes lines are observed in the rotational Raman spectrum. Both these equations indicate that these lines are equally spaced and successive difference between the two lines is 4B. On the other hand the difference between the first Stokes and anti-Stokes lines is 12B.

Similar to classical theory, rotational Raman transitions are also a function of change in the polarisability of the molecule and thus one can observe rotational Raman scattering in homonuclear diatomic molecules, for example, H_2, N_2 ...... The molecular parameters can thus be calculated for these kinds of molecules from the value of B.

Example 5.25: If $^{1}H^{35}Cl$ molecule is irradiated with 404.7 nm mercury line, calculate the first two Stokes as well as anti-Stokes lines observed in the rotational Raman spectrum. The bond length of $^{1}H^{35}Cl$ is 0.1275 nm.

Solution: $$\mu = \frac{(1 \text{ g mol}^{-1})\ (35 \text{ g mol}^{-1})}{(1 + 35) \text{ g mol}^{-1}\ (6.02 \times 10^{23} \text{mol}^{-1})} = 1.615 \times 10^{-24} \text{ g}$$

$$= 1.615 \times 10^{-27} \text{ kg}$$

$$I = (1.615 \times 10^{-27} \text{ kg})\ (0.1275 \times 10^{-9} \text{ m})^2 = 2.625 \times 10^{-47} \text{ kg m}^2$$

$$B = \frac{h \times 10^{-2}}{8\pi^2 I c} = \frac{(6.62 \times 10^{-34} \text{ J s})\,(10^{-2} \text{ m cm}^{-1})}{(8\pi^2)\,(2.625 \times 10^{-47} \text{ kg m}^2)\,(3.0 \times 10^8 \text{ m s}^{-1})}$$

$$= 10.65 \text{ cm}^{-1}$$

$$\bar{\nu}_1 \text{ (J = 0 to J = 2)} = (4)\ (10.65 \text{ cm}^{-1}) \left(0 + \frac{3}{2}\right) = 63.9 \text{ cm}^{-1}$$

$$\bar{\nu}_2 \text{ (J = 1 to J = 3)} = (4)\ (10.65 \text{ cm}^{-1}) \left(1 + \frac{3}{2}\right) = 106.5 \text{ cm}^{-1}$$

$$\bar{\nu}_{inc} = \frac{1}{404.7 \times 10^{-9} \text{ m}} = 2470966 \text{ m}^{-1}$$

The first Stokes line will be observed at = 2470966 – 6390 = 2464576 m^{-1} = 405.75 nm

IInd Stokes line will be observed at = 2470966 – 10650 = 2460316 = 406.45 nm

Similarly Ist anti-Stokes line will be observed at = 2470966 + 6390 = 2477356 m^{-1} = 403.66 nm

IInd anti-Stokes line will be observed at = 2470966 + 10650 = 2481616 m^{-1} = 402.96 nm

5.6.5 ROTATIONAL-VIBRATIONAL RAMAN SCATTERING

Similar to infrared spectrum one can also observe the rotational-vibrational Raman spectrum in diatomic molecules. The selection rules are the combination of Raman rotational and Raman vibrational spectra, *i.e.*, $\Delta v = \pm 1$ and $\Delta J = 0, \pm 2$. $\Delta v = \pm 1$ corresponds to the frequency of the vibrational transition when the rotational quantum number of the two states do not change. Similar to R and P branches, one can get O, Q and S branches which corresponds to $\Delta J = 0$, $\Delta J = +2$ and $\Delta J = -2$ respectively.

$$\bar{\nu}_O = \bar{\nu}_0 + B_{v'} J' (J' + 1) - B_{v''} J'' (J'' + 1)$$

$$= \bar{\nu}_0 + B_{v'} (J'' - 2)\ (J'' - 1) - B_{v''} J'' (J'' + 1)$$

$$= \bar{\nu}_0 + 2B_{v'} - (3B_{v'} + B_{v''})\, J + (B_{v'} - B_{v''})\, J^2, \quad J = 2, 3, \ldots \qquad \ldots(5.155)$$

$$\bar{\nu}_Q = \bar{\nu}_0 + B_{v'} J' (J' + 1) - B_{v''} J'' (J'' + 1)$$

$$= \bar{\nu}_0 + (B_{v'} - B_{v''})\, J + (B_{v'} - B_{v''})\, J^2, \quad J = 0, 1, 2, \ldots \qquad \ldots(5.156)$$

$$\bar{\nu}_S = \bar{\nu}_0 + B_{v'} J' (J' + 1) - B_{v''} J'' (J'' + 1)$$

$$= \bar{\nu}_0 + 6B_{v'} + (5B_{v'} - B_{v''})\, J + (B_{v'} - B_{v''})\, J^2, \quad J = 0, 1, 2, \ldots \qquad \ldots(5.157)$$

Thus the Stokes lines will occur at wave number given as:

$$\bar{\nu}_{RQ} = \bar{\nu}_{inc} - \bar{\nu}_Q$$

$$= \bar{\nu}_{inc} - \bar{\nu}_0 - (B_{v'} - B_{v''})\,J - (B_{v'} - B_{v''})\,J^2 \qquad ...(5.158)$$

$$\bar{\nu}_{RO} = \bar{\nu}_{inc} - \bar{\nu}_O$$

$$= \bar{\nu}_{inc} - \bar{\nu}_0 - 2B_{v'} + (3B_{v'} + B_{v''})\,J - (B_{v'} - B_{v''})\,J^2 \qquad ...(5.159)$$

$$\bar{\nu}_{RS} = \bar{\nu}_{inc} - \bar{\nu}_O$$

$$= \bar{\nu}_{inc} - \bar{\nu}_0 - 6B_{v'} - (5B_{v'} - B_{v''})\,J - (B_{v'} - B_{v''})\,J^2 \qquad ...(5.160)$$

In all these expressions, similar to rotational-vibrational spectrum, $\Delta\bar{\nu}_0 = \bar{\nu}_{inc} - \bar{\nu}_0$ corresponds to null or zero line and the centrifugal distortion term D, is neglected. The above expressions can be simplified further if we assume that $B_{v'} = B_{v''}$.

In general, increase in the vibrational quantum number is only 1, *i.e.,* $v = 0$ to $v = 1$. During this vibrational excitation, the difference between $B_{v'}$ and $B_{v''}$ is very small and thus the lines of Q branch are not very well resolved. Due to this the intensity of Q branch seems to be very high. On the other hand, the lines of O and S branches are nicely resolved but are very weak. The successive difference between these lines is double the amount observed for rotational-vibrational spectrum, *i.e.,* 4B. Calculations of the rotational constants are nearly on the same lines as has been carried out for rotational-vibrational spectrum. Similar to vibrational Raman, one can again find the rotational constants of the molecules which do not possess permanent dipole moment.

Anti-Stokes lines are observed on the higher frequency side of Rayleigh line at the same distance from the exciting line. The intensity of the anti-Stokes lines is very low. Further the resolution of Raman spectra is not very good and thus the effect of centrifugal distortion can be neglected.

Lastly, for large molecules, the resolution between the rotational lines is further reduced and thus the analysis of rotational structure is ignored. Even the O and S branches of rotational-vibrational band, *i.e.,* their contours, are rarely observed, because these branches are very weak as compared to Q branch.

5.6.6 VIBRATIONAL RAMAN SPECTRUM OF POLYATOMIC MOLECULES

Like infrared spectra, vibrational Raman spectra of polyatomic molecules are also complicated. Some times it is easy to predict whether a particular vibrational mode will be Raman active or not and in other cases one has to involve a detailed consideration, whether or not polarisability changes during vibration. For example, if the molecule has little or no symmetry, all the vibrational modes, in general, are Raman active, whereas if the molecule has a considerable symmetry, it is not always easy to predict. But the detailed study has given a very important and general rule. It is known as *rule of mutual exclusion. If a molecule has a center of symmetry, Raman active vibrations are infrared inactive and infrared active vibrations will be Raman inactive.* If there is no centre of symmetry, then some (but not necessarily all) vibrations may be both Raman and infrared active. Water molecule has no centre of symmetry and all its vibrational modes are Raman as well as infrared active. On the other hand, CO_2 molecule has a centre of symmetry. So the symmetrical stretching vibration ($\bar{\omega}_2$) is infrared inactive but Raman active. On other hand the asymmetric stretching ($\bar{\omega}_1$) and bending ($\bar{\omega}_3$) modes are Raman inactive, but infrared active.

The converse of this is also true, *i.e.,* if vibrational modes of molecule are active both in Raman and infrared spectrum, it is certain that the molecule has no center of symmetry. However if the vibrational modes which are infrared active but Raman inactive and vice versa, it implies that the molecules are having centre of symmetry. But one should be very careful because as we have learnt earlier, some of Raman lines are very weak and we may not be able to observe them. Thus, extremely useful information can be obtained by using the results of Raman and infrared spectroscopy.

Example 5.26: The molecular formula of a compound is XY_2. It shows two vibrational bands in infrared and only one in vibrational Raman. Further Raman active band is infrared inactive. Speculate the structure of the molecule.

Solution: Since this is a triatomic molecule, it will be having four vibrational degrees of freedom for linear molecule and three for the non-linear molecule, *i.e.,* two stretching mode and two or one bending modes. The data have also shown that the molecule is having centre of symmetry. So the molecule can be neither bent nor linear like Y – X – X. So the structure will be Y– X – Y, linear. In the infrared, symmetrical stretching will be inactive and two bending modes will be degenerate. In Raman spectrum, symmetric stretching will be active and other two will be inactive.

5.7 ELECTRONIC SPECTRUM

Similar to the discussion in the earlier sections, in this section we will also describe first the electronic spectrum of diatomic molecules and extending it to polyatomic molecules.

5.7.1 ELECTRONIC SPECTRUM OF DIATOMIC MOLECULES

To start with let us write the expression for Born-Oppenheimer approximation as written in section 4.1.

$$E_T = E_{elc} + E_{vib} + E_{rot} \quad ...(5.161)$$

which means that each form of energy is completely independent of each other. Although we will not discuss in this book, the finer details of electronic spectrum do indicate the presence interaction of each motion with one another. These interactions do affect the spectrum. On shining radiation or supplying energy, the change in the total energy can be written as:

$$\Delta E_T = \Delta E_{elc} + \Delta E_{vib} + \Delta E_{rot} \text{ in Joules}$$

As mentioned earlier, rotational quantum is much smaller than vibrational and vibrational quantum is even much smaller than electronic. This means that during electronic excitation, excitation to vibrational and rotational energy levels of higher electronic state will also be involved. The combination of electronic and vibrational energy levels leads to coarse spectrum, whereas further combination of rotational motion leads to fine structure of electronic spectrum. In this book, we will only concentrate on the coarse electronic spectrum of the molecules. Unlike pure rotational and vibrational spectra, where the molecules must either possess a permanent dipole moment or oscillating dipole moment should be finite, electronic spectrum is observed for all the molecules (howsoever weak it may be), because a change in dipole moment is observed in all the molecules on excitation. Thus, one can get the information regarding rotational and vibrational constants for all the molecules, even for homonuclear diatomic molecules (like O_2, N_2 etc.,) from the electronic spectrum, which was not possible from the pure rotational and vibrational spectrum. Further it is also possible to calculate the rotational and vibrational constants for the excited electronic states, even though molecules exist for very short duration in these states. The above discussion does not mean that every electronic transition will be observed with equal intensity. Some will be weak ($\varepsilon \sim 10$ to $100\ M^{-1}\ cm^{-1}$), some moderate ($\varepsilon \sim 100$ to $1000\ M^{-1}\ cm^{-1}$) and some strong ($\varepsilon \sim > 10^3\ M^{-1}\ cm^{-1}$). The intensity of the transition is controlled by the value of transition moment integral, which in turn can be evaluated qualitatively with the help of selection rules or quantitatively by actually determining these relations.

5.7.2 VIBRATIONAL COARSE SPECTRUM

Neglecting the contribution from the rotational motion, the total energy can be written as:

$$E_T = E_{elc} + E_{vib} \text{ in J} \qquad ...(5.162)$$

or

$$T = T_e + G(v) \text{ in cm}^{-1} \qquad ...(5.163)$$

where T_e is the term value for electronic energy and $G(v)$ for vibrational motion. For vibrational motion of the molecules in different electronic states, we use the model of anharmonic oscillator and substitute the value of $G(v)$ in equation (5.163).

$$T = T_e + \bar{\omega}_e\ (v + 1/2) - \bar{\omega}_e\ x_e\ (v + 1/2)^2 + ... \qquad ...(5.164)$$

The energy level diagram is depicted in Fig. 5.22. As usual the single prime designates the upper state (T_e', v') and double prime refers to lower state (T_e'', v''). Spacing between the vibrational energy levels in the lower electronic state is more than that shown in upper state. This is because the excited electronic state corresponds to a weaker bond in the molecule and hence a smaller vibrational frequency.

Using the equation (5.163) for the term values of a molecule, we find that wave numbers of the spectral lines corresponding to the transitions between the electronic states are given by:

$$\bar{\nu} = T' - T'' = T_e' - T_e'' + G(v') - G(v'') \qquad ...(5.165)$$

$$= \bar{\nu}_{elc} + \left[\bar{\omega}_e'\left(v' + \frac{1}{2}\right) - \bar{\omega}_e'\ x_e'\left(v' + \frac{1}{2}\right) + ...\right]$$

$$- \left[\bar{\omega}_e''\left(v'' + \frac{1}{2}\right) - \bar{\omega}_e''\ x_e''\left(v'' + \frac{1}{2}\right)^2 ...\right] \qquad ...(5.166)$$

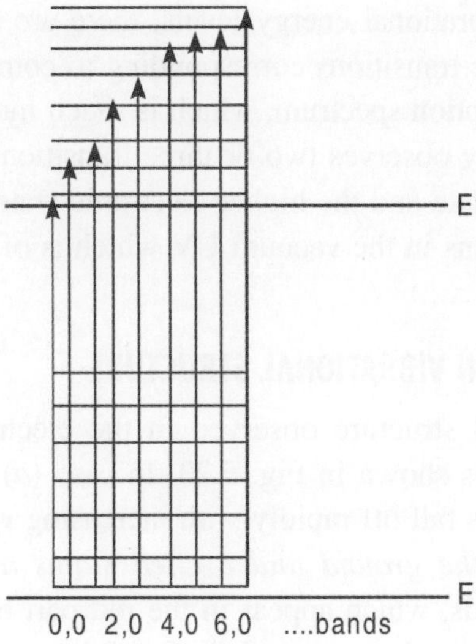

Figure 5.22: *Schematic representation of the electronic energy levels E″ and E′. Vibrational levels have been drawn in between the electronic states. Different vibrational bands are also drawn starting from v″ = 0 from ground electronic state*

where $\bar{\nu}_{elc}$ corresponds to the electronic transition from $v'' = 0$ to $v' = 0$. There are some selection rules for combining the two electronic states, but there are virtually no selection rules for vibrational levels when a molecule undergoes electronic transition. In other words every combination of v' and v'' is possible theoretically and thus one would expect a number of lines in a given electronic transition. In reality this does not happen. As mentioned in Table 5.4, the vibrational levels other than $v'' = 0$ in a given ground electronic state are hardly populated and so the only transitions which are observed with reasonable intensity will originate from $v'' = 0$. These lines are shown in Fig. 5.22. According to the convention followed it can be represented as (v', v''), *i.e.,* the upper state is mentioned first, followed by the lower state, for example, (0, 0), (1, 0), (2, 0)...... A set of such lines is called a band, because under high resolution each line is composed of many rotational lines, where as under low resolution it appears broad and diffuse. In spectroscopic language the set of lines is also called $v'' = 0$ progression since the values of v' increases by unity for each line in the set. It is also clear from Fig. 5.22 that the gap between the lines observed in the electronic transition keeps on decreasing as the value of v' increases and in some cases the gap becomes so small that it cannot be resolved and appears to be continuous. It is due to the consequence of anharmonic oscillation, which causes the excited vibrational levels to converge.

Equation (5.166) indicates that it is possible to get a number of lines. From equation (5.166), it is possible to calculate the values of $\bar{\omega}_e$, $\bar{\omega}_e x'_e$ and $\bar{\omega}''_e$, $\bar{\omega}''_e x''_e$ as well as separation between the electronic states $\bar{\nu}_e$ (*i.e.,* when $v' = v'' = 0$). The knowledge of excited state vibrational constants can provide information about the nature of bonding in the excited state, even though we know that these states are very unstable and the molecules may exist in these states for very short duration.

Similar to rotational and vibrational energy levels, there are also many electronic states and one can observe many electronic transitions corresponding to combination of lower state with the upper states. This leads to absorption spectrum, which is much more complex than what is shown in Fig. 5.22. In general, one only observes two or three transitions because the energy difference between the ground electronic state and the higher electronic states higher than 3, is so large that one only observes these transitions in the vacuum UV which is of no practical interest for normal purposes.

5.7.3 INTENSITY DISTRIBUTION IN VIBRATIONAL STRUCTURE

The intensity of vibrational structure observed in the electronic spectrum can broadly be classified into three categories as shown in Fig. 5.23. In case (*a*), the most intense band is (0,0) and the intensities of other bands fall off rapidly with increasing v'. Such a case is observed *when equilibrium bond distance in the ground and excited states are nearly equal,* for example, for the atmospheric oxygen bands, which appear in the red part of solar spectrum. In the second case (*b*), the intensities of bands in the progression $v'' = 0$ increases with increase of v' and then starts decreasing slowly. CO is an example, which shows this behaviour. *In this case, equilibrium bond distance of the molecules in the upper state is greater than that in the lower state.* The

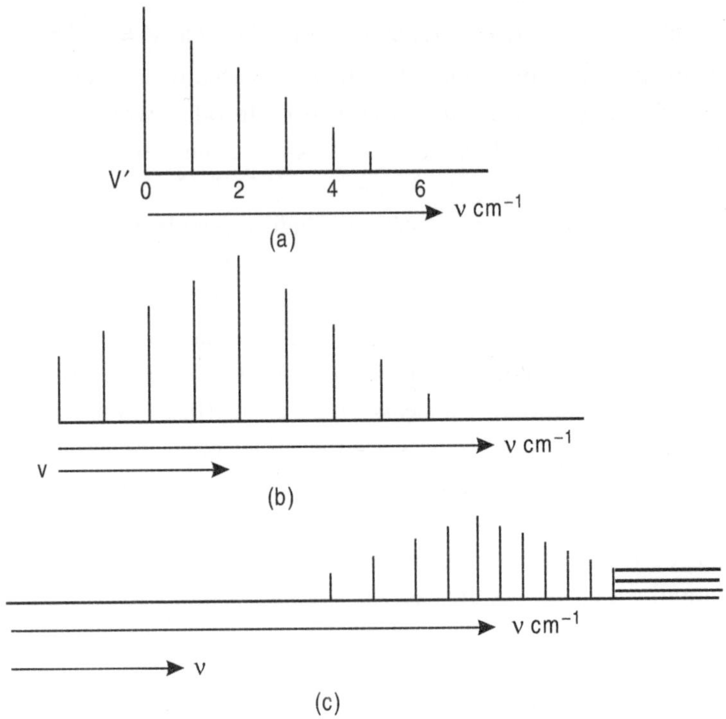

Figure 5.23: *Three typical cases of intensity distribution in absorption spectral band series. For simplicity, the separation between the bands is shown to be the same in all the three cases. Naturally these cases would be observed in different band systems, which would in general not have the same band separation*

third case is represented by the electronic spectrum of I_2 molecule, which is similar to as shown in Fig. 5.23*c* and the potential energy curve or the upper state is displaced towards the right side of the lower state potential energy curve. The structure of this band system is similar to that represented by 5.23*b* except (0, 0) band is not observed and the gap between the bands keep on decreasing with increase of v' till it joins the continuum toward short wave length. The maximum of the intensity either lies at very high value of v' or it may lie in the continuum. The intensity distribution thus observed in the vibrational structure of the electronic spectrum of various molecules in general lies between these three categories.

All these kinds of intensity distribution in spectra can be explained with the help of *Franck-Condon principle,* stated as follows: *The electron jump in a molecule takes place so rapidly* ($\sim 10^{-15}$ *s*) *in comparison to the vibrational motion* ($\sim 10^{-12}$ *s*) *that immediately after ward the nuclei still have nearly the same relative position and velocity as before jump.* All this can be explained very easily if we take the help of potential energy curves as shown in Fig. 5.24. In this diagram we have considered both ground and excited state potential energy curves as stable ones, *i.e.,* possessing minima. As explained in section 2.7, the most probable position of a particle in $v'' = 0$ state is in the middle (Fig. 5.25), which is different from the classical mechanics (at the end positions).

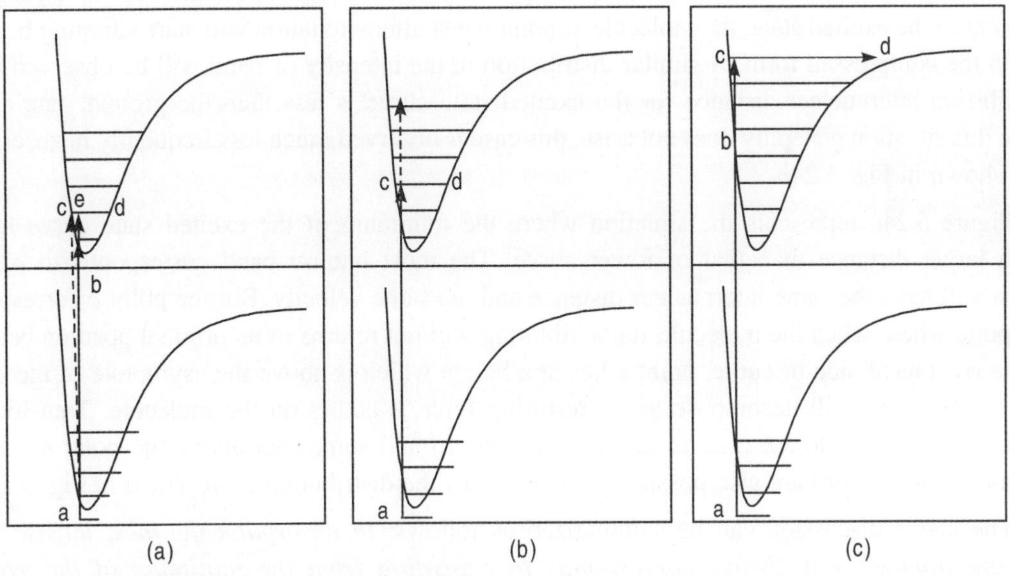

Figure 5.24: *Depiction of Franck-Condon Principle according to classical mechanics. (a), (b) and (c) depict the excited state potential energy curve at equal distance, slightly shifted and large shifted towards right as compared to ground electronic state*

In Fig. 5.24*a*, potential energy curves of two electronic states are so chosen that their minima are placed very nearly one above the other (*i.e.,* equal internuclear distance). If we neglect the zero point energy, the molecule will be at the minimum of the potential energy curve. In absorption the transition which is following the Franck-Condon principle is the (0, 0) band, *i.e.*, the electron is moving from point *a* to *b*. On the other hand if we consider the transition to higher vibrational state, say *c d* ($v' = 1$), it is possible only if during the electronic excitation, either the position (*a* to *c*) or velocity (*a* to *e*) or both are not altered to an appreciable extent. At the point *e*, where the position is the same, the electron is possessing kinetic energy of magnitude *e b* and thus it (or velocity) can be zero only at the turning points *c* or *d*. Thus, according to Franck-Condon principle this kind of transition is highly improbable if not forbidden. To explain the observation of transition (1, 0), decrease in the intensity is due to the fact that change in the position or velocity or both is quite small for $v' = 1$. For (2, 0), (3, 0)….. bands, the changes in the position or velocity or both are quite large and thus one would expect a large decrease in the intensity distribution observed in the electronic spectrum as shown in Fig. 5.24*a*.

Let us consider the Fig. 5.24 *b*. In this case, equilibrium internuclear distance of the excited state potential energy curve is at longer distance as compared to that in the ground state. Therefore the 0–0 band would not be stronger band because this does involve a larger change in distance or velocity. On the other hand let us consider the transition *a* to *c*. At point *c*, the distance between the nuclei is the same and at the turning point *c*, it will have zero kinetic energy, equal to the point *a*, before excitation. Thus this transition will be strongest. If the higher vibrational levels are considered, an appreciable change in position and velocity will be noticed and it will reduce the intensity, based on the Franck-Condon principle. Since the equilibrium internuclear distance is different in the excited state, the molecule at point *c* just after excitation will start vibrating because it is in the compressed form. A similar distribution in the intensity of band will be observed if the equilibrium internuclear distance for the excited state curve is less than the ground state curve. Since this situation generally does not arise, this case is observed much less frequently in absorption than shown in Fig. 5.24*b*.

Figure 5.24*c* represents the situation where the minimum of the excited state curve lies at much larger distance than that of lower curve. The most intense band corresponds to *a* to *c*, which will have the same internuclear distance and the same velocity. But the point *c* corresponds to a point where when the molecule starts vibrating will not returns to its original position because on the right hand side of curve, point *c* lies at a height which is above the asymptote of the curve and the molecule will decompose as no restoring force is acting on the molecule. Transition to points somewhat, below *b* (*i.e.*, in the discrete region) and somewhat above the point *b* (*i.e.*, in the continuum region) are also possible. This explains the distribution represented in Fig. 5.24*c*.

The above discussion can be summarized as follows: *In absorption the most intense band starting from* $v'' = 0$ *always corresponds to transition from the minimum of the ground potential energy curve to the vertically upward.*

5.7.4 WAVE MECHANICAL DESCRIPTION OF FRANCK-CONDON PRINCIPLE

Better insight of the Franck-Condon principle, and the basic assumptions involved in it, *i.e.*, transitions vertically upwards (in absorption) or downwards (in emission) in the potential energy diagrams corresponding to most intense, can be better understood from wave mechanical treatment. As we have learnt earlier that according to wave mechanical treatment, the oscillator will never be quite at rest, but will be oscillating in the ground state and will be described by the probability density function as given in section 2.6.7. This suggests that the vertical transitions from the ground state will be taking place over a certain range of distance *r* and will also be ending over a small range of distance *r* in the excited state (Fig. 5.25). It is quite clear from Fig. 5.25 that instead of getting only one vertical intense band, a number of bands with v' progression are observed with $v'' = 0$ in absorption spectrum with a more or less broad intense maximum. Observation of different kinds of bands (Fig. 5.24*a*, and *b*) depend on the nature of potential energy curves, *e.g.*, the extent of band series and the continuum which may possible join on to it, is greater the steeper the upper potential energy curve.

As mentioned earlier, the probability of transition is given by the transition moment integral, and intensity is proportional to the square of the transition moment integral, *i.e.*,

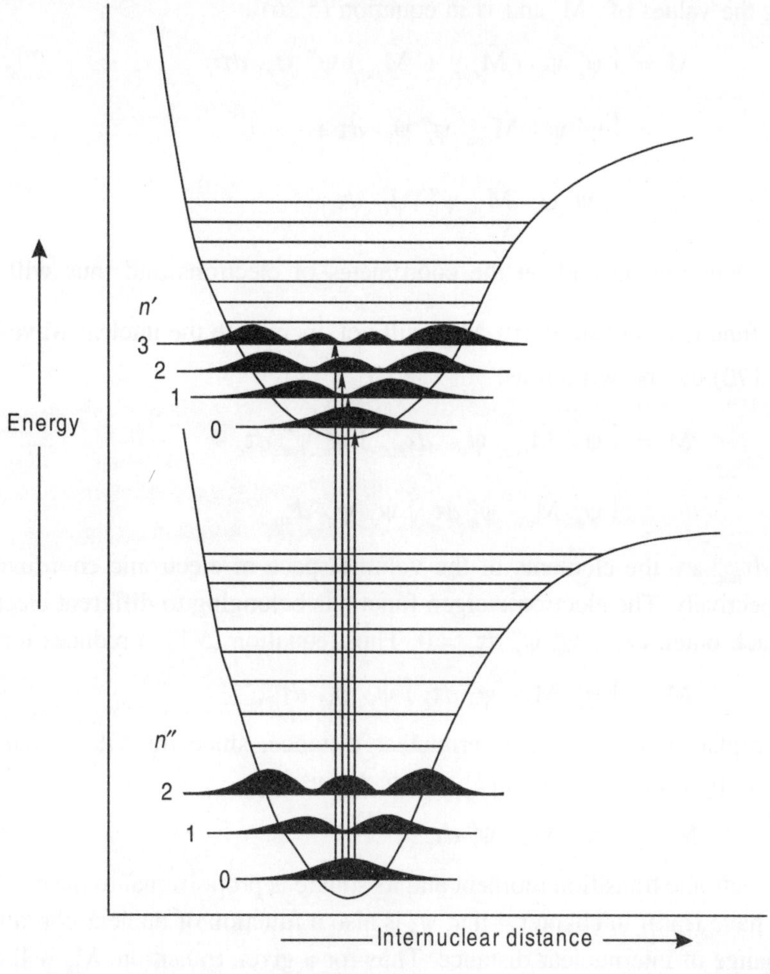

Figure 5.25 a: *Figure depicts the potential energy curves for the two electronic states, having nearly same equilibrium internuclear distance. Shaded areas show the probability density functions of the few vibrational states of harmonic oscillator, also drawn on the potential energy curves. The vertical lines indicates the vibronic transition starting from n″ = 0*

$$M = \int \psi' \hat{M} \psi'' d\tau \qquad ...(5.26)$$

where each term has the usual meaning. Considering only the electronic and vibration motions, the total wave function can be written as:

$$\psi = \psi_e \psi_v \qquad ...(5.167)$$

where ψ_e and ψ_v are the electronic and vibrational wave functions respectively. Further M can be resolved into two parts, one part depends on the nuclear coordinates and the other part depends on the electronic coordinates.

$$\hat{M} = \hat{M}_{nuc} + \hat{M}_{elc} \qquad ...(5.168)$$

Substituting the values of $\hat{M}$ and ψ in equation (5.26)

$$M = \int \psi'_e \psi_{v'} (\hat{M}_{nuc} + \hat{M}_{elc}) \psi''_e \psi_{v''} d\tau \qquad ...(5.169)$$

$$= \int \psi'_e \psi_{v'} \hat{M}_{nuc} \psi''_e \psi_{v''} d\tau + \int \psi'_e \psi_{v'} \hat{M}_{elc} \psi''_e \psi_{v''} d\tau \qquad ...(5.170)$$

Since $\hat{M}_{nuc}$ does not depend on the coordinates of electrons and thus will not operate on electronic wave functions and similarly $\hat{M}_{elc}$ will not operate on the nuclear wave functions. Thus the equation (5.170) can be written as:

$$M = \int \psi_{v'} \hat{M}_{nuc} \psi_{v''} d\tau_{nuc} \int \psi'_e \psi''_e d\tau_e + \int \psi'_e \hat{M}_{elc} \psi''_e d\tau_e \int \psi_{v'} \psi_{v''} d\tau_{nuc} \qquad ...(5.171)$$

where $d\tau_e$ and $d\tau_{nuc}$ are the elements of the volume space of electronic coordinates and nuclear coordinates respectively. The electronic eigen functions belonging to different electronic states are orthogonal to each other, *i.e.*, $\int \psi'_e \psi''_e d\tau_e = 0$. Thus, equation (5.171) reduces to:

$$M = \int \psi'_e \hat{M}_{elc} \psi''_e d\tau_e \int \psi_{v'} \psi_{v''} d\tau_r \qquad ...(5.172)$$

where $d\tau_{nuc}$ is replaced by $d\tau_r$, the internuclear distance, since the vibrational wave function depends on r. The first integral in (5.172) is depicted as M_e:

$$M_e = \int \psi'_e \hat{M}_{elc} \psi''_e d\tau_e \qquad ...(5.173)$$

is called as the electronic transition moment and its square is proportional to the electronic transition probability. We have learnt in chapter 4 that ψ_e is also a function of nuclear coordinates and varies slightly with change of internuclear distance. Thus for a given transition, $\mathbf{M}_e$ will depend to some extent on the internuclear distance. Assumption made in the Franck-Condon principle is that the variation of $\mathbf{M}_e$ with r is so small that one can take the average value $\langle \mathbf{M}_e \rangle$ in the equation (5.172).

$$\mathbf{M} = \langle \mathbf{M}_e \rangle \int \psi_{v'} \psi_{v''} d\tau_r \qquad (5.174)$$

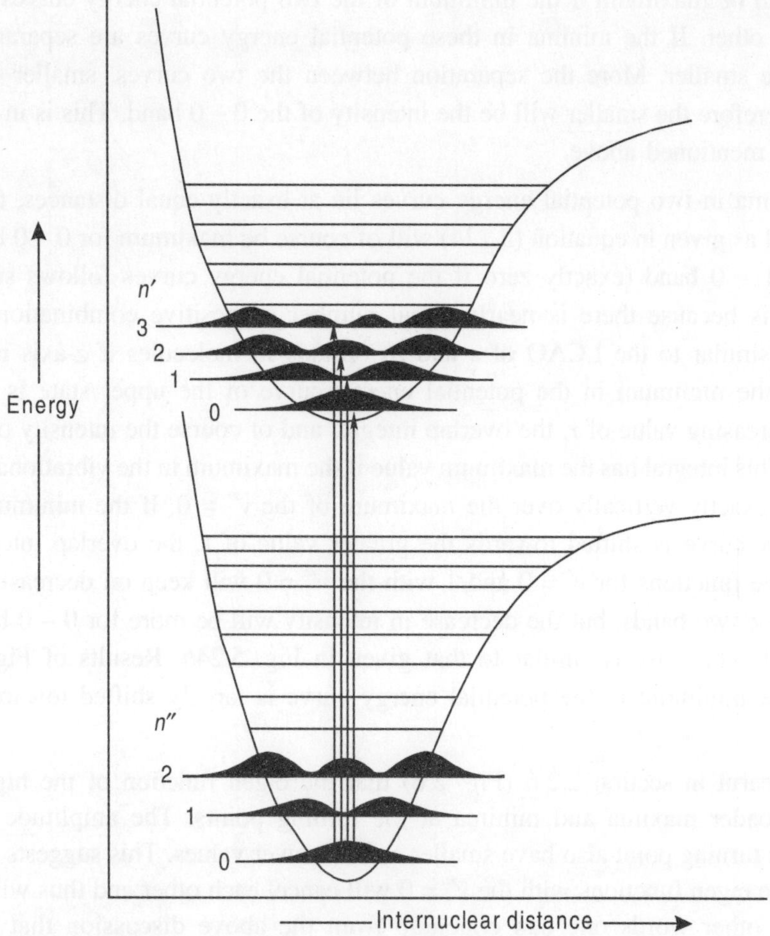

Figure 5.25 b: *Figure depicts the potential energy curves for the two electronic states, having nearly different equilibrium internuclear distance. Shaded areas show the probability density functions of the few vibrational states of harmonic oscillator, also drawn on the potential energy curves. The vertical lines indicates the vibronic transition starting from* $v'' = 0$

The transition probability and therefore the intensity is proportional to the square of the expression given in equation (5.174), *i.e.*, proportional to the square of the integral over the product of the vibrational eigen function of the two states (also known as overlap integral). The first part $(\langle \mathbf{M}_e \rangle)$ indicates whether the particular electronic transition is allowed or not and gives rise to the selection rules for the electronic transition. The second part gives rise to the distribution of the band intensity or shape of the particular electronic band, also called as *Franck-Condon factors*. Now this expression can be used to explain the qualitative features of the spectrum described in the previous section.

Let us consider first the 0 – 0 band. As given in Fig. 5.25, it is clear that the vibrational wave functions given in both the electronic states are bell shaped. Naturally the product of two eigen functions for each value of *r* and hence also the integral of this product (overlap integral) over all the r values will be maximum if the minimum of the two potential energy curves will lie exactly one above the other. If the minima in these potential energy curves are separated, the overlap integral will be smaller. More the separation between the two curves, smaller will be overlap integral and therefore the smaller will be the intensity of the 0 – 0 band. This is in agreement with what has been mentioned above.

If the minima in two potential energy curves lie at exactly equal distances, the value of the overlap integral as given in equation (5.174) will of course be maximum for 0 – 0 band but will be minimum for 1 – 0 band (exactly zero if the potential energy curves follows simple harmonic motion). This is because there is nearly equal number of positive combinations and negative combinations (similar to the LCAO of s and p_x orbitals in molecules if z-axis is the molecular axis). Now if the minimum in the potential energy curve of the upper state is shifted slightly towards the increasing value of r, the overlap integral and of course the intensity of 1 – 0 band, at first increase. This integral has the maximum value if the maximum in the vibrational wave function for $v' = 1$ lies exactly vertically over the maximum of the $v'' = 0$. If the minimum of the upper potential energy curve is shifted towards the greater value of r, the overlap integral of the two vibrational wave functions for $v' = 0$ and 1 with the $v'' = 0$ will keep on decreasing and thus the intensity of these two bands, but the decrease in intensity will be more for 0 – 0 band rather than for 1 – 0 band. This case is similar to that given in Fig. 5.24*b*. Results of Fig. 5.24*c* can be explained if the minimum in the potential energy curve is largely shifted towards increasing r, value.

We have learnt in section 2.2.6 (Fig. 2.6) that the eigen function of the higher vibrational levels have broader maxima and minima at the turning points. The amplitude of these eigen functions at the turning point also have smaller and narrower values. This suggests that the overlap integral of these eigen functions with the $v'' = 0$ will cancel each other and thus will lead to nearly zero value. In other words one can conclude from the above discussion that for a series of absorption bands for $v'' = 0$, the overlap integral (and thus intensity) of $v'' = 0$ series will be largest if the maximum of the vibrational eigen function of the upper state (v') lies vertically above the maximum of the ground state vibrational eigen function ($v'' = 0$) and the values of the overlap integral decreases for greater and smaller values of v'. Thus one can see from the above discussion that the results discussed above are in agreement with those mentioned in section 5.7.3. On the same lines one can also explain the results for other series starting with other values of v' and v''.

5.7.5 SELECTION RULES

We have already learnt in section 5.2.2 that intensity of any transition is proportional to the square of the transition moment integral and the transition is *allowed* only if this integral (equation 5.26) is non-zero and *forbidden* if the integral is zero. Further selection rule is the statement, which defines whether the transition moment integral is zero or non-zero, *i.e.*, either the transition is forbidden or allowed. These rules have been discussed by, many workers. Here these rules will be summarized only.

Spin Forbidden: The transition is only allowed if the spin of the two states are the same, *i.e.*, $\Delta S = 0$. This rule is vigorously obeyed by the light molecules but breaks down in case of polyatomic molecules containing heavy atoms. In case of polyatomic molecules containing heavy atoms, spin orbital angular momentum and orbital angular momentum interact with each other. This interaction introduces a term in the Hamiltonian operator, which operates both on spin and spatial part of the wave function. In the presence of spin-orbital interaction, there is mixing of states, the singlet state has a little character of the triplet state and the triplet state has some contribution from the singlet state. This makes the integral (5.26) non-zero if the spin wave functions are also introduced. The mixed orbital wave function of triplet state is given by:

$$\Psi_T = \psi_T^0 + \lambda_{TS}\,\psi_S^0 \qquad ...(5.175)$$

where ψ_T, ψ_T^0 and ψ_S^0 are the wave functions of the mixed triplet state, pure triplet state and pure singlet state respectively. Singlet-triplet mixing coefficient λ_{TS}, is inversely proportional to the energy difference between singlet and triplet states. Smaller the value of $(E_S - E_T)$, greater is the mixing and stronger the singlet triplet transition.

Symmetry Forbidden: *The integral (5.26) will be non-zero if $\psi' \hat{M} \psi''$ is totally symmetric.* Since $\hat{M} = \Sigma e_i r_i$, is always anti-symmetric with respect to centre of inversion, the product of $\psi' \psi''$ has to be anti-symmetric. To make the complete integral symmetric this is possible only if one of the states is symmetric and the other anti-symmetric. This leads to selection rules:

$$g \leftrightarrow u \text{ is allowed and } u \leftrightarrow u \text{ and } g \leftrightarrow g \text{ are forbidden.} \qquad ...(5.176)$$

The permissibility of the transition between opposite symmetries is allowed and is known as **Parity (Laporte) rules.** According to this rule the parity, **P**, of a system should change for an allowed transition:

$$P = (-1)^l, \text{ where } l \text{ is the orbital quantum number.} \qquad ...(5.177)$$

Example 5.27: Calculate the parity for permissible transition $\Delta l/\Delta L = \pm 1$. (*a*) Carbon ($1s^2\,2s^2\,2p^2$), (*b*) Nitrogen ($1s^2\,2s^2\,2p^3$) and (*c*) oxygen ($1s^2\,2s^2\,2p^4$).

Solution: Using equation (5.177),

(*a*) For carbon, $L = \Sigma\, l_i = 2$, $\therefore$ $P = (-1)^2 = +1$, even

(*b*) For nitrogen, $L = \Sigma\, l_i = 3$, $\therefore$ $P = (-1)^3 = -1$, odd

(*c*) For oxygen, $L = \Sigma\, l_i = 4$, $\therefore$ $P = (-1)^4 = +1$, even

One thing that is to be kept in mind while applying parity selection rule is that it is based only on one of the symmetry elements (centre of inversion) of the symmetry point group whereas the electronic states of polyatomic molecules are assigned on the basis when all the symmetry elements are used. So sometimes the parity allowed transition are forbidden on the basis of complete symmetry point group, for example, the first and second electronic transitions in benzene are allowed on the basis of parity selection rules but are forbidden when complete symmetry of the states are taken into account.

In some cases the weak transitions are observed though they are symmetry forbidden. This is due to the coupling of vibrational motion of the nuclei with the electronic motion. Coupling of vibrational motion changes the symmetry of the molecule and the assignment of the electronic state is based on symmetry consideration. Hence above rule is not rigorously followed. These types of transitions are called *Vibronic Transitions.*

5.7.6 ELECTRONIC SPECTRUM OF POLYATOMIC MOLECULES

Structure of molecules (Chapter 4) has clearly shown that in general three kinds of molecular orbitals are present in polyatomic molecules and their relative energies are shown in Fig. 5.26. These are:

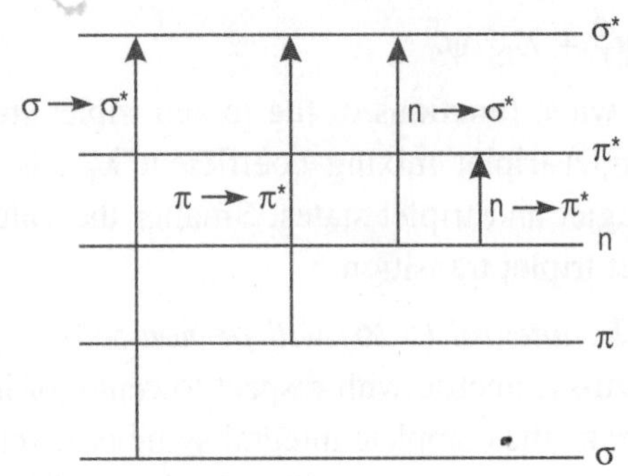

Figure 5.26: *Schematic energy level diagram of different kind of molecular orbitals present in molecules. Various transitions are also shown*

(*i*) **σ and σ* molecular orbitals:** These are present in saturated bonds as well as in multiple bonds. Since the overlap of the atomic orbitals of the combining atoms is very strong, σ and σ* m.o's., are of much lower and higher energies respectively, with respect to atomic orbitals.

(*ii*) **π and π* molecular orbitals:** These are present in double and triple bonds. Since the formation of these molecular orbitals involve lateral overlap of atomic orbitals, these are less stabilized or destabilized in comparison to σ and σ* orbitals. The common examples are: >C=C<, —C ≡ C—, >C=O etc.,

(*iii*) Non-bonding (*n*) orbitals or lone pair of electrons, which do not take part in the bond formation and more or less are localized on the atom itself. The energies of these electrons are nearly equal to the energy of the atomic orbitals, for example, lone pair of electrons present on nitrogen, oxygen and halogen etc.,

Theoretically there may be more transitions, which can occur in polyatomic molecules but we will concentrate only on these three kinds of transitions.

(*i*) **σ* ← σ transition:** As the name suggests an electron from σ bonding m.o. is excited to σ* anti-bonding m.o. From the Fig. 5.26, it is evident that this transition is of highest energy and thus corresponds to a very high frequency or short wavelength. These transitions in general lie in the vacuum UV region. For example, in saturated hydrocarbons, these transitions are present at 60 nm and are allowed transitions. Due to their presence in vacuum UV, these transitions are of no practical use.

(*ii*) **$\pi^* \leftarrow \pi$ transitions:** In this case an electron from π bonding m.o. is excited to π^* anti-bonding m.o. and these kinds of transitions are present in the multiple bonds. In simple ethylene, $\pi^* \leftarrow \pi$ transition is an allowed one and is observed at ~ 170 nm in the vacuum UV region. Similar to $\sigma^* \leftarrow \sigma$ transition, this transition is also of no practical interest. But it has been observed that the introduction of one more double bond conjugated to the first one, shifts the $\pi^* \leftarrow \pi$ transition to longer wavelength by ~ 30 nm. For example, the first member of the conjugated diene system is 1,3-butadiene and the longest wavelength transition is observed at ~ 200 nm. As shown in section 4.18, if the double bonds in a molecule are present conjugated to each other, the π electrons are delocalized over the whole molecule. The resonance structures show that each carbon-carbon bond along the chain has appreciable double bond character. For example, presence of cis and trans butadiene confirms the free movement of π electrons along the chain. Besides using the more complicated theories simple free electron model can be combined with the experimental data to get many useful information regarding the structure of the molecules.

For example, consider the molecule β-carotene, which is highly conjugated and $\pi^* \leftarrow \pi$ transition is observed at 451 nm in the visible region. This molecule possesses 22–π electrons. Free movement of these electrons suggest that skeleton can be considered as a roughly uniform region of low potential bounded at the end of the molecule by region of infinitely high potential. The resulting square well potential leads to the model, which is close to a particle in one-dimensional box. The energies of the levels are given by the expression as:

$$E_n = \frac{h^2}{8\,m\,a^2} n^2 \quad n = 1, 2, \ldots \qquad \ldots(5.178)$$

where a is the effective length of the molecule. Two electrons with opposite spins can be accommodated in one energy level. Thus the highest occupied molecular orbital (called as HOMO) and lowest unoccupied molecular orbital (called as LUMO) are 11 and 12 respectively. The lowest energy transition given by the expression:

$$\Delta E = E_{12} - E_{11} = \frac{h^2}{8\,m\,a^2}[12^2 - 11^2] = \frac{23\,h^2}{8\,m\,a^2} \qquad \ldots(5.179)$$

$$\Delta E = h\,\nu = \frac{h\,c}{\lambda} = \frac{23\,h^2}{8\,m\,a^2} \qquad \ldots(5.180)$$

$$a = \left[\frac{23\,h\,\lambda}{8\,m\,c}\right]^{\frac{1}{2}} \qquad \ldots(5.181)$$

Substituting the values of constants and λ = 451 nm, a, is found to be 17.8 × 10^{-10} m. This value is very close to that which would be expected for the extended conjugated carbon chain of β-carotene molecule.

The other and very important class of extended conjugation is present in aromatic molecules. The simplest molecule in this class of compounds is benzene. The explanation of the absorption spectrum of benzene is not as simple as given above for linear conjugated systems. But qualitatively it can be explained by considering the free electron moving on a circular ring model. In this class of compounds, the transitions are of $\pi^* \leftarrow \pi$ type but the first two transitions (*i.e.,* 257 nm and 200 nm) are forbidden. These are observed, though weakly, because of the failure of certain selection rules based on symmetry. This failure is due to the interaction of vibrational motion with the electronic motion.

(*iii*) **$\pi^* \leftarrow n$ Transitions:** In this case, the electron from non-bonded m.o. is excited to π^* anti-bonding m.o. This type of transition is present in molecules containing elements of group V, VI and VII (*i.e.,* N, P, O, S, F, etc.). In general these kinds of transitions are very weak, with molar absorption coefficient of the order of ~ 10 M^{-1} cm^{-1}. It is clear from the shape of this kind of molecules. For example, consider a carbonyl group:

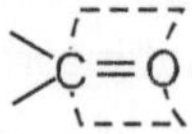

The dotted lines show the π and π^* orbitals, which are perpendicular to the lone pair of electrons. Thus the overlaps of n and π^* orbitals will be nearly zero and will reduce the transition moment integral to zero.

The effect of extended conjugation on $\pi^* \leftarrow n$ transition will have the same effect as observed in case of simple ethylene molecule, *i.e.*, presence of further double bonds will increase the energy of HOMO and decrease that of LUMO, whereas that of nonbonding orbitals are not affected very much. Due to this a red shift is observed in $\pi^* \leftarrow n$ transition or band maximum is shifted to longer wavelength as shown in Fig. 5.26. Table 5.8 lists few of the electronic transitions and molecular extinction coefficients of the transition present in ethylenic and carbonyl double bonds. Tables 5.9 and 5.10 give the λ_{max} and ε_{max} of the extended double bonds in ethylenes and molecules containing carbonyl group.

5.7.7 EFFECT OF SUBSTITUENTS ON ELECTRONIC TRANSITIONS

The most common substituents, which are present and can affect the electronic spectra of polyatomic molecules, are the ones, which contain lone pair of electrons. Most common groups are $-NH_2$, –OH, X (halogens), etc. In general these substituents can have two kinds of effects.

(*i*) **Inductive Effect:** This effect can only change the charge densities at the neighbouring atoms of the substituents and decreases sharply as one moves away from the substituent. If the charge density decreases on the neighbouring atom, it is called negative inductive effect (–I) and if charge density increases on the neighbouring atom, it is called positive effect (+I). Naturally the former is due to the more electronegative atoms of the substituents and the latter is because of electropostive atoms of the substituents in comparison to the adjacent atoms. Since this effect only changes the charge densities on the atoms of the parent molecules, it only modifies the Coulomb integrals (α). In other words either the energy of the molecular orbitals are raised or lowered in reference to

Table 5.8: *Band maxima of $\pi^* \leftarrow n$ and $\pi^* \leftarrow \pi$ transitions of few multiple bonds*

Chromophore	$\pi^* \leftarrow \pi$ (strong) (nm)	$\pi^* \leftarrow n$ (weak) (nm)
>C=C<	170	–
—C≡C—	170	–
>C=O	166	280
>C=N—	190	300
Ph—N=N—Ph	260	350
>C=S	230 – 310	599

Table 5.9: *Band maxima of $\pi^* \leftarrow \pi$ transition and molecular extinction coefficient of ethylelne and few conjugated ethylenes*

Chromophore	λ_{max} (nm)	ε_{max} (M^{-1} cm^{-1})
Ethylene —C=C—	170	16,000
1, 3 Butadiene —C=C—C=C—	210	20,900
1, 3, 5-Hexatriene —C=C—C=C—C=C—	268	42,700
1, 3, 5, 7-Octatetrene	304	
1, 3, 5, 7, 9-Decapentaene	334	121,000
1, 3, 5, 7, 9, 11-Dodecahexaene	364	138,000
1, 3, 5, 7, 9, 11, 13-Tetradecaheptaene	390	
1, 3, 5, 7, 9, 11, 13, 15-Hexadecaoctaene	410	
1, 3, 5, 7, 9, 11, 13, 15, 17, 19-Eicosadecaene	447	
All trans-ß-carotene (17-double bonds)	480	139,000

Table 5.10: *Effect of conjugation on the transition of carbonyl group*

Chromophore	$\pi^* \leftarrow \pi$ (strong) (nm)	$\pi^* \leftarrow n$ (weak) (nm)
H_2C = CH—CH=O	208	328, 13
>C=C—C=O	240	320
>C=C—C=C—C=O	270	350

the parent molecule containing no substituent. Since the difference between HOMO and LUMO does not change, the band maximum of the transition is hardly affected due to this effect, but intensity of the respective band increases because of the increase in the dipole length.

(*ii*) **Resonance Effect or Mesomeric Effect:** In this case the lone pair of electrons takes part in the bonding and tries to behave like extended conjugation. Detailed calculations have shown that the energies of both HOMO and LUMO are raised but the increase in the energy of HOMO is more than that of LUMO. Due to this energy difference between HOMO and LUMO decreases and thus leads to shift of the band maximum towards longer wavelength. The red shift in the band maximum depends on the interaction of the lone pair of electrons with HOMO and LUMO. This depends on three factors:

(a) **Energy difference between the Lone Pair of Electrons and HOMO and LUMO:** Since the energy difference between HOMO and lone pairs of electrons is less than that between the LUMO and lone pair of electrons, the energy of HOMO will be raised more than that of LUMO. The interaction of the non-bonding electrons with the σ orbital will be less than that with the π orbital. Thus, the resonance effect is more pronounced on the $\pi^* \leftarrow \pi$ transition than on $\sigma^* \leftarrow \sigma$ transition.

(b) **Electronegativity of Heteroatom:** Greater the electronegativity of the heteroatom containing the lone pair, the more tightly it will be bound to that atom. This will reduce the interaction of the lone pair of electrons with the parent molecular orbitals. In other words, the resonance effect will be small, *i.e.,* the λ_{max} of the original transition will only be slightly shifted from its original position. The effect of the substituent on the transition (λ_{max}) can be arranged as: $-NR_2 \gg NH_2 > -OH > I > Br > Cl > F$.

(c) **Arrangement of the Lone Pair in the Space:** As we have learnt in the last chapter, the interaction of the orbitals is the maximum if these are properly arranged in space. For example, if n and π orbitals are parallel to each other their interaction is the maximum and if these are perpendicular, it is the minimum. Because of this reason, the energies of π and π^* orbitals will be affected accordingly. In other words, the red shift in the $\pi^* \leftarrow \pi$ transition will be maximum in the former case and minimum in the latter case.

The effect of substituents can be discussed by taking few molecules. For example, consider methylamine (CH_3NH_2). Presence of $-NH_2$ group containing lone pair of electrons allows the possibility of $\sigma^* \leftarrow n$ transition at 213 nm in addition to and also tends to increase the wave length of the $\sigma^* \leftarrow \sigma$ absorption band (170 nm) instead of 122 nm in methane and 135 nm in ethane.

Table 5.11: *Absorption spectrum of Carbonyl Compounds*

Compounds		λ_{max} (nm)	ε_{max}
Formaldehyde $HCH = O$	Vapour	304 175	18 18,000
Acetaldehyde $CH_3CH = O$	Vapour	289 182	12.5 10,000
Propionaldehyde $C_2H_5CH = O$	Vapour	290	15
Acetone $(CH_3)_2\ C = O$	Vapour	274	13.6
Methylpropylketone $MePrC = O$	Hexane	278	15.5
Cyclopentanone	Isooctane	300	18
Cyclohexanone	Isooctane	291	15
Acetylacetone, $CH_3COCH_2COCH_3$	Ethanol	272	110
Acetylchloride CH_3COCl	Heptane	240	34
Acetyl bromide, CH_3COBr	Heptane	250	93
Ethyl acetate, $CH_3COC_2H_5$	Vapour	211	47.5
Acetamide, CH_3CONH_2	Water	214	100
Acetic anhydride,$(CH_3CO)_2O$	Vapour	220	73
Acetic acid, CH_3COOH	Vapour	203	depends on pressure

$\pi^* \leftarrow n$ transition of the carbonyl group in formaldehyde lies at ~310 nm. Replacing hydrogen atom by hydroxyl group (—OH, formic acid) and amino group (—NH_2, formamide) has pronounced effect on this transition and large blue shifts are observed (204 and 214 nm respectively). Both —NH_2 and —OH groups are having –I effect and will decrease the energy of the non-bonding electron. The resonance effect of lone pair of electron will increase the energy of the π^* anti-bonding orbital. Both the effects are acting in the same direction, *i.e.,* increase the energy between the non-bonding and π^* anti-bonding m.o. Table 5.11 lists the long wavelength band of formaldehyde and few compounds derived by replacing H- atom by the substituents.

5.7.8 TRANSITION METAL COMPLEXES

The absorption spectra of transition metal complexes are more complex as compared to the organic chromophores. This is because of the unpaired electrons in *d*-orbitals and environment around the metal ion. That is, it is not only the electronic configuration interactions or spin-orbit coupling which leads to number of electronic states, but the presence of ligand field also splits them further. The details of these more complex interactions and thus diagrams can only be obtained by quantum mechanical calculations, which are beyond the scope of this book. In this section, only qualitative picture of two main types of transitions, which are present in transition metal complexes, will be given.

1. ***d–d* Transitions:** We have learnt in Chapter 4 that all the 5–*d* orbitals of a given shell having the same principal quantum number are degenerate in a free atom. The degeneracy is maintained if the metal atom or ion is, surrounded by a spherical field. In the transition metal complexes, the ligands are placed at the corners of tetrahedron, octahedron or square plane. In other words, the immediate environment of the atom or ion is no longer spherical symmetrical and thus degeneracy of *d*-orbitals is removed. In Fig. 5.27, the *d*-orbitals are arranged in increasing order of energy in different environments. For example, in octahedral complexes, the *d*-orbitals are arranged into two sets, one comprising of three *d*-orbitals (t_{2g}) of lower energy and the other of two *d*-orbitals (e_g) of higher energy. The energy gap between these set of orbitals is denoted by either 10 D*q* or by Δ and the gap increases with the increase of field strength of the ligand, as shown in Fig. 5.28. The energy of t_{2g} level is lowered by 2/5 Δ or 4 D*q* amount and that of e_g is increased by an amount of 3/5 Δ or 6 D*q*. The electronic configuration of d^1 and d^9 metal ions can be written as $t_{2g}^1\, e_g^0$ or simply t_{2g}^1 and $t_{2g}^6\, e_g^3$ respectively in the ground state, wheras those of the excited states are $t_{2g}^0\, e_g^1$ or e_g^1 and $t_{2g}^5\, e_g^4$. In other words, in case of Cu^{2+} ion, it may be said that there is a hole in the e_g orbital in the ground state and a hole in the t_{2g} orbital in the excited state. On the left side of the Fig. 5.28, the spectroscopic term symbol is ^2D in the ground states of d^1 and d^9 ions in the absence of a ligand field, whereas in the presence of ligand field the degeneracy of *d*-orbitals is removed and thus leads to a transition which involves the promotion of an electron from *d*-orbital (t_{2g}, t_2 for Ti^{3+}) to *d*-orbital (e_g, *e* for Ti^{3+}). These transitions are generally present at quite a longer wavelength and also very weak. These bands are always broad and usually without any vibrational structure.

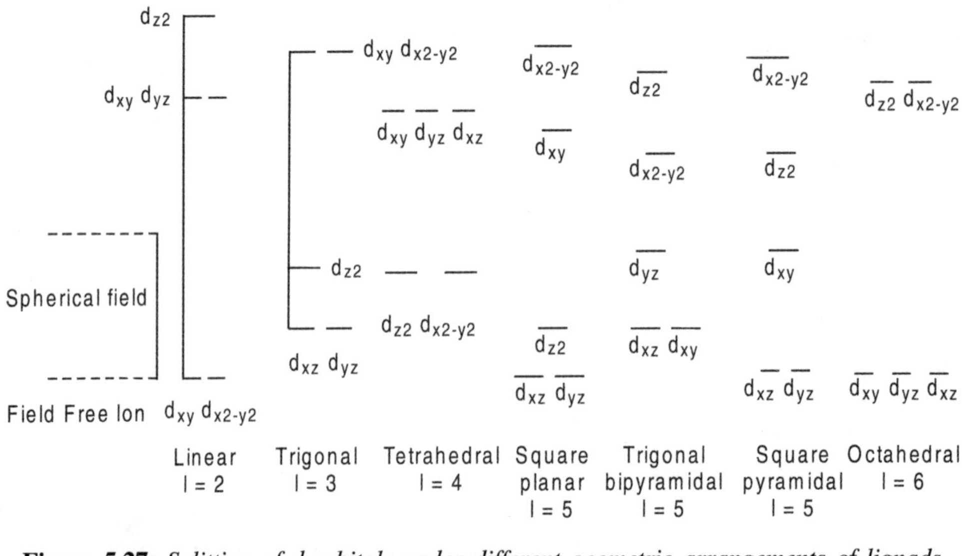

Figure 5.27: *Splitting of d-orbitals under different geometric arrangements of lignads*

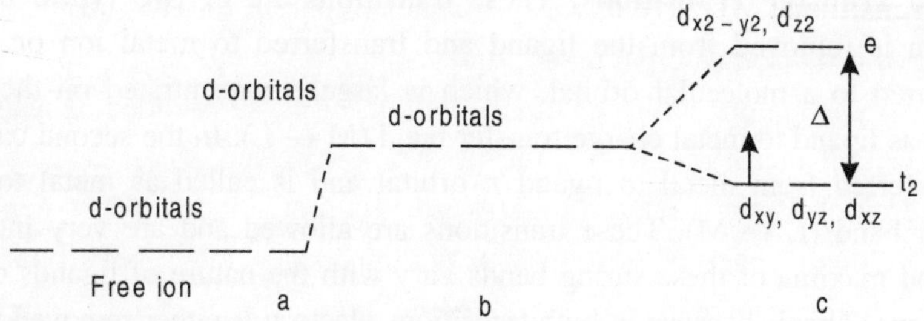

Figure 5.28: *Energy level diagram of metal d-orbitals. (a) Free metal ion (Ti^{3+}) in presence of spherical field. (b) Energies of d-orbitals is raised due to close proximity of negative charge. (c) Splitting of the d-orbitals of a metal ion (Ti^{3+}) in presence of octahedral field and gap between them depends on the ligand filed strength*

Sometimes these transitions are hidden under other strong transitions or appear as shoulder towards the red side of the strong transitions. Based on parity (or Laporte) selection rules *d–d* transitions are forbidden, because allowed transitions do involve the distribution of electrons over the quantum shell which changes by unity, *i.e.,* $\Delta l = \pm 1$. But we do observe these transitions and as mentioned earlier these are weak, the molecular absorption coefficient is less than, 50 M^{-1} cm^{-1}. The break down of parity selection rules, are because of the following reasons:

(*i*) In case of certain transition metal complexes, the environment around the metal ion lacks centre of symmetry. Because of this, mixing of *d* and *p* (or *f*) orbitals does take place to certain extent, *i.e.,* states arising because of this mixing are no longer pure *d*-states and possesses small contribution from *p*- or *f*-orbitals in *d*-orbitals. Now a *d–d* transition which is forbidden in a free atom or ion, now involves a small amount of transfer from *d*-orbital to either *p*- or *f*-orbital when the ion is a part of the complex. Since transition *d* to *p* or *f* is allowed, the normally forbidden transitions may occur with low intensity. The intensity of this transition is proportional to the extent of mixing of *d* and *p* or *f* orbitals. Based on above arguments, *d–d* transitions, in tetrahedral complexes should be substantially more intense than the octahedral complexes since the metal ion is not at a centre of symmetry.

(*ii*) Second reason for the observation of weak *d–d* transition is due to the involvement of molecular vibrations. In case of transition metal complexes, where a metal atom or ion is present at the centre of symmetry, the mixing of *d* and *p* orbitals is possible. But during molecular vibrational motions, it may happen that environments around the metal ion may depart from the symmetrical equilibrium position for some time and thus leads to the mixing of *d* and *p* orbitals. Because of this reason break down of parity selection rules takes place, which can be justified in detail by quantum mechanical calculations. These kinds of transitions, which gain intensity because of mixing of vibrational motions with electronic ones, are called *vibronic bands.*

2. **Charge Transfer Transitions:** These transitions are of two types. In one case, an electron is removed from the ligand and transferred to metal ion or more correctly transferred to a molecular orbital, which is largely concentrated on the metal. This is known as ligand to metal charge transfer band ($M \leftarrow L$). In the second case, the electron is transferred from metal to ligand π orbital and is called as metal to ligand charge transfer band ($L \leftarrow M$). These transitions are allowed and are very intense. However the band maxima of these strong bands vary with the nature of ligands or the nature of metal ions. This is because in both transitions electron is either removed from the ligand or from the metal and thus can be related to the oxidation potentials of these species. For example, in the first case ($M \leftarrow L$) the absorption band maximum shifts to the longer wavelength if the ligand is more easily oxidized. Consider the complex of Co^{3+} $[Co(NH_3)_5X]^{2+}$, where $X = F^-$, Br^- or I^-. In case of fluoride ion the band maximum appears at nearly the same place where $[Co(NH_3)_6]^{3+}$ absorbs, whereas it shifts to longer wavelength when F^- is replaced by Cl^- or Br^-. Finally when $X = I^-$, the $L \rightarrow M$ charge transfer band is shifted towards longer wavelength to such an extent that it obscures the weak *d–d* transition.

   In the second case although the energy of the first charge transfer band, in special cases, depends on the detailed electronic structure of the metal ion, it is roughly true that the bands move to longer wavelengths as the oxidizing power of the metal ion increases. Further more, for the corresponding complexes of metal ions of different transition series the bands move to shorter wavelength as we go from the first to the latter series. The direction of this change is in agreement with the increase in stability of high valences in the latter series.

5.8 BEHAVIOUR OF ELECTRONICALLY EXCITED MOLECULES

Depending on the wavelength of excitation the molecule can be raised to one of the several different electronically excited states. The various ways in which this excited molecule can behave are shown in Fig. 5.29, known as *Jablonski diagram.* The processes by which excited molecule can decay to the ground state can be divided into following categories.

(*i*) **Photophysical Process:** In this process the molecule returns to ground state without change in the structure of molecule. This includes fluorescence, phosphorescence, non-radiative processes etc.

(*ii*) **Photochemical Processes:** The excited molecule decomposes into atoms or radicals and these intermediates may take part in further reactions. Unlike the above processes this is generally accompanied either with the change of mass in the reactants or change in the molecular structure.

Some of these processes are discussed in the following sections briefly.

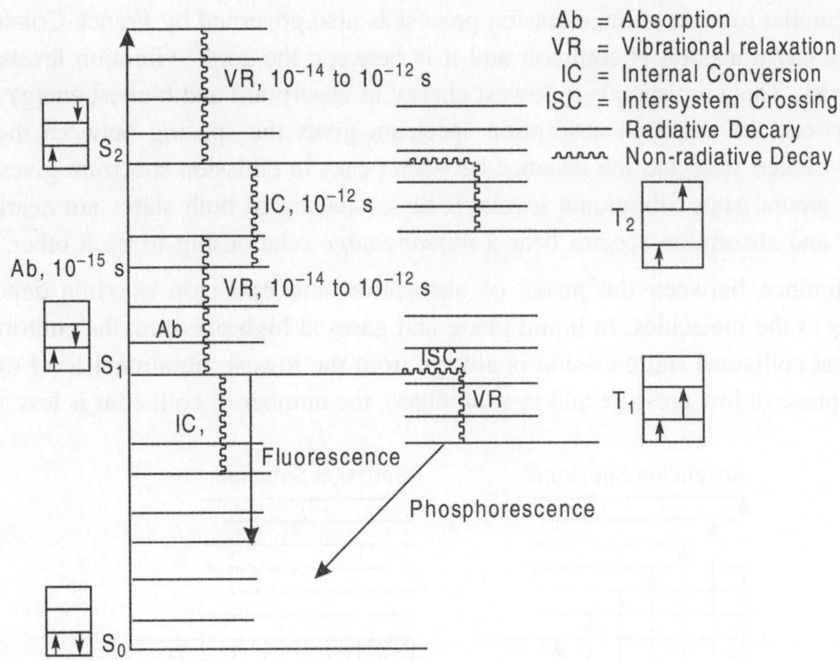

Figure 5.29: *Jablonski diagram showing the fate of polyatomic molecule upon photo-excitation*

5.8.1 FLUORESCENCE

If the excited state has the same spin multiplicity as the ground state ($\Delta S = 0$), the radiation emitted during the transition is known as fluorescence. Generally both the states are singlet but both the states can be doublet or triplet as well.

The simplest example of fluorescence is given by, the monoatomic vapours at low pressure. Sodium atom in its ground state ($^2S_{1/2}$) absorbs radiation to give $^2P_{1/2}$ and $^2P_{3/2}$ states. The atoms revert back to ground state with a mean lifetime of ~ 10^{-8} s. In this case the exciting and the emitting radiation have the same wavelength. This kind of fluorescence is called *resonance fluorescence.*

Molecules in their ground electronic state are generally present in their lowest vibrational levels. When radiation is absorbed, the molecules are excited both electronically and vibrationally according to Franck-Condon principle. As a result instead of a single line a broad band is obtained and the intensity distribution is governed by Franck-Condon principle. As we know the probability of transition between different levels is not equal, the intensity of each transition, can be represented by the arrow of unequal heights in the spectrum, Fig. 5.30.

Vibrational relaxation (energy loss) or de-excitation in the excited electronic state is also a rapid process (10^{-12} s and particularly in the condensed phase). Molecules revert to the lowest vibrational level of the electronically excited state during this period. The emission thus occurs at a rate depending on the environments and structure of the molecule. As in absorption the molecules after emission return to ground state vibrational level, which is different from where it is started. The intensity of transition (emission) in the spectra is again represented by the height of the arrow

(Fig. 5.30). Similar to absorption, emission process is also governed by Franck-Condon principle. In both cases one transition is common and it is between the zeroth vibration levels of both the electronic states. This corresponds to lowest energy in absorption and highest energy in emission. Distance between the peaks in absorption spectrum gives the spacing between the vibrational levels of the excited state and the distance between peaks in emission spectrum gives the spacing between the ground state vibrational levels. If these spacing in both states are nearly equal, the fluorescence and absorption spectra bear a *mirror image* relationship to each other.

The difference between the peaks of absorption and emission maxima depends on the environments of the molecules. In liquid phase and gases at high pressure, the emitting molecules suffer frequent collisions and emission is always from the lowest vibrational level of the excited state. In gas phase at low pressure and in solid phase, the number of collisions is less and emission

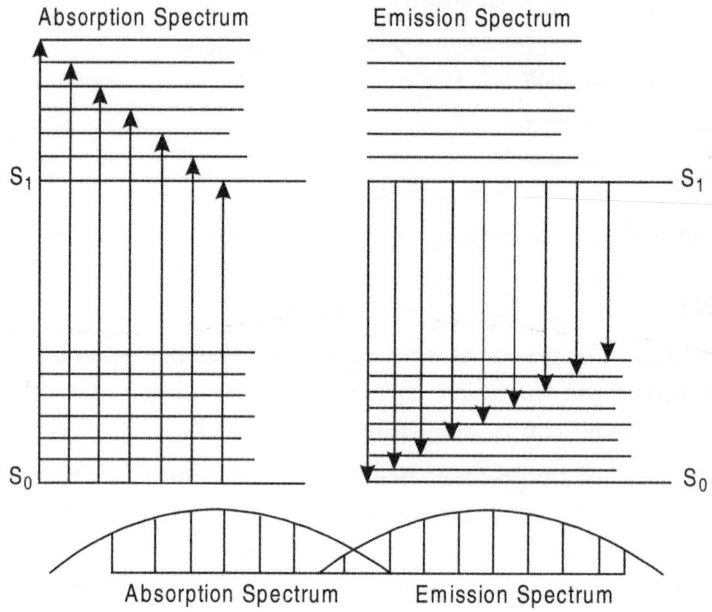

Figure 5.30: *Absorption and emission spectrum and their vibrational distribution*

collisions is less and emission can occur directly from the level of excitation. In latter case difference between absorption and emission maxima is small, whereas in former case difference between absorption and emission maxima is large. Thus resonance fluorescence cannot be observed from molecules under normal conditions.

It is generally observed that in liquid solutions or in gas phase at high pressure, the fluorescence spectrum and fluorescence quantum yield are independent of wavelength of excitation. This is because vibrational relaxation between same electronic states and relaxation from upper electronic states ($S_2 \rightarrow S_1$ etc.) are very fast in comparison to emission from S_1. Thus emission always takes place from the lowest vibrational level of the first electronic state (S_1) except in few cases where emission occurs from S_2. Well-known example is azulene.

5.8.2 PHOSPHORESCENCE

Emission of radiation from states having different spin multiplicity ($\Delta S \neq 0$) is known as phosphorescence. The states generally involved are singlet and triplet, but other combinations are also possible. Phosphorescence is red shifted with respect to and has a longer lifetime as compared to fluorescence. This phenomenon has been known for a long time, but the designation of name, the triplet state, to this metastable state was given by, Kasha and Lewis in late forties. Since two states have different spin multiplicities, the direct absorption to this state is highly improbable. It can be seen from Jablonski diagram (Fig. 5.29) that the triplet state is populated indirectly. The molecule is first excited to S_1 or after relaxation from S_2, S_3 ....to S_1 state. Molecules in S_1 state crosses over to triplet state and the phenomenon is known as intersystem crossing. The crossing over process is in competition with the other processes of the excited singlet state (*e.g.*, fluorescence, internal conversion etc.). The population of triplet states depends on the rate constants of each process. Following factors are important in increasing the triplet state population.

(*a*) **Mixing of Singlet and Triplet States:** Mixing of two wave functions or energy levels is inversely proportional to the energy difference between the two states. Smaller the energy gap, greater will be the population of triplet state. For example, in carbonyl group, difference between S_1 and T_1 states is only ~20 kJ mol^{-1} and thus, very weak fluorescence is observed. In benzene, this difference is 125.41 kJ mol^{-1}, both fluorescence and phosphorescence are observed.

(*b*) **Heavy Atom Effect:** Presence of heavy atom in the molecule or in a system with orbital, which can penetrate more deeply into the neighbouring atom, increases the population of the triplet state. This is due to the interaction of spin and orbital angular momentum. The effect of a heavy atom (in going from fluorine to iodine), present with in the molecule or/and in the solvent is to decrease the singlet state lifetime of the molecule but to increase the ratio of phosphorescence to fluorescence quantum yield.

(*c*) **Introduction of Paramagnetic Centre:** Paramagnetic species has the spin angular momentum and thus are associated with strong inhomogeneous magnetic field. This field is similar, to that created by a high atomic number atom. This increases the spin-orbit coupling leading to an increase in absorption and emission between singlet and triplet states. Like singlet state, triplet state also undergoes the following processes:

(*a*) Emission of radiation (phosphorescence) which is red shifted to fluorescence, by an amount equivalent to energy $\Delta E = E(S_1) - E(T_1)$, provided that emission takes place from zero vibrational level.

(*b*) Triplet $\rightarrow$ triplet annihilation,

$$T_1 + T_1 \rightarrow S_1^* + S_0 \qquad ...(5.182)$$

and emission from S_1 state is similar to fluorescence but it is long lived.

(*c*) Intersystem crossing to highly excited vibrational levels of ground electronic state.

(*d*) Collisional quenching due to impurities.

In view of the above processes, phosphorescence quantum yield depends on the relative rate constant of each process. Due to long lifetime of triplet state the process (*d*) is more efficient than process (*b*) and phosphorescence in solution is generally not observed at room temperature. The dissolved molecular oxygen is very efficient in collisional quenching of phosphorescence. Many of the phosphorescence studies are thus carried out on molecules trapped in solid matrix or at low temperatures where the process (*d*) is less efficient. Although the natural lifetime of triplet state is independent of temperature, rates of other processes mentioned above do vary with temperature and thus affects the lifetime.

5.8.3 DELAYED FLUORESCENCE

Besides fluorescence and phosphorescence there is another kind of emission with a longer lifetime but with a spectrum similar to that of fluorescence. This is called *delayed fluorescence.* There are two ways by which this fluorescence can occur.

1. **E-Typed Delayed Fluorescence:** This is produced by the thermal activation of molecules from triplet states. Intensity of this emission, relative to that of triplet-singlet, emission decreases exponentially with decrease of temperature and activation energy of this process is equal to energy difference between first singlet and first triplet state. The intensity is proportional to the first power of the rate of absorption of radiation. Lifetime is the same as that of the triplet state under the same conditions.
2. **P-Typed Delayed Fluorescence:** This is due to triplet-triplet annihilation, *i.e.,* two triplet molecules collide to form S_1 and S_0 molecules. Intensity relative to phosphorescence emission does not follow the exponential temperature law, but is proportional the square of absorption of exciting radiation. For low rates of absorption this emission decays exponentially. The lifetime is expressed to be half of the lifetime of triplet state in the same solution.

5.8.4 EXCIMER FORMATION

Emission spectral distribution and its characteristics have been found to be concentration dependent and in many cases new features appear in the emission. Pyrene emission spectrum in UV region is structured at concentration up to 10^{-4} M. As the concentration increases, a new structureless weak emission band appears at the expense of structured band. Its intensity is proportional to the square of the pyrene concentration. This has been explained as the emission from a dimer (excimer) of pyrene.

$$\text{Pyrene}^* \ (A^*) + \text{Pyrene}_0 \ (A_0) \rightarrow (\text{Pyrene-Pyrene})^* \rightarrow 2 \text{ Pyrene } (A_0) + h\nu \qquad ...(5.183)$$

The excimer formation is described by the union of a molecule in the excited state with a molecule in the ground state. Its formation depends on:

(*i*) Whether lifetime of the excited state is long enough for excimer formation.

(*ii*) Whether the excimer dissociates rapidly.

(*iii*) Whether intermolecular interaction energy is sufficient to cause quenching of dimer fluorescence.

Exact interpretation of excimer formation varies from excimer to excimer but bonding in excimer probably involves contribution from both charge transfer states and delocalization of singlet energy between two systems. Similar types of emissions are observed in crystalline hydrocarbons and in particular those systems having lattices in which there is high molecular overlap, *e.g.*, pyrene, perylene, benz-pyrenes, coronene etc.,

5.8.5 RADIATIONLESS TRANSITIONS

In fluorescence and phosphorescence a molecule goes from higher state to lower state by emission of radiation. *There are certain processes in which a molecule can change its state, with the same or different spin multiplicity without emitting radiation. Such processes are called radiationless transitions or processes.* These are mainly of two types.

1. **Internal Conversion:** In this process, like fluorescence there is no change of spin between two states, *e.g.*, $S_1 \rightsquigarrow S_0$, $S_2 \rightsquigarrow S_1$ etc., The wiggly arrow is used to indicate a process not involving radiation as shown in Jablonski diagram. If potential energy curve of ground state crosses or comes close to the vibrational level lower than that to which the molecule is excited, there is a finite probability that the molecule will cross over to the other intersecting state at this point. The molecule acquires vibrational energy in the lower electronic state and comes to thermal equilibrium after collisions with other molecules. Conversions from higher to lower singlet sates ($S_n \rightsquigarrow S_{n-1}$) are more common than from $S_1 \rightsquigarrow S_0$. This is particularly true of polyatomic molecules because of the presence of numerous potential energy curves and their intersections. Rate of this process in higher singlet states is ~ 10^{12} s^{-1} and is completed before emission starts from S_1 state. On the other hand, rate of intersystem conversion in S_1 state is ~ 10^8 s^{-1}. That is the reason why fluorescence always occurs from first singlet state although molecules are excited to higher singlet states.

2. **Intersystem Crossing:** This is similar to internal conversion but the cross-over takes place between the two states having different spin multiplicities (*e.g.*, $S_1 \rightsquigarrow T_1$ or $T_1 \rightsquigarrow S_0$). Recent theories have been developed to explain the radiationless transitions. The most important restriction is of spin, which may decrease the rate of such process by a factor as high as 10^6 over the equivalent states. Rate of intersystem crossing is ~ 10^7 s^{-1}, whereas that of internal conversion is 10^{12} s^{-1} in S_2, S_3 ..... and 10^8 s^{-1} in S_1 state.

   Other factors affecting the rate of radiative or nonradiaitve processes are the considerations of overlap of orbitals and their symmetries. If two orbitals do not overlap, the probability of a transition is low, for example, in $\pi^* \leftrightarrow n$ transitions, n and π^* orbitals are present in planes at right angle to each other and their overlap is small as compared to π and π^* orbitals. That is why $\pi^* \leftarrow n$ transitions are weaker than $\pi^* \leftarrow \pi$ transition. As said earlier, the symmetry of the molecule should also be maintained during the transition. This restriction also applies to the $\pi^* \leftarrow \pi$ transition of carbonyl group.

5.9 NUCLEAR MAGNETIC RESONANCE

Magnetic resonance spectroscopy is quite different from the rotational, vibrational and electronic spectroscopy in the sense that in the latter three branches of spectroscopy the transitions are *stimulated by the interaction of electrical dipole of the molecules with the electrical vector of the radiation,* whereas in the former branch of spectroscopy transition is stimulated by the magnetic field generated by the nuclei and electronic spinning motion. These are called *magnetic dipole induced transitions.* Thus in this technique the transition takes place between the states of molecules, which are obtained when molecules are placed in the external magnetic field and radiation of frequency either in the radiowave (Nuclear Magnetic Resonance, NMR) or microwave region (Electron Spin Resonance, ESR) are applied. Although the NMR spectroscopy can be used for any nucleus having finite nuclear spin angular momentum, the proton nucleus is simple and very well studied. In this section only proton NMR will be discussed. Local environments of a nucleus determine the frequency of absorption. Thus, NMR spectrum contains much more information regarding the interactions and motion of these nuclei. Further the frequency of these resonance absorption can be measured to an accuracy of about 1 part per million, different from one another. This characteristic of high resolution NMR makes this branch more popular and useful over other spectroscopic techniques discussed earlier, where the absorption of different chromophores often overlap and make it more difficult to assign the absorption of frequency to any particular group or chromophore.

It is well known that any spinning charge behaves like a magnet and possesses a magnetic moment. It has been shown earlier that electrons and nuclei spin around their own axis and thus behaves like small magnets. The energy of these magnetic moments depends on their orientation around the external magnetic field just as the energy of the electrical dipole depends on its orientation with the electrical field. The common example is that of a compass needle. If this needle is pointed toward east and is allowed to have free rotation, it will point itself towards north. This is because the energy is lowered in this direction. The other example can be molecules present in the sample tube when placed in the external magnetic field. The only difference between the two systems described above is that the first one can have any value of energy, being a macro-system. Whereas the latter being microscopic system, can have only discrete values of energies. This means that the magnetic moment can orient itself with the external magnetic field only at certain angle. Now if the proper resonance frequency is applied to this sample placed in the external magnetic field, a transition will take place, bringing the molecule to an excited state, with the spin of the proton flipped in the opposite direction.

It is clear from the above discussion that only those nuclei will have NMR spectrum, which have finite nuclear spin. Similar to electron, each proton and neutron is having angular momentum given by $\hbar$ units. As in the case of multi-electron atoms, the nuclei in the atoms, which are having many protons and neutrons, can be represented by the resultant total angular momentum, I and thus can have magnetic moment. *Unlike multi-electron atoms the resultant I value for nuclei can only be determined experimentally.* But with lot of results available one can predict the following values of I.

Table 5.12: *Some characteristics of nuclear spin of the few nuclei*

Isotope radioactive	Natural abundance (%)	Spin I	Nuclear G factor	Magnetic moment[1] μ/μ_N	NMR frequency at 1 T (ν/MHz)
$^{1}n^{*}$		1/2	−3.8260	−1.9130	29.167
^{1}H	99.9844	1/2	5.5857	2.79285	42.576
^{2}H	0.0156	1	0.85745	0.85745	6.536
$^{3}H^{*}$		1/2	−4.2553	−2.12765	32.4338
^{13}C	1.108	1/2	1.4046	0.7023	10.705
^{14}N	99.635	1	0.40356	0.40356	3.076
^{17}O	0.037	5/2	−0.7572	−1.893	5.772
^{19}F	100	1/2	5.2567	2.62835	40.054
^{31}P	100	1/2	2.2634	1.1317	17.238
^{33}S	0.74	3/2	0.4289	0.6434	3.266
^{35}Cl	75.4	3/2	0.5479	0.8218	4.171
^{37}Cl	24.6	3/2	0.4561	0.6841	3.472

1. The nuclear magneton has the value $\mu_N = eh/2\pi m_p = 5.05082 \times 10^{-27}$ J T^{-1}; μ is the magnetic moment of the spin state with the largest values of m_I: $\mu = g_I \, \mu_N I$.

(*i*) All the nuclei with odd mass number have a finite resultant spin and is given by the integral multiple of $\frac{1}{2}$ (*i.e.*, $\frac{1}{2}\hbar$, $\frac{3}{2}\hbar$, $\frac{5}{2}\hbar$ ....).

(*ii*) All the nuclei with even mass number but having odd number of protons possess integral multiple of spin (*i.e.*, 1, 2, 3.........).

(*iii*) All the nuclei with even mass number and even number of protons possess **zero spin.**

The spins of some nuclei are given in Table 5.12. The eigen values of the resultant spin angular momentum of nuclei I, similar to the total S value of electrons, is given by the relation:

$$\mathbf{I} = [I(I+1)]^{\frac{1}{2}}\hbar \qquad ...(5.184)$$

and similarly the eigen values of the components of the spin angular momentum along the *z*-direction are given by:

$$I_Z = m_I \hbar \qquad ...(5.185)$$

where $$m_I = I.\ I-1,\ I-2,........0..........-I+2,\ -I+1,\ -I \qquad ...(5.186)$$

That is each state associated with I, spin angular momentum, are having 2I + 1 degeneracy.

Quantitatively, the values of the energies of these states are due to the interaction of the magnetic moments induced in the nuclei with external magnetic field B_0, applied in any direction. The values are given by the equation (5.187).

$$E_{mag} = -\mu B_0 = -\mu B_0 \cos\theta = -\mu_z B_0 \quad ...(5.187)$$

where E_{mag} is the energy of the level in presence of magnetic field, **μ** is the magnetic moment and μ_z is the component of the magnetic moment along the direction of the external magnetic field strength B_0, θ is the angle between magnetic field and the magnetic moment. As can be seen from the Appendix 3, the magnetic moment, can be given by the relation:

$$\mu = \frac{g_N e}{2 m_p} I \quad ...(5.188)$$

where g_N is a dimensionless quantity called nuclear *g* factor, *e* is the charge on the electron and m_p is the mass of the proton. The nuclear *g* factor can have positive and negative values and are compiled in Table 5.12. The positive value of g_N indicates that the magnetic moment and angular momentum are parallel. Negative value of g_N indicates that these are anti-parallel. The component of the magnetic moment along the *z*-axis is given by:

$$\mu_Z = \frac{g_N e}{2 m_p} m_Z = \frac{g_N eh}{4\pi m_p} m_I \quad ...(5.189)$$

$$= g_N \mu_N m_I \quad ...(5.190)$$

where μ_N is the basic unit of nuclear dipole moment, given by equation (5.191) and is called as Bohr's nuclear magneton.

$$\mu_N = \frac{eh}{4\pi m_p} \quad ...(5.191)$$

Substituting the values of each quantity in equation (5.191):

$$\mu_N = \frac{(1.602\times10^{-19}\ C)(6.626\times10^{-34}\ Js)}{(4\pi)(1.6726\times10^{-27}\ kg)}$$

$$= 5.0508 \times 10^{-27}\ J\ kg^{-1}\ C^{-1}\ s^{-1} = 5.0508 \times 10^{-27}\ JT^{-1}$$

where T is the electromagnetic SI unit of magnetic field, called *Tesla*. The dimensions of Tesla are kg s^{-2} C^{-1}. In cgs units (which differ in dimension of charge) Bohr magneton is related as:

$$\mu_N = \frac{eh}{4\pi m_p c} \quad ...(5.192)$$

where *c* is the velocity of light. The old unit of magnetic field is Guass (G) and 10,000 G = 1 T. Substituting the values of m_Z in equation (5.187):

$$E_{mag} = -g_N \mu_Z m_I B_0 \quad ...(5.193)$$

Figure 5.31 shows the plot of E_{mag} with B_0 for a proton. Unlike the energy of the rotational, vibrational and electronic motions, the energies of the magnetic states depend on the field strength of the external magnetic field. The gap between the two magnetic states increases with increase of magnetic field. After defining the energy levels we are now in a position to calculate the frequency at which the resonance absorption will occur. The specific selection rules are:

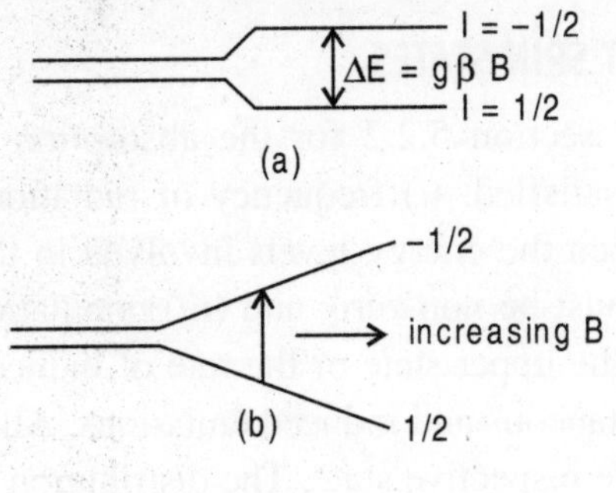

Figure 5.31: *Energy level of protons in a magnetic field. (b) Energy levels in increasing magnetic filed*

$$\Delta m_I = \pm 1 \qquad \text{...(5.194)}$$

In case of PMR, there can be only two energy levels corresponding to $m_I = +\frac{1}{2}$ and $m_I = -\frac{1}{2}$. The former is of lower energy and the latter of higher energy. The transition will occur between these states as the selection rule is satisfied. The energy separation is:

$$\Delta E = -\mu_N g_N \left(-\frac{1}{2} - \frac{1}{2}\right) B_0 = \mu_N g_N B_0 \qquad \text{...(5.195a)}$$

$$\nu = \frac{\Delta E}{h} = \frac{g_N \mu_N B_0}{h} \qquad \text{...(5.195b)}$$

For $I > \frac{1}{2}$, there will be more than two energy levels (Fig. 5.32). According to equations (5.186) and (5.190), these levels will be equally spaced and thus there will be only *one* NMR frequency for each nucleus. The frequency of absorption, given by (5.195*b*) is called as *Larmor frequency.*

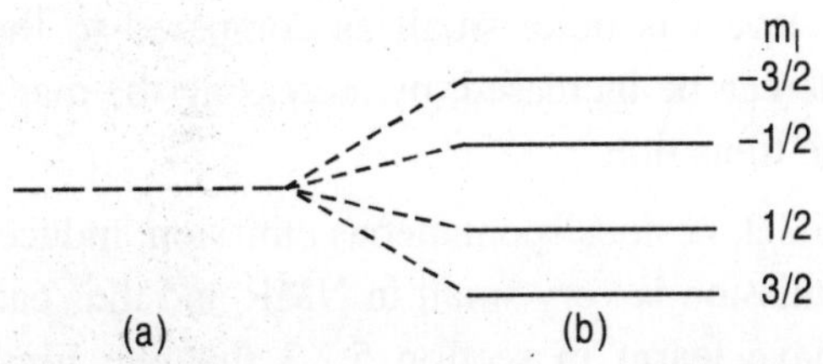

Figure 5.32: *Energy level diagram of a nucleus with I = 3/2 in (a) absence and (b) presence of magnetic field*

Example 5.28: Calculate the lowest and highest value of μ_Z for deuterium atom. The g_N value for deuterium is 0.857.

Solution: The nuclear spin for D is 1. The maximum and minimum values of I will be +1 and –1. Thus substituting the respective values of each quantity in equation (5.193), we get

$$\mu_Z = \pm (0.857)\,(5.0508 \times 10^{-27}\ \mathrm{J\,T^{-1}})\,(1) = \pm 4.32 \times 10^{-27}\ \mathrm{J\,T^{-1}}$$

5.9.1 POPULATION OF DIFFERENT SPIN STATES

Recalling the discussion of section 5.2.2 for the absorption of radiation by a sample, the following conditions are to be satisfied: (*i*) frequency of radiation must match with the natural frequency (*i.e.,* separation between the energy levels involved in the transition of the molecules), (*ii*) transition moment integral must be non-zero, and (*iii*) population of the lower state should be considerably larger than that in the upper state or the rate of induced absorption should be slower than the sum of the rates of spontaneous and induced emissions. All the three rates are proportional to the number of molecules in the respective state. The distribution of molecules in different states can be calculated with the help of Boltzmann equation.

$$\frac{n_1}{n_0} = \frac{e^{-\frac{E_1}{k_b T}}}{e^{-\frac{E_0}{k_b T}}} = e^{-\frac{\Delta E}{k_b T}} = e^{-\frac{g N_\mu B_0}{k_b T}} \qquad \text{...(5.196)}$$

Here n_1 and n_0 are the number of molecules in the excited and ground states respectively, k_b is the Boltzmann constant and T is the temperature in degree Kelvin. For proton at 1 Tesla

$$\frac{\Delta E}{k_b T} = \frac{g_N\,\mu_N\,B_0}{k_b\,T} = \frac{(5.585)\,(5.0508 \times 10^{-27}\ \mathrm{J\,T^{-1}})\,(1\ \mathrm{T})}{(1.38 \times 10^{-23}\ \mathrm{J\,K^{-1}})\,(298\ \mathrm{K})} = 6.8 \times 10^{-6}$$

Since the exponent in the equation (5.196) is so small that the exponential term can be expanded and the higher terms can be neglected. Thus, equation (5.196) reduces to:

$$\frac{n_1}{n_0} = 1 - 6.8 \times 10^{-6} \qquad \text{...(5.197)}$$

This means that the ground state contains only few nuclei per million more than the excited state. This is unlike the other branches of spectroscopy discussed earlier. The reason is that the separation between the two levels is quite small as compared to the thermal energy k_bT. The population in the ground state can be increased, by increasing the magnetic field strength, thereby increasing the intensity of the transition.

As discussed in section 5.2.2, besides spontaneous emission, induced emission also takes place in NMR. The spontaneous emission is very small in NMR and thus can be neglected. This can be explained as follows. We have learnt in section 5.2.3 that the Einstein's coefficient (A_{21}) of spontaneous emission is proportional to ν^3 (5.24), *i.e.,* it is very strongly dependent on frequency. Thus, its value for the spontaneous emission (fluorescence) in the UV/Visible region (proportional to ~10^{42}) will be very large as compared to the similar emission in radio wave region (proportional to 10^{18}) of the NMR spectrum. If the intensity of radiation is very large, *i.e.,* if the rate of induced absorption is very higher than rate of emission (called as relaxation), the population of the upper level may become equal to that of the lower state. This phenomenon is called *saturation.* Under such condition net absorption is zero. Two kinds of relaxation are observed in NMR experiments.

(*i*) **Spin-Lattice Relaxation:** In this case the energy released during emission is transmitted to the environments, called *lattice vibrational system.* In other words this process requires interactions of spins of the nuclei and lattice system. This is also called as *longitudinal relaxation.* Thus, the rate of longitudinal or spin lattice relaxation $\left(\frac{1}{T_1}\right)$ controls this relaxation process.

(*ii*) **Spin-Spin Relaxation:** This is also called *transverse relaxation.* In this process the relaxation can proceed with and without the release of energy. If the process is accompanied with the relaxation of energy, it is due to spin-lattice relaxation. If it is with out the relaxation of energy it is due to spin-spin relaxation. In general longitudinal relaxation rate is much smaller than the transverse relaxation rate $\left(\frac{1}{T_2}\right)$, *i.e.,* $T_2 \ll T_1$ and thus, the thermal equilibrium is achieved by transverse relaxation.

Example 5.29: Calculate the frequency of the radiation for the absorption in NMR signal if proton is placed in a magnetic field of 1 T.

Solution: Using equation (5.195*b*):

$$\nu = \frac{g_N \mu_N B_0}{h} = \frac{(5.585)(5.0508\times10^{-27}\ \mathrm{J\,T^{-1}})(1\mathrm{T})}{6.62\times10^{-27}\ \mathrm{Js}} = 42.61 \times 10^6\ \mathrm{s^{-1}}$$

$$= 42.61\ \mathrm{MHz}$$

(1 MHz corresponds to 10^6 cycles per second). Thus, the energy difference that corresponds to the transition of nuclear spins is of the size of radio waves.

5.9.2 NUCLEAR MAGNETIC RESONANCE SPECTROMETER

NMR spectrometer consists of mainly three components. Homogeneous magnetic field to produce energy levels, the radio wave frequency transmitter for exciting the nuclei from lower state to upper state and the radio frequency receiver to receive the frequencies emitted by the excited nuclei. The resolution and accuracy of the spectrum thus depends on the homogeneity of the magnetic field and the accuracy of the radio wave transmitter. Since the chemical shift (section 5.9.4) produced in proton NMR is few parts per million, the variation in magnetic field should not be more than a part in 10^8 or 10^9 and similarly the radio wave frequency wave generator should be constant over a part in 10^9. This can be achieved in a crystal-controlled transmitter. As shown in Fig. 5.33 the sample tube is placed in between the electromagnet and let the lines of force are represented along the *z*-axis. Radio wave transmitter coil and radio wave receiver coil are wrapped around the sample tube in such a manner that these are mutually, as well as, perpendicular to the direction of the magnetic field. This is because the magnetic field produced by the transmitter coil is perpendicular to the external magnetic field produced by the electromagnet and the radio wave receiver should not receive any signal from the transmitter. It is clear from the equation (5.195*b*) that NMR spectrum can be observed either by the variation of the transmitter frequency or by varying the magnetic field. In the first case the sample is placed between the poles of the electromagnet of constant field strength and signal to sample is sent through the radio wave oscillation. When the frequency of the oscillator matches with the Larmor frequency, the system absorbs the radiation sent by the oscillator. This absorption changes the impedance of the radio wave frequency circuit. The change in impedance is measured by, a bridge circuit, amplified and displayed on a charge recorder. The time axis of the recorder reflects the frequency range scanned. When the frequency of the oscillator passes through the resonance and displaed on a

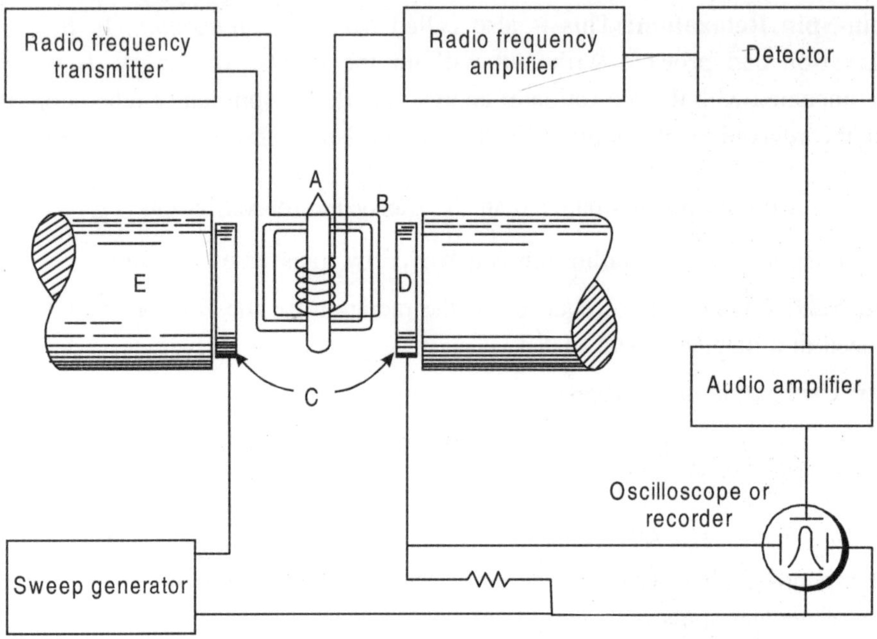

Figure 5.33: *The schematic diagram of NMR spectrometer*

charge recorder. The time axis of the recorder reflects the frequency range scanned. When the frequency of the oscillator passes through the resonance frequency, the impedance of the oscillator comes back to the normal value. Such a trace is recorded in the Fig. 5.34.

In practice it is difficult to design a variable frequency oscillator, which can have the desired, frequency range. In general second technique is followed. In this case, the radio wave frequency oscillator is designed to have a fixed frequency and the most common oscillator can generate frequency of 60 MHz. Better accuracy and easy interpretation of spectra can be achieved with higher frequency oscillator. Spectrometers with frequency up to 750 MHz are available. The magnet is designed to provide high field. The small coils, known as sweep generator, which can vary the field over a small range, *e.g.,* are added. The sweep generator produces a steadily increasing voltage that drops to zero when certain voltage is reached. This also controls the sweep of the oscilloscope and chart recorder. In the presence of magnetic field the nuclei in the sample start precessing around the direction of the electromagnet. When the magnetic field reaches a value at which the nuclei of the sample are precessing with frequency that matches the Larmor frequency, the radio frequency from the transmitter is absorbed by the nuclei and are excited to the upper state. This can only take place when there is a resonance. At the same time the excited molecules emit the same frequency and the signals are picked up by the radio wave, receiver coils. This can be displayed either on the oscilloscope or on the chart recorder. In this case the time axis depicts the strength of the magnetic field. Even if the magnetic field is homogeneous a small inhomogeniety in the magnetic field is developed when it passes through the sample tube. In order to avoid this or to improve the resolution, the sample tube is rotated around its axis so that the nuclei experience a field that is averaged over the volume of the sample.

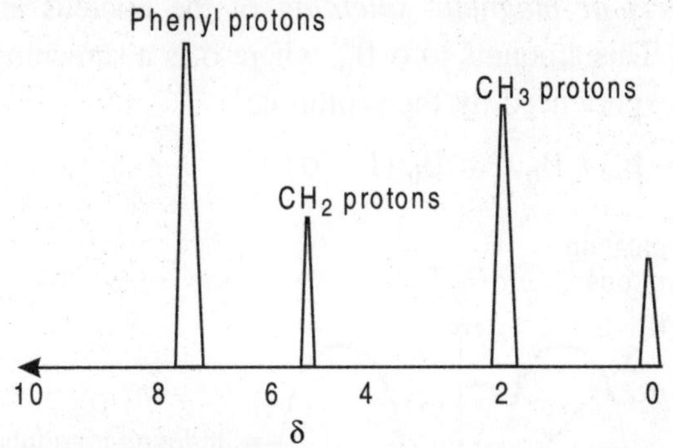

Figure 5.34: *NMR trace from NMR spectrophotometer under low resolution. Line at zero is due to reference compound. Height represents the number of equivalent protons*

Similar to other branches of spectroscopy, the transitions in NMR also have finite line width. This is because of the interactions of magnetic moment with the surroundings. NMR lines observed in case of solutions are much narrower than those observed in solids. This is because in case of solutions, molecules are having random motions and thus the dipole-dipole coupling of the nuclei with other molecules is averaged to zero. On the other hand in case of solids, position of nuclei is fixed and direct interactions by a dipole-dipole coupling are possible. Even though this leads to broadening of NMR lines, but still lot of information regarding the distance between the nuclei with magnetic moment and their orientation in the crystal can be obtained. These-a-days methods have been developed, which have made possible to get a high-resolution NMR spectra of solids.

5.9.3 CHEMICAL SHIFT

The material discussed so far is relevant only to a nuclear physicist, because it tells us about the calculation of nuclear *g* factor or it tells us how one can detect magnetically different nuclei in a compound. For at least a couple of reasons listed below, NMR is not used for this purpose. Firstly to carry out the experiments to detect the magnetically different nuclei one has to modify the equipment so that resonance lines can be observed. Although it is not impossible, it is quite difficult also. Secondly the elemental composition of organic compounds can be determined more easily and accurately by other techniques such as elemental analysis or mass spectrometrically. Thus, the importance of NMR is not to detect merely the magnetically different nuclei but to find out how the environments in a molecule around a particular nucleus affect the NMR resonance line. In other words resonance line of individual nucleus is influenced by the distribution of electron density in the chemical bond of the molecule. Thus, NMR frequency of a proton nucleus in a molecule is dependent on its structure.

Since the bare nucleus or isolated nucleus is not feasible, consider hydrogen atom, which is having a nucleus with spin 1/2 and an electron in 1*s* orbital. Unperturbed charge cloud has a spherical distribution. When an external magnetic field is applied, it induces circulations in the electron cloud surrounding the nucleus such that following Lenz's law a magnetic moment μ is induced and this is opposite to the B_0, as shown in Fig. 5.35. Because of this reason, the external magnetic field felt by the proton will not be equal to B_0, but will be less than B_0. This effect is known as *diamagnetic effect or magnetic shielding* of the nucleus and is proportional to the external magnetic field B_0. This amounts to σB_0, where σ is a screening or shielding constant of a proton. The actual field experienced by the proton is:

$$B = B_0 - B_0 \sigma = B_0 (1 - \sigma) \quad ...(5.198)$$

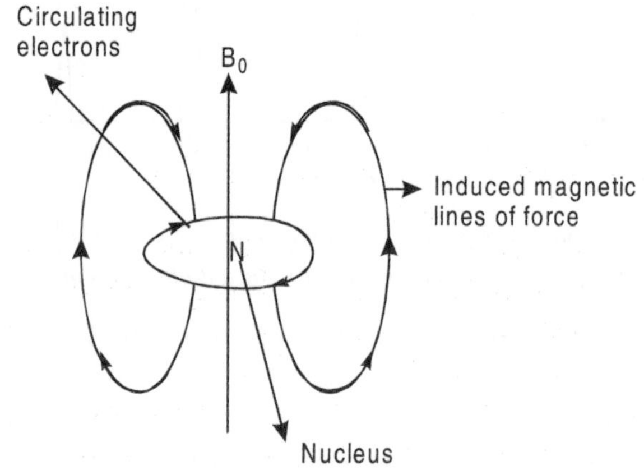

Figure 5.35: *Diamagnetical shielding of nucleus by the rotation of electrons around the nucleus*

The magnitude of screening constant σ in case of proton, which possesses unperturbed spherical electron distribution, can be calculated using Lamb's formula. But the situation in molecules is much more complex, because in these cases one has to consider the electronic circulation within the entire molecules. Due to this reason it may happen that induced magnetic field produced by the motion of electronical charge may be aligned along the direction of B_0 field and thus strengthens B_0 field. This is known as *deshielding or paramagnetic effect.* The value of shielding constant is less than nearly by a factor of 10^{-6} for the protons and nearly 10^{-3} for most of the other nuclei. The magnitude of the shielding constant depends on the charge density around the nucleus, as well as, on the orientation of the molecule with respect to the magnetic field. Since in liquid or gas phase the directional part of σ averages out and thus can be treated as a scalar quantity or pure number. Effect of this actual magnetic field (5.198) on the energy levels of the proton is depicted in Fig. 5.36. It clearly indicates that the gap between the two energy levels will decrease and thus the frequency given by the relation (5.195*b*) will also be less. To match this frequency with the frequency of the radio wave transmitter one has to increase the strength of the magnetic field. In other words the signal observed will be shifted to higher field in comparison to the isolated proton. This shift is called *chemical shift*. Since the isolated proton is not available chemical shift is always compared with respect to some standard reference. Commonly used standard reference is tetramethyl silane (TMS), if organic solvent is used and water if the aqueous medium is used for taking NMR. These are known as internal reference standards. The chemical shift defined in the following manner is a unitless quantity.

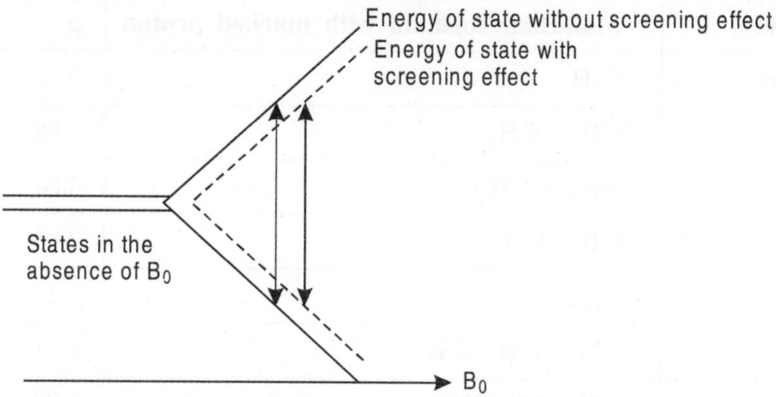

Figure 5.36: *Splitting of m_I states of proton as a function of external magnetic field in the absence and presence of screening effect*

$$\delta \text{ (parts per million)} = \frac{(B_{sample} - B_{ref})\,(10^6)}{B_{ref}} \qquad ...(5.199)$$

Substituting the values of magnetic field from equation (5.198):

$$\delta = \frac{(\sigma_{ref} - \sigma_{sample})\,(10^6)}{1-\sigma_{ref}} \approx (\sigma_{ref} - \sigma_{sam}) \times 10^6 \qquad ...(5.200)$$

because $\sigma_{ref} \ll 1$. The advantage of using δ scale is that it is independent of the magnetic field strength, where as if δ is defined in terms of field strength or frequency it will be a function of frequency or field strength respectively used in the spectrometer. The chemical shift (5.200) is the difference between the shielding constants of the reference and sample, expressed as part per million. This will in general have positive value for most of the organic compounds if TMS is used as internal reference, because the protons in TMS are highly shielded. Another chemical shift scale, which is used by some workers is defined as:

$$\Gamma = 10 - \delta \qquad ...(5.201)$$

Table 5.13 gives the chemical shifts of a proton under different environments as compared to TMS and chemical shift as large as 16 ppm has been observed for proton. Thus, it is clear from Table 5.13 that for the methyl proton the chemical shift is different under different environments, *e.g.*, CH_3—C— (saturated), CH_3—C—X (X = halogen atom), CH_3O, CH_3—N<, the values of δ are ~1, ~2, ~4, and ~3 respectively. The important characteristic about shielding factors is that they have different values for different environments of proton. Thus, the position of the resonance line can be used to determine the nature of the group (*e.g.*, CH_3, —NH_2, —OH) in which proton is located in different environments. Now if one combines it with area under the curve (because the intensity of line is proportional to the number of equivalent nuclei), it gives the best information

Table 5.13: *Proton resonances (δ and τ values) of few selected organic compounds*

Compounds	Molecular formula with marked proton	μ	τ
Cyclopropane	C_3H_6	0.22	9.78
Ethane	CH_3—CH_3	0.88	9.12
Ethylene	CH_2=CH_2	4.84	5.16
Ethyne	CH≡CH	2.88	7.12
Benzene	C_6H_6	7.27	2.73
Propene	CH=CH—CH_3	1.71	8.29
Propyne	CH≡CH CH_3	1.80	8.20
Acetone	CH_3—CO—CH_3	2.17	7.83
Cyclohexane	C_6H_{12}	1.44	8.56
Methyl chloride	CH_3Cl	3.10	6.90
Methylene chloride	CH_2Cl_2	5.30	4.70
Chloroform	$CHCl_3$	7.27	2.73
Ethanol	CH_3CH_2OH	1.22	8.78
	CH_3CH_2OH	3.70	6.30
	CH_3CH_2OH	2.58	7.42
Acetic acid	CH_3COOH	2.10	7.90
	CH_3COOH	8.63	1.37
Acetaldehyde	CH_3CHO	2.20	7.80
	CH_3CHO	9.80	0.2
Diethyl ether	$(CH_3CH_2)_2O$	1.16	8.84

	$(CH_3CH_2)_2O$	3.36	6.64
Ethyl acetate	$CH_3COOCH_2CH_3$	2.03	7.97
	$CH_3COOCH_2CH_3$	1.25	8.75
	$CH_3COOCH_2CH_3$	4.12	6.88
Trimethylamine	$N(CH_3)_3$	2.12	
Triethylamine	$N(CH_2CH_3)_3$	2.42	
Toluene	$C_6H_5CH_3$	2.32	
Benzaldehyde	C_6H_5CHO	9.96	

about the number of protons around the particular nucleus and environments in which the concerned nucleus is present. That is why it known as the best tool to diagnose the structure of molecules.

5.9.4 ORIGIN OF PROTON CHEMICAL SHIFTS

As mentioned in the last section, chemical shift is because of shielding of proton, which in turn depends on the charge distribution around the particular nucleus and structure of the molecule. The chemical shift in turn is reflected by, the shielding or screening constant of the nucleus. The contributions to the shielding constants are due to many factors, which are discussed below briefly.

(*a*) **Influence of the Electron Density at the Proton:** As discussed earlier the shielding constant of neutral hydrogen atom can be calculated using Lamb's equation, because of the spherical symmetry of proton. This value is found to be nearly 17.8 ppm. If the charge density at the proton is reduced because of the inductive effect, the value of the shielding constant will be reduced and thus NMR line will be observed at lower field. For example, consider the proton in the following groups: —CH_3, OCH_3, OH, —NH_2 and —COOH. It has been observed that the shielding is maximum in the first group and minimum in the last group. Based on this and from other data in Table 5.13, it can be concluded that the shielding of the proton is minimum if the acidity of the group is quite large. Similarly the screening constants in hydrogen halides expectedly fall in the order of HF < HCl < HBr < HI.

(*b*) **Influence of the Electron Density at Neighbouring Carbon Atoms:** In organic chemistry generally the proton is not directly connected to the electronegative atom, but it is through the carbon atom. Nevertheless the effect of the electronegative atom is similar. The only difference is that the charge density at the neighbouring carbon atom becomes a determining factor and this factor keeps on decreasing as one moves away from the electronegative atom. Because of this, chemical shift decreases as the distance from the electronegative elements increases. For example, the chemical shifts (δ values) observed in case of CH_3F, CH_3Cl, CH_3Br, CH_3I and CH_3H are respectively 4.13, 2.84,

2.45, 1.98 and 0.13 ppm. Similarly the changes in chemical shifts ($\Delta\delta$) noted for protons *a*, *b* and *c* in CH_3 (*c*) —CH_2 (*b*) —CH_2 (*a*) NO_2 are 3.45, 0.72, and 0.12 ppm respectively, if the proton resonance frequencies of propane are used as reference.

In case of metal derivatives the proton attached, to the carbon atom is shielded much more than the proton of TMS. Thus, the chemical shift observed has the negative value, but the influence of metal atom on the neighbouring carbon atoms decreases sharply in comparison to the electronegative element. For examples, the δ for the protons attached to the carbon atom immediate to the metal atom is having a negative δ with respect to TMS. But these values become positive for the protons attached to the second carbon atom, as shown below:

CH_3Li	CH_3—CH_2Li	$(CH_3$—$CH_2)_2Mg$
–1.3	1.33 –0.99	1.26 –0.64

(*c*) **Influence of Induced Magnetic Moments of Neighbouring Atoms and Bonds:** Besides the change in the charge density at the respective nucleus due to the electronegativity of the neighbouring atom, the presence of magnetic dipole at the neighbouring atoms or group can also alter the local magnetic field responsible for the resonance frequency and hence the chemical shift. The one thing, which should be kept in mind, is that these kinds of effects are having directional property and thus depend on the structure of the molecule or its orientation in the external magnetic field. For example, consider a proton B and we are going to see the effect of induced magnetic field generated by the electron cloud around A on B. As we have already learnt (Fig. 5.35) that the induced diamagnetic circulation of the charge cloud around any nucleus produces a magnetic field, which is opposite to the direction of external magnetic field B_0. In other words the magnetic field experienced by the nucleus is less than that of B_0 and thus the nucleus is shielded. Now the effect of this induced magnetic field of nucleus A depends on the orientation of the bond B—A to the B_0. As seen from Fig. 5.37(*b*) and 5.37(*c*), where B_0 is perpendicular to B—A bond, the direction of the induced magnetic field acting on B is parallel to B_0. Whereas from Fig. 5.37(*a*), the direction of the induced magnetic field acting on B is opposite to B_0. In other words in the former two cases the total magnetic field felt by the nucleus B is more than that of B_0, we call it as *deshielding or paramagnetic effect* and in the last case the total magnetic field felt by B is less than that of B_0, we call it *shielding or diamagnetic effect*. Similarly the paramagnetic circulation of charge cloud which supports the external magnetic field at the respective nucleus A, may however act diamagnetically at the particular proton B if the bond B—A is perpendicular to the magnetic field B_0 and paramagnetically if the molecular axis is along the B_0 axis, *i.e., the diamagnetic effect reduces the effective external field at the proton and consequently increases the shielding whereas the paramagnetic effect increases the magnetic field and thus decreases the shielding at the proton.* Couple of generalization can be made:

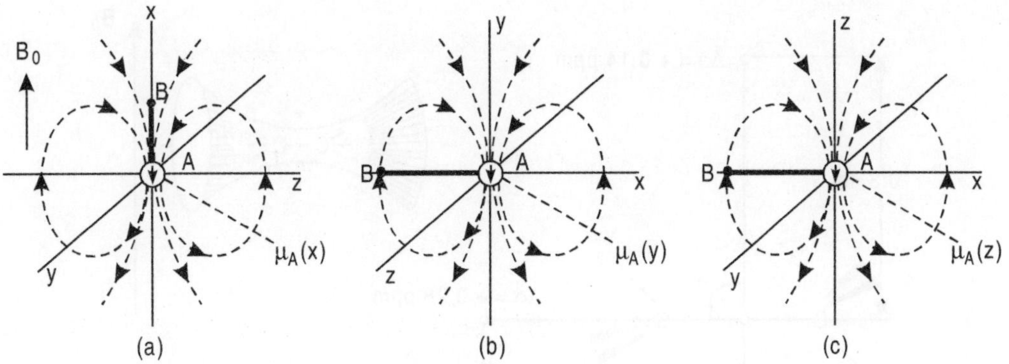

Figure 5.37: *Effect of the direction of the induced magnetic field on the proton due the adjacent atom A. In (c) and (b), it is parallel to the B_0 and in (a) it is opposite to B_0*

(*i*) If the symmetry axis of the charge distribution around a nucleus coincides with the direction of the external field, only diamagnetic currents occur. But it is not always possible because of thermal motions. In all other cases both diamagnetic and paramagnetic currents are present.

(*ii*) The charge distribution surrounding the protons has approximately spherical symmetry irrespective of the presence of magnetic field or chemical bonding, even if a molecule of which the proton forms a part, is changing its orientation with respect to the external magnetic field direction due to rotation and vibration. Because of this mainly diamagnetic currents will flow around proton due to almost axial symmetry of the charge cloud with respect to the field direction.

(*iii*) Diamagnetic circulation of charge around a nucleus or group exert paramagnetic effect or deshield the proton if the proton is perpendicular or away from the field direction and diamagnetic effect on the proton if it coincides with the B_0 direction, whereas the paramagnetic circulation of charge cloud has the opposite effects.

In the gas phase or liquid phase, because of random motion these effects average out to be zero if the neighbouring atom or group is magnetically isotropic, *i.e.*, components of the magnetic susceptibility of nucleus A along *x*, *y* and *z* directions (χ_x, χ_y, χ_z) are equal. But most of the times these groups are not magnetically isotropic and thus can have large effects on the local field.

Consider the molecules CH_3—CH_3, CH_2=CH_2 and CH≡CH. Based on the acidity criterion, the acetylene protons are deshielded maximum and ethane minimum, *i.e.*, chemical shift should be minimum for ethane and maximum for acetylene protons. But the experimental sequence is CH_3—CH_3 (δ = 0.88), CH≡CH, (δ = 2.88) and CH_2=CH_2 (δ = 5.84). All these systems are anisotropic. The results can be explained as follows, by taking two extreme orientations of the molecules, *i.e.*, either parallel or perpendicular to external magnetic field (Fig. 5.38), (Fig. 5.39) and (Fig. 5.40) respectively.

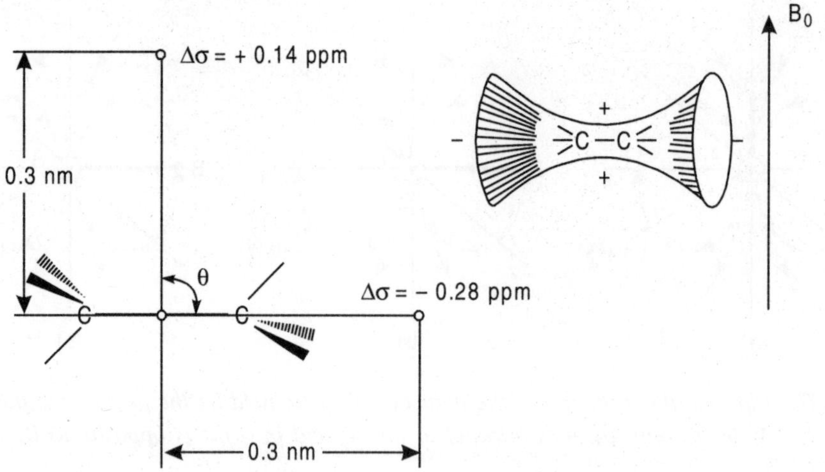

Figure 5.38: *Ethane molecule*

When ethane molecule is oriented parallel to B_0 (Fig. 5.38*a*), mainly diamagnetic currents flow and thus have weak paramagnetic effects on the proton because the C—H bond in ethane is nearly perpendicular (~109°) to the C—C bond axis, whereas in case of Fig. 5.38*b*, when C—C axis is perpendicular to B_0, paramagnetic currents flow around the carbon and thus have paramagnetic effects on the proton. The total resultant paramagnetic effect from neighbouring carbon reduces the shielding of the protons by their shielding effects.

Consider ethylene molecule, as depicted in Fig. 5.39*a* and 5.39*b*. The paramagnetic currents produce the diamagnetic effects in the former and paramagnetic effects in the latter case. The net effect of carbon on the proton is that the protons are better shielded in ethylene, but they will be less shielded as compared to the protons of ethane. Thus the overall effect is that δ ($CH_2{=}CH_2$) > δ (CH—CH).

Similarly the acetylene molecule can be represented either parallel to B_0 (Fig. 5.40*a*) or perpendicular to B_0 (Fig. 5.40*b*). In the former case diamagnetic currents flowing around carbon produces diamagnetic effects on the adjoining protons, whereas, in latter case paramagnetic currents, around carbon produces diamagnetic effects on protons. Since the protons of acetylene are acidic than those of ethylene than those of ethane, thus the net effect is that the acetylene protons are better shielded in comparison to the protons of ethane or ethylene. But owing to greater acidity of acetylene protons, these are less shielded by, their own diamagnetic currents, than the proton of ethane and ethylene. The net result of these effects is that order of chemical shifts among these molecules is:

δ ($CH_2{=}CH_2$) > δ ($CH{\equiv}CH$) > δ ($CH_3{-}CH_3$).

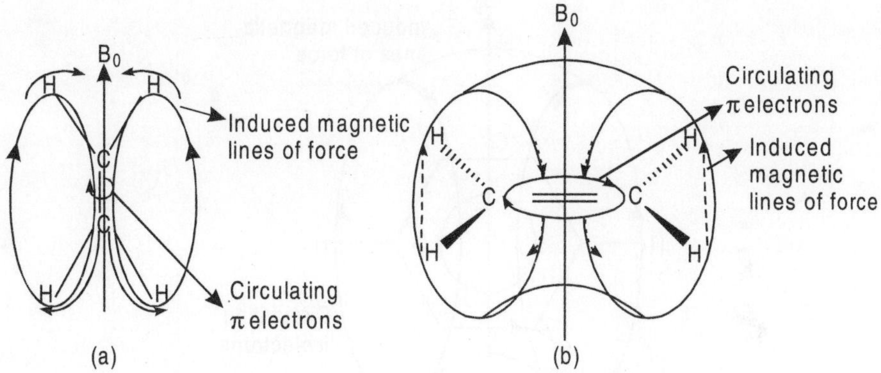

Figure 5.39: *Deshielding of ethylene protons where magnetic filed is: (a) parallel to the bond axis (b) perpendicular to the bond axis*

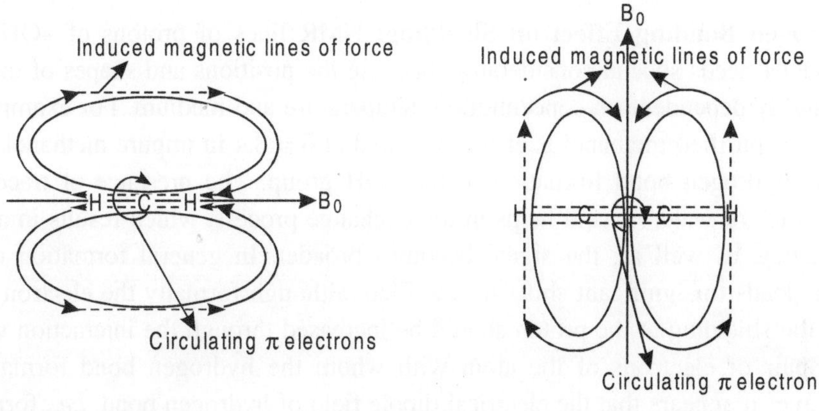

Figure 5.40: *Shielding effect of ethyne protons where magnetic field is: (a) parallel to the bond axis, (b) perpendicular to the bond axis*

(*d*) **Ring Current Effect in Cyclic Conjugated π System:** Benzene molecule forms the simplest example. Unlike ethylenic protons (δ = 4.84), the chemical shift observed for benzene is at δ = 7.27, *i.e.*, the benzene protons are more deshielded in comparison to ethylene ones. In the simple model, an aromatic molecule can be visualized as a current loop where π electrons are free to move in a circular framework formed by the σ bonds. When these kinds of molecules are subjected to external magnetic field B_0, the diamagnetic ring currents are introduced. Because of motion of π electrons, a magnetic field having a centre at aromatic ring and opposite in direction to B_0, as shown in Fig. 5.41, is induced. As a result protons in the molecular plane and outside the ring are deshielded, whereas protons in the region above and below the plane of aromatic system are shielded. This effect is called ring current.

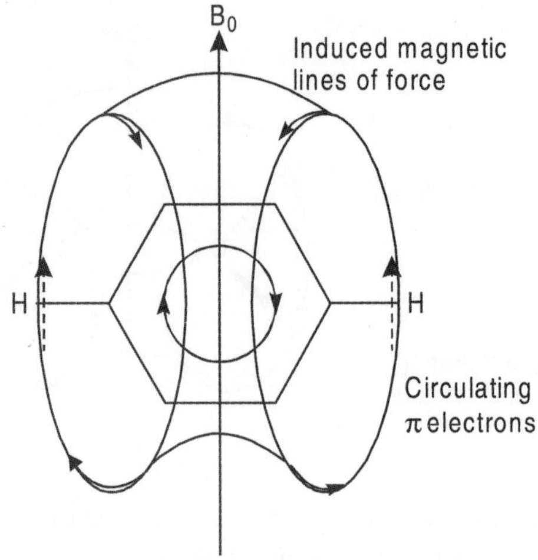

Figure 5.41: *Ring current effects in benzene*

(*e*) **Hydrogen Bonding Effect on Shielding:** NMR lines of protons of –OH, $–NH_2$ and —COOH needs special consideration because the positions and shapes of the transitions are highly dependent on concentration, temperature and medium. For example, the –OH signal in purified methanol is at $\delta = 8.14$ and at $\delta = 8.4$ in impure methanol. This is due to the hydrogen bond formation of the –OH group. The presence of trace amount of impurities of water or acid helps in the exchange process, which results in a shift in the frequency, as well as, the signal becomes broader. In general formation of hydrogen bonds leads to significant shifts to low filed, although formally the electron density and thus the shielding at the proton should be increased through the interaction with the free lone pair of electrons of the atom with whom the hydrogen bond formation occurs. However it appears that the electrical dipole field of hydrogen bond, *i.e.*, formulated as a pure electrostatic bond affects a deshielding. Like infrared spectroscopy the intra and intermolecular hydrogen bonding can also be detected with NMR, because the latter one is concentration dependent.

(*f*) **Paramagnetic Materials:** Paramagnetic materials containing unpaired electrons do possess a magnetic moment. The spin magnetic moment can align with the field and thus lowers the energy of the state by doing so. In this case the interaction of the magnetic moment and magnetic field becomes more favourable when field is strong. So the paramagnetic material is attracted into the magnetic field. The size of the force can be used to calculate the number of unpaired electrons.

(*g*) **Lastly** the effect of medium or solvent surrounding the nucleus also plays a major role if the solvents are interacting one. These interactions can be hydrogen bonding or chemical exchange or creating a significant magnetic field. The first two are the chemical interactions and the last one is the magnetic interactions.

The chemical shifts observed for protons in most of the organic compounds lies in the range of 0 to 10 on a δ scale. Whereas for other nuclei, the chemical shifts are much larger than those noticed for proton. This is because the other nuclei are surrounded by more number of electrons. For example in case of ^{13}C NMR, the chemical shift observed for this nuclei, measured with respect to TMS, range from 200 ppm for $\genfrac{}{}{0pt}{}{R}{R'}\rangle C=O$ to nearly zero for methyl group. Chemical shifts for ^{14}N and ^{15}N using NH_3 as a reference are in the range of 800 ppm. On the other hand chemical shifts for phosphorus vary between ~ 100 to ~ 200 ppm when measured with respect to P_4O_6 as a reference.

NMR spectrum represented in terms of chemical shift depends on the kind of spectrometer used. For example, in case of constant frequency NMR spectrometer since the magnetic field strength increases from left to right, the chemical shift scale (δ) will increase from right to left, *i.e.,* NMR lines towards left represent those nuclei which are deshielded and shielding increases from left to right, as shown in Fig. 5.42. Spectrum on the right of TMS line depicts those protons, which are shielded more than those of TMS.

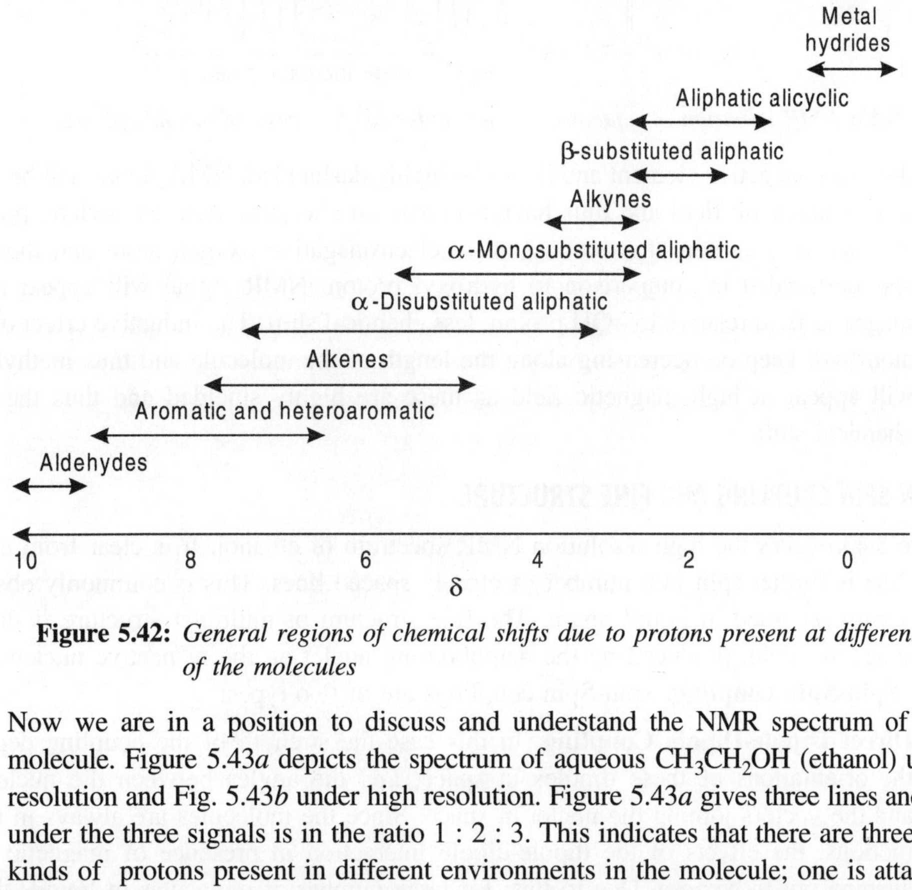

Figure 5.42: *General regions of chemical shifts due to protons present at different positions of the molecules*

Now we are in a position to discuss and understand the NMR spectrum of a simple molecule. Figure 5.43*a* depicts the spectrum of aqueous CH_3CH_2OH (ethanol) under low resolution and Fig. 5.43*b* under high resolution. Figure 5.43*a* gives three lines and the area under the three signals is in the ratio 1 : 2 : 3. This indicates that there are three different kinds of protons present in different environments in the molecule; one is attached to –OH group, the other is to methylene (>CH_2) group and the third to methyl (–CH_3) group. Ratio of the areas under these signals also corresponds to the number of protons attached to three groups respectively. Further the chemical shifts observed for each proton can also be explained on the principles discussed earlier. The hydroxyl proton is attached to

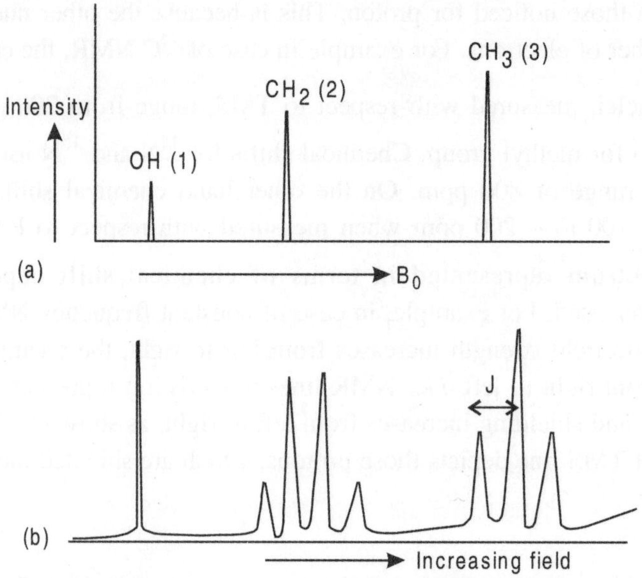

Figure 5.43: *NMR spectrum of aqueous ethanol under (a) low resolution and (b) high resolution*

the electronegative element and it will be highly deshielded. NMR signal will be observed at low magnetic field and thus having maximum chemical shift. Methylene protons are attached to a carbon atom connected to electronegative oxygen atom and thus will be less deshielded in comparison to hydroxyl proton. NMR signal will appear at higher magnetic field relative to –OH proton, less chemical shift. The inductive effect of oxygen atom will keep on decreasing along the length of the molecule and thus methyl protons will appear at high magnetic field as these are highly shielded and thus the smallest chemical shift.

5.9.5 SPIN-SPIN COUPLING AND FINE STRUCTURE

Figure 5.43*b* gives the high resolution NMR spectrum of ethanol. It is clear from the figure that each line is further split into number of closely spaced lines. This is commonly observed in NMR spectrum recorded in liquid phase. The fine structure or multiplet structure is due to the effect of magnetic field, produced by the neighbouring nuclei on the respective nucleus. This is known as **Spin-Spin coupling.** Spin-Spin couplings are of two types:

(*i*) **Direct Dipole-Dipole Coupling.** In this case the strength of the coupling depends on the orientations of these dipoles in space, *i.e.,* the angles between the nuclear spins and the vectors joining the nuclei in space. Since the molecules are always in tumbling motions, the effects of the dipole-dipole interaction in presence of magnetic field B_0 average out to be zero. Due to this, for large number of molecules in liquids the direct interactions have nearly no effect on the NMR signals. On the other hand in case of solids where the nuclei are fixed in space, the direct interactions have large effects and leads to broadening of lines.

(*ii*) **Dipole-Dipole Coupling Through Electrons:** Ramsey and Parcel suggested a mechanism for dipole-dipole interactions involving electrons that form a chemical bond. Consider, for example, two nuclei A and X, each having I = 1/2. Suppose nucleus A has its spin oriented, parallel to B_0. An electron near nucleus A will tend to orient its spin anti-parallel to that of A, because of the tendency of magnetic moments to pair in anti-parallel fashion. If this electron is in an orbital with other electron, then by Pauli Exclusion Principle

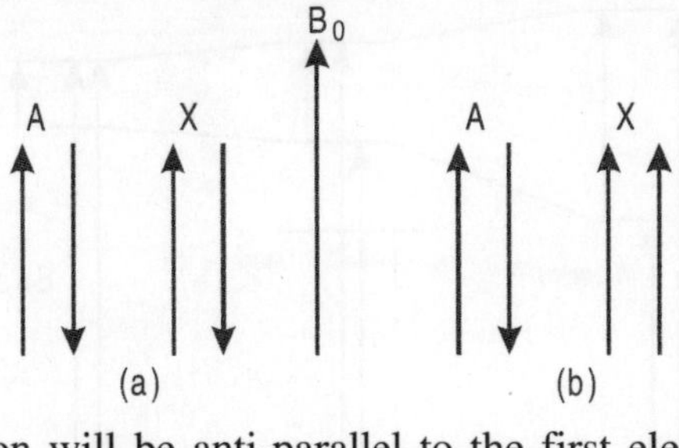

the spin of this electron will be anti-parallel to the first electron but parallel to that of nucleus A. When the second electron is near the nucleus X it will tend to orient the spin of the nucleus X anti-parallel to itself, which is anti-parallel to the nucleus A. *Thus, the information about the spin of the nucleus of A is transmitted to nucleus X via bonding electrons*. In this situation, which is represented by (*a*) is the most favourable (*i.e.,* the lowest energy) situation that occurs when the nucleus A and nucleus X are anti-parallel to each other. The other situation where the spins of the nuclei are parallel to each other (represented in *b*) is a state of only slightly higher energy than that of situation *a*. In other words, the coupling of this kind will lead to energy level of anti-parallel orientation of nuclear spins of A and X of slightly lower energy than for the parallel spin of the nuclei. This is very simple explanation, whereas in reality these interactions are more complicated and some times may lead to reverse results. But the results of coupling discussed above are more common.

Now when the nucleus A is placed in the magnetic field B_0 it will not have only the splitting of energy levels (Zeeman effect) because of B_0, but also experience the effects of nearby nucleus X. There will be two energy levels; one of little lower energy when the spins of nuclei A and X are anti-parallel and the other of higher energy when the spins of the nuclei A and X are parallel to each other. Because of these interactions and splitting of energy levels, the NMR line of A will split into two lines. Similarly the nucleus X will also experience the effect of external magnetic field B_0 and the effect of orientations of the nucleus A. The NMR line of X will also split into two lines. The NMR spectrum of AX system is shown in Fig. 5.44.

For AX system four states of spin orientations are:

1. α (A) α (X) = $\alpha\ \alpha$ = M_I = 1
2. α (A) β (X) = $\alpha\ \beta$ = M_I = 0
3. β (A) α (X) = $\beta\ \alpha$ = M_I = 0
4. β (A) β (X) = $\beta\ \beta$ = M_I = –1

From the above discussion it is clear that spin-spin coupling does not depend on the external magnetic field. Unlike the chemical shift, which is induced by and hence proportional to the applied field, spin-spin interactions is characteristic of the molecule. The magnitude of the interaction between nuclei A and B is given by spin-spin coupling constant J_{AB}, which is always expressed in hertz (Hz).

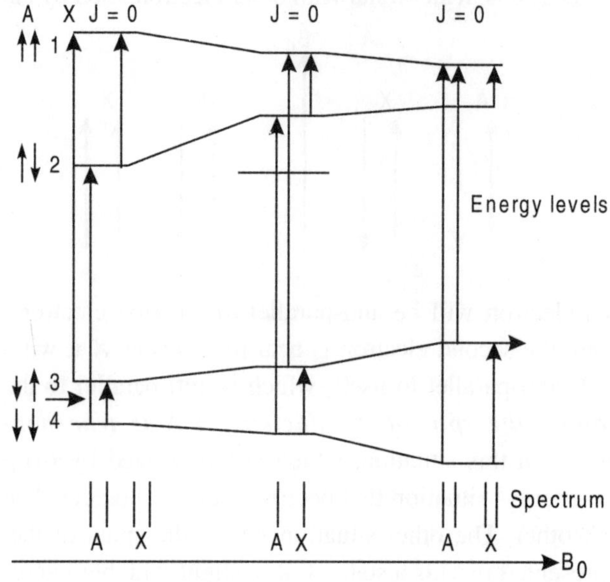

Figure 5.44: *Schematic representation of the energy levels and spectrum of AX system under, different conditions of spin couplings*

Spin-spin coupling can occur when two nuclei are bonded together, such as $^{13}C–^{1}H$, $^{31}P^{1}H$ or where several bonds intervene, such as $^{1}H_A{}^{12}C^{1}H_B$. In general the spin-spin interaction information is carried by, the electrons through chemical bonds and not through space. Possible exceptions, especially with nuclei other than hydrogen have been suggested. Due to this, spin-spin splitting constant keeps on decreasing as the distance between the interacting nuclei increases. *At this level it is proper to consider the interaction between the immediate neighbouring nuclei.*

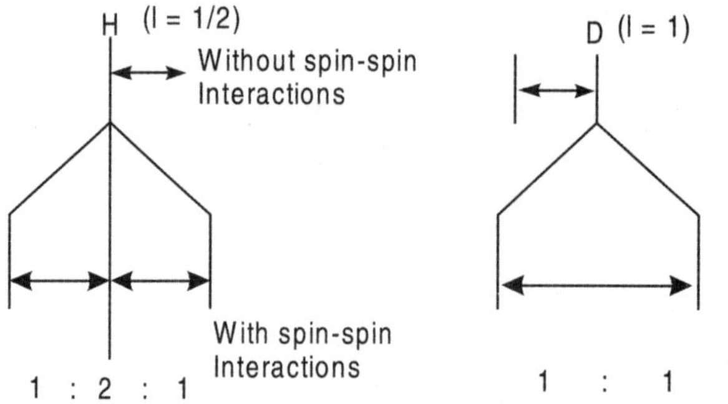

Figure 5.45: *Schematic representation of the splitting of H and D lines in presence of each other*

For nucleus with, spin > 1/2, more number of lines results from spin-spin interactions, since there are 2I +1 possible orientations of nuclear spin I relative to the applied field. Thus, for the molecule HD (deuterium has I = 1), the proton resonance spectrum consists of three lines while deuterium consists of two lines, as shown in Fig. 5.45. Typical values of spin-spin coupling constants for proton are given in Table 5.14.

5.9.6 COUPLING BETWEEN EQUIVALENT NUCLEI

NMR spectrum of CH_3CH_2OH under high resolution is shown in Fig. 5.43*b*. It indicates that NMR lines due the methylene and methyl protons are further split into four (with intensities in the ratio of 1 : 3 : 3 : 1) and three lines (with intensity ratio of 1 : 2 : 1) respectively. These results can be interpreted as follows. Splitting of one line of methyl protons into three lines is due to the presence of methylene protons, which can also generate small magnetic field. Since each of the

↑↑ ↑↓ ↓↑ ↓↓

methylene proton spin can have two orientations (+ 1/2 or –1/2), there can be four possible orientations of the two nuclei. Out of which two are equivalent. Upward arrow represents a component of nuclear magnetic moment along the direction of B_0 and the downward arrow in opposite to B_0. Because of this the desired nucleus (methyl protons) will experience three distinct magnetic fields and thus will give rise to three lines, each corresponding to three local fields. The intensity of the middle line is twice that of the wings. This is due to the fact that there are two possible orientations of the nuclear spin of the methylene protons, which lead to same resultant magnetic moments. Similarly the four peaks of methylene protons are due to the presence of three methyl protons, which can be arranged as:

↑↑↑ ↑↑↓ ↑↓↑ ↑↓↓ ↓↓↑ ↓↑↓ ↑↓↓ ↓↓↓

Since all the three arrangements of the two central combinations give rise to the same magnetic field, there will be four distinct arrangements and thus producing four lines. The intensity ratio of arranging spins of the three protons to get the distinct values of spin magnetic moments is given as 1 : 3 : 3 : 1.

If CH_3CH_2OH is 100 % pure and the hydroxyl proton does not involve in hydrogen bonding, the NMR signal of hydroxyl proton will also be split into three lines of intensities 1 : 2 : 1 and the NMR signals of methylene protons will be more complicated than that described in the above paragraph. The latter part will be discussed later on in the next section, but the observance of one line in case of hydroxyl proton is due to hydrogen bonding and the fast exchange of hydroxyl proton with some trace impurities like water or acid.

At this point one thing definitely comes into mind is the failure to observe more number of lines because of spin-spin coupling between the protons belonging to either methylene or methyl groups. Here we would like to specify clearly that the coupling between the equivalent protons does exist, but it is does not lead to any observable splitting of lines. This can be shown, by carrying out quantum mechanical calculations but it is beyond the scope of this book. Couple of things we will define here which are very useful and will help in understanding the NMR spectrum.

Chemically Equivalent Nuclei

Nuclei are said to be chemically equivalent when they have the same chemical shift. Usually this occurs as a result of molecular symmetry, for example, 2- and 6- protons or 3- and 5- protons of phenol. Sometimes the similar chemical shifts are also observed because of accidental coincidence shielding effects.

Magnetically Equivalent Nuclei

Nuclei are said to be magnetically equivalent if all of them are having the same chemical shift and all of the nuclei in the set are coupled equally with any other single nucleus in the molecule. For example, in case of difluoromethane (I) which is having a tetrahedral structure, due to symmetry H_a and H_b are magnetically equivalent because these are coupled equally to F_a

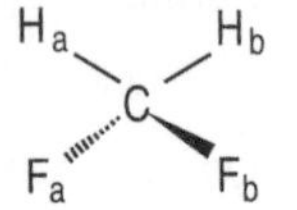

I

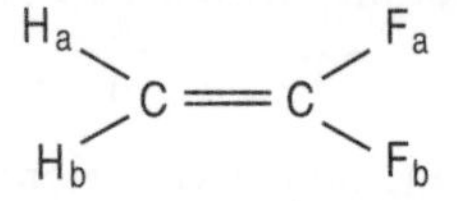

II

and F_b. But the situation is different in case of 1,1-difluorethene (II). The protons H_a and fluorine F_a are cis to each other, whereas proton H_a and F_b are trans to each other. Since the coupling constant J (cis) and J (trans) are different, the two protons H_a and H_b are magnetically different.

When there is a rapid internal motion, *i.e.,* internal rotation or inversion in a molecule, the equivalence of nuclei should be determined on an overall average basis, rather than in one of the individual conformation. For example, in CH_3CH_2OH, the three-methyl protons are magnetically equivalent, because these are coupled equally on the average with each of the methylene protons, even though in anyone of the three stable conformations if frozen would be magnetically non-equivalent, as shown in Fig. 5.46.

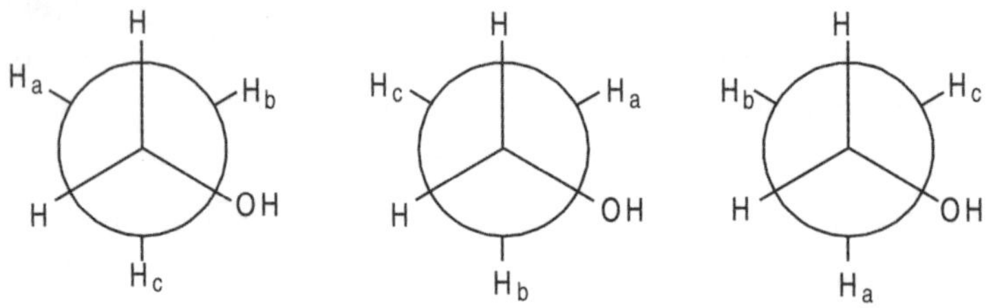

Figure 5.46: *Three different staggered conformations of ethanol in Newman's projection*

5.9.7 FIRST ORDER ANALYSIS

So far in the last two sections emphasis has been laid on how the NMR spectrum of particular molecule is affected by the presence of other nucleus or nuclei having finite spin value. The kind of analysis carried out in the last section about NMR spectrum of CH_3CH_2OH, is known as first order analysis and this is applicable only under certain conditions mentioned below:

(*a*) *The chemical shift difference between the nuclei (or groups of nuclei) must be much larger than the spin-spin coupling between them.* Since the spin coupling constant J_{AB} is expressed in Hz, the chemical shift difference between the nuclei AB and *i.e.,* $(\nu_A - \nu_B)$

should also be expressed in the same units, *i.e.,* Hz. Chemical shifts when expressed in frequency units increases with the increase of field strength, B_0 (radio frequency); the NMR spectrum recorded at high field strength are more likely to follow this condition than those recorded at low magnetic field strength.

(*b*) *Coupling must involve groups of nuclei that are magnetically equivalent not just chemically equivalent.* The criterion of the equivalency in nuclei is discussed in the last section.

When first order analysis is applicable, the number of components in a multiplet, their spacing and their relative intensities can be determined easily from the following rules.

(*i*) A nucleus or group of nuclei coupled to set of n nuclei with spin I will have its resonance split into $2n\text{I} + 1$ lines. For the case of proton where I = 1/12, the number of lines for n equivalent protons will be $n + 1$.

(*ii*) Relative intensities of $(2n\text{I} + 1)$ lines can be determined from the number of ways each spin state can be formed. For the case of I = 1/2, the intensities of $(n + 1)$ lines corresponds to the coefficients of the binomial theorem, as given in the following Table, known as Pascle triangle.

(*iii*) $(2n\text{I} + 1)$ lines are equally spaced, with the frequency separation between adjacent lines being equal to J, the coupling constant.

(*iv*) Coupling between nuclei with magnetically equivalent set of molecules does not affect the spectrum. When two or more nuclei are present near and couple with the respective nucleus the effect can be considered by a repetitive procedure. For example, if the number of neighbouring protons around the respective nucleus is n_A and n_B, the total number of lines observed in a multiplet will be $(n_A + 1)(n_B + 1)$. Unless a very high resolution NMR spectrometer is used, the theoretical number of lines is never observed.

This could be because: (*i*) lines are so closely spaced that these cannot be resolved and (*ii*) frequencies of some of these lines may match each other and thus only one line is observed. Usually it is convenient to consider the largest coupling first but it is immaterial to the final result.

Pascle Triangle

n								Relative Intensity							
0								1							
1							1		1						
2						1		2		1					
3					1		3		3		1				
4				1		4		6		4		1			
5			1		5		10		10		5		1		
6		1		6		15		20		15		6		1	
7	1		7		21		35		35		21		7		1

For example: consider acrylonitrile molecule.

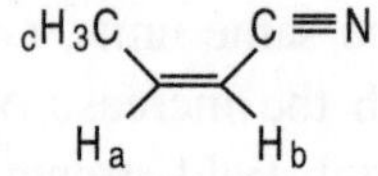

H_a lines will split into two lines because of H_b proton of equal intensities. Each line will further split into four lines of intensities 1 : 3 : 3 : 1, because of three methyl protons. Same will be the case for proton H_b. CH_3 (*c*) protons will have two lines because of proton H_a (1 : 1) and then each will be further split into two each (of equal intensities) Fig. 5.47.

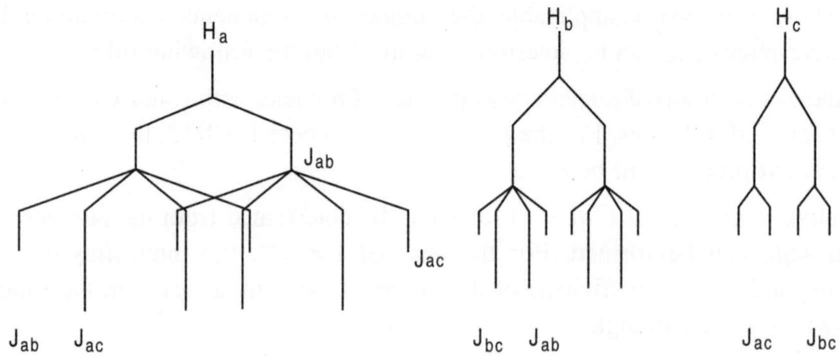

Figure 5.47: *Use of repetitive first order analysis (on arbitrary scale) of* ^1H *spectrum of cis* $CH_3CH = CHCN$. *The olefin coupling constant is 11 Hz, while the 3-bond and 4-bond couplings to the methyl protons are 6.7 and 1.5 Hz respectively. The intensity ratio may also be little off from the true ones*

In the end of this section it is concluded that we have only given very basic and simple treatment to NMR spectrum. For more details some advanced level books concerning only NMR may be studied. One thing may be mentioned here that since spin-spin splitting is due to the effect of neighbouring groups, it imparts very important additional structural information.

APPENDIX 5.1: EVALUATION OF TRANSITION MOMENT INTEGRAL FOR ROTATIONAL SPECTRUM

For rotational spectrum, the transition moment integral **M** should be non-zero. For simplicity we will only consider the *z*-component $\hat{M}_z$ in detail. $\hat{M}_x$ and $\hat{M}_y$ components can also be solved in a similar manner. The *z*-component M_z (J′, *m*′; J″, *m*″, *i.e.*, *m*′ and *m*″ are the quantum numbers representing the non-degenerate rotational levels in the presence of external magnetic field) can be written as:

$$\mathbf{M}_z (J', m'; J'' m'') = \int \psi'_{rot} (J', m') \; \hat{M}_z \, \psi''_{rot} \quad (J'', m'')\, d\tau \qquad \text{...(A 5.1)}$$

Substituting the values ψ'_{rot}, and ψ''_{rot}, $\hat{M}_z = \hat{M}_0 \cos\theta$ and $d\tau = \sin\theta \, d\theta \, d\varphi$

$\mathbf{M}_z$ (J′, *m*′; J″, *m*″)

$$= N'_r \, N''_r \, M_0 \int_0^{\pi} P_{J'}^{|m'|}(\cos\theta) \, \cos\theta \, P_{J''}^{|m''|}(\cos\theta) \, \sin\theta \, d\theta \int_0^{2\pi} e^{im'\varphi} \, e^{-im''\varphi} \, d\varphi \qquad \text{...(A 5.2)}$$

where N'_r and N'_r are the normalizing constants and M_0 is the dipole moment vector. The second integral of the equation (A 5.2) will be zero except for $m' = m''$ and its value is 2π. For the evaluation of first integral, certain mathematical relations existing for spherical harmonics have to be used, *i.e.*,

$$\cos\theta \, P_J^{|m|}(\cos\theta) = \frac{J + |m|}{2J + 1} P_{J-1}^{|m|}(\cos\theta) + \frac{J - |m| + 1}{2J + 1} P_{J+1}^{|m|} (\cos\theta) \qquad \text{...(A 5.3)}$$

$$\sin\theta \, P_J^{|m|}(\cos\theta) = \frac{J + |m|}{2J + 1} P_{J-1}^{|m|}(\cos\theta) + \frac{J - |m| + 1}{2J + 1} P_{J+1}^{|m|} (\cos\theta) \qquad \text{...(A 5.4)}$$

Substituting the first recursion relation in equation (A 5.2) and rearranging for $m' = m''$,

$$\hat{M}_z (J', m'; J'', m'') = 2\pi N'_r \, N''_r \, M_0 \frac{J + |m|}{2J + 1} \int_0^{\pi} P_{J'}^{|m|}(\cos\theta) \, P_{J''-1}^{|m|}(\cos\theta) \, \sin\theta \, d\theta \; +$$

$$2\pi N'_r \, N''_r \, M_0 \frac{J + |m| + 1}{2J + 1} \int_0^{\pi} P_{J'}^{|m|}(\cos\theta) \, P_{J''+1}^{|m|}(\cos\theta) \sin\theta \, d\theta \qquad \text{...(A 5.5)}$$

Other useful relations about spherical harmonics are:

$$\int_0^{\pi} P_{J'}^{|m|}(\cos\theta) \, P_{J''-1}^{|m|}(\cos\theta) \, \sin\theta \, d\theta = 0 \qquad \text{if } J_1 \neq J_2 \qquad \text{...(A 5.6)}$$

$$\int_0^{\pi} P_{J'}^{|m|}(\cos\theta) \, P_{J''-1}^{|m|}(\cos\theta) \, \sin\theta \, d\theta = \frac{2}{2J + 1} \frac{(J + |m|)}{(J - |m|)} \qquad \text{if } J_1 = J_2 \qquad \text{...(A 5.7)}$$

If the above relations are used to evaluate the integrals of the equations (A 5.5), it can be seen that $\hat{M}_z$ (J′, m'; J″, m'') will be non-zero for J′ = J″ – 1 or J′ = J″ + 1 and also $m' = m''$. Similarly it can also be shown that $\mathbf{M}_x$ (J′, m'; J″, m'') and $\mathbf{M}_y$ (J′, m'; J″, m''), *i.e.*, x and y components of the transition moment integral will be non-zero for J′ = J″ – 1 or J′ = J″ + 1 and at the same time $m' = m'' \pm 1$. This proves that the selection rules $\Delta J = \pm 1$. The selection rule $\Delta m = 0, \pm 1$ are valid when external magnetic field is applied.

APPENDIX 5.2 : EVALUATION OF TRANSITION MOMENT INTEGRAL FOR THE VIBRATIONAL SPECTRUM

The expression for the transition moment integral in case of vibrational transitions is:

$$\mathbf{M}\ (v', v'') = \int \psi_{v'}\ \hat{M}\ \psi_{v''}\ d\tau \qquad \text{...(A 5.8)}$$

where $\psi_{v'}$ and $\psi_{v''}$ are the wave functions of the final and initial states. $\hat{M}$ is the oscillating dipole moment operator. For diatomic molecule and considering that molecular axis is the x-axis, the oscillating dipole moment μ is given as:

$$\mu = \mu_0 + \left(\frac{\partial \mu}{\partial x}\right)_{x=0} x + \ldots. \qquad \text{...(A 5.9)}$$

where μ_0 is the permanent dipole moment and other terms depict the change of dipole moment with the displacement of the nuclei (x) from their mean positions. For simple harmonic oscillator, the higher terms in equation (A 5.9) can be neglected. Substituting the value of M in equation (A 5.8), we get

$$\mathbf{M}_x\ (v', v'') = \int \psi_{v'}\ \mu_0\ \psi_{v''}\ dx + \left(\frac{\partial \mu}{\partial x}\right)_{x=0} \int \psi_{v'}\ x\ \psi_{v''}\ dx \qquad \text{...(A 5.10)}$$

The first term is zero because μ_0 is constant and $\psi_{v'}$ and $\psi_{v''}$ are orthogonal to each other. For finite value of first integral $\Delta v = 0$ ($v' = v''$). It will only mean pure rotational spectrum. Further $\psi_{v'}$ and $\psi_{v''}$ are Hermite polynomial and following recursion relation exist for Hermite polynomial.

$$x\ H_v\ (x) = 1/2\ H_{v+1}\ (x) + v\ H_{v-1}\ (x) \qquad \text{...(A 5.11)}$$

and using

$$\Psi_v = N_v\ e^{-\frac{\alpha x^2}{2}}\ H_v\ (\sqrt{\alpha}x) \qquad \text{...(A 5.12)}$$

where

$$\alpha = \text{constant} = \frac{4\pi^2 \mu\ \nu}{h}$$

$$\mathbf{M_x}\ (v'\ v'') = \left(\frac{\partial \mu}{\partial x}\right)_{x=0} N_{v'}\ N_{v''} \int e^{-\frac{\alpha x^2}{2}}\ H_{v'}\ (\sqrt{\alpha}x)\ x e^{-\frac{\alpha x^2}{2}}\ H_{v''}(\sqrt{\alpha}x)\ dx \qquad \text{...(A 5.13)}$$

$$= \left(\frac{\partial \mu}{\partial x}\right)_{x=0} N_{v'} N_{v''} [\frac{1}{2}\int e^{-\alpha x^2} H_{v'} (\sqrt{\alpha}x) H_{v''+1} (\sqrt{\alpha}x)\, dx$$

$$+ \int ve^{-\alpha x^2} H_{v'} (\sqrt{\alpha}x) H_{v''-1} (\sqrt{\alpha}x)\, dx] \quad \text{...(A 5.14)}$$

Both these integrals will be zero unless $v' = v'' + 1$ or $v' = v'' - 1$. So the transition will be observed between the vibrational levels if $\Delta v = \pm 1$.This is called fundamental transition.

For an-harmonic oscillator, the higher terms in the expression (A 5.10) will also be included *i.e.*,

$$\mu = \mu_0 + \left(\frac{\partial \mu}{\partial x}\right)_{x=0} x + \left(\frac{\partial^2}{\partial x^2}\right)_{x=0} x^2 + \left(\frac{\partial^3}{\partial x^3}\right)_{x=0} x^3 + \quad \text{....(A 5.15)}$$

Due to this, the transition moment integral will be non-zero for $\Delta v = \pm 2$, $\Delta v = \pm 3$ ... besides $\Delta v = \pm 1$. The $\Delta v = \pm 2$ is called first overtone, $\Delta v = \pm 3$, the second overtone and so on.

APPENDIX 5.3: EVALUATION OF BOHR NUCLEAR MAGNETON

In this section we will derive the relation between the orbital angular momentum of the electron and its orbital magnetic moment. Figure A 5.1 depicts the motion of an electron in a circular orbit of radius r with speed v. The revolving electron is equivalent to a single-turn current loop. The magnetic moment in a current loop is given as:

$$\boldsymbol{\mu} = i\,A \quad \text{...(A 5.16)}$$

where i is the amount of charge that passes through any point in the orbit divided by the time required, in amperes (coulomb/second) and A is area of the orbit (in square meters). The charge on the electron is e and time T required for one revolution of the electron is (circumference of the circle divided by velocity):

$$T = \frac{2\pi r}{v} \quad \text{...(A 5.17)}$$

So the value of i will be given as:

$$i = \frac{e}{2\pi r / v} = \frac{ev}{2\pi r} \quad \text{...(A 5.18)}$$

The area of the single turn loop (*i.e.*, of orbit) $= \pi r^2$

Substituting the values of i and A in equation (A 5.16) and generating to the vector formulation,

$$\boldsymbol{\mu} = -\frac{ev}{2\pi r}\pi r^2 = -\frac{evr}{2} = -\frac{e\,(m_e vr)}{2m_e} = -\frac{e}{2m_e}\mathbf{L} \quad \text{...(A 5.19)}$$

where **L** is the orbital angular momentum of the electron in an orbit. Thus equation (A 5.19) gives the relation between the magnetic moment (**μ**) of the electron in an orbit and its orbital angular momentum **L**. The minus sign arises because the revolving electron carries a negative charge. This tells us that the magnetic moment and orbital angular momentum vectors are oppositely aligned.

Bohr's postulates tell that the orbital angular momentum of an electron is quantized and is given by the relation:

$$\mathbf{L} = \frac{nh}{2\pi}, \quad n = 1, 2, 3\ldots\ldots \qquad \text{...(A 5.20)}$$

The smallest value of orbital angular momentum is $\frac{h}{2\pi}$. Substituting this value of L in equation (A 5.19) and considering magnitude only:

$$\boldsymbol{\mu}\ (\text{Bohr magneton}) = \frac{e}{2m_e}\frac{h}{2\pi} = \frac{eh}{4\pi m_e} \qquad \text{...(A 5.21)}$$

This value of magnetic moment is called as *Bohr's magneton,* the unit in which magnetic moments are expressed at the level of the electron or proton or neutron, depending on the mass of the particle used. Thus, *Bohr magneton is equal to the orbital magnetic moment of a particle (electron, proton and neutron depending on the mass of the particle) revolving in an orbit with smallest allowed (non-zero) value of orbital angular momentum.*

The expression of magnetic moment given in equation (A 5.19) does not agree with that obtained experimentally. The reason being that here we have used the classical mechanic principles to get the value of magnetic moment of a charged particle moving in circle. We know very well that the small particles like electron, proton, neutron etc., do not follow the classical principles. Thus, to have the proper agreement between the above relation and the experimentally determined values, a *fudge factor* g_e or g_p (also know as *Lande g factor*) have been used. Thus, the value of magnetic moment given in equation (A 5.19) can be written as:

$$\boldsymbol{\mu} = -\frac{g_e e}{2m_e}\mathbf{L} \qquad \text{...(A 5.22)}$$

and similarly for the motion of proton and neutron, the relation between the magnetic moment and spin angular momentum in the nucleus is given as:

$$\boldsymbol{\mu}_N = -\frac{g_N e}{2m_p}\mathbf{I}. \qquad \text{...(A 5.23)}$$

PROBLEMS

1. Show that the rotational transitions of a diatomic molecule occur in the microwave region or far infrared region of the spectrum.
2. The successive difference between the rotational lines of the rotational spectrum of $^1H^{79}Br$ is 16.72 cm^{-1}. Calculate the value of moment of inertia and internuclear distance.
3. If it is assumed that the internuclear distance between the diatomic molecule does not depend on the mass of the nucleus, calculate the values of B for: (*i*) $^2H^{79}Br$, (*ii*) $^1H^{81}Br$ and (*iii*) $^2H^{81}Br$.
4. Using the data obtained in problem 3, calculate the (*i*) ΔE, (*ii*) ν of transition and (*iii*) transition frequency in wave numbers when molecule goes from J = 4 to J = 5.
5. Calculate the value of J_{max} in each case, *i.e.,* $^2H^{79}Br$, $^1H^{81}Br$, $^2H^{81}Br$ and $^1H^{79}Br$ at: (*i*) 300 K (*ii*) 1000 K and (*iii*) 3000 K.
6. The internuclear distance between $^{12}C^{16}O$ is 113 pm. Calculate the value of: (*i*) moment of inertia, (*ii*) rotational constant B, (*iii*) transition between J_{max} to J_{max} + 1 at 300 K, 1000 K and 3000 K.
7. Calculate the population ratio of $\frac{J_{max}}{J_0}$ for $^{12}C^{16}O$ at 300 K, 1000 K and 3000 K using r = 113 pm.
8. The equilibrium internuclear distance between $^{14}N^{16}O$ is 115.08 pm. Calculate the value of B in Hz as well as in cm^{-1}. Also calculate transition frequency between J = 5 to J = 6.
9. The equilibrium internuclear distance between $^{35}C^{19}F$ is 162.81 pm. Plot $\frac{N_J}{N_0}$ versus J for this molecule at 300 K. Find out the value of J_{max} from this plot and compare it with that obtained from equation (5.58).
10. If we assume that rotation of $^1H^{35}Cl$ molecule in J = 5 state can be approximated by classical mechanics, calculate the number of revolutions made by HCl in this state at 300 K. B (HCl) = 10.2 cm^{-1}.
11. Plot $\frac{N_J}{N_0}$ versus J for $^{14}N^{16}O$ using the data of problem 8. Find J_{max} from this plot. Compare it with that obtained by using equation (5.58). Also compare the J_{max} obtained for NO and ClF and comment over the results.
12. Show that the vibrational spectrum of diatomic molecules undergoing simple harmonic motion depicts only one line and this generally lies in infrared region.
13. Force constant of $^{12}C^{16}O$ is 1906 N m^{-1}. Calculate the vibrational frequency of the molecule.
14. The vibrational frequency of the molecule $^{23}Na^{35}Cl$ and $^{39}K^{35}Cl$ are at 378 and 278 cm^{-1} respectively. Calculate the respective force constants and compare the bond strengths of these molecules among themselves qualitatively.
15. The vibrational frequencies observed for $^1H^{35}Cl$, $^1H^{79}Br$ and $^1H^{127}I$ are at 2886, 2559 and 2230 cm^{-1} respectively. Calculate the respective force constants and compare the bond strengths qualitatively.

16. Using the data of problem 15, calculate the value of $\frac{N_1}{N_0}$ for each molecule at (*i*) 300 K, (*ii*) 1000 K and (*iii*) 3000 K. Explain why the value of $\frac{N_1}{N_0}$ obtained is different.

17. Using the data of problem 14, calculate the values of $\frac{N_1}{N_0}$ for each molecule at (*i*) 300 K, (*ii*) 1000 K and (*iii*) 3000 K. Explain why the value of $\frac{N_1}{N_0}$ obtained for NaCl is smaller than that obtained for KCl.

18. Plot $\frac{N_v}{N_0}$ versus v for $^{14}N^{16}O$ molecule at: (*i*) 300 K, (*ii*) 1000 K and (*iii*) 3000 K. Compare these plots with $\frac{N_J}{N_0}$ obtained for the same molecule in problem 11. Why these two plots are different. Carry out the similar plot for $^{79}Br^{19}F$ at these temperatures. Force constants for NO and BrF are 1598 and 129 Nm^{-1} respectively.

19. The force constant for $^{14}N^{16}O$ is 1598 Nm^{-1}. Construct the potential energy curve for NO using Hook's law. Potential energy curve for NO can also be constructed using Morse potential (equation 5.99). Using $\beta = 1.75 \times 10^{10}\ m^{-1}$ and $D_0 = 626.1\ kJ\ mol^{-1}$, draw the potential energy curve on the same plot and compare the two curves.

20. Repeat the problem 19 for HCl. Given, $\beta = 1.82 \times 10^{10}\ m^{-1}$, $D_e = 7.31 \times 10^{-19}$ J molecule^{-1}. Compare the two curves.

21. Using equations (5.108), (5.109) and (5.110), calculate the vibrational frequency and an-harmonic constants for HCl molecule from the following data.

v	0	1	2	3	4	5
$\hat{w}_e$ (cm^{-1})	0	2885.9	5668.0	8347.0	10,923.1	13,196.5

22. Using the data of vibrational frequency and an-harmonic constants of HCl obtained in problem 21, calculate E_0, E_1, E_2, E_3, E_4 and E_5. Compare these values with those obtained in exercise 5.19 from the experimental data.

23. Repeat the problem 22 for NO.

24. Prove that the fraction, $f_i\left(\frac{N_v}{N}\right)$, where N_v and N are the number of molecules in the vth state and total number of molecules respectively) of molecules in any vth state is given by: $f_i = (1 - e^{-\frac{h\nu}{k_b T}})$.

25. Calculate the fraction of CO molecules in the ground vibrational state at 300 K, (*ii*) 1000 K and (*iii*) 3000 K, $\bar{w}_e$ (cm^{-1}) = 2143.

26. Repeat the problem 25 for $^{127}I^{127}I$, $\bar{w}_e$ (cm^{-1}) = 213. Compare the fractions of two molecules.

27. Construct rotation-vibrational spectrum of NO assuming that rotational constants are same for all the vibrational states. r = 115.08 pm and k = 1598 N m^{-1}.

28. Repeat the problem 27 for $^{79}Br^{19}F$. $r = 175.6$ pm, $k = 129$ N m^{-1}.

29. Using equations (5.50) and (5.64), derive equation (5.66), *i.e.*, $\langle \bar{w}^2 e \rangle = \dfrac{4B^3}{D}$.

30. Calculate the value of $\bar{w}_e$ for $^1H^{35}Cl$, given $B_e = 10.59$ cm^{-1}, D = 0.005 cm^{-1}.

31. Calculate the value of B_e for $^7Li^7Li$, given D = 0.009 cm^{-1}, $\bar{w}_e = 35.14$ cm^{-1}.

32. Calculate the value of D for NO, given $B_e = 1.705$ cm^{-1}. $\bar{w}_e = 1904$ cm^{-1}.

33. The $\bar{w}_e$ and $\bar{w}_e\, x_e$ for NO are 1904 cm^{-1} and 13.97 cm^{-1} respectively. Calculate the fundamental and first three overtones. See whether the overtones are integral multiple of the fundamental.

34. Consider $^{35}C^{19}F$ as a rigid rotator-harmonic oscillator, having $r_e = 162.81$ pm, $\alpha_e = 0.004$ cm^{-1}, $k =$ 457.6 N m^{-1}. Construct the first five rotational levels for $v = 0$ and $v = 1$ vibrational states. Show the allowed absorption transitions and calculate the frequencies of first four transitions of R and P branches.

35. Consider equation (5.63). Show that one can determine B and D by graphical method. Using the following data for $^1H^{35}Cl$ molecule, calculate B and D by this method.

Transition	3–4	4–5	5–6	6–7	7–8	8–9	9–10	10–11
($\bar{\nu}$ cm^{-1})	83.03	104.1	124.3	145.03	165.51	185.86	206.38	226.50

36. Using the values of B and D obtained in problem 37, calculate the following frequencies (cm^{-1}) for 0–1, 1–2, and 2–3 rotational lines for the ground state of HCl in the non-rigid rotator approximation.

37. Using the graphical method developed in problem 37, calculate B and D for CO from the following data.

Transitions	0–1	1–2	2–3	3–4	4–5	5–6
$\bar{\nu}$ (cm^{-1})	3.8454	7.6906	11.5355	15.3799	19.2238	23.06685

38. (*a*) Using Morse potential energy curve for a diatomic molecule, calculate the value of force constant and an-harmonic oscillator constants. (*b*) Show that $\nu = \dfrac{\beta}{\pi}\left(\dfrac{D_e}{2\mu}\right)^{\frac{1}{2}}$. Calculate the value of ν for H_2. $\beta = 1.963 \times 10^{10}$ m^{-1}, $D_e = 456$ kJ mol^{-1}. Compare this value with experimental 4.40×10^3 cm^{-1}.

39. Calculate the value of D_e for CO from the data. $D_0 = 1071.6$ kJ mol^{-1}, $\bar{w}_e = 2170.2$ cm^{-1}. Also calculate vibrational constants.

40. Given $D_0 = 431.8$ kJ mol^{-1} and $\bar{w}_e = 4395.2$ cm^{-1}, calculate the value of D_e as well as the vibrational constants.

41. Consider that a particle is oscillating between the limits of potential energy curve (Fig. 5.14) and if we assume that the energy levels are given by one dimensional model, show that the graph between the successive levels decreases as v increases.

42. Vibrational transitions in the an-harmonic oscillator can be depicted as $\bar{\nu} = \bar{w}_e\, v - \bar{w}_e\, x_e\, v\,(v + 1)$. Suggest a graphical procedure to determine $\bar{w}_e$ and $\bar{w}_e\, x_e$.

43. The infrared spectrum of $^{127}I^{35}Cl$ shows the following data.

Transition	0–1	0–2	0–3	0–4	0–5
$\bar{\nu}$ (cm^{-1})	381.2	759.6	1135.00	1507.4	1877.00

Calculate $\bar{w}_e$ and $\bar{w}_e\, x_e$ using graphical method and analytical methods.

44. First few P and R lines in the rotational-vibrational spectrum of $^{1}H^{127}I$ are given below. Calculate B_e, B_0, B_1 and α_e. Also find out the value of J where the band head formation takes place.

Line	R (0)	R(1)	R(2)	P(1)	P(2)	P(3)
Frequency (cm^{-1})	2242.6	2254.8	2266.6	2217.7	2203.8	2190.2

45. The electronic spectrum of $^{12}C^{32}S$ gives the following value of rotational constants in different vibrational states. Calculate B_e and α_e and J value for the band head formation. Also use equation developed in problem 48 to see whether band head will be observed.

v	0	1	2	3
B_v (cm^{-1})	0.81708	0.81116	0.80524	0.79932

46. For a given molecule, the value of B′ and B″ are 8.80 and 11.25 cm^{-1} respectively. Calculate the value of J for the band head formation. If the relation depicting the total number of molecules in all the rotational levels, $N = \Sigma N_J = \dfrac{8\pi^2 c\, I\, k_b\, T}{h} = \dfrac{k_b T}{B}$, then find out whether we will observe the band head or not.

47. Draw the NMR spectrum for each of the following compounds in $CDCl_3$. It may be assumed that first order principle is applicable.

(*i*) $CH_3CH_2\,CH_2\,Cl$

(*ii*) $CH_3CH_2\,COCH_2CH_3$

(*iii*) 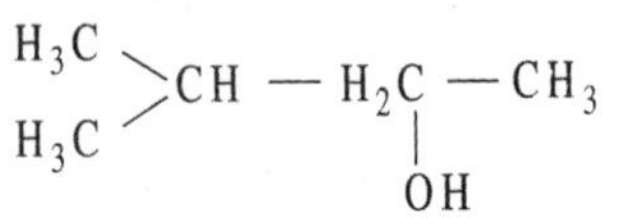

(*iv*) 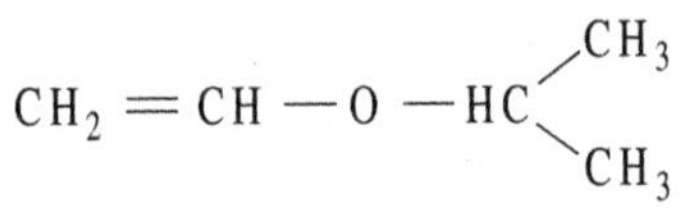

(*v*) $Cl—CH_2\,C\,(H) = O$

(*vi*) $(CH_3)_2 — N — CH\,(OH)$

Appendix

Table A-1: Basic Units.

Physical quantity	Unit	Symbol
Length	Metre	M
Mass	Kilogram	Kg
Time	Second	S
Electric current	Ampere	A
Temperature	Kelvin	K
Luminous intensity of light	Candela	Cd
Plane angle	Radian	Rad
Solid angle	Steradian	Srn

Table A-2: Symbols for some derived SI units.

Physical quantity	Name of SI unit	Symbol of SI unit	Definition of SI unit
Force	Newton	N	$kg\ m\ s^{-2}$
Pressure	Pascal	Pa	$Kg\ m^{-1}\ s^{-2}\ (N\ m^{-2})$
Energy	Joule	J	$kg\ m^2\ s^{-2}$
Power	Watt	W	$kg\ m^2\ s^{-3}\ (J\ s^{-1})$
Electric charge	Coulomb	C	As
Electric potential difference	Volt	V	$kg\ m^2\ s^{-2}\ A^{-2}\ (J\ A^{-1}\ s^{-1})$
Electric resistance	Ohm	Ω	$kg\ m^2\ s^{-3}\ A^{-2}\ (V\ A^{-1})$

Electric conductance	Seimens	s	Kg^{-1} m^{-2}
Electric capacitance	Farad	F	A^2 s^2 kg^1 $m^{-2}(A\ s\ V^{-1})$
Magnetic flux	Weber	W	kg $m^2s^{-2}A^{-1}(A\ s\ V^{-1})$
Magnetic flux density	Tesla	T	kg s^{-2} A^{-1} $(V\ s\ m^{-2})$
Luminousa flux	Lumen	Lm	Cd sr
Illumination	Lux	Ix	Cd sr m^{-2}
Frequency	Hertz	Hz	s^{-1}
Inductance	Henry	H	kg m^2 $s^{-2}A^{-2}(A\ s\ V^{-1})$

Table A-3: Prefixes for fractions and multiples of SI units.

Prefix	Symbol	Factor	Prefix	Symbol	Factor
tera	T	10^{12}	centi*	c	10^{-2}
giga	G	10^9	mili	m	10^{-3}
mega	M	10^6	micro	μ	10^{-6}
kilo	k	10^3	nano	n	10^{-9}
hecto*	h	10^2	pico	P	10^{-2}
Deca*	da	10	femto	f	10^{-15}
Deci*	d	10^{-1}	atto	a	10^{-18}

** An asterisk indicates that the use of these prefixes is limited to occasions where recommended prefixes are inconvenient.

Table A-4: Conversion factor from normal units to SI units.

Physical quantity	Unit	SI equivalent
Length	Angstrom (A)	10^{-10} m
	Inch (in)	0.0254 m
	mile	1.60934 km
	Nautical mile*, international	1.852 km
Area	barn*	10^{-28} m^2
	square inch* (in^2)	645.16 m^2
Volume	cubic inch	1.63871×10^{-5} m^3
Mass	pound*	0.45359237 kg

Force	dyne* (dyn)	10^{-5} N
	poundal	0.138255 N
	pound force	4.44822 N
Pressure	atmosphere* (atm)	101.325 kNm^{-2}
	bar* (bar)	105 N m^{-2}
	torr (torr)	133.322 Nm^{-2}
	p. s .i.	6894.76 N m^{-2}
Energy	calorie international*	4.1868 J
	calorie 15ºC (calrr)	4.1855 J
	calorie thermochemical (cal_{th})	4.184 J
	B. t. u.	1055.06 J
	electron volt (eV)	1.6021 × $10–^{19}$J
	erg* (erg)	10^{-7} J
Power	horse power	745.700 W
Radio activity	curie* (Ci)	3.7 × 10^{10} s^{-1}
Exposure	roentgen (R)	2.5 × 10^{-4} C kg^{-1}
Magnetic flux density	gauss* (G)	10^{-4}T
Kinematic viscosity	stoke* (St)	10^{-4} m^2 s^{-1}
Dynamic viscosity	poise (P)	10^{-1} kg m^{-1} s^{-1}

* An asterisk indicates that the conversions are exact.

Table A-5: Revised list of general physical constants and their values.

Constant	Symbol	Value	Units (SI)	Units (cgs)
Speed of light	c	2.997923	10^8 m s^{-1}	10^{10}cm s^{-f}
Elementary charge	e	1.602191	10^{-19} C	10^{-20}cm $g^{-1/2}$
		4.803250		10^{-10} cm c^{-1} g^{-1} s
Avogadro constant	N_{av}	6.022169	10^{23} mol^{-1}	10^{23} mol^{-1}
Atomic mass unit	u	1.660531	10^{-27}kg	10^{-24} g
Electron rest mass	m_e	9.109558	10^{-31} kg	10^{-28} g
Proton rest mass	m_p	1.672614	10^{-27} kg	10^{-24} g

Faraday constant	F	9.648670	10^4 C mol^{-1}	10^{-3} cm g^{-1} mol^{-1}
Planck constant	h	6.626196	10^{-34} J s	10^{-27} erg s
Boltzmann constant	k_b	1.380622	10^{-23} J K^{-1}	10^{-16} erg K^{-1}
Fine structure constant	α	7.29096	10^{-3}	10^{-3}
Bohr magneton	μB	9.274095	10^{-24} J T^{-1}	10^{-21} erg G^{-1}
Nuclear magneton	μN	5.05404	10^{-27} J T^{-1}	10^{-24} erg G^{-1}
Permittivity	ε_0	8.85419	10^{-12} C^2 N^{-1} m^{-2}	
Permeability of vacuum	μ_0	4π	10^{-7} kg s^{-2} A^{-2}	
Gas constant	R	8.31433	J K^{-1} mol^{-1}	8.31433×10^7 erg K^{-1} mol^{-1}; 1.98717 cal K^{-1} mol^{-1}; 0.052056 atm L K^{-1} mol^{-1}
Stefan-Boltzmann constant	σ	5.66961	10^{-8}W m^{-2}K^{-4}	10^{-5}erg cm^{-2} K^{-4}
Gravitational	G	6.6732	10^{-11} Nm^2 kg	10^{-8} dyn cm^2 g^{-2}
Bohr radius	a_o	0.529	10^{-10}m	10^{-8} cm, (Å)

Table A-6: Various branches of spectroscopy and the molecular phenomenon associated with them

Branch	Phenomenon	Wavelength	Information
γ-rays	Nuclear transitions	0.3 – 0.003 Å	Oxidation sates
X-rays	Electronic transitions (inner electrons)	100 – 0.3 Å	Core electrons energies < Å
Visible/UV	Electronic transitions (outer transitions	1μm – 300 Å	Electron configuration
Infrared	Molecular vibration, Molecular rotation	300 – 1μm	Surface bonds
Microwaves, Far infrared	Electron spin, molecular rotation	30 – 0.3 m	Bond lengths and bond angles
Radio waves	Nuclear magnetic resonance, nuclear quadrupole resonance	300 – 3 m	Electronic structure near the nucleus

Index

A

AB molecules, 355
Absorbance, 482
Absorption coefficient, 482
Absorption cross section, 482
Absorption spectra, 483
Addition of operators, 87
Allyl systems, 434
Angular momentum operators, 98
  equivalent electron, 306
  one electron, 293
  multi-electron, 300
  non-equivalent electrons, 300
  spin, 265
  total, 296, 301
Angular plots, 223
  energy levels, 526
  Morse potential, 522
  Spectrum, 527-529
Approximate methods, 240
  perturbation method, 149
  variation method, 250
Atomic orbital, 212
  angular plots, 213
  Aufbau principle, 282
Atomic radii, 292

B

Beer-Lambert's law, 480
BeH_2, 386
Benzene, 442
  HMO's of, 444-447
Black body radiation, 16
Bohr's theory, 30
  correspondence principle, 42, 129
  Draw backs, 50
  finite mass in, 38
  uncertainty principle in, 61,70
Bond order, and bond length, 346
Born-Oppenheimer approximation, 322
Butadiene, 437

C

Carbon compounds
  saturated, 404
  unsturated, 407
Carbon dioxide, 390

Carbonyl group, 411
Commutation of operators, 88
Compton effect, 27
Coulomb integral, 331, 362
Correction for finite mass, 38
Correspondence principle, 42, 129
Cyclobutadiene, 439
Crystal field theory, 458

D

d-d transition, 570
de Broglie hypothesis, 55
Degree of freedom, 5
Delayed fluorescence, 576
Delocalization energy, 438
Diatomic molecules
  LCAO-M.O. energy levels of, 324, 341, 345
  electronic structure of, 346, 349
  term symbols, 375
  MO wave functions, 334
  VB treatment of, 367, 368
Dipole length, 490
Dipole moment, 378
Dynamical variables, 83, 119

E

Eigen functions, 84
Eigen values, 84
Einstein coefficients of,
  induced absorption, 488
  induced emission, 488
  spontaneous emission, 488
Electronic configuration, 346
Electron affinitily, 291
Electronegativity, 380
  Mullikan's method, 383
  Pauling's method, 380
Electron spin resonance (ESR), 495
Electronic spectrum, 553, 564
  Franck-Condon Principle, 556, 558
  vibration coarse structure, 554
Emision spectra, 485
  selection ruls, 562
Energies levels of
  harmonic oscillator, 148
  hydrogen atom, 201
  m.o.s, 341, 345
  particle in a box, 124, 134
  particle in a ring, 138
Energy operator, 98
Equivalent bonds, 389
Equivalent electron, 306
Equivalent orbitals, 389
Ethylene, 407, 432
Ethyne, 409
Excimer, 576
Exchange degeneracy, 248
Exchange integral, 363

F

Field concept, 12
Fluorescence spectrum, 574
Franck-Condon principle,
  Classical, 557
  Quantum mechanical, 556, 558
Free particle, 122
Functions, 83

General coordinates, 5
Group velocity, 54

H

Hamilton's equation, 11
Hamilton's function, 2
Hamilton' principle, 2
Hamiltonian operator, 91
symmetry, 239
Harmonic oscillator, 143
eigen functions, 151
eigen values, 151
Hartree method, 274
Hartree-Fock SCF method, 272
Heat capacity of solids, 22
Heisenberg uncertainty principle, 57, 61, 70
Heitler-London theory, 321
Helium atom, 248, 258
Hermite polynomial, 151
Hermitian operator, 84
Hermiticity, 111
Heteronuclear diatomic molecules, 352, 368
Homonuclear diatomic molecules 349, 367
Huckel molecular orbital theory, 427
ethylene, 432
allyl system, 434
benzene, 442-447
butadiene, 437
cyclobutadiene, 439
ethylene, 432
Hund's rule, 314
Hybridization, 413
angular dependence, 423
strength of hybrid orbitals, 426
Hydrogen atom, 170, 228, 255
radial part, 191
Φ part, 176
Θ part, 180
Hydrogen like atoms, 201, 206
geometrical description, 212
Hydrogen molecule, 337, 358
Hydrogen
H_2^+, M.O. treatment, 328
bonding, 35

I

Ioic radii, 292
Infrared spectra, application 541
Inductive effect 556
Intrinsic lifetime, 480
Internal conversion, 574
Ionization energy 290
Intersystem crossing, 575

J

Jablonski diagram, 572

L

Ladder operators, 116, 118
LCAO-MO theory, 325
Laguerre polynomial, 194
Laplacian operator, 73
Laporte selection rule, 563
Legendre polynomial, 182
Lengrangian function, 4
Ligand field theory, 465
Linear momentum, 92
Lowering operators, 116-119

M

Matter, wave particle duality, 52
Mean value theorem, 121
Mesomeric effect, 568
M.O. and VBT, comparison, 370
Molecular orbital theory, 324, 464
Molecular absorvity, 482
Molecular absorption coeff, 480
Multi-electron atoms, 272, 300
Mutual exclusion principle, 550

N

Newtonian mechanics, 4
Nodes, 127, 156, 218
  radial, 218
Non-interacting particles, 132
Normalization constant, 77
Nuclear Magnetic Resonance, 494, 578
  chemical shift, 585
  chemically equivalent proton, 600
  deshielding effect, 586
  diamagnetic effect, 586
  hydrogen bond effect, 594
  magnetically equivalent proton, 600
  population of states, 582
  ring current, 593
  spectrometer, 583
  spin-spin coupling, 593
  spin-spin relaxation, 580

O

Old quantum theory, 16
Operators, 83
Operator algebra,
  addition, 87
  angular momentum, 98
  associative multiplication, 88
  commutator, 88
  energy, 98
  Hamiltonian, 91
  Heisenberg rule, 108
  Hermiticity, 111
  Hermitian quantum mechanical, 89
  Laplacian, 93
  Cartesian to polar coordinates of ∇, 94
  Linear momentum, 92
Orbital,
  angular momemtum, 293
Orthogonal relation, 78
Oscillator strength, 484

P

Particle in
  one dimensional box, 122,124
  on a ring, 138
  three dimensional box, 134
Pascal tringle, 601
Pauli exclusion principle, 268
Perturbation method, 240
Phosphorescence, 575
Photoelectric effect, 25
Planck's constant, 16
Planck's law of radiation, 16
Periodic table, 284
Postulates of quantum mechanics, 119
Power series method, 140

Product of operators, 87
Probability function, 195

Q

Quantum, 16
  mechanical operator, 91
  mechanical calculations, 233
  molecular, 324, 340
Quantum numbers, 121, 140, 148, 162, 202
Quantum theory, 16

R

Radial distribution, 192
Radial equation, 187
Radial probability function, 196
Radial wave function, 214
Radial plots, 199
Radiative lifetime, 576
Radiation less transition, 577
Raman scattering, 495, 543
  Classical theory, 544
  rotational, 549
  rotational-vibration, 551
  vibrational spectrum, 548, 552
  quantum theory, 547
Reduced mass 39
Resonance, 373
Resonance effect, 568
Rigid rotator, 156
Rotational spectrum, 495
  angular momentum, 496
  effect of nuclear spin, 512
  non-rigid rotator, 508
  polyatomic molecules, 505
  quantum restriction, 496
  selection rules, 498
  spectrum, 501
  spectrum intensity, 503
  radial equation, 187
Rotational-vibrations spectrum, 529-535
  band origin, 532
  energy levels, 529
  Null line, 532
  selection rules, 530
  spectrum, 531
  vibrational rotator, 529
  Rydberg constant, 36

S

Schrodinger wave equation, 74, 91, 122, 124, 135 139, 145, 157, 172, 322, 328, 357
Screening effect, 279
SCF theory, 272
Selection rules, 492
Secular equation, 254
Separation of variables, 85
Spin, 265, 490
Angular momentum, 295
Spin forbidden transctions, 563
Standing waves, 20
Superposition of functions, 85

T

Term symbols
  aromic, 303
  molecular, 375

Theories of
  absorption and emission, 483
  classical, 484
  quantum mechanical, 485
Transition metal complexes, 567
  bonding, 454
  VBT, 455
  CFT, 458, 461
  LFT, 465
  MOT, 466, 474
  Π bonding, 470
Transaction metal complexes spectra, 565
  d-d-transitions, 570
Triatomic molecules, 385
  beryllium hydride, 386, 388
  carbon dioxide, 390, 395, 396
  water, 399, 402
Tunneling quantum effect, 232
Thin wall, 232
Two electron atoms, 246

U

Unsaturated hydrocarbons, 283

V

Valence bond theory
  diatomic molecule, 358
  triatomic molecules, 384, 388, 395, 402
Virial Theorm, 229
Variation method, 250
Vibrational spectrum, 495
  anharmonic oscillator, 522
  combination bands, 539
  energy levels, 516, 526
  fermi resonance, 540
  overtones, 539, 540
  polyatomic molecules, 537
  selection rules, 515
  spectrum, 518, 527
  thermal distribution, 519
Vib-Raman spectrum, 552
Vib-Rot-spectrum, 537

W

Wave function, 76
  normalization of, 77
  orthoganlization, 78
  physical meaning of, 76
  angular, 221
Wave mechanics, 73
Water, 399, 402
Wein's law, 19, 21
Werner's theory, 452

Z

Zero-point energy, 149